T0205754

Lecture Notes in Computer Science   **13807**

More information about this series at https://link.springer.com/bookseries/558

Leonid Karlinsky · Tomer Michaeli ·
Ko Nishino (Eds.)

Computer Vision – ECCV 2022 Workshops

Tel Aviv, Israel, October 23–27, 2022
Proceedings, Part VII

Springer

Editors
Leonid Karlinsky
IBM Research - MIT-IBM Watson AI Lab
Massachusetts, USA

Tomer Michaeli ⓘD
Technion – Israel Institute of Technology
Haifa, Israel

Ko Nishino ⓘD
Kyoto University
Kyoto, Japan

ISSN 0302-9743 ISSN 1611-3349 (electronic)
Lecture Notes in Computer Science
ISBN 978-3-031-25081-1 ISBN 978-3-031-25082-8 (eBook)
https://doi.org/10.1007/978-3-031-25082-8

This Springer imprint is published by the registered company Springer Nature Switzerland AG
The registered company address is: Gewerbestrasse 11, 6330 Cham, Switzerland

Foreword

Organizing the European Conference on Computer Vision (ECCV 2022) in Tel-Aviv during a global pandemic was no easy feat. The uncertainty level was extremely high, and decisions had to be postponed to the last minute. Still, we managed to plan things just in time for ECCV 2022 to be held in person. Participation in physical events is crucial to stimulating collaborations and nurturing the culture of the Computer Vision community.

There were many people who worked hard to ensure attendees enjoyed the best science at the 17th edition of ECCV. We are grateful to the Program Chairs Gabriel Brostow and Tal Hassner, who went above and beyond to ensure the ECCV reviewing process ran smoothly. The scientific program included dozens of workshops and tutorials in addition to the main conference and we would like to thank Leonid Karlinsky and Tomer Michaeli for their hard work. Finally, special thanks to the web chairs Lorenzo Baraldi and Kosta Derpanis, who put in extra hours to transfer information fast and efficiently to the ECCV community.

We would like to express gratitude to our generous sponsors and the Industry Chairs Dimosthenis Karatzas and Chen Sagiv, who oversaw industry relations and proposed new ways for academia-industry collaboration and technology transfer. It's great to see so much industrial interest in what we're doing!

Authors' draft versions of the papers appeared online with open access on both the Computer Vision Foundation (CVF) and the European Computer Vision Association (ECVA) websites as with previous ECCVs. Springer, the publisher of the proceedings, has arranged for archival publication. The final version of the papers is hosted by SpringerLink, with active references and supplementary materials. It benefits all potential readers that we offer both a free and citeable version for all researchers, as well as an authoritative, citeable version for SpringerLink readers. Our thanks go to Ronan Nugent from Springer, who helped us negotiate this agreement. Last but not least, we wish to thank Eric Mortensen, our publication chair, whose expertise made the process smooth.

October 2022

Rita Cucchiara
Jiří Matas
Amnon Shashua
Lihi Zelnik-Manor

Preface

Welcome to the workshop proceedings of the 17th European Conference on Computer Vision (ECCV 2022). This year, the main ECCV event was accompanied by 60 workshops, scheduled between October 23–24, 2022. We received 103 workshop proposals on diverse computer vision topics and unfortunately had to decline many valuable proposals because of space limitations. We strove to achieve a balance between topics, as well as between established and new series. Due to the uncertainty associated with the COVID-19 pandemic around the proposal submission deadline, we allowed two workshop formats: hybrid and purely online. Some proposers switched their preferred format as we drew near the conference dates. The final program included 30 hybrid workshops and 30 purely online workshops. Not all workshops published their papers in the ECCV workshop proceedings, or had papers at all. These volumes collect the edited papers from 38 out of the 60 workshops. We sincerely thank the ECCV general chairs for trusting us with the responsibility for the workshops, the workshop organizers for their hard work in putting together exciting programs, and the workshop presenters and authors for contributing to ECCV.

October 2022

Tomer Michaeli
Leonid Karlinsky
Ko Nishino

Organization

General Chairs

Rita Cucchiara   University of Modena and Reggio Emilia, Italy
Jiří Matas   Czech Technical University in Prague,
     Czech Republic
Amnon Shashua   Hebrew University of Jerusalem, Israel
Lihi Zelnik-Manor   Technion – Israel Institute of Technology, Israel

Program Chairs

Shai Avidan   Tel-Aviv University, Israel
Gabriel Brostow   University College London, UK
Giovanni Maria Farinella   University of Catania, Italy
Tal Hassner   Facebook AI, USA

Program Technical Chair

Pavel Lifshits   Technion – Israel Institute of Technology, Israel

Workshops Chairs

Leonid Karlinsky   IBM Research - MIT-IBM Watson AI Lab, USA
Tomer Michaeli   Technion – Israel Institute of Technology, Israel
Ko Nishino   Kyoto University, Japan

Tutorial Chairs

Thomas Pock   Graz University of Technology, Austria
Natalia Neverova   Facebook AI Research, UK

Demo Chair

Bohyung Han   Seoul National University, South Korea

Social and Student Activities Chairs

Tatiana Tommasi   Italian Institute of Technology, Italy
Sagie Benaim   University of Copenhagen, Denmark

Diversity and Inclusion Chairs

Xi Yin   Facebook AI Research, USA
Bryan Russell   Adobe, USA

Communications Chairs

Lorenzo Baraldi   University of Modena and Reggio Emilia, Italy
Kosta Derpanis   York University and Samsung AI Centre Toronto,
                                Canada

Industrial Liaison Chairs

Dimosthenis Karatzas   Universitat Autònoma de Barcelona, Spain
Chen Sagiv   SagivTech, Israel

Finance Chair

Gerard Medioni   University of Southern California and Amazon,
                                USA

Publication Chair

Eric Mortensen   MiCROTEC, USA

Workshops Organizers

W01 - AI for Space

Tat-Jun Chin   The University of Adelaide, Australia
Luca Carlone   Massachusetts Institute of Technology, USA
Djamila Aouada   University of Luxembourg, Luxembourg
Binfeng Pan   Northwestern Polytechnical University, China
Viorela Ila   The University of Sydney, Australia
Benjamin Morrell   NASA Jet Propulsion Lab, USA
Grzegorz Kakareko   Spire Global, USA

W02 - Vision for Art

Alessio Del Bue   Istituto Italiano di Tecnologia, Italy
Peter Bell   Philipps-Universität Marburg, Germany
Leonardo L. Impett   École Polytechnique Fédérale de Lausanne
                                (EPFL), Switzerland
Noa Garcia   Osaka University, Japan
Stuart James   Istituto Italiano di Tecnologia, Italy

W03 - Adversarial Robustness in the Real World

Angtian Wang	Johns Hopkins University, USA
Yutong Bai	Johns Hopkins University, USA
Adam Kortylewski	Max Planck Institute for Informatics, Germany
Cihang Xie	University of California, Santa Cruz, USA
Alan Yuille	Johns Hopkins University, USA
Xinyun Chen	University of California, Berkeley, USA
Judy Hoffman	Georgia Institute of Technology, USA
Wieland Brendel	University of Tübingen, Germany
Matthias Hein	University of Tübingen, Germany
Hang Su	Tsinghua University, China
Dawn Song	University of California, Berkeley, USA
Jun Zhu	Tsinghua University, China
Philippe Burlina	Johns Hopkins University, USA
Rama Chellappa	Johns Hopkins University, USA
Yinpeng Dong	Tsinghua University, China
Yingwei Li	Johns Hopkins University, USA
Ju He	Johns Hopkins University, USA
Alexander Robey	University of Pennsylvania, USA

W04 - Autonomous Vehicle Vision

Rui Fan	Tongji University, China
Nemanja Djuric	Aurora Innovation, USA
Wenshuo Wang	McGill University, Canada
Peter Ondruska	Toyota Woven Planet, UK
Jie Li	Toyota Research Institute, USA

W05 - Learning With Limited and Imperfect Data

Noel C. F. Codella	Microsoft, USA
Zsolt Kira	Georgia Institute of Technology, USA
Shuai Zheng	Cruise LLC, USA
Judy Hoffman	Georgia Institute of Technology, USA
Tatiana Tommasi	Politecnico di Torino, Italy
Xiaojuan Qi	The University of Hong Kong, China
Sadeep Jayasumana	University of Oxford, UK
Viraj Prabhu	Georgia Institute of Technology, USA
Yunhui Guo	University of Texas at Dallas, USA
Ming-Ming Cheng	Nankai University, China

W06 - Advances in Image Manipulation

Radu Timofte	University of Würzburg, Germany, and ETH Zurich, Switzerland
Andrey Ignatov	AI Benchmark and ETH Zurich, Switzerland
Ren Yang	ETH Zurich, Switzerland
Marcos V. Conde	University of Würzburg, Germany
Furkan Kınlı	Özyeğin University, Turkey

W07 - Medical Computer Vision

Tal Arbel	McGill University, Canada
Ayelet Akselrod-Ballin	Reichman University, Israel
Vasileios Belagiannis	Otto von Guericke University, Germany
Qi Dou	The Chinese University of Hong Kong, China
Moti Freiman	Technion, Israel
Nicolas Padoy	University of Strasbourg, France
Tammy Riklin Raviv	Ben Gurion University, Israel
Mathias Unberath	Johns Hopkins University, USA
Yuyin Zhou	University of California, Santa Cruz, USA

W08 - Computer Vision for Metaverse

Bichen Wu	Meta Reality Labs, USA
Peizhao Zhang	Facebook, USA
Xiaoliang Dai	Facebook, USA
Tao Xu	Facebook, USA
Hang Zhang	Meta, USA
Péter Vajda	Facebook, USA
Fernando de la Torre	Carnegie Mellon University, USA
Angela Dai	Technical University of Munich, Germany
Bryan Catanzaro	NVIDIA, USA

W09 - Self-Supervised Learning: What Is Next?

Yuki M. Asano	University of Amsterdam, The Netherlands
Christian Rupprecht	University of Oxford, UK
Diane Larlus	Naver Labs Europe, France
Andrew Zisserman	University of Oxford, UK

W10 - Self-Supervised Learning for Next-Generation Industry-Level Autonomous Driving

Xiaodan Liang	Sun Yat-sen University, China
Hang Xu	Huawei Noah's Ark Lab, China

Fisher Yu	ETH Zürich, Switzerland
Wei Zhang	Huawei Noah's Ark Lab, China
Michael C. Kampffmeyer	UiT The Arctic University of Norway, Norway
Ping Luo	The University of Hong Kong, China

W11 - ISIC Skin Image Analysis

M. Emre Celebi	University of Central Arkansas, USA
Catarina Barata	Instituto Superior Técnico, Portugal
Allan Halpern	Memorial Sloan Kettering Cancer Center, USA
Philipp Tschandl	Medical University of Vienna, Austria
Marc Combalia	Hospital Clínic of Barcelona, Spain
Yuan Liu	Google Health, USA

W12 - Cross-Modal Human-Robot Interaction

Fengda Zhu	Monash University, Australia
Yi Zhu	Huawei Noah's Ark Lab, China
Xiaodan Liang	Sun Yat-sen University, China
Liwei Wang	The Chinese University of Hong Kong, China
Xiaojun Chang	University of Technology Sydney, Australia
Nicu Sebe	University of Trento, Italy

W13 - Text in Everything

Ron Litman	Amazon AI Labs, Israel
Aviad Aberdam	Amazon AI Labs, Israel
Shai Mazor	Amazon AI Labs, Israel
Hadar Averbuch-Elor	Cornell University, USA
Dimosthenis Karatzas	Universitat Autònoma de Barcelona, Spain
R. Manmatha	Amazon AI Labs, USA

W14 - BioImage Computing

Jan Funke	HHMI Janelia Research Campus, USA
Alexander Krull	University of Birmingham, UK
Dagmar Kainmueller	Max Delbrück Center, Germany
Florian Jug	Human Technopole, Italy
Anna Kreshuk	EMBL-European Bioinformatics Institute, Germany
Martin Weigert	École Polytechnique Fédérale de Lausanne (EPFL), Switzerland
Virginie Uhlmann	EMBL-European Bioinformatics Institute, UK

Peter Bajcsy   National Institute of Standards and Technology,
                                          USA

Erik Meijering   University of New South Wales, Australia

W15 - Visual Object-Oriented Learning Meets Interaction: Discovery, Representations, and Applications

Kaichun Mo   Stanford University, USA
Yanchao Yang   Stanford University, USA
Jiayuan Gu   University of California, San Diego, USA
Shubham Tulsiani   Carnegie Mellon University, USA
Hongjing Lu   University of California, Los Angeles, USA
Leonidas Guibas   Stanford University, USA

W16 - AI for Creative Video Editing and Understanding

Fabian Caba   Adobe Research, USA
Anyi Rao   The Chinese University of Hong Kong, China
Alejandro Pardo   King Abdullah University of Science and
                                          Technology, Saudi Arabia
Linning Xu   The Chinese University of Hong Kong, China
Yu Xiong   The Chinese University of Hong Kong, China
Victor A. Escorcia   Samsung AI Center, UK
Ali Thabet   Reality Labs at Meta, USA
Dong Liu   Netflix Research, USA
Dahua Lin   The Chinese University of Hong Kong, China
Bernard Ghanem   King Abdullah University of Science and
                                          Technology, Saudi Arabia

W17 - Visual Inductive Priors for Data-Efficient Deep Learning

Jan C. van Gemert   Delft University of Technology, The Netherlands
Nergis Tömen   Delft University of Technology, The Netherlands
Ekin Dogus Cubuk   Google Brain, USA
Robert-Jan Bruintjes   Delft University of Technology, The Netherlands
Attila Lengyel   Delft University of Technology, The Netherlands
Osman Semih Kayhan   Bosch Security Systems, The Netherlands
Marcos Baptista Ríos   Alice Biometrics, Spain
Lorenzo Brigato   Sapienza University of Rome, Italy

W18 - Mobile Intelligent Photography and Imaging

Chongyi Li   Nanyang Technological University, Singapore
Shangchen Zhou   Nanyang Technological University, Singapore

Ruicheng Feng	Nanyang Technological University, Singapore
Jun Jiang	SenseBrain Research, USA
Wenxiu Sun	SenseTime Group Limited, China
Chen Change Loy	Nanyang Technological University, Singapore
Jinwei Gu	SenseBrain Research, USA

W19 - People Analysis: From Face, Body and Fashion to 3D Virtual Avatars

Alberto Del Bimbo	University of Florence, Italy
Mohamed Daoudi	IMT Nord Europe, France
Roberto Vezzani	University of Modena and Reggio Emilia, Italy
Xavier Alameda-Pineda	Inria Grenoble, France
Marcella Cornia	University of Modena and Reggio Emilia, Italy
Guido Borghi	University of Bologna, Italy
Claudio Ferrari	University of Parma, Italy
Federico Becattini	University of Florence, Italy
Andrea Pilzer	NVIDIA AI Technology Center, Italy
Zhiwen Chen	Alibaba Group, China
Xiangyu Zhu	Chinese Academy of Sciences, China
Ye Pan	Shanghai Jiao Tong University, China
Xiaoming Liu	Michigan State University, USA

W20 - Safe Artificial Intelligence for Automated Driving

Timo Saemann	Valeo, Germany
Oliver Wasenmüller	Hochschule Mannheim, Germany
Markus Enzweiler	Esslingen University of Applied Sciences, Germany
Peter Schlicht	CARIAD, Germany
Joachim Sicking	Fraunhofer IAIS, Germany
Stefan Milz	Spleenlab.ai and Technische Universität Ilmenau, Germany
Fabian Hüger	Volkswagen Group Research, Germany
Seyed Ghobadi	University of Applied Sciences Mittelhessen, Germany
Ruby Moritz	Volkswagen Group Research, Germany
Oliver Grau	Intel Labs, Germany
Frédérik Blank	Bosch, Germany
Thomas Stauner	BMW Group, Germany

W21 - Real-World Surveillance: Applications and Challenges

| Kamal Nasrollahi | Aalborg University, Denmark |
| Sergio Escalera | Universitat Autònoma de Barcelona, Spain |

Radu Tudor Ionescu | University of Bucharest, Romania
Fahad Shahbaz Khan | Mohamed bin Zayed University of Artificial Intelligence, United Arab Emirates
Thomas B. Moeslund | Aalborg University, Denmark
Anthony Hoogs | Kitware, USA
Shmuel Peleg | The Hebrew University, Israel
Mubarak Shah | University of Central Florida, USA

W22 - Affective Behavior Analysis In-the-Wild

Dimitrios Kollias | Queen Mary University of London, UK
Stefanos Zafeiriou | Imperial College London, UK
Elnar Hajiyev | Realeyes, UK
Viktoriia Sharmanska | University of Sussex, UK

W23 - Visual Perception for Navigation in Human Environments: The JackRabbot Human Body Pose Dataset and Benchmark

Hamid Rezatofighi | Monash University, Australia
Edward Vendrow | Stanford University, USA
Ian Reid | University of Adelaide, Australia
Silvio Savarese | Stanford University, USA

W24 - Distributed Smart Cameras

Niki Martinel | University of Udine, Italy
Ehsan Adeli | Stanford University, USA
Rita Pucci | University of Udine, Italy
Animashree Anandkumar | Caltech and NVIDIA, USA
Caifeng Shan | Shandong University of Science and Technology, China
Yue Gao | Tsinghua University, China
Christian Micheloni | University of Udine, Italy
Hamid Aghajan | Ghent University, Belgium
Li Fei-Fei | Stanford University, USA

W25 - Causality in Vision

Yulei Niu | Columbia University, USA
Hanwang Zhang | Nanyang Technological University, Singapore
Peng Cui | Tsinghua University, China
Song-Chun Zhu | University of California, Los Angeles, USA
Qianru Sun | Singapore Management University, Singapore
Mike Zheng Shou | National University of Singapore, Singapore
Kaihua Tang | Nanyang Technological University, Singapore

W26 - In-Vehicle Sensing and Monitorization

Jaime S. Cardoso	INESC TEC and Universidade do Porto, Portugal
Pedro M. Carvalho	INESC TEC and Polytechnic of Porto, Portugal
João Ribeiro Pinto	Bosch Car Multimedia and Universidade do Porto, Portugal
Paula Viana	INESC TEC and Polytechnic of Porto, Portugal
Christer Ahlström	Swedish National Road and Transport Research Institute, Sweden
Carolina Pinto	Bosch Car Multimedia, Portugal

W27 - Assistive Computer Vision and Robotics

Marco Leo	National Research Council of Italy, Italy
Giovanni Maria Farinella	University of Catania, Italy
Antonino Furnari	University of Catania, Italy
Mohan Trivedi	University of California, San Diego, USA
Gérard Medioni	Amazon, USA

W28 - Computational Aspects of Deep Learning

Iuri Frosio	NVIDIA, Italy
Sophia Shao	University of California, Berkeley, USA
Lorenzo Baraldi	University of Modena and Reggio Emilia, Italy
Claudio Baecchi	University of Florence, Italy
Frederic Pariente	NVIDIA, France
Giuseppe Fiameni	NVIDIA, Italy

W29 - Computer Vision for Civil and Infrastructure Engineering

Joakim Bruslund Haurum	Aalborg University, Denmark
Mingzhu Wang	Loughborough University, UK
Ajmal Mian	University of Western Australia, Australia
Thomas B. Moeslund	Aalborg University, Denmark

W30 - AI-Enabled Medical Image Analysis: Digital Pathology and Radiology/COVID-19

Jaime S. Cardoso	INESC TEC and Universidade do Porto, Portugal
Stefanos Kollias	National Technical University of Athens, Greece
Sara P. Oliveira	INESC TEC, Portugal
Mattias Rantalainen	Karolinska Institutet, Sweden
Jeroen van der Laak	Radboud University Medical Center, The Netherlands
Cameron Po-Hsuan Chen	Google Health, USA

Diana Felizardo   IMP Diagnostics, Portugal
Ana Monteiro   IMP Diagnostics, Portugal
Isabel M. Pinto   IMP Diagnostics, Portugal
Pedro C. Neto   INESC TEC, Portugal
Xujiong Ye   University of Lincoln, UK
Luc Bidaut   University of Lincoln, UK
Francesco Rundo   STMicroelectronics, Italy
Dimitrios Kollias   Queen Mary University of London, UK
Giuseppe Banna   Portsmouth Hospitals University, UK

W31 - Compositional and Multimodal Perception

Kazuki Kozuka   Panasonic Corporation, Japan
Zelun Luo   Stanford University, USA
Ehsan Adeli   Stanford University, USA
Ranjay Krishna   University of Washington, USA
Juan Carlos Niebles   Salesforce and Stanford University, USA
Li Fei-Fei   Stanford University, USA

W32 - Uncertainty Quantification for Computer Vision

Andrea Pilzer   NVIDIA, Italy
Martin Trapp   Aalto University, Finland
Arno Solin   Aalto University, Finland
Yingzhen Li   Imperial College London, UK
Neill D. F. Campbell   University of Bath, UK

W33 - Recovering 6D Object Pose

Martin Sundermeyer   DLR German Aerospace Center, Germany
Tomáš Hodaň   Reality Labs at Meta, USA
Yann Labbé   Inria Paris, France
Gu Wang   Tsinghua University, China
Lingni Ma   Reality Labs at Meta, USA
Eric Brachmann   Niantic, Germany
Bertram Drost   MVTec, Germany
Sindi Shkodrani   Reality Labs at Meta, USA
Rigas Kouskouridas   Scape Technologies, UK
Ales Leonardis   University of Birmingham, UK
Carsten Steger   Technical University of Munich and MVTec,
                          Germany
Vincent Lepetit   École des Ponts ParisTech, France, and TU Graz,
                          Austria
Jiří Matas   Czech Technical University in Prague,
                          Czech Republic

W34 - Drawings and Abstract Imagery: Representation and Analysis

Diane Oyen	Los Alamos National Laboratory, USA
Kushal Kafle	Adobe Research, USA
Michal Kucer	Los Alamos National Laboratory, USA
Pradyumna Reddy	University College London, UK
Cory Scott	University of California, Irvine, USA

W35 - Sign Language Understanding

Liliane Momeni	University of Oxford, UK
Gül Varol	École des Ponts ParisTech, France
Hannah Bull	University of Paris-Saclay, France
Prajwal K. R.	University of Oxford, UK
Neil Fox	University College London, UK
Ben Saunders	University of Surrey, UK
Necati Cihan Camgöz	Meta Reality Labs, Switzerland
Richard Bowden	University of Surrey, UK
Andrew Zisserman	University of Oxford, UK
Bencie Woll	University College London, UK
Sergio Escalera	Universitat Autònoma de Barcelona, Spain
Jose L. Alba-Castro	Universidade de Vigo, Spain
Thomas B. Moeslund	Aalborg University, Denmark
Julio C. S. Jacques Junior	Universitat Autònoma de Barcelona, Spain
Manuel Vázquez Enríquez	Universidade de Vigo, Spain

W36 - A Challenge for Out-of-Distribution Generalization in Computer Vision

Adam Kortylewski	Max Planck Institute for Informatics, Germany
Bingchen Zhao	University of Edinburgh, UK
Jiahao Wang	Max Planck Institute for Informatics, Germany
Shaozuo Yu	The Chinese University of Hong Kong, China
Siwei Yang	Hong Kong University of Science and Technology, China
Dan Hendrycks	University of California, Berkeley, USA
Oliver Zendel	Austrian Institute of Technology, Austria
Dawn Song	University of California, Berkeley, USA
Alan Yuille	Johns Hopkins University, USA

W37 - Vision With Biased or Scarce Data

Kuan-Chuan Peng	Mitsubishi Electric Research Labs, USA
Ziyan Wu	United Imaging Intelligence, USA

W38 - Visual Object Tracking Challenge

Matej Kristan   University of Ljubljana, Slovenia
Aleš Leonardis   University of Birmingham, UK
Jiří Matas   Czech Technical University in Prague,
                               Czech Republic
Hyung Jin Chang   University of Birmingham, UK
Joni-Kristian Kämäräinen   Tampere University, Finland
Roman Pflugfelder   Technical University of Munich, Germany,
                               Technion, Israel, and Austrian Institute of
                               Technology, Austria
Luka Čehovin Zajc   University of Ljubljana, Slovenia
Alan Lukežič   University of Ljubljana, Slovenia
Gustavo Fernández   Austrian Institute of Technology, Austria
Michael Felsberg   Linköping University, Sweden
Martin Danelljan   ETH Zurich, Switzerland

Contents – Part VII

W29 - Computer Vision for Civil and Infrastructure Engineering

W30 - AI-Enabled Medical Image Analysis: Digital Pathology and Radiology/COVID-19

W28 - Computational Aspects of Deep Learning

W28 - Computational Aspects of Deep Learning

Deep Learning has been one the most significant breakthroughs in computer science in the last ten years. It has achieved significant progress in terms of the effectiveness of prediction models in many research topics and fields of application. This paradigm shift has radically changed the way research is conducted; AI is becoming a computational science where gigantic models with billions of parameters are trained on large-scale supercomputers. While this transition is leading to better and more accurate results by accelerating scientific discovery and technology advance, the availability of such computational power and the ability to harness it is a key success factor. In this context, optimization and careful design of neural architectures play an increasingly important role that directly affects the pace of research, the effectiveness of state-of-the-art models, their applicability at production scale and, last but not least, the reduction of energy consumed to train and evaluate models. Architectural choices and strategies to train models, in fact, have an exceptional impact on run-time and discovery time, thus ultimately affecting the speed of progress of many research areas. The need for effective and efficient solutions is important in most research areas and essential to help researchers even in those situations where the availability of computational resources is scarce or severely restricted. This workshop presented novel research works that focus on the development of deep neural network architectures in computationally challenging domains.

October 2022

Iuri Frosio
Sophia Shao
Lorenzo Baraldi
Claudio Baecchi
Frederic Pariente
Giuseppe Fiameni

EdgeNeXt: Efficiently Amalgamated CNN-Transformer Architecture for Mobile Vision Applications

Muhammad Maaz[1(✉)], Abdelrahman Shaker[1], Hisham Cholakkal[1], Salman Khan[1,2], Syed Waqas Zamir[3], Rao Muhammad Anwer[1,4], and Fahad Shahbaz Khan[1,5]

[1] Mohamed bin Zayed University of AI, Abu Dhabi, UAE
muhammad.maaz@mbzuai.ac.ae
[2] Australian National University, Canberra, Australia
[3] Inception Institute of Artificial Intelligence (IIAI), Abu Dhabi, UAE
[4] Aalto University, Espoo, Finland
[5] Linköping University, Linköping, Sweden

Abstract. In the pursuit of achieving ever-increasing accuracy, large and complex neural networks are usually developed. Such models demand high computational resources and therefore cannot be deployed on edge devices. It is of great interest to build resource-efficient general purpose networks due to their usefulness in several application areas. In this work, we strive to effectively combine the strengths of both CNN and Transformer models and propose a new efficient hybrid architecture EdgeNeXt. Specifically in EdgeNeXt, we introduce split depth-wise transpose attention (STDA) encoder that splits input tensors into multiple channel groups and utilizes depth-wise convolution along with self-attention across channel dimensions to implicitly increase the receptive field and encode multi-scale features. Our extensive experiments on classification, detection and segmentation tasks, reveal the merits of the proposed approach, outperforming state-of-the-art methods with comparatively lower compute requirements. Our EdgeNeXt model with 1.3M parameters achieves 71.2% top-1 accuracy on ImageNet-1K, outperforming MobileViT with an absolute gain of 2.2% with 28% reduction in FLOPs. Further, our EdgeNeXt model with 5.6M parameters achieves 79.4% top-1 accuracy on ImageNet-1K. The code and models are available at https://t.ly/_Vu9.

Keywords: Edge devices · Hybrid model · Convolutional neural network · Self-attention · Transformers · Image classification · Object detection and segmentation

1 Introduction

Convolutional neural networks (CNNs) and the recently introduced vision transformers (ViTs) have significantly advanced the state-of-the-art in several mainstream computer

---

M. Maaz and A. Shaker—Equal contribution.

© The Author(s), under exclusive license to Springer Nature Switzerland AG 2023
L. Karlinsky et al. (Eds.): ECCV 2022 Workshops, LNCS 13807, pp. 3–20, 2023.
https://doi.org/10.1007/978-3-031-25082-8_1

vision tasks, including object recognition, detection and segmentation [20,37]. The general trend is to make the network architectures more deeper and sophisticated in the pursuit of ever-increasing accuracy. While striving for higher accuracy, most existing CNN and ViT-based architectures ignore the aspect of computational efficiency (*i.e.*, model size and speed) which is crucial to operating on resource-constrained devices such as mobile platforms. In many real-world applications *e.g.*, robotics and self-driving cars, the recognition process is desired to be both accurate and have low latency on resource-constrained mobile platforms.

Most existing approaches typically utilize carefully designed efficient variants of convolutions to achieve a trade-off between speed and accuracy on resource-constrained mobile platforms [19,28,36]. Other than these approaches, few existing works [16,39] employ hardware-aware neural architecture search (NAS) to build low latency accurate models for mobile devices. While being easy to train and efficient in encoding local image details, these aforementioned light-weight CNNs do not explicitly model global interactions between pixels.

The introduction of self-attention in vision transformers (ViTs) [8] has made it possible to explicitly model this global interaction, however, this typically comes at the cost of slow inference because of the self-attention computation [24]. This becomes an important challenge for designing a lightweight ViT variant for mobile vision applications.

The majority of the existing works employ CNN-based designs in developing efficient models. However, the convolution operation in CNNs inherits two main limitations: First, it has local receptive field and thereby unable to model global context; Second, the learned weights are stationary at inference times, making CNNs inflexible to adapt to the input content. While both of these issues can be alleviated with Transformers, they are typically compute intensive. Few recent works [29,46]

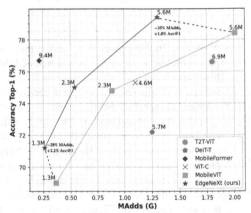

Fig. 1. Comparison of our proposed EdgeNeXt models with SOTA ViTs and hybrid architecture designs. The x-axis shows the multiplication-addition (MAdd) operations and y-axis displays the top-1 ImageNet-1K classification accuracy. The number of parameters are mentioned for each corresponding point in the graph. Our EdgeNeXt shows better compute (parameters and MAdds) versus accuracy trade-off compared to recent approaches.

have investigated designing lightweight architectures for mobile vision tasks by combining the strengths of CNNs and ViTs. However, these approaches mainly focus on optimizing the parameters and incur higher multiply-adds (MAdds) operations which restricts high-speed inference on mobile devices. The MAdds are higher since the complexity of the attention block is quadratic with respect to the input size [29]. This becomes further problematic due to multiple attention blocks in the network architecture. Here, we argue that the model size, parameters, and MAdds are all desired to be small with respect to the resource-constrained devices when designing a unified mobile architecture that effectively combines the complementary advantages of CNNs and ViTs (see Fig. 1).

Contributions. We propose a new light-weight architecture, named *EdgeNeXt*, that is efficient in terms of model size, parameters *and* MAdds, while being superior in accuracy on mobile vision tasks. Specifically, we introduce split depth-wise transpose attention (SDTA) encoder that effectively learns both local and global representations to address the issue of limited receptive fields in CNNs without increasing the number of parameters and MAdd operations. Our proposed architecture shows favorable performance in terms of both accuracy *and* latency compared to state-of-the-art mobile networks on various tasks including image classification, object detection, and semantic segmentation. Our EdgeNeXt backbone with 5.6M parameters and 1.3G MAdds achieves 79.4% top-1 ImageNet-1K classification accuracy which is superior to its recently introduced MobileViT counterpart [29], while requiring 35% less MAdds. For object detection and semantic segmentation tasks, the proposed EdgeNeXt achieves higher mAP and mIOU with fewer MAdds and a comparable number of parameters, compared to all the published lightweight models in literature.

2   Related Work

In recent years, designing lightweight hardware-efficient convolutional neural networks for mobile vision tasks has been well studied in literature. The current methods focus on designing efficient versions of convolutions for low-powered edge devices [17,19]. Among these methods, MobileNet [17] is the most widely used architecture which employs depth-wise separable convolutions [5]. On the other hand, ShuffleNet [47] uses channel shuffling and low-cost group convolutions. MobileNetV2 [36] introduces inverted residual block with linear bottleneck, achieving promising performance on various vision tasks. ESPNetv2 [31] utilizes depth-wise dilated convolutions to increase the receptive field of the network without increasing the network complexity. The hardware-aware neural architecture search (NAS) has also been explored to find a better trade-off between speed and accuracy on mobile devices [16,39]. Although these CCNs are faster to train and infer on mobile devices, they lack global interaction between pixels which limits their accuracy.

Recently, Desovitskiy *et al.* [8] introduces a vision transformer architecture based on the self-attention mechanism [41] for vision tasks. Their proposed architecture utilizes large-scale pre-training data (e.g. JFT-300M), extensive data augmentations, and a longer training schedule to achieve competitive performance. Later, DeiT [40] proposes to integrate distillation token in this architecture and only employ training on ImageNet-1K [35] dataset. Since then, several variants of ViTs and hybrid architectures are proposed in the literature, adding image-specific inductive bias to ViTs for obtaining improved performance on different vision tasks [9,11,38,42,48].

ViT models achieve competitive results for several visual recognition tasks [8,24]. However, it is difficult to deploy these models on resource-constrained edge devices because of the high computational cost of the multi-headed self-attention (MHA). There has been recent work on designing lightweight hybrid networks for mobile vision tasks that combine the advantages of CNNs and transformers. MobileFormer [4] employs parallel branches of MobileNetV2 [36] and ViTs [8] with a bridge connecting both branches for local-global interaction. Mehta *et al.* [29] consider transformers as convolution and propose a MobileViT block for local-global image context fusion. Their approach achieves superior performance on image classification surpassing previous light-weight CNNs and ViTs using a similar parameter budget.

Although MobileViT [29] mainly focuses on optimizing parameters and latency, MHA is still the main efficiency bottleneck in this model, especially for the number of MAdds and the inference time on edge devices. The complexity of MHA in Mobile-ViT is quadratic with respect to the input size, which is the main efficiency bottleneck given their existing nine attention blocks in MobileViT-S model. In this work, we strive to design a new light-weight architecture for mobile devices that is efficient in terms of both parameters and MAdds, while being superior in accuracy on mobile vision tasks. Our proposed architecture, EdgeNeXt, is built on the recently introduced CNN method, ConvNeXt [25], which modernizes the ResNet [14] architecture following the ViT design choices. Within our EdgeNeXt, we introduce an SDTA block that combines depth-wise convolutions with adaptive kernel sizes along with transpose attention in an efficient manner, obtaining an optimal accuracy-speed trade-off.

3   EdgeNeXt

The main objective of this work is to develop a lightweight hybrid design that effectively fuses the merits of ViTs and CNNs for low-powered edge devices. The computational overhead in ViTs (e.g., MobileViT [29]) is mainly due to the self-attention operation. In contrast to MobileViT, the attention block in our model has linear complexity with respect to the input spatial dimension of $\mathcal{O}(Nd^2)$, where N is the number of patches, and d is the feature/channel dimension. The self-attention operation in our model is applied across channel dimensions instead of the spatial dimension. Furthermore, we demonstrate that with a much lower number of attention blocks (3 versus 9 in Mobile-ViT), we can surpass their performance mark. In this way, the proposed framework can model global representations with a limited number of MAdds which is a fundamental criterion to ensure low-latency inference on edge devices. To motivate our proposed architecture, we present two desirable properties.

a) **Encoding the global information efficiently.** The intrinsic characteristic of self-attention to learn global representations is crucial for vision tasks. To inherit this advantage efficiently, we use cross-covariance attention to incorporate the attention operation across the feature channel dimension instead of the spatial dimension within a relatively small number of network blocks. This reduces the complexity of the original self-attention operation from quadratic to linear in terms of number of tokens and implicitly encodes the global information effectively.

b) **Adaptive kernel sizes.** Large-kernel convolutions are known to be computationally expensive since the number of parameters and FLOPs quadratically increases as the kernel size grows. Although a larger kernel size is helpful to increase the receptive field, using such large kernels across the whole network hierarchy is expensive and sub-optimal. We propose an adaptive kernel sizes mechanism to reduce this complexity and capture different levels of features in the network. Inspired by the hierarchy of the CNNs, we use smaller kernels at the early stages, while larger kernels at the latter stages in the convolution encoder blocks. This design choice is optimal as early stages in CNN usually capture low-level features and smaller kernels are suitable for this purpose. However, in later stages of the network, large convolutional kernels are required to capture high-level features [45]. We explain our architectural details next.

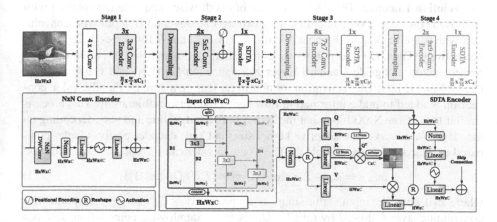

Fig. 2. Top Row: The overall architecture of our framework is a stage-wise design. Here, the first stage downsamples the input image to $1/4^{th}$ resolution using 4×4 strided convolution followed by three 3×3 Convolution (Conv.) encoders. In stages 2–4, 2×2 strided convolutions are used for downsampling at the start, followed by $N \times N$ Convolution and the Split depth-wise Transpose Attention (SDTA) encoders. **Bottom Row:** We present the design of the Conv. encoder (Left) and the SDTA encoder (right). The Conv. encoder uses $N \times N$ depth-wise convolutions for spatial mixing followed by two pointwise convolutions for channel mixing. The SDTA Encoder splits the input tensor into B channel groups and applies 3×3 depth-wise convolutions for multi-scale spatial mixing. The skip connections between branches increase the overall receptive field of the network. The branches $B3$ and $B4$ are progressively activated in stages 3 and 4, increasing the overall receptive field in the deeper layers of the network. Within the proposed SDTA, we utilize Transpose Attention followed by a light-weight MLP, that applies attention to feature channels and has linear complexity with respect to the input image.

Overall Architecture. Figure 2 illustrates an overview of the proposed EdgeNeXt architecture. The main ingredients are two-fold: **(1)** adaptive $N \times N$ Conv. encoder, and **(2)** split depth-wise transpose attention (SDTA) encoder. Our EdgeNeXt architecture builds on the design principles of ConvNeXt [25] and extracts hierarchical features at four different scales across the four stages. The input image of size $H \times W \times 3$ is passed through a patchify stem layer at the beginning of the network, implemented using a 4×4 non-overlapping convolution followed by a layer norm, which results in

$\frac{H}{4} \times \frac{W}{4} \times C1$ feature maps. Then, the output is passed to 3×3 Conv. encoder to extract local features. The second stage begins with a downsampling layer implemented using 2×2 strided convolution that reduces the spatial sizes by half and increases the channels, resulting in $\frac{H}{8} \times \frac{W}{8} \times C2$ feature maps, followed by two consecutive 5×5 Conv. encoders. Positional Encoding (PE) is also added before the SDTA block in the second stage only. We observe that PE is sensitive for dense prediction tasks (e.g., object detection and segmentation) as well as adding it in all stages increases the latency of the network. Hence, we add it only once in the network to encode the spatial location information. The output feature maps are further passed to the third and fourth stages, to generate $\frac{H}{16} \times \frac{W}{16} \times C3$ and $\frac{H}{32} \times \frac{W}{32} \times C4$ dimensional features, respectively.

Convolution Encoder. This block consists of depth-wise separable convolution with adaptive kernel sizes. We can define it by two separate layers: **(1)** depth-wise convolution with adaptive $N \times N$ kernels. We use $k = 3, 5, 7$, and 9 for stages 1, 2, 3, and 4, respectively. Then, **(2)** two point-wise convolution layers are used to enrich the local representation alongside standard Layer Normalization [2] (LN) and Gaussian Error Linear Unit [15] (GELU) activation for non-linear feature mapping. Finally, a skip connection is added to make information flow across the network hierarchy. This block is similar to the ConvNeXt block but the kernel sizes are dynamic and vary depending on the stage. We observe that adaptive kernel sizes in Conv. encoder perform better compared to static kernel sizes (Table 8). The Conv. encoder can be represented as follows:

$$x_{i+1} = x_i + Linear_G(Linear(LN(Dw(x_i)))), \tag{1}$$

where x_i denotes the input feature maps of shape $H \times W \times C$, $Linear_G$ is a point-wise convolution layer followed by GELU, Dw is $k \times k$ depth-wise convolution, LN is a normalization layer, and x_{i+1} denotes the output feature maps of the Conv. encoder.

SDTA Encoder. There are two main components in the proposed split depth-wise transpose attention (SDTA) encoder. The first component strives to learn an adaptive multi-scale feature representation by encoding various spatial levels within the input image and the second part implicitly encodes global image representations. The first part of our encoder is inspired by Res2Net [12] where we adopt a multi-scale processing approach by developing hierarchical representation into a single block. This makes the spatial receptive field of the output feature representation more flexible and adaptive. Different from Res2Net, the first block in our SDTA encoder does not use the 1×1 pointwise convolution layers to ensure a lightweight network with a constrained number of parameters and MAdds. Also, we use adaptive number of subsets per stage to allow effective and flexible feature encoding. In our STDA encoder, we split the input tensor $H \times W \times C$ into s subsets, each subset is denoted by x_i and has the same spatial size with C/s channels, where $i \in \{1, 2, ..., s\}$ and C is the number of channels. Each feature maps subset (except the first subset) is passed to 3×3 depth-wise convolution, denoted by d_i, and the output is denoted by y_i. Also, the output of d_{i-1}, denoted by y_{i-1}, is added to the feature subset x_i, and then fed into d_i. The number of subsets s is adaptive based on the stage number t, where $t \in \{2, 3, 4\}$. We can write y_i as follows:

$$y_i = \begin{cases} x_i & i = 1; \\ d_i(x_i) & i = 2, t = 2; \\ d_i(x_i + y_{i-1}) & 2 < i \leq s, t. \end{cases} \tag{2}$$

Each depth-wise operation d_i, as shown in SDTA encoder in Fig. 2, receives feature maps output from all previous splits $\{x_j, j \leq i\}$.

As mentioned earlier, the overhead of the transformer self-attention layer is infeasible for vision tasks on edge-devices because it comes at the cost of higher MAdds and latency. To alleviate this issue and encode the global context efficiently, we use transposed query and key attention feature maps in our SDTA encoder [1]. This operation has a linear complexity by applying the dot-product operation of the MSA across channel dimensions instead of the spatial dimension, which allows us to compute cross-covariance across channels to generate attention feature maps that have implicit knowledge about the global representations. Given a normalized tensor Y of shape $H \times W \times C$, we compute query (Q), key (K), and value (V) projections using three linear layers, yielding $Q = W^Q Y$, $K=W^K Y$, and $V=W^V Y$, with dimensions $HW \times C$, where W^Q, W^K, and W^V are the projection weights for Q, K, and V respectively. Then, L2 norm is applied to Q and K before computing the cross-covariance attention as it stabilizes the training. Instead of applying the dot-product between Q and K^T along the spatial dimension i.e., $(HW \times C) \cdot (C \times HW)$, we apply the dot-product across the channel dimensions between Q^T and K i.e., $(C \times HW) \cdot (HW \times C)$, producing $C \times C$ softmax scaled attention score matrix. To get the final attention maps, we multiply the scores by V and sum them up. The transposed attention operation can be expressed as follows:

$$\hat{X} = Attention(Q, K, V) + X, \tag{3}$$

$$s.t., \quad Attention(Q, K, V) = V \cdot \texttt{softmax}(Q^T \cdot K) \tag{4}$$

where X is the input and $\hat{X}$ is the output feature tensor. After that, two 1×1 pointwise convolution layers, LN and GELU activation are used to generate non-linear features. Table 1 shows the sequence of Conv. and STDA encoders with the corresponding input size at each layer with more design details about extra-extra small, extra-small and small models.

4 Experiments

In this section, we evaluate our EdgeNeXt model on ImageNet-1K classification, COCO object detection, and Pascal VOC segmentation benchmarks.

4.1 Dataset

We use ImageNet-1K [35] dataset in all classification experiments. The dataset provides approximately 1.28M training and 50K validation images for 1000 categories. Following the literature [17,29], we report top-1 accuracy on the validation set for all experiments. For object detection, we use COCO [22] dataset which provides approximately 118k training and 5k validation images respectively. For segmentation, we use Pascal VOC 2012 dataset [10] which provides almost 10k images with semantic segmentation masks. Following the standard practice as in [29], we use extra data and annotations from [22] and [13] as well.

4.2 Implementation Details

We train our EdgeNeXt models at an input resolution of 256×256 with an effective batch size of 4096. All the experiments are run for 300 epochs with AdamW [27] optimizer, and with a learning rate and weight decay of 6e-3 and 0.05 respectively. We use cosine learning rate schedule [26] with linear warmup for 20 epochs. The data augmentations used during training are Random Resized Crop (RRC), Horizontal Flip, and RandAugment [6], where RandAugment is only used for the EdgeNeXt-S model. We also use multi-scale sampler [29] during training. Further stochastic depth [18] with a rate of 0.1 is used for EdgeNeXt-S model only. We use EMA [32] with a momentum of 0.9995 during training. For inference, the images are resized to 292×292 followed by a center crop at 256×256 resolution. We also train and report the accuracy of our EdgeNeXt-S model at 224×224 resolution for a fair comparison with previous methods. The classification experiments are run on eight A100 GPUs with an average training time of almost 30 h for the EdgeNeXt-S model.

Table 1. EdgeNeXt Architectures. Description of the models' layers with respect to output size, kernel size, and output channels, repeated n times, along with the models MAdds and parameters. The number of the output channels for small, extra-small, and extra-extra small models is chosen to match the number of parameters with the counterpart MobileViT model. We use adaptive kernel sizes in Conv. Encoder to reduce the model complexity and capture different levels of features. Also, we pad the output size of the last stage to be able to apply the 9×9 filter.

Layer	Output size	n	Kernel	Output channels		
				XXS	XS	S
Image	256×256	1	–	–	–	–
Stem	64×64	1	4×4	24	32	48
Conv. Encoder	64×64	3	3×3	24	32	48
Downsampling	32×32	1	2×2	48	64	96
Conv. Encoder	32×32	2	5×5	48	64	96
STDA Encoder	32×32	1	–	48	64	96
Downsampling	16×16	1	2×2	88	100	160
Conv. Encoder	16×16	8	7×7	88	100	160
STDA Encoder	16×16	1	–	88	100	160
Downsampling	8×8	1	2×2	168	192	304
Conv. Encoder	8×8	2	9×9	168	192	304
STDA Encoder	8×8	1	–	168	192	304
Global Average Pooling	1×1	1	–	–	–	–
Linear	1×1	1	–	1000	1000	1000
Model MAdds				0.3G	0.5G	1.3G
Model Prameters				1.3M	2.3M	5.6M

For detection and segmentation tasks, we finetune EdgeNeXt following similar settings as in [29] and report mean average precision (mAP) at IOU of 0.50–0.95 and mean intersection over union (mIOU) respectively. The experiments are run on four A100 GPUs with an average training time of ~36 and ~7 h for detection and segmentation respectively.

We also report the latency of our models on NVIDIA Jetson Nano[1] and NVIDIA A100 40 GB GPU. For Jetson Nano, we convert all the models to TensorRT[2] engines and perform inference in FP16 mode using a batch size of 1. For A100, similar to [25], we use PyTorch v1.8.1 with a batch size of 256 to measure the latency.

Table 2. Comparisons of our proposed EdgeNeXt model with state-of-the-art lightweight fully convolutional, transformer-based and hybrid models on ImageNet-1K classification task. Our model achieves a better trade-off between accuracy and compute (i.e., parameters and multiplication-addition (MAdds)).

Frameworks	Models	Date	Input	Params↓	MAdds↓	Top1↑
	MobileNetV2	CVPR2018	224^2	6.9M	585M	74.7
ConvNets	ShuffleNetV2	ECCV2018	224^2	5.5M	597M	74.5
	MobileNetV3	ICCV2019	224^2	5.4M	219M	75.2
ViTs	T2T-ViT	ICCV2021	224^2	6.9M	1.80G	76.5
	DeiT-T	ICML2021	224^2	5.7M	1.25G	72.2
	MobileFormer	CoRR2021	224^2	9.4M	214M	76.7
Hybrid	ViT-C	NeurIPS2021	224^2	4.6M	1.10G	75.3
	CoaT-Lite-T	ICCV2021	224^2	5.7M	1.60G	77.5
	MobileViT-S	ICLR2022	256^2	5.6M	2.01G	78.4
	EdgeNeXt-S	Ours	224^2	5.6M	965M	78.8
	EdgeNeXt-S	Ours	256^2	5.6M	1.30G	79.4

4.3 Image Classification

Table 2 compares our proposed EdgeNeXt model with previous state-of-the-art fully convolutional (ConvNets), transformer-based (ViTs) and hybrid models. Overall, our model demonstrates better accuracy versus compute (parameters and MAdds) trade-off compared to all three categories of methods (see Fig. 1).

Comparison with ConvNets. EdgeNeXt surpasses ligh-weight ConvNets by a formidable margin in terms of top-1 accuracy with similar parameters (Table 2). Normally, ConvNets have less MAdds compared to transformer and hybrid models because of no attention computation, however, they lack the global receptive field. For instance,

---

[1] https://developer.nvidia.com/embedded/jetson-nano-developer-kit.

[2] https://github.com/NVIDIA/TensorRT.

Table 3. Comparison of different variants of EdgeNeXt with the counterpart models of Mobile-ViT. The last two columns list the latency in ms and μs on NVIDIA Jetson Nano and A100 devices, respectively. Our EdgeNext models provide higher accuracy with lower latency for each model size, indicating the flexibility of our design to scale down to as few as 1.3M parameters.

Model	Date	Input	Params↓	MAdds↓	Top1↑	Nano↓	A100↓
MobileViT-XXS			1.3M	364M	69.0	21.0 ms	216 μs
MobileViT-XS	ICLR2022	256^2	2.3M	886M	74.8	35.1 ms	423 μs
MobileViT-S			5.6M	2.01G	78.4	53.0 ms	559 μs
EdgeNeXt-XXS			1.3M	261M	71.2	19.3 ms	142 μs
EdgeNeXt-XS	Ours	256^2	2.3M	538M	75.0	31.6 ms	227 μs
EdgeNeXt-S			5.6M	1.30G	79.4	48.8 ms	332 μs

EdgeNeXt-S has higher MAdds compared to MobileNetV2 [36], but it obtains 4.1% gain in top-1 accuracy with less number of parameters. Also, our EdgeNeXt-S outperforms ShuffleNetV2 [28] and MobileNetV3 [16] by 4.3% and 3.6% respectively, with comparable number of parameters.

Comparison with ViTs. Our EdgeNeXt outperforms recent ViT variants on ImageNet1K dataset with fewer parameters and MAdds. For example, EdgeNeXt-S obtains 78.8% top-1 accuracy, surpassing T2T-ViT [44] and DeiT-T [40] by 2.3% and 6.6% absolute margins respectively.

Comparison with Hybrid Models. The proposed EdgeNeXt outperforms Mobile-Former [4], ViT-C [42], CoaT-Lite-T [7] with less number of parameters and fewer MAdds (Table 2). For a fair comparison with MobileViT [29], we train our model at an input resolution of 256×256 and show consistent gains for different models sizes (i.e., S, XS, and XXS) with fewer MAdds and faster inference on the edge devices (Table 3). For instance, our EdgeNeXt-XXS model achieves 71.2% top-1 accuracy with only 1.3M parameters, surpassing the corresponding MobileViT version by 2.2%. Finally, our EdgeNeXt-S model attains 79.4% accuracy on ImageNet with only 5.6M parameters, a margin of 1.0% as compared to the corresponding MobileViT-S model. This demonstrates the effectiveness and the generalization of our design.

Further, we also train our EdgeNeXt-S model using knowledge distillation following [34] and achieves 81.1% top-1 ImageNet accuracy.

4.4   ImageNet-21K Pretraining

To further explore the capacity of EdgeNeXt, we designed EdgeNeXt-B model with 18.5M parameters and 3.8MAdds and pretrain it on a subset of ImageNet-21K [35] dataset followed by finetuning on standard ImageNet-1K dataset. ImageNet-21K (winter'21 release) contains around 13M images and 19K classes. We follow [33] to pre-process the pretraining data by removing classes with fewer examples and split it into training and validation sets containing around 11M and 522K images respectively over

10,450 classes. We refer this dataset as ImageNet-21K-P. We strictly follow the training recipes of [25] for ImageNet-21K-P pretaining. Further, we initialize the ImageNet-21K-P training with ImageNet-1K pretrained model for faster convergence. Finally, we finetune ImageNet-21K model on ImageNet-1K for 30 epochs with a learning rate of $7.5e^{-5}$ and an effective batch size of 512. The results are summarized in Table 4.

4.5 Inference on Edge Devices

We compute the inference time of our EdgeNeXt models on the NVIDIA Jetson Nano edge device and compare it with the state-of-the-art MobileViT [29] model (Table 3). All the models are converted to TensorRT engines and inference is performed in FP16 mode. Our model attains low latency on the edge device with similar parameters, fewer MAdds, and higher top-1 accuracy. Table 3 also lists the inference time on A100 GPU for both MobileViT and EdgeNeXt models. It can be observed that our EdgeNeXt-XXS model is ~34% faster than the MobileViT-XSS model on A100 as compared to only ~8% faster on Jetson Nano, indicating that EdgeNeXt better utilizes the advanced hardware as compared to MobileViT.

Table 4. Large-scale ImageNet-21K-P pretraining of EdgeNeXt-B model. Our model achieves better accuracy vs compute trade-off compared to SOTA ConvNeXt [25] and MobileViT-V2 [30].

Model	Pretraining	Input	Params↓	MAdds↓	Top1↑
ConvNeXt-T	None	224^2	28.6M	4.5G	82.1
ConvNeXt-T	ImageNet-21K	224^2	28.6M	4.5G	82.9
MobileViT-V2	None	256^2	18.5M	7.5G	81.2
MobileViT-V2	ImageNet-21K-P	256^2	18.5M	7.5G	82.4
EdgeNeXt-B	None	256^2	18.5M	3.8G	82.5
EdgeNeXt-B	ImageNet-21K-P	256^2	18.5M	3.8G	83.3

4.6 Object Detection

We use EdgeNeXt as a backbone in SSDLite and finetune the model on COCO 2017 dataset [22] at an input resolution of 320×320. The difference between SSD [23] and SSDLite is that the standard convolutions are replaced with separable convolutions in the SSD head. The results are reported in Table 5. EdgeNeXt consistently outperforms MobileNet backbones and gives competitive performance compared to

Table 5. Comparisons with SOTA on COCO object detection. EdgeNeXt improves over previous approaches.

Model	Params↓	MAdds↓	mAP↑
MobileNetV1	5.1M	1.3G	22.2
MobileNetV2	4.3M	800M	22.1
MobileNetV3	5.0M	620M	22.0
MobileViT-S	5.7M	3.4G	27.7
EdgeNeXt-S (ours)	6.2M	2.1G	27.9

MobileVit backbone. With less number of MAdds and a comparable number of parameters, EdgeNeXt achieves the highest 27.9 box AP, ~38% fewer MAdds than MobileViT.

4.7 Semantic Segmentation

We use EdgeNeXt as backbone in DeepLabv3 [3] and finetune the model on Pascal VOC [10] dataset at an input resolution of 512×512. DeepLabv3 uses dilated convolution in cascade design along with spatial pyramid pooling to encode multi-scale features which are useful in encoding objects at multiple scales. Our model obtains 80.2 mIOU on the validation dataset, providing a 1.1 points gain over MobileViT with $\sim$36% fewer MAdds (Table 6).

Table 6. Comparisons with SOTA on VOC semantic segmentation. Our model provides reasonable gains.

Model	Params↓	MAdds↓	mIOU↑
MobileNetV1	11.1M	14.2G	75.3
MobileNetV2	4.5M	5.8G	75.7
MobileViT-S	5.7M	13.7G	79.1
EdgeNeXt-S (ours)	6.5M	8.7G	80.2

5 Ablations

In this section, we ablate different design choices in our proposed EdgeNeXt model.

SDTA Encoder and Adaptive Kernel Sizes. Table 7 illustrates the importance of SDTA encoders and adaptive kernel sizes in our proposed architecture. Replacing SDTA encoders with convolution encoders degrades the accuracy by 1.1%, indicating the usefulness of SDTA encoders in our design. When we fix kernel size to 7 in all four stages of the network, it further reduces the accuracy by 0.4%. Overall, our proposed design provides an optimal speed-accuracy trade-off.

We also ablate the contributions of SDTA components (e.g., adaptive branching and positional encoding) in Table 7. Removing adaptive branching and positional encoding slightly decreases the accuracy.

Table 7. Ablation on different components of EdgeNeXt and SDTA encoder. The results show the benefits of SDTA encoders and adaptive kernels in our design. Further, adaptive branching and positional encoding (PE) are also required in SDTA module.

	Model	Top1↑	Latency↓
Base	EdgeNeXt-S	79.4	332 μs
Different	w/o SDTA Encoders	78.3	265 μs
Components	+ w/o Adaptive Kernels	77.9	301 μs
SDTA	w/o Adaptive Branching	79.3	332 μs
Components	+ w/o PE	79.2	301 μs

Hybrid Design. Table 8 ablates the different hybrid design choices for our EdgeNeXt model. Motivated from MetaFormer [43], we replace all convolutional modules in the last two stages with SDTA encoders. The results show superior performance when all blocks in the last two stages are SDTA blocks, but it increases the latency (row-2 vs 3). Our hybrid design where we propose to use an SDTA module as the last block in the last three stages provides an optimal speed-accuracy trade-off.

Table 8. Ablation on the hybrid architecture. Using one SDTA encoder as the last block in the last three stages provides an optimal accuracy-latency trade-off.

Model Configuration	Top1↑	Latency↓
1: Conv=[3, 3, 9, 0], SDTA=[0, 0, 0, 3]	79.3	303 μs
2: Conv=[3, 3, 0, 0], SDTA=[0, 0, 9, 3]	79.7	393 μs
3: Conv=[3, 2, 8, 2], SDTA=[0, 1, 1, 1]	79.4	332 μs

Table 9 provides an ablation of the importance of using SDTA encoders at different stages of the network. It is noticeable that progressively adding an SDTA encoder as the last block of the last three stages improves the accuracy with some loss in inference latency. However, in row 4, we obtain the best trade-off between accuracy and speed where the SDTA encoder is added as the last block in the last three stages of the network. Further, we notice that adding a global SDTA encoder to the first stage of the network is not helpful where the features are not much mature.

We also provide an ablation on using the SDTA module at the start of each stage versus at the end. Table 10 shows that using the global SDTA encoder at the end of each stage is more beneficial. This observation is consistent with the recent work [21].

Activation and Normalization. EdgeNeXt uses GELU activation and layer normalization throughout the network. We found that the current PyTorch implementations of GELU and layer normalization are not optimal for high speed inference. To this end, we replace GELU with Hard-Swish and layer-norm with batch-norm and retrain our models. Figure 3 indicates that it reduces the accuracy slightly, however, reduces the latency by a large margin.

Table 9. Ablation on using SDTA encoder at different stages of the network. Including SDTA encoders in the last three stages improves performance, whereas a global SDTA encoder is not helpful in the first stage.

Model Configuration	Top1↑	Latency↓
1: Conv=[3, 3, 9, 3], SDTA=[0, 0, 0, 0]	78.3	265 μs
2: Conv=[3, 3, 9, 2], SDTA=[0, 0, 0, 1]	78.6	290 μs
3: Conv=[3, 3, 8, 2], SDTA=[0, 0, 1, 1]	79.1	310 μs
4: Conv=[3, 2, 8, 2], SDTA=[0, 1, 1, 1]	79.4	332 μs
5: Conv=[2, 2, 8, 2], SDTA=[1, 1, 1, 1]	79.2	387 μs

Table 10. Ablation on using SDTA at the start and end of each stage in EdgeNeXt. The results show that it is generally beneficial to use SDTA at the end of each stage.

SDTA Configuration	Top1↑	Latency↓
Start of Stage (SDTA=[0, 1, 1, 1])	79.0	332 μs
End of Stage (SDTA=[0, 1, 1, 1])	79.4	332 μs

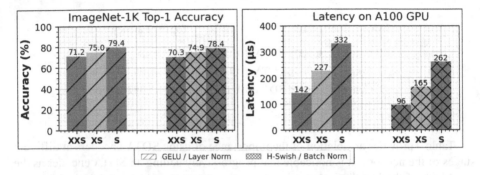

Fig. 3. Ablation on the effect of using different activation functions and normalization layers on accuracy and latency of our network variants. Using Hard Swish activation and batch normalization instead of GELU and layer normalization significantly improves the latency at the cost of some loss in accuracy.

6 Qualitative Results

Figures 4 and 5 shows the qualitative results of EdgeNeXt detection and segmentation models respectively. Our model can detect and segment objects in various views.

Fig. 4. Qualitative results of our EdgeNeXt detection model on COCO validation dataset. The model is trained on COCO dataset with 80 detection classes. Our model can effectively localize and classify objects in diverse scenes.

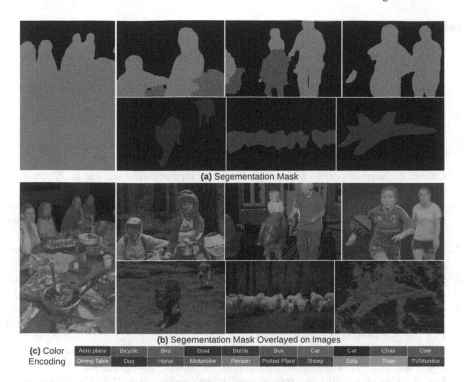

Fig. 5. Qualitative results of our EdgeNeXt segmentation model on unseen COCO validation dataset. The model is trained on Pascal VOC dataset with 20 segmentation classes. **(a)** shows the predicted semantic segmentation mask where 'black 'color represents the background pixels. **(b)** displays the predicted masks on top of original images. **(c)** represents the color encodings for all Pascal VOC classes for the displayed segmentation masks. Our model provides high-quality segmentation masks on unseen COCO images. (Color figure online)

7   Conclusion

The success of the transformer models comes with a higher computational overhead compared to CNNs. Self-attention operation is the major contributor to this overhead, which makes vision transformers slow on the edge devices compared to CNN-based mobile architectures. In this paper, we introduce a hybrid design consisting of convolution and efficient self-attention based encoders to jointly model local and global information effectively, while being efficient in terms of both parameters and MAdds on vision tasks with superior performance compared to state-of-the-art methods. Our experimental results show promising performance for different variants of EdgeNeXt, which demonstrates the effectiveness and the generalization ability of the proposed model.

References

1. Ali, A., et al.: XCiT: cross-covariance image transformers. In: Advances in Neural Information Processing Systems (2021)
2. Ba, J.L., Kiros, J.R., Hinton, G.E.: Layer normalization. arXiv preprint arXiv:1607.06450 (2016)
3. Chen, L.C., Papandreou, G., Schroff, F., Adam, H.: Rethinking atrous convolution for semantic image segmentation. arXiv preprint arXiv:1706.05587 (2017)
4. Chen, Y., et al.: Mobile-former: bridging mobilenet and transformer. In: Proceedings of the IEEE/CVF Conference on Computer Vision and Pattern Recognition (2022)
5. Chollet, F.: Xception: deep learning with depthwise separable convolutions. In: Proceedings of the IEEE/CVF Conference on Computer Vision and Pattern Recognition (2017)
6. Cubuk, E.D., Zoph, B., Shlens, J., Le, Q.: RandAugment: practical automated data augmentation with a reduced search space. In: Advances in Neural Information Processing Systems (2020)
7. Dai, Z., Liu, H., Le, Q., Tan, M.: CoAtNet: marrying convolution and attention for all data sizes. In: Advances in Neural Information Processing Systems (2021)
8. Dosovitskiy, A., et al.: An image is worth 16×16 words: transformers for image recognition at scale. arXiv preprint arXiv:2010.11929 (2020)
9. d'Ascoli, S., Touvron, H., Leavitt, M.L., Morcos, A.S., Biroli, G., Sagun, L.: ConViT: improving vision transformers with soft convolutional inductive biases. In: International Conference on Machine Learning (2021)
10. Everingham, M., Van Gool, L., Williams, C.K., Winn, J., Zisserman, A.: The pascal visual object classes (VOC) challenge. Int. J. Comput. Vis. **88**(2), 303–338 (2010)
11. Fan, H., et al.: Multiscale vision transformers. In: Proceedings of the IEEE/CVF International Conference on Computer Vision (2021)
12. Gao, S.H., Cheng, M.M., Zhao, K., Zhang, X.Y., Yang, M.H., Torr, P.: Res2Net: a new multiscale backbone architecture. IEEE Trans. Pattern Anal. Mach. Intell. **43**(2), 652–662 (2019)
13. Hariharan, B., Arbeláez, P., Bourdev, L., Maji, S., Malik, J.: Semantic contours from inverse detectors. In: Proceedings of the IEEE/CVF International Conference on Computer Vision (2011)
14. He, K., Zhang, X., Ren, S., Sun, J.: Deep residual learning for image recognition. In: Proceedings of the IEEE/CVF Conference on Computer Vision and Pattern Recognition (2016)
15. Hendrycks, D., Gimpel, K.: Gaussian error linear units (GELUs). arXiv preprint arXiv:1606.08415 (2016)
16. Howard, A., et al.: Searching for MobileNetV3. In: Proceedings of the IEEE/CVF International Conference on Computer Vision (2019)
17. Howard, A.G., et al.: MobileNets: efficient convolutional neural networks for mobile vision applications. CoRR abs/1704.04861 (2017)
18. Huang, G., Sun, Yu., Liu, Z., Sedra, D., Weinberger, K.Q.: Deep networks with stochastic depth. In: Leibe, B., Matas, J., Sebe, N., Welling, M. (eds.) ECCV 2016. LNCS, vol. 9908, pp. 646–661. Springer, Cham (2016). https://doi.org/10.1007/978-3-319-46493-0_39
19. Iandola, F.N., Han, S., Moskewicz, M.W., Ashraf, K., Dally, W.J., Keutzer, K.: SqueezeNet: AlexNet-level accuracy with $50\times$ fewer parameters and <0.5 mb model size. arXiv preprint arXiv:1602.07360 (2016)
20. Khan, S., Naseer, M., Hayat, M., Zamir, S.W., Khan, F.S., Shah, M.: Transformers in vision: a survey. ACM Comput. Surv. (CSUR) **54**, 1–41 (2021)
21. Li, Y., et al.: Improved multiscale vision transformers for classification and detection. In: Proceedings of the IEEE/CVF Conference on Computer Vision and Pattern Recognition (2022)

22. Lin, T.-Y., et al.: Microsoft COCO: common objects in context. In: Fleet, D., Pajdla, T., Schiele, B., Tuytelaars, T. (eds.) ECCV 2014. LNCS, vol. 8693, pp. 740–755. Springer, Cham (2014). https://doi.org/10.1007/978-3-319-10602-1_48

23. Liu, W., et al.: SSD: single shot multibox detector. In: Leibe, B., Matas, J., Sebe, N., Welling, M. (eds.) ECCV 2016. LNCS, vol. 9905, pp. 21–37. Springer, Cham (2016). https://doi.org/10.1007/978-3-319-46448-0_2

24. Liu, Z., et al.: Swin transformer: hierarchical vision transformer using shifted windows. In: Proceedings of the IEEE/CVF International Conference on Computer Vision (2021)

25. Liu, Z., Mao, H., Wu, C.Y., Feichtenhofer, C., Darrell, T., Xie, S.: A convnet for the 2020s. In: Proceedings of the IEEE/CVF Conference on Computer Vision and Pattern Recognition (2022)

26. Loshchilov, I., Hutter, F.: SGDR: stochastic gradient descent with warm restarts. In: International Conference on Learning Representations (2017)

27. Loshchilov, I., Hutter, F.: Decoupled weight decay regularization. In: International Conference on Learning Representations (2019)

28. Ma, N., Zhang, X., Zheng, H.-T., Sun, J.: ShuffleNet V2: practical guidelines for efficient CNN architecture design. In: Ferrari, V., Hebert, M., Sminchisescu, C., Weiss, Y. (eds.) Computer Vision – ECCV 2018. LNCS, vol. 11218, pp. 122–138. Springer, Cham (2018). https://doi.org/10.1007/978-3-030-01264-9_8

29. Mehta, S., Rastegari, M.: MobileViT: light-weight, general-purpose, and mobile-friendly vision transformer. In: International Conference on Learning Representations (2022)

30. Mehta, S., Rastegari, M.: Separable self-attention for mobile vision transformers. arXiv preprint arXiv:2206.02680 (2022)

31. Mehta, S., Rastegari, M., Shapiro, L., Hajishirzi, H.: ESPNetv2: a light-weight, power efficient, and general purpose convolutional neural network. In: Proceedings of the IEEE/CVF Conference on Computer Vision and Pattern Recognition (2019)

32. Polyak, B.T., Juditsky, A.B.: Acceleration of stochastic approximation by averaging. SIAM J. Control Optim. **30**, 838–855 (1992)

33. Ridnik, T., Ben-Baruch, E., Noy, A., Zelnik-Manor, L.: ImageNet-21k pretraining for the masses. arXiv preprint arXiv:2104.10972 (2021)

34. Ridnik, T., Lawen, H., Ben-Baruch, E., Noy, A.: Solving imagenet: a unified scheme for training any backbone to top results. arXiv preprint arXiv:2204.03475 (2022)

35. Russakovsky, O., et al.: ImageNet large scale visual recognition challenge. Int. J. Comput. Vis. **115**(3), 211–252 (2015). https://doi.org/10.1007/s11263-015-0816-y

36. Sandler, M., Howard, A., Zhu, M., Zhmoginov, A., Chen, L.C.: MobileNetv 2: inverted residuals and linear bottlenecks. In: Proceedings of the IEEE/CVF Conference on Computer Vision and Pattern Recognition (2018)

37. Schmidhuber, J.: Deep learning in neural networks: an overview. Neural Netw. **61**, 85–117 (2015)

38. Srinivas, A., Lin, T.Y., Parmar, N., Shlens, J., Abbeel, P., Vaswani, A.: Bottleneck transformers for visual recognition. In: Proceedings of the IEEE/CVF Conference on Computer Vision and Pattern Recognition (2021)

39. Tan, M., et al.: MnasNet: platform-aware neural architecture search for mobile. In: Proceedings of the IEEE/CVF Conference on Computer Vision and Pattern Recognition (2019)

40. Touvron, H., Cord, M., Douze, M., Massa, F., Sablayrolles, A., Jégou, H.: Training data-efficient image transformers & distillation through attention. In: International Conference on Machine Learning (2021)

41. Vaswani, A., et al.: Attention is all you need. In: Advances in Neural Information Processing Systems (2017)

42. Xiao, T., Dollar, P., Singh, M., Mintun, E., Darrell, T., Girshick, R.: Early convolutions help transformers see better. In: Advances in Neural Information Processing Systems (2021)

43. Yu, W., et al.: MetaFormer is actually what you need for vision. In: Proceedings of the IEEE/CVF Conference on Computer Vision and Pattern Recognition (2021)
44. Yuan, L., et al.: Tokens-to-token ViT: training vision transformers from scratch on imagenet. In: Proceedings of the IEEE/CVF International Conference on Computer Vision (2021)
45. Zeiler, M.D., Fergus, R.: Visualizing and understanding convolutional networks. In: Fleet, D., Pajdla, T., Schiele, B., Tuytelaars, T. (eds.) ECCV 2014. LNCS, vol. 8689, pp. 818–833. Springer, Cham (2014). https://doi.org/10.1007/978-3-319-10590-1_53
46. Zhang, H., Hu, W., Wang, X.: EdgeFormer: improving light-weight convnets by learning from vision transformers. arXiv preprint arXiv:2203.03952 (2022)
47. Zhang, X., Zhou, X., Lin, M., Sun, J.: ShuffleNet: an extremely efficient convolutional neural network for mobile devices. In: Proceedings of the IEEE/CVF Conference on Computer Vision and Pattern Recognition (2018)
48. Zhou, J., Wang, P., Wang, F., Liu, Q., Li, H., Jin, R.: ELSA: enhanced local self-attention for vision transformer. arXiv preprint arXiv:2112.12786 (2021)

Continual Inference: A Library for Efficient Online Inference with Deep Neural Networks in PyTorch

Lukas Hedegaard$^{(\boxtimes)}$ and Alexandros Iosifidis

Department of Electrical and Computer Engineering,
Aarhus University, Aarhus, Denmark
{lhm,ai}@ece.au.dk

Abstract. We present *Continual Inference*, a Python library for implementing Continual Inference Networks (CINs), a class of Neural Networks designed for redundancy-free online inference. This paper offers a comprehensive introduction and guide to CINs and their implementation, as well as best-practices and code examples for composing basic modules into complex neural network architectures that perform online inference with an order of magnitude less floating-point operations than their non-CIN counterparts. Continual Inference provides drop-in replacements of PyTorch modules and is readily downloadable via the Python Package Index and at www.github.com/lukashedegaard/continual-inference.

Keywords: Online inference · Continual Inference Network · Deep Neural Network · Python · PyTorch · Library

1 Introduction

Designing and implementing Deep Neural Networks, which offer good performance in online inference scenarios, is an important but overlooked discipline in Deep Learning and Computer Vision. Research in areas such as Human Activity Recognition focuses heavily on improving accuracy on select benchmark datasets with limited focus on computational complexity and still less on efficient online inference capabilities. Yet, important real-life applications such as human monitoring [13,18], driver assistance [3], and autonomous vehicles depend on performing predictions on a continual input stream with low latency and low energy consumption.

Continual Inference Networks (CINs) [7–9], are a recent family of Deep Neural Networks, which can accelerate a wide range of architectures for time-series processing (e.g., CNNs and Transformers) during online inference, even though source networks may have been trained exclusively for offline processing.

This paper provides comprehensive introduction to CINs (Sect. 2), the guiding principles of their design and implementation via the Continual Inference library (Sect. 4), and summarizes and compares achieved reductions in stepwise computational complexity and memory-usage using the library (Sect. 4).

© The Author(s), under exclusive license to Springer Nature Switzerland AG 2023
L. Karlinsky et al. (Eds.): ECCV 2022 Workshops, LNCS 13807, pp. 21–34, 2023.
https://doi.org/10.1007/978-3-031-25082-8_2

2   Continual Inference Networks

Originally introduced in [9] and subsequently elaborated in [7,8], Continual Inference Networks denote a variety of Neural Network, which can operate without redundancy during online inference on a continual input stream, as well as offline during batch inference. Specifically, CINs comply with Definition 1 [7]:

Definition 1 (Continual Inference Network). *A Continual Inference Network is a Deep Neural Network, which*

- *is capable of continual step inference without computational redundancy,*
- *is capable of batch inference corresponding to a non-continual Neural Network,*
- *produces identical outputs for batch inference and step inference given identical receptive fields,*
- *uses one set of trainable parameters for both batch and step inference.*

Many prior networks can be viewed as CINs. This includes networks, which perform their task within a single time-step (e.g., object detection and image recognition models), or which inherently process temporal data step-by-step (e.g., Recurrent Neural Networks such as LSTMs [10] and GRUs [2]). Some network types, however, are limited to batch inference exclusively. These include Convolutional Neural Networks (CNNs) with temporal convolutional components (e.g., 3D CNNs), as well as Transformers with tokens spanning the temporal dimension. While they can in principle be used for online inference, it is an inefficient process, where input steps are assembled to full (spatio-)temporal batches and fed to the network in a sliding window fashion, with many redundant intermediary computations as a result.

While some specialty architectures have been devised to let 3D convolutional network variants make predictions step by step [11,16], and accordingly also qualify as CINs, these were not weight-compatible with regular 3D CNNs. Recently, *Continual* 3D Convolutions [9] changed this. Through a reformulation of the 3D convolution to compute outputs for each time-step individually rather than for the whole spatio-temporal input at once, well-performing 3D CNNs such as X3D [4], Slow [5], and I3D [1] trained for Trimmed Activity Recognition were re-implemented to execute step by step without any re-training. Likewise, Spatio-temporal Graph Convolutional Networks for Skeleton-based Action Recognition [14,15,20], which originally operated only on complete sequences of skeleton graphs, were transformed to perform stepwise inference as well though a continual formulation of their Spatio-temporal Graph Convolution blocks [8]. Temporal Transformer networks had also been restricted to operate on batches until a *Continual* Multi-head Attention (*Co*MHA) [7] was introduced, which is weight-compatible with the original MHA [19], while being able to compute updated outputs for each time step.

With these innovations, many existing DNNs can be converted to operate efficiently during online inference. In general, non-continual networks, which are transformed to continual ones attain reductions in per-step computational complexity in proportion to the temporal receptive field of the network. In some cases, these savings can amount to multiple orders of magnitude [8]. Still, the implementation of Continual Inference Networks with temporal convolutions and Multi-head Attention in frameworks such as PyTorch [12] requires deep knowledge and practical experience with CINs. With the Continual Inference library described in the next section, we hope to change this.

3   Library Design

3.1   Principles

The fundamental feature of CINs, that networks are flexible and perform well on both online inference and batch inference, is a guiding principle in the design of the Continual Inference library as well: Refactoring of existing implementations in pure PyTorch should be straightforward. In the following, we will adopt the Python import abbreviations `import continual as co` and `from torch import nn`. The library follows Principle 1 to ensure that `co` modules can be used as drop-in replacements for `nn` modules without behavior change:

Principle 1 (Compatibility with PyTorch). *`co` modules with identical names to `nn` modules also have:*

1. *identical* `forward`*,*
2. *identical model weights,*
3. *identical or extended constructors,*
4. *identical or extended supporting functions.*

Before proceeding to the enhanced functionality of `co` modules, let us state our assumption to the input format:

Assumption 1 (Order of input dimensions). *Inputs to `co` modules use the order* $(B, C, T, S_1, S_2, ...)$ *for multi-step inputs and* $(B, C, S_1, S_2, ...)$ *for single-step inputs, where B is the batch size, C is the input channel size, T is the temporal size, and S_n are additional optional dimensions.*

The core difference between Continual Inference Networks and regular networks is their ability to efficiently compute results for each time-step. Besides the regular `forward` function found in `nn` modules, `co` modules add multiple call modes that allow for continual inference with a simple interface:

Principle 2 (Call modes). `co` *modules provide three forward operations:*

1. `forward`*: takes a (spatio-) temporal input and operates identically to the* `forward` *of an* `nn` *module,*
2. `forward_step`*: takes a single time-step as input without a time-dimension and produces an output corresponding to* `forward`*, had it's input been shifted by one time-step, given identical prior inputs.*
3. `forward_steps`*: takes multiple time-steps as input and produces outputs identical to applying* `forward_step` *the number of times corresponding to the temporal size of the input.*

Furthermore, the __call__ method of co *modules can be changed to use any of the three by either setting the* call_mode *attribute of the module or applying the* co.call_mode() *context with a string spelling out the wanted forward type.*

Let us exemplify Principle 2 in practice. Example 1.1 shows how the different forward functions introduced in Principle 2.1 can be used. Principle 2.2 is illustrated in Example 1.2.

```
import torch
import continual as co

con = co.Conv3d(in_channels=4,
                out_channels=8,
                kernel_size=3)
assert con.delay == 2
assert con.receptive_field == 3

reg = torch.nn.Conv3d(in_channels=4,
                      out_channels=8,
                      kernel_size=3)
# Reuse weights
con.load_state_dict(reg.state_dict())

x = torch.randn((2, 3, 5, 6, 7))   # B,C,T,H,W
y = con.forward(x)
assert torch.equal(y, reg.forward(x))

# Multiple steps
firsts = con.forward_steps(x[:, :, :4])
assert torch.allclose(firsts, y[:, :, : con.delay])

# Single step
last = con.forward_step(x[:, :, 4])
assert torch.allclose(last, y[:, :, con.delay])
```

Example 1.1. Definition and usage of co.Conv3d and its forward modes.

```
net(x)            # Invokes 'forward' by default

net.call_mode = "forward_step"
net(x[:, :, 0])   # Invokes 'forward_step'

with co.call_mode("forward_steps"):
    net(x)        # Invokes 'forward_steps'

net(x[:, :, 0])   # Invokes 'forward_step' again
```

Example 1.2. Changing the call_mode for a continual module net.

Continual modules, which use information from multiple time-steps, are inherently stateful. Whenever forward_step or forward_steps is invoked, intermediary results needed for future step results are optimistically computed and stored. Principle 3 states the rules for state-manipulation and updates.

Principle 3 (State). *Module state is updated according to the following rules:*

- forward_step *and* forward_steps *use and update state by default.*
- *Step results may be computed without updating internal state by passing* update_state=False *to either* forward_step *or* forward_steps.
- forward *neither uses nor updates state.*
- *Module state can be wiped by invoking the* clean_state() *method.*
- *A module produces non-empty outputs after its has conducted a number of stateful forwards steps corresponding to its* delay.

Regular nn modules predominantly operate on input batches in an offline setting and do not have a built-in concept of delay. co modules on the other hand are designed to operate on time-series. Since co modules often integrate information over multiple time-steps and online operation is causal by nature, some modules produce the output corresponding to a given input only after observing additional steps. For instance, a co.Conv1d module with kernel_size = 3 produces an output from the third input step as illustrated in Fig. 1. The delay of a module is calculated according to Principle 4:

Principle 4 (Delay). co *modules produce step outputs that are delayed by*

$$d = f - p - 1 \tag{1}$$

steps relative to the earliest input step used in the computation, where f is the receptive field and p is the temporal padding.

While padding is used in regular networks to retain the size of feature-maps in consecutive layers, this interpretation of temporal padding does not make sense in the context of an infinite, continual input, as handled by CINs. Instead, we may interpret padding as a reduction in delay. For instance, a co.Conv1d module with kernel_size = 3 and padding = 2 has a delay of zero, because the padded zeros already "saturated" the state before-hand. This is illustrated in Fig. 2. Considering, that co modules expect an infinite and continual input stream, end-padding padding is omitted by default. If an end-padding is required, the library supports its use by either passing manually defined zeros as steps or by setting pad_end = True for an invocation of the forward_steps function.

Similar to padding, the stride of a co module impacts the timing of the outputs. Specifically, stride results in empty outputs every $(s - 1)/s$ outputs, as well as larger delays for downstream network modules through increased receptive fields. This is stated in Principles 5 and 6.

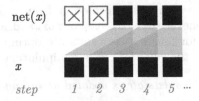

Fig. 1. Sketch of delay and receptive field. Here, the stepwise operation of a co module net with `receptive_field = 3` is illustrated. ■ are non-zero step-features and ⊠ are empty outputs.

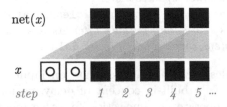

Fig. 2. Sketch of how padding reduces delay. Here, the stepwise operation of a co module net with `receptive_field = 3`, `padding = 2` is illustrated. ⊡ are padded zeros and ■ are non-zero step-features.

Principle 5 (Stride and prediction rate). *For neural network of N modules with strides $s^{(i)}, i \in \{1..N\}$, the accumulated stride at any given layer is*

$$s_{acc}^{(i)} = s^{(i)} \cdot s_{acc}^{(i-1)} \quad i \in 1..N \tag{2}$$

$$s_{acc}^{(0)} = s^{(0)}. \tag{3}$$

Equivalently, the resulting network stride is

$$s_{NN} = \prod_{i=1}^{N} s^{(i)}, \tag{4}$$

and the network prediction rate is

$$r_{NN} = 1/s_{NN}. \tag{5}$$

Accordingly, the outputs of a co network are empty every $(s_{NN} - 1)/s_{NN}$ steps.

Principle 6 (Accumulated delay). *The accumulated receptive field of a downstream module i in a network of N modules is given by:*

$$f_{acc}^{(i)} = f^{(i)} + (f_{acc}^{(i-1)} - 1)s^{(i)}, \quad i \in 1..N \tag{6}$$

$$f_{acc}^{(0)} = f^{(0)}. \tag{7}$$

The accumulated delay of layer i in a network is

$$d^{(i)} = f_{acc}^{(i)} - p_{acc}^{(i)} - 1, \tag{8}$$

where the accumulated padding p_{acc} is given by

$$p_{acc}^{(i)} = p^{(i)} \cdot s_{acc}^{(i-1)}, \quad i \in 1..N, \tag{9}$$

$$p_{acc}^{(0)} = p^{(0)}. \tag{10}$$

Figure 3 illustrates a mixed example, where the first layer of a two-layer network has **padding** = 2 and **stride** = 2. Noting layer attributes in consecutive order, and using Eqs. 2 to 10, the example has the following network attributes:

$$
\begin{aligned}
s &= \{2, \quad 1\} \\
p &= \{2, \quad 0\} \\
s_{acc} &= \{2, \quad 2 \cdot 1 = 2\} \\
p_{acc} &= \{2, \quad 2 + 2 \cdot 0 = 2\} \\
f_{acc} &= \{3, \quad 3 + (3-1) \cdot 2 = 7\} \\
d_{acc} &= \{3 - 2 - 1 = 0, \quad 7 - 2 - 1 = 4\} \\
s_{NN} &= s_{acc}^{(1)} = 2 \\
r_{NN} &= 1/s_{NN} = 1/2 \\
d_{NN} &= d_{acc}^{(1)} = 4.
\end{aligned}
$$

Before continuing onto the specific modules, we have to discuss a final principle of CINs, namely that of parallel modules.

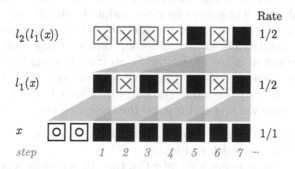

Fig. 3. A mixed example of delay and outputs under padding and stride. Here, we illustrate the stepwise operation of two co module layers, l_1 with **receptive_field** = 3, **padding** = 2, and **stride** = 2 and l_2 with **receptive_field** = 3, no padding and **stride** = 1. ◎ denotes a padded zero, ■ is a non-zero step-feature, and ⊠ is an empty output.

Principle 7 (Parallel modules). *Modules can be arranged in parallel to execute on each their separate stream of data under the following rules:*

- *Parallel modules follow the same global clock.*
- *The delay of a collection of parallel modules is the maximum delay of any module in the collection.*
- *If the merger of parallel step values includes an empty value, then the resulting step output of the merger is also empty.*

A discussion of residual connections provides a practical example for Principle 7.

Residual Connections. The residual connection is a simple but crucial tool for avoiding vanishing and exploding gradients; by adding the input of a module to its output, gradients can flow freely through models with hundreds of layers. Without exaggeration, we can state that almost all recent deep architectures at the time of writing use some form of residual connection [5,6,19,20]. Yet, their implementation in Continual Inference Networks may not follow common intuition in all cases. Let us first consider the residual connection during regular forward operation as found in a non-continual residual shown in Fig. 4a. Here, the wrapped module will almost always use padding to ensure equal input and output shapes (known as "equal padding"). For a module with receptive field three, we would thus have a padding of one. In this case, the forward computation of the residual amounts to adding the input to the output of the convolution. However, the implementation of forward_step illustrated in Fig. 4b is different. Since the first output uses information from the second step, the module has a delay of one. Accordingly, the residual connection requires a delay of one as well.

Now consider the same scenario but without padding. This will be quite foreign to many Deep Learning practitioners, and it is not clear how exactly to align residuals. We will use a separate module to shrink the residual by an equivalent amount as the wrapped module. Of the possible alignment choices, a sensible approach is to discard the border values to align the feature maps on *center*. Contrary to other alignment forms, this has the benefit of weight-compatibility between the no-padding case and the case with equal padding described in the former paragraph. The outputs of step 3 in Figs. 4b and 5b are equal given the same weights and inputs. However, two issues arise:

1. Delay mismatch: While the residual connection has a delay of one, the wrapped module has a delay of two.
2. Mix of empty and non-empty results: C.f. the differences in delay, the residual will start producing non-empty outputs before the wrapped module.

Principle 7 helps us navigate this. Despite the internal delay mismatch, the delay of the whole residual module corresponds to the largest delay, in this case two. Consequently, the whole residual module produces outputs from the third step, despite the fact that the delayed input already has non-empty outputs from the second step. Both of these issues can also be avoided if we force residuals to employ the same delay as the wrapped module. This corresponds to a *lagging* alignment. However, using such a strategy breaks weight compatibility between the same residual modules with and without padding.

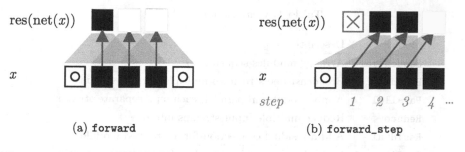

Fig. 4. Residual connections ↑ over a module with receptive field of size ▲ and padding one ("equal padding") ⊡. ⊠ are empty outputs.

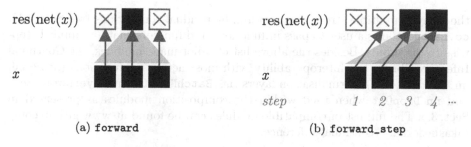

Fig. 5. Centered residual connections ↑ over a module with receptive field of size ▲ and no padding. ⊠ are empty outputs.

3.2   Core Modules

Designed as an augmentation of PyTorch, the Continual Inference library provides a collection of basic building blocks for composing neural networks. Following Principle 1, we use the same public interfaces as PyTorch, i.e. class constructor, function names and arguments, and attribute names, to ensure that co modules can be used as drop-in replacements for nn modules. The basic modules can be categorized as follows:

- Convolutions [9]: co.Conv1d, co.Conv2d, ...
- Pooling: co.AvgPool1d, co.MaxPool1d, ...
- Linear: co.Linear.
- Transformer [7]: co.TransformerEncoder, ...
- Shape: co.Delay, co.Reshape.
- Arithmetic: co.Lambda, co.Add, ...

Here, the MultiheadAttention implementation is a special case, which features two distinct versions of continual operation: 1) "single-output", where only the attention output corresponding to the latest input is produced, and 2) "retrospective", where updates to prior outputs are also produced retrospectively. The details of this are explained in greater detail in [7]. Linear co modules follow the nn modules closely, but ensure compatibility of dimension c.f. Assumption 1. co.Delay adds a specified delay to the input stream. This is handy for aligning

Table 1. Composition modules.

Module	Description
Sequential	Arrange modules sequentially
Broadcast	Broadcast one stream to multiple parallel streams
Parallel	Apply modules in parallel, each on a separate stream
Reduce	Reduce multiple input streams into one
Residual	Add a residual connection for a wrapped module
Conditional	Conditionally invoke a module (or another) at runtime

the delay of multiple streams as required by residual connections (see Sect. 3.1). co.Lambda allows a user to pass in functions and functors that are applied stepwise to the inputs. Besides the above list of tailor-made modules, the Continual Inference library has interoperability with most activation functions (nn.ReLU, nn.Softmax, etc.), normalisation layers (nn.BatchNorm1d, nn.LayerNorm, etc.), and nn.DropOut when used within the composition modules as presented in Sect. 3.3. The full list of compatible modules can be found at www.github.com/lukashedegaard/continual-inference.

3.3 Composition Modules

In PyTorch, modules are composed by either by using the nn.Sequential container or by creating a new class which inherits from nn.Module and manually controls data flow within the forward function. While the latter is commonly used to handle complex modules in a simple and easily debuggable manner, it is not necessarily the simplest approach for implementing complex Continual Inference Networks. In addition to defining the basic forward flow, a CIN implementation also needs to handle stepwise computations, which require meticulous alignment of delays if Principle 2 is to be kept.

Instead, we expand the container interface of PyTorch to include modules for parallel and conditional processing (Table 1). While each module is simple in nature, they can be used to compose complex neural network architectures, which retain all the principles in Sect. 3.1 without explicitly needing to consider them. A brief overview and description of each co container module is given in Sect. 3.3. To get a practical understanding of these, we will give implementation examples of two common architecture blocks, the residual connection as discussed in Sect. 3.1 and an Inception module [17].

Example 1.3 shows three equivalent implementations of a residual 3D convolution block. res1 is the verbose version, in which co.Broadcast is used to split a single input into two parallel stream, co.Parallel specifies that conv handles the first stream, while a delay is used on the second. co.Reduce merges the streams via an add reduce operation. Due to the commonality of broadcast-apply-reduce operations, the library features a co.BroadcastReduce shorthand to specify such composition more succinctly. Even shorter, co.Residual can

```
conv = co.Conv3d(1, 1, kernel_size=3, padding=1)

res1 = co.Sequential(co.Broadcast(2),
                     co.Parallel(conv, co.Delay(1)),
                     co.Reduce("sum"))

res2 = co.BroadcastReduce(conv, co.Delay(1))

res3 = co.Residual(conv)
```

Example 1.3. Equivalent implementations of a residual block.

automatically infer the needed delay from the module it wraps. Other reduction functions can be specified in co.BroadcastReduce and co.Residual using the reduce argument, which is "sum" by default. The code in Example 1.3 correspond to Fig. 4. The centered residual module in Fig. 5 is easily specified as co.Residual(conv, residual_shrink=True) where conv has padding = 0.

```
def norm_relu(conv):
    return co.Sequential(conv,
                         nn.BatchNorm3d(conv.out_channels),
                         nn.ReLU())

inception_module = co.BroadcastReduce(
    co.Conv3d(192, 64, 1),
    co.Sequential(
        norm_relu(co.Conv3d(192, 96, 1)),
        norm_relu(co.Conv3d(96, 128, 3, padding=1)),
    ),
    co.Sequential(
        norm_relu(co.Conv3d(192, 16, 1)),
        norm_relu(co.Conv3d(16, 32, 5, padding=2))
    ),
    co.Sequential(
        co.MaxPool3d(kernel_size=(1, 3, 3),
                     padding=(0, 1, 1),
                     stride=1),
        norm_relu(co.Conv3d(192, 32, 1)),
    ),
    reduce="concat",
)
```

Example 1.4. *Continual* Inception module using a mix of co and nn modules.

We can showcase a more advanced application of parallel streams by considering an Inception module [17]. An Inception module broadcasts the input into four streams and applies convolution of varying kernel sizes in parallel before concatenating the channels to produce one output. Without the co container modules, it would be complicated to keep track of and align delays of the different branches to create valid `forward`, `forward_step`, and `forward_steps` methods. Using `co.Sequential`, which automatically sums up delays, and `co.BroadcastReduce`, which automatically adds delays to match the branch with highest inherent delay, the implementation becomes simple as shown in Example 1.4.

Table 2. Dataset performance, parameter count, maximum allocated memory (Max mem.), and floating-point operations (FLOPs) of continual and non-continual models on video and spatio-temporal graph classification datasets. Subscript$_{xx}$ denotes expanded temporal average pooling, b1 and b2 denote one and two block transformer decoders, and superscript* indicates architectures where network stride was reduced to one. Parentheses show the improvement/deterioration of the continual model relative to the corresponding non-continual model. The noted metrics were originally presented in [7–9].

Model	Dataset performace (%)		Params (M)	Max mem. (MB)	FLOPs (G)
	Kinetics-400 (Acc.)				
X3D-L	69.3		06.2	240.7	19.17
CoX3D-L$_{64}$	71.6 (+2.3)		06.2	184.4 0(75%)	01.25 0($\downarrow$ 15.34$\times$)
X3D-M	67.2		03.8	126.3	04.97
CoX3D-M$_{64}$	71.0 (+3.8)		03.8	069.0 0(55%)	00.33 0($\downarrow$ 15.06$\times$)
X3D-S	64.7		03.8	061.3	02.06
CoX3D-S$_{64}$	67.3 (+2.6)		03.8	042.0 0(69%)	00.17 0($\downarrow$ 12.12$\times$)
Slow-8$\times$8	67.4		32.5	266.0	54.87
CoSlow$_{64}$	73.1 (+5.7)		32.5	176.4 0(66%)	06.90 00($\downarrow$ 7.95$\times$)
I3D	64.0		28.0	191.6	28.61
CoI3D$_8$	59.6 (−4.4)		28.0	235.9 (123%)	05.68 00($\downarrow$ 5.04$\times$)
	THUMOS14 (mAP)	TVSeries (mcAP)			
OadTR-b2	64.2	89.0	15.9	067.6	01.08
CoOadTR-b2	64.4 (+0.2)	88.2 (−0.8)	15.9	071.7 (106%)	00.41 00($\downarrow$ 2.61$\times$)
OadTR-b1	64.4	89.1	09.6	043.3	00.67
CoOadTR-b1	64.5 (+0.1)	88.0 (−1.1)	09.6	045.1 (104%)	00.01 0($\downarrow$ 63.49$\times$)
	NTU RGB+D 60 (Acc.)				
	X-Sub	X-View			
ST-GCN	86.0	93.4	03.1	045.3	16.73 $\downarrow$ 000.0$\times$
CoST-GCN*	86.3 (+0.3)	93.8 (+0.4)	03.1	036.1 0(80%)	00.16 ($\downarrow$ 107.7$\times$)
AGCN	86.4	94.3	03.5	048.4	18.69 $\downarrow$ 000.0$\times$
CoAGCN*	84.1 (−2.3)	92.6 (−1.7)	03.5	037.4 0(77%)	00.17 ($\downarrow$ 108.8$\times$)
S-TR	86.8	93.8	03.1	074.2	16.14 $\downarrow$ 000.0$\times$
CoS-TR*	86.3 (−0.3)	92.4 (−1.4)	03.1	036.1 0(49%)	00.15 ($\downarrow$ 107.6$\times$)

4 Performance Comparisons

Using the basic co modules and composition building blocks, continual versions of advanced neural networks have been implemented in multiple recent works with manyfold speedups and significant reductions in memory consumption during online inference [7–9]. Specifically, the 3D-CNNs *Co*X3D, *Co*I3D, and *Co*Slow for video-based Human Activity Recognition were proposed in [9]; the Transformer *Co*OadTR for Online Action Detection in [7]; and Spatio-temporal Graph Convolutional Networks *Co*ST-GCN, *Co*AGCN, and *Co*S-TR for Skeleton-based Action Recognition in [8]. While direct conversion from regular to continual versions of the above noted architectures works well in accelerating inference in itself, further improvements can be achieved by exploiting some core characteristics of CINs: in [9], accuracy was improved by increasing model receptive fields through expansions of temporal global average pooling to 64 steps, and in [8], the stride of temporal convolutions was reduced to one to increase prediction rates. Table 2 presents a summary of benchmark performance, computational complexity, and maximum allocated memory on GPU for each of these networks alongside with their non-continual counterparts [7–9].

5 Conclusion

We presented Continual Inference, an easy-to-use library for implementing Continual Inference Networks in Python. Following interfaces closely, the components provided in the library are backwards-compatible drop-in replacements for PyTorch modules, which add the capability of redundancy-free online inference without the need for intimate knowledge of CINs nor their meticulous low-level implementation. Having shown the vast computational advantages of CINs over regular neural networks in multiple settings of video and spatio-temporal graph classification, we hope that this library will contribute to the adoption of CINs and the advancement of use-cases requiring low-latency online inference under recourse constraints in general.

Acknowledgement. This work has received funding from the European Union's Horizon 2020 research and innovation programme under grant agreement No 871449 (OpenDR).

References

1. Carreira, J., Zisserman, A.: Quo Vadis, action recognition? A new model and the kinetics dataset. In: IEEE/CVF Conference on Computer Vision and Pattern Recognition (CVPR), pp. 4724–4733 (2017)
2. Cho, K., van Merriënboer, B., Bahdanau, D., Bengio, Y.: On the properties of neural machine translation: encoder-decoder approaches. In: Proceedings of SSST-8, Eighth Workshop on Syntax, Semantics and Structure in Statistical Translation, pp. 103–111 (2014)

3. Enkelmann, W.: Video-based driver assistance-from basic functions to applications. Int. J. Comput. Vis. (IJCV) **45**(3), 201–221 (2001)
4. Feichtenhofer, C.: X3D: expanding architectures for efficient video recognition. In: IEEE/CVF Conference on Computer Vision and Pattern Recognition (CVPR) (2020)
5. Feichtenhofer, C., Fan, H., Malik, J., He, K.: SlowFast networks for video recognition. In: IEEE/CVF International Conference on Computer Vision (ICCV), pp. 6201–6210 (2019)
6. He, K., Zhang, X., Ren, S., Sun, J.: Deep residual learning for image recognition. In: IEEE Conference on Computer Vision and Pattern Recognition (CVPR), pp. 770–778 (2016)
7. Hedegaard, L., Bakhtiarnia, A., Iosifidis, A.: Continual transformers: redundancy-free attention for online inference. In: International Conference on Learning Representations (ICLR) (2023)
8. Hedegaard, L., Heidari, N., Iosifidis, A.: Online skeleton-based action recognition with continual spatio-temporal graph convolutional networks. Preprint arXiv:2203.11009 (2022)
9. Hedegaard, L., Iosifidis, A.: Continual 3D convolutional neural networks for real-time processing of videos. In: Avidan, S., Brostow, G., Cissé, M., Farinella, G.M., Hassner, T. (eds.) ECCV 2022. LNCS, vol. 13664, pp. 369–385. Springer, Cham (2022). https://doi.org/10.1007/978-3-031-19772-7_22
10. Hochreiter, S., Schmidhuber, J.: Long short-term memory. Neural Comput. **9**, 1735–1780 (1997)
11. Köpüklü, O., Hörmann, S., Herzog, F., Cevikalp, H., Rigoll, G.: Dissected 3D CNNs: temporal skip connections for efficient online video processing. Preprint arXiv:2009.14639 (2020)
12. Paszke, A., et al.: PyTorch: an imperative style, high-performance deep learning library. In: Advances in Neural Information Processing Systems, vol. 32, pp. 8024–8035. Curran Associates, Inc. (2019)
13. Pigou, L., van den Oord, A., Dieleman, S., Van Herreweghe, M., Dambre, J.: Beyond temporal pooling: recurrence and temporal convolutions for gesture recognition in video. Int. J. Comput. Vis. (IJCV) **126**(2), 430–439 (2018)
14. Plizzari, C., Cannici, M., Matteucci, M.: Skeleton-based action recognition via spatial and temporal transformer networks. Comput. Vis. Image Underst. **208**, 103219 (2021)
15. Shi, L., Zhang, Y., Cheng, J., Lu, H.: Two-stream adaptive graph convolutional networks for skeleton-based action recognition. In: IEEE Conference on Computer Vision and Pattern Recognition, pp. 12026–12035 (2019)
16. Singh, G., Cuzzolin, F.: Recurrent convolutions for causal 3D CNNs. In: IEEE/CVF International Conference on Computer Vision Workshop (ICCVW), pp. 1456–1465 (2019)
17. Szegedy, C., et al.: Going deeper with convolutions. In: IEEE Conference on Computer Vision and Pattern Recognition (CVPR), pp. 1–9 (2015)
18. Tavakolian, M., Hadid, A.: A spatiotemporal convolutional neural network for automatic pain intensity estimation from facial dynamics. Int. J. Comput. Vis. (IJCV) **127**(10), 1413–1425 (2019)
19. Vaswani, A., et al.: Attention is all you need. In: Advances in Neural Information Processing Systems (NeurIPS), vol. 30, pp. 5998–6008 (2017)
20. Yan, S., Xiong, Y., Lin, D.: Spatial temporal graph convolutional networks for skeleton-based action recognition. In: AAAI Conference on Artificial Intelligence, pp. 7444–7452 (2018)

Hydra Attention: Efficient Attention with Many Heads

Daniel Bolya[1,2]([✉]), Cheng-Yang Fu[2], Xiaoliang Dai[2], Peizhao Zhang[2], and Judy Hoffman[1]

[1] Georgia Tech, Atlanta, USA
{dbolya,judy}@gatech.edu
[2] Meta AI, Menlo Park, USA
{chengyangfu,xiaoliangdai,stzpz}@fb.com

Abstract. While transformers have begun to dominate many tasks in vision, applying them to large images is still computationally difficult. A large reason for this is that self-attention scales quadratically with the number of tokens, which in turn, scales quadratically with the image size. On larger images (e.g., 1080p), over 60% of the total computation in the network is spent solely on creating and applying attention matrices. We take a step toward solving this issue by introducing Hydra Attention, an extremely efficient attention operation for Vision Transformers (ViTs). Paradoxically, this efficiency comes from taking multi-head attention to its extreme: by using as many attention heads as there are features, Hydra Attention is computationally linear in both tokens and features with no hidden constants, making it significantly faster than standard self-attention in an off-the-shelf ViT-B/16 by a factor of the token count. Moreover, Hydra Attention retains high accuracy on ImageNet and, in some cases, actually *improves* it.

Keywords: Vision Transformers · Attention · Token efficiency

1 Introduction

Because of their generality and high capacity to learn from large amounts of data, transformers [32] have been a dominant force in natural language processing (NLP) for the last couple of years [6,17,25]. And now, with the introduction of Vision Transformers (ViTs) [10], the same takeover is happening in vision.

Yet, unlike in NLP, the pure instantiation of transformers that can be seen in NLP with BERT [17] or in vision with ViT [10] are not the force dominating computer vision tasks. Instead, much more vision-specialized attention-based

---

D. Bolya—This work was done under an internship at Meta AI.

Supplementary Information The online version contains supplementary material available at https://doi.org/10.1007/978-3-031-25082-8_3.

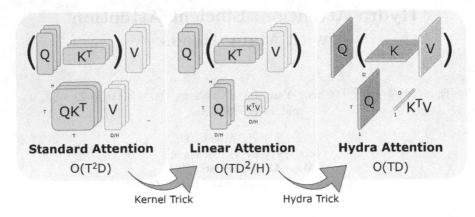

Fig. 1. Hydra Attention. Standard attention [32] scales with the square of the number of tokens T. Using a decomposable kernel, we can rearrange the order of operations as in [16] such that attention scales with the square of features D instead. Our Hydra attention goes one step further by maximizing the number of attention heads H, resulting in an $O(TD)$ operation in both space and time.

architectures such as Swin [21] or MViT [11,20] or attention-conv mixtures like LeViT [13] are being used instead.

The primary reason behind this discrepancy is efficiency: specialized vision transformers can perform better with less compute—either by adding conv layers, by using vision-specific local window attention, or by using some other way to cheaply add visual inductive bias. While pure ViTs can perform well at scale (90.45% top-1 on ImageNet [38]), the primary mechanism of a pure transformer—multihead self-attention [32]—can be an extreme bottleneck when applying a model on the large images required by several downstream tasks.

In fact, when applying an off-the-shelf ViT on 1080p images, common for benchmark tasks such as segmentation (e.g., CityScapes [7]), 60% of the total computation in the network (see Table 4) is spent simply on creating and applying attention matrices for self-attention, compared to 4% on 224 × 224 ImageNet [9] images. In a pure transformer, these attention matrices scale computationally with the square of tokens, which can already be prohibitively expensive (such as with long sentences in NLP). But in a ViT, the problem is compounded further by the tokens scaling with the square of the image size, meaning doubling the image size increases the computation in attention by a factor of 16.

There are already a wealth of techniques that have been explored to address this problem in the NLP space. Several works have introduced "linear" attention (in terms of tokens) either by re-arranging the order of computation using a "kernel trick" [5,16,24,28] or projecting to a token-independent low-rank space [5,24,34], some doing both. However, most of these "linear" attention methods trade computation across the tokens for computation across the features, making them rather expensive. In fact, recently, Flash Attention [8] has shown that an

IO efficient implementation of multihead self-attention can outperform most of these "linear" attention methods even with token counts in the thousands.

A few works have attempted efficient attention in the vision space, too, but none have been explored on their own in a traditional ViT shell. PolyNL [2] treats attention as an efficient third-order polynomial, but this hasn't yet been explored in a ViT architecture. Attention Free Transformer [37] has an AFT-Simple variant that is similarly efficient, but it performs poorly in a pure ViT and requires extra support from convs and position encodings. We test both of these methods in a standard DeiT [31] shell (see Table 1), and find that both methods, while efficient, result in a significant accuracy drop. Thus, there is room in the literature for a truly efficient, accurate, and general replacement for multihead self-attention.

To that extent, we introduce Hydra Attention (see Fig. 1). Our method results from a somewhat paradoxical behavior in linear attention: with standard multi-head self-attention, adding more heads to the model keeps the amount of computation the same. However, after changing the order of operations in linear attention, adding more heads actually *reduces* the compute cost of the layer. We take this observation to its extreme by setting the number of heads in the model to be equal to the number of features, thereby creating an attention module that's computationally linear with respect to both tokens and features.

Not only is Hydra Attention a more general formulation of previous efficient attention works (see Sect. 3.5), but when using the right kernel, it can be significantly more accurate (see Table 1). In fact, when mixed with standard multi-head attention, Hydra Attention can actually *increase* the accuracy of a baseline DeiT-B model while being faster (see Fig. 4). And by being derived from multihead attention, our method retains several of attention's nice properties, such as explainability (see Fig. 3) and generality to different tasks.

However, while Hydra Attention is general and efficient for large images, in this paper we focus solely on ImageNet [9] classification using DeiT-B [31], which traditionally uses smaller 224×224 and 384×384 images. While the efficiency gains aren't as much here (10–27% based on image size), other efficient attention methods (e.g., [2,37]) already suffer from huge accuracy drops in this regime (see Table 1), whereas Hydra Attention does not. We hope Hydra Attention can become a stepping stone for general, pure transformers with large numbers of tokens in the future.

Our contributions are as follows: we perform a study to validate how many heads a transformer can have (Fig. 2) and find that 12 is the limit for softmax attention, but with the right kernel, any number is feasible. Then we use that observation to introduce Hydra Attention (Sect. 3) for pure transformers by increasing the number of heads in multihead self-attention. We then analyze the action of Hydra Attention mathematically (Sect. 3.4) and introduce a method to visualize its focus (Fig. 3). Finally, we find that by replacing specific attention layers with Hydra Attention (Fig. 4), we can either *improve* accuracy by 1% or match the accuracy of the baseline, while producing a strictly faster model using DeiT-B [31] on ImageNet-1k [9].

2   Related Work

In this paper, our goal is to speed up the inference time of a transformer by removing the token squared computation bottleneck in multihead self-attention.

Efficient Attention. Multihead Self-Attention [32] is a notoriously slow operation, and there have been plenty of works trying to address its computational shortcomings in different domains.

In NLP, several works approximate attention with a decomposable kernel function [5,16,24,28]. This "kernel trick" allows them to reorder the matrix multiplications to be quadratic in terms of features instead of tokens. Some of these methods go further and reduce the dimensionality of this matrix multiplication through a projection to a low rank space [5,24,34]. However, these "linear" attention methods trade computation across the tokens for computation across the features, which can make them expensive. In fact, in the domain of this paper (ImageNet classification), there aren't enough tokens to justify these approaches and most of them produce a *slower* model. And even with thousands of tokens, Flash Attention [8] has shown that an IO-aware implementation of multihead self-attention can actually outspeed even the fastest of these methods.

But reordering operations isn't the only way to speed up attention. In fact, the most common way to "linearize" attention in vision is by using local window attention (e.g., [3,19,21]). This is indeed computationally linear with respect to the number of tokens, but local window attention can be difficult to compute (especially in the case of Swin [21]) and this is only possible with dense, spatially ordered modalities such as images and videos.

Our goal is instead to produce a linear attention method that is efficient, fast to compute, and general across several different modalities.

Efficient Transformers. Replacing the attention module is not the only way to speed up the inference time of a transformer. In fact, depending on the task and the number of tokens, other efficient transformer methods can be more desirable. For instance, attention only accounts for 4% of the total network computation on ImageNet [9] classification, meaning 4% is the maximum obtainable speed-up if only attention is modified.

There are several efficient vision transformers that mix convs and attention together to create a more efficient end product, such as LeViT [13], MobileViT [22], Mobile-Former [4], and LVT [35]. All of these are a valid strategy for images, and we view them as adjacent techniques. Other vision-specific attention papers such as [2,37] use convolutions in addition to their efficient attention, making it difficult to discern whether the improvement comes from the attention method or the introduction of convolution.

In this paper, we make no modifications to the underlying ViT architecture except to swap multihead self-attention for Hydra Attention in order to clearly isolate its impact on performance.

Multihead Attention. Hydra Attention relies on increasing the number of heads used in multihead attention. Interestingly enough, since its introduction in [32], the number of heads used for multihead attention has not been explored in much depth. Some studies have been done on pruning attention heads [23,33], however all studies have been in the direction in reducing the number of heads. In fact, even with ViT-G, the largest ViT models explored in [38], the authors only use 16 attention heads. Thus, we conduct this study ourselves in Fig. 2.

3   Hydra Attention

Standard multihead self-attention [32] scales quadratically with the number of tokens in an image. More concretely, if T is the number of tokens and D is the number of feature dimensions, then creating and applying an attention matrix are both $O(T^2 D)$. This poses a problem, then, when T is large (as it is the case with large images), as this operation can quickly become computationally infeasible.

3.1   The Kernel Trick

As discussed in Sect. 2, many works [5,16,24,28] have already attempted to address this by introducing "linear" attention. Given queries Q, keys K, and values V in $\mathbb{R}^{T \times D}$, standard softmax self-attention is computed as

$$A(Q, K, V) = \text{softmax}\left(\frac{QK^T}{\sqrt{D}}\right) V \qquad (1)$$

Computing QK^T is $O(T^2 D)$ and creates a $T \times T$ matrix, which scales poorly with T. As in [16], we can generalize this operation by treating softmax($\cdot$) as a pairwise similarity between Q and K. That is, for some similarity function sim($\cdot$), we can write

$$A(Q, K, V) = \text{sim}(Q, K)V \qquad (2)$$

If we then choose a decomposable kernel with feature representation $\phi(\cdot)$ such that $\text{sim}(x, y) = \phi(x)\phi(y)^T$, we can obtain

$$A(Q, K, V; \phi) = \left(\phi(Q)\phi(K)^T\right) V \qquad (3)$$

Then by associativity, we can change the order of computation such that

$$A(Q, K, V; \phi) = \phi(Q)\left(\phi(K)^T V\right) \qquad (4)$$

This allows us to compute $\phi(K)^T V$ first, leading to an operation that is $O(TD^2)$ and that creates a D^2 matrix instead of a T^2 one. Note this formulation differs slightly from [16], in that we leave the normalization to the similarity function rather than make it explicit.

3.2 Multi-head Attention

Despite being linear with respect to T, the result in Eq. 4 is still undesirable: D is typically large (≥ 768) and so creating a $D \times D$ matrix and performing $O(TD^2)$ operations can still be quite costly. However, Eq. 1 through Eq. 4 assume that we create one attention matrix, and thus have one "head".

In practice, most vision transformers use H heads (typically between 6 and 16), where each head creates and applies its own attention matrix. Following [32], each of heads operate on their own D/H subset of features from Q, K, and V. Thus Eq. 1 becomes

$$A(Q_h, K_h, V_h) = \text{softmax}\left(\frac{Q_h K_h^T}{\sqrt{D}}\right) V_h \qquad \forall h \in \{1, \ldots, H\} \tag{5}$$

where $Q_h, K_h, V_h \in \mathbb{R}^{T \times \frac{D}{H}}$. This keeps the total number of operations the same:

$$O(HT^2 D/H) = O(T^2 D) \tag{6}$$

The same is not true, however, for linear attention. Equation 4 becomes

$$A(Q_h, K_h, V_h; \phi) = \phi(Q_h)\left(\phi(K_h)^T V_h\right) \qquad \forall h \in \{1, \ldots, H\} \tag{7}$$

By computing attention in this way, adding heads actually *decreases* the number of operations:

$$O(HT(D/H)^2) = O(TD^2/H) \tag{8}$$

3.3 Adding Heads

Given Eq. 8, the more heads we add to the network, the faster multihead linear attention becomes. That begs the question, how many heads can we reasonably add, anyway? Most transformers in the wild use between 6 and 16 heads [10,17, 32,38] depending on the number of features D, but what happens if you increase the number of heads beyond that?

To find out, we train DeiT-B [31] on ImageNet-1k [9] and vary the number of heads H using either standard multi-head self-attention (Eq. 5, MSA) with softmax or multi-head linear attention (Eq. 7, MLA) with cosine similarity, plotting the results in Fig. 2. In terms of memory usage, MSA runs out of memory when $H > 96$ and MLA runs out of memory when $H < 3$.

In terms of performance, while the accuracy for MSA tanks for $H > 12$, the accuracy for MLA with cosine similarity stays quite consistent all the way up to $H = 768$. Amazingly, at this number of heads, H is equal to D, meaning each head has only a scalar features to work with!

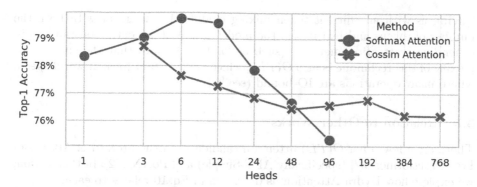

Fig. 2. Varying Heads. We train a DeiT-B model on ImageNet-1k with different numbers of heads using either standard self-attention (blue) using softmax or multi-head linear attention (red) using cosine similarity. Results for standard self-attention ran out of memory for $H > 96$ and multi-head linear attention for $H < 3$. Softmax attention seems to crash in accuracy as we add more heads, while multi-head linear attention stays consistent. Note that H must divide $D = 768$.

3.4 The Hydra Trick

As shown in Fig. 2, it's feasible to scale H up arbitrarily as long as the similarity function $\text{sim}(x, y)$ is not softmax. To exploit this, we introduce the "hydra trick", where we set $H = D$:

$$A(Q_h, K_h, V_h; \phi) = \phi(Q_h)\left(\phi(K_h)^T V_h\right) \qquad \forall h \in \{1, \ldots, D\} \qquad (9)$$

In this case, each Q_h, K_h, V_h is a column vector in $\mathbb{R}^{T \times 1}$. If we then vectorize the operation across the heads, we end up with

$$\text{Hydra}(Q, K, V; \phi) = \phi(Q) \odot \sum_{t=1}^{T} \phi(K)^t \odot V^t \qquad (10)$$

where $\odot$ denotes element-wise multiplication. Note there is a subtle difference between this vectorization and Eq. 9: ϕ is applied to the entirety of Q and K, rather than to individual column vectors Q_h and K_h. This is important because for each token, Q_h and K_h are scalars, and taking the similarity between two scalars is very restrictive (e.g., cosine similarity can only output -1, 0, or +1).

Also, while the derivation of Eq. 10 comes from multihead attention, it actually ends up performing something quite different: it first creates a global feature vector $\sum_{t=1}^{T} \phi(K)^t \odot V^t$ that aggregates information across all the tokens in the image. Then each $\phi(Q)$ gates the importance of this global feature for each output token. Thus, Hydra Attention mixes information through a global bottleneck, rather than doing explicit token-to-token mixing as in standard self-attention.

This results in a computational complexity of

$$O(TD(D/H)) = O(TD) \qquad (11)$$

leaving us with an efficient token mixing module that is linear with both the number of tokens and features in the model, and with no extra constants as in other linear attention methods (such as [5,16,34]). Note that the space complexity of this technique is also $O(TD)$, which is important for real-world speed, where many operations are IO-bound (see [8]).

3.5   Relation to Other Works

There are a few other $O(TD)$ attention candidates in the literature: Attention-Free Transformer [37] (specifically AFT-Simple) and PolyNL [2]. In this section, we explore how Hydra Attention as described in Eq. 10 relates to each.

AFT-Simple [37] is described as

$$\text{AFT-Simple}(Q, K, V) = \sigma(Q) \odot \sum_{t=1}^{T} \text{softmax}(K)^t \odot V^t \tag{12}$$

where $\sigma(\cdot)$ denotes sigmoid. If we allow ϕ to vary between Q and K, this is a direct specilization of Eq. 10 with $\phi(Q) = \sigma(Q)$ and $\phi(K) = \text{softmax}(K)$.

PolyNL [2], on the other hand, is described as

$$\text{PolyNL}(X; W_1, W_2, W_3) = \left(X \odot \frac{1}{T} \sum_{t=1}^{T} XW_1 \odot XW_2 \right) W_3 \tag{13}$$

If we denote $K = XW_1$ and $V = XW_2$, and let $\phi_{\text{mean}}(x) = x/\sqrt{T}$, we can write

$$\text{PolyNL}(X; W_1, W_2, W_3) = \text{Hydra}(X, K, V; \phi_{\text{mean}})W_3 \tag{14}$$

Thus, Hydra attention can be seen as a more general form of other $O(TD)$ attention methods.

4   Experiments

For all experiments, unless otherwise noted, we use DeiT-B [31] with default settings trained on ImageNet-1k [9] reported as Top-1 accuracy on the validation set. When not specified, the function used for $\phi(\cdot)$ in Eq. 10 is L2 normalization such that $\text{sim}(\cdot, \cdot)$ is cosine similarity. To compute throughput, we sweep over several batch sizes and report the highest average throughput on 30 batches after 10 discarded warm-up iterations.

4.1   The Choice of Kernel

In most of our experiments, following [16] we use cosine similarity as our kernel function for Eq. 10. In Table 1, we explore other possible kernels, including those used by other candidate attention replacement methods as discussed in Sect. 3.5. Yet, no kernel we test outperforms simple cosine similarity.

Table 1. Kernel Choice. Here we vary the choice of kernel function through its feature representation $\phi(\cdot)$ in Eq. 10. We also compare against AFT and PolyNL here as mentioned in Sect. 3.5. Note that some kernels can be asymmetric, with different $\phi(Q)$ and $\phi(K)$. See the appendix for more kernels.

Method	Kernel	$\phi(Q)$	$\phi(K)$	Accuracy
Hydra	Cosine Similarity	$x/\|x\|_2$		**76.37**
Hydra	Mean	$x/\sqrt{T}$		75.95
Hydra	Tanh Softmax	$\tanh(x)$	$\text{softmax}(x)$	74.18
AFT-Simple [37]	Sigmoid Softmax	$\sigma(x)$	$\text{softmax}(x)$	74.02
PolyNL [2]	Mean	$x/\sqrt{T}$		73.96

This might be because cosine similarity changes the nature of attention. With MSA (Eq. 5), attention exclusively mixes information contained in V, as the mixing weights $\text{sim}(Q, K)$ must sum to 1. That's not the case when using cosine similarity or other unrestricted dot-product kernels like mean. And it turns out, these weights summing to 1 might not be a desirable property in the first place: AFT-Simple [37] as described in Eq. 12 sets $\phi(Q) = \sigma(Q)$ and $\phi(K) = \text{softmax}(K)$, which is closer to a strict mixing of V, but the performance suffers as a result (see Table 1).

We also test using $\tanh(Q)$ instead of $\sigma(Q)$ to see if cosine similarity allowing the result to be negative was the reason, but that performs only slightly better than AFT-Simple. Thus, in this computationally constrained environment, it seems that leaving the kernel to be as unrestricted as possible while normalizing it in some way is important. We test several other kernels and note them in the appendix, but none outperform this simple technique.

4.2 Visualizing Hydra Attention

One of the most desirable qualities of self-attention is its explainability: visualizing the focus of an attention-based model (e.g. with attention rollout [1]) is typically straightforward. The same is less true for Hydra attention.

In order to visualize the focus of a Hydra attention module, we could construct attention matrices $\phi(Q)_h \phi(K)_h^T$ for $h \in \{1, \ldots, D\}$, but each would be rank 1 and it isn't clear how to combine D different attention matrices when each is responsible for a different feature dimension. Simply averaging the heads together produces a meaningless result because each feature dimension encodes different information.

Instead, let's look at the information that each token contributes to the output for the class token. If we sample just the class token c's output from Eq. 10, we get

$$\phi(Q)^c \odot \sum_{t=1}^{T} \phi(K)^t \odot V^t = \sum_{t=1}^{T} \phi(Q)^c \odot \phi(K)^t \odot V^t \tag{15}$$

Fig. 3. Hydra Attention Visualization. Visualization of the class token's Hydra attention in the last layer as specified in Sect. 4.2. The 4 images on the left are predicted correctly, while the two examples on the right are misclassified. In the top right image, the network focuses on the head of the wrong dog, guessing the wrong breed. Then on the bottom right, the network misses the bird completely. See the Appendix for more examples.

Thus, each token t has a contribution to the output of the class token c given by

$$\phi(Q)^c \odot \phi(K)^t \odot V^t \tag{16}$$

To tell how this relates to the final prediction, we can use a method similar to Grad-CAM [27]: set the loss to be the logit for the predicted class, then obtain the gradient g with respect to the output of the Hydra attention layer. Then the contribution of each token along the direction of that gradient is

$$(\phi(Q)^c \odot \phi(K)^t \odot V^t)^T g \tag{17}$$

We plot this quantity for several different images in Fig. 3 and the Appendix. For these visualizations, we normalize Eq. 17 along the tokens and show the positive values. These focus maps show that while the math might be different, Hydra attention is performing much the same function as standard self-attention.

4.3 Which Layers Can We Replace?

As discussed in Sect. 3.4 and Sect. 4.1, Hydra Attention with a cosine similarity kernel mixes information between tokens in a different way to standard MSA [32]. Thus, it is perhaps unreasonable to replace every attention layer in the network with Hydra attention. In fact, Hydra attention creates a global feature from the tokens and applies that to each token weighted by Q. Because this is a global operation, it would make more sense in the later layers of the network, as at that point information has already been mixed locally. We test this in Fig. 4, where we progressively replace the MSA attention layers in DeiT-B with Hydra attention following different strategies.

In this experiment, we observe that if we start replacing from the first layer of the network, the performance of the model quickly degrades. However, as it

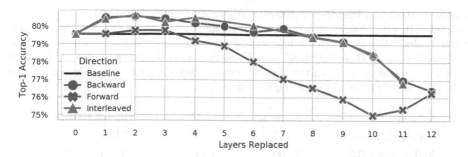

Fig. 4. Which layers can we replace? Replacing softmax self-attention with Hydra attention using different replacement strategies: from the front, from the back, or by interleaving the layers. In all cases, 0 indicates no layers replaced (the baseline), and 12 indicates that all layers were replaced. Surprisingly, with the right layer replacement strategy, Hydra attention can actually *improve* accuracy on ImageNet by 1%, while being faster. Alternatively, we can replace up to 8 layers with no accuracy drop.

turns out, if we replace the layers in reverse starting with the last layer, we can actually *improve* the accuracy of the model. And this improvement is so great that we can replace the last 8 layers of the network and still match the accuracy of the baseline DeiT-B model.

Then, if Hydra attention can be complementary with standard softmax attention, perhaps the best way to combine the two is to interleave them. In Fig. 4, we also attempt to alternate MSA and Hydra layers following the principle that Hydra attention layers should follow MSA layers. However, we don't observe much tangible benefit to this interleaving strategy over starting at the back, suggesting that the number, not necessary the placement, of Hydra layers is what's important.

Note that other efficient attention methods such as AFT [37] and UFO-ViT [29] add conv layers instead of interspersing regular attention layers. Adding these convs serves much the same purpose as using self-attention to perform local mixing, but it's not clear whether the benefit of these prior methods come from the conv layers or their proposed attention layer. In this case, we've clearly isolated that Hydra attention can not only benefit the speed of the model, but also its performance. Future work may be interested in using convs instead.

4.4 Results

We present our final accuracy and FLOP count using Hydra attention in Table 2 compared to standard $O(T^2D)$ attention and other $O(TD)$ methods on ImageNet-1k. Hydra attention achieves 2.4% higher accuracy compared to other $O(TD)$ methods when replacing all layers. And when replacing fewer layers, Hydra attention can strictly outperform the baseline standard attention model: with 2 layers, accuracy increases by 1.1% at 0.7% reduced FLOPs and 2.3% increase in throughput, and with 8 layers, accuracy stays the same with 2.7% reduced FLOPs and 6.3% faster throughput. Interestingly enough, the actual

Table 2. Results. Results for different attention methods in a DeiT-B [31] shell on ImageNet-1k [9] val trained on 224px images along with throughput measured on a V100. Hydra attention results in less accuracy drop than other $O(TD)$ attention methods (AFT-Simple [37] and PolyNL [2]). Moreover, if we don't replace every attention layer in the network, Hydra attention can improve accuracy or keep it the same while still reducing FLOPs and increasing throughput.

Method	Accuracy (%)		FLOPs (G)		Speed (im/s)	
Standard Attention [32]	79.57		17.58		314.8	
AFT-Simple [37]	74.02	(-5.55)	**16.87**	(-4.0%)	346.1	(+9.9%)
PolyNL [2]	73.96	(-5.61)	**16.87**	(-4.0%)	353.8	(+12.4%)
Hydra (2 layers)	**80.64**	(+1.1)	17.46	(-0.7%)	321.9	(+2.3%)
Hydra (8 layers)	79.45	(-0.12)	17.11	(-2.7%)	334.8	(+6.3%)
Hydra (12 layers)	76.40	(-3.17)	**16.87**	(-4.0%)	346.8	(+10.2%)

Table 3. 384px Fine-Tuning. Results for the models in Table 2 fine-tuned with 384px images for 30 epochs. Even with more tokens, Hydra attention can still improve the accuracy over the baseline by 0.59% with 2 layers and increase throughput by 15.4% with 7 layers while matching the baseline's accuracy.

Method	Accuracy (%)		FLOPs (G)		Speed (im/s)	
Standard Attention [32]	81.33		55.54		92.5	
Hydra (2 layers)	**81.92**	(+0.59)	54.52	(-1.8%)	96.3	(+4.1%)
Hydra (7 layers)	81.26	(-0.07)	51.96	(-6.4%)	106.8	(+15.4%)
Hydra (12 layers)	77.85	(-3.48)	**49.40**	(-11.0%)	117.6	(+27.1%)

throughput increase outpaces the flops reduction substantially. This could be due to the observation in [8] that attention is memory-bound and because Hydra Attention uses less memory than standard attention.

Larger Images. To explore whether Hydra Attention retains these gains with more tokens, in Table 3 we fine-tune the backwards replacement models from Fig. 4 at a 384px resolution for 30 epochs using the hyperparameters suggested in [31]. This results in a model with almost 3 times the number of tokens, which should both accentuate the difference between $O(TD)$ and $O(T^2D)$ attention and indicate whether the global information propogation strategy of Hydra Attention is effective at these higher token counts. And indeed, in Table 3, we see the same trend as with 224px images: Hydra Attention can increase accuracy by 0.59% and throughput by 4.1% with 2 layers or keep accuracy the same and increase throughput by 15.4% with 7 layers this time.

Limitations. Okay, but Hydra attention is 197x faster than standard attention (with $T = 197$), so why is the maximum FLOP count reduction only 4%? Well, it turns out that with ViT-B/16 224×224 images ($T = 197, D = 768$), only 4.10% of total model FLOPs reside in creating and applying attention matrices.

Table 4. FLOP Count vs Image Size. FLOP count scaling of a ViT-B/16 model across different attention methods as image size increases. We also list the percent of total computation taken by creating and applying attention matrices. While Hydra attention significantly improves the FLOP count of the model at large image sizes, so does local window attention, which has already been shown effective on large images [19]. A limitation of Hydra attention is that it can only be 4% faster than local window attention, though it's more general and can lead to proportionally higher throughputs.

Image Size	T	Baseline		Hydra		Local Window	
		GFLOPs	Attn	GFLOPs	Attn	GFLOPs	Attn
224	197	17.6	4.10%	16.8	0.02%	17.6	4.10%
384	577	55.1	11.13%	49.0	0.02%	51.1	4.10%
448	785	78.0	14.56%	66.7	0.02%	69.5	4.10%
1024	4097	657.3	47.06%	348.1	0.02%	362.8	4.10%
1280	6401	1298.9	58.14%	543.8	0.02%	566.9	4.10%

With Hydra attention, this is reduced down to 0.02%, essentially eliminating the cost of attention in the model. While this does result in a raw throughput increase of up to 10.2% (see Table 2), we can clearly do better.

Of course, the story changes as you increase the image size: in Table 4, we repeat this computation for different image sizes, and the computation of standard attention balloons all the way up to 58% with 1280px images, while Hydra attention remains negligible at 0.02%. We test 384px images ourselves in Table 3, and the speed-up for Hydra Attention is much more pronounced (up to a 27.1% throughput increase). However, further work needs to be done to validate Hydra Attention on tasks that use more tokens (e.g. instance segmentation [15]). Though in those tasks, we'd be comparing against the local window attention used in ViTDet [19], which has already been shown to be effective for large token regimes in images. Compared to local window attention, Hydra attention uses only 4% fewer FLOPs at any image size, though its throughput would likely be proportionally higher (due to less memory usage).

In general, the usefulness of Hydra attention lies in its generality. Local window attention is a powerful solution for dense image prediction, but quickly becomes cumbersome with token sparsity (e.g., with masked pretraining [12,14,30] or token pruning [18,26,36]). We leave this for future work to explore.

5   Conclusion and Future Directions

In this paper, we introduce Hydra Attention, an efficient attention module with many heads. We show that Hydra Attention outperforms other $O(TD)$ attention methods in Table 1 and can even work in tandem with traditional multihead self-attention to improve the accuracy of a baseline DeiT-B model in Fig. 4. However, while Hydra attention works well on ImageNet classification (Table 2, Table 3), its real potential for speed-up lies in larger images (Table 4).

We've taken the first step in showing that Hydra attention can work at all and hope that future work can explore its use in other, more token-intensive domains such as detection, segmentation, or video. Moreover, Hydra attention is a general technique that doesn't make any assumptions about the relationships between tokens, so it can be applied to further improve the speed of token-sparse applications such as masked pretraining [12,14,30] or token pruning [18,26,36]. We hope Hydra attention can be used as a step toward more powerful, efficient, and general transformers in the future.

References

1. Abnar, S., Zuidema, W.: Quantifying attention flow in transformers. arXiv:2005.00928 [cs.LG] (2020)
2. Babiloni, F., Marras, I., Kokkinos, F., Deng, J., Chrysos, G., Zafeiriou, S.: Poly-NL: linear complexity non-local layers with 3rd order polynomials. In: ICCV (2021)
3. Chen, B., Wang, R., Ming, D., Feng, X.: ViT-P: rethinking data-efficient vision transformers from locality. arXiv:2203.02358 [cs.CV] (2022)
4. Chen, Y., et al.: Mobile-former: bridging mobilenet and transformer. In: CVPR (2022)
5. Choromanski, K., et al.: Rethinking attention with performers. arXiv:2009.14794 [cs.LG] (2020)
6. Chowdhery, A., et al.: PaLM: scaling language modeling with pathways. arXiv:2204.02311 [cs.CL] (2022)
7. Cordts, M., et al.: The cityscapes dataset for semantic urban scene understanding. In: CVPR (2016)
8. Dao, T., Fu, D.Y., Ermon, S., Rudra, A., Ré, C.: FlashAttention: fast and memory-efficient exact attention with IO-awareness. arXiv:2205.14135 [cs.LG] (2022)
9. Deng, J., Dong, W., Socher, R., Li, L.J., Li, K., Fei-Fei, L.: ImageNet: a large-scale hierarchical image database. In: CVPR (2009)
10. Dosovitskiy, A., et al.: An image is worth 16×16 words: transformers for image recognition at scale. In: ICLR (2020)
11. Fan, H., et al.: Multiscale vision transformers. In: ICCV (2021)
12. Feichtenhofer, C., Fan, H., Li, Y., He, K.: Masked autoencoders as spatiotemporal learners. arXiv:2205.09113 [cs.CV] (2022)
13. Graham, B., et al.: LeViT: a vision transformer in convnet's clothing for faster inference. In: ICCV (2021)
14. He, K., Chen, X., Xie, S., Li, Y., Dollár, P., Girshick, R.: Masked autoencoders are scalable vision learners. In: CVPR (2022)
15. He, K., Gkioxari, G., Dollár, P., Girshick, R.: Mask R-CNN. In: ICCV (2017)
16. Katharopoulos, A., Vyas, A., Pappas, N., Fleuret, F.: Transformers are RNNs: fast autoregressive transformers with linear attention. In: ICML (2020)
17. Kenton, J.D.M.W.C., Toutanova, L.K.: BERT: pre-training of deep bidirectional transformers for language understanding. In: NAACL-HLT (2019)
18. Kong, Z., et al.: SPViT: enabling faster vision transformers via soft token pruning. arXiv:2112.13890 [cs.CV] (2021)
19. Li, Y., Mao, H., Girshick, R., He, K.: Exploring plain vision transformer backbones for object detection. arXiv:2203.16527 [cs.CV] (2022)

20. Li, Y., et al.: MViTv2: improved multiscale vision transformers for classification and detection. In: CVPR (2022)
21. Liu, Z., et al.: Swin transformer: hierarchical vision transformer using shifted windows. In: ICCV (2021)
22. Mehta, S., Rastegari, M.: MobileViT: light-weight, general-purpose, and mobile-friendly vision transformer. arXiv:2110.02178 [cs.CV] (2021)
23. Michel, P., Levy, O., Neubig, G.: Are sixteen heads really better than one? In: NeurIPS (2019)
24. Peng, H., Pappas, N., Yogatama, D., Schwartz, R., Smith, N.A., Kong, L.: Random feature attention. arXiv:2103.02143 [cs.CL] (2021)
25. Radford, A., et al.: Language models are unsupervised multitask learners. OpenAI Blog (2019)
26. Rao, Y., Zhao, W., Liu, B., Lu, J., Zhou, J., Hsieh, C.J.: DynamicViT: efficient vision transformers with dynamic token sparsification. In: NeurIPS (2021)
27. Selvaraju, R.R., Cogswell, M., Das, A., Vedantam, R., Parikh, D., Batra, D.: Grad-CAM: visual explanations from deep networks via gradient-based localization. In: ICCV (2017)
28. Shen, Z., Zhang, M., Zhao, H., Yi, S., Li, H.: Efficient attention: attention with linear complexities. In: WACV (2021)
29. Song, J.: UFO-ViT: high performance linear vision transformer without softmax. arXiv:2109.14382 [cs.CV] (2021)
30. Tong, Z., Song, Y., Wang, J., Wang, L.: VideoMAE: masked autoencoders are data-efficient learners for self-supervised video pre-training. arXiv:2203.12602 [cs.CV] (2022)
31. Touvron, H., Cord, M., Douze, M., Massa, F., Sablayrolles, A., Jégou, H.: Training data-efficient image transformers & distillation through attention. arXiv:2012.12877 [cs.CV] (2020)
32. Vaswani, A., et al.: Attention is all you need. In: NeurIPS (2017)
33. Voita, E., Talbot, D., Moiseev, F., Sennrich, R., Titov, I.: Analyzing multi-head self-attention: specialized heads do the heavy lifting, the rest can be pruned. arXiv:1905.09418 [cs.CL] (2019)
34. Wang, S., Li, B.Z., Khabsa, M., Fang, H., Ma, H.: Linformer: self-attention with linear complexity. arXiv:2006.04768 [cs.LG] (2020)
35. Yang, C., et al.: Lite vision transformer with enhanced self-attention. In: CVPR (2022)
36. Yin, H., Vahdat, A., Alvarez, J., Mallya, A., Kautz, J., Molchanov, P.: AdaViT: adaptive tokens for efficient vision transformer. In: CVPR (2022)
37. Zhai, S., et al.: An attention free transformer. arXiv:2105.14103 [cs.LG] (2021)
38. Zhai, X., Kolesnikov, A., Houlsby, N., Beyer, L.: Scaling vision transformers. arXiv:2106.04560 [cs.CV] (2021)

BiTAT: Neural Network Binarization with Task-Dependent Aggregated Transformation

Geon Park[1(✉)], Jaehong Yoon[1], Haiyang Zhang[3], Xing Zhang[3],
Sung Ju Hwang[1,2], and Yonina C. Eldar[3]

[1] KAIST, Daejeon, South Korea
{geon.park,jaehong.yoon}@kaist.ac.kr
[2] AITRICS, Seoul, South Korea
[3] Weizmann Institute of Science, Rehovot, Israel

Abstract. Neural network quantization aims to transform high-precision weights and activations of a given neural network into low-precision weights/activations for reduced memory usage and computation, while preserving the performance of the original model. However, extreme quantization (1-bit weight/1-bit activations) of compactly-designed backbone architectures (e.g., MobileNets) often used for edge-device deployments results in severe performance degeneration. This paper proposes a novel Quantization-Aware Training (QAT) method that can effectively alleviate performance degeneration even with extreme quantization by focusing on the inter-weight dependencies, between the weights within each layer and across consecutive layers. To minimize the quantization impact of each weight on others, we perform an orthonormal transformation of the weights at each layer by training an input-dependent correlation matrix and importance vector, such that each weight is disentangled from the others. Then, we quantize the weights based on their importance to minimize the loss of the information from the original weights/activations. We further perform progressive layer-wise quantization from the bottom layer to the top, so that quantization at each layer reflects the quantized distributions of weights and activations at previous layers. We validate the effectiveness of our method on various benchmark datasets against strong neural quantization baselines, demonstrating that it alleviates the performance degeneration on ImageNet and successfully preserves the full-precision model performance on CIFAR-100 with compact backbone networks.

Keywords: Neural network binarization · Quantization-aware training

1 Introduction

Over the past decade, deep neural networks have achieved tremendous success in solving various real-world problems, such as image/text generation [5,14], unsu-

---

G. Park and J. Yoon—Equal contibution.

Supplementary Information The online version contains supplementary material available at https://doi.org/10.1007/978-3-031-25082-8_4.

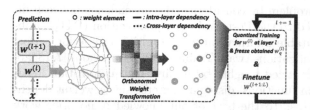

METHOD	BRECQ [16]	DBQ [7]	ReActNet [18]	Ours
BIT$_w$ / BIT$_a$	2/4	4/8	1/1	1/1
CORRELATION	block	N/A	N/A	block
TASK-BASED Q	✓	✗	✗	✓
STRUCTURED	node	✗	✗	dynamic
APPROACH	PTQ[1]	QAT[2]	QAT	QAT
T1@IMGNET	66.60%	70.5%	68.26%	**68.51%**
FLOPs ×10^7	3.31	3.60	1.2	1.2

[1] Post-training Quantization
[2] Quantization-aware Training

Fig. 1. Left: An Illustration of our proposed method. Weight elements in a layer is highly correlated to each other along with the weights in other layers. Our BiTAT sequentially obtains quantized weights of each layer based on the importance of disentangled weights to others using a trainable orthonormal rotation matrix and importance vector. **Right:** Categorization of relevant and strong quantization methods to ours.

pervised representation learning [4,10,34], and multi-modal training [26,28,33]. Recently, network architectures that aim to solve target tasks are becoming increasingly larger, based on the empirical observations of their improved performance. However, as the models become larger, it is increasingly difficult to deploy them on resource-limited edge devices with limited memory and computational power. Therefore, many recent works focus on building resource-efficient deep neural networks to bridge the gap between the scale of deep neural networks and actual permissible computational complexity/memory-bounds for on-device model deployments. Some of these works consider designing computation- and memory-efficient modules for neural architectures, while others focus on compressing a given neural network by either pruning its weights [6,11,17,32] or reducing the bits used to represent the weights and activations [3,7,16]. The last approach, *neural network quantization,* is beneficial for building on-device AI systems since the edge devices oftentimes only support low bitwidth-precision parameters and/or operations. However, it inevitably suffers from the non-negligible forgetting of the encoded information from the full-precision models. Such loss of information becomes worse with extreme quantization into binary neural networks with 1-bit weights and 1-bit activations [3,24,27,35].

How can we then effectively preserve the original model performance even with extremely low-precision deep neural networks? To address this question, we focus on the somewhat overlooked properties of neural networks for quantization: the weights in a layer are highly correlated with each other and weights in the consecutive layers. Quantizing the weights will inevitably affect the weights within the same layer, since they together comprise a transformation represented by the layer. Thus, quantizing the weights and activations at a specific layer will adjust the correlation and relative importance between them. Moreover, it will also largely impact the next layer that directly uses the output of the layer, which together comprise a function represented by the neural network.

Despite their impact on neural network quantization, such inter-weight dependencies have been relatively overlooked. As shown in Fig. 1 Right, although

BRECQ [16] addresses the problem by considering the dependency between filters in each block, it is limited to the Post-Training Quantization (PTQ) problem, which suffers from inevitable information loss, resulting in inferior performance. The most recent Quantization-Aware Training (QAT) methods [7,18] are concerned with obtaining quantized weights by minimizing quantization losses with parameterized activation functions, disregarding cross-layer weight dependencies during training process. To the best of our knowledge, no prior work explicitly considers dependencies among the weights for QAT.

To tackle this challenging problem, we propose a new QAT method, referred to as Neural Network **B**inarization with **T**ask-dependent **A**ggregated **T**ransformation (**BiTAT**), as illustrated in Table 1 Left. Our method sequentially quantizes the weights at each layer of a pre-trained neural network based on chunk-wise input-dependent weight importance by training orthonormal dependency matrices and scaling vectors. While quantizing each layer, we fine-tune the subsequent full-precision layers, which utilize the quantized layer as an input for a few epochs while keeping the quantized weights frozen. we aggregate redundant input dimensions for transformation matrices and scaling vectors, significantly reducing the computational cost of the quantization process. Such consideration of inter-weight dependencies allows our BiTAT algorithm to better preserve the information from a given high-precision network, allowing it to achieve comparable performance to the original full-precision network even with extreme quantization, such as binarization of both weights and activations. The main contributions of the paper can be summarized as follows:

- We demonstrate that weight dependencies within each layer and across layers play an essential role in preserving the model performance during quantized training.
- We propose an input-dependent quantization-aware training method that binarizes neural networks. We disentangle the correlation in the weights from across multiple layers by training rotation matrices and importance vectors, which guides the quantization process to consider the disentangled weights' importance.
- We empirically validate our method on several benchmark datasets against state-of-the-art neural network quantization methods, showing that it significantly outperforms baselines with the compact neural network architecture.

2   Related Work

Minimizing the Quantization Error. Quantization methods for deep neural networks can be broadly categorized into several strategies [25]. We first introduce the methods by minimizing the quantization error. Many existing neural quantization methods aim to minimize the weight/activation discrepancy between quantized models and their high-precision counterparts. XNOR-Net [27] aims to minimize the least-squares error between quantized and full-precision weights for each output channel in layers. DBQ [7] and QIL [13] perform layerwise quantization with parametric scale or transformation functions optimized to the task.

Yet, they quantize full-precision weight elements regardless of the correlation between other weights. While TSQ [29] and Real-to-Bin [20] propose to minimize the ℓ_2 distance between the quantized activations and the real-valued network's activations by leveraging intra-layer weight dependency, they do not consider cross-layer dependencies. Recently, BRECQ [16] and the work in a similar vein on post-training quantization [22] consider the interdependencies between the weights and the activations by using a Taylor series-based approach. However, calculating the Hessian matrix for a large neural network is often intractable, and thus they resort to strong assumptions such as small block-diagonality of the Hessian matrix to make them feasible. BiTAT solves this problem by training the dependency matrices alongside the quantized weights while grouping similar weights together to reduce the computational cost.

Modifying the Task Loss Function. A line of methods aims to achieve better generalization performance during quantization by taking sophisticatedly-designed loss functions. BNN-DL [9] adds a distributional loss term that enforces the distribution of the weights to be quantization-friendly. Apprentice [21] uses knowledge distillation (KD) to preserve the knowledge of the full-precision teacher network in the quantized network. However, such methods only put a constraint on the distributional properties of the weights, not the dependencies and the values of the weight elements. CI-BCNN [30] parameterizes bitcount operations by exploring the interaction between output channels using reinforcement learning and quantizes the floating-point accumulation in convolution operations based on them. However, reinforcement learning is expensive, and it still does not consider cross-layer dependencies.

Reducing the Gradient Error. Bi-Real Net [19] devises a better gradient estimator for the sign function used to binarize the activations and a magnitude-aware gradient correction method. It further modifies the MobileNetV1's architecture to better fit the quantized operations. ReActNet [18] achieves state-of-the-art performance for binary neural networks by training a generalized activation function for compact network architecture used in [19]. However, the quantizer functions in these methods conduct element-wise unstructured compression without considering the change in other correlated weights during quantization training. This makes the search process converge to the suboptimal solutions since task loss is the only guide for finding the optimal quantized weights, which is often insufficient for high-dimensional and complex architectures. However, our proposed method can obtain a better-informed guide that compels the training procedure to spend more time searching in areas that are more likely to contain high-performing quantized weights.

3   Weight Importance for Quantization-Aware Training

We first introduce the problem of Quantization-Aware Training (QAT) in Sect. 3.1 and show that the dependency among the neural network weights plays a crucial role in preserving the performance of a quantized model obtained with QAT in Sect. 3.2. We further show that the dependency between consecutive layers critically affects the performance of the quantized model in Sect. 3.3.

3.1 Problem Statement

We aim to quantize a full-precision neural network into a binary neural network (BNN), where the obtained quantized network is composed of binarized 1-bit weights and activations, which preserves the performance of the original full-precision model. Let $f(\cdot; \mathcal{W})$ be a L-layered neural network parameterized by a set of pre-trained weights $\mathcal{W} = \{w^{(1)}, \ldots, w^{(L)}\}$, where $w^{(l)} \in \mathbb{R}^{d_{l-1} \times d_l}$ is the weight at layer l and d_0 is the dimensionality of the input. Given a training dataset $\mathcal{X}$ and corresponding labels $\mathcal{Y}$, existing QAT methods [2,7,13,23,27,31] search for optimal quantized weights by solving for the optimization problem that can be generally described as follows:

$$\underset{\mathcal{W}, \phi}{\text{minimize}} \ \mathcal{L}_{task}\left(f\left(\mathcal{X}; Q\left(\mathcal{W}; \phi\right)\right), \mathcal{Y}\right), \tag{1}$$

where $\mathcal{L}_{task}$ is a standard task loss function, such as cross-entropy loss, and $Q(\cdot; \phi)$ is the weight quantization function parameterized by ϕ which transforms a real-valued vector to a discrete, binary vector. The quantization function used in existing works typically minimize loss terms based on the Mean Squared Error (MSE) between the full-precision weights and the quantized weights at each layer:

$$Q(w) := \alpha^* b^*, \qquad \text{where} \qquad \alpha^*, b^* = \underset{\alpha \in \mathbb{R}, b \in \{-1,1\}^m}{\arg \min} \ \|w - \alpha b\|_2^2, \tag{2}$$

where m is the dimensionality of the target weight. For inference, QAT methods use $w_q = Q(w)$ as the final quantized parameters. They iteratively search for the quantized weights based on the task loss with a stochastic gradient descent optimizer, where the model parameters converge into the ball-like region around the full-precision weights w.

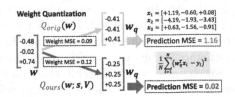

Fig. 2. A simple experiment that cross-layer weight correlation is critical to find well-performing quantized weights during QAT.

However, the region around the optimal full-precision weights may contain suboptimal solutions with high errors. We demonstrate such inefficiency of the existing quantizer formulation through a simple experiment in Fig. 2. Suppose we have three input points, x_1, x_2, and x_3, and full-precision weights w. Quantized training of the weight using Eq. 2 successfully reduces MSE between the quantized weight and the full-precision, but the task prediction loss using w_q is nonetheless very high.

We hypothesize that the main source of error comes from the independent application of the quantization process to each weight element. However, neural network weights are not independent, but highly correlated and thus quantizing a set of weights will largely affect the others. Moreover, after quantization, the relative importance among weights could also largely change. Both factors lead to high quantization errors in the pre-activations. On the other hand, our proposed QAT method *BiTAT*, described in Sect. 4, achieves a quantized model with much smaller MSE. This results from the consideration of the inter-weights dependencies, which we describe in the next subsection.

3.2 Disentangling Weight Dependencies via Input-dependent Orthornormal Transformation

How can we then find the low-precision subspace, which contains the best-performing quantized weights on the task, by exploiting the inter-weight dependencies? The properties in the input distribution give us some insights into this question. Let us consider a task composed of centered N training samples $\{x_1, \ldots, x_N\} = \mathcal{X} \in \mathbb{R}^{N \times d_0}$. We can obtain principal components of the training samples $v_1, \ldots, v_{d_0} \in \mathbb{R}^{d_0}$ and the corresponding coefficients $\lambda_1, \ldots, \lambda_{d_0} \geq 0$, in a descending order. Let us further suppose that we optimize a single-layered neural network parameterized by $w^{(1)}$. Neurons corresponding to the columns of $w^{(1)}$ are oriented in a similar direction to the principal components with higher variances (i.e., v_i than v_j, where $i < j$) that is much more likely to get activated than the others. We apply a change of basis to the column space of the weight matrix $w^{(1)}$ with the bases $(v_1, \ldots, v_{d_0})$:

$$V \widetilde{w}^{(1)} = w^{(1)} \tag{3}$$

$$\widetilde{w}^{(1)} = V^\top w^{(1)}, \tag{4}$$

where $V = [v_1 \mid \cdots \mid v_{d_0}] \in \mathbb{R}^{d_0 \times d_0}$ is an orthonormal matrix. The top rows of the transformed weight matrix $\widetilde{w}^{(1)}$ will contain more important weights, whereas the bottom rows will contain less important ones. Therefore, the accuracy of the model will be more affected by the perturbations of the weights at top rows than ones at the bottom rows. Note that this transformation can also be applied to the convolutional layer by "unfolding" the input image or feature map into a set of patches, which enables us to convert the convolutional weights into a matrix (The detailed descriptions of the orthonormal transformations for convolutional layers is provided in the supplementary file).

We can also easily generalize the method to multi-layer neural networks, by taking the inputs for the l-th layer as the "training set", assuming that all of the previous layer's weights are fixed, as follows:

$$\left\{ x_i^{(l)} = \sigma \left(w^{(l)\top} x_i^{(l-1)} \right) \right\}_{i=1}^N, \tag{5}$$

where $\sigma(\cdot)$ is the nonlinear transformation defined by both the non-linear activations and any layers other than linear transformation with the weights, such

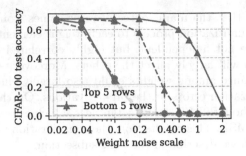

Fig. 3. Solid lines: Test accuracy of a MobileNetV2 model on CIFAR-100 dataset, after adding Gaussian noise to the top 5 rows and the bottom 5 rows of $\hat{\boldsymbol{w}}^{(l)}$ for all layers, considering the dependency on the lower layers. **Dashed lines:** Not considering the dependency on the lower layers. The x axis is in log scale. (Color figure online)

as a average-pooling or Batch Normalization. Then, we straightforwardly obtain the change-of-basis matrix $V^{(l)}$ and $\boldsymbol{s}^{(l)}$ for layer l. The impact of transformed weights is shown in Fig. 3. We compute the principal components of each layer in the initial pre-trained model and measure the test accuracy when adding the noise to the top-5 highest-variance *(dashed red)* or lowest-variance components *(dashed blue)* per layer. While a model with perturbed high-variance components degenerates the performance as the noise scale increases, a model with perturbed low-variance components consistently obtains high performance even with large perturbations. This shows that preserving the important weight components that correspond to high-variance input components is critically important for effective neural network quantization that can preserve the loss of the original model.

3.3 Cross-Layer Weight Correlation Impacts Model Performance

So far, we only described the dependency among the weights within a single layer. However, dependencies between the weights across different layers also significantly impact the performance. To validate that, we perform layerwise sequential training from the bottom layer to the top. At the training of each layer, the model computes the principal components of the target layer and adds noise to its top-5 high/low components. As shown in Fig. 3, progressive training with the low-variance components *(solid blue)* achieves significantly improved accuracy over the end-to-end training counterpart *(dashed blue)* with a high noise scale, which demonstrates the beneficial effect of modeling weight dependencies in earlier layers. We describe further details in the supplementary file.

4 Task-Dependent Weight Transformation for Neural Network Binarization

Our objective is to train binarized weights $\boldsymbol{w}_q$ given pre-trained full-precision weights. We effectively mitigate performance degeneration from the loss of information incurred by binarization by focusing on the inter-weight dependencies within each layer and across consecutive layers. We first reformulate the quantization function Q in Eq. 2 with the weight correlation matrix V and the importance vector s so that each weight is disentangled from the others while allowing larger quantization errors on the unimportant disentangled weights:

$$Q(w; s, V) = \arg\min_{w_q \in \mathbb{Q}} \left\| \text{diag}(s) \left(V^\top \otimes w - V^\top \otimes w_q \right) \right\|_F^2 + \gamma \left\| w_q \right\|_1, \quad (6)$$

where $s \in \mathbb{R}^{d_o}$ is a scaling term that assigns different importance scores to each row of $\hat{w}$. We denote $\mathbb{Q} = \{\alpha \odot b : \alpha \in \mathbb{R}^{d_{out}}, b \in \{-1, 1\}^{d_{in} \times d_{out}}\}$ as the set of possible binarized values for $w \in \mathbb{R}^{d_{in} \times d_{out}}$ with a scalar scaling factor for each output channel. The operation $\otimes$ denotes permuted matrix multiplication with index replacement, where we detail the computation in the following subsection. We additionally include ℓ_1 norm adjusted by a hyperparameter γ. At the same time, we want our quantized model to minimize the empirical task loss (e.g., cross-entropy loss) for a given dataset. Thus we formulate the full objective in the form of a bilevel optimization problem to find the best quantized weights which minimizes the task loss by considering the cross-layer weight dependencies and the relative importance among weights:

$$w^*, s^*, V^* = \arg\min_{w, s, V} \mathcal{L}_{task} \left(f \left(\mathcal{X}; w_q \right), \mathcal{Y} \right), \qquad \text{where } w_q = Q(w; s, V).$$

$$(7)$$

After the quantized training, the quantized weights w_q^* are determined by $w_q^* = Q(w^*; s^*, V^*)$.

In practice, directly solving the above bilevel optimization problem is impractical due to its excessive computational cost. We therefore instead consider the following relaxed problem:

$$w^*, s^*, V^* = \arg\min_{w, s, V} \left\{ \mathcal{L}_{task} \left(f(\mathcal{X}; \text{sgn}(w)), \mathcal{Y} \right) \right.$$

$$\left. + \lambda \left\| \text{diag}(s) V^\top \otimes (w - \text{sgn}(w)) \right\|_F^2 + \gamma \left\| \text{sgn}(w) \right\|_1 \right\}, \quad (8)$$

where λ is a hyperparameter to balance between the quantization objective and task loss. Since it is impossible to compute the gradients for discrete values in quantized weights, we adopt the straight-through estimator [1] that is broadly used across QAT methods: $\text{sgn}(w)$ indicates the sign function applied elementwise to w. We follow [18] for the derivative of $\text{sgn}(\cdot)$. Finally, we obtain the desired quantized weights by $w_q^* = \text{sgn}(w^*)$. In order to obtain the off-diagonal parts of the cross-layer dependency matrix V, we minimize Eq. 8 with respect to s and V to dynamically determine the values (we omit $\mathcal{X}$ and $\mathcal{Y}$ from the arguments of $\mathcal{L}_{train}(\cdot)$ for readability):

$$\mathcal{L}_{train}(w, s, V) = \mathcal{L}_{task} \left(f \left(\mathcal{X}; \text{sgn}(w) \right), \mathcal{Y} \right) + \lambda \left\| \text{diag}(s) V^\top \otimes (w - \text{sgn}(w)) \right\|_F^2$$

$$+ \gamma \left\| \text{sgn}(w) \right\|_1 + Reg(s, V),$$

$$(9)$$

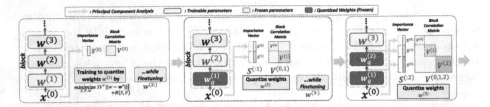

Fig. 4. Quantization-aware Training with BiTAT: We perform a sequential training process: quantization training of a layer - rapid finetuning for upper layers. At each layerwise quantization, we also train the importance vector and orthonormal correlation matrix, which are initialized by PCA components of the current and lower layer inputs in the target block, and guide the quantization to consider the importance of disentangled weights.

where $Reg(\boldsymbol{s}, \boldsymbol{V}) := \|\boldsymbol{V}\boldsymbol{V}^\top - \boldsymbol{I}\|^2 + |\sigma - \sum_i \log(s_i)|^2$ is a regulariztion term which enforces $\boldsymbol{V}$ to be orthogonal and keeps the scale of $\boldsymbol{s}$ constant. Here, σ is the constant initial value of $\sum_i \log(s_i)$, which is a non-negative importance score.

4.1 Layer-Progressive Quantization with Block-Wise Weight Dependency

While we obtain the objective function in Eq. 9, it is inefficient to perform quantization-aware training while considering the complete correlations of all weights in the given neural network. Therefore, we only consider cross-layer dependencies between only few consecutive layers (we denote it as a *block*), and initialize $\boldsymbol{s}$ and $\boldsymbol{V}$ using Principal Component Analysis (PCA) on the inputs to those layers within each block.

Formally, we define a weight correlation matrix in a neural network block $V^{(block)} \in \mathbb{R}^{(\sum_{i=1}^k d_i) \times (\sum_{i=1}^k d_i)}$, where k is the number of layers in a block, similarly to the block-diagonal formulation in [16] to express the dependencies between weights across layers in the off-diagonal parts. We initialize $\boldsymbol{s}^{(l)}$ and in-diagonal parts $V^{(l)}$ by applying PCA on the input covariance matrix:

$$\boldsymbol{s}^{(l)} \leftarrow (\boldsymbol{\lambda}^{(l)})^{\frac{1}{2}}, \quad \boldsymbol{V}^{(l)} \leftarrow \boldsymbol{U}^{(l)},$$
$$\text{where } \boldsymbol{U}^{(l)}\boldsymbol{\lambda}^{(l)}(\boldsymbol{U}^{(l)})^\top := \frac{1}{N}\sum_{i=1}^N \boldsymbol{o}_i^{(l-1)}\boldsymbol{o}_i^{(l-1)\top}, \quad (10)$$

where $\boldsymbol{o}^{(l)}$ is the output of l-th layer and $\boldsymbol{o}^{(0)} = \boldsymbol{x}$. This allows the weights at l-th layer to consider the dependencies on the weights from the earlier layers within the same neural block, and we refer to this method as *sequential quantization*. Then, instead of having one set of $\boldsymbol{s}$ and V for each layer, we can keep the previous layer's $\boldsymbol{s}$ and V and expand them. Specifically, when quantizing layer l which is a part of the block that starts with the layer m, we first apply PCA on the input covariance matrix to obtain $\boldsymbol{\lambda}^{(l)}$ and $U^{(l)}$. We then expand the existing

$s^{(m:l-1)}$ and $V^{(m:l-1)}$ to obtain $s \in \mathbb{R}^{D+d_{l-1}}$ and $s \in \mathbb{R}^{D+d_{l-1}}$ as follows[1]:

$$[s^{(m:l)}]_i := \begin{cases} [s^{(m:l-1)}]_i, & i \le D, \\ [(\lambda^{(l)})^{\frac{1}{2}}]_{i-D}, & D < i, \end{cases} \quad [V^{(m:l)}]_{i,j} := \begin{cases} [V^{(m:l-1)}]_{i,j}, & i,j \le D, \\ [U^{(l)}]_{i-D,j-D}, & D < i,j, \\ 0, & \text{otherwise}, \end{cases} \quad (11)$$

where $D = \sum_{i=m}^{l-2} d_i$, as illustrated in Fig. 4. The weight dependencies between different layers (i.e., off-diagonal areas) are trainable and zero-initialized. To enable the matrix multiplication of the weights with the expanded s and V, we define the expanded block weights as follows[2]:

$$w^{(m:l)} = \left[\text{PadCol}(w^{(m:l-1)}, d_l); w^{(l)}\right], \quad (12)$$

where $\text{PadCol}(\cdot, c)$ zero-pads the input matrix to the right by c columns. Then, our final objective from Eq. 9 with cross-layer dependencies is given as follows:

$$\begin{aligned} \mathcal{L}_{train}&(w^{(l:L)}, s^{(m:l)}, V^{(m:l)}) \\ &= \mathcal{L}_{task}\left(f\left(\mathcal{X}; \{\text{sgn}(w^{(l)}), w^{(l+1:L)}\}\right), \mathcal{Y}\right) \\ &\quad + \lambda \left\|\text{diag}(s^{(m:l)})V^{(m:l)^\top}\left(w^{(m:l)} - \text{sgn}(w^{(m:l)})\right)\right\|_F^2 \\ &\quad + \gamma \left\|\text{sgn}(w^{(m:l)})\right\|_1 + Reg(s^{(m:l)}, V^{(m:l)}). \end{aligned} \quad (13)$$

Given the backbone architecture with L layers, we minimize $\mathcal{L}_{train}(w^{(l)}, s^{(l)}, V^{(l)})$ with respect to $w^{(l)}, s^{(l)}$, and $V^{(l)}$ to find the desired binarized weights $w_q^{*(l)}$ for layer l while keeping the other layers frozen. Next, we finetune the following layers using the task loss function a few epochs before performing QAT on following layers, as illustrated in Fig. 4. This sequential quantization proceeds from the bottom layer to the top and the obtained binarized weights are frozen during the training.

4.2   Cost-Efficient BiTAT via Aggregated Weight Correlation Using Reduction Matrix

We derived a QAT formulation which focus on the cross-layer weight dependency by learning block-wise weight correlation matrices. Yet, as the number of inputs to higher layers is often large, the model constructs higher-dimensional $V^{(l)}$ on upper blocks, which is costly. In order to reduce the training memory footprint as well as the computational complexity, we aggregate the input dimensions into several small groups based on functional similarity using k-means clustering.

First, we take feature vectors, the outputs of the l-th layer $o_1^{(l)}, \ldots, o_N^{(l)} \in \mathbb{R}^{d_l}$ for each output dimension, to obtain d_l points $p_1, p_2, \ldots, p_{d_l} \in \mathbb{R}^N$, then aim

---

[1] $[\cdot]_i$ indicates the i-th element of the object inside the brackets.
[2] $[A; B]$ indicates vertical concatenation of the matrices A and B.

Algorithm 1. Neural Network Binarization with Task-dependent Aggregated Transformation

---

1: **Input:** Pre-trained weights $w^{(1)}, \ldots, w^{(L)}$ for L layers, task loss function $\mathcal{L}$, Maximum size of input-dimension group k, quantization epochs per layer N_{ep}.
2: **Output:** Quantized weights $w^{*(1)}, \ldots, w^{*(L)}$.
3: $\mathcal{B}_1, \ldots, \mathcal{B}_n$ ←Divide the neural network into n blocks
4: **for each** block $\mathcal{B}$ **do**
5:     $s = [], V = []$
6:     **for each** layer l in $\mathcal{B}$ **do**
7:         $o^{(l-1)} \leftarrow$ inputs for layer l
8:         $P \leftarrow$ **if** $d_{l-1} > k$ **then** K-MEANS$(X^{(l)}, k)$ **else** $I_{d_{l-1}}$ ▷ Grouping permutation matrix
9:         $U \operatorname{diag}(\lambda) U^{\top} = \operatorname{PCA}(P o^{(l-1)})$ ▷ Initialization values for the expanded part
10:         $s \leftarrow [s; \lambda^{\frac{1}{2}}], V \leftarrow \begin{bmatrix} V & 0 \\ 0 & U \end{bmatrix}$ ▷ expand s and V
11:         $w^{(l:L)}, s, V \leftarrow \arg\min_{w,s,V} \mathcal{L}_{train}(w^{(l:L)}, s, V)$ ▷ Iterate for N_{ep} epochs
12:         $w_q^{(l)} \leftarrow \operatorname{sgn}(w^{(l)})$

---

to cluster the points to k groups using k-means clustering, each containing N/k points. Let $g_i \in \{1, 2, \ldots, k\}$ indicate the group index of p_i, for $i = 1, \ldots, d_l$. We construct the reduction matrix $P \in \mathbb{R}^{k \times d_l}$, where $P_{ij} = \frac{1}{N/k}$ if $g_j = i$, and 0 otherwise. Each group corresponds to a single row of the reduced $\widehat{V}^{(l+1)} \in \mathbb{R}^{k \times k}$ instead of the original dimension $d_l \times d_l$. In practice, this **significantly reduces the memory consumption** of the V (down to **0.07%**). Now, we replace s and $V^{\top}$ in Eq. 13 to $\widehat{s}$ and $\widehat{V}^{\top} P$, respectively, initializing $\widehat{s}$ and $\widehat{V}$ with the grouped input covariance $\frac{1}{N} \sum_{i=1}^{N} (P o_i^{(l)})(P o_i^{(l)})^{\top}$. We describe the full training process of our proposed method in Algorithm 1. The total number of training epochs taken in training is $O(L N_{ep})$, where L is the number of layers, and N_{ep} is the number of epochs for the quantizing step for each layer.

5   Experiments

We validate a new quantization-aware training method, BiTAT, over multiple benchmark datasets; CIFAR-10, CIFAR-100 [15], and ILSVRC2012 ImageNet [8] datasets. We use MobileNet V1 [12] backbone network, which is a compact neural architecture designed for mobile devices. We follow overall experimental setups from prior works [18,31].

Baselines and Training Details. While our method aims to solve the QAT problem, we extensively compare our *BiTAT* against various methods; Post-training Quantization (PTQ) method: BRECQ [16], and Quantization-aware Training (QAT) methods: DBQ [7], EBConv [3], Bi-Real Net [19], Real-to-Bin [20], LCQ [31], MeliusNet [2], ReActNet [18]. Note that DBQ, LCQ, and MeliusNet, which keep some crucial layers, such as 1×1 downsampling layers, in full-precision, leading to inefficiency at evaluation time. Due to the page limit,

Table 1. Performance comparison of BiTAT with baselines. We report the averaged test accuracy across three independent runs. The best results are highlighted in bold, and results of cost-expensive models ($10^8 \uparrow$ ImgNet FLOPs) are de-emphasized in gray. We refer to several results reported from their own papers, denoted as †.

METHODS	ARCHITECTURE	BITWIDTH WEIGHT/ ACTIV.	IMGNET FLOPs ($\times 10^7$)	IMGNET ACC (%)	CIFAR-10 ACC (%)	CIFAR-100 ACC (%)
Full-precision	ResNet-18	32/32	200.0	69.8	93.02	75.61
	MobileNet V2	32/32	31.40	71.9	94.43	68.08
BRECQ [16]	MobileNet V2	4/4	3.31	66.57†	–	–
DBQ [7]	MobileNet V2	4/8	3.60	70.54†	93.77	73.20
LCQ [31]	ResNet-18	2/2	15.00	68.9†	–	–
	MobileNet V2	4/4	3.31	70.8†	–	–
MeliusNet59 [2]	N/A	1/1	24.50	70.7†	–	–
Bi-Real Net [19]	ResNet-18	1/1	15.00	56.4†	–	–
Real-to-Bin [20]	ResNet-18	1/1	15.00	65.4†	–	76.2†
EBConv [3]	ResNet-18	1/1	11.00	71.2†	–	76.5†
ReActNet-C [18]	MobileNet V1	1/1	14.00	71.4†	90.77	–
ReActNet-A [18]	MobileNet V1	1/1	**1.20**	68.26	89.73	65.51
BiTAT (Ours)	MobileNet V1	1/1	**1.20**	**68.51**	**90.21**	**68.36**

we provide the details on baselines, and the training and inference phase during QAT including hyperparameter setups in the Supplementary file. We also discuss about the limitations and societal impacts of our work in the Supplementary file.

5.1   Quantitative Analysis

We compare our BiTAT against various PTQ and QAT-based methods in Table 1 on multiple datasets. BRECQ introduces an adaptive PTQ method by focusing on the weight dependency via hessian matrix computations, resulting in significant performance deterioration and excessive training time. DBQ and LCQ suggest QAT methods, but the degree of bitwidth compression for the weights and activations is limited to 2- to 8-bits, which is insufficient to meet our interest in achieving neural network binarization with 1-bits weights and activations. MeliusNet only suffers a small accuracy drop, but it has a high OP count. DBQ and LCQ restrict the bit-width compression to be higher at 4 bits so that they cannot enjoy the XNOR-Bitcount optimization for speedup. Although Bi-Real Net, Real-to-Bin, and EBConv successfully achieve neural network binarization, over-parameterized ResNet is adopted as backbone networks, resulting in higher OP count. Moreover, except EBConv, these works still suffer from a significant accuracy drop. ReActNet binarizes all of the weights and activations (except the first and last layer) in compact network architectures while preventing model

Method	Intra-layer Transform	Cross-layer Transform	Accuracy (%)	Train Time (hours)
ReActNet [18]	N/A	N/A	65.51 ± 0.74	10.75
	✗	✗	68.17 ± 0.07	3.49
BiTAT	✓	✗	67.82 ± 0.22	3.66
(Ours)	✓	✓	68.21 ± 0.24	8.50
	w/ Filter-wise Transform		67.86 ± 0.11	3.01
	w/ Aggregated Transform		**68.36 ± 0.45**	**3.11**

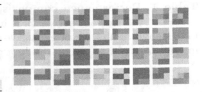

Fig. 5. Left: Ablation study for analyzing core components in our method. We report the averaged performance and 95% confidence interval across 3 independent runs and the complete BiTAT result is highlighted in gray background. **Right: Visualization of the weight grouping** during sequential quantization of BiTAT. Each 3×3 square represents a convolutional filter of the topmost layer (26th, excluding the classifier) of our model, and each unique color represents each group to which weight elements belong.

convergence failure. Nevertheless, the method still suffers from considerable performance degeneration of the binarized model. On the other hand, our BiTAT prevents information loss during quantized training up to 1-bits, showing a superior performance than ReActNet, 0.37 % ↑ for ImageNet, 0.53% ↑ for CIFAR-10, and 2.31% ↑ for CIFAR-100. Note that BiTAT further achieves on par performance of the MobileNet backbone for CIFAR-100. The results support our claim on layer-wise quantization from the bottom layer to the top, reflecting the disentangled weight importance and correlation with the quantized weights at earlier layers.

Ablation Study. We conduct ablation studies to analyze the effect of salient components in our proposed method in Fig. 5 Left. BiTAT based on layer-wise sequential quantization without weight transformation already surpasses the performance of ReActNet, demonstrating that layer-wise progressive QAT through an implicit reflection of adjusted importance plays a critical role in preserving the pre-trained models during quantization. We adopt intra-layer weight transformation using the input-dependent orthonormal matrix, but no significant benefits are observed. Thus, we expect that only disentangling intra-layer weight dependency is insufficient to fully reflect the adjusted importance of each weight due to a binarization of earlier weights/activations. This is evident that BiTAT considering both intra-layer and cross-layer weight dependencies achieves improved performance than the case with only intra-layer dependency. Yet, this requires considerable additional training time to compute with a chunk-wise transformation matrix. In the end, BiTAT with aggregated transformations, which is our full method, outperforms our defective variants in both terms of model performance and training time by drastically removing redundant correlation through reduction matrices. We note that using k-means clustering for aggregated correlation is also essential, as another variant, BiTAT with filter-wise transformations, which filter-wisely aggregates the weights instead, results in deteriorated performance.

5.2   Qualitative Analysis

Visualization of Reduction Matrix. We visualize the weight grouping for BiTAT in Fig. 5 Right to analyze the effect of the reduction matrix, which groups the weight dependencies in each layer based on the similarity between the input dimensions. Each 3×3 square represents a convolutional filter, and each unique color in weight elements represents which group each weight is assigned to, determined by the k-means algorithm, as described in Sect. 4.2. We observe that weight elements in the same filter do not share their dependencies; rather, on average, they often belong to four-five different weight groups. Opposite to these observations, BRECQ regards the weights in each filter as the same group for computing the dependencies in different layers, which is problematic since weight elements in the same filter can behave differently from each other.

Visualization of Cross-Layer Weight Dependency. In Fig. 6, we visualize learned transformation matrices V (*top row*), which shows that many weight elements at each layer are also dependent on other layer weights as highlighted in darker colors, verifying our initial claim. Further, we provide visualizations for their multiplications with corresponding importance vectors diag$(s)V^\top$ (*bottom row*). Here, the row of $V^\top$ is sorted by the relative importance in increasing order at each layer. We observe that important weights in a layer affect other layers, demonstrating that cross-layer weight dependency impacts the model performance during quantized training.

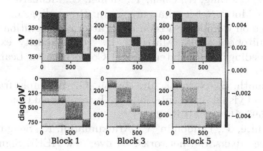

Fig. 6. Visualization of the learned V matrix and the diag$(s)V^\top$ of three blocks of the network, with the CIFAR-100 dataset. Notice the off-diagonal parts which represent cross-layer dependencies.

6   Conclusion

In this work, we explored long-overlooked factors that are crucial in preventing the performance degeneration with extreme neural network quantization: the inter-weight dependencies. That is, quantization of a set of weights affect the weights for other neurons within each layer, as well as weights in consecutive layers. Grounded by the empirical analyses of the node interdependency, we propose a Quantization-Aware Training (QAT) method for binarizing the weights and activations of a given neural network with minimal loss of performance. Specifically, we proposed orthonormal transformation of the weights at each layer to disentangle the correlation among the weights to minimize the negative impact of quantization on other weights. Further, we learned scaling term to allow varying degree of quantization error for each weight based on their measured importance, for layer-wise quantization. Then we proposed an iterative algorithm to perform the layerwise quantization in a progressive manner. We demonstrate the effectiveness of our method in neural network binarization on multiple benchmark datasets with compact backbone networks, largely outperforming state-of-the-art baselines.

Acknowlegements. This work was supported by Institute of Information & communications Technology Planning & Evaluation (IITP) grant funded by the Korea government (MSIT) (No. 2019-0-00075, Artificial Intelligence Graduate School Program (KAIST)). This work is supported by Samsung Advanced Institute of Technology.

References

1. Bengio, Y., Léonard, N., Courville, A.C.: Estimating or propagating gradients through stochastic neurons for conditional computation. arXiv preprint arXiv:1308.3432 (2013)
2. Bethge, J., Bartz, C., Yang, H., Chen, Y., Meinel, C.: MeliusNet: can binary neural networks achieve MobileNet-level accuracy? In: Proceedings of the IEEE International Conference on Computer Vision and Pattern Recognition (CVPR) (2020)
3. Bulat, A., Tzimiropoulos, G., Martinez, B.: High-capacity expert binary networks. In: Proceedings of the International Conference on Learning Representations (ICLR) (2021)
4. Chen, T., Kornblith, S., Norouzi, M., Hinton, G.: A simple framework for contrastive learning of visual representations. In: International Conference on Machine Learning, pp. 1597–1607. PMLR (2020)
5. Creswell, A., White, T., Dumoulin, V., Arulkumaran, K., Sengupta, B., Bharath, A.A.: Generative adversarial networks: an overview. IEEE Signal Process. Mag. **35**, 53–65 (2018)
6. Dai, B., Zhu, C., Guo, B., Wipf, D.: Compressing neural networks using the variational information bottleneck. In: Proceedings of the International Conference on Machine Learning (ICML) (2018)
7. Dbouk, H., Sanghvi, H., Mehendale, M., Shanbhag, N.: DBQ: a differentiable branch quantizer for lightweight deep neural networks. In: Vedaldi, A., Bischof, H., Brox, T., Frahm, J.-M. (eds.) ECCV 2020. LNCS, vol. 12372, pp. 90–106. Springer, Cham (2020). https://doi.org/10.1007/978-3-030-58583-9_6
8. Deng, J., Dong, W., Socher, R., Li, L.J., Li, K., Fei-Fei, L.: ImageNet: a large-scale hierarchical image database. In: Proceedings of the IEEE International Conference on Computer Vision and Pattern Recognition (CVPR) (2009)
9. Ding, R., Chin, T., Liu, Z., Marculescu, D.: Regularizing activation distribution for training binarized deep networks. In: Proceedings of the IEEE International Conference on Computer Vision and Pattern Recognition (CVPR) (2019)

10. Gidaris, S., Singh, P., Komodakis, N.: Unsupervised representation learning by predicting image rotations. arXiv preprint arXiv:1803.07728 (2018)
11. He, Y., Ding, Y., Liu, P., Zhu, L., Zhang, H., Yang, Y.: Learning filter pruning criteria for deep convolutional neural networks acceleration. In: Proceedings of the IEEE International Conference on Computer Vision and Pattern Recognition (CVPR) (2020)
12. Howard, A.G., et al.: MobileNets: efficient convolutional neural networks for mobile vision applications. arXiv preprint arXiv:1704.04861 (2017)
13. Jung, S., et al.: Learning to quantize deep networks by optimizing quantization intervals with task loss. In: Proceedings of the IEEE International Conference on Computer Vision and Pattern Recognition (CVPR) (2019)
14. Karras, T., et al.: Alias-free generative adversarial networks. In: Advances in Neural Information Processing Systems (NeurIPS) (2021)
15. Krizhevsky, A., Hinton, G., et al.: Learning multiple layers of features from tiny images (2009)
16. Li, Y., et al.: BRECQ: pushing the limit of post-training quantization by block reconstruction. In: Proceedings of the International Conference on Learning Representations (ICLR) (2021)
17. Lin, M., et al.: HRank: filter pruning using high-rank feature map. In: Proceedings of the IEEE International Conference on Computer Vision and Pattern Recognition (CVPR) (2020)
18. Liu, Z., Shen, Z., Savvides, M., Cheng, K.-T.: ReActNet: towards precise binary neural network with generalized activation functions. In: Vedaldi, A., Bischof, H., Brox, T., Frahm, J.-M. (eds.) ECCV 2020. LNCS, vol. 12359, pp. 143–159. Springer, Cham (2020). https://doi.org/10.1007/978-3-030-58568-6_9
19. Liu, Z., Wu, B., Luo, W., Yang, X., Liu, W., Cheng, K.-T.: Bi-real net: enhancing the performance of 1-bit CNNs with improved representational capability and advanced training algorithm. In: Ferrari, V., Hebert, M., Sminchisescu, C., Weiss, Y. (eds.) ECCV 2018. LNCS, vol. 11219, pp. 747–763. Springer, Cham (2018). https://doi.org/10.1007/978-3-030-01267-0_44
20. Martinez, B., Yang, J., Bulat, A., Tzimiropoulos, G.: Training binary neural networks with real-to-binary convolutions. In: Proceedings of the International Conference on Learning Representations (ICLR) (2020)
21. Mishra, A., Marr, D.: Apprentice: using knowledge distillation techniques to improve low-precision network accuracy. In: Proceedings of the International Conference on Learning Representations (ICLR) (2018)
22. Nagel, M., Amjad, R.A., van Baalen, M., Louizos, C., Blankevoort, T.: Up or down? Adaptive rounding for post-training quantization. In: Proceedings of the International Conference on Machine Learning (ICML) (2020)
23. Park, E., Yoo, S.: PROFIT: a novel training method for sub-4-bit MobileNet models. In: Vedaldi, A., Bischof, H., Brox, T., Frahm, J.-M. (eds.) ECCV 2020. LNCS, vol. 12351, pp. 430–446. Springer, Cham (2020). https://doi.org/10.1007/978-3-030-58539-6_26
24. Qin, H., Gong, R., Liu, X., Bai, X., Song, J., Sebe, N.: Binary neural networks: a survey. Pattern Recogn. (2020)
25. Qin, H., Gong, R., Liu, X., Bai, X., Song, J., Sebe, N.: Binary neural networks: a survey. In: Proceedings of the International Conference on Pattern Recognition (ICPR) (2020)
26. Radford, A., et al.: Learning transferable visual models from natural language supervision. arXiv preprint arXiv:2103.00020 (2021)

27. Rastegari, M., Ordonez, V., Redmon, J., Farhadi, A.: XNOR-Net: ImageNet classification using binary convolutional neural networks. In: Leibe, B., Matas, J., Sebe, N., Welling, M. (eds.) ECCV 2016. LNCS, vol. 9908, pp. 525–542. Springer, Cham (2016). https://doi.org/10.1007/978-3-319-46493-0_32

28. Su, W., et al.: VL-BERT: pre-training of generic visual-linguistic representations. arXiv preprint arXiv:1908.08530 (2019)

29. Wang, P., Hu, Q., Zhang, Y., Zhang, C., Liu, Y., Cheng, J.: Two-step quantization for low-bit neural networks. In: Proceedings of the IEEE International Conference on Computer Vision and Pattern Recognition (CVPR) (2018)

30. Wang, Z., Lu, J., Tao, C., Zhou, J., Tian, Q.: Learning channel-wise interactions for binary convolutional neural networks. In: Proceedings of the IEEE International Conference on Computer Vision and Pattern Recognition (CVPR) (2019)

31. Yamamoto, K.: Learnable companding quantization for accurate low-bit neural networks. In: Proceedings of the IEEE International Conference on Computer Vision and Pattern Recognition (CVPR) (2021)

32. Yoon, J., Hwang, S.J.: Combined group and exclusive sparsity for deep neural networks. In: Proceedings of the International Conference on Machine Learning (ICML) (2017)

33. Zareian, A., Rosa, K.D., Hu, D.H., Chang, S.F.: Open-vocabulary object detection using captions. In: Proceedings of the IEEE/CVF Conference on Computer Vision and Pattern Recognition, pp. 14393–14402 (2021)

34. Zbontar, J., Jing, L., Misra, I., LeCun, Y., Deny, S.: Barlow twins: self-supervised learning via redundancy reduction. arXiv preprint arXiv:2103.03230 (2021)

35. Zhuang, B., Shen, C., Tan, M., Liu, L., Reid, I.: Structured binary neural networks for accurate image classification and semantic segmentation. In: Proceedings of the IEEE International Conference on Computer Vision and Pattern Recognition (CVPR) (2019)

Power Awareness in Low Precision Neural Networks

Nurit Spingarn Eliezer[1,2]($\boxtimes$) (ID), Ron Banner[2](ID), Hilla Ben-Yaakov[2](ID),
Elad Hoffer[2](ID), and Tomer Michaeli[1](ID)

[1] Technion–Israel Institute of Technology, Haifa, Israel
nurits@campus.technion.ac.il, tomer.m@ee.technion.ac.il
[2] Habana labs, Intel, Haifa, Israel
{rbanner,hbenyaakov,ehoffer}@habana.ai

Abstract. Existing approaches for reducing DNN power consumption rely on quite general principles, including avoidance of multiplication operations and aggressive quantization of weights and activations. However, these methods do not consider the precise power consumed by each module in the network and are therefore not optimal. In this paper we develop accurate power consumption models for all arithmetic operations in the DNN, under various working conditions. We reveal several important factors that have been overlooked to date. Based on our analysis, we present PANN (power-aware neural network), a simple approach for approximating any full-precision network by a low-power fixed-precision variant. Our method can be applied to a pre-trained network and can also be used during training to achieve improved performance. Unlike previous methods, PANN incurs only a minor degradation in accuracy w.r.t. the full-precision version of the network and enables to seamlessly traverse the power-accuracy trade-off at deployment time.

Keywords: Quantization · Power consumption of DNNs · Multiplier-free DNNs

1 Introduction

With the ever increasing popularity of deep neural networks (DNNs) for tasks like face detection, voice recognition, and image enhancement, power consumption has become one of the major considerations in the design of DNNs for resource-limited end-devices. Over the last several years, a plethora of approaches have been introduced for achieving power efficiency in DNNs. These range from specialized architectures [16,40,42,45], to hardware oriented methods like multiplier-free designs and low-precision arithmetic.

Multiplier aware methods attempt to reduce power consumption by avoiding the costly multiplication operations, which dominate the computations in

---

Supplementary Information The online version contains supplementary material available at https://doi.org/10.1007/978-3-031-25082-8_5.

a DNN. Several works replaced multiplications by additions [6,7,27] or by bit shift operations [8] or both [48]. Others employed efficient matrix multiplication operators [26,46]. However, most methods in this category introduce dedicated architectures, which require training the network from scratch. This poses a severe limitation, as different variants of the network need to be trained for different power constraints.

Low-precision DNNs reduce power consumption by using low-precision arithmetic. This is done either via quantization-aware training (QAT) or with post-training quantization (PTQ) techniques. The latter avoid the need for re-training the network but often still require access to a small number of calibration samples in order to adapt the network's weights. These techniques include approaches like re-training, fine-tuning, calibration and optimization [4,18,28,37]. All existing methods in this category suffer from a large drop in accuracy with respect to the full-precision version of the network, especially when working at very low bit widths. Moreover, similarly to the multiplier-free approaches, they do not provide a mechanism for traversing the power-accuracy trade-off without actually changing the hardware (e.g., replacing an 8×8 bits multiplier by a 4×4 bits one).

In this work, we introduce a *power-aware neural network* (PANN) approach that allows to dramatically cut down the power consumption of DNNs. Our method can be applied at post-training to improve the power efficiency of a pre-trained model, or in a QAT setting to obtain even improved results. Our approach is based on careful analysis of the power consumed by additions and multiplications, as functions of several factors. We rely on bit toggling activity, which is the main factor affecting dynamic power consumption, and support our theoretical analysis with accurate gate-level simulations on a 5 nm process.

Our first important observation is that a major portion of the power consumed by a DNN is due to the use of signed integers. We therefore present a simple method for converting any pre-trained model to use unsigned arithmetic. This conversion does not change the functionality of the model and, as can be seen in Fig. 1(a), dramatically reduces power consumption on common hardware configurations.

Our second observation is that the multiplier's power consumption is dominated by the larger bit width among its two inputs. Therefore, although high accuracy can often be achieved with quite drastic quantization of only the weights, this common practice turns out to be ineffective in terms of power consumption. To take advantage of the ability to achieve high accuracy with drastic weight quantization, here we propose a method to remove the multiplier altogether. Our approach can work in combination with any activation quantization method. We show theoretically and experimentally that this method is far advantageous over existing quantization methods at low power budgets, both at post-training and in QAT settings (see Fig. 1(a) and Sect. 6).

Our method allows working under any power constraint by tuning the average number of additions used to approximate each multiply-accumulate (MAC) operation. This is in contrast to regular quantization methods, which are limited to particular bit-width values. This allows traversing the power-accuracy trade-off without changing the architecture (e.g., multiplier bit width), as required by existing methods.

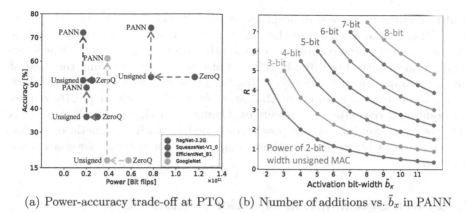

(a) Power-accuracy trade-off at PTQ   (b) Number of additions vs. $\tilde{b}_x$ in PANN

Fig. 1. (a) For each pre-trained full-precision model, we used ZeroQ [5] to quantize the weights and activations to 4 bits at post-training. Converting the quantized models to work with unsigned arithmetic (←), already cuts down 33% of the power consumption (assuming a 32 bit accumulator). Using our PANN approach to quantize the weights (at post-training) and remove the multiplier, further decreases power consumption and allows achieving higher accuracy for the same power level (↑). See more examples in SM Sect. 6.1. (b) Each color represents the power of an unsigned b_x-bit MAC for some value of b_x. In PANN, we can move on a constant power curve by modifying the number of additions per element R (vertical axis) on the expense of the activation bit width $\tilde{b}_x$ (horizontal axis).

2 Related Work

2.1 Avoiding Multiplications

In fixed point (integer) representation, additions are typically much more power-efficient than multiplications [14,15]. Some works suggested to binarize or ternarize the weights to enable working with additions only [7,27,29]. However, this often severely impairs the network's accuracy. Recent works suggested to replace multiplications by bit shifts [8] or additions [6] or both [48]. Other methods reduce the number of multiplications by inducing sparsity [32,47], decomposition into smaller intermediate products [23], Winograd based convolutions [26], or Strassen's matrix multiplication algorithm [46]. Some of these methods require internal changes in the model, a dedicated backpropagation scheme, or other modifications to the training process.

2.2 Quantization

DNN quantization approaches include post-training quantization (PTQ), which is applied to a pre-trained model, and quantization-aware training (QAT), where the network's weights are adapted to the quantization during training [2,9,12,31]. PTQ methods are more flexible in that they do not require access to the training set. These methods show optimal results for 8-bit quantization, but tend to incur a large drop in accuracy at low bit widths. To battle this effect, some PTQ methods minimize the quantization errors of each layer individually by optimizing the parameters over a calibration set [17, 35, 37]. Others use nonuniform quantization [10, 30]. Effort is also invested in avoiding the need of any data sample for calibration [5, 13, 36, 43]. These methods, however, still show a significant drop in accuracy at the lower bit widths, while frequently requiring additional computational resources. Common to all quantization works is that they lack analysis of the power consumed by each arithmetic operation as a function of bit-width, and thus cannot strive for optimal power-accuracy trade-offs.

3   Power Consumption of a Conventional DNN

The power consumption associated with running a DNN on a processing unit can be broadly attributed to two sources: memory movement, and compute. Recent hardware architectures effectively battle the power consumption associated with data movement to/from the memory by increasing the size of the local memory [1, 19, 44] or even using on-chip memory (as in the Graphcore[1] accelerator). Furthermore, modern accelerators reuse activations [11, 20, 25, 34], bringing them from memory only once and using them for all required computations. With these approaches, the dominant power consumer becomes the compute, on which we focus here (see additional discussion in SM Sec. 9).

The compute power is composed of a static power component and a dynamic one. The static power is due to a constant leakage current, and therefore does not depend on the circuit's activity. The dynamic power consumed by each node in the circuit is given by $P = CV^2 f \alpha$, where C is the node capacitance, V is the supply voltage, f is the operating frequency, and α is the switching activity factor (the average number of bit flips per clock) [38]. Here we focus on dynamic power, which is a major contributor to the overall power consumption (see SM Sect. 1 and [22, 24]) and is the only factor affected by how computations are performed on a given hardware.

Most of the computations in a forward pass of a DNN correspond to MAC operations. Here we focus on the popular implementation of MACs, which involves a multiplier that accepts two b-bit numbers and outputs a b_{acc}-bit result ($b_{\mathrm{acc}} = 2b$ to account for the largest possible product), and an accumulator with a large bit width B to which the multiplier's output is added repeatedly (see Fig. 2).

3.1   Dynamic Power vs. Simplistic Power Consumption Indicators

Before providing an analysis of the average dynamic power consumed by a MAC operation, it is important to note that dynamic power can be very different from other simplistic approximations of power consumption, like those relying

---

[1] https://www.graphcore.ai/.

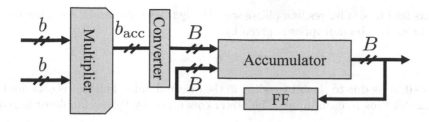

Fig. 2. Multiply-accumulate. The multiplier accepts two b-bit inputs. The product is then summed with the previous B bit sum, which awaits in the flip-flop (FF) register.

Table 1. Average number of bit flips per signed MAC. The b-bit multiplier inputs are drawn uniformly from $[-2^{b-1}, 2^{b-1})$ and its $b_{acc} = 2b$ bit output is summed with the B-bit number in the FF.

Element	Toggles
Multiplier (b-bit) inputs	$0.5b+0.5b$
Multiplier's internal units	$0.5b^2$
Accumulator (B-bit) input	$0.5B$
Accumulator sum & FF	$0.5b_{acc}+0.5b_{acc}$

on total gate counts or the number of units participating in a specific computation. Specifically, the dynamic power depends on the bit status of consecutive operations. Thus, the number of toggles per multiplication or addition is affected by all inner components, many of which are typically not at all related to the current product or addition, but rather to the previous one. For example, consider the sequence of MACs $2 \times 96 + 3 \times 2 + 1 \times 111$. Here, many bits are toggled in the 2nd MAC even though it involves small numbers. This is because the 1st and 3rd MACs involve large numbers (see SM Fig. 2 and Sect. 2 for more details).

3.2   Bit Toggling Simulation

To understand how much power each of the components in Fig. 2 consumes, we simulated them in Python. For the multiplier, we used the Booth-encoding architecture, which is considered efficient in terms of bit toggling [3]. For the accumulator, we simulated a serial adder. Our Python simulation allows measuring the total number of bit flips in each MAC operation, including at the inputs, at the outputs, in the flip-flop (FF) register holding the previous sum, and within each of the internal components (*e.g.*, the full-adders) of the multiplier.

Table 1 shows the average number of bit flips per MAC when both inputs to the multiplier are drawn uniformly at random from $[-2^{b-1}, 2^{b-1})$ (Gaussian inputs lead to similar results; please see SM Figs. 3–4). As can be seen, the power consumed by the multiplier is given by[2]

$$P_{\text{mult}} = 0.5b^2 + b, \tag{1}$$

where $0.5b^2$ is due to the bit toggling in the internal units, and $0.5b$ is contributed by the bit flips in each input. The power consumed by the accumulator is given by

$$P_{\text{acc}} = 0.5B + 2b, \tag{2}$$

where $0.5B$ is due to the bit toggling in its input coming from the multiplier, $0.5b_{\text{acc}} = b$ (recall $b_{\text{acc}} = 2b$) to the bit flips at the output, and an additional $0.5b_{\text{acc}} = b$ to the bit flips in the FF. To verify our model, we compared it with an accurate physical gate-level simulation on a 5 nm process, and found good agreement between the two (see Figs. 3(a) and 3(b)). This leads us to the following observation.

Observation 1. *A dominant source of power consumption is the bit toggling at the input of the accumulator (0.5B).*

Suppose, for example, we use $b = 4$ bits for representing the weights and activations, and employ a $B = 32$ bit accumulator, as common in modern architectures [21,41]. Then the toggling at the input of the accumulator ($0.5B = 16$) is responsible for 44.4% of the total power consumption ($P_{\text{mult}} + P_{\text{acc}} = 36$). At lower bit widths, this percentage is even larger.

Unfortunately, existing quantization methods and multiplier-free designs do not battle this source of power consumption. It has recently been shown in [39] that the bit-width B of the accumulator can be somewhat reduced by explicitly accounting for overflows. However, this approach requires dedicated training, and degrades the network's classification accuracy at low values of B. As we now show, it is possible to drastically reduce the bit toggles at the input of the accumulator at post-training *without changing the model's functionality* (and thus its classification accuracy).

4   Switching to Unsigned Arithmetic

Since the output of the multiplier has only $2b$ bits, one could expect to experience no more than b bit flips on average at the accumulator's input. Why do we have $0.5B$ bit flips instead? The reason is rooted in the use of signed arithmetic. Specifically, negative numbers are represented using two's complement, and thus switching between positive and negative numbers results in flipping of many of the higher bits. For example, when using a 32 bit accumulator,

---

[2] The power consumed by a single bit flip may vary across platforms (*e.g.*, between a 5 nm and a 45 nm fabrication), but the number of bit flips per MAC does not change. We therefore report power in units of bit-flips, which allows comparing between implementations while ignoring the platform.

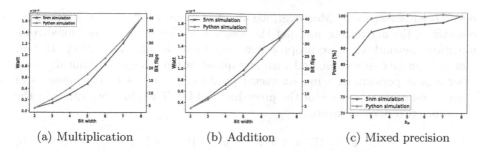

(a) Multiplication   (b) Addition   (c) Mixed precision

Fig. 3. Bit-flip simulation vs. 5 nm gate level simulation. (a) In red we plot the power consumed by a multiplication operation as measured in our Python simulation (bit flips). This curve agrees with the theoretical power model in Eq. (3), i.e. $P_{\text{mult}} = 0.5b^2 + b$ and with the power measurements on a 5-nm silicon process (in blue). (b) In this experiment, we measured the power consumed by a b-bit accumulator without the FF. In red is the power measured in our Python simulation, which is very close to our theoretical model $P_{\text{acc}} = 0.5b + 0.5b = b$. The power measurements for the 5 nm silicon process (blue) nicely agree with our Python simulation. (c) Here we measured the power consumed by 8×8 multiplier for $b_x = 8$ and different values of b_w. We plot the ratio between that power and the power consumed when $b_w = 8$. See more examples in SM Sect. 5.

if the output of the multiplier switches from $+2$ to -2, then the bits at the input of the accumulator switch from 00000000000000000000000000000010 to 11111111111111111111111111111110. Note that this effect is dominant only at the accumulator's input simply because sign changes at the output are rare.

If we could work with unsigned integers, then the higher bits at the accumulator's input would always remain zero, which would lead to a substantial reduction in power consumption without any performance degradation. To quantify this, we repeated the experiment of Sect. 3, but with the b-bit inputs to the multiplier now drawn uniformly from $[0, 2^b)$ (see SM Sec. 2 for details). In this case, the average number of bit flips at the input of the accumulator reduced from $0.5B$ to $0.5b_{\text{acc}} = b$. Specifically, the average power consumption of an unsigned MAC operation was measured to be

$$P_{\text{mult}}^{\text{u}} = 0.5b^2 + b \tag{3}$$

due to the multiplier and

$$P_{\text{acc}}^{\text{u}} = 3b \tag{4}$$

due to the accumulator. In (4), $2b$ bit flips occur at the accumulator's output and the FF, and b bit flips occur at the accumulator's input coming from the multiplier. Thus, although the mutliplier's power (3) turns out to be the same as in the signed setting (1), the accumulator's power (4) is substantially reduced w.r.t. the signed case (2).

The concept of unsigned arithmetic in quantization is not new (*e.g.*, asymmetric quantization [18]). However, its consequences on power consumption have never been pointed out. Moreover, asymmetric quantization is usually used for quantizing the activations and not the weights. It also requires additional computations beyond what we propose here (see more details in SM Sect. 4). Our method can be easily combined with any quantization scheme. Specifically, consider a layer performing a matrix-vector product, $y = Wx + b$. The elements of x are non-negative because of the preceding ReLU[3]. Therefore, we can split the layer into two parallel layers as

$$y^+ = W^+x + b^+, \qquad y^- = W^-x + b^-, \tag{5}$$

where $W^+ = \text{ReLU}(W)$, $b^+ = \text{ReLU}(b)$, $W^- = \text{ReLU}(-W)$, $b^- = \text{ReLU}(-b)$, and compute

$$y = y^+ - y^-. \tag{6}$$

This way, all MACs are converted to unsigned ones in (5), and only a single subtraction per output element is needed in (6). This one subtraction is negligible w.r.t. the MACs, whose number is usually in the thousands. Please see SM Fig. 7(b) for a schematic illustration.

Figure 1(a) shows the effect that this approach has on the power consumption of several pretrained networks for ImageNet classification. With a 32 bit accumulator, merely switching to unsigned arithmetic cuts 58% of the power consumption of these networks. We verified the power save on a real gate level simulation (see SM Fig. 7(a)). Please see more experiments with other accumulator bit widths in SM Sect. 3.2.

5   Removing the Multiplier

Having reduced the power consumed by the accumulator, we now turn to treat the multiplier. A common practice in quantization methods is to use different bit widths for the weights and the activations. This flexibility allows achieving good classification accuracy with quite aggressive quantization of one of them (typically the weights), but a finer quantization of the other. An interesting question is whether this approach is beneficial in terms of power consumption.

We repeated the experiment of Sect. 3, this time with the multiplier inputs having different bit widths, b_w and b_x. We focused on the standard setting of signed numbers, which we drew uniformly from $[-2^{b_w-1}, 2^{b_w-1})$ and $[-2^{b_x-1}, 2^{b_x-1})$. Interestingly, we found that the average number of bit flips in the multiplier's internal units is affected only by the larger among b_w and b_x. Accounting also for the bit flips at the inputs, we obtained that the multiplier's total power is

$$P_{\text{mult}} = 0.5 \left(\max\{b_w, b_x\}\right)^2 + 0.5(b_w + b_x). \tag{7}$$

Using our Python simulation, we found this surprising behavior to be characteristic of both the Booth multiplier and the simple serial multiplier (see SM Figs. 5–6.). We also verified this observation with accurate simulations on a 5 nm silicon

---

[3] Batch-norm layers should first be absorbed into the weights and biases.

process gate level synthesis. Specifically, we used an 8×8 multiplier and measured the power when one of the inputs was drawn uniformly from $[0, 2^8)$ and the other from $[0, 2^{b_w})$. For all values of $b_w < 8$, we got $88\% - 100\%$ of the power that was measured when both inputs were drawn from $[0, 2^8)$ (See Fig. 3(c) and more examples in SM Sect. 5). This leads us to our second important observation.

Observation 2. *There is marginal benefit in the common practice of decreasing the bit width of only the weights or only the activations, at least in terms of the power consumed by the multiplier.*

It should be noted that in the case of unsigned numbers, there exists some power save when reducing one of the bit widths, especially for the serial multiplier (see SM Fig. 6). This highlights again the importance of unsigned arithmetic. However, in our experiments we do not take this extra benefit of our approach into account when computing power consumption, so that our reports are conservative.

To benefit from the ability to achieve high precision with drastic quantization of only the weights, we now explore a solution that removes the multiplier altogether. Unlike other multiplier-free designs, our method allows converting any full-precision pre-trained model into a low-precision power-efficient one without changing the architecture.

5.1   Power Aware Weight Quantization

Consider the computation

$$y = \sum_{i=1}^{d} w_i \cdot x_i, \tag{8}$$

which involves d MACs. Here, $\{w_i, x_i\}$ are the weights and activations of a convolution or a fully-connected layer. Given $\{w_i, x_i\}$ in full precision, our goal is to accurately approximate (8) in a power-efficient manner. When quantizing the weights and activations we obtain the approximation

$$y \approx \sum_{i=1}^{d} \gamma_w \mathcal{Q}_w(w_i) \cdot \gamma_x \mathcal{Q}_x(x_i), \tag{9}$$

where the quantizers $\mathcal{Q}_w(\cdot)$ and $\mathcal{Q}_x(\cdot)$ map $\mathbb{R}$ to $\mathbb{Z}$, and γ_w and γ_x are their quantization steps[4]. To make the computation (9) power efficient, we propose to *implement multiplications via additions*. Specifically, assume $\mathcal{Q}_w(w_i)$ is a non-negative integer (as in Sect. 4). Then we can implement the term $\mathcal{Q}_w(w_i) \cdot \mathcal{Q}_x(x_i)$ as

$$\mathcal{Q}_w(w_i) \cdot \mathcal{Q}_x(x_i) = \underbrace{\mathcal{Q}_x(x_i) + \cdots + \mathcal{Q}_x(x_i)}_{\mathcal{Q}_w(w_i) \text{ times}}, \tag{10}$$

---

[4] In quantized models MAC operations are always performed on integers and rescaling is applied at the end.

so that (9) is computed as

$$y \approx \gamma_w \gamma_x \sum_{i=1}^{d} \sum_{j=1}^{\mathcal{Q}_w(w_i)} \mathcal{Q}_x(x_i). \tag{11}$$

This is the basis for our power-aware neural network (PANN) design.

It may seem non-intuitive that repeated additions can be more efficient than using a multiplier. Seemingly, if that were the case then multipliers would have been designed to work this way in the first place. However, recall that conventional multipliers use equal bit widths for both inputs, and do not consume less power when only one of their inputs is fed with small numbers (corresponding to a smaller bit width). By contrast, in PANN we do enjoy from taking $\mathcal{Q}_w(w_i)$ to be very small. As we will see, in this setting repeated additions do become advantageous.

Let $\boldsymbol{w} = (w_1, \ldots, w_d)^T$ and $\boldsymbol{x} = (x_1, \ldots, x_d)^T$ denote the full precision weights and activations, and denote their quantized versions by $\boldsymbol{w}_q = (\mathcal{Q}_w(w_1), \ldots, \mathcal{Q}_w(w_d))^T$ and $\boldsymbol{x}_q = (\mathcal{Q}_x(x_1), \ldots, \mathcal{Q}_x(x_d))^T$, respectively. In contrast to conventional quantization methods, here we do not need the quantized weights be confined to any particular range of the form $[0, 2^{b_w})$. Indeed, what controls our approximation accuracy is not the largest possible entry in $\boldsymbol{w}_q$, but rather the number of additions per input element, which is $\|\boldsymbol{w}_q\|_1/d$. Therefore, given a budget of R additions per input element, we propose to use a quantization step of $\gamma_w = \|\boldsymbol{w}\|_1/(Rd)$ in (9), so that

$$\mathcal{Q}(w_i) = \text{round}(w_i/\gamma_w). \tag{12}$$

This quantization ensures that the number of additions per input element is indeed as close as possible to the prescribed R. We remark that although we assumed unsigned weights, this quantization procedure can also be used for signed weights (after quantization, the positive and negative weights can be treated separately in order to save power, as in Sect. 4).

5.2   Power Consumption

We emphasize that in PANN, we would not necessarily want to use the same bit width for the activations as in regular quantization. We therefore denote the activation bit width in PANN by $\tilde{b}_x$ to distuinguish it from the b_x bits we would use with a regular quantizer. To estimate the power consumed by our approach, note that we have approximately $\|\boldsymbol{w}\|_1$ additions of $\tilde{b}_x$ bit numbers. On average, each such addition leads to $0.5\tilde{b}_x$ bit flips at the accumulator's output and $0.5\tilde{b}_x$ bit flips in the FF register (see Table 1). The input to the accumulator, however, remains fixed for $\mathcal{Q}_w(w_i)$ times when approximating the ith MAC and therefore changes a total of only d times throughout the entire computation in (11), each time with $0.5\tilde{b}_x$ bit flips on average. Thus, overall, the average power per element consumed by PANN is

$$P_{\text{PANN}} = \frac{\|\boldsymbol{w}\|_1 \tilde{b}_x + 0.5\tilde{b}_x d}{d} = (R + 0.5)\tilde{b}_x. \tag{13}$$

Algorithm 1. Determining the optimal parameters for PANN

1: **Input:** Power budget P
2: **Output:** Optimal $\tilde{b}_x, R$
3: **for each** $\tilde{b}_x \in [\tilde{b}_x^{\min}, \tilde{b}_x^{\max}]$ **do**
4:     Set $R = P/\tilde{b}_x - 0.5$ (Eq. (13))
5:     Quantize the weights using Eq. (12) with $\gamma_w = \|w\|/(Rd)$
6:     Quantize the activations to $\tilde{b}_x$ bits using any quantization method
7:     Run the network on a validation set, with multiplications replaced by additions
        using Eq. (10)
8:     Save the accuracy to $\text{Acc}(\tilde{b}_x)$.
9: **end for**
10: set $\tilde{b}_x \leftarrow \arg\max_{\tilde{b}_x} \text{Acc}(\tilde{b}_x), \quad R \leftarrow P/\tilde{b}_x - 0.5$

This implies that to comply with a prescribed power budget, we can either increase the activation bit width $\tilde{b}_x$ on the expense of the number of additions R, or vice versa.

Figure 1(b) depicts the combinations of $\tilde{b}_x$ and R that lead to the same power consumption as that of a b_x bit unsigned MAC, $P_{\text{MAC}}^{\text{u}} = 0.5b_x^2 + 4b_x$ (see (3), (4)), for several values of b_x (different colors). When we traverse such an equal-power curve, we also change the quantization error. Thus, the question is whether there exist points along each curve, which lead to lower errors than those obtained with regular quantization at the bit-width corresponding to that curve.

A theoretical analysis of the quantization error incurred by our method in comparison to regular uniform quantization, for a given power consumption budget, can be found in SM Sec. 10. This analysis provides insight into the difference between the approaches, however it is valid only for uniformly distributed weights and activations, which is often not an accurate enough assumption for DNNs. In practice, the best way to determine the optimal bit width $\tilde{b}_x$ is by running the quantized network on a validation set, as summarized in Algorithm 1.

6   Experiments

We now examine PANN in DNN classification experiments. We start by examining its performance at post training, and then move on to employ it during training. Here we focus only on the effect of removing the multiplier (vertical arrows in Fig. 1(a)). Namely, we assume all models have already been converted to unsigned arithmetic (recall this by itself reduces a lot of the power consumption).

6.1   PANN at Post Training

We illustrate PANN's performance in conjunction with a variety of post training methods for quantizing the activations, including the data free approaches GDFQ [43] and ZeroQ [5], the small calibration set method ACIQ [4], and the optimization based approach BRECQ [28], which is currently the state-of-the-art for post training quantization at low bit widths. Table 2 reports results with ResNet-50 on ImageNet (see SM Sect. 6.1 for results with other models). For the baseline methods, we always use equal bit widths for the weights and activations. Each row also shows our PANN variant, which works at the precise same power budget, where we choose the optimal $\tilde{b}_x$ and R using Algorithm 1. As can be seen, PANN exhibits only a minor degradation w.r.t. the full-precision model, even when working at the power budget of 2 bit networks. This is while all existing methods completely fail in this regime. Beyond the advantage in accuracy, it is important to note that the regular MAC approach requires changing the multiplier when changing the power budget (it uses a $b_x \times b_x$ multiplier for a b_x bit-width power budget). PANN, on the other hand, uses no multiplier and thus requires no hardware changes. Namely, to move between different equal-power curves (Fig 1(b)), all we need is to change one of the parameters ($\tilde{b}_x$ or R).

Table 2. PTQ: Classification accuracy [%] of ResNet-50 on ImageNet (FP 76.11%). The baselines (Base.) use equal bit widths for weights and activations. This bit width determines the power P, reported in first column in units of Giga bit-flips. The power is calculated as $P^u_{mult} + P^u_{acc}$ (Eqs. (3), (4)) times the number of MACs in the network. In each row, our variant PANN is tuned to work at the same power budget, for which we choose the optimal $\tilde{b}_x$ and R using Algorithm 1.

Power (Bits)				ACIQ		ZeroQ		GDFQ		BRECQ	
	A Mem.	W Mem.	Latency	Base.	**Our**	Base.	**Our**	Base.	**Our**	Base.	**Our**
265 (8)	1×	0.625×	7.5×	76.02	76.10	75.90	75.77	76.17	76.05	76.10	76.05
217 (6)	1.3×	0.83×	4.7×	75.41	76.05	73.57	74.65	76.05	76.02	75.86	76.01
134 (5)	2×	0.8×	3.5×	74.02	75.50	58.62	74.32	71.40	75.96	75.75	75.96
99 (4)	2.3×	0.75×	2.9×	66.12	75.10	3.53	68.24	50.81	75.20	75.42	75.80
68 (3)	2×	1×	2.2×	7.73	74.16	1.51	68.12	0.24	74.85	68.12	74.62
41 (2)	3×	1.5×	1.1×	0.20	71.55	0.10	62.96	0.13	74.32	18.80	73.21

6.2   PANN for Quantization Aware Training

To use PANN during training, we employ a straight-through estimator for back-propagation through the quantizers. Table 3 compares our method to LSQ [9], which is a state-of-the-art QAT approach, where in PANN we use LSQ for quantizing the activations. As can be seen, PANN outperforms LSQ for various models and power budgets. In Table 4 we compare our method to the multiplication-free approaches AdderNet [6] and ShiftAddNet [48], which are also training-based techniques. For each method, we report the *addition factor*, which is the ratio between its number of additions per layer and a regular layer. For example, AdderNet uses no multiplications but twice as many additions, so that its addition factor is 2. ShiftAddNet, on the other hand, uses one addition and one shift

Table 3. QAT: Comparison with LSQ. Imagenet classification accuracy [%] of various models. We report the bit width of LSQ and power in Giga bit-flips.

Bits (Power), Net	LSQ	PANN
18 (2), ResNet-18	67.32	70.83
30 (3), ResNet-18	69.81	71.12
41 (2), ResNet-50	71.36	76.65
68 (3), ResNet-50	73.54	76.78
155 (2), VGG-16bn	71.15	73.30

Table 4. QAT: Comparison with multiplier-free methods. Classification accuracy [%] of ResNet-20 on CIFAR-10. The top row specifies weight/activation bit widths, and the addition factor is specified in parentheses.

Method	6/6	5/5	4/4	3/3
Our (1×)	91.15	91.05	89.93	85.62
Our (1.5×)	91.52	91.50	90.05	86.12
Our (2×)	91.63	91.61	90.10	86.84
ShiftAddNet (1.5×)	87.72	87.61	86.76	85.10
AdderNet (2×)	67.39	65.53	64.31	63.50

operation. According to [48], a shift operation costs between 0.2 (on FPGA) and 0.8 (on a 45 nm ASIC) an addition operation. Therefore ShiftAddNet's addition factor is between 1.2 and 1.8, and for simplicity we regard it as 1.5. In PANN, we can choose any addition factor R, and therefore examine our method for $R = 1, 1.5, 2$. We can see in the table that PANN outperforms both AdderNet and ShiftAddNet for all bit widths, even when using a smaller addition factor. Please see more QAT comparisons in SM Sect. 6.2.

We now analyze the effect of PANN on other inference aspects besides power. One important aspect is *runtime memory footprint*. When working with batches, the runtime memory consumption is dominated by the activations [33]. The optimal number of bits $\tilde{b}_x$ we use for the activations is typically larger than the bit width b_x used in regular quantization. The second column of Table 2 reports the factor $b\tilde{b}_x/b_x$ by which the runtime memory of PANN exceeds that of the baseline model. As can be seen, this factor never exceeds 3, however it can be lowered on the expense of accuracy or latency. Please see an example in Table 11 in SM. The third column shows the increase in the runtime memory footprint of the weights (see discussion in SM Sect. 8). In the comparisons with the multiplier-free methods (Table 4 and Tables 7–8 in SM), there is no change in the memory footprint since we keep the same bit width for the activations. A second important factor is *latency*. Recall we remove the multiplier and remain only with the accumulator. Since addition is faster than multiplication, one could potentially use a higher clock-rate and thus gain speed. However, if we conservatively assume the original clock-rate, then the latency is increased by R (each multiplication is replaced by R additions). As can be seen in Table 2 fourth column, the increase in latency is quite small at the lower power budgets. For the multiplier-free methods, we obtain improvement in accuracy even for $R = 1$. In that case, our latency is smaller than that of AdderNet (2×) and ShiftAddNet (1.5×). Please refer to SM Sec. 8 for more analyses. Note that in all experiments, we constrained PANN to the precise same power budget as the regular MAC nets. This is the reason for the small increase in latency (especially at the power budgets corresponding to 5–8 bits). However, we can also aim for a slightly smaller power budget, which often significantly improves the latency on the expense of only a slight degradation in accuracy. For example, at an 8 bit power budget with BRECQ (top-right cell in Table 2), reducing the addition factor from 7.5 to 2 (which reduces the power from 265 to 83 Giga bit-flips) leads to only a slight drop in classification accuracy (from 76.05% to 75.65%). This ability to easily traverse the power-accuracy-latency trade-off is one of the strengths of PANN.

7   Conclusion

We presented an approach for reducing the power consumption of DNNs. Our technique relies on a detailed analysis of the power consumption of each arithmetic module in the network, and makes use of two key principles: switching to unsigned arithemtic, and employing a new weight quantization method that allows removing the multiplier. Our method substantially improves upon existing approaches, both at post-training and when used during training, leading to a higher accuracy at any power consumption budget.

Broader implications State-of-the-art AI techniques are not only data-hungry at training time, but also energy-hungry at inference time. This may limit their potential use in mobile devices which have relatively short battery life and tend to warm fast. We believe that energy efficient deep learning, of the type we propose here can bridge this gap and bring AI to mobile end-users, even allowing to run multiple AI applications in parallel on the same device.

Acknowledgements. This research was partially supported by the Ollendorff Miverva Center at the Viterbi Faculty of Electrical and Computer Engineering, Technion.

References

1. Abts, D., et al.: Think fast: a tensor streaming processor (tsp) for accelerating deep learning workloads. In: 2020 ACM/IEEE 47th Annual International Symposium on Computer Architecture (ISCA), pp. 145–158. IEEE (2020)
2. Achterhold, J., Koehler, J.M., Schmeink, A., Genewein, T.: Variational network quantization. In: International Conference on Learning Representations (2018)

3. Asif, S., Kong, Y.: Performance analysis of wallace and radix-4 booth-wallace multipliers. In: 2015 Electronic System Level Synthesis Conference (ESLsyn), pp. 17–22. IEEE (2015)

4. Banner, R., Nahshan, Y., Soudry, D.: Post training 4-bit quantization of convolutional networks for rapid-deployment. In: Advances in Neural Information Processing Systems, pp. 7950–7958 (2019)

5. Cai, Y., Yao, Z., Dong, Z., Gholami, A., Mahoney, M.W., Keutzer, K.: Zeroq: a novel zero shot quantization framework. In: Proceedings of the IEEE/CVF Conference on Computer Vision and Pattern Recognition, pp. 13169–13178 (2020)

6. Chen, H., et al.: Addernet: do we really need multiplications in deep learning? In: Proceedings of the IEEE/CVF Conference on Computer Vision and Pattern Recognition, pp. 1468–1477 (2020)

7. Courbariaux, M., Bengio, Y., David, J.P.: Binaryconnect: training deep neural networks with binary weights during propagations. arXiv preprint arXiv:1511.00363 (2015)

8. Elhoushi, M., Chen, Z., Shafiq, F., Tian, Y.H., Li, J.Y.: Deepshift: towards multiplication-less neural networks. arXiv preprint arXiv:1905.13298 (2019)

9. Esser, S.K., McKinstry, J.L., Bablani, D., Appuswamy, R., Modha, D.S.: Learned step size quantization. In: International Conference on Learning Representations (2019)

10. Fang, J., Shafiee, A., Abdel-Aziz, H., Thorsley, D., Georgiadis, G., Hassoun, J.H.: Post-training piecewise linear quantization for deep neural networks. In: Vedaldi, A., Bischof, H., Brox, T., Frahm, J.-M. (eds.) ECCV 2020. LNCS, vol. 12347, pp. 69–86. Springer, Cham (2020). https://doi.org/10.1007/978-3-030-58536-5_5

11. Gudaparthi, S., Narayanan, S., Balasubramonian, R., Giacomin, E., Kambalasubramanyam, H., Gaillardon, P.E.: Wire-aware architecture and dataflow for CNN accelerators. In: Proceedings of the 52nd Annual IEEE/ACM International Symposium on Microarchitecture, pp. 1–13 (2019)

12. Gupta, S., Agrawal, A., Gopalakrishnan, K., Narayanan, P.: Deep learning with limited numerical precision. In: International Conference on Machine Learning, pp. 1737–1746 (2015)

13. Haroush, M., Hubara, I., Hoffer, E., Soudry, D.: The knowledge within: Methods for data-free model compression. In: Proceedings of the IEEE/CVF Conference on Computer Vision and Pattern Recognition, pp. 8494–8502 (2020)

14. Horowitz, M.: Computing's energy problem (and what we can do about it). In:2014 IEEE International Solid-State Circuits Conference Digest of Technical Papers (ISSCC), pp. 10–14 (2014)

15. Horowitz, M.: Energy table for 45nm process. In: Stanford VLSI wiki (2014)

16. Huang, N.C., Chou, H.J., Wu, K.C.: Efficient systolic array based on decomposable mac for quantized deep neural networks (2019)

17. Hubara, I., Nahshan, Y., Hanani, Y., Banner, R., Soudry, D.: Improving post training neural quantization: Layer-wise calibration and integer programming. arXiv preprint arXiv:2006.10518 (2020)

18. Jacob, B., et al.: Quantization and training of neural networks for efficient integer-arithmetic-only inference. In: Proceedings of the IEEE Conference on Computer Vision and Pattern Recognition, pp. 2704–2713 (2018)

19. Jiao, Y., et al.: 7.2 a 12nm programmable convolution-efficient neural-processing-unit chip achieving 825tops. In: 2020 IEEE International Solid-State Circuits Conference-(ISSCC), pp. 136–140. IEEE (2020)

20. Jouppi, N.P., et al.: In-datacenter performance analysis of a tensor processing unit. In: Proceedings of the 44th Annual International Symposium on Computer Architecture, pp. 1–12 (2017)

21. Kalamkar, D.,et al.: A study of bfloat16 for deep learning training. arXiv preprint arXiv:1905.12322 (2019)

22. Karimi, N., Moos, T., Moradi, A.: Exploring the effect of device aging on static power analysis attacks. UMBC Faculty Collection (2019)

23. Kim, Y., Park, E., Yoo, S., Choi, T., Yang, L., Shin, D.: Compression of deep convolutional neural networks for fast and low power mobile applications. In: Bengio, Y., LeCun, Y. (eds.) 4th International Conference on Learning Representations, ICLR 2016, San Juan, Puerto Rico, May 2–4, 2016, Conference Track Proceedings (2016). http://arxiv.org/abs/1511.06530

24. Kim, Y., Kim, H., Yadav, N., Li, S., Choi, K.K.: Low-power RTL code generation for advanced CNN algorithms toward object detection in autonomous vehicles. Electronics 9(3), 478 (2020)

25. Kwon, H., Chatarasi, P., Pellauer, M., Parashar, A., Sarkar, V., Krishna, T.: Understanding reuse, performance, and hardware cost of DNN dataflow: a data-centric approach. In: Proceedings of the 52nd Annual IEEE/ACM International Symposium on Microarchitecture, pp. 754–768 (2019)

26. Lavin, A., Gray, S.: Fast algorithms for convolutional neural networks. In: Proceedings of the IEEE Conference on Computer Vision and Pattern Recognition, pp. 4013–4021 (2016)

27. Li, F., Zhang, B., Liu, B.: Ternary weight networks. arXiv preprint arXiv:1605.04711 (2016)

28. Li, Y., et al.: BRECQ: pushing the limit of post-training quantization by block reconstruction. In: International Conference on Learning Representations (2021). https://openreview.net/forum?id=POWv6hDd9XH

29. Lin, Z., Courbariaux, M., Memisevic, R., Bengio, Y.: Neural networks with few multiplications. arXiv preprint arXiv:1510.03009 (2015)

30. Liu, X., Ye, M., Zhou, D., Liu, Q.: Post-training quantization with multiple points: Mixed precision without mixed precision. In: Proceedings of the AAAI Conference on Artificial Intelligence, vol. 35, pp. 8697–8705 (2021)

31. Louizos, C., Reisser, M., Blankevoort, T., Gavves, E., Welling, M.: Relaxed quantization for discretized neural networks. arXiv preprint arXiv:1810.01875 (2018)

32. Mahmoud, M.: Tensordash: Exploiting sparsity to accelerate deep neural network training and inference (2020)

33. Mishra, A., Nurvitadhi, E., Cook, J.J., Marr, D.: WRPN: wide reduced-precision networks. arXiv preprint arXiv:1709.01134 (2017)

34. Mukherjee, A., Saurav, K., Nair, P., Shekhar, S., Lis, M.: A case for emerging memories in dnn accelerators. In: 2021 Design, Automation & Test in Europe Conference & Exhibition (DATE), pp. 938–941. IEEE (2021)

35. Nagel, M., Amjad, R.A., van Baalen, M., Louizos, C., Blankevoort, T.: Up or down? adaptive rounding for post-training quantization. arXiv preprint arXiv:2004.10568 (2020)

36. Nagel, M., Baalen, M.v., Blankevoort, T., Welling, M.: Data-free quantization through weight equalization and bias correction. In: Proceedings of the IEEE International Conference on Computer Vision, pp. 1325–1334 (2019)

37. Nahshan, Y., et al.: Loss aware post-training quantization. arXiv preprint arXiv:1911.07190 (2019)

38. Nasser, Y., Prévotet, J.C., Hélard, M., Lorandel, J.: Dynamic power estimation based on switching activity propagation. In: 2017 27th International Conference on Field Programmable Logic and Applications (FPL), pp. 1–2. IEEE (2017)

39. Ni, R., Chu, H.m., Castaneda Fernandez, O., Chiang, P.V., Studer, C., Goldstein, T.: Wrapnet: Neural net inference with ultra-low-precision arithmetic. In: 9th International Conference on Learning Representations (ICLR 2021) (2021)

40. Radosavovic, I., Kosaraju, R.P., Girshick, R., He, K., Dollár, P.: Designing network design spaces. In: Proceedings of the IEEE/CVF Conference on Computer Vision and Pattern Recognition, pp. 10428–10436 (2020)

41. Rodriguez, A., et al.: Lower numerical precision deep learning inference and training. Intel White Paper **3**, 1–19 (2018)

42. Sandler, M., Howard, A., Zhu, M., Zhmoginov, A., Chen, L.C.: Mobilenetv 2: inverted residuals and linear bottlenecks. In: Proceedings of the IEEE Conference on Computer Vision and Pattern Recognition, pp. 4510–4520 (2018)

43. Xu, S., et al.: Generative low-bitwidth data free quantization. In: Vedaldi, A., Bischof, H., Brox, T., Frahm, J.-M. (eds.) ECCV 2020. LNCS, vol. 12357, pp. 1–17. Springer, Cham (2020). https://doi.org/10.1007/978-3-030-58610-2_1

44. Tam, E., et al.: Breaking the memory wall for AI chip with a new dimension. In: 2020 5th South-East Europe Design Automation, Computer Engineering, Computer Networks and Social Media Conference (SEEDA-CECNSM), pp. 1–7. IEEE (2020)

45. Tan, M., et al.: Mnasnet: platform-aware neural architecture search for mobile. In: Proceedings of the IEEE/CVF Conference on Computer Vision and Pattern Recognition, pp. 2820–2828 (2019)

46. Tschannen, M., Khanna, A., Anandkumar, A.: StrassenNets: deep learning with a multiplication budget. In: International Conference on Machine Learning. pp. 4985–4994. PMLR (2018)

47. Venkatesh, G., Nurvitadhi, E., Marr, D.: Accelerating deep convolutional networks using low-precision and sparsity (2016)

48. You, H., et al.: ShiftaddNet: a hardware-inspired deep network. In: Advances in Neural Information Processing Systems, vol. 33 (2020)

Augmenting Legacy Networks for Flexible Inference

Jason Clemons[(✉)] [iD], Iuri Frosio, Maying Shen, Jose M. Alvarez,
and Stephen Keckler

NVIDIA Corporation, Santa Clara, CA, USA
`jclemons@nvidia.com`

Abstract. Once deployed in the field, Deep Neural Networks (DNNs)
run on devices with widely different compute capabilities and whose computational load varies over time. Dynamic network architectures are one
of the existing techniques developed to handle the varying computational load in real-time deployments. Here we introduce LeAF (Legacy
Augmentation for Flexible inference), a novel paradigm to augment the
key-phases of a pre-trained DNN with alternative, trainable, shallow
phases that can be executed in place of the original ones. At run time,
LeAF allows changing the network architecture without any computational overhead, to effectively handle different loads. LeAF-ResNet50 has
a storage overhead of less than 14% with respect to the legacy DNN; its
accuracy varies from the original accuracy of 76.1% to 64.8% while requiring 4 to 0.68 GFLOPs, in line with state-of-the-art results obtained with
non-legacy and less flexible methods. We examine how LeAF's dynamic
routing strategy impacts the accuracy and the use of the available computational resources as a function of the compute capability and load
of the device, with particular attention to the case of an unpredictable
batch size. We show that the optimal configurations for a given network
can indeed vary based on the system metrics (such as latency or FLOPs),
batch size and compute capability of the machine.

Keywords: Dynamic architecture · Real-time systems · Legacy · Fast
inference

1 Introduction

Deep Neural Networks (DNNs) deliver state-of-the-art results in a wide range
of applications. Very often, however, they are characterized by a high inference
cost [2,28]; when pushed to their limit, DNNs use a large amount of computational resources for small gains in model performance such as accuracy. These
high resource costs pose a barrier to the adoption of state-of-the-art DNNs in
the field.

The reasons for these inefficiencies are varied. Large or deep DNNs require
high compute at inference time. Efficient, static architectures like residual connections [6,9,13,14], pixel shuffle [23] or pruning [1,10,17,26] partially alleviate

© The Author(s), under exclusive license to Springer Nature Switzerland AG 2023
L. Karlinsky et al. (Eds.): ECCV 2022 Workshops, LNCS 13807, pp. 84–98, 2023.
https://doi.org/10.1007/978-3-031-25082-8_6

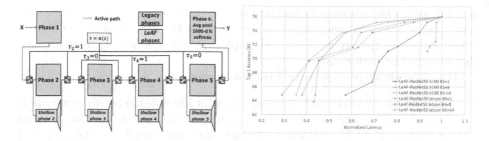

Fig. 1. The left panels shows LeAF-ResNet50, where shallow phases augment the legacy DNN. The external routing policy, $\tau = \pi(s)$, is disentangled from LeAF-ResNet50: it can be any user-defined function of the state of the compute device, s; here, the shallow phases 2 and 4 are executed together with the legacy phases 3 and 5. The right panel compares LeAF-ResNet50 performance on an A100 GPU(solid lines) and Jetson Xavier (dashed lines) as the batch size BS is varied between 1,8 and 64. The latency is normalized to the latency for the given batch size when using the base network. Notice that the curves are different for different batch sizes.

this issue, but low utilization of computational resources is still possible when a DNN *overthinks* about easy cases [25,32]. The complexity of dealing with inefficient DNNs is further increased when taking into account deploying models onto a wide range of devices, as different computing systems require diverse optimal implementations of the same DNN [7,25]. Even the computational load on the same device changes over time [11]. These issues can be alleviated by affording a DNN the ability to vary its architecture *on the fly* based on current system state. Thus, providing this capability is highly desirable for practical implementation and field deployment.

A variety of solutions to create DNNs with *variable* architecture exists. *Dynamic* DNNs are models that incorporate a gating policy function to route the input towards complementary paths with varying compute complexity [21,25,27], each specialized during training to handle different classes of data, thus making dynamic DNNs highly accurate and efficient. *Anytime prediction* DNNs [12,24] use early exits that allow selecting among a limited set of compute/accuracy compromises. All-for-one [2] DNNs are initially trained to be *flexible, i.e.,* such that sub-networks can be run while keeping a high accuracy; a single DNN configuration is then selected for deployment, based on the target device.

The analysis of networks with dynamic routing reveals interesting aspects in terms of real performance and efficiency on deployed systems. Different paths are characterized by different compute requirements and thus performances, that also change with the device and its current load status. In other words, when optimizing performance for deployment, all factors including not only the target model performance, but also the batch size and the device compute capability must be taken into account.

We introduce a DNN augmentation paradigm that provides full real-time control of the DNN configuration, and preserve the legacy network. Our technique produces a *flexible* DNN whose architecture can be changed on the fly accordingly to the constraints imposed by the state of the compute device, and achieving accuracy/compute cost compromises in the same ballpark of that of existing state-of-the-art methods. We explore using this flexible inference and characterize the performance.

In this work we provide several contributions:

- We introduce LeAF(Legacy Augmentation for Flexible inference), a paradigm to transform pre-trained, static DNNs into *flexible* ones, while also allowing execution of the legacy DNN. LeAF networks disentangle the gating policy problem from the computational aspects, allowing selection of the optimal accuracy/compute cost compromise on the fly.
- We perform a thorough analysis of the computational aspects of LeAF-DNNs for different tasks, network architectures and compute devices. We analyze the relation between FLOPs and latency and highlight the importance of the batch size for the deployment of effective DNNs. The validity of these insights extends beyond LeAF, showing for instance that results reported in literature for batch size 1 do not generalize naively to larger batch sizes.
- We demonstrate that DNNs with variable architectures allow many possible configurations, but few of them belong to the Pareto set in the accuracy / compute cost plane; we leverage this by fine-tuning LeAF-DNNs only for those configurations, but the insight applies again to a larger class of variable architecture networks.

2 Related Work

There are a few key differences in the attempts to speed up and reduce the energy consumption of DNNs. The first major difference is the one between *static* and *variable* network architectures. These can be further subdivided into *anytime prediction networks* with early exits and architectures with variable data-flow paths (generally referred to as *dynamic networks* in literature). An orthogonal classification can be made between methods that disentangle the (or "use an *external*") routing policy from the task at hand (like LeAF) and those that do not (or have an *internal* routing policy).

2.1 Static Architectures

DNN training can be made faster or less energy hungry by the adoption of skip/residual connections that help gradient propagation [6,9,11,14], but this option does little to save energy at inference time - on the contrary, it may favour the adoption of deeper networks characterized by costly inference. To reduce latency or energy consumption at inference time, software implementations making an effective use of the underlying hardware (like pixel shuffle [23]) or reducing the computational cost of the DNN operations (as for reduced precision [3,33]) exist. Another widely used possibility when deploying DNNs is offered by pruning. Weight pruning methods such as SNIP [15] or N:M pruning [26] remove individual parameters to induce sparsity, but need hardware support to get the actual acceleration, whereas channel pruning methods are more practical on modern hardware (e.g., GPUs). Typically, pruning is formulated as the resource (memory, computes, latency, etc.) constrained problem of selecting an optimal sub-network. Some channel pruning methods remove the least salient channels until reaching the desired cost in terms of FLOPs [16,17,30], but these same method hardly achieves optimal latency because of the documented discrepancy between the FLOPs and latency on complex devices. Platform aware methods achieves better latency reduction performances. HALP [22] estimates the channel latency cost using prior knowledge of the target platform, and consequently minimizes the loss drop under the user-defined latency constraints. NetAdapt [29] iteratively removes channels until reaching the latency goal with empirical latency measurements. However, these result in a single sub-networks that are device-specific by definition, and are also constraint-specific. Thus, storing all of the sets of parameters associated with different network training, each aimed at satisfying a different constraint, is possible but impractical because of the large memory footprint and the overhead associated with moving such a large amount of data in case of an architecture switch.

2.2   Variable Architectures

DNNs with a variable execution path are implemented in different flavors. One is to have the input automatically routed towards different blocks of the DNN, each specialized during training to handle different classes (*e.g.,* to classify pets or vehicles). These DNNs are referred to as *dynamic* networks. One example is the outrageously large neural network [21] that can be seen as a large mixture of experts each with small inference computational cost, but overall characterized by an huge memory footprint. Other dynamically routed networks, like SkipNet [27] or ConvNet-AIG [25], are designed such that, for a given input, different set of filters or layers are skipped. Slimmable networks [31] also allow skipping set of filters, but do not include any automatically routing mechanism, and therefore cannot be referred to as a dynamic network. Dynamic DNNs can surpass the original network in terms of accuracy at a lower computational cost, but training is often non trivial as the gating function is non-differentiable, and it also introduces a small computational overhead. Furthermore, the network architecture and training does not support the legacy preservation of the original DNN, contrary to LeAF.

A different type of DNNs with variable architecture is represented by *anytime prediction* networks, where early exit blocks can be added on top of an existing architecture as in the case of Branchynet [24] or Patience-based Early Exit [32]. Measuring the confidence of the network output at an early exit allows preventing the network overthinking and thus getting a high confidence result in a short amount of time. Some anytime prediction DNNs adopt an ad-hoc architecture that leverages information at multiple scales (see the Multi-Scale DenseNet (MSDNet) [12]), while other authors propose changing the cost function for anytime prediction networks to handle the different levels of prediction noise generated at different depths [11]. Dual dynamic Inference DDI [28] mixes anytime prediction networks and dynamic bypasses, using an LSTM network for gating functions.

Overall, variable architecture DNNs achieves better performance / compute cost compromises than static ones, but they have their own drawback. Legacy is often not considered, as we do in LeAF. Anytime prediction networks have limited flexibility, as the number of exits scales less than linearly with the network depth, while in LeAF it is proportional to the number of combinations of the shallow phases although only a fraction of these is useful in practice as we show in Sect. 3.2. Furthermore, early exits are not easily integrated into non-purely sequential architectures, like a U-nets. Lastly, the advantage claimed for variable architecture DNNs often refer to edge devices working with batch size 1. This is however only one of the interesting cases for latency reduction and/or energy saving: in some applications, like stereo vision, the same DNN may process image pairs, while servers are often required to handle large and possibly unpredictable batch sizes.

2.3   Internal vs. External Routing Policy

Dynamic networks include the computation of the data routing path into their architecture [7] - we say that they have an *internal routing policy*. Anytime prediction DNNs offer the possibility to automatically stop the execution based on the estimated confidence at a given exit (internal routing policy), but the user also has the possibility to decide a-priori the desired exit and the resulting performance/compute cost compromise, thus adopting an *external routing policy*. In LeAF, we disentangle the problem of creating a variable network architecture from that of selecting the best compute path for a given batch of any size. This is something that we have in common with slimmable networks [31], probably the closest work to ours, and one of the few aimed at creating a *flexible* network architecture, where the routing policy can be explicitly controlled by the user and changed on the fly to meet the time-varying constraints of the compute device. The One-For-All approach [2] partially achieves the same level of flexibility by adopting a network with an ad-hoc architecture that is trained to maximize the accuracy of any of the possible sub-networks (10^{19}) that can be extracted from it. This offers the flexibility of easily picking the sub-networks with the best performance/compute cost compromise for any target device. However, since the sub-network has a static architecture with no routing option, any form of flexibility is eventually lost once the DNN is fine-tuned and deployed in the field. None of the aforementioned methods consider the legacy aspect in any way, as LeAF-networks do.

3   Legacy Augmentation for Flexible Inference (LeAF)

3.1   LeAF Overview

Many DNNs include sets of layers with similar characteristics, that we call *phases*. For example, Resnet [8] has 6 phases (Fig. 1), Mobilenetv2 [20] has 12.

Table 1. LeAF training parameters for ResNet-50 and Mobilenetv2. We use SGD with step scheduler for the learning rate and batch size 32

Network	Training step	Learning Rate	Parameters	Epochs	Shallow phases
LeAF-ResNet-50	base-training	.1	lr_step=10, lr_gamma=.1	60	$2^4 = 16$
LeAF-ResNet-50	fine-tuning	.1	lr_step=10, lr_gamma=.1	60	6
LeAF-Mobilenetv2	base-training	4.5e–3	lr_step=1,lr_gamma=.96,weight_decay=4e-6	80	$2^5 = 32$
LeAF-Mobilenetv2	fine-tuning	4.5e–4	lr_step=1,lr_gamma=.97, weight_decay=4e-6	100	10

LeAF augments the legacy DNN with alternative, user-selectable *shallow phases* that can be executed in place of the original ones. Each shallow phase is designed to respect the original size of the input and output tensors, whereas internally the number of channels is a fraction of the original one. Thus, LeAF can be applied to a variety of network shapes such U-nets, ResNets and GANs. For a flexbile, LeAF-DNN with n phases, the number of possible execution paths is then 2^n, including the original one. During execution, any policy function $\tau = \pi(s)$ can be used to control the set of active shallow phases (Fig. 1), where s is the current state of the system (and could, for instance, include the current load or target accuracy). As τ is a binary vector, reconfiguration overhead is minimal allowing configuration updates in real time for any processed batch.

3.2   Training LeAF

To create a LeAF-DNN, we start from a pre-trained DNN with parameters θ that are frozen while we train the parameters ω of the shallow phases. This significantly decreases the overhead of training a DNN with multiple paths, while also allowing to continue to execute the legacy network. To preserve the accuracy of the legacy DNN, we also freeze any statistics in the batch normalization layers.

We train LeAF-DNNs using the adaptive loss in Eq. 1 based on [11] [31] where $J(\omega, \tau_i; \theta, X, y)$ is the traditional loss (*e.g.*, cross entropy) computed for the execution profile τ_i, whereas α_i is a multiplicative factor corresponding to the fraction of time a path is expected to run. $E[J(\omega; \theta, \tau_i; X, y)]$ is the sample average computed over the last 100 training iterations and is treated as a constant re-weighting factor during training. In our experiments we set $\alpha_i = 1/|\tau|$ (therefore assigning to any path has the same probability of being executed).

$$J_{tot\_adaptive}(\omega; \theta, X, y) = \sum_{i=0}^{|\tau|} \frac{J(\omega, \tau_i; \theta, X, y)}{E[J(\omega, \tau_i; \theta, X, y)]} \cdot \alpha_i, \qquad (1)$$

Before training, we initialize the weights ω of the shallow phases by sampling from the original weights θ, as we found this to be more stable. Training is then performed in two steps, using the parameters in Table 1. In the base-training step, we use all the 2^n configurations of the LeAF-DNN, and set $\alpha_i = 1/(2^n) \ \forall i$. In the forward pass, for each batch we loop over all the configurations and accumulate the loss in Eq. 1 before doing the optimization step[1].

---

[1] We note that if the number of configurations becomes large, it may be more efficient to randomly sample the configuration for each mini-batch.

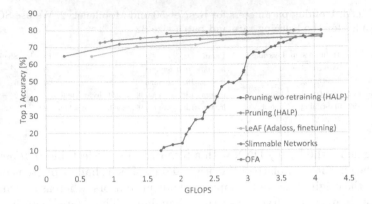

Fig. 2. Accuracy vs. FLOPs for various dynamic and flexible network methods including pruning and LeAF.

Once the base-training step is complete, we measure the accuracy and compute cost of each configuration, including a target batch size[2,3]. Figure 3 shows the accuracy vs. FLOPs for LeAF-Resnet-50 on Imagenet for batch size 256. Some of the network configurations achieve a sub-optimal accuracy/FLOPs ratio: including these in the cost function in Eq. 1 is detrimental of the final accuracy. Therefore, we identify the configurations that reside in the Pareto set (Fig. 3), set $\alpha_i = 0$ in Eq. 1 for all those configurations that do not, and proceed with the fine-tuning training phase. To identify that a configuration is in the Pareto set we iterate over all the configurations and determine if there another configuration with a lower system cost and higher model performance/accuracy. If no such other configuration exists then the this configuration is considered in the Pareto set. We eventually store the Pareto set configurations with their final accuracy and costs in a look-up-table that can be easily used to implemented the policy gating function $\tau = \pi(s)$.

Figure 2 shows the results of various techniques for producing flexible networks. It can be seen that LeAF results have similar performance to other techniques. Thus we choose to use LeAF as a proxy to understand the impact of various system characteristics on the execution of models. We believe the similarities of the techniques makes the insights applicable to other dynamic network executions.

---

[2] The compute cost can be derived analytically in case of FLOPs, or experimentally in case of latency, energy or power consumption.

[3] When the compute cost is measured in FLOPS, the batch size will be normalized away. Other systems cost metrics (*e.g.*, latency) may be a function of the batch size, as detailed in the Results Section.

4   Evaluation

LeAF provides a novel solution to augmenting pre-trained models to generate flexible networks. The LeAF-DNNs with variable architecture can be used to adapt the network working point based on system performance and achieve performance/compute cost compromises similar to those of similar techniques. For LeAF-ResNet-50 we show a top-1 accuracy of 64.8% using approximately .69 GFLOPs with the same model that contains the weights of original ResNet-50, and 76.146% at 4 GFLOPs (legacy network). This comes at a cost of 13.8% more memory for the parameters in the model. We show that we can apply LeAF to efficient networks such as MobileNetV2 for as little as 7.7% memory overhead to store the weights.

4.1   Methodology

We apply LeAF to different network architectures to measure its effectiveness and to study in detail the computational behavior of dynamic neural networks. In particular, we consider as representative case studies the widely used ResNet-50 and MobileNetV2 for image classification on Imagenet [5], each augmented with LeAF.

To train, we use NVIDIA DGX systems with either 8 T V100-DGXS-16GB GPUs or 8 A100-SXM-80GB GPUs, with PyTorch 1.10 [19], CUDA 11.3 and cuDNN 8.2.2. We use pretrained DNNs from Torchvision [19] as legacy networks for ResNet-50 and MobileNetV2.

For each network we grouped the layers into the phases as shown in Fig. 1. For ResNet we create shallow phases by taking the ResNet residual layers in *conv2_x*, *conv3_x*, *conv4_x* and *conv5_x* and using a fraction of the original channel count. We add a 1×1 convolutional layer to the end of each shallow phase to return the channel count back to its size in the original phase, and allow switching between the shallow and original phase. We adopt a similar grouping strategy for MobileNetV2 where we augment each of the 7 bottleneck groups (but the first one) with shallow phases. In our evaluation we use shallow phases containing 25% of the channels in the original phases. The shallow phases require an additional 13.8% parameters in ResNet-50 and only 7.7% in MobileNetV2. The lower amount in MobileNetV2 is due to the impact of the scaling factor on the expansion factors in the inverted residuals. We then train each LeAF-network with the procedure in the former Section and the meta parameters in Table 1.

To measure the latency on a high-end GPU, we use a system with an AMD EPYC 7742, 512 GB RAM and Ampere based A100-SXM-80GB GPU. We measure the average GPU execution time through the pyTorch profiler and report it as latency. We ensure the clocks are locked to stock values.

Beyond evaluation on the A100 GPU, we also measure the performances of LeAF-ResNet-50 on a Jetson Xavier NX system with 384 CUDA cores, maintaining the FP32 precision of the network. We lock the clocks of the GPU to 1.1GHz for this evaluation. We vary the batch size from 1 to 256 by powers of 2.

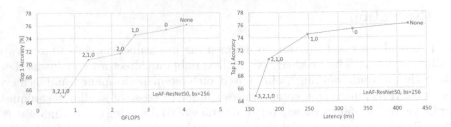

Fig. 3. Left: Accuracy vs. GFLOPs on a A100 GPU for LeAF-ResNet-50 for a batch size of 256. Points are labeled with the active set of shallow phases. Right: Accuracy Vs Latency on the same GPU for LeAF-ResNet-50 for a batch size of 256.

In the following, we perform a detailed analysis on the relation between FLOPs, latency and batch size for LeAF-ResNet-50 and then move to the evaluation of LeAFon MobileNetV2 for image classification.

4.2 LeAF-ResNet-50: On High End GPU

Figure 3 shows the Top-1 accuracy vs. FLOPs for LeAF-Resnet-50 in the left plot and Top-1 accuracy vs. latency in the right plot for a batch size of 256 on an A100 GPU. It can be seen that the plots have a similar but distinctly different shape. This demonstrates that while GFLOPs can be used as a proxy for latency it is better to use latency directly. This can be further illustrated by examining the x-axis distance of the configuration with {0} to the neighboring configurations of {None} and {1, 0}. In the GFLOPs plot (left) configuration {0} is closer to {None} while in the latency plot (right) it is closer to configuration {1, 0}. Furthermore the GFLOPs plot includes configuration {2, 0} on the Pareto frontier while that point is not included when measuring latency directly.

This reinforces the idea, already emerged in previous works [4,22,29], that latency and FLOPs are not completely interchangeable on complex, parallel system like GPUs. For example, an underutilized GPU can actually provide more parallel FLOPs without increasing the execution time by utilizing more of the hardware. Furthermore, the scheduling of work on a GPU is complex and can have unexpected interactions. This has an impact on the development of truly flexible architectures for DNN: depending on the system constraints to satisfy, one has to build the correct Pareto set for the problem in hand.

We bring to the attention of the reader another phenomenon that, to the best of our knowledge, has not been highlighted before. Figure 4 shows that, for varying batch size, the Pareto set in the accuracy vs. latency space varies as well, that may be (at least at first thought) unexpected. We believe the reason is again the complexity of a GPU system, whose behavior in terms of latency is nonlinear with respect to the occupancy of the computational cores. As for a given batch size a shallow phase may saturate the compute capability of a GPU while another may not, increasing the batch size may sometime (but not always) lead to zero latency overhead when switching from the shallow to the

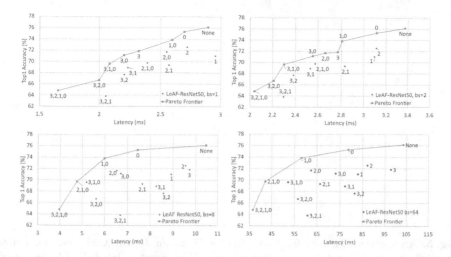

Fig. 4. Accuracy vs. latency for LeAF-ResNet-50 for batch size (bs) of 1, 2, 8, and 64, after base-training. Each point is labeled with the set of active shallow phases.

legacy phase. Considering that the GPU task scheduler is not under direct control, predicting the latency based on the configuration τ and batch size becomes a hard exercise (that is one of the reasons why in OFA [2] latency is predicted through an additional DNN). Therefore, when deploying a LeAF-DNN in the field and optimizing for latency (or power, energy, and so on), one should select the working batch size *a priori* and estimate the Pareto set accordingly to it. Alternatively, (although we did not explore this possibility here), one could also create a LeAF-DNN that is trained to deliver the highest accuracy for the configurations and batch sizes with the highest system performance.

4.3 LeAF-ResNet-50 Latency on Mobile

Figure 5 shows the Pareto sets in the accuracy vs. latency space at batch sizes of 1, 8, 16 and 64 when running on the Jetson Xavier NX. Similar to what was already reported for the high-end A100 GPU, we notice that in this case the Pareto set also varies as a function of the batch size for smaller batch sizes. Figure 5 shows that the relative positions of the configurations in the accuracy vs. latency space are less stable for lower batch sizes as demonstrated by the difference in the Pareto sets and relative configuration locations for the batch sizes 1 and 8 when compared to batch sizes 16 and 64. However, it can be seen that the relative position of the configurations stays consistent between batch sizes 16 and 64. Though not shown here, our experiments show that this relative positioning continues for higher batch sizes as well. Thus, for low batch size on this platform, it would be best to use different configurations for different batch sizes when optimizing for the latency. For larger batch size that fully utilize the GPU, the relative system performance stabilizes.

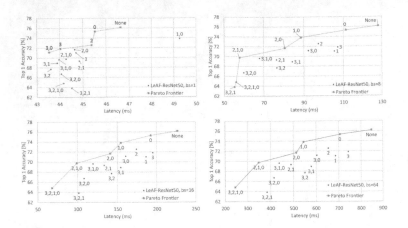

Fig. 5. Accuracy vs. latency for LeAF-ResNet-50 on an Jetson Xavier NX for batch size (bs) of 1, 8, 16, and 64 after base-training. Each point is labeled with the set of active shallow phases.

The upper right panel of Fig. 5 shows an example of the complex interactions of scheduling work on these systems. The configuration $\{1, 0\}$ repeatably took longer to execute than the base configuration. We believe this to be caused by an interaction between the kernel sizes, run-time framework and work scheduling.

To investigate the computational aspects of LeAF-ResNet-50 in more in detail on this device, we used the nvprof profiling tool [18] and analyzed the load on the GPU on Jetson Xavier NX system when running LeAF-ResNet-50. We found that increasing the batch size of the GPU leads to higher utilization metrics. The achieved *occupancy* is a measure of the utilization of the GPU's full compute capability, whereas the simultaneous multiprocessor (SM) *efficiency* is a measure of how many of the SMs are running a given kernel. For batch size 8, these two metrics are 46% and 89% respectively, meaning that almost 90% of the SMs are active, while they are lower for smaller batch sizes (< 40% and < 85%, respectively). For batch sizes larger than 8 the processing becomes more serialized leading to a latency increase that is more predictable and the Pareto set no longer changes. We find that at batch size 64 and above, 95% of the SMs are active further showing that more compute will be serialized. This is consistent with our hypothesis that the Pareto set stops changing as the GPU compute capability is saturated.

Overall, we found that the Pareto set is a function of the batch size both on high-end GPUs (like the A100 tested in the main paper) and on GPUs with smaller compute capabilities (like the Jetson Xavier NX tested here). Thus, when trying to deploy a network without having to retrain, it is valuable to be able adjust the network's architecture dynamically based not only on the specific device and its load conditions, but also considering the specific application and the expected distribution of the batch size so that we can optimize the system performance. For example, mobile platforms may use a single camera or stereo camera as inputs. Depending on which is used, the optimal flexible network configuration can change. This can be extended easily to 8 cameras in an autonomous vehicle application where there may be different LeAF configurations used to ensure optimal system performance.

4.4 Analysis as Function of Device

Our analysis of the LeAF ResNet-50 network shows different Pareto sets based on the device that is being used to run the network as noted by comparing Fig. 4 and Fig. 5. For instance, when running on the Jetson Xavier NX and increasing the batch size from 1 to 256 by powers of 2, we found that above a batch size of 8 the shallow phase set $\{2,0\}$ is present in the Pareto set. However, when running the same experiment on the A100 GPU, we found that $\{2,0\}$ is only present in the Pareto set for a batch size of 2. Furthermore, we can see that in Fig. 5 the Pareto set reaches a steady state with 6 shallow phase configurations for the mobile system as the batch size increases. However, for the A100 GPU in the main paper, the Pareto set reaches a steady state with 5 configurations as the batch size increases. The difference is the shallow phase configuration $\{2,0\}$. It can also be seen that the points that do not belong to the Pareto set form a different relative shape between the two devices. This shows that optimal configuration for the same network and batch size can vary based on the device. This is due to the fact that different phases saturate the capabilities of different devices at different thresholds.

4.5 LeAF on MobileNetV2

We apply LeAF to MobileNetV2 to investigate its adoption on NAS-based DNNs that are already targeted at efficient inference. As shown in Fig. 6, before fine-tuning the top-1 accuracy of LeAF-MobileNetV2 goes from 71.79% for the legacy configuration at 9.6 GFLOPs to 50.05% at 4.75 GFLOPs when all the five shallow phases are active. The Pareto set contains 10 configurations (out of the 32 possible execution paths) confirming that, in a DNN with variable architecture, not all the paths are equally important. After fine-tuning the Pareto set configurations, we observe an average increase of 2.4% in terms of accuracy (green triangles in Fig. 6), that is more pronounced for low-compute configurations.

Figure 7 shows the normalized latency and FLOPs for the different configurations of LeAF-MobileNetV2 for batch size 1 and 256 on an A100 GPU. The figures shows a stronger latency / FLOPs correlation (64% vs. 56%) for a larger batch size. The fact that, for batch size 1, a significant decrease in FLOPs does not correspond to a significant decrease in latency suggests that the system is likely underutilized, while for larger batch sizes the occupancy is higher and, since operations begin to be serialized even on a GPU, the scaling between latency and FLOPs tend to become more linear.

The results on LeAF-Mobilenetv2 show that applying LeAF to DNNs designed to be already fairly efficient led to a larger degradation in performances (when compared to the results obtained for ResNet-50) as more shallow phases

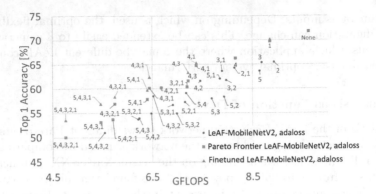

Fig. 6. Accuracy vs. FLOPs for all the compute configurations of LeAF-MobileNetV2, after base-training and after fine-tuning, for batch size 32. (Color figure online)

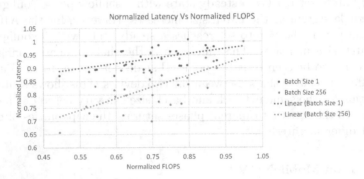

Fig. 7. MobileNetV2 Latency vs. FLOPs for batch sizes 1 and 256. The relationship between FLOPs and latency is batch size dependent.

are activated. This is expected, given the design space exploration performed to generate such networks. Nonetheless, LeAF continues to generate a single model that can be tuned in real time to satisfy the system constraints.

5   Discussion and Conclusion

In LeAF, we disentangle the selection and the execution of the compute path in flexible DNNs, to increase their versatility. This turns out to be a critical feature for the deployment of many real-time systems, where the currently available system resources determine the operating point to use. For example, a mostly idle system could use a highly accurate but costly DNN configuration, and switch to a lower performing, less costly one in case of high load scenario. LeAF provides flexible networks by augmenting a legacy network with lower system resource paths. This technique allows us to convert any network into a flexible one while preserving the ability to run the original network at full accuracy. Our results show that this technique produces models that are competitive with other state of the art techniques for dynamic networks that, however, do not have this same capability. We have shown the performance and capability of our technique while varying the system compute capability, batch size and network models.

While our experiments were performed using LeAF networks, we believe the results are applicable to other dynamic and flexible models. We have demonstrated the need to directly use the target system metric in our comparison of FLOP count and latency performance on the actual systems. We have shown that as the compute capability of the system is varied, the target configurations for running with a constrained resources will probably vary as well. Our analysis shows the importance of the batch size on the performance of DNNs with variable architecture - something that so far has not been highlighted enough. The fact that beyond a target device one should also specify a target batch size adds an additional axis to the problem of creating a flexible DNN. Optimizing performance in real-time systems with time varying constraints is a complex problem. LeAF helps to provide a way to maximize model performance as system resource constraints are varied during run-time.

References

1. Alvarez, J.M., Salzmann, M.: Learning the number of neurons in deep networks. In: NeurIPS (2016)
2. Cai, H., Gan, C., Wang, T., Zhang, Z., Han, S.: Once for all: train one network and specialize it for efficient deployment. In: ICLR (2020)
3. Cai, Y., Yao, Z., Dong, Z., Gholami, A., Mahoney, M.W., Keutzer, K.: ZeroQ: a novel zero shot quantization framework. In: CVPR (2020)
4. Dai, X., et al.: ChamNet: Towards efficient network design through platform-aware model adaptation. In: CVPR (2019)
5. Deng, J., Dong, W., Socher, R., Li, L.J., Li, K., Fei-Fei, L.: ImageNet: a large-scale hierarchical image database. In: 2009 IEEE Conference on Computer Vision and Pattern Recognition, pp. 248–255. IEEE (2009)
6. Ding, X., Zhang, X., Ma, N., Han, J., Ding, G., Sun, J.: RepVGG: Making VGG-style convnets great again. In: CVPR (2021)
7. Han, Y., Huang, G., Song, S., Yang, L., Wang, H., Wang, Y.: Dynamic neural networks: a survey. IEEE Trans, PAMI (2021)
8. He, K., Zhang, X., Ren, S., Sun, J.: Deep residual learning for image recognition. In: CVPR (2016)
9. He, K., Zhang, X., Ren, S., Sun, J.: Identity mappings in deep residual networks. In: Leibe, B., Matas, J., Sebe, N., Welling, M. (eds.) ECCV 2016. LNCS, vol. 9908, pp. 630–645. Springer, Cham (2016). https://doi.org/10.1007/978-3-319-46493-0_38
10. He, Y., Zhang, X., Sun, J.: Channel pruning for accelerating very deep neural networks. In: ICCV (2017)
11. Hu, H., Dey, D., Hebert, M., Bagnell, J.: Learning anytime predictions in neural networks via adaptive loss balancing. In: AAAI (2019)
12. Huang, G., Chen, D., Li, T., Wu, F., van der Maaten, L., Weinberger, K.: Multi-scale dense networks for resource efficient image classification. In: ICLR (2018)
13. Huang, G., Sun, Yu., Liu, Z., Sedra, D., Weinberger, K.Q.: Deep networks with stochastic depth. In: Leibe, B., Matas, J., Sebe, N., Welling, M. (eds.) ECCV 2016. LNCS, vol. 9908, pp. 646–661. Springer, Cham (2016). https://doi.org/10.1007/978-3-319-46493-0_39

14. Jastrzebski, S., Arpit, D., Ballas, N., Verma, V., Che, T., Bengio, Y.: Residual connections encourage iterative inference. In: ICLR (2018)
15. Lee, N., Ajanthan, T., Torr, P.H.: Snip: Single-shot network pruning based on connection sensitivity. CoRR abs/1810.02340 (2018)
16. Li, B., Wu, B., Su, J., Wang, G.: EagleEye: fast sub-net evaluation for efficient neural network pruning. In: Vedaldi, A., Bischof, H., Brox, T., Frahm, J.-M. (eds.) ECCV 2020. LNCS, vol. 12347, pp. 639–654. Springer, Cham (2020). https://doi. org/10.1007/978-3-030-58536-5_38
17. Molchanov, P., Mallya, A., Tyree, S., Frosio, I., Kautz, J.: Importance estimation for neural network pruning. In: CVPR (2019)
18. NVIDIA: CUDA Toolkit Documentation. http://www.docs.nvidia.com/cuda/ profiler-users-guide/index.html. Accessed 30 Oct 2021
19. Paszke, A., et al.: PyTorch: an imperative style, high-Performance deep learning library. In: NEURIPS (2019)
20. Sandler, M., Howard, A., Zhu, M., Zhmoginov, A., Chen, L.C.: MobileNetV2: inverted residuals and linear bottlenecks. In: CVPR (2018)
21. Shazeer, N., Mirhoseini, A., Maziarz, K., Davis, A., Le, Q.V., Hinton, G.E., Dean, J.: Outrageously large neural networks: the sparsely-gated mixture-of-experts layer. In: ICLR (2017)
22. Shen, M., Yin, H., Molchanov, P., Mao, L., Liu, J., Alvarez, J.M.: HALP: hardware-aware latency pruning. CoRR abs/2110.10811 (2021)
23. Shi, W., et al.: Real-time single image and video super-resolution using an efficient sub-pixel convolutional neural network. In: CVPR (2016)
24. Teerapittayanon, S., McDanel, B., Kung, H.: BranchyNet: fast inference via early exiting from deep neural networks. In: ICPR (2016)
25. Veit, A., Belongie, S.: Convolutional networks with adaptive inference graphs. Int. J. Comput. Vis. **128**(3), 730–741 (2019). https://doi.org/10.1007/s11263-019-01190-4
26. Wang, W., et al.: Accelerate CNNs from three dimensions: a comprehensive pruning framework. In: ICML (2021)
27. Wang, X., Yu, F., Dou, Z.-Y., Darrell, T., Gonzalez, J.E.: SkipNet: learning dynamic routing in convolutional networks. In: Ferrari, V., Hebert, M., Sminchisescu, C., Weiss, Y. (eds.) ECCV 2018. LNCS, vol. 11217, pp. 420–436. Springer, Cham (2018). https://doi.org/10.1007/978-3-030-01261-8_25
28. Wang, Y., et al.: Dual dynamic inference: enabling more efficient, adaptive, and controllable deep inference. IEEE J. Selected Top. Sig. Process. **14**(4), 623–633 (2020)
29. Yang, T.-J., et al.: NetAdapt: platform-aware neural network adaptation for mobile applications. In: Ferrari, V., Hebert, M., Sminchisescu, C., Weiss, Y. (eds.) ECCV 2018. LNCS, vol. 11214, pp. 289–304. Springer, Cham (2018). https://doi.org/10. 1007/978-3-030-01249-6_18
30. Yu, J., Huang, T.S.: Network slimming by slimmable networks: towards one-shot architecture search for channel numbers. CoRR abs/1903.11728 (2019). http:// arxiv.org/1903.11728
31. Yu, J., Yang, L., Xu, N., Yang, J., Huang, T.: Slimmable neural networks. In: ICLR (2019)
32. Zhou, W., Xu, C., Ge, T., McAuley, J.J., Xu, K., Wei, F.: Bert loses patience: fast and robust inference with early exit. In: NeurIPS (2020)
33. Zhu, C., Han, S., Mao, H., Dally, W.J.: Trained ternary quantization. In: ICLR (2017)

Deep Neural Network Compression for Image Inpainting

Soyeong Kim[✉] [ID], Do-Yeon Kim[ID], and Jaekyun Moon[ID]

Korea Advanced Institute of Science and Technology, Daejeon, Republic of Korea
{best004,dy.kim}@kaist.ac.kr, jmoon@kaist.edu

Abstract. Image inpainting techniques have recently been developed leveraging deep neural networks and have seen many real-world applications. However, image inpainting networks, which are typically based on generative adversarial network (GAN), suffer from high parameter complexities and long inference time. While there are some efforts to compress image-to-image translation GAN, compressing image inpainting networks has rarely been explored. In this paper, we aim to create a small and efficient GAN-based inpainting model by compressing the generator of the inpainting model without sacrificing the quality of reconstructed images. We propose novel channel pruning and knowledge distillation techniques that are specialized for image inpainting models with mask information. Experimental results demonstrate that our compressed inpainting model with only one-tenth of the model size achieves similar performance to the full model.

Keywords: Image inpainting · Network pruning · Knowledge distillation

1 Introduction

Image inpainting is one area that is receiving significant attention nowadays. A primary goal of image inpainting is to fill the missing masked regions with semantically relevant image patches. Image inpainting technique can be used in various applications such as removing undesirable obstacles and restoring damaged objects. To generate the missing parts having desired structure and texture, consistent with surrounding regions, recent works utilize the concept of Generative Adversarial Network (GAN) [3]. GAN based image inpainting models thus far have shown a decent quality for reconstructed patches by making use of the ability to generate a naturally looking image. However, despite its success in utilizing GAN, none of the existing works takes into account the computational side of the inpainting model, which may hinder real-world deployment in resource-limited environments. Since GAN is notorious for requiring high computational complexity and memory usage, it is necessary to create a small inpainting model to utilize it in resource-constrained applications such as mobile devices. Moreover, in situation where a large number of images need to be processed within a limited time, fast and lightweight inpainting networks become more valuable.

L. Karlinsky et al. (Eds.): ECCV 2022 Workshops, LNCS 13807, pp. 99–114, 2023.
https://doi.org/10.1007/978-3-031-25082-8_7

Motivations. Unfortunately, several studies such as [22] have shown that most neural network compression techniques, which are usually used in regression and classification problems, have limited performance when applied to GAN. GAN has more complex structures and unstable training processes because of having to train two neural networks at the same time. This characteristic leads to high instability. Nevertheless, several researchers have developed techniques to compress image-to-image translation GAN models [14,23]. They mainly adopt techniques such as channel pruning, quantization and knowledge distillation which are designed for compressing GAN. However, there was still no attempt to compress the GAN-based image inpainting model. Since the generative models used for image inpainting have different characteristics from image-to-image translation GAN, it is desired to have a new model compression technique specialized for image inpainting.

Contributions. In this paper, we propose new channel pruning and knowledge distillation methods tailored to image inpainting. In our channel pruning method, we first define the gradients with respect to reconstructed region as the sensitivity to the masked region. Based on the sensitivity, we prune the unimportant channels in the masked region to only keep the channels that are sensitive to the masked region. Regarding the knowledge distillation, we propose three types of distillation to transfer the knowledge: one pixel-level distillation and two image-level distillation losses. For pixel-level distillation, we utilize partial convolution to focus only on the valid pixels when matching the features of two generators. For image-level distillation, we use both an extra pre-trained network and a pre-trained discriminator to extract information in the feature maps.

Our scheme based on channel pruning and knowledge distillation tailored to image inpainting achieves the best performance on globally and locally consistent image completion (GLCIC) [9], which is a standard model for image inpainting. We can remove 70% of channels through entire layers with only a small degradation in image resolution.

2   Related Works

Image Inpainting. Recent works on image inpainting exploit GAN to directly predict pixel values inside the hole [9,15,20,27,28]. Context Encoder [20] is the first deep inpainting model which generates the missing regions using encoder-decoder architecture. Iizuka et al. [9] proposed local and global discriminators and applied dilated convolution layers [26] instead of fully connected layers. [27] takes a two-steps approach using a coarse-to-fine network and introduces a contextual attention layer which can borrow features from distant locations. Liu et al. [15] devised partial convolution layers that operate on only valid and uncorrupted pixels. Yu et al. [28] proposed gated convolution layers which generalize partial convolution layers by learning the feature selection for each layer. While the above works on inpainting focus on developing high quality image patches, they do not focus on reducing the size of the inpainting models for speed and lightness.

Network Pruning. Most existing network pruning methods can be broadly categorized into unstructured [6] and structured pruning [12]. Unstructured pruning prunes the network at the level of parameter individually, so that it can produce fine-grained sparse neural network. As a result, unstructured pruning shows a high sparsity ratio. However, it requires a special hardware [5] to accelerate training or inference. On the other hand, structured pruning does not require any particularly designed hardware specialized for operating pruned network efficiently, since it prunes the network in the level of filter or layer. Wen et al. [24] proposed group-LASSO regularization which turns out to induce unnecessary filter weights to be zero and Liu et al. [17] regularizes the parameters of batch normalization and removes the corresponding filters based on the magnitude of its scaling factor during training. To exploit the advantage of being able to implement a fast and lightweight model at software level, we make use of structured pruning in our method.

Knowledge Distillation. The history of knowledge distillation dates back to Hinton el al. [8] where the authors proposed distillation loss which allows the small student network to mimic the knowledge from a large teacher network. A recent survey [4] divides the knowledge distillation techniques into three categories: response-based, feature-based and relation-based knowledge distillation. In response-based distillation, the student model directly refers to the final output of the teacher model. Hinton el al. [8] used the last prediction, namely the soft labels, of the teacher model which is more informative compared to one-hot labels. Feature-based distillation uses not only the output of the last layer but also of intermediate layers. Romero et al. [21] defined the output of hidden layers as hints which are responsible for guiding a student's learning process. Relation-based methods identify the relation between the different activations or neurons. Yim et al. [25] capture the correlation between the feature maps and use it in the role of distilled knowledge. In this paper, we adopt response-base and feature-based distillation methods at each pixel-level distillation and image-level distillation step.

GAN Compression. Recently, image-to-image translation GAN compression methods have been actively developed [10,14,23]. Shu et al. [22] proposed a channel pruning by using co-evolutionary algorithms and reduced memory usage and FLOPs. However, their method is restrictive because it can be applied only to the CycleGAN algorithm. Chen et al. [2] and Aguinaldo et al. [1] used knowledge distillation to train a student generator by inheriting knowledge from a pre-trained teacher generator. While these methods use only a single approach, there are several works that combine multiple techniques to improve the efficiency. Li et al. [14] automatically found the channels using neural architecture search (NAS) and transferred the knowledge from the intermediate layers of the teacher generator. Jin et al. [10] directly pruned the trained teacher generator using one-step pruning and proposed a novel knowledge distillation with kernel alignment. Wang et al. [23] suggested a unified optimization form that integrated channel pruning, network quantization, and knowledge distillation. However, the above works do not focus on compressing the image inpainting GAN networks.

To the best of our knowledge, our work is the first attempt to compress the image inpainting network for efficient inference.

3   Proposed Algorithm

Given a pre-trained generator G_T of the image inpainting model, our goal is to create a smaller generator G_S. We proceed with channel pruning in the first step and re-train the pruned student model with distillation losses, as shown in Fig. 1. In the following, we describe our pruning and knowledge distillation methods tailored to the image inpainting model.

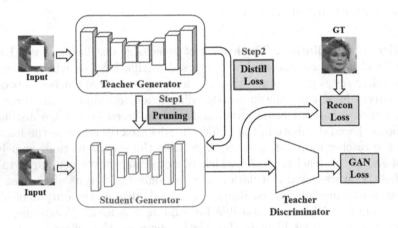

Fig. 1. Proposed compression scheme for image inpainting

3.1   Channel Pruning for Image Inpainting

Forward Weight Magnitude Information. We first describe our channel pruning method, which prunes channels according to their importance. Existing works on channel pruning [7,12,19] use only the forward information (incoming weight magnitude) connected to a specific channel to calculate the importance. More specifically, the forward score (FS) of the k-th channel in the i-th layer is defined as the sum of the incoming weight magnitudes:

$$\text{FS}_{W_k^i} = \sum_{l=1}^{n_{i-1}} |W_{l,k}^i| \tag{1}$$

where $W_{l,k}^i$ is the weight between the l-th channel in the $(i-1)$-th layer, the k-th channel in the i-th layer and n_{i-1} is the number of channels in the $(i-1)$-th layer. Our pruning method uses not only the forward weight magnitude but also the backward information from the generated image to compute the importance, as described below.

Backward Gradient Information. We describe how we calculate the backward information tailored to image inpainting. Since the final output of the inpainting model contains reconstructed masked region, we aim to prevent performance degradation by keeping the channels that are sensitive to masked regions. In order to determine which channels are sensitive to masked regions, we propose a new loss $\mathcal{L}_S$ which is used for scoring the sensitivity of the channels. We apply the input mask M_g to both the ground-truth input image x and the generated image $G_T(x, M_g)$ to define $\mathcal{L}_S$ as the $\ell 1$ norm of the difference of these two values:

$$\mathcal{L}_S = \|G_T(x, M_g) \odot M_g - x \odot M_g\|_1 \tag{2}$$

Intuitively, channels with high gradients with respect to $\mathcal{L}_S$ can be interpreted as important channels. As Fig. 2 shows, we propose an importance scoring method of the channels based on the gradient of outgoing weights to capture the change from output images. The backward score (BS) of the k-th channel in the i-th layer W_k^i is defined as the sum of the gradient of weights between the i-th layer and the $(i+1)$-th layer:

$$\mathrm{BS}_{W_k^i} = \sum_{t=1}^{n_{i+1}} \left| \frac{d\mathcal{L}_S}{dW_{k,t}^i} \right| \tag{3}$$

where $W_{k,t}^i$ is the weight between the k-th channel in the i-th layer, the t-th channel in the $(i+1)$-th layer and n_{i+1} is the number of channels in the $(i+1)$-th layer. The above score reflects the information of the output images, in particular, sensitive to the masked area. If we keep the channels having high sensitivity, the model can converge rapidly to a good model that generates masks close to the ground truth, when fine-tuning the student generator G_S.

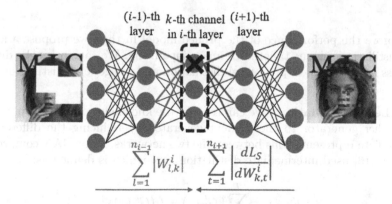

Fig. 2. Proposed pruning method considering both the forward magnitude information and the backward gradient information

Proposed Pruning Method. To make the network leverage both the forward and backward information, one naive approach is using the sum of two scores as the importance scores. However, when this method is applied, forward information is ignored because the backward score is dominant over the forward score. To balance two sets of information, we use the following method. We first prune a certain portion of channels based on Eq. (3) and then prune the remaining channels based on Eq. (1). For example, if we want to keep 30% of channels in each layer, we can first select 50% of channels based on the outgoing gradients, and then choose the final 30% of the channels with the large weight magnitudes among the selected channels. Figure 2 summarizes our algorithm.

3.2   Knowledge Distillation for Image Inpainting

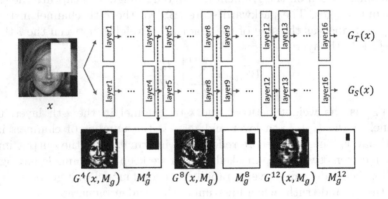

Fig. 3. Pixel-level distillation using partial convolution

To improve the performance of the pruned model further, we propose a knowledge distillation scheme tailored to image inpainting. We categorize the distillation loss into two types: pixel-level distillation and image-level distillation.

Pixel-Level Distillation. We can transfer the knowledge of feature space from the teacher generator to the student generator by reducing the difference of intermediate representations between the two networks. Prior GAN compression works [14,16] used intermediate distillation loss, which is defined as:

$$\mathcal{L}_{\text{distill}} = \sum_{t=1}^{T} \left\| G_T^t(x) - f_t(G_S^t(x)) \right\|_2 \tag{4}$$

where $G_T^t(x)$ and $G_S^t(x)$ are the intermediate activations of the t-th chosen layers in the teacher and student generators. There is one problem here: the intermediate activations of the teacher and student have different depth dimensions. To match the dimension, [14, 16] proposed a learnable convolution layer f_t. f_t is the 1×1 convolution layer that matches the depth dimension of the features of teacher and student generators.

While the convolution layer f_t operates on the whole region of activations of each layer, the features in the inpainting model have both valid pixels and invalid pixels. Matching the features without distinguishing between two types of pixels can degrade performance. To prevent performance degradation, partial convolution [15] updates the mask and re-normalizes the feature map on only valid pixels. By removing the effect of invalid pixels, partial convolution can achieve better image inpainting results than using normal convolution. We substitute the original f_t to the partial convolution layer g_t to focus only on the valid value when matching two activations. We apply partial convolution at chosen layers and match only the valid region of intermediate features, as in Fig. 3. Our first type distillation loss $\mathcal{L}_{\text{distill1}}$ is defined as follows:

$$\mathcal{L}_{\text{distill1}} = \sum_{t=1}^{T} \left\| G_T^t(x, M_g) \odot M_g^t - g_t(G_S^t(x, M_g), M_g^t) \right\|_2 \tag{5}$$

where g_t is the partial convolution layer and M_g^t is the updated mask after the t-th partial convolution layer. In detail, the partial convolution layer g_t is defined as:

$$X^{t'} = \begin{cases} W^t(X^t \odot M^t)\frac{\text{sum}(1)}{\text{sum}(M^t)} + b, & \text{if } \text{sum}(M^t) > 0 \\ 0, & \text{otherwise} \end{cases} \tag{6}$$

where W^t and b are the convolution kernel and bias of the t-th layer, and X^t and M^t are input feature values and input mask values for the convolution window of t-th layer. After the partial convolution process, the new input mask is described as follows:

$$M^{t'} = \begin{cases} 1, & \text{if } \text{sum}(M^t) > 0 \\ 0, & \text{otherwise} \end{cases} \tag{7}$$

Image-Level Distillation. We use two types of image-level distillation losses. For the first loss, we utilize both perceptual loss and style loss. Different from prior works [13, 15] that compared the output images to the ground truth images, we compared the output images from G_S to the output images from G_T. We input two images into the VGG-16 network pre-trained on ImageNet and measured the euclidean distance between the feature maps of VGG-16. Our second type distillation loss $\mathcal{L}_{\text{distill2}}$ is the sum of the perceptual loss and style loss:

$$\mathcal{L}_{\text{distill2}} = \mathcal{L}_{\text{perceptual}} + \mathcal{L}_{\text{style}} \tag{8}$$

where $\mathcal{L}_{\text{perceptual}}$ is the perceptual loss and $\mathcal{L}_{\text{style}}$ is the style loss as defined in [11].

For the last distillation loss, we propose another method to use the intermediate features of the teacher discriminator. Chen et al. [2] proposed the distillation method for a discriminator to guide the objective function of generators. Since the teacher discriminator was already trained to distinguish between true and fake images, it can measure the manifold of the target domain. Our third type distillation loss $\mathcal{L}_{distill3}$ is defined as the distance between the feature maps of teacher discriminator:

$$\mathcal{L}_{distill3} = \frac{1}{n} \sum_{i=1}^{n} \left\| \hat{D}_T \left(G_T \left(x_i, M_g \right) \right) - \hat{D}_T \left(G_S \left(x_i, M_g \right) \right) \right\|_1 \qquad (9)$$

where $\hat{D}_T$ is the selected layers of the teacher discriminator.

Proposed Distillation Loss. Overall, our final objective to train the student model is written as follows:

$$\begin{aligned} \mathcal{L}_{total} =& \lambda_{adv} \mathcal{L}_{adv} + \lambda_{recon} \mathcal{L}_{recon} + \lambda_{distill1} \mathcal{L}_{distill1} \\ & + \lambda_{distill2} \mathcal{L}_{distill2} + \lambda_{distill3} \mathcal{L}_{distill3} \end{aligned} \qquad (10)$$

where $\mathcal{L}_{adv}$ is the loss to solve the min-max optimization for GAN training. $\mathcal{L}_{recon}$ is the loss to bring the output of the generator closer to the ground truth in the pixel-level. We used $\mathcal{L}_{adv}$ and $\mathcal{L}_{recon}$ as described in [9].

4   Experiments

4.1   Training Setting

We conducted experiments on a standard image inpainting model, GLCIC [9]. We basically followed the training setting of GLCIC [9], but we changed a few settings for the proposed losses. We trained the student generator and pre-trained teacher discriminator at the same time with a batch size of 16 and a learning rate of 1e-5. GLCIC [9] has two discriminators: global discriminator and local discriminator. Therefore, when we find $\mathcal{L}_{distill3}$ in Eq. (9), we use both discriminators of teacher network. We use the sum of the loss from the two discriminators as $\mathcal{L}_{distill3}$.

Datasets. CelebA dataset [18] contains over 180,000 human face images. We divided the dataset by 8:2 and used 162,080 data as training data, and 40521 data sets as test data. The dataset has images of 178×218 pixel. We cropped images to 178×178 pixel and resized them to 160×160 pixel images. We used 96×96 pixel image patches as the input of the local discriminator. For mask datasets, we generated a random rectangular hole in the [48, 96] pixel range and filled it with the mean pixel value of the training dataset.

Evaluation Metrics. We evaluated the performance in terms of PSNR, SSIM and $\ell1$ norm. PSNR is the peak signal-to-noise ratio that is commonly used to measure the sum over squared pixel value difference. SSIM is the structural similarity index measure to evaluate structural information differences and similarities. The hypothesis of this metric is that the pixels that are close spatially have strong dependencies. $\ell1$ norm is the distance between the pixel value of ground truth images and completed images.

4.2   Pruning Effectiveness

Qualitative Results Without Fine-tuning. We first show the effectiveness of our channel pruning method. We removed 25% of channels from each layer in the teacher generator by our pruning method and other methods. We compared qualitative results of pruned generator that uses the three different pruning methods: (1) Random; (2) Magnitude; (3) Ours. (1) Random pruning removes random channels and (2) magnitude pruning retains channels which have large forward weight magnitude as described in Eq. (1). (3) Our scheme used the proposed score $\mathcal{L}_S$ as described in Eq. (2). We removed 20% of channels based on the outgoing gradients and eliminated the final 5% of channels with the small weight magnitudes among remaining channels.

As shown in Fig. 4, (1) Random and (2) Magnitude destroy the image quality and cannot reconstruct the missing regions. These results imply that important channels for generating missing regions are removed. On the other hand, (3) our scheme preserves the structures and textures in the estimated pixels. Even before fine-tuning, the qualitative performance of our student generator outperformed the other baselines. The results prove that our pruning method can search the efficient network architectures in the full network.

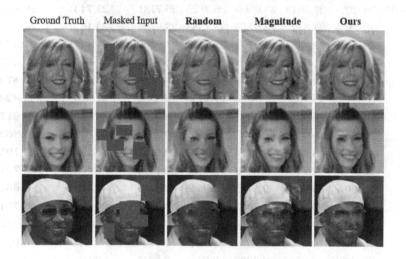

Fig. 4. Qualitative results without fine-tuning.

Quantitative Results with Fine-tuning. We obtained the student genera-tor by one-shot pruning and removed 70% of channels in each layer from the teacher generator. We fine-tuned the student generator with a teacher discrim-inator and evaluated the performance metrics. We compare 5 different pruning methods: (1) Random; (2) Mag; (3) Large Grad; (4) Small Grad; (5) Ours. To examine the effect of our proposed sensitivity measurement, Large Grad and

Small Grad only consider the backward information. Large Grad keeps channels with high scores proposed in the Eq. (3). Conversely, Small Grad keeps the channels with low scores proposed in the Eq. (3). Our scheme considers both forward and backward information as proposed in Subsect. 3.1. In the training process, we used adversarial loss, reconstruction loss and simple perceptual loss. Quantitative compression results on CelebA dataset are summarized in Table 1. Here, the Full refers to the teacher generator. We randomly sampled 5000 generated images and generated 5000 masks for evaluation. We found that randomly pruned generator degraded all metrics and Mag methods improved the results slightly. As can be seen by comparing Large Grad and Small Grad, the proposed sensitivity measurement has successfully removed unimportant channels. Small Grad cannot even achieve the performance of Mag. Our pruning method tailored to inpainting outperforms other methods in all metrics.

Table 1. Pruning effectiveness results over CelebA. We fine-tuned the student generator created by 5 different pruning methods using adversarial loss, reconstruction loss and simple perceptual loss. *Higher is better. †Lower is better.

	Mask	Full	Rand	Mag	Large Grad	Small Grad	Ours
PSNR*	10–20%	27.5018	25.6509	26.3825	26.7181	25.7434	**26.8523**
	20–30%	24.842	23.0216	23.7383	23.9755	23.1611	**24.0889**
	30–40%	22.7353	20.9753	21.7172	21.9002	21.1171	**21.9393**
	40–50%	21.025	19.4135	20.1277	20.2976	19.6076	**20.3714**
	50–60%	19.3974	17.9646	18.6092	18.7838	18.1553	**18.8727**
	Average	23.1003	21.4050	22.1150	22.3350	21.5569	**22.4249**
SSIM*	10–20%	0.9244	0.9041	0.9127	0.9166	0.9056	**0.9176**
	20–30%	0.8727	0.8429	0.8565	0.8612	0.8463	**0.8621**
	30–40%	0.8111	0.7714	0.7903	0.7951	0.7764	**0.7959**
	40–50%	0.7407	0.6962	0.7189	0.7244	0.7050	**0.7278**
	50–60%	0.6605	0.6169	0.6422	0.6482	0.6261	**0.6518**
	Average	0.8019	0.7663	0.7841	0.7891	0.7719	**0.7910**
$\ell 1$†	10–20%	0.0119	0.0154	0.0139	0.0133	0.0154	**0.0131**
	20–30%	0.0202	0.0259	0.0234	0.0227	0.0258	**0.0225**
	30–40%	0.0308	0.0393	0.0354	0.0347	0.0392	**0.0344**
	40–50%	0.0425	0.0535	0.0483	0.0474	0.0526	**0.0469**
	50–60%	0.0567	0.0700	0.0638	0.0626	0.0689	**0.0617**
	Average	0.0324	0.0408	0.0370	0.0361	0.0404	**0.0357**

4.3   Knowledge Distillation Effectiveness

We compared knowledge distillation results based on different pruning methods. As in Subsect. 4.2, we removed 70% channels in each layer according to

Table 2. Quantitative results of knowledge distillation over CelebA. We fine-tuned the student generator created by 5 different pruning methods using adversarial loss, reconstruction loss and proposed distillation loss. *Higher is better. †Lower is better.

	Mask	Full	Rand	Mag	Large Grad	Small Grad	Ours
PSNR*	10–20%	27.5018	26.5566	27.1204	27.1747	26.5310	**27.2856**
	20–30%	24.842	23.8173	24.3517	24.4302	23.7501	**24.5819**
	30–40%	22.7353	21.7146	22.2404	22.3506	21.5752	**22.4038**
	40–50%	21.025	20.1202	20.5830	20.6846	19.8501	**20.7880**
	50–60%	19.3974	18.6009	19.0189	19.1431	18.2083	**19.2875**
	Average	23.1003	22.1619	22.6629	22.7567	21.9830	**22.8694**
SSIM*	10–20%	0.9244	0.9144	0.9214	0.9218	0.9150	**0.9231**
	20–30%	0.8727	0.8580	0.8685	0.8697	0.8584	**0.8713**
	30–40%	0.8111	0.7907	0.8050	0.8073	0.7907	**0.8088**
	40–50%	0.7407	0.7188	0.7368	0.7397	0.7179	**0.7426**
	50–60%	0.6605	0.6420	0.6627	0.6661	0.6393	**0.6702**
	Average	0.8019	0.7848	0.7989	0.8009	0.7843	**0.8031**
$\ell 1$†	10–20%	0.0119	0.0137	0.0127	0.0125	0.0140	**0.0124**
	20–30%	0.0202	0.0234	0.0217	0.0214	0.0239	**0.0210**
	30–40%	0.0308	0.0355	0.0330	0.0323	0.0366	**0.0322**
	40–50%	0.0425	0.0483	0.0454	0.0445	0.0504	**0.0439**
	50–60%	0.0567	0.0632	0.0602	0.0589	0.0673	**0.0578**
	Average	0.0324	0.0369	0.0346	0.0340	0.0384	**0.0335**

5 different one-shot pruning methods: (1) Random; (2) Mag; (3) Large Grad; (4) Small Grad; (5) Ours. To fine-tune the student generator, we used our proposed distillation method as described in Eq. (10), instead of the simple perceptual loss. The parameters are set as $\lambda_{adv} = 0.01$, $\lambda_{recon} = 20$, $\lambda_{distill1} = 10$, $\lambda_{distill2} = 10$, $\lambda_{distill3} = 1$. We show quantitative distillation results on CelebA dataset in Table 2. Compared to Table 1, overall values of PSNR, SSIM and $\ell 1$ loss are improved in every pruning method. The overall trend of the results is similar to Table 1, but one difference is that Small Grad has lower performance than random pruning. It means that non-sensitive channels are more detrimental than random channels. Our final result of applying the proposed pruning and distillation method show similar performance to the full model. The proposed distillation method helps compressed network to generate high quality images. We show several completed images for a large rectangular mask in Fig. 5.

4.4 Ablation Study on the Proposed Pruning Method

In Subsect. 3.1, we proposed the pruning method considering both sides. We pruned channels based on the magnitudes of incoming weights and gradients of outgoing weights (Mag + Grad). To prove the effectiveness of our method,

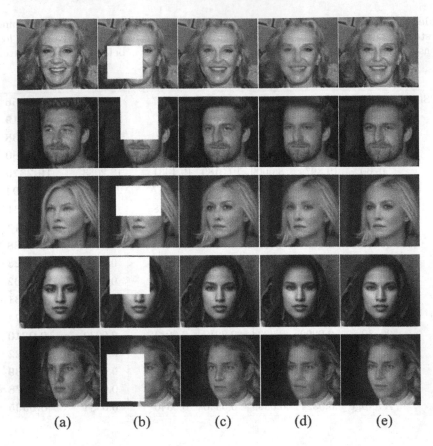

Fig. 5. Qualitative results of teacher and student generator. (a) Ground truth (b) Masked input (c) Output of teacher generator (d) Output of student generator with pruning method (e) Output of student generator with pruning and distillation methods

we analyze the combination of pruning methods on both sides. We compared four different pruning combination: the magnitudes of incoming weights and gradients of outgoing weights (Mag + Grad), the magnitudes of incoming weights and magnitudes of outgoing weights (Mag + Mag), the gradients of incoming weights and magnitudes of outgoing weights (Grad + Mag) and the gradients of incoming weights and gradients of outgoing weights (Grad + Grad). As shown in Table 3, our method (Mag + Grad) is better than other pruning methods in PSNR, SSIM and $\ell 1$.

4.5   Ablation Study on the Proposed Distillation Method

We further analyze the effectiveness of knowledge distillation. For all experiments, we fix the setup for channel pruning as in the proposed pruning method. We show the quantitative results in Table 4. Base corresponds to the results of

Table 3. Ablation study of different pruning bases. We fine-tuned the student generator created by 4 different pruning combinations using adversarial loss, reconstruction loss and simple perceptual loss. *Higher is better. †Lower is better.

	Mask Ratio	Mag + Grad	Mag + Mag	Grad + Mag	Grad + Grad
PSNR*	10–20%	**26.8523**	26.0584	26.3906	26.7376
	20–30%	**24.0889**	23.4293	23.7302	24.0828
	30–40%	21.9393	21.4481	21.6850	**21.9572**
	40–50%	**20.3714**	19.8875	20.1191	20.3414
	50–60%	**18.8727**	18.3646	18.6291	18.8446
	Average	**22.4249**	21.8376	22.1108	22.3927
SSIM*	10–20%	**0.9176**	0.9088	0.9130	0.9163
	20–30%	0.8621	0.8509	0.8565	**0.8624**
	30–40%	**0.7959**	0.7832	0.7898	0.7959
	40–50%	**0.7278**	0.7114	0.7188	0.7263
	50–60%	0.6518	0.6336	0.6418	**0.6519**
	Average	**0.7910**	0.7776	0.7840	0.7905
ℓ1†	10–20%	**0.0131**	0.0145	0.0140	0.0133
	20–30%	0.0225	0.0246	0.0236	**0.0224**
	30–40%	0.0344	0.0370	0.0358	**0.0343**
	40–50%	**0.0469**	0.0502	0.0448	0.0469
	50–60%	**0.0617**	0.0662	0.0643	0.0619
	Average	**0.0357**	0.0385	0.0373	0.0358

the last column in Table 1 where only a simple perceptual loss is used. In the next two columns, we compare the results of type 1 distillation loss that does not use partial convolution with the results of type 1 loss that used the partial convolution. The results of additional image-level distillation losses, type1+2 and type1+2+3, are shown in the last two columns. We choose the activation maps of 4-th layer, 8-th layer and 12-th layer in the student generator to construct the loss in Eq. (5) and select the 2-nd and 4-th activation maps of the teacher discriminator to construct the loss in Eq. (9). The performance of type 1+2+3 distillation loss outperforms other losses and is close to the full model.

4.6   Memory Reduction and Speedup

Finally, we evaluated the number of parameters, model size, MACs and inference time per one image between teacher and student generators. Table 5 shows the details. The teacher model has about 6 M parameters and 24.3 MB model size, but the student model only has about 0.5 M parameters and 2.2 MB. Therefore, the student model can run on devices having limited resources. We also measured the computation cost of the inpainting model by using MACs. MACs treat mutiply-add $(a + b \times c)$ operation as one tensor operation, so 1 MACs is equal

Table 4. Ablation study of knowledge distillation over CelebA. We fine-tuned the student generator using adversarial loss, reconstruction loss and different types of distillation loss. *Higher is better. †Lower is better.

	Mask	Full	Base	Type1 (w/o pconv)	Type 1	Type 1+2	Type 1+2+3
PSNR*	10–20%	27.5018	26.8523	27.2650	**27.3455**	27.3444	27.2856
	20–30%	24.842	24.0889	24.4813	**24.5982**	24.5397	24.5819
	30–40%	22.7353	21.9393	22.3819	22.3802	22.3465	**22.4038**
	40–50%	21.025	20.3714	20.7607	20.7253	20.7492	**20.7880**
	50–60%	19.3974	18.8727	19.2282	19.2199	19.2306	**19.2875**
	Average	23.1003	22.4249	22.8234	22.8538	22.8421	**22.8694**
SSIM*	10–20%	0.9244	0.9176	0.9222	0.9230	**0.9235**	0.9231
	20–30%	0.8727	0.8621	0.8697	**0.8713**	0.8710	**0.8713**
	30–40%	0.8111	0.7959	0.8072	0.8079	0.8075	**0.8088**
	40–50%	0.7407	0.7278	0.7404	0.7421	0.7423	**0.7426**
	50–60%	0.6605	0.6518	0.6678	0.6697	0.6685	**0.6702**
	Average	0.8019	0.7910	0.8015	0.8029	0.8026	**0.8031**
$90\ell1$†	10–20%	0.0119	0.0131	0.0127	0.0125	**0.0124**	0.0124
	20–30%	0.0202	0.0225	0.0216	0.0213	0.0212	**0.0210**
	30–40%	0.0308	0.0344	0.0328	0.0327	0.0325	**0.0322**
	40–50%	0.0425	0.0469	0.0448	0.0447	0.0442	**0.0439**
	50–60%	0.0567	0.0617	0.0590	0.0589	0.0583	**0.0578**
	Average	0.0324	0.0357	0.0342	0.340	0.0337	**0.0335**

to 2 FLOPs. The teacher model requires 17.75 G MACs, but the student model only needs 1.60 G, reduced by 10x. We also measured the inference time per image on GPU (GeForce RTX 3090) and CPU. The inference time on the student model is reduced by 3x on GPU and 4.5x on CPU compared to the teacher model. Even with small computational burden, we can inpaint the image holes without the loss of visual fidelity.

Table 5. Memory reduction and speedup of the student model.

Model	Teacher model	Student model
# Parameters	6.072 M	0.538 M
Model size	24.3 MB	2.2 MB
MACs	17.75 G	1.60 G
Inference time (GPU)	15.75 ms	5.08 ms
Inference time (CPU)	125.94 ms	28.13 ms

5   Conclusion

In this paper, we proposed a novel compression framework for image inpainting for the first time. By introducing channel pruning and knowledge distillation methods specialized for image inpainting, our method can effectively compress the inpainting model while generating similar quality of images compared to the original model. Our scheme enables improvement in image inpainting tasks in various resource-constrained applications where reducing the size of the model is of paramount importance.

Acknowledgments. This work was supported by Samsung Electronics Co., Ltd. and by National Research Foundation of Korea (No. 2019R1I1A2A02061135).

References

1. Aguinaldo, A., Chiang, P.Y., Gain, A., Patil, A., Pearson, K., Feizi, S.: Compressing gans using knowledge distillation. arXiv preprint arXiv:1902.00159 (2019)
2. Chen, H., et al.: Distilling portable generative adversarial networks for image translation. In: Proceedings of the AAAI Conference on Artificial Intelligence, vol. 34, pp. 3585–3592 (2020)
3. Goodfellow, I., et al.: Generative adversarial nets. Adv. Neural Inf. Process. Syst. **27** (2014)
4. Gou, J., Yu, B., Maybank, S.J., Tao, D.: Knowledge distillation: a survey. Int. J. Comput. Vis. **129**(6), 1789–1819 (2021)
5. Han, S., et al.: EIE: efficient inference engine on compressed deep neural network. ACM SIGARCH Comput. Archit. News **44**(3), 243–254 (2016)
6. Han, S., Pool, J., Tran, J., Dally, W.: Learning both weights and connections for efficient neural network. Adv. Neural Inf. Process. Syst. **28** (2015)
7. He, Y., Zhang, X., Sun, J.: Channel pruning for accelerating very deep neural networks. In: Proceedings of the IEEE International Conference on Computer Vision, pp. 1389–1397 (2017)
8. Hinton, G., Vinyals, O., Dean, J.: Distilling the knowledge in a neural network (2015). arXiv preprint arXiv:1503.02531 2 (2015)
9. Iizuka, S., Simo-Serra, E., Ishikawa, H.: Globally and locally consistent image completion. ACM Trans. Graph. (ToG) **36**(4), 1–14 (2017)
10. Jin, Q., et al.: Teachers do more than teach: compressing image-to-image models. In: Proceedings of the IEEE/CVF Conference on Computer Vision and Pattern Recognition, pp. 13600–13611 (2021)
11. Johnson, J., Alahi, A., Fei-Fei, L.: Perceptual losses for real-time style transfer and super-resolution. In: Leibe, B., Matas, J., Sebe, N., Welling, M. (eds.) ECCV 2016. LNCS, vol. 9906, pp. 694–711. Springer, Cham (2016). https://doi.org/10.1007/978-3-319-46475-6_43
12. Li, H., Kadav, A., Durdanovic, I., Samet, H., Graf, H.P.: Pruning filters for efficient convnets. arXiv preprint arXiv:1608.08710 (2016)
13. Li, J., Wang, N., Zhang, L., Du, B., Tao, D.: Recurrent feature reasoning for image inpainting. In: Proceedings of the IEEE/CVF Conference on Computer Vision and Pattern Recognition, pp. 7760–7768 (2020)

14. Li, M., Lin, J., Ding, Y., Liu, Z., Zhu, J.Y., Han, S.: Gan compression: efficient architectures for interactive conditional gans. In: Proceedings of the IEEE/CVF Conference on Computer Vision and Pattern Recognition, pp. 5284–5294 (2020)

15. Liu, G., Reda, F.A., Shih, K.J., Wang, T.C., Tao, A., Catanzaro, B.: Image inpainting for irregular holes using partial convolutions. In: Proceedings of the European Conference on Computer Vision (ECCV), pp. 85–100 (2018)

16. Liu, Y., Shu, Z., Li, Y., Lin, Z., Perazzi, F., Kung, S.Y.: Content-aware gan compression. In: Proceedings of the IEEE/CVF Conference on Computer Vision and Pattern Recognition, pp. 12156–12166 (2021)

17. Liu, Z., Li, J., Shen, Z., Huang, G., Yan, S., Zhang, C.: Learning efficient convolutional networks through network slimming. In: Proceedings of the IEEE International Conference on Computer Vision, pp. 2736–2744 (2017)

18. Liu, Z., Luo, P., Wang, X., Tang, X.: Deep learning face attributes in the wild. In: Proceedings of the IEEE International Conference on Computer Vision, pp. 3730–3738 (2015)

19. Luo, J.H., Wu, J., Lin, W.: ThiNet: a filter level pruning method for deep neural network compression. In: Proceedings of the IEEE International Conference on Computer Vision, pp. 5058–5066 (2017)

20. Pathak, D., Krahenbuhl, P., Donahue, J., Darrell, T., Efros, A.A.: Context encoders: feature learning by inpainting. In: Proceedings of the IEEE Conference on Computer Vision and Pattern Recognition, pp. 2536–2544 (2016)

21. Romero, A., Ballas, N., Kahou, S.E., Chassang, A., Gatta, C., Bengio, Y.: Fitnets: Hints for thin deep nets. arXiv preprint arXiv:1412.6550 (2014)

22. Shu, H., et al.: Co-evolutionary compression for unpaired image translation. In: Proceedings of the IEEE/CVF International Conference on Computer Vision, pp. 3235–3244 (2019)

23. Wang, H., Gui, S., Yang, H., Liu, J., Wang, Z.: GAN slimming: all-in-one GAN compression by a unified optimization framework. In: Vedaldi, A., Bischof, H., Brox, T., Frahm, J.-M. (eds.) ECCV 2020. LNCS, vol. 12349, pp. 54–73. Springer, Cham (2020). https://doi.org/10.1007/978-3-030-58548-8_4

24. Wen, W., Wu, C., Wang, Y., Chen, Y., Li, H.: Learning structured sparsity in deep neural networks. Adv. Neural Inf. Process. Syst. **29** (2016)

25. Yim, J., Joo, D., Bae, J., Kim, J.: A gift from knowledge distillation: Fast optimization, network minimization and transfer learning. In: Proceedings of the IEEE Conference on Computer Vision and Pattern Recognition, pp. 4133–4141 (2017)

26. Yu, F., Koltun, V.: Multi-scale context aggregation by dilated convolutions. arXiv preprint arXiv:1511.07122 (2015)

27. Yu, J., Lin, Z., Yang, J., Shen, X., Lu, X., Huang, T.S.: Generative image inpainting with contextual attention. In: Proceedings of the IEEE Conference on Computer Vision and Pattern Recognition, pp. 5505–5514 (2018)

28. Yu, J., Lin, Z., Yang, J., Shen, X., Lu, X., Huang, T.S.: Free-form image inpainting with gated convolution. In: Proceedings of the IEEE/CVF International Conference on Computer Vision, pp. 4471–4480 (2019)

QFT: Post-training Quantization via Fast Joint Finetuning of All Degrees of Freedom

Alex Finkelstein[✉], Ella Fuchs, Idan Tal, Mark Grobman, Niv Vosco, and Eldad Meller

Hailo, Tel-Aviv, Israel
{alexf,eldadm}@hailo.ai
http://www.hailo.ai

Abstract. The post-training quantization (PTQ) challenge of bringing quantized neural net accuracy close to original has drawn much attention driven by industry demand. Many of the methods emphasize optimization of a specific per-layer *degree of freedom* (DoF), such as grid step size, preconditioning factors, nudges to weights and biases, often chained to others in multi-step solutions. Here we rethink quantized network parameterization in HW-aware fashion, towards a unified analysis of all quantization DoF, permitting for the first time their joint end-to-end finetuning. Our single-step simple and extendable method, dubbed quantization-aware finetuning (QFT), achieves 4b-weights quantization results on-par with SoTA within PTQ constraints of speed and resource.

1 Introduction

Quantization is a standard step in the deployment of Deep Neural Networks inference to embedded or otherwise resource-constrained environments [1–5]. Using less bits for weights and intermediates helps reduce power consumption and increase inference speed. However, accuracy is lost, increasingly so for lower bit-widths, unless applying an optimizing 'quantization method' [6] for the choice of quantized weights and other deployment parameters. It is standard to divide quantization methods into two broad groups - post-training quantization (PTQ) and quantization-aware training (QAT). Broadly speaking, QAT methods involve at least a few epochs of quantization-augmented training on the original labeled training set while PTQ methods are fast and require a small amount of unlabeled data. PTQ methods are also valued for their robustness and ease of use, e.g. minimal per-network handcrafted settings. This is of special interest to HW vendors designing AI accelerators. Their hardware-tailored quantization tools try to provide an automated network-agnostic *compilation* of quantized network from a pre-trained one, that can be quickly iterated across deployment settings (e.g., compression levels).

Supplementary Information The online version contains supplementary material available at https://doi.org/10.1007/978-3-031-25082-8_8.

Curiously, it seems there's not much published about how well QAT methods do in the PTQ regime - i.e., when they are constrained to use a small amount of compute resources and unlabeled data, but otherwise adhering to end-to-end training of all weights. For conciseness, we term this stripped-down QAT the Quantization-aware Finetuning (QFT) regime. In this paper, we explore the potential of this setting to provide a SoTA PTQ baseline while using a small amount of data, no labels, and a quick single-GPU run.

We emphasize joint training of all quantization parameters, including auxiliary ones such as re-coding factors and scales, bringing under that umbrella also the cross-layer factorization [7,8] formerly treated as pre-quantization step but here re-framed as a per-channel scale of activations. To that end, we address the issue of HW constraints and relations among parameters through the concept of the *offline subgraph* fed by their *independent subset*. In Sect. 3, we lay a foundation to that in terms of a principled and HW-aware analysis of the arithmetic pipeline, also called for by deployability concerns [9] oft arising for works in this field. Curiously, even the standard layer-wise/channel-wise 'flag' can yet benefit from a HW-anchored definition, which we supply, and then proceed to uncover and rigorously analyze a *doubly-channelwise* kernel quantization mode. In Sect. 4, we provide extensive experimentation to show that our simple QFT method achieves accuracy on-par with SoTA PTQ methods while having less steps and hand-crafted decisions, and leaves much room to further extension. *To sum up, our contributions are as follows:*

– We show that QAT scaled down to the PTQ regime can give results on-par with SoTA method when all degrees of freedom (DoF) are jointly fine-tuned.
– We outline a recipe for setting up such *all-dof trainability* for any network and deployment arithmetic. Applied to convolutions, it supersedes the layerwise/channelwise spec language by more explicit *vector scales* semantics, covering the DoF of [7,8] and an under-explored *doubly-channelwise* mode.

2   Prior Work

Quantization Aware Training (QAT): Straight-through Estimator (STE) was introduced in [10] as a generic enabler of backprop through non-differentiable ops, and applied to trained binarization [11] and integer quantization [1,2] by injecting round&clip ops into the training graph. In [12] 4b quantization with accuracy matching or exceeding full-precision were demonstrated. Drop-in alternatives for STE were suggested by [13,14] among others. Starting from [15–17] quantization step sizes have been trained alongside weights and biases. This enabled better results and/or reduced training [17] which however still entails at least a few epochs on the full labeled training set. On top of that, the setup requires careful adherence to original training procedure, often creating complex interplay with quantization, e.g., around BatchNorm [9] - just folding it away gives inferior convergence (if any).

Post-training Quantization (PTQ): Many architectures, e.g. ResNets, can quantize well to 8-bit by the trivial "round to nearest" approach, choosing the range for each tensor by a naive $max(|.|)$ applied on weights and data sampled on a calibration set. For 4-bit weights, range-finding by an optimal Clipping, in, e.g., *MMSE* (minimum mean-square error) sense, is necessary [13,18], often used as an initialization for further tuning. Some architectures s.a. MobileNet still presented a challenge for 8b quantization especially in "layerwise" setting, solved by heuristics utilizing two other degrees of freedom, bias correction [8,19] for removal of systematic quantization error and inverse-proportional cross-layer factorization [7,8]. The latter pre-conditions network weights so that the kernels and/or activations suffer less quantization error, by "equalizing" the dynamic ranges of different channels.

While PTQ originally [1,2] implied no training, some works partially relaxed that self-limitation. Massively down-scaled QAT for a subset of params was proposed in [20] for scale factors and in [19] for biases. In [14,21] the teacher-student ("reconstruction") training was performed layer-by-layer, avoiding overfit concerns at the expense of using a local proxy for the loss. In [14,22], the STE was replaced by a specially designed constraints relaxation procedure, progressively penalizing deviation of weights from grid. In [22] a block-by-block reconstruction was adopted, better approximating the network loss, at the expense of some block-structure hyperparameters to be set per network.

3 Methods

An overview of our analysis applied to convolutions can be found in Fig. 1. The notation in the figure and text below is as follows. $W^l_{..,m,n}$ is the pre-trained kernel for conv layer #l where input (output) channels are indexed by m (n), respectively ('..' indicate spatial dimensions which can be safely ignored). Tensors part to the "online", HW-simulating part of the graph are denoted with $[\hat{.}]$; that includes inter-layer data ("activations") $\hat{A}$, and compile-time precomputed tensors s.a. quantized-weights $\hat{W}$ and HW rescale factors $\hat{F}$. Scale tensors, denoted by $(S_x)_{i,j}$, are drawn as same shape as the tensor they divide, even if only changing along some of the dimensions.

3.1 Quantization Description by Over-Parameterized Scale Tensors

We start by suggesting a quantization language that properly models the arithmetic hardware (HW) and its constraints. This allows us to exploit all degrees-of-freedom (DoF) and verify the quantization result can be deployed. The typical approach defines the quantizer as a stand-alone operation to be "injected" into otherwise full-precision graph:

$$Q_{b,s}(x) = s * clip\left(\lfloor x/s \rceil, \pm(2^{b-1} - 1)\right) \tag{1}$$

with the quantization scale s having the direct interpretation as the "quantization bin/step size" for a certain tensor or slice thereof, hence "per-channel" or

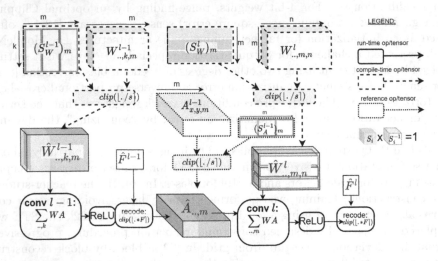

(A) Layerwise quantization: scalar rescale factors.
Cross-layer factorization as a single vector DoF, impacting quantized kernels of **two** layers.
Reformulated as vector scale of activation - kernels are same, only their "scale tensor" changes.

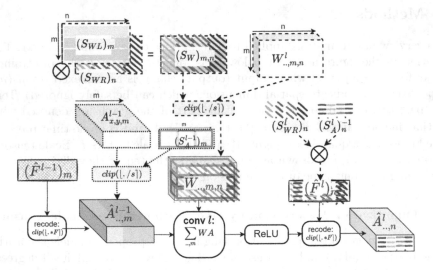

(B) Channelwise Quantization - per-channel rescale factors. Under some conditions, they can compensate for both per-input-channel and per-output-channel kernel scaling, enabling a **doubly-channelwise** quantization with 2 vector DoF per layer.

Fig. 1. Unified picture of layerwise/[doubly-]channelwise quantization and cross-layer factorization, achieved by postulating per-channel ("vector") scale of activation. Building on the analysis, we use a DL framework to infer and backprop (using STE [10] where appropriate) thru the full graph including both "online" (run-time) and "offline" (compile-time) parts of computation. All DoF per HW spec, including Weights and HW-specifics such as factors and vector scales are trained on the same footing.

"per-tensor" quantization. These scales (and offsets for asymmetric case) are seen as constants or, in case of trainable scales [16], as variables, implicitly assumed to be independently trainable. We suggest instead looking at them as *tensors*, nodes in the *offline part* of the computational graph. This will enable a training aware of their relations which express the HW constraints and freedoms.

We propose the following HW-aware analysis and simulation recipe :

1. Start with over-parameterized scales, e.g. S_x shaped as the respective x.
2. Analyze constraints of actual arithmetic pipeline to arrive at an under-determined system of equations describing the relations of scales.
3. Re-parameterize more compactly by finding the degrees of freedom (DoF) - some (non-unique) subset of independent variables, determining the rest.
4. Implement this inference in differentiable way as the *offline* part of network simulation graph to enable train-ability of all DoF.

Deferring its further discussion to Sect. 3.2 and Appendix A, we now exemplify the process for a simple convolution, putting Fig. 1 on a mathematical basis.

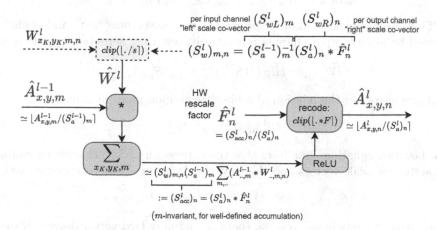

Fig. 2. Analyzing quantization parameters constraints and relations for a simple convolution. Can parameterize all via either $\{S_a, \hat{F}\}$ or $\{S_{wL}, S_{wR}\}$, at most two vector DoF, less if additional HW constraints apply (layerwise rescale, special activation, etc.).

Degrees-of-freedom Analysis for a Simple Convolution. In Fig. 2 we present an exhaustive analysis of the fully-integer deployment arithmetic for a single layer, taking for brevity of exposition a bias-less convolution with symmetric quantization. Quantized weights $\hat{W}$, derived from full-precision ones by scaling with an over-parameterized scale tensor $(S_w)_{m,n}$ and round&clip op, are convolved with activations $\hat{A}$, produced by their respective layers via an activation function followed by a HW re-code, for a symmetric case comprising just a multiplication by factor $\hat{F}_n$. Note that such op always exists, even if implicitly as a composite "dequantize+requantize" for non-fully-integer platforms restoring full FP32 representation between layers.

We begin by a rigorous re-definition of channelwise/layerwise HW spec as $\hat{F}_n$ being vector/scalar (with/without n-variation). The HW activations (inter-layer data) are in approximate correspondence to activations in full-precision run, via a scale $(S_a)_m^l$, which, *key to our exposition*, is a per-channel vector.

We now introduce the constraint of all terms in the partial sum having the same scale. This constraint is derived from hardware use of multiply-and-accumulate primitive. This is formalized as a m-invariant encoding of the addends, by a well-defined "accumulator scale" that can be factored out as:

$$\sum_{m,..}(\hat{A}_{..,m}^{l-1} * \hat{W}_{..,m,n}^l) = (S_{acc}^l)_n^{-1}\sum_{m,..}(A_{..,m}^{l-1} * W_{..,m,n}^l)$$

$$(S_{acc}^l)_n := (S_w^l)_{m,n}(S_a^{l-1})_m \tag{2}$$

This constraint mandates that weight-scale tensor is decomposed as follows, defining accumulator scale vector and left/right weight scale co-vectors:

$$(S_w^l)_{m,n} := (S_a^{l-1})_m^{-1}(S_{acc}^l)_n := (S_{wL}^l)_m * (S_{wR}^l)_n \tag{3}$$

Thus, weight scale is an outer-product of two co-vectors ("left" and "right"), indexed by input/output channel respectively, rewriting kernel quantization as:

$$\hat{W}_{m,n}^l = clip\left(\lfloor(S_{wL}^l)_m^{-1}W_{m,n}^l(S_{wR}^l)_n^{-1}\rceil\right) \tag{4}$$

Next, we introduce the relation of scale change upon multiplicative recode:

$$(S_a^l)_n = (S_{acc})_n/\hat{F}_n^l \tag{5}$$

The last two equations finally enable the expression of kernel scale co-vectors via activation scale vectors and the HW rescale factor (be it vector or scalar):

$$(S_{wL}^l)_m = (S_a^{l-1})_m^{-1} \; ; \quad (S_{wR}^l)_n = (S_a^l)_n * \hat{F}_n^l \tag{6}$$

Thus, given "channelwise" rescale, there are actually **two** vector degrees of freedom per layer, which can be parameterized via Eq. 6 or, on some condition, via the "left/right" scales (see next subsection).

This formulation treats input/output channels symmetrically. Our "right" scale vector corresponds to the standard per-[output-]channel quantization step-size. However, the "left" scale vector, novel to this exposition, is equivalent to cross-layer factorization DoF, formerly [7,8] described "externally" to quantization analysis, as preconditioning factors applied to kernel rows, with inverse factors applied to columns of previous layer's kernel. We thus rethink this DoF within quantization, as one of two scale co-vectors, constrained by relation to the preceding layer - to its rescale factors (Fig. 1, B) or kernel scales (Fig. 1, A), as per channelwise/layerwise rescale capability of the arithmetic pipeline.

Note that the two "scale co-vectors" are not directly interpretable as "quantization step size" of a certain tensor slice, thus generalizing this abstraction, apparently somewhat insufficient to fully describe *standard* quantization (not a non-uniform or vector-quantization, for which it is obvious). Now, is there an optimization opportunity in quantizing kernels with both per-input-channel and per-output-channel granularity, not yet explicitly discussed as it doesn't fit well into the simple "well-defined step-size" quantization picture?

Doubly-channelwise Kernel Quantization. It might be possible to re-parameterize the equations with the left/right kernel scale co-vectors as independent variables, determining the rest of quantization parameters - scale-vectors for activations, and the rescale factors.

$$(S_a^{l-1})_m = (S_{wL}^l)_m^{-1} \; ; \quad (S_a^l)_n = (S_{wL}^{l+1})_n^{-1} \tag{7}$$

$$\hat{F}_n^l = (S_{wR}^l)_n/(S_a^l)_n = (S_{wR}^l)_n(S_{wL}^{l+1})_n \tag{8}$$

For a scheme with a free choice of "doubly-channelwise" kernel quantization parameters to be successfully deployed (Fig. 1 B), we therefore need the following:

1. Hardware must support per-channel $(F = F_n)$ rescale factors in the re-coding block - which we take as a HW-aware redefinition of "channelwise" quantization. On top of that, in case of fan-out from layer l, with "next" $l + 1$ layer not uniquely defined, a separate rescale factors vector has to be applied for each outgoing edge. Yet other nuances may apply to special layers.
2. Optionally - with activation scale vectors fully determined (up to scalar) by next layer's "left" scale co-vector, the activation encoding might need more bits than that of kernel's so that info loss by activation quantization is less of a concern. This usually holds for 4b weight and 8b activation setting.

For a standard test of the error-reduction benefit to be had assuming the above, we experiment with Minimum Mean Square Error (MMSE) quantization of MobileNetV2 convolutional kernels, see Fig. (3). Layerwise, channelwise, doubly-channelwise MMSE are defined as follows:

$$MMSE(W) := \min_s \|W - s * clip(\lfloor W/s \rceil)\| \tag{9}$$

$$MMSE_{Ch}(W) := \min_{S_{wR}} \left\|W_{m,n} - (S_{wR})_n * clip\left(\left\lfloor \frac{W_{m,n}}{(S_{wR})_n}\right\rceil\right)\right\| = \sqrt{\textstyle\sum_n MMSE^2(W_{n,..})} \tag{10}$$

$$MMSE_{dCh}(W) := \min_{S_{wL},S_{wR}} \left\|W_{m,n} - (S_{wL})_m(S_{wR})_n * clip\left(\left\lfloor \frac{W_{m,n}}{(S_{wL})_m(S_{wR})_n}\right\rceil\right)\right\| \tag{11}$$

Note that channelwise MMSE problem is separable into per-kernel-slice standard MMSE subproblems, while double-channelwise MMSE is non-trivially different. We optimize using the 'progressive projection' algorithm adopted from [13], which we extend for the doubly-channelwise problem by alternating rows/columns projections (Appendix B). In Fig. (3) we can see consistent MMSE reduction by virtue of adding any vector DoF, as expected. We then draw the local error after application of our global optimization (see next chapter).

We see that the end-to-end optimization consistently utilizes the (double) granularity but only partially drives local error reduction. This shows that alignment of local vs. global (in layer/network sense) optimization objectives might be quite partial and motivates end-to-end training, in-line with observations of [16] for their full-length scale-training QAT.

To recap, given a favorable deployment arrangement (which, in light of the result, might be worth designing for), convolutional kernels could be quantized in doubly-channelwise fashion, improving kernel representation - but better yet, placing an overlooked DoF at the disposal of a global fine-tuning.

Activation's Vector Scale Degree of Freedom. Now we look into the opposite case - "layerwise quantization", for an arithmetic pipeline not supporting per-channel rescale. In that case, out of the two possible DoF in Eq. 6 just one DoF is available, the activation's per-channel vector scale, related to kernel scale left/right co-vectors - of next/previous layer respectively. This DoF previously [7,8] formalized as pre-quantization modification of kernel. The preconditioning view provides for a modular tool but on the other hand leaves untapped potential for holistic co-optimization with other DoFs. Here we re-frame the weight-factorization as modifications to quantization parameters, rather that to the weights themselves, for a unified view of Clipping and Equalization as manifestations of 'vector scales'.

Within the 'local' view of improving tensor representation, this single DoF impacts three "range granularity desirables": (A) per-channel A^{l-1} activation range/scale (B) per-output-channel W_{l-1} scaling (C) per-input-channel W_l scaling. Channel-equalizing heuristics balancing (A),(B) were proposed in [7], while [8] focused on balancing (B) against (C), better serving the 4bW 8bA case when adapted by replacing *naive-max* with *MMSE*. However, ideally we'd like to balance all three, while also replacing local heuristic with data-driven end-to-end training towards improving downstream network objective. This is what we set to do next.

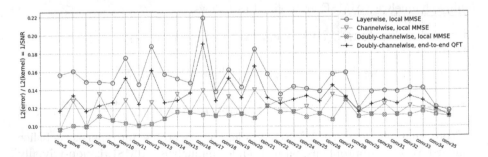

Fig. 3. MobileNetV2 kernels quantization error norm, across scale-tensor granularity, optimization procedure. For this net there seems to be gain from every extra vector degree of freedom, at least in terms of local error reduction

3.2   QFT: Downscaled all-DoF QAT as a Simple PTQ Baseline

We now describe our Quantization-aware Fine-tuning procedure. We avoid the use of labels, instead relying on a pure knowledge-distillation (KD) [23] method, with the 'teacher' being the full-precision pre-trained network [24–26]. The student to be trained is the quantized network, or more precisely, a deployment-aware graph simulating both the 'online' (HW run-time) and 'offline' (compile-time) computations. For the training loss, we explored usage of internal layers [26], taking the norm of difference of Teacher's and Student's (properly decoded) activations. The last layer of backbone alone was found to give good baseline results, 'covering' all layers and making use of spatially-rich information available before the global pooling. Results are further improved by mixing-in a classic KD [23] cross-entropy loss on logits.

Offline Subgraph. Central to our analysis is the overparameterized quantization and its resolution from degrees-of-freedom (DoF). If the HW constraints and relations among the quantization parameters comprise a system of equations, the offline subgraph is its symbolic "solution", a computation inferring all quantization parameters, both virtual "scales" and deployment constants such as rescale-factors, from a maximal unconstrained subset that is the *quantization DoF*. For simple convolutions described thus far, the offline subgraph amounts to applying Eqs. (6) and (4) to weights, activations' vector scales and rescale factors:

$$Vars := \{W^l_{..,m,n}, (S^l_A)_n, \hat{F}^l_n | l = 1..L\}$$

our suggested parameterization of independently trainable subset of quantization parameters. Layerwise quantization setting is supported by demoting the rescale factor to scalar $\hat{F}^l$. The output of the offline subgraph is the 'exports' $\hat{W}, \hat{F}$ fed into the online, HW-emulating part. Graph is differentiable by virtue of Straight-through Estimator (STE) [10] applied to augment any $clip(\lfloor . \rceil)$ op, representing the lossy part of the online/offline quantization of activations/weights, respectively. Both the pre-trained weights and all quantization params are end-to-end fine-tuned on the same footing, while train-deploy compatibility is maintained by enforcing all constraints in the forward pass of the offline subgraph.

   This approach replaces explicit quantization-parameter gradient definitions [15–17] by a native gradient flow through the offline subgraph. This enables for the first time an end-to-end training of all quantization parameters that are interlinked or not localized to a single layer / single Eq.(1)-style expression as required by LSQ [16] and similar methods. This includes the activation-scale vector degree of freedom, now chosen in a trainable, not heuristic way.

   For application of the offline-graph concept in real deployments, the equations have to be generalized in a few directions which we broadly outline:

1. Additive arithmetics (bias, ew-add), optional 'zero-point' parameters joining the scales for asymmetric encoding, and their respective relations.

2. Other layer types - e.g. *depthwise* convs not having L/R scale but only one, *elementwise* add/multiply introducing more sets of rescale-factors, *non-arithmetic* layers s.a. Maxpool,Concat,ResizeNN introducing relations between vector scales of their input and output, etc.
3. More specialized constraints - quantization of rescale-factors, non-homogenous activation functions precluding cross-layer factorization, etc.

These call for simulation approach centered around a repeated application of same 2 principles: (A) Explicitly model all HW arithmetics (B) Express all constraints in the offline subgraph. See Appendix A for further detail.

4   Experiments

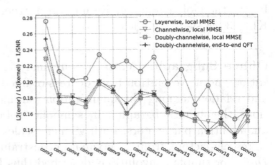

We apply our all-DoF QFT to classic ImageNet-1K [27] classification CNNs [28–31], pre-trained by courtesy of [22]. BatchNorm ops are folded, and no regularization or augmentation is applied. We run most experiments on a setup with per-channel rescale factors, and weight-only quantization. For this setup we utilize the doubly-channelwise quantization, training two vector DoF per layer. We also experiment with a deployment-oriented setup, with 8b (unsigned) activations and layerwise rescale factors. Here, only the cross-layer activation vector scale DoF is available to be fine-tuned jointly with weights and biases.

Fig. 4. ResNet18 kernels quantization error norm, across scale-tensor granularity, optimization procedure. This net is well served by just a simple per-output-channel MMSE scale, which however is also easily recovered by the end-to-end training

The hyperparameters of the Quantization-aware Finetuning are uniform across the networks we use: Distillation from the FP net serving as teacher, using a sum of classic [23] cross-entropy loss at the logits layer, and a L2 'reconstruction' loss at the last (before average-pool) convolutional layer output, weighted x4. We train for 12 epochs of 8K images each, with Adam optimizer and cosine learning rate, decaying across 4 epochs

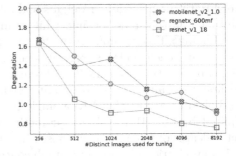

Fig. 5. Effect of dataset size.

starting from 1e-4 and reloading at /2 (i.e. 5e-5@ep4, 2.5e-5@ep8). Batch size is set at 16 to enable easily reproducing on smaller GPUs (e.g. RTX2080 8G).

Dataset size ablation is presented in Fig. 5. We scale down distinct images used but increase epochs to keep the total images fed constant at 32K, so convergence properties are roughly the same. We see that results deteriorate gracefully down to 1K images and well below. Beyond a few K images there seem to be diminishing returns at least for our "safe" LR=1e–4 regime. We therefore set 8K images (0.7% of the training set) as our working point for the rest of the section.

Overall effect of fine-tuning all degrees-of-freedom vs. "plain" training of weights and biases alone (fixed scales) is explored in Table 1. In both cases, we start by plain uniform ("scalar") MMSE initialization (solving Eq.(9) via repeated projection algorithm [13]) of all vector scales.

Table 1. Full/restricted QFT ablation: effect of optimizing all degrees-of-freedom by joint end-to-end fine-tuning, vs. only training weights and biases (fixed scalar scales)

Weight compression PTQ benchmark: ImageNet-1K accuracy (-*degradation*).							
Methods	Bits (W/A)	ResNet18	MobileNetV2	RegNet0.6G	MnasNet2	ResNet50	RegNet3.2G
Full Precision	32/32	71.25	72.8	73.8	76.66	76.8	78.5
fixed scales	4/32, chw	69.75 (−1.5)	71.0 (−1.8)	71.4 (−2.4)	75.9 (−0.75)	75.65 (−1.15)	77.05 (−1.45)
trained scales	4/32, chw	70.5 (−0.75)	71.75 (−1.05)	72.8 (−1.0)	76.1 (−0.55)	76.3 (−0.5)	77.9 (−0.6)
fixed scales	4/8, lw	69.85 (−1.4)	71.1 (−1.7)	71.4 (−2.4)	75.85 (−0.8)	75.6 (−1.2)	77.05 (−1.45)
trained scales	4/8, lw	70.05 (−1.2)	71.5 (−1.3)	72.45 (−1.35)	76.0 (−0.65)	76.05 (−0.75)	77.5 (−1.0)

We can see major improvement for full vs. restricted QFT across nets. The gain is still significant but smaller for the constrained *4/8,lw* where just the cross-layer DoF is available (vs. double input/output channelwise). This again testifies to the added value of utilizing either DoF or both.

Local Optimization. We now note that this experiment (Table 1) deprives the "fixed scales" setting of any usage of the vector scales DoF, and the full QFT of possibly better-positioned initialization. In Table 2 we test QFT initialized with fully locally-optimized doubly-channelwise scales, which are then either left fixed or trained jointly as in Table 1. We observe net-dependent behavior. In ResNet18 case, fixed-scales training gains strongly from the local optimization which indeed seems (Fig. 4) well-aligned with the global, but the holistic approach recovers those as well and if equipped with the better init, even yields a slight further gain. However, for MobileNetV2 the alignment is weak, as seen from Table 2 and by-layer quantization-error analysis in Fig. 3,

Table 2. Effect of locally-optimal scale-vectors initialization on post-QFT degradation. The two init options are solutions of Eq. 9/Eq. 11 respectively

Initialization	scalar-mmse		2-vector-mmse	
scales in QFT	fixed	trained	fixed	trained
ResNet18	−1.3	−0.8	−0.8	−0.57
MobilenetV2	−1.85	−1.05	−1.7	−1.1
RegNet600	−2.5	−1.0	−3.8	−1.5

and for RegNet the local pre-optimization is even detrimental to global objective. We thus use a "no bells & whistles" scalar init in the rest of this section, and fully rely on the end-to-end training to optimize all vector DoF.

Comparison to State-of-the-Art Methods on ImageNet-1K. In Table 3, we compare our results obtained by QFT to SoTA PTQ methods, in layerwise and channelwise regime. While it is common to leave first layer not quantized for such experiments, we note following [32] that this is quite an arbitrary choice. Instead, for a 'flat overhead rate' across nets, we quantize in 8b a few smallest layers, added-up by increasing size till their cumulative size is 1% of total backbone footprint. Impact on resource and deployment is equally negligible.

For the *4/8,lw* regime the degradation is mostly outperforming other method, and is never away from the 1% ballpark. For the *4/32,chw* case we achieve sub-1% degradation results, mostly on-par with SoTA [22]. In all experiments QFT doesn't use any per-network modification. Further gains per net may be had by trying: (A) Block-by-block reconstruction as in [22] (B) Local mmse pre-optimization (Table 2) (C) Scan for best learning rate (and other hyperparameters) - or any other standard train tuning technique. We stress that QFT is a single-stage method performing holistic optimization of DoF to which separate PTQ stages are often dedicated [3] - clipping, equalization, bias-correction, etc.

Table 3. Comparing QFT to SoTA PTQ methods. chw/lw means channelwise/layerwise configuration. Results of other methods are quoted from respective papers and compared to respective full-precision evaluations for fair degradation values in (.).

ImageNet-1K accuracy (*-degradation*); results within 0.2 of best method in **bold**							
Methods	Bits (W/A)	ResNet18	MobileNetV2	RegNet0.6G	MnasNet2	ResNet50	RegNet3.2G
Full Precision	32/32	71.2	72.8	73.8	76.7	76.8	78.5
Adaround [14]	4/32, lw	68.7 (**−1.0**)	69.8(−1.9)	72.0 (−1.7)	74.9 (−1.8)	75.2 (−0.9)	77.1 (−1.3)
Adaround [14]	4/8, lw	68.6 (**−1.1**)	69.3 (−2.4)	–	–	75.0 (−1.1)	–
QFT (ours)	4/8, lw	70.1 (**−1.1**)	71.6 (**−1.2**)	72.5 (**−1.3**)	76.1 (**−0.6**)	76.2 (**−0.6**)	77.5 (**−1.0**)
BRECQ [22]	4/32, chw	70.7 (**−0.4**)	71.7 (**−0.8**)	73.0 (**−0.7**)	76.0 (**−0.7**)	76.3 (−0.7)	78.0 (**−0.4**)
QFT (ours)	4/32, chw	70.4 (−0.8)	71.8 (**−1.0**)	72.8 (−1.0)	76.2 (**−0.5**)	76.4 (**−0.4**)	77.8 (−0.7)

QFT is also a speedy method. Run times for experiments here vary between 20–90min depending on network size; for some net can be pushed yet down by batch size hike and x3 epochs drop, e.g., achieving FP-1% accurate ResNet18 in **5 min** in one experiment. Speed can be credited to a standard end-to-end usage of GPUs and SW stack, in contrast to layer-by-layer methods s.a. [14,21,22] which may hit IO bottleneck, making high GPU utilization a challenge.

5   Conclusions and Outlook

In this work we bridge the gap between two long-standing complementary approached to DNN quantization - PTQ and QAT. We show that a successful

PTQ method can be constructed from QAT operating in quick fine-tuning mode (hence "Quantization-aware Fine-tuning" - QFT), stripped of massive data and labels usage, BatchNorm-quantization issues and any per-network configuration. Our QFT baseline is underpinned by jointly training all possible degrees-of-freedom and as such it optimally adapts to HW configuration - layerwise, channelwise, etc. QFT introduces into the PTQ domain: (a) an end-to-end holistic optimization approach and (b) the application of ideas from the QAT literature. We show that QFT can be competitive for 4bW PTQ, while kept simple and without many of the bells and whistles of SoTA PTQ methods. We thus suggest it as a principled PTQ baseline, further extendable by ideas from other methods, and by a scale-up across the PTQ-QAT continuum. We leave exploration of those directions to future work.

References

1. Jacob, B., et al.: Quantization and training of neural networks for efficient integer-arithmetic-only inference. CoRR abs/1712.05877 (2017)
2. Krishnamoorthi, R.: Quantizing deep convolutional networks for efficient inference: a whitepaper. CoRR abs/1806.08342 (2018)
3. Nagel, M., Fournarakis, M., Amjad, R.A., Bondarenko, Y., van Baalen, M., Blankevoort, T.: A white paper on neural network quantization. CoRR abs/2106.08295 (2021)
4. Wu, H., Judd, P., Zhang, X., Isaev, M., Micikevicius, P.: Integer quantization for deep learning inference: principles and empirical evaluation. CoRR abs/2004.09602 (2020)
5. Kozlov, A., Lazarevich, I., Shamporov, V., Lyalyushkin, N., Gorbachev, Y.: Neural network compression framework for fast model inference. CoRR abs/2002.08679 (2020)
6. Gholami, A., Kim, S., Dong, Z., Yao, Z., Mahoney, M.W., Keutzer, K.: A survey of quantization methods for efficient neural network inference. CoRR abs/2103.13630 (2021)
7. Meller, E., Finkelstein, A., Almog, U., Grobman, M.: Same, same but different - recovering neural network quantization error through weight factorization. CoRR abs/1902.01917 (2019)
8. Nagel, M., van Baalen, M., Blankevoort, T., Welling, M.: Data-free quantization through weight equalization and bias correction. In: 2019 IEEE/CVF International Conference on Computer Vision, ICCV 2019, pp. 1325–1334. Seoul, Korea (South), 27 October - 2 November 2019. IEEE (2019)
9. Li, Y., et al.: MQBench: towards reproducible and deployable model quantization benchmark. In: Vanschoren, J., Yeung, S., (eds.): Proceedings of the Neural Information Processing Systems Track on Datasets and Benchmarks, vol. 1 (2021)
10. Bengio, Y., Léonard, N., Courville, A.C.: Estimating or propagating gradients through stochastic neurons for conditional computation. CoRR abs/1308.3432 (2013)
11. Hubara, I., Courbariaux, M., Soudry, D., El-Yaniv, R., Bengio, Y.: Binarized neural networks. In: Lee, D.D., Sugiyama, M., von Luxburg, U., Guyon, I., Garnett, R., (eds.): Advances in Neural Information Processing Systems 29: Annual Conference on Neural Information Processing Systems 2016, pp. 4107–4115, 5-10 December 2016, Barcelona, Spain (2016)

12. McKinstry, J.L., et al.: Discovering low-precision networks close to full-precision networks for efficient embedded inference. CoRR abs/1809.04191 (2018)
13. Liu, Z.G., Mattina, M.: Learning low-precision neural networks without straight-through estimator(ste). CoRR abs/1903.01061 (2019)
14. Nagel, M., Amjad, R.A., van Baalen, M., Louizos, C., Blankevoort, T.: Up or down? adaptive rounding for post-training quantization. In: Proceedings of the 37th International Conference on Machine Learning, ICML 2020, pp. 7197–7206, 13-18 July 2020, Virtual Event. Volume 119 of Proceedings of Machine Learning Research, PMLR (2020)
15. Choi, J., Wang, Z., Venkataramani, S., Chuang, P.I., Srinivasan, V., Gopalakrishnan, K.: PACT: parameterized clipping activation for quantized neural networks. CoRR abs/1805.06085 (2018)
16. Esser, S.K., McKinstry, J.L., Bablani, D., Appuswamy, R., Modha, D.S.: Learned step size quantization. In: 8th International Conference on Learning Representations, ICLR 2020, Addis Ababa, Ethiopia, April 26–30 (2020), OpenReview.net (2020)
17. Jain, S.R., Gural, A., Wu, M., Dick, C.: Trained quantization thresholds for accurate and efficient fixed-point inference of deep neural networks. In: Dhillon, I.S., Papailiopoulos, D.S., Sze, V., (eds.): Proceedings of Machine Learning and Systems 2020, MLSys 2020, Austin, TX, USA, 2–4 March 2020, mlsys.org (2020)
18. Banner, R., Nahshan, Y., Hoffer, E., Soudry, D.: ACIQ: analytical clipping for integer quantization of neural networks. CoRR abs/1810.05723 (2018)
19. Finkelstein, A., Almog, U., Grobman, M.: Fighting quantization bias with bias. CoRR abs/1906.03193 (2019)
20. Choukroun, Y., Kravchik, E., Yang, F., Kisilev, P.: Low-bit quantization of neural networks for efficient inference. In: 2019 IEEE/CVF International Conference on Computer Vision Workshops, ICCV Workshops 2019, pp. 3009–3018. Seoul, Korea (South), 27-28 October 2019. IEEE (2019)
21. Hubara, I., Nahshan, Y., Hanani, Y., Banner, R., Soudry, D.: Improving post training neural quantization: layer-wise calibration and integer programming. CoRR abs/2006.10518 (2020)
22. Li, Y., et al.: BRECQ: pushing the limit of post-training quantization by block reconstruction. In: 9th International Conference on Learning Representations, ICLR 2021, Virtual Event, Austria, 3–7 May 2021. OpenReview.net (2021)
23. Hinton, G.E., Vinyals, O., Dean, J.: Distilling the knowledge in a neural network. CoRR abs/1503.02531 (2015)
24. Mishra, A.K., Marr, D.: Apprentice: using knowledge distillation techniques to improve low-precision network accuracy. In: 6th International Conference on Learning Representations, ICLR 2018, Vancouver, BC, Canada, 30 April - 3 May 2018. Conference Track Proceedings, OpenReview.net (2018)
25. Polino, A., Pascanu, R., Alistarh, D.: Model compression via distillation and quantization. In: 6th International Conference on Learning Representations, ICLR 2018, Vancouver, BC, Canada, 30 April - 3 May 2018. Conference Track Proceedings, OpenReview.net (2018)
26. Zhuang, B., Shen, C., Tan, M., Liu, L., Reid, I.: Towards effective low-bitwidth convolutional neural networks. In: Proceedings of the IEEE Conference on Computer Vision and Pattern Recognition (CVPR) (2018)
27. Russakovsky, O., et al.: Imagenet large scale visual recognition challenge. Int. J. Comput. Vis. **115**(3), 211–252 (2015)

28. He, K., Zhang, X., Ren, S., Sun, J.: Deep residual learning for image recognition. In: 2016 IEEE Conference on Computer Vision and Pattern Recognition, CVPR 2016, pp. 770–778. Las Vegas, NV, USA, 27-30 June 2016. IEEE Computer Society (2016)

29. Radosavovic, I., Kosaraju, R.P., Girshick, R.B., He, K., Dollár, P.: Designing network design spaces. In: 2020 IEEE/CVF Conference on Computer Vision and Pattern Recognition, CVPR 2020, pp. 10425–10433. Seattle, WA, USA, 13–19 June 2020. Computer Vision Foundation / IEEE (2020)

30. Tan, M., et al.: MnasNet: platform-aware neural architecture search for mobile. In: IEEE Conference on Computer Vision and Pattern Recognition, CVPR 2019, pp. 2820–2828. Long Beach, CA, USA, 16–20 June 2019. Computer Vision Foundation / IEEE (2019)

31. Sandler, M., Howard, A.G., Zhu, M., Zhmoginov, A., Chen, L.: Mobilenetv 2: Inverted residuals and linear bottlenecks. In: 2018 IEEE Conference on Computer Vision and Pattern Recognition, CVPR 2018, pp. 4510–4520. Salt Lake City, UT, USA, 18–22 June 2018. Computer Vision Foundation / IEEE Computer Society (2018)

32. Gluska, S., Grobman, M.: Exploring neural networks quantization via layer-wise quantization analysis. CoRR abs/2012.08420 (2020)

Searching for N:M Fine-grained Sparsity of Weights and Activations in Neural Networks

Ruth Akiva-Hochman[1(✉)] , Shahaf E. Finder[1] , Javier S. Turek[2] ,
and Eran Treister[1]

[1] Ben-Gurion University of the Negev, Be'er Sheva, Israel
{ruthho,finders}@post.bgu.ac.il, erant@cs.bgu.ac.il
[2] Intel Labs, Hillsboro, OR, USA
javier.turek@intel.com

Abstract. Sparsity in deep neural networks has been extensively studied to compress and accelerate models for environments with limited resources. The general approach of pruning aims at enforcing sparsity on the obtained model, with minimal accuracy loss, but with a sparsity structure that enables acceleration on hardware. The sparsity can be enforced on either the weights or activations of the network, and existing works tend to focus on either one for the entire network. In this paper, we suggest a strategy based on Neural Architecture Search (NAS) to sparsify both activations and weights throughout the network, while utilizing the recent approach of *N:M* fine-grained structured sparsity that enables practical acceleration on dedicated GPUs. We show that a combination of weight and activation pruning is superior to each option separately. Furthermore, during the training, the choice between pruning the weights of activations can be motivated by practical inference costs (e.g., memory bandwidth). We demonstrate the efficiency of the approach on several image classification datasets.

Keywords: Neural architecture search · Weight pruning · Activation pruning · N:M fine-grained Sparsity

1 Introduction

Deep Neural Networks (DNNs) have achieved remarkable success in many fields, including computer vision, natural language processing, and more [10]. However, many state-of-the-art networks require a vast amount of computation and memory and are difficult to apply in environments with low computational capabilities. In an attempt to overcome this issue various approaches have been proposed, including quantization [1,8,18], designing and searching for efficient model architectures [5,16,34,35], knowledge distillation [11], and pruning [13,14,44]. In this work, we focus on a combination of pruning and Neural Architecture Search (NAS).

R. Akiva-Hochman and S. E. Finder—Contributed equally.

L. Karlinsky et al. (Eds.): ECCV 2022 Workshops, LNCS 13807, pp. 130–143, 2023.
https://doi.org/10.1007/978-3-031-25082-8_9

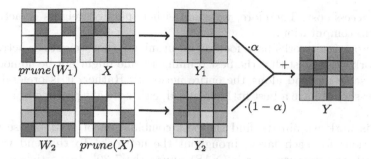

$prune(W_1)$ X Y_1 $\cdot\alpha$ $+$

W_2 $prune(X)$ Y_2 $\cdot(1-\alpha)$ Y

Fig. 1. Illustration of a single pruning search-layer. During training, the layer calculates the two pruning options and returns a weighted sum of their outputs.

The general approach of pruning aims at enforcing sparsity on the obtained model, with minimal accuracy loss. The most common pruning methods take a trained model and gradually zero out certain entries in it [44]. However, the resulting sparsity structure is typically irregular, and is hard to exploit in hardware for actual speedup. Some works have tackled this issue by zeroing groups of weights [27], but practice shows that doing so does not achieve comparable results to the unstructured versions.

Recently, advances in hardware architectures led to support accelerated 2:4 structured fine-grained sparsity [26]. Here, $N{:}M$ sparsity indicates that only N weights are non-zero for every M continuous-in-memory weights. This can be applied in several common operations including linear, convolutional, transformer blocks, etc. Generally, all these operations are based on matrix-matrix multiplications, which is the main computational component in DNNs. The main advantage of $N{:}M$ sparsity is its fine-grained structure. That is, within each block of M numbers we have N non-zeros in unstructured locations. As M is larger and N is smaller, we get closer to a fully unstructured pruning. Hence, on the one hand this approach is effective to maintain competitive accuracy, and on the other hand it can be exploited in hardware for speedup.

Previous literature has focused on weight pruning, either given trained weights or during training [28]. In particular, that is the case in both [26,42] that consider $N{:}M$ sparsity. Alternatively, sparsity in DNNs can also be enforced on the hidden feature maps, also referred to as activations [21], because typically the weight matrices are multiplied with activation matrices, using a matrix-matrix multiplication kernel. This is intriguing because at least when the ReLU activation is used, we can expect a significant portion of the activation entries to be zeros, hence in principle, we can exploit this natural sparsity for speedup without loosing accuracy. The issue is that the natural sparsity of the activations is not structured, and again, cannot be fully exploited for speedup as it is.

At this point we note that a matrix-matrix multiplication is symmetric in terms of the computations. That is, the multiplication and its potential speedup does not depend on the roles of the matrices in the network. It may depend on the dimensions of the matrices, i.e., the amount of floating point multiplications, and memory access cost. Therefore, we could either sparsify weights or activations to speedup computation.

While previous works have focused on pruning either weights or activations, in this work we suggest that the best pruning scheme might not be homogeneous (i.e., the same method along the entire network). Rather, it is preferable that each layer should have a best fitting method, either weight pruning or activation pruning[1].

In this work we aim to find the best combination of $N{:}M$ sparse weights or activations for each layer throughout the network. To this end we adopt the Neural Architecture Search (NAS) approach [7,30]. In particular, we use the differential NAS [41], where the architectural choice in every search-layer is determined by a softmax function, and is optimized together with the network's weights—see Fig. 1 for an illustration of our differential NAS for a single search-layer. To accommodate both activation or weight pruning in a unified training scheme within the differential NAS, we employ a gradient-based training for the fine-grained structured pruning of the weights or activations, using the Straight Through Estimator (STE) [40]. That is, during the forward pass, the top-k entries in the weights or activations in each convolutional layer are chosen according to their absolute value, while the other entries are zeroed out. During the backward pass, the STE allows to optimize for all the weight parameters or propagate gradient information through the pruned activation entries. In addition, we propose a new bandwidth loss term to account for the change in computational performance that is related to memory access. We use this term to augment the loss function for training the network, to allow us to direct the NAS towards certain choices of weights or activation pruning.

To summarize, in this work we propose a novel framework for optimizing the use of $N{:}M$ fine-grained sparsity of the weights or activations, depending on the layer. We demonstrate how such mix of choices is preferable over a homogeneous pruning scheme for image classification.

2   Related Work

2.1   Unstructured and Structured Weight Pruning

The most popular approaches for learning a sparse neural networks are greedy pruning methods, which generally train a dense neural network, perform a pruning routine, and retrain the sparse network. The pruning routine can be either single-shot [22] or iterative [12,13,36,38]. Most of these methods choose the pruned weights according to a score function, which is based on weight magnitude [12,13], gradient [38], or trained importance [36]. While these methods can achieve high sparsity with minimal performance impact, the sparsity is generally unstructured, making it difficult to translate to actual speedup on hardware [42].

---

[1] To the best of our knowledge, only sparse-dense matrix-matrix products can be efficiently applied in hardware, and the speedup of sparse-sparse products is much more complicated to achieve. Hence, we consider the pruning of either the weights or the activations. However, the pruning of both can also be considered in our framework.

One possible approach to enforce structure is to give a score to groups of weights [23,27], but practice shows that doing so does not achieve comparable results to the unstructured versions. This result is not too surprising considering the randomness of the initial state in the training and the resulting non-grouped nature of the weights. Without some intervention (e.g., adding regularization or modifying the architecture), weights do not naturally group for significant structured pruning.

2.2 *N:M* Fine-grained Structured Weight Pruning

NVIDIA Amper's latest architecture introduces a new sparse tensor core design with a fixed pruning purpose of 50% weight. Meaning that a model with $N{:}M$ structured fine-grained sparsity, where within every group of M continuous weights only N are non-zero, achieves better accuracy and performance on the GPU [43]. To achieve $N{:}M$ structured fine-grained sparsity, [26] proposes greedy pruning with retraining. The work [42] improve on that by first utilizing the straight-through estimator (STE) for training sparse $N{:}M$ weights:

$$W_{t+1} \leftarrow W_t - \gamma_t \nabla_{W_t} \mathcal{L}(\tilde{W}_t; X_t), \tag{1}$$

where $\tilde{W} = prune(W)$ is the pruned weights. Then, they further improved the training routine by adding weight-decay for the zeroed weights (referred to as SR-STE). Furthermore, [17] suggests $N{:}M$ sparsity to accelerate DNN training, by using a transposable fine-grained sparsity mask, where the same mask can be used for both forward and backward passes. Lastly, [33] suggests fixing M and using a different N per layer, which is chosen by an efficient heuristic. This method has the advantage of flexibility in choosing a different N for each layer, assuming that dedicated hardware can support such variety of pruning ratios and still can achieve practical acceleration.

2.3 Activation Pruning

Another direction to make the computation more efficient would be to encourage structured sparsity in each layer's input (feature maps). The motivation here is that the operations we want to prune are at their core a matrix-matrix products between weights and feature maps. Hence, in this sense the need for sparsity is symmetric between feature maps and weights.

One attempt for inducing activation sparsity is through ℓ_1 penalty [9]. Another attempt is the use of Hoyer regularization in addition to Forced Activation-Threshold Rectified Linear Unit activation function suggested by [21],

$$FATReLU(x) = \begin{cases} 0 & x < T \\ x & \text{otherwise} \end{cases}. \tag{2}$$

This function is not only non-differentiable at T but also non-continuous. The authors suggests that this activation function should be used only on a pretrained model, due to the difficulty of learning a model with such activation function. Another property of this approach is that there is no structure guarantee on the sparsity of the activations. To the best of our knowledge, no work has explored N:M fine-grained structured sparsity for activations yet.

2.4 Neural Architecture Search

In recent years there has been an increasing interest in designing neural network architecture automatically. Previous NAS algorithms automatically discover the topology structure of the network or the network size [25]. Many different search strategies can be used to explore the space of neural architectures. A prevailing search strategy is to use evolutionary algorithm [29] or reinforcement-learning based methods [34,45]. However, they require huge computational resources and cannot be directly used on large-scale target datasets. Alternatively, differentiable methods dramatically reduce the computation costs. Through these approaches, the discrete search strategy can become continuous by relaxing the selection of candidate operations to variations of softmax (e.g., Gumbel softmax) of all possible operations, which enables the use of gradients to efficiently optimize the architecture search space [39].

In the context of compressed networks, [5] searches for the depth and width of a small network using a differentiable search process, aiming to minimize the computational cost. NAS was also recently applied to finding optimal mixed precision quantization [3,32,41].

3   Method

In this work, our goal is to find the optimal choice between weight and activation pruning for each layer, for a given network and a specific N:M configuration. To this extent, we model this task as a NAS problem as detailed below.

3.1   Pruning Search-Layer

To simplify our search space, we start with a given architecture (e.g., ResNet), and replace each convolutional layer with a simple search-layer, which is constructed of weighted average of weight pruning and activation pruning layers:

$$Y = \alpha(prune(W_1) \cdot X) + (1 - \alpha)(W_2 \cdot prune(X)), \tag{3}$$

where W_1 and W_2 are the set of weights for the weights and activations pruning, respectively. The term $prune()$ denotes an operator that keeps the N parameters with the highest absolute values in a group of M parameters, while setting the others to zero, and α is a learnable parameter indicating the importance of each pruning type, similar to [24,41]. This configuration is presented visually in Fig. 1, and in Fig. 2b.

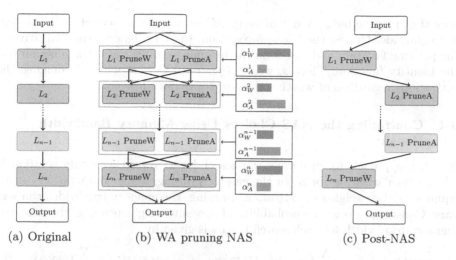

(a) Original   (b) WA pruning NAS   (c) Post-NAS

Fig. 2. Illustration of our framework. (a) the original model, (b) the same model with pruning search-layers replacing the original layers, (c) the final model after training and finalizing each search-layer's choice.

In (3), the parameter α can be optimized through gradient decent, and one can use softmax to bound it between 0 and 1. However, empirically, this tends to keep α at the value 0.5. We discuss two different approaches to avoid this, Gumbel-softmax (Subsect. 3.2) and memory bandwidth penalty (Subsect. 3.4).

3.2   Gumbel-Softmax

Gumbel-softmax [19] is a popular approach [5], that replaces softmax with

$$\alpha_w = \frac{\exp((\log(p_w) + g_w)/\tau)}{\sum_{j \in \{w,a\}} \exp((\log(p_j) + g_j)/\tau)}, \quad \alpha_a = 1 - \alpha_w \quad (4)$$
$$\text{s.t.} \quad g_j = -\log(-\log(u)), \quad j \in \{w,a\}, \quad u \sim U(0,1),$$

where $U(0,1)$ indicates the uniform distribution between zero and one, p_w, p_a are the learnable probabilities that represent the two choices of pruning (weights and activations) and τ is the softmax temperature. During training, the temperature is smoothly reduced to 0, where $\tau = 0$ causes Gumbel-Softmax to return a one hot vector. That is, it ensures that the decision for pruning weights or pruning activations is certain.

3.3   Training for N:M Sparse Activations Using STE

When training the $N{:}M$ pruned layers within the pruning search-layers, we use STE [2] to perform online magnitude-based pruning, as discussed in Sect. 2 for the weights. We note that to our knowledge, we are the first to utilize it for activation pruning. During the forward pass, we apply a pruning function to keep only the sparse subset (N out of every M) of weights or activations according to magnitude. During the back-propagation, the STE ignores the derivative of the pruning function and passes on the incoming gradient as if the function was the identity function. That is, we set $\partial prune/\partial X = 1$ (where X denotes the activation) regardless of whether the entry was zeroed or not.

3.4 Controlling the NAS Choices Using Memory Bandwidth Penalty

Another approach to encourage the search-layer toward a certain solution is adding regularization for α. In the case of pruning, while computation can be equal for either weights or activations pruning, the memory bandwidth can still vary. Considering α as the probability of choosing weight-pruning, the expected memory bandwidth for each search-layer l is given by

$$\mathbb{E}(\text{Mem}_l) = \alpha_l(\frac{N}{M}\mathcal{N}(W_l) + \mathcal{N}(X_l)) + (1 - \alpha_l)(\mathcal{N}(W_l) + \frac{N}{M}\mathcal{N}(X_l)), \quad (5)$$

where W and X are the weight and activations tensors respectively, and $\mathcal{N}(T)$ is the number of elements in tensor T. The memory bandwidth penalty for the entire network is then given by

$$\mathcal{L}_{mem} = \frac{\sum_{l=1}^{L} \mathbb{E}(\text{Mem}_l)}{\sum_{l=1}^{L} (\mathcal{N}(W_l) + \mathcal{N}(X_l))}. \quad (6)$$

The training loss is then given by

$$\mathcal{L} = \mathcal{L}_{obj} + \lambda_{mem}\mathcal{L}_{mem}, \quad (7)$$

where $\mathcal{L}_{obj}$ is the original objective loss, and $\mathcal{L}_{mem}$ is weighted by a regularization parameter λ_{mem}. We note that Eq. (5)–(6) do not consider a batch size, but in principle, batched inference can be useful in certain scenarios where multiple input streams are used. Considering batched inference, the memory bandwidth of the activations can become much more significant in terms of cost, favoring the pruning of activations.

3.5 Finalizing the Search-Layers Choices

When the training cycle is over, we finalize it by selecting a pruning method for each layer according the resulting α. That is, we replace the search-layer with the chosen pruned layer. This step is illustrated in Fig. 2c.

When using Gumbel-softmax, α tends to converge to 0 or 1 and therefore requires no fine-tuning. However, when using the penalty term alone, the parameter α might converge to a number in $[0, 1]$, usually close to zero or close to one. Therefore, changing this to a hard assignment between weights or activations pruning (i.e., setting $\alpha \in \{0, 1\}$) might affect the result of the network. To this end, after choosing each pruning option, we fine-tune for a few epochs to finalize the training.

4   Results

We show our performance of different $N{:}M$ sparse patterns: 2:4, 4:8 (50%), 2:8, 4:16 (75%) and 2:16 (87.5%) sparsity. For each $N{:}M$ option, we compare between three methods of pruning: weight pruning (STE-W), activation pruning (STE-A) and a combination of both using NAS (WA-NAS). We ran our experiments on NVIDIA 24GB RTX 3090 GPUs.

4.1   Image Classification

We evaluate our approach by training ResNet-20 on the datasets CIFAR-10 and CIFAR-100 [20], and ResNet-18 and ResNet-50 on ImageNet [4]. CIFAR-10 and CIFAR-100 have 10 and 100 classes respectively, both containing $5 \cdot 10^4$ training images and 10^4 test images. ImageNet contains $1.28 \cdot 10^6$ training images and $5 \cdot 10^4$ test images of 10^3 different classes.

CIFAR-10/100. We trained ResNet-20 from scratch for each of the pruning methods. We used SGD with momentum of 0.9 and weight decay of $2 \cdot 10^{-4}$ for 300 epochs with batch size 128. The learning rate is initialized with 0.1 and reduced by a cosine scheduler. This training setup was used for all our experiment, denoted by STE-W and STE-A for weight and activation pruning using STE, respectively. For the pruning search (WA-NAS) we set $\lambda_{mem} = 0.01$, and τ in Gumbel-Softmax is linearly decayed from 5 to 0.01.

In Table 1, we observe that for the same $N{:}M$ structural sparse patterns that STE-W surpasses STE-A (although not by a lot) and that WA-NAS leads consistently to better performance than STE-W and STE-A. In addition, our WA-NAS 4:8 and 2:4 methods outperform the state-of-the-art unstructured weight sparsity methods, including LCCL [6], SFP [14] and FPGM [15]. We note that in these experiments we did not observe any advantage for SR-STE-W over STE-W.

Table 1. Results for ResNet-20 pruning trained on CIFAR-10 and CIFAR-100.

Method	CIFAR-10			CIFAR-100		
	Acc	Acc Drop	FLOPs ($\cdot 10^7$)	Acc	Acc Drop	FLOPs ($\cdot 10^7$)
LCCL [6] [†]	91.68%	1.06%	2.61 (36.0%)	64.66%	2.87%	2.73 (33.1%)
SFP [14] [†]	90.83%	1.37%	2.43 (42.2%)	64.37%	3.25%	2.43 (42.2%)
FPGM [15] [†]	91.09%	1.11%	2.43 (42.2%)	66.86%	0.76%	2.43 (42.2%)
STE-W 4:8	92.13%	0.54%	2.05 (50.05%)	66.87%	1.51%	2.05 (50.05%)
STE-A 4:8	91.54%	1.13%	2.05 (50.05%)	66.48%	1.9%	2.05 (50.05%)
WA-NAS 4:8	92.78%	-0.11%	2.05 (50.05%)	67.69%	0.69%	2.05 (50.05%)
STE-W 2:4	91.77%	0.9%	2.05 (50.05%)	66.4%	1.98%	2.05 (50.05%)
STE-A 2:4	91.07%	1.6%	2.05 (50.05%)	65.44%	2.94%	2.05 (50.05%)
WA-NAS 2:4	92.15%	0.52%	2.05 (50.05%)	66.63%	1.47%	2.05 (50.05%)

[†] Unstructured pruning method

ImageNet. For Imagenet, we trained ResNet-18 and ResNet-50 from scratch for each of the pruning methods. We used SGD with momentum of 0.9 and weight decay of 10^{-4} for 120 epochs with batch size 256, distributed over four GPUs. The learning rate is initialized with 0.05 and reduced by a cosine scheduler. For the pruning search (WA-NAS) we set $\lambda_{mem} = 0.01$, and τ in Gumbel-Softmax is linearly decayed from 5 to 0.01.

The results for ResNet-18 and ResNet-50 are reported in Table 2 and Table 3 respectively. We again see that STE-A by itself is inferior to the two STE-W variants with lower accuracy by approximately 1%, hence we show STE-A for 2:4 only. We also reached the same conclusion of [42], that SR-STE training performs consistently better than STE-W, hence we show only partial results for STE-W, and omit the NAS option when trained with STE alone for the weights pruning. We also note that in Domino [33], the sparsity is distributed unevenly by using a different N per layer, this approach can also be combined with our method.

In Table 2, for ResNet-18, WA-NAS achieves better accuracy than SR-STE except for the 2:4 configuration. We have not been able to replicate the accuracy of [42] using the authors' code, and when training SR-STE using our training configuration, WA-NAS does surpass its score. For 2:16, our method surpasses the accuracy reported in [33] for SR-STE, however, Domino's [33] flexibility in N choice leads to better results than WA-NAS that uses a fixed N. In Table 3, for ResNet-50, our method is able to surpass or achieve the same accuracy of SR-STE. Similar to ResNet-18, our method is slightly outperformed by Domino, but can be combined with it if the framework is beneficial in hardware.

Table 2. Results for ResNet-18 pruning trained on ImageNet dataset.

Method	N:M	Top-1 Acc	FLOPs ($\cdot 10^8$)
Baseline		70.55	18.1(0%)
STE-A	2:4	68.3	9.7 (47%)
STE-W	2:4	69.3	9.7 (47%)
Nvidia [26]	2:4	70.7	9.7 (47%)
SR-STE [42]	2:4	71.2	9.7 (47%)
SR-STE[†]	2:4	70.3	9.7 (47%)
WA-NAS (ours)	2:4	70.9	9.7 (47%)
SR-STE[†]	2:8	68.9	5.4 (70%)
WA-NAS (ours)	2:8	69.7	5.4 (70%)
SR-STE[†]	4:16	68.9	5.4 (70%)
WA-NAS (ours)	4:16	69.9	5.4 (70%)
SR-STE[‡]	2:16	66.65	3.3 (82%)
Domino [33]	N:16	68.0	3.3 (82%)
WA-NAS (ours)	2:16	67.5	3.3 (82%)

[†] Our training of SR-STE
[‡] SR-STE results as reported by [33]

Table 3. Results for ResNet-50 pruning trained on ImageNet dataset.

Method	$N:M$	Top-1 Acc	FLOPs ($\cdot 10^8$)
Baseline		77.22	40.89 (0%)
STE-A	2:4	75.9	21 (49%)
STE-W	2:4	76.4	21 (49%)
Nvidia [26]	2:4	76.8	21 (49%)
SR-STE [42]	2:4	77.0	21 (49%)
WA-NAS (ours)	2:4	77.1	21 (49%)
SR-STE [42]	2:8	76.2	11 (73%)
WA-NAS (ours)	2:8	76.2	11 (73%)
SR-STE [42][†]	4:16	76.3	11 (73%)
SR-STE [42][‡]	4:16	76.5	11 (73%)
Domino [33]	N:16	76.7	11 (73%)
WA-NAS (ours)	4:16	76.5	11 (73%)
SR-STE [42][‡]	2:16	74.4	6.2 (85%)
Domino [33]	N:16	75.4	6.2 (85%)
WA-NAS (ours)	2:16	75.0	6.2 (85%)

[†] Our training of SR-STE
[‡] SR-STE results as reported by [33]

4.2 Ablation Study

To investigate the contribution of each component, we trained multiple configurations of pruned ResNet-18 trained on ImageNet50 [31,37], a subset of 50 classes from the ImageNet dataset. The configurations include 2:4 and 4:8 pruning and the α regulation terms: Gumbel-Softmax, memory bandwidth penalty, and a combination of both. The results are shown in Table 4. We make two observations: 1. WA-NAS consistently achieves better results than pruning only weights or activations. 2. The proposed regulation terms achieve comparable accuracy.

4.3 NAS Pruning Choice Analysis

We demonstrate the NAS pruning choice of each layer in Table 5. The model shown is ResNet-18 with a 2:4 sparsity pattern, trained on ImageNet50 using our NAS framework with Gumbel-Softmax. We observe, in this model and others, that the NAS often chooses to prune activations in down-sampling layers. In addition, our method tends to prune the weights in the first few layers. In our opinion, that is because pruning the first few activation maps may cause a loss of critical initial feature information that undergoes all the processing throughout the network. Lastly, the majority of the parameters in the ResNet models are located in the last layers, which have small activation maps for the ImageNet dataset, so it is not surprising that the choice would be to prune the weights.

Table 4. Ablation Study, ResNet-18 train on ImageNet50 using NAS pruning with different regularization configurations.

Method	$\mathcal{L}_{mem}$	Gumbel-softmax	$N{:}M$	Top-1 Acc
Baseline	–	–	–	88.04
STE-W	–	–	2:4	87.6
SR-STE [42]	–	–	2:4	87.6
STE-A	–	–	2:4	87.32
WA-NAS	✓	✗	2:4	87.76
WA-NAS	✗	✓	2:4	87.72
WA-NAS	✓	✓	2:4	87.86
STE-W	–	–	2:8	87.24
SR-STE [42]	–	–	2:8	87.32
STE-A	–	–	2:8	86.68
WA-NAS	✓	✗	2:8	87.5
WA-NAS	✗	✓	2:8	87.6
WA-NAS	✓	✓	2:8	87.56

Table 5. WA-NAS 2:4 pruning choices for ResNet-18 trained on ImageNet50 with Gumbel-Softmax regularization.

Block	Conv	Pruning Choice
1.1	conv1	pruneW
1.1	conv2	pruneW
1.2	conv1	pruneW
1.2	conv2	pruneA
2.1	conv1	pruneA
2.1	conv2	pruneA
2.1	downsample	pruneW
2.2	conv1	pruneA
2.2	conv2	pruneA
3.1	conv1	pruneW
3.1	conv2	pruneW
3.1	downsample	pruneA
3.2	conv1	pruneW
3.2	conv2	pruneW
4.1	conv1	pruneW
4.1	conv2	pruneW
4.1	downsample	pruneA
4.2	conv1	pruneW
4.2	conv2	pruneW

5   Conclusion

In this paper, we propose a strategy based on differential NAS to sparsify both activations and weights in neural networks, while utilizing the recent approach of N:M fine-grained structured sparsity that enables practical acceleration on dedicated hardware. While most of the existing works focus either on sparsification of weights or activations for the entire network, we show that a combination of weight and activation sparsities is superior to each option separately. The motivation here is that the core operations that we wish to accelerate are the matrix-matrix products between weights and activation maps, so the need for sparsity is symmetric between both. Our results show that it can be utilized to increase the network performance. Furthermore, during the training, the choice between sparsifying the weights of activations can be motivated by practical inference costs (e.g., memory bandwidth). We demonstrate the efficiency of the approach on image classification datasets.

Acknowledgements. This work was supported in part by the Israel Innovation Authority through the Avatar consortium. The authors also thank the Israeli Council for Higher Education (CHE) via the Data Science Research Center and the Lynn and William Frankel Center for Computer Science at BGU. SF is also supported by Kreitman High-tech scholarship.

References

1. Banner, R., Hubara, I., Hoffer, E., Soudry, D.: Scalable methods for 8-bit training of neural networks. In: Advances in Neural Information Processing Systems, pp. 5145–5153 (2018)
2. Bengio, Y., Léonard, N., Courville, A.: Estimating or propagating gradients through stochastic neurons for conditional computation. preprint arXiv:1308.3432 (2013)
3. Cai, Z., Vasconcelos, N.: Rethinking differentiable search for mixed-precision neural networks. In: Proceedings of the IEEE/CVF Conference on Computer Vision and Pattern Recognition (CVPR), pp. 2349–2358 (2020)
4. Deng, D., Socher, L., K-Li, F.F.L.: Imagenet: a large-scale hierarchical image database. In: 2009 IEEE Conference on Computer Vision and Pattern Recognition, pp. 248–255. IEEE (2009)
5. Dong, Y.: Network pruning via transformable architecture search. Advances in Neural Information Processing Systems (NeurIPS) (2019)
6. Dong, X., Huang, J., Yang, Y., Yan, S.: More is less: a more complicated network with less inference complexity. In: CVPR, pp. 5840–5848 (2017)
7. Elsken, T., Metzen, J.H., Hutter, F.: Neural architecture search: a survey. J. Mach. Learn. Res. **20**(1), 1997–2017 (2019)
8. Finder, S.E., Zohav, Y., Ashkenazi, M., Treister, E.: Wavelet feature maps compression for image-to-image CNNs. arXiv preprint arXiv:2205.12268 (2022)
9. Georgiadis, G.: Accelerating convolutional neural networks via activation map compression. In: Proceedings of the IEEE/CVF Conference on Computer Vision and Pattern Recognition, pp. 7085–7095 (2019)

10. Goodfellow, I., Bengio, Y., Courville, A., Bengio, Y.: Deep Learning, vol. 1. MIT Press, Cambridge (2016)

11. Gou, J., Yu, B., Maybank, S.J., Tao, D.: Knowledge distillation: a survey. Int. J. Comput. Vision **129**, 1789–1819 (2021)

12. Guo, Y., Yao, A., Chen, Y.: Dynamic network surgery for efficient DNNs. In: Advances in Neural Information Processing Systems (NIPS), pp. 1379–1387 (2016)

13. Han, S., Pool, J., Tran, J., Dally, W.: Learning both weights and connections for efficient neural network. In: Advances in Neural Information Processing Systems, pp. 1135–1143 (2015)

14. He, Y., Kang, G., Dong, X., Fu, Y., Yang, Y.: Soft filter pruning for accelerating deep convolutional neural networks. In: IJCAI International Joint Conference on Artificial Intelligence (2018)

15. He, Y., Liu, P., Wang, Z., Hu, Z., Yang, Y.: Filter pruning via geometric median for deep convolutional neural networks acceleration. In: Proceedings of the IEEE/CVF Conference on Computer Vision and Pattern Recognition, pp. 4340–4349 (2019)

16. Howard, A., et al.: Searching for MobileNetV3. In: IEEE/CVF International Conference on Computer Vision, pp. 1314–1324 (2019)

17. Hubara, I., Chmiel, B., Island, M., Banner, R., Naor, J., Soudry, D.: Accelerated sparse neural training: a provable and efficient method to find n: m transposable masks. In: Advances in Neural Information Processing Systems (NeurIPS), vol. 34, pp. 21099–21111 (2021)

18. Hubara, I., Courbariaux, M., Soudry, D., El-Yaniv, R., Bengio, Y.: Quantized neural networks: training neural networks with low precision weights and activations. J. Mach. Learn. Res. **18**(1), 6869–6898 (2017)

19. Jang, G., Poole: categorical reparameterization with gumbel-softmax. In: International Conference on Learning Representations (ICLR) (2017)

20. Krizhevsky, H.: Learning multiple layers of features from tiny images. Technical report, Citeseer (2009)

21. Kurtz, M., et al.: Inducing and exploiting activation sparsity for fast inference on deep neural networks. In: International Conference on Machine Learning (ICML), pp. 5533–5543. PMLR (2020)

22. Lee, N., Ajanthan, T., Torr, P.H.: Snip: Single-shot network pruning based on connection sensitivity. In: International Conference on Learning Representations (ICLR) (2019)

23. Li, H., Kadav, A., Durdanovic, I., Samet, H., Graf, H.P.: Pruning filters for efficient convnets. In: International Conference on Learning Representations (ICLR) (2017)

24. Liu, H., Simonyan, K., Yang, Y.: Darts: differentiable architecture search. In: International Conference on Learning Representations (ICLR) (2019)

25. Liu, C., et al.: Progressive neural architecture search. In: Ferrari, V., Hebert, M., Sminchisescu, C., Weiss, Y. (eds.) ECCV 2018. LNCS, vol. 11205, pp. 19–35. Springer, Cham (2018). https://doi.org/10.1007/978-3-030-01246-5_2

26. Mishra, A., et al.: Accelerating sparse deep neural networks. arXiv preprint arXiv:2104.08378 (2021)

27. Molchanov, P., Mallya, A., Tyree, S., Frosio, I., Kautz, J.: Importance estimation for neural network pruning. In: Proceedings of the IEEE Conference on Computer Vision and Pattern Recognition, pp. 11264–11272 (2019)

28. Raihan, M.A., Aamodt, T.: Sparse weight activation training. In: Advances in Neural Information Processing Systems (NeurIPS), vol. 33, pp. 15625–15638 (2020)

29. Real, E., et al.: Large-scale evolution of image classifiers. In: International Conference on Machine Learning, pp. 2902–2911. PMLR (2017)

30. Ren, P., et al.: A comprehensive survey of neural architecture search: challenges and solutions. ACM Computing Surveys (CSUR), pp. 1–34 (2021)

31. Ronen, M., Finder, S.E., Freifeld, O.: Deepdpm: Deep clustering with an unknown number of clusters. In: Conference on Computer Vision and Pattern Recognition (2022)

32. Sun, Q., Li, X., Jiao, L., Ren, Y., Shang, F., Liu, F.: fast and effective: a novel sequential single-path search for mixed-precision-quantized networks. IEEE Trans. Cybern. 1–13 (2022)

33. Sun, W., et al.: Dominosearch: Find layer-wise fine-grained n: M sparse schemes from dense neural networks. In: Advances in Neural Information Processing Systems (NeurIPS), vol. 34, pp. 20721–20732 (2021)

34. Tan, C., Pang, V., Sandler, H., Le, V.: MnasNet: platform-aware neural architecture search for mobile. In: Proceedings of the IEEE/CVF Conference on Computer Vision and Pattern Recognition, pp. 2820–2828 (2019)

35. Tan, M., Le, Q.: EfficientNet: rethinking model scaling for convolutional neural networks. In: International Conference on Machine Learning, pp. 6105–6114 (2019)

36. Tanaka, H., Kunin, D., Yamins, D.L., Ganguli, S.: Pruning neural networks without any data by iteratively conserving synaptic flow. preprint arXiv:2006.05467 (2020)

37. Van Gansbeke, W., Vandenhende, S., Georgoulis, S., Proesmans, M., Van Gool, L.: SCAN: learning to classify images without labels. In: Vedaldi, A., Bischof, H., Brox, T., Frahm, J.-M. (eds.) ECCV 2020. LNCS, vol. 12355, pp. 268–285. Springer, Cham (2020). https://doi.org/10.1007/978-3-030-58607-2_16

38. Wang, C., Zhang, G., Grosse, R.: Picking winning tickets before training by preserving gradient flow. In: International Conference on Learning Representations (2020)

39. Wu, D., Zhang, W., Sun, W., Tian, V., Jia, K.: FBNet: hardware-aware efficient convnet design via differentiable neural architecture search. In: Proceedings of the IEEE/CVF Conference on Computer Vision and Pattern Recognition, pp. 10734–10742 (2019)

40. Yin, P., Lyu, J., Zhang, S., Osher, S., Qi, Y., Xin, J.: Understanding straight-through estimator in training activation quantized neural nets. In: International Conference on Learning Representations (ICLR) (2019)

41. Yu, H., Han, Q., Li, J., Shi, J., Cheng, G., Fan, B.: Search what you want: barrier panelty NAS for mixed precision quantization. In: Vedaldi, A., Bischof, H., Brox, T., Frahm, J.-M. (eds.) ECCV 2020. LNCS, vol. 12354, pp. 1–16. Springer, Cham (2020). https://doi.org/10.1007/978-3-030-58545-7_1

42. Zhou, M., Zhu, L., Zhang, Y., Sun, L.: Learning N: M fine-grained structured sparse neural networks from scratch. In: International Conference on Learning Representations (2021)

43. Zhu, M., Zhang, T., Gu, Z., Xie, Y.: Sparse tensor core: algorithm and hardware co-design for vector-wise sparse neural networks on modern GPUs. In: Proceedings of the 52nd Annual IEEE/ACM International Symposium on Microarchitecture, pp. 359–371 (2019)

44. Zhu, M., Gupta, S.: To prune, or not to prune: exploring the efficacy of pruning for model compression. arXiv preprint arXiv:1710.01878 (2017)

45. Zoph, V., Shlens, L.: Learning transferable architectures for scalable image recognition. In: Proceedings of the IEEE Conference on Computer Vision and Pattern Recognition, pp. 8697–8710 (2018)

W29 - Computer Vision for Civil and Infrastructure Engineering

W29 - Computer Vision for Civil and Infrastructure Engineering

Civil and infrastructure engineering are corner stones in modern society, and as the world population continues to grow, the infrastructure and built environment has to keep up. This has led to significant interest in utilizing computer vision to assist and contribute with the inspection processes and in contributing to the built environment, both during and after construction. There is huge potential for computer vision in many aspects of the civil and infrastructure domain which has yet to be realized, and this workshop aims at bringing practitioners and researchers from both domains together to realize this potential.

October 2022

Joakim Bruslund Haurum
Mingzhu Wang
Ajmal Mian
Thomas B. Moeslund

Image Illumination Enhancement for Construction Worker Pose Estimation in Low-light Conditions

Xinyu Chen and Yantao Yu[✉]

Department of Civil and Environmental Engineering,
The Hong Kong University of Science and Technology, Hong Kong, China
xchengl@connect.ust.hk, ceyantao@ust.hk

Abstract. Many construction scenes feature low-light work, such as nighttime construction and tunnel construction. Poor lighting and low visibility will increase the risk of site accidents. One of the leading causes of construction accidents is unsafe worker behavior, which can be predicted via worker posture estimation. Therefore, this study proposes an Unsupervised Illumination Reflectance Estimation (UIRE-Net) framework for estimating the dark worker pose. On the basis of lightness-color consistency, in spite of ungratified illumination conditions, the "true color" of objects depends on the illumination reflectance only. The illumination reflectance estimation is monotonous to neighboring pixel differences, making the extracted features robust for worker pose estimation. In addition, the proposed UIRE-Net restores image brightness without relying on image pairs. A testing experiment based on nighttime construction workers is conducted to validate the veracity.

Keywords: Low light enhancement · Construction worker pose estimation · Occupational safety and health · Illumination reflection estimation

1 Introduction

Low-light construction is a common activity in many construction scenes, including nighttime pavement maintenance, nighttime railway construction, and tunnel construction. Night construction can bring some efficiency improvement, for example, outdoor construction activities at night (such as pavement projects) can reduce the interference of traffic jams caused by vehicles. In summer, the temperature at night is much lower than during the day, so the cool air can speed up the material delivery cycle and shorten the idling time of the machine. According to Arditi's [1] study, workers under low light have a higher fatigue index and are more prone to risk and skeletal muscle strain. With the development of deep learning, human pose estimation is an important method for fatigue estimation of workers in the construction field.

Supplementary Information The online version contains supplementary material available at https://doi.org/10.1007/978-3-031-25082-8_10.

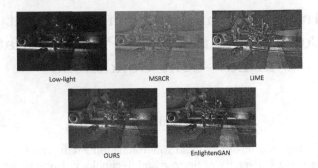

Fig. 1. Visual comparisons on a typical low light image with various enhancement methods.

Under the condition of low illumination, due to dim light and insufficient exposure, the captured image has high noise, low contrast, and low brightness, which brings severe challenges to pose estimation. Under extremely low light conditions, it is difficult to distinguish with human eyes, and CNN loses the ability of feature extraction. Therefore, it is difficult to maintain robust pose estimation without target features. Results of dark worker pose estimation benchmarks show that even the current top estimator tries to keep its SOTA performance in low light hardly. Therefore, it is urgent to develop a more robust framework for recognizing a workers' posture in low light. Existing studies have suggested enhancing the quality of images before feeding them into the detection algorithms, but this strategy gives rise to some intractable inadequacies. In most cases, low light image enhancement methods are optimized to improve the visual perception of images rather than detection performance. Figure 1 shows an example of improving a lowlight image with various enhancement methods. UIRE-Net brightens the image while keeping the natural color and features, compared to state-of-the-art technologies. According to Retinex theory, the perceived brightness of the human eye depends on the combination of illumination reflectance and light field intensity [15]. The image is determined by the lighting distribution and the illumination reflectance. Thus, the illumination of color describes the physical properties of the object and is independent of the flux of radiant energy [14]. Therefore, low light enhancement can be expressed as an estimation problem of illumination reflectance. The reflectance of illumination intensity is obtained by iterating the illumination distribution in Retinex theory. By solving the reflectivity illumination, the low-light image and downstream workers' pose estimation are improved. The objectives of this research include:

1) To propose an Unsupervised Illumination Reflectance Estimation method for low-light pose estimation, to detect and estimate worker behavior under nonuniform and poor lighting conditions.

2) To train the image enhancement network by unsupervised novel designed loss functions, It doesn't need for paired images.

3) To propose a training method of low light workers' pose model. By darking normal light workman' pose dataset, the low-light dataset is expanded and reduce the gap between the normal light and the low light dataset. It could solve the lack of low-light human pose data set.

2 Related Work

2.1 Human Pose Estimation Methods in Construction

Worker pose estimation based on the motion sensor of a wearable device uses a wearable device to collect human body signals. The most widely used wearable sensor is the inertial measurement units(IMUS). IMUS uses accelerometers, gyroscopes, and compasses to collect real-time positions of body parts signals. Many researchers use IMUS to capture the movement of construction workers. For example, Chen et al. [3] used IMU-based data to predict the uncomfortable postures of construction workers. Valero et al. [28] described the unsafe posture of construction Work-related Musculoskeletal Disorders(WMSD). The IMUS can not be applied to the attitude estimation of construction workers because of the output error [29] and the burden of wearing [21]. Because of the problems posed by these intrusive sensors [7], people are forced to use computer vision for attitude estimation.

3D human pose estimation based on an RGB-D camera is also widely used in architecture. For instance, Yu et al. [32] used RGB-D cameras to model various parts of workers' bodies. By analyzing these biomechanical models of human postures, workers were assessed for fatigue. Martinez et al. [20] and Chen et al. [2] used a simple neural network trained to estimate 3D pose and get a good performance. However, the study of Y, Yu et al. [33] showed that if 3D pose estimation is used for outdoor pose estimation of construction workers, it is easy to fail due to RGB-D camera errors.

Yang et al. [31] proposed a hand-made feature extraction algorithm and trained in construction worker datasets. This method has poor generalization performance on other datasets [27]. In order to increase the generalization ability and improve the accuracy of human pose estimation, Meiyin Liu et al. [18] proposed to use the deep neural network model for 2D pose estimation of construction workers. X.Yan et al. [30] simultaneously used three posture classifiers to identify worker backs, arms, and legs. It performed 2D workers posture estimation and fatigue prediction for construction workers. Computer vision-based methods mostly rely on datasets for training. Existing data sets lack images of human pose estimation in low light. There is a large gap between low light and normal light, thus the human pose estimation models trained under normal light are difficult to show their good performance in low light.

2.2 Image Illumination Enhancement

low-light image enhancement methods are mainly divided into traditional methods and deep learning-based methods. Earlier traditional methods mainly rely on local statistics or intensity mapping. For instance, the histogram-based solutions expand the enhanced dynamic range by adjusting the DIS distribution of global [4,11] or local [16,25] images. Subsequently, solutions are based on the Retinex theory. For instance, Multiscale Retinex with Color Restoration (MSRCR) [13] uses reflection mapping as the final enhancement result. Fu et al. [5] used a weighted variational model to estimate the reflectivity and illuminance of images to improve image lighting conditions. LIME [8] estimates the rough illumination map by searching for the maximum intensity of each pixel in the RGB channel and refines the map by a structural prior. The disadvantages of based on the Retinex theory are that colors are prone to local distortion and excessive exposure.

In recent years, with the development of deep neural networks, many enhancement methods have begun to use deep learning techniques. For example, Enlighter-GAN [12] is the first enhanced network in the GAN model. It designs discriminator and generator without unpaired datasets to simulate normal lighting images. But Enlighter-GAN requires the selection of multi-exposure training data. Zero-DCE [6] further addresses the problem by establishing multiple specific image curves grounded on zero-inference. However, the model based on light recovery curves has uneven image recovery capability under various dark conditions. Considering this, image features need to be estimated for robustness enhancement. In this paper, we try to integrate Retinex theory into an unsupervised end-to-end training framework, which is based on a deep neural network model based on luminance-color constancy. It can effectively solve the image enhancement problem under multiple light conditions.

2.3   Construction Field Data Set Expansion

Algorithms based on computer vision have a good performance in construction. However, it is very important to build a high-quality and huge training image database for training high-performance deep learning models. These models often require tens of thousands of manually annotated data sets and cost a lot of manpower and resources. To solve this problem, Soltani et al. [24] used 3D modeling software (3DsMax) to generate annotated images. A 3D excavator model was used to create composite images from different perspectives of the building scene. Seo and Lee [22] used virtual human models to build human body postures such as standing and bending. Hong et al. [9] used building Information Modeling (BIM) to construct a synthetic dataset containing annotated information on infrastructure.

Hwang et al. [10] also developed a synthetic data set to simulate excavator posture for training the model. However, the training image database developed by 3D virtual modeling tools also has some limitations. Compared with the real datasets, the training image database lacks real background, light, and image texture. In addition, these models are difficult to deal with problems such as light conversion, and cannot well simulate the scene at night. And building such 3D models often takes a lot of time and resources. To solve these problems, the low light training image database not only needs to use real images to build but also needs to contain the required scene characteristics such as light intensity, noise, etc. Therefore, we propose an expansion method of the human pose database under low light. This paper simulates human pose under low light by adding Gaussian noise and reducing light intensity in the normal light human pose datasets. There is no need to collect pictures and label human posture in a low-light environment, which greatly reduces the labeling cost.

2.4   Research Challenges and Objectives

In this section, we review the human pose estimation algorithms of construction workers, low-light image enhancement, and construction datasets expansion. The computer vision-based pose estimation for workers under low light is a basic step for assessing workers' fatigue degree and monitoring worker behavior in low light. A posture estimation method in low light can be used in many monitoring tasks to make workers more secure and reduce fatigue injuries. However, there are three main challenges to the existing worker attitude estimation methods in low light.

Firstly, detectors cannot estimate human pose accurate results in low light, because most of the images for model training are collected during the daytime. Existing methods are difficult to detect human key points from low-light images since the low-light intensity is easy to cause the disappearance of human key points and trigger motion blur issues. Secondly, the collection and labeling of low-light datasets is a very difficult problem, because of the influence of light factors, it is easy to make mistakes in labeling the information of human joints in low-light pictures. Finally, the existing image enhancement technology can not produce a good performance in many low-light construction scenes. To solve the above research problems, an image enhancement deep neural network based on Retinex theory is developed in this paper, which preprocesses the input images with the enhancement network. By doing that, some features of construction workers that are not visible in low-light images will become recognizable and detectable. To solve the problem of lack of datasets and difficulty in labeling human pose in low light, this .paper proposes a method of darkening normal human pose datasets. Darken normal light datasets by adjusting light brightness and adding noise. The darken human pose datasets help estimate human pose under low-light conditions.

3   Methodology

In this section, firstly the overview framework of methodology is introduced. Then, the three main modules involved in the vision-based approach, low light dataset extension, image light enhancement based on Retinex theory, and human pose estimation are described in detail.

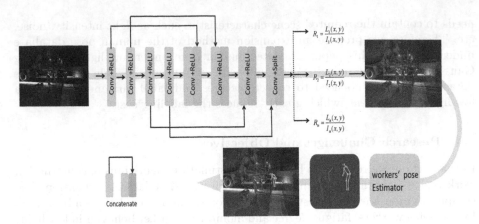

Fig. 2. Architecture of the proposed UIRE-Net. Upper row present the Unsupervised Illumination Reflection method and lower row denotes the Estimator.

3.1   Overall Framework

The overall framework of the proposed method is shown in Fig. 2. First, the images collected at night are fed into the light enhancement network, which is a deep neural network based on Retinex theory to repair low-light images. Then, the human pose model is used to estimate the human pose of the workers. Finally, the estimation results of worker body posture under low light are obtained. To sum up, the technical innovation of the modified method is mainly reflected in 2 aspects: (a) The deep neural network based on Retinex is used to enhance and restore images under low light. UIRE-Net solves the problems of low light visibility and poor detection accuracy. (b) To simulate the low-light data set of human body posture, We reduce brightness and add noise. Our method resolve the lack of low light datasets.

3.2   Proposed UIRE-Net

Main Idea of UIR Model. The unsupervised illumination reflection module is mainly to solve the problem of insufficient light intensity in low light construction. When estimating human pose from images, the insufficient light source could reduce the visual quality and the performance of the detection algorithm. If we directly increase the contrast of low-light image, it will lead to overexposure and color distortion. And details are lost in the brightest and darkest areas of the image. To this end, we specially designed a deep learning method based on Retinex theory to enhance low-light images, enhance image lighting and restore color without color distortion and retain details to the maximum extent.

On the basis of Retinex theory, a given image $I(x,y) \in R^{W \times H \times C}$ can be decomposed as source illumination and scene reflectance, which is,

$$I(x,y) = R(x,y) \cdot L(x,y), \tag{1}$$

where $L(x, y)$ is the spatial distribution of source illumination, $R(x, y)$ denotes the distribution of scene reflectance. In light of Retinex theory, lightness-color constancy refers to the resilience of perceived color and lightness to spatial and spectral illumination variations. Based on this, we attempt to estimate the reflectance as guidance for automatic dark image enhancement, with the merits of a simple and differentiable expression relying on the input images only and monotonous to preserve the differences of neighboring pixels. Assuming that the illumination map under normal light condition L is an identity matrix, according to the effective formulation of Retinex, the reflectance for enhancement can be obtained. Arguably, the formulation of $R(x, y)$ is an ill-posed problem, a direct decomposition results in unnatural artifacts. So we design the enhanced net and use iterative algorithms to gradually eliminate the impact, which is

$$\log R_i(x, y) = \log I_i(x, y) - \log L_i(x, y), \quad \forall i = 1 \ldots n. \tag{2}$$

where n is the number of iterations. Here we set $n = 8$ empirically, which will be discussed in the ablation study in detail.

The detailed architecture of the UIR model is demonstrated in TABLE I. As we can see, the proposed UIR model consists of 8 convolutional layers with skip connections. Specifically, the first 7 layers are with 32 convolutional kernels of size 3×3 and stride 1 followed by the ReLU activation function, and the last one has 8 convolutional kernels of size 3×3 and stride 1 followed by the Tanh activation function. Down-sampling and batch normalization are weeded out for retaining neighboring context among pixels. Essentially, the last convolutional layer splits the estimated lighting reflectance Ri, and the given image Ii is enhanced iteratively in terms of the parameter maps R_1 to R_n.

Table 1. Detailed architecture of UIR model

Input	Dimensions	Operator	Output
Image	$256 \times 256 \times 3$	Conv+ReLU	Conv1
Conv1	$256 \times 256 \times 32$	Conv+ReLU	Conv2
Conv2	$256 \times 256 \times 32$	Conv+ReLU	Conv3
Conv3	$256 \times 256 \times 32$	Conv+ReLU	Conv4
Conv1+Conv4	$256 \times 256 \times 64$	Conv+ReLU	Conv5
Conv2+Conv5	$256 \times 256 \times 64$	Conv+ReLU	Conv6
Conv3+Conv6	$256 \times 256 \times 64$	Conv+ReLU	Conv7
Conv4+Conv7	$256 \times 256 \times 64$	Conv+ReLU	Conv8
Conv7+Conv8	$256 \times 256 \times 8$	Split	$R_1 \cdots R_n$
Image $\times R_n$	–	–	Enhanced image

Unsupervised Loss Functions. Several differentiable loss functions are devised to facilitate the estimated reflectance enhancing images for boosting detection performance so that the ULR model can be trained unsupervised.

(1)Information Difference Loss. In worker pose estimation tasks, the quality of feature extraction is critical. Thereby, an information difference loss is devised to quantify the differences between the improved image and the original image, which is,

$$L_{idl} = \|V(I_n) - V(I_0)\|_2^2, \tag{3}$$

where $V(I)$ is the feature extraction operator based on VGG-16 [23], I_0 is the original raw image and I_n is the corresponding enhanced image after n time iterations. The VGG network is leveraged here for its simple and concise architecture for the effective computation of information difference.

(2)Exposure Control Loss. Because the enhancement sometimes results in under and over-exposure, an exposure control loss is required to limit the undesired phenomenon. To obtain an average intensity Y, the improved pictures are split into 16×16 non-overlapping local regions. According to Exposure fusion theory, a well-exposedness level E is defined as the gray level in RGB space. As a result, the exposure control loss, which measures the distance between average intensity Y and well-exposedness level E, is calculated as follows:

$$L_{ecl} = \frac{1}{M} \sum_{m=1}^{M} |Y_m - E|, \tag{4}$$

where M represents the number of non-overlapping local regions, E is set as 0.6 empirically, for barely performance nuance is witnessed within the small value range.

(3)Color Constancy Loss. By introducing a color constancy loss, which confines the improved RGB channels at an approximate similar level by connecting the three adjusted channels, the potential color discrepancies caused by enhancement approaches are corrected, which is,

$$L_{ccl} = \frac{1}{w \times h} \left[\left(I^R - I^G\right)^2 + \left(I^R - I^B\right)^2 + \left(I^B - I^G\right)^2 \right], \tag{5}$$

where I^R, I^G, and I^B describe the average intensity values of RGB channels of enhanced images, w and h stand for the width and height of the images, respectively.

(4)Illumination Smoothness Loss. An illumination smoothness loss is applied to the estimated reflectance parameter map Ri to ensure the smoothness of the illumination component in RGB space for maintaining the monotonicity of pixel-level surrounding context during iteration., which is expressed as:

$$L_{isl} = \frac{1}{N} \sum_{i=1}^{N} \left(|\nabla_x R_i^{c \in \{R,G,B\}}| + |\nabla_y R_i^{c \in \{R,G,B\}}| \right)^2, \tag{6}$$

where N represents the number of iterations, ∇ is the first-order differential operator, accordingly ∇_x and ∇_y denote the horizontal and vertical gradient operations, respectively.

To sum up, the total loss is expressed as,

$$L = \omega_a L_{idl} + \omega_b L_{ecl} + \omega_c L_{ccl} + \omega_d L_{isl}, \qquad (7)$$

where ω_a, ω_b, ω_c, and ω_d are the weights controlling the significance of losses.

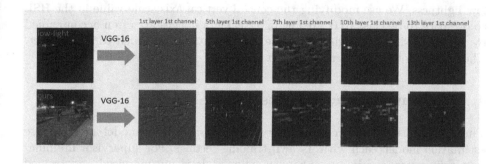

Fig. 3. Visualization of features extracted by the VGG-16. The features taken by the VGG-16 are visualized. The features from the original image are in the first row, and the images enhanced by the UIR model are in the second row. It is obvious that features derived from the improved patch are more conspicuous and distinguishable than those extracted from the original patch.

UIRE-Net for Worker Pose Estimation. In this paper, we choose HRNet as the backbone network for worker pose estimation. HRNet [26] even exceeds all Top-down Methods on COCO Test (77 % AP), Suggesting its robustness in a crowded scene. The HRNET method has high key point detection accuracy and high efficiency in computational complexity and parameters without the supervision of an intermediate heat map. We adjust the input size and structure of the network according to the characteristics of workers in human body detection photos on construction sites so that it can be better applied to the construction field. Specifically, The specific adjustments are to change the original input size from 384×288 to 512×384 and increase the network depth. This is because in the estimation of construction workers, the posture size of the human body collected and focus is a large size, and the small workers' body at a distance is not applicable. Therefore, we changed the input size and depth to reduce the parameters contained in the network, so that the results can be obtained batter in training and testing.

As shown in Fig. 3, a qualitative comparison is done between the attributes of the original patch and the upgraded patch. The characteristics taken from the original patch appear jumbled and indistinct, however, the augmented patch's features are significant and distinct. As a result, more discriminative features are used to increase the tracker's dark tracking ability.

Data Expansion. Due to the difficulty of collecting and labeling human posture data under low light, there is a lack of such large databases. In this paper, a data expansion method is proposed to convert the data set collected under normal optical fiber to low light. The low-light conversion method designed in this paper is inspired by [19]. RGB (red, green, and blue) images are converted into HSL (hue, saturation, and brightness) images to adjust brightness. To get low lightness, We use modifying the central vertical axis scale value in the HSL color space to be close to 0. At low light, camera images are often accompanied by noise. To simulate nighttime noise, Gaussian noise is added to the dataset. The specific formula is in the support material.

4   Experiments and Results

The proposed UIRE-Net consists of two modules: the UIR model and the pose estimation model, each with its own training settings. For supervisor illumination reflectance, the image enhancement module is unsupervised and trained on SICE [33] and construction datasets. SICE [33] is a dataset comprising 589 image sequences and 4,413 high-resolution samples with considerable variability in exposure levels from both indoor and outdoor scenes. The training set for the proposed unsupervised illumination reflectance estimation is made up of 220 image sequences and 2,000 low-contrast photos and associated high-quality correspondences. The construction dataset contains 300 different construction site images, including day and night, indoor and outdoor. This data set contains the most common construction lighting levels. To reduce the influence of different lighting environments, the UIR model was trained by using SICE and multi-exposure construction site datasets to ensure the generalization performance of UIRE-Net under various lighting conditions of construction scenes and enhance the ability of the estimator to meet the lighting variation problem.

The input resolution is set to 256×256, and the network is optimized using the ADAM optimizer with default parameters and a fixed learning rate of 0.001. The weights of loss functions: w_a, w_b, w_c, w_d are set as 0.005, 40, 10, and 1,500, empirically. All the results presented in the experiments are based on the test set from Crowdpose and our collected construction workers, and they are evaluated in Object keypoint similarity (OKS) via the official evaluation tool: .

4.1   Data Sets

For the performance evaluation, worker pose estimation is set as the benchmark for comparison. Consequently, Crowdpose [17] and our collected construction worker dataset is utilized for experiments. Crowdpose consists of 20,000 images with 80,000 annotated human bodies in a diversity of appearance, pose, scale, and occlusion, while most of them are captured under normal light conditions. Our collected construction worker dataset has 100 images with 580 workers' bodies, which are collected from several construction sites in dark environments, as shown in Fig. 4.

Fig. 4. Example of our collected and labeled construction worker dataset .

4.2 Comparison with State-of-the-art Methods

In this experiment, HRNet [26] is selected as the benchmark for the worker pose estimation, 16,000 images in Crowdpose are used as training and 4,000 images are used as test images. Meanwhile, the low-light images are compared with the UIRE-net. Finally, it is validated on a real-site data set. Three different training strategies are progressively applied to improve results. For the first phase, the estimator performs pre-training only on the Crowdpose dataset. In the second stage, the model is pre-trained on Crowdpose and fine-tuned on DARK Crowdpose. In the third stage, all test baselines are pre-trained on WIDER FACE, trained again on the DARK Crowdpose, and fine-tuned on the enhanced DARK Crowdpose to get the final test results, as shown in Fig. 5. The proposed UIRE-Net is compared with state-of-the-art methods for estimation in a dark environment, as shown in Table 2. Model2 is trained directly on the original low-light dataset, so it is not applicable to other enhancement test. --- means not applicable.

Table 2. Results on darken CrowdPose test set. Test model is divided into three part and we report the results respectively.

Model	Darken CrowdPose data set (mAP)				
	Raw	MSRCR	LIME	Enlighter-GAN	Ours
Model 1	0.602	0.641	0.647	0.651	0.674
Model 2	0.706	---	---	---	---
Model 3	---	0.712	0.724	0.747	0.785

In general, the Enlighter-GAN approach based on the deep learning method performs better than traditional image enhancement methods LIME and MSRCR. However, the performance of UIRE-net is superior to all competitors. As can be seen from Fig. 6, the qualitative comparison is consistent with the quantitative comparison. By visualizing the worker pose in the test images,

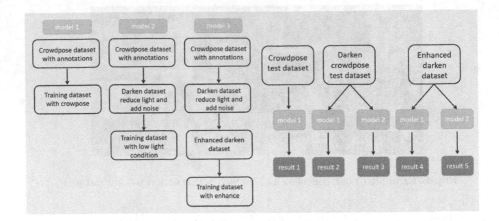

Fig. 5. Process of model training and testing.

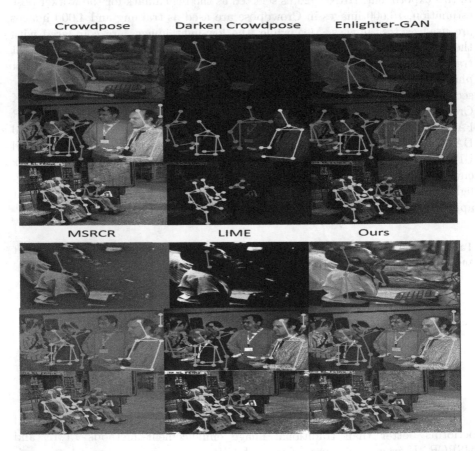

Fig. 6. Qualitative comparison of pose estimation performance with state-of-the-art methods in low light.

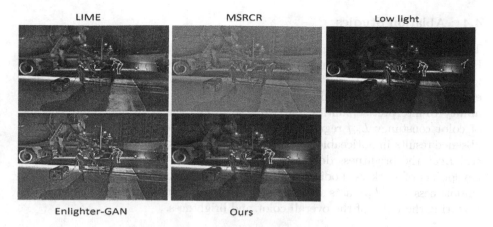

Fig. 7. Example of pose estimation in construction data set.

the proposed method detects more human key points than the most advanced methods and reduces false estimates. UIRE-net shows a performance increase of approximately 17% over dark images without enhancements. This is mainly because the enhanced image can better display the human pose information in the image.

4.3   Real-world Evaluation

Table 3. Results on low-light construction data set.

Model	Low-light construction data set (mAP)				
	Raw	MSRCR	LIME	Enlighter-GAN	Ours
Model 1	0.581	0.612	0.658	0.649	0.697
model 2	0.632	---	---	---	---
model 3	---	0.704	0.783	0.778	0.815

We tested UIRE-Net and other comparison algorithms in real low-light construction scenarios to verify the effectiveness and generalization of our method. Real scenes include night construction sites, night road maintenance, low light tunnel construction, etc. As you can see in Fig. 7, in the low-light scene, it is estimated that the network will mistakenly detect the human figure in the bright place as a human body, and there will also be the problem that the dark place cannot detect. In Table 3, UIRE-Net shows the highest accuracy in different training model modes. Observing the examples, UIRE-Net lightens up the workers in the extremely dark areas and preserves the well-lit areas, thus improving the performance of estimators in the dark.

4.4   Ablation Studies

This subsection delves deeper into the contribution of each loss in UIRE-Net. Figure 8 depicts the influence of loss functions, with various combinations of losses being used to train the findings. The loss of information difference L_{idl} enhanced the features produced from the input image, improving the model's ability to interpret semantics, particularly around the key point area. The loss of color constancy L_{ccl} regulates color restoration over RGB channels, and its absence results in noticeable color distortion. Without the loss of exposure control $Lecl$, the brightness decreases, and the low-light area fails to recover. the key points of workers' bodies cannot be detected accurately. The illumination smoothness loss L_{isl} acts as a link between surrounding pixel-level changes, Ensuring the unity of the overall color and brightness.

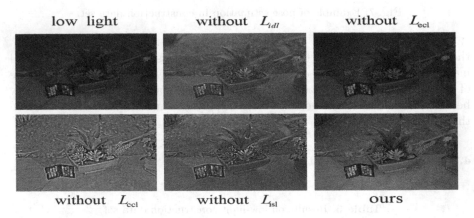

Fig. 8. Qualitative comparison of pose estimation performance with state-of-the-art methods in low light.

5   Conclusions

This work establishes a framework for low-light worker posture estimation. Inspired by the Retinex theory, we use invariable reflectivity characteristics to enhance the light intensity and characteristics of the image. UIRE-Net is dedicated to mitigating the interference of attitude estimation due to brightness effects. Compared with the SOTA low light enhancement methods, the superiority of UIRE-NET in low light human pose estimation is proved. The reliability and applicability of our method are further verified by testing on a real low-light construction data set. In conclusion, we forcefully believe that this study can help promote construction protection and supervision for construction workers in low light.

References

1. Arditi, D., Lee, D.E., Polat, G.: Fatal accidents in nighttime vs daytime highway construction work zones. J. Safety Res. **38**(4), 399–405 (2007)
2. Chen, J., Qiu, J., Ahn, C.: Construction worker's awkward posture recognition through supervised motion tensor decomposition. Autom. Constr. **77**, 67–81 (2017)
3. Cheng, T., Teizer, J., Migliaccio, G.C., Gatti, U.C.: Automated task-level activity analysis through fusion of real time location sensors and worker's thoracic posture data. Autom. Constr. **29**, 24–39 (2013)
4. Coltuc, D., Bolon, P., Chassery, J.M.: Exact histogram specification. IEEE Trans. Image Process. **15**(5), 1143–1152 (2006)
5. Fu, Y., Lam, A., Sato, I., Okabe, T., Sato, Y.: Separating reflective and fluorescent components using high frequency illumination in the spectral domain. In: Proceedings of the IEEE International Conference on Computer Vision, pp. 457–464 (2013)
6. Guo, C., et al.: Zero-reference deep curve estimation for low-light image enhancement. In: Proceedings of the IEEE/CVF Conference on Computer Vision and Pattern Recognition, pp. 1780–1789 (2020)
7. Guo, H., Yu, Y., Skitmore, M.: Visualization technology-based construction safety management: a review. Autom. Constr. **73**, 135–144 (2017)
8. Guo, X., Li, Y., Ling, H.: Lime: low-light image enhancement via illumination map estimation. IEEE Trans. Image Process. **26**(2), 982–993 (2016)
9. Hong, Y., Park, S., Kim, H., Kim, H.: Synthetic data generation using building information models. Autom. Constr. **130**, 103871 (2021)
10. Hwang, J., Kim, J., Chi, S., Seo, J.: Development of training image database using web crawling for vision-based site monitoring. Autom. Constr. **135**, 104141 (2022)
11. Ibrahim, H., Kong, N.S.P.: Brightness preserving dynamic histogram equalization for image contrast enhancement. IEEE Trans. Consum. Electron. **53**(4), 1752–1758 (2007)
12. Jiang, Y.: EnlightenGAN: deep light enhancement without paired supervision. IEEE Trans. Image Process. **30**, 2340–2349 (2021)
13. Jobson, D.J., Rahman, Z.u., Woodell, G.A.: A multiscale retinex for bridging the gap between color images and the human observation of scenes. IEEE Trans. Image Process. **6**(7), 965–976 (1997)
14. Land, E.H.: The retinex theory of color vision. Sci. Am. **237**(6), 108–129 (1977)
15. Land, E.H., McCann, J.J.: Lightness and retinex theory. Josa **61**(1), 1–11 (1971)
16. Lee, C., Lee, C., Kim, C.S.: Contrast enhancement based on layered difference representation of 2D histograms. IEEE Trans. Image Process. **22**(12), 5372–5384 (2013)
17. Li, J., Wang, C., Zhu, H., Mao, Y., Fang, H.S., Lu, C.: CrowdPose: efficient crowded scenes pose estimation and a new benchmark. In: Proceedings of the IEEE/CVF Conference on Computer Vision and Pattern Recognition, pp. 10863–10872 (2019)
18. Liu, M., Han, S., Lee, S.: Potential of convolutional neural network-based 2D human pose estimation for on-site activity analysis of construction workers. In: Computing in Civil Engineering 2017, pp. 141–149 (2017)
19. Lv, F., Li, Y., Lu, F.: Attention guided low-light image enhancement with a large scale low-light simulation dataset. arXiv preprint arXiv:1908.00682 (2019)
20. Martinez, J., Hossain, R., Romero, J., Little, J.J.: A simple yet effective baseline for 3d human pose estimation. In: Proceedings of the IEEE International Conference on Computer Vision, pp. 2640–2649 (2017)

21. Premerlani, W., Bizard, P.: Direction cosine matrix IMU: Theory. Diy Drone: USA 1 (2009)
22. Seo, J., Lee, S.: Automated postural ergonomic risk assessment using vision-based posture classification. Autom. Constr. **128**, 103725 (2021)
23. Simonyan, K., Zisserman, A.: Very deep convolutional networks for large-scale image recognition. In: International Conference on Learning Representations (2015)
24. Soltani, M.M., Zhu, Z., Hammad, A.: Automated annotation for visual recognition of construction resources using synthetic images. Autom. Constr. **62**, 14–23 (2016)
25. Stark, J.A.: Adaptive image contrast enhancement using generalizations of histogram equalization. IEEE Trans. Image Process. **9**(5), 889–896 (2000)
26. Sun, K., Xiao, B., Liu, D., Wang, J.: Deep high-resolution representation learning for human pose estimation. In: Proceedings of the IEEE/CVF Conference on Computer Vision and Pattern Recognition, pp. 5693–5703 (2019)
27. Tompson, J., Goroshin, R., Jain, A., LeCun, Y., Bregler, C.: Efficient object localization using convolutional networks. In: Proceedings of the IEEE Conference on Computer Vision and Pattern Recognition, pp. 648–656 (2015)
28. Valero, E., Sivanathan, A., Bosché, F., Abdel-Wahab, M.: Musculoskeletal disorders in construction: a review and a novel system for activity tracking with body area network. Appl. Ergon. **54**, 120–130 (2016)
29. Yan, X., Li, H., Li, A.R., Zhang, H.: Wearable IMU-based real-time motion warning system for construction workers' musculoskeletal disorders prevention. Autom. Constr. **74**, 2–11 (2017)
30. Yan, X., Li, H., Wang, C., Seo, J., Zhang, H., Wang, H.: Development of ergonomic posture recognition technique based on 2D ordinary camera for construction hazard prevention through view-invariant features in 2D skeleton motion. Adv. Eng. Inform. **34**, 152–163 (2017)
31. Yang, J., Shi, Z., Wu, Z.: Vision-based action recognition of construction workers using dense trajectories. Adv. Eng. Inform. **30**(3), 327–336 (2016)
32. Yu, Y., Guo, H., Ding, Q., Li, H., Skitmore, M.: An experimental study of real-time identification of construction workers' unsafe behaviors. Autom. Constr. **82**, 193–206 (2017)
33. Yu, Y., Li, H., Yang, X., Kong, L., Luo, X., Wong, A.Y.: An automatic and non-invasive physical fatigue assessment method for construction workers. Autom. Constr. **103**, 1–12 (2019)

Towards an Error-free Deep Occupancy Detector for Smart Camera Parking System

Tung-Lam Duong$^{(\boxtimes)}$ ⓘ, Van-Duc Le ⓘ, Tien-Cuong Bui, and Hai-Thien To

Seoul National University, Seoul, Korea
{dtlam26,levanduc,cuongbt91,haithienld}@snu.ac.kr

Abstract. Although the smart camera parking system concept has existed for decades, a few approaches have fully addressed the system's scalability and reliability. As the cornerstone of a smart parking system is the ability to detect occupancy, traditional methods use the classification backbone to predict spots from a manual labeled grid. This is time-consuming and loses the system's scalability. Additionally, most of the approaches use deep learning models, making them not error-free and not reliable at scale. Thus, we propose an end-to-end smart camera parking system where we provide an autonomous detecting occupancy by an object detector called OcpDet. Our detector also provides meaningful information from contrastive modules: training and spatial knowledge, which avert false detections during inference. We benchmark OcpDet on the existing PKLot dataset and reach competitive results compared to traditional classification solutions. We also introduce an additional SNU-SPS dataset, in which we estimate the system performance from various views and conduct system evaluation in parking assignment tasks. The result from our dataset shows that our system is promising for real-world applications.

1 Introduction

According to the 2018 UN media, 68% of the world's population will move to urban areas by 2050 [20]. This dense population in towns and cities directly leads to an increase in the number of cars and other vehicles, which raised a major concern on the parking management on their capacities and efficiency. Letting drivers wander in the city to find an appropriate parking slot in a tight city space causes significant air pollution and wastes drivers' time and energy. It also leaves empty spaces in parking lots and varies statistic measurements on parking occupancy rate, which trouble operators to exploit their facility for revenue. In addition, these factors may worsen, particularly during peak hours when the flow density is at its maximum. For real concrete evidence, a recent report by INRIX [9] shows that on average, a typical American driver spends 17 h a year looking for a parking space, which can go up to 107 h when addressing a dense population city like New York. From [24] analysis, the exceeding of CO2 emissions can rise

© The Author(s), under exclusive license to Springer Nature Switzerland AG 2023
L. Karlinsky et al. (Eds.): ECCV 2022 Workshops, LNCS 13807, pp. 163–178, 2023.
https://doi.org/10.1007/978-3-031-25082-8_11

nearly three times due to this problem. Therefore, a stable future city needs a Smart Parking System (SPS) that can serve as a link between drivers and parking operators and benefit both sides. By suggesting optimal parking places to drivers and managing their destination, a future SPS not only minimizes vehicle emissions (via decremented delays in finding the vacant parking spot [2]), but also provides operators a reliable number of customers to boost revenues through e.g. dynamic pricing [25].

Regardless of potential promises, most SPS functionalities are strictly bounded by the performance of correctly determining the occupancy of a parking lot. Hence, our current parking system relies heavily on sensors as the first layer of the system [2]. However, despite its high precision, this turns out to be pricey when scaling up the parking lot size for future perspective, as each sensor (magnetometer/ultrasonic sensor) [25] is designed to operate solely on a single parking spot. An effective solution for this drawback is applying computer vision(CV) to occupancy detection. A single camera can cover a multiple of parking locations and eliminate the need for a sensor per parking spot [3]. Furthermore, because most of nowadays parking lots have security cameras, it reduces installation and maintenance costs and supports multiple additional tasks for better parking management, such as wrong parking placement, abnormal behavior, and theft detection [29] with which sensors fail to cope.

Although computer vision is a promising approach, no datasets exist for a full CV-SPS intention. Most popular datasets PKLot [10], and CNRPark-Ext [4] and their solutions [4,5,22,28,29] are constrained to a small number of parking lots and treat each parking spot as a binary classification image. Regarding performance, there are three main drawbacks to this type of dataset and their following research. First, it limits solutions to operate in the classification scheme solely. Second, when the number of slots increases, reliable deep-learning classification solutions [22,28] require multiple forward passes and are slow to run in real-time feedback to drivers or to stack more additional tasks. Lastly, a parking operator using solutions from this dataset must reannotation every parking spot for new installation. For example, an operator is in charge of three parking lots with at least 300 parking spots in each facility. He must perform 900 annotations to use the solution, and this procedure will be conducted again when the positions of the cameras change. Therefore, an automatic parking space localization and classification solution for a scalable CV-SPS is needed to deal with the future urban population. Moreover, when we take into the functionalities of SPS in these datasets, there is no information on the parking location or surrounding traffic in this dataset as they only focus on occupancy results. It creates a big gap between existing data and the SPS's scope in the CV paradigm.

Aware of those flaws in current CV approaches and the importance of CV in future SPS, we proposed a complete CV-SPS with a new SPS-based dataset called SNU-SPS. We provide a different solution to the first layer of SPS by treating the parking lot occupancy as an object-detection task and propose a new variant model architecture called OcpDet. Changing the scope to object-detection lifts the system's performance to real-time operation and produces results frame by frame. However, most object detector is not error-free and can be potentially wrong in a real-world inference. Hence, to maintain a reliable SPS, we provide SPS with a result filter in the second layer by addressing results from two modules of our object detector: the spatial estimator module and the training error module. While the spatial estimator module creates a predicted parking region and compares it with the model output to form the spatial error, the training error module catches the training difference of an inference frame and assigns an error. Incorrect inferences are recorded, marked as unusable information, and collected for fine-tuning and retraining the detection model. Then finally, only correct/believable detection results are stored and analyzed in the middleware layer and transparent to operators and drivers. From this layer, we can support optimal routing for drivers and alert operators about upcoming occupied parking slots at the top layer. SNU-SPS dataset is created to support this idea of the system. It contains parking slots captured from multiple parking lots at various angles, ranges, and positions with different light and contrast settings to train the detectors. Furthermore, we provide parking lot GPS with its occupancy rate and surrounding traffic information for system performance analysis.

We extensively test our proposed system on our dataset for efficiency evaluation and conduct detection measurements with the popular parking datasets PKLot for a detailed benchmark. The results from our experiment raise a competitive performance compared to exhaustive classification methods.

2   Related Work

2.1   Automatic Parking Occupancy Detection

Because most previous work focuses on solving the SPS occupancy task as image classification from datasets [4,10] with manual label mask/grids location of a parking lot, none of available datasets could be found for automatic parking space detection. It leads to a small amount of effort on this topic. To the best of our knowledge, there are currently two main approaches to this topic: a mask-based method and a detector-based method.

A mask-based method aims to provide parking patches directly from captures and perform binary classification for the occupancy. The perspective transformation method [7,8,21] is usually used in this scheme to bring the parking lot to a 2D grid presentation. Therefore, it can save time for self-annotating parking locations and exploit the classification machine-learning and deep-learning backbones. However, since the perspective projection process is highly dependent on the camera setting to the parking lot, classification models need to be retrained for different camera settings, which questions the scalability of those methods. Notice this behavior, [18] has introduced a GAN approach that generates the parking place's masks from a team of drones, but there is no comprehensive measurement of the correctness of these masks. In addition, this method requires a top-view capture of the lot, making it unrealistic for indoor parking facilities.

In contrast to mask-based solutions, detector-based approaches perform detection and classification tasks in a single process by a CNN architecture instead of separating them into two processes, which maintains the flexibility and fast inference for CV-SPS infrastructure. The CNN architecture in this realm regresses a parking slot as a foreground or a region of interest and optimizes its classification score. This procedure can be classified into two-stage and one-stage detectors. While the two-stage detector, such as Faster RCNN [26] focuses on the first stage to propose the regions of interest and performs classification on those regions in the second stage, the one-stage detector combines both tasks by grid-anchor regressions. However, because an empty or small parking slot is easily confused as a part of the image's background, both of these architectures face a lot of flaws [23]. Most recent works [16] only used detectors to find a parked car in the parking lot and determine the occupancy rate by a preowned parking lot's capacity and location. This approach relaxes the problems into the well-known car detection, but it limits the extension of the SPS for letting drivers know the location of the parking slot. Recent developments in new architectures such as YOLO [6], and RetinaNet [19] have opened some flexibilities in the small object capture. The idea of using a drone's captures is also used by [14]. The author performs the car detection at top-views by Faster RCNN and YOLO and combines it with the layout proposal. This method faces the same drawback as [18] and is restricted for car detections. For a complete occupancy detection from a detector, only [23] has conducted on a RetinaNet on PKLot [10] dataset. However, the results show much confusion between moving cars and occupied parking slots. While the main reason for this inefficiency is the nature of the PKLot dataset itself (partial area of the parking lot is annotated), the method's performance can be improved if there is an attention mechanism on the parking lot region. In addition, there is a potential non-optimized model design as there is no information on the grid-anchor feature selection.

2.2   Deep Model Uncertainty

As a model performance is a reflection of the coverage of the training dataset, recent research tends to capture model performance in the wild by measuring its uncertainty/stability or contradiction. While [1,15] aimed to create this uncertainty from the outputs by comparing an image and its noise version, [12,13] measured the stability of the prediction of the same image from different models' inferences. However, these methods are not designed for online inference and require some delay for a model judgment. Only [30] can achieve the model uncertainty within a single forward pass inference. Instead of replicating the contradiction from inferences, the author trained the model with a different head to predict this uncertainty directly. Even though it sounds naive and simple, his approach has shown promising results on general object detection datasets.

2.3   Qualitative Comparison Between Existing Approaches and the Proposed System

Aware of the lack of discovery in the current approaches for the CV-SPS system and the need for an efficient detector-based method, we provide a compact detector-based solution with its corresponding dataset. To our knowledge, our

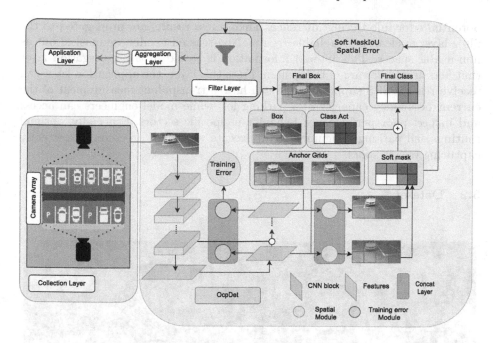

Fig. 1. Our CV-SPS Overall Architecture

CV-SPS is the first end-to-end solution for automatic parking occupancy detection. Our detector-based solution inherits the feature pyramid structure from RetinaNet [19] with two additional module heads: the spatial module and the training error module. While we inherit [30] as the training error module for training error information, we design the spatial estimator module as a parking region proposal to compare the spatial error. It can be understood as a foreground filter, aiming to filter out detection not covered in the predicted foreground region. Hence, this module provides the model with better localization attention, which [23] did not address.

3   Proposed Method and Dataset

3.1   Overall System Architecture

As shown in Fig. 1, we divide a parking lot into sectors to create a scalable and efficient CV-SPS. Each sector is controlled and well-observed by a camera and has non-overlapped observing areas among cameras. This constraint reduces the complexity of the problem by duplicating observation or high occlusion. Assuming the parking lot can be set up with this requirement, our overall system architecture consists of 4 layers: a collection layer, a filter layer, an aggregation layer, and an application layer. The collection layer is responsible for gathering the detection results from distributed cameras in the parking lot as well as their potential error info during inference. Then, these results are propagated to the filter layer to cleanse for reliable results. Non-trusted results are masked out as non-usable spaces. This filtered information is stored in the aggregation layer that acts as middleware of the system. From this layer, the application layer can receive reliable SPS support. Users can have a transparent measurement of the current occupancy capacity of a parking lot, while model engineers can access and inspect poor performance behavior in specific sectors. Especially, optimal routings and parking assignments can execute with high precision by correctly capturing vacant spots.

3.2   Dataset

Fig. 2. SNU-SPS dataset images representation from various indoor and outdoor views. Annotation colors (Red: Occupied, Green: Available, Blue: Restricted and Yellow: Illegal) (Color figure online)

We introduced our SNU-SPS dataset containing nearly 3500 images to support our system. Those images are captured from various views, heights (1–3m), and light conditions in indoor and outdoor parking lots. Each parking lot has different parking spot background colors. The total images were manually checked, labeled, and attached by GPS to the corresponding parking slot. The protocol used to construct the SNU-SPS dataset is composed as follows:

Image Acquisition: All images are captured with a full-HD resolution. For the training set, it is captured randomly for one month in 15 parking lots. Meanwhile, the test set is captured consecutively in 6 parking lots from 3–6pm through 5 working days. It should be noted that none of the 6 parking lots are in the training set. Moreover, test samples contains various weather conditions (sun/rain/cloudy) and has corresponding surrounding traffic measurements from the open government website.

Labeling: For each parking sector, images were labeled as *available/ occupied/ illegal/ restricted* of each parking space. Each annotation is covered by four keypoints that specify for the localization of a parking lot. We formulate the wrapping bounding boxes for the detector from these key points. Especially, we provide optional image masks for the test set to filter out overlapping areas and non-important localization among capture among parking lots. The intention is to maintain the system's constraints and preserve a better parking assignment benchmark.

Table 1. Training and Testing Sets

Set Type	Total Images	Total Labels	Classes			
			Available	Occupied	Illegal	Restricted
Train	2848	18263	7229	10596	396	42
Test	574	2747	1291	1336	36	84

3.3 Automatic Parking Occupancy Detector

When addressing parking occupancy as an object detector, the most arousing problem is the confusion of the parking slots with the image background information such as moving cars or blank spots. The only meaningful visual information is the thin lines separating spots. However, it is usually missed at the lower level of deep neural networks.

RetinaNet: is a promised solution due to its feature pyramid network (FPN). In short, the FPN backbone combines standard convolutional network lower features with lateral connections of early-level features. Hence, the network can construct rich, multi-scale object features, which maintain the impact of the line features in the network. However, from [23] results, despite capturing good center localization, traditional RetinaNet could not expand the parking space tightly when the mAP dropped dramatically from 63.64 to 4.75 when raising from 0.5 to 0.75 precision. This might be non-optimized anchor grid features and lack of a location attention mechanism. Moreover, the Resnet backbone is quite heavy for computation and may not be able to scale up with other additional SPS tasks, limiting the scope of CV-SPS.

$$L_{size} = \sum_{i=0}^{N} D_{p,k}^{i} / D_{p,c}^{i} \tag{1}$$

OcpDet: Noticing this limitation, we replace the heavy Resnet backbone with a lightweight Mobilenet for faster inference and build up our model from this called OcpDet. We intend to give OcpDet more reference points rather than just centers and sizes of parking slot boxes. Therefore, instead of solely detecting the bounding boxes by their centers and sizes, our detector also predicts the key points of parking slots in the localization head. From this scope, we design a new loss function L_{size} which aims to maximize the predicted coverage boxes to their corresponding predicted key points. We treat these N keypoints as anchors which pull the box corners closer to them, where N is less than 4. We denoted the distance between a box corner p to the its corresponding keypoint k as $D_{p,k}$ and the distance between a box corner p to the center c as $D_{p,c}$. When we set the keypoints as our dataset keypoints, the loss implicitly guides the model to focus more on the border of a parking slot.

Spatial Estimator Module: Increasing solely reference points is not enough to make our OcpDet robust with background and foreground conflict in a parking lot. Cars vs. parked places or empty spots vs. walls can be highly confused. Because our model is a single-stage object detector, it divides an image into anchor patches and represents them at different scale levels for classification prediction and localization regression. As the parking lot layout is usually aligned, equally separated, and non-overlapped among spots, it is very convenient to ideally wrap a parking slot in one single anchor patch representation. Therefore, we create a soft head estimator for the active anchors that can be considered as parking slots from each level feature generator of FPN.

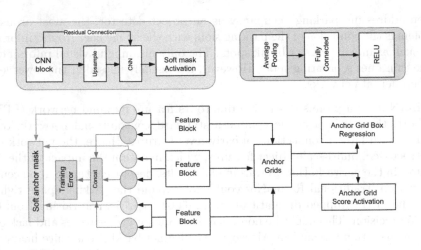

Fig. 3. Module Architectures for Spatial Estimatior Module and Training Error Module

We attach a residual convolution block for each of the N feature levels and average the last channel to get a 2D map of anchor patches. Noted that, we avoid flattening the map to avoid insufficient computations as the fully connected layers can go up to huge connections for early-level features. For example, a 256×256 anchor can cost 65k connection and additional 65 parameters for a dense layer. With that intuition, as parking slots are in the foreground, we trigger this 2D prediction by a sigmoid activation and treat it as a binary classification (Fig. 3). From this activation map, we can create a soft mask for parking slot locations as a reference for the model spatial outputs. To train this activation map, we compare each map M from N feature levels with its corresponding classification target map C in the classification head by a ϵ difference. This loss can act as a regularizer for the model to provide attention to the foreground region of the model. Moreover, as the soft mask head is not directly connected to the localization head prediction, it gives the model another degree of freedom to operate while implicitly improving the localization through top-level features. To deeply understand the role of the spatial module, we investigate the way of predicting the foreground from multi-level anchor-grid features through the FPN. Regardless of the grid's anchor box size, it will be considered a foreground if the classification activation is switched on. Therefore, we look into the pixel activation map through the class activation map (HiRes-GradCam) [11] to understand which pixel contributes to the final decision. As demonstrated in Fig. 4, there is a high density of pixel activations at each middle of parking slots whenever the model predicts the location class, which is reasonable as a class determination belongs to what having inside a parking border. The class confidence is reduced when the number of surrounding activation maps is dimmed Fig. 4-a and disappears in Fig. 4-b. This means the activation map only produces a reliable decision when it falls inside a parking spot. We notice that the spatial module's heatmap can evade this problem by adding the activation of the soft mask prediction. Following this idea, we injected the predicted classification parking-grid information into the classification head by adding its prediction to

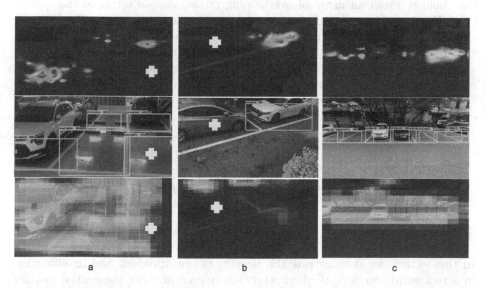

Fig. 4. HiRes-GradCam activations following by Image Detections (green: "occupied", blue: "available") and their coresponding soft mask heatmaps. White arrows stand for weak detection area: (a) low confidence score, (b) zero detection. c) is an exmaple of our model inference (Color figure online)

the ground confidence. We avoid multiplication for overfitting results with the predicted soft-mask.

Training Error Module: as mentioned above, we use [30] method as a model error predictor from training knowledge. In short, [30] represents this error as predicted loss during training. The method selects pairs of training batch samples in each iteration to represent the model inference stability during training. A good pair of samples should have a low loss during training, while a poor inference will raise the loss. Hence, during training, a predictor module will try to estimate this loss by accessing multi-level features, allowing it to choose necessary information between layers to capture the loss behavior correctly. In other words, this method can also act like an abstract regularizer for the model to avoid conflict between batches. Compare to the Spatial Estimator Module in Fig. 3, it shares the same feature map, but average out each feature by an average pooling before aggregation for loss prediction. Therefore, this module can operate independently from the model output performance.

3.4 Result Filter

Therefore, after obtaining the detection results with two additional predictions from the spatial and training error modules, the foreground N detections of each sector D are compared with its soft spatial activation map S. We compute the overlap value for each detection and select the strong-belief region by a threshold γ. From an array of overlapping ratios, we can estimate the spatial error $Err_{spatial}$ for each frame capture detection as demonstrated in Eq. 2. Noted that, if $\mathbb{1}(R)$ from a detection reach near the zero bottoms of the overlap ratio, the spatial module will get rid of this detection. We combine this spatial error with the training error $Err_{training}$ from the training error module by averaging aggregation as a final error score Err_{total} for the inference. From this score, we can eliminate whether to believe these detections results or not by thresholding our belief.

$$Err_{spatial} = 1 - \frac{1}{N}\sum_{i=0}^{N} \mathbb{1}(\frac{D_i \cap S}{D_i}) \quad \text{s.t} \quad \mathbb{1}(R) = \begin{cases} 1, & \text{if } R > \gamma \\ R, & \text{otherwise} \end{cases} \quad (2)$$

$$Err_{total} = \alpha Err_{training} + (1 - \alpha)Err_{spatial} \quad (3)$$

4 Experiment

In this section, we demonstrate the benefits of our approach and its efficiency in a real-world scenario. We first study the impact of using the spatial module and reference keypoints for improving parking localization and compare it with the existing solutions. Second, we demonstrate the effectiveness of filtering out unstable inference to the model's actual performance. Finally, we measure the impact of OcpDet's performance with the optimal routing and parking assignment task.

4.1 Experimental Setup

Training settings: For our experiments, with the scope of an efficient and quick response in the first system layer, we only addressed the single-stage object detector: SSD-Mobilenet (denoted MBN), Mobilenet-FPN (our model backbone, denoted MBN-FPN), and our OcpDet. The training engine for these models is Tensorflow Object Detection API. We trained OcpDet for both datasets in 25000 iterations with batchsize 48 by SGD optimizer. To keep the detection robust to the small parking spots, we selected with high-resolution 896-pixel input instead of the traditional 300-pixel or 640-pixel.

Benchmark settings: We used PKLot dataset [10] and our SNU-SPS dataset for the solutions benchmark. We did not use CNRPark-Ext because this dataset's parking spot border lines are faded and not consistently visible. PKLot dataset contains three sub-datasets capturing from high view: PUCPR, UFPR05, and UFPR04, which leads to a small scale of parking spots. All of these sub-dataset parking locations are partially annotated. Thus, we provide masks for each sub-dataset to clip out non-annotated parking regions to avoid false positives during training and make it suitable for the detection benchmark. We split the dataset from each sub-dataset in half for training and testing, as the PKLot authors suggested. For our SNU-SPS, we have a separate test set. However, due to a small label of *illegal* and *restricted* classes in our dataset, our test set is only addressed with two classes: *occupied* and *available*. In addition, because the scope of our dataset is to detect correctly in a sector, we only address medium and large ground truths in the test set. During the assignment application test, the detection will be filtered out overlapping areas by our provided masks.

4.2 Detection Performance

In this part, we conduct our experiment on the efficiency of our spatial module and the localization improvement of additional reference keypoints. Our main metric for the detection evaluation is mAP(mean average of precision) and recall ranging from 0.5 to 0.95 IoU(intersection over union). Meanwhile, as the classification task is the side task to benchmark with a classification approach, we address the mAP(0.5) from each class for the comparison. As results are summarized in Table 2, the performance of the localization attention leads in all data settings. We did not compare with [23] solution because of his insufficient model's performance.

In the PKLot dataset, MBN struggles to learn the features because it is lack of top feature generation from the grid and lines to capture the small objects. Meanwhile, thank to the FPN, both MBN-FPN and our model outperform MBN in this dataset. Due to the fixed location of the parking lot's captures, OcpDet can strongly overfit the position of each parking space and turn into a grid classifier. As demonstrated in Fig. 5, the localization guidance from the soft mask generations helps the anchor patches avoid negative samples that do not belong to the parking area. Our model can boost the performance to near perfection using the soft mask head during training. In addition, by using additional

Table 2. PKLot and SNU-SPS Detection Benchmark

Method	Test Set	Recall (0.5:0.95)	mAP			Classification Score	
			0.5	0.75	(0.5:0.95)	Occupied	Available
OcpDet	PUCPR	0.88	0.98	0.98	0.84	0.98	0.98
MBN-FPN		0.76	0.86	0.85	0.72	0.83	0.90
MBN [27]		0.41	0.48	0.35	0.31	0.49	0.46
Classifier [28]		–	–	–	–	0.99	0.99
OcpDet	UFPR05	0.98	0.99	0.99	0.97	0.99	0.99
MBN-FPN		0.84	0.93	0.90	0.82	0.93	0.94
MBN [27]		0.42	0.51	0.42	0.37	0.50	0.53
Classifier [28]		–	–	–	–	0.99	0.99
OcpDet	UFPR04	0.96	0.99	0.99	0.93	0.99	0.99
MBN-FPN		0.83	0.95	0.90	0.79	0.95	0.96
MBN [27]		0.43	0.52	0.44	0.36	0.51	0.53
Classifier [28]		–	–	–	–	0.99	0.99
OcpDet	SNU-SPS	0.56	0.81	0.48	0.47	0.83	0.80
OcpDet_{spatial}		0.56	0.83	0.50	0.47	0.85	0.82
OcpDet_{both}		0.56	0.82	0.49	0.47	0.84	0.81
OcpDet_{ll} [30]		0.55	0.82	0.49	0.47	0.83	0.81
MBN-FPN		0.54	0.77	0.46	0.45	0.80	0.74
MBN [27]		0.51	0.71	0.48	0.44	0.73	0.69

keypoints, our approach improves the localization detection and preserves its tightness among scales.

In contrast to the unchanged parking layout of the PKLot dataset, our SNU-SPS creates more challenges for the model to select the correct anchor patch due to its various capture positions. This makes not only our model but also other detectors struggle. Therefore, we omit the classifier method for a fair comparison. In this dataset, both the reference keypoints and the soft mask generator are not much efficient as the model grid is not stationary. Despite the challenge of adaptability, when looking into the heatmap of the soft mask through different scales of the spatial module in Fig. 5, the obtained mask on the parking lot still has denser attention than other foreground predictions. Combined with the impact of reference keypoints, OcpDet leaves a gap of nearly 5% on mAP(0.5) to original MBN-FPN and 10% to MBN. Moreover, in the first row of Fig. 5, it shows that our model is not sensitive with car appearance. The prediction only activates inside in the parking zone where the lines are visible.

4.3 Result Filter Performance

To evaluate the result filter efficiency, we solely address OcpDet on SNU-SPS as there is still room for model improvement. We calculate the error of an inference sample from the formula Eq 2 and remove a maximum of 100 samples (20% of the test set) from the test set. For the purpose of testing out the benefit from

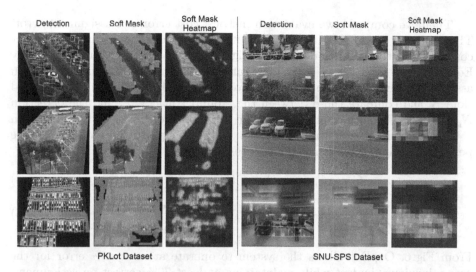

| Detection | Soft Mask | Soft Mask Heatmap | Detection | Soft Mask | Soft Mask Heatmap |

PKLot Dataset | SNU-SPS Dataset

Fig. 5. Visualization occupancy detections, soft mask predictions and soft mask predictions heatmap on PKLot dataset and SNU-SPS dataset

the spatial module and the training error module, we lock the γ by 0.7 and set α from a set of $[0, 0.4, 1]$ with respect to $OcpDet_{ll}$, $OcpDet_{both}$ and $OcpDet_{spatial}$ in Table 2. According to the experiment, the result filter has boosted the model's overall accuracy, proving that the filter can ensure a better quality from the detector regardless of α assignment. Because the spatial filter can score and remove unreliable results from detections by overlapping with the soft mask, its results are slightly better than the [30] approach on the training filter.

4.4 Optimal Routing and Parking Assignment

To make a comprehensive benchmark on the impact of detection results on the assignment application layer, we collect the traffic information over ten days from the government website and associate that information with the test set to form a close loop simulation. Each day from 3 to 6pm, there will be about 100 requirements for booking a vacant spot to 6

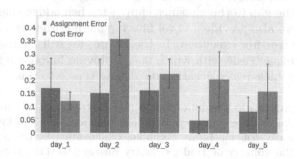

Fig. 6. The cost errors and the assignment errors: averaging simulation for 5 d at 6 parking lots

parking lots. The suggested optimal road for each request will be assigned from the MapQuest API. We consider the Hungarian assignment [17] from the masked-out test label as the ground truth for assigning vacant spots.

Then, we compute two evaluation metrics: cost error and assignment error. The cost error is computed by the absolute error of the ground truth assignment cost $C_{g,i}$ and the vacancy detection assignment cost $C_{p,i}$, which is designed by Eq. 4. The assignment cost is computed after getting the assignments. From each assigned parking lot in 6 parking lots, we compute the total number of booked slots N and compare those between the ground truth $N_{g,i,j}$ and the detection $N_{p,i,j}$ by an absolute error. We perform normalization to keep the value from 0 to 1. The simulation will repeat ten times to capture the model's average performance due to different traffic statuses.

$$C = \gamma C_{price} + (1 - \gamma)(C_{travel} + C_{distance}) \tag{4}$$

$$Err_{cost} = \sum_{i=3}^{6} \frac{|C_{g,i} - C_{p,i}|}{C_{g,i}} \qquad Err_{assign} = \sum_{i=3}^{6} \sum_{j=1}^{6} \frac{|N_{g,i,j} - N_{g,i,j}|}{N_{g,i}} \tag{5}$$

From Fig. 6, OcpDet allows the system to operate at most 40% error for the cost-minimizing budget while maintaining at least 70% correct on assignment. Because the cost error is related to the distance travel, wrong assignments placed on far-distance drives can cause a huge gap to the optimal cost. But this error is not much in terms of matching the number of assignments. From these metrics, we can overview the system's benefits to users. Operators will benefit the most as their vacant spaces will automatically be assigned with minima error. In contrast, some drivers may get some disadvantages from the system assignment.

5   Conclusion & Future Work

This paper proposes a novel end-to-end CV-SPS with a detailed benchmark on both old and new datasets. Even though our dataset is small, it shows challenging factors, and it is the first dataset for computer vision with full CV-SPS scope. Our method has proved its efficiency to some extent and can potentially close the gap to the classifier approach when addressing stationary views. Moreover, we also provide a novel filtering method that can help the system approach an error-free execution. In the future, we will continue building our dataset to a bigger scale with which it can provide functional information for SPS, such as vehicle reidentification or parking type selection.

Acknowledgement. Dataset and experiment in this work were supported by the Automotive Industry Building Program (1415177436, Building an open platform ecosystem for future technology innovation in the automotive industry) funded by the Ministry of Trade, Industry Energy (MOTIE, Korea).

References

1. Aghdam, H.H., Gonzalez-Garcia, A., Weijer, J.v.d., López, A.M.: Active learning for deep detection neural networks. In: Proceedings of the IEEE/CVF International Conference on Computer Vision, pp. 3672–3680 (2019)

2. Al-Turjman, F., Malekloo, A.: Smart parking in IoT-enabled cities: a survey. Sustain. Cities Soc. **49**, 101608 (2019)
3. Lisboa de Almeida, P.R., Honório Alves, J., Stubs Parpinelli, R., Barddal, J.P.: A systematic review on computer vision-based parking lot management applied on public datasets. arXiv e-prints pp. arXiv-2203 (2022)
4. Amato, G., et al.: A wireless smart camera network for parking monitoring. In: 2018 IEEE Globecom Workshops (GC Wkshps), pp. 1–6. IEEE (2018)
5. Amato, G., Carrara, F., Falchi, F., Gennaro, C., Meghini, C., Vairo, C.: Deep learning for decentralized parking lot occupancy detection. Expert Syst. Appl. **72**, 327–334 (2017)
6. Bochkovskiy, A., Wang, C.Y., Liao, H.Y.M.: YOLOv4: optimal speed and accuracy of object detection. arXiv preprint arXiv:2004.10934 (2020)
7. Bohush, R., Yarashevich, P., Ablameyko, S., Kalganova, T.: Extraction of image parking spaces in intelligent video surveillance systems (2019)
8. Bura, H., Lin, N., Kumar, N., Malekar, S., Nagaraj, S., Liu, K.: An edge based smart parking solution using camera networks and deep learning. In: 2018 IEEE International Conference on Cognitive Computing (ICCC), pp. 17–24. IEEE (2018)
9. Cookson, G.: Parking pain-inrix offers a silver bullet. INRIX-INRIX http://inrix.com/blog/2017/07/parkingsurvey/. Accessed 21 Nov (2017)
10. De Almeida, P.R., Oliveira, L.S., Britto, A.S., Jr., Silva, E.J., Jr., Koerich, A.L.: PKLot-a robust dataset for parking lot classification. Expert Syst. Appl. **42**(11), 4937–4949 (2015)
11. De Almeida, P.R., Oliveira, L.S., Britto, A.S., Jr., Silva, E.J., Jr., Koerich, A.L.: PKLot-a robust dataset for parking lot classification. Expert Syst. Appl. **42**(11), 4937–4949 (2015)
12. Feng, D., Wei, X., Rosenbaum, L., Maki, A., Dietmayer, K.: Deep active learning for efficient training of a lidar 3D object detector. In: 2019 IEEE Intelligent Vehicles Symposium (IV), pp. 667–674. IEEE (2019)
13. Haussmann, E., et al.: Scalable active learning for object detection. In: 2020 IEEE Intelligent Vehicles Symposium (IV), pp. 1430–1435. IEEE (2020)
14. Hsieh, M.R., Lin, Y.L., Hsu, W.H.: Drone-based object counting by spatially regularized regional proposal network. In: Proceedings of the IEEE International Conference on Computer vision, pp. 4145–4153 (2017)
15. Kao, C.-C., Lee, T.-Y., Sen, P., Liu, M.-Y.: Localization-aware active learning for object detection. In: Jawahar, C.V., Li, H., Mori, G., Schindler, K. (eds.) ACCV 2018. LNCS, vol. 11366, pp. 506–522. Springer, Cham (2019). https://doi.org/10.1007/978-3-030-20876-9_32
16. Kirtibhai Patel, R., Meduri, P.: Faster R-CNN based automatic parking space detection. In: 2020 The 3rd International Conference on Machine Learning and Machine Intelligence, pp. 105–109 (2020)
17. Kuhn, H.W.: The hungarian method for the assignment problem. Naval Res. Logistics Q. **2**(1–2), 83–97 (1955)
18. Li, X., Chuah, M.C., Bhattacharya, S.: UAV assisted smart parking solution. In: 2017 International Conference on Unmanned Aircraft Systems (ICUAS), pp. 1006–1013. IEEE (2017)
19. Lin, T.Y., Goyal, P., Girshick, R., He, K., Dollár, P.: Focal loss for dense object detection. In: Proceedings of the IEEE International Conference on Computer Vision, pp. 2980–2988 (2017)
20. Media, U.N.: 68% of the world population projected to live in urban areas by 2050, says un. https://www.un.org/development/desa/en/news/population/2018-revision-of-world-urbanization-prospects.html (2018). Accessed 08 Jun 2022

21. Nieto, R.M., García-Martín, Á., Hauptmann, A.G., Martínez, J.M.: Automatic vacant parking places management system using multicamera vehicle detection. IEEE Trans. Intell. Transp. Syst. **20**(3), 1069–1080 (2018)
22. Nyambal, J., Klein, R.: Automated parking space detection using convolutional neural networks. In: 2017 Pattern Recognition Association of South Africa and Robotics and Mechatronics (PRASA-RobMech), pp. 1–6. IEEE (2017)
23. Padmasiri, H., Madurawe, R., Abeysinghe, C., Meedeniya, D.: Automated vehicle parking occupancy detection in real-time. In: 2020 Moratuwa Engineering Research Conference (MERCon), pp. 1–6. IEEE (2020)
24. Paidi, V., Håkansson, J., Fleyeh, H., Nyberg, R.G.: CO2 emissions induced by vehicles cruising for empty parking spaces in an open parking lot. Sustainability **14**(7), 3742 (2022)
25. Polycarpou, E., Lambrinos, L., Protopapadakis, E.: Smart parking solutions for urban areas. In: 2013 IEEE 14th International Symposium on" A World of Wireless, Mobile and Multimedia Networks"(WoWMoM), pp. 1–6. IEEE (2013)
26. Ren, S., He, K., Girshick, R., Sun, J.: Faster R-CNN: towards real-time object detection with region proposal networks. In: Advances in Neural Information Processing Systems, vol. 28 (2015)
27. Sandler, M., Howard, A., Zhu, M., Zhmoginov, A., Chen, L.C.: Mobilenetv 2: Inverted residuals and linear bottlenecks. In: Proceedings of the IEEE Conference on Computer Vision and Pattern Recognition, pp. 4510–4520 (2018)
28. Valipour, S., Siam, M., Stroulia, E., Jagersand, M.: Parking-stall vacancy indicator system, based on deep convolutional neural networks. In: 2016 IEEE 3rd World Forum on Internet of Things (WF-IoT), pp. 655–660. IEEE (2016)
29. Varghese, A., Sreelekha, G.: An efficient algorithm for detection of vacant spaces in delimited and non-delimited parking lots. IEEE Trans. Intell. Transp. Syst. **21**(10), 4052–4062 (2019)
30. Yoo, D., Kweon, I.S.: Learning loss for active learning. In: Proceedings of the IEEE/CVF Conference on Computer Vision and Pattern Recognition, pp. 93–102 (2019)

CrackSeg9k: A Collection and Benchmark for Crack Segmentation Datasets and Frameworks

Shreyas Kulkarni[1(✉)], Shreyas Singh[1], Dhananjay Balakrishnan[1],
Siddharth Sharma[1], Saipraneeth Devunuri[2],
and Sai Chowdeswara Rao Korlapati[3]

[1] Indian Institute of Technology, Madras, India
shreyas.kulkarni@smail.iitm.ac.in
[2] University of Illinois Urbana-Champaign, Champaign, USA
[3] Rizzo International, Inc., Pittsburgh, USA

Abstract. The detection of cracks is a crucial task in monitoring structural health and ensuring structural safety. The manual process of crack detection is time-consuming and subjective to the inspectors. Several researchers have tried tackling this problem using traditional Image Processing or learning-based techniques. However, their scope of work is limited to detecting cracks on a single type of surface (walls, pavements, glass, etc.). The metrics used to evaluate these methods are also varied across the literature, making it challenging to compare techniques. This paper addresses these problems by combining previously available datasets and unifying the annotations by tackling the inherent problems within each dataset, such as noise and distortions. We also present a pipeline that combines Image Processing and Deep Learning models. Finally, we benchmark the results of proposed models on these metrics on our new dataset and compare them with state-of-the-art models in the literature.

Keywords: Deep learning applications · Image processing · Semantic segmentation · Crack detection · Datasets · DeepLab · CFD

1 Introduction

Cracks are common building distress, which is a potential threat to the safety and integrity of the buildings. Localizing and fixing the cracks is a major responsibility in maintaining the building. However, the task of detecting cracks is both tedious and repetitive. To expedite this process and alleviate the workload on experts, it is necessary to achieve automation in crack detection and segmentation.

For a long time, such crack detection has been done manually. Of late, much research has gone into developing automated techniques. Initial work focused on

Supplementary Information The online version contains supplementary material available at https://doi.org/10.1007/978-3-031-25082-8_12.

L. Karlinsky et al. (Eds.): ECCV 2022 Workshops, LNCS 13807, pp. 179–195, 2023.
https://doi.org/10.1007/978-3-031-25082-8_12

using image thresholding and edge detection techniques [1,11,17,45]. Recently, focus has shifted to using Deep Learning for classification [12,23,41,42], object detection [11,27,30,33] and semantic segmentation [26,40,46,50] of cracks. Attempts have been made to use various techniques, including feature pyramid networks [48] and segmentation models like U-Net [24]. Some experiments have been conducted using pre-trained Conv-Nets [52], with limited crack-detection abilities. Better results were obtained when crack segmentation was attempted using models from the R-CNN family [20].

The Deep Learning approaches are highly dependent on the data that has been used for training. The study by [13] makes use of 3 datasets: AigleRN [4] (38 images), CFD dataset [21] (118 images), and the HTR dataset (134 images, not publicly available). Another work by [25] performs analysis on the CFD dataset [21] and Crack500 dataset [47]. [31] have evaluated their pipeline on two datasets containing 56 grayscale and 166 RGB images respectively. SDDNet [6] (200 images) manually created a dataset to check their model's performance. Deep-Crack [28] too highlights the lack of a benchmark dataset and makes an effort to create a dataset of 537 RGB images, which has been included in our dataset.

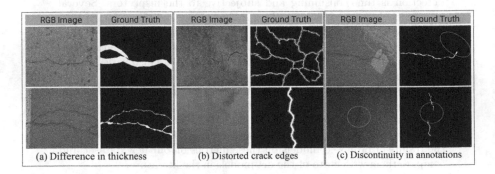

Fig. 1. Inconsistent annotations among sub-datasets

All the attempts mentioned above have fallen short of requirements primarily due to the quality and quantity of publicly available data sets. Obtaining an annotated data set with segmented masks for cracks is a challenge. Most crack data sets have less than 500 images, with several having less than 100 images. Further, the data set being used is not generalizable, and this, combined with the quality of the data sets, leads to non-reproducible results. The datasets used in all cases have different resolutions of the images and different types of annotations leading to inconsistencies as shown in Fig. 1. This makes it more difficult to compare with other works. We work towards unifying all the sub-datasets involved to deal with such issues.

We aim to create an efficient Crack segmentation model using the latest Computer Vision and Deep Learning advances. While we did perform bounding box prediction to localize crack regions in images of arbitrary resolutions, our primary focus has remained on binary segmentation of the corresponding cracks. To overcome the limiting factor of the quality of data sets, we combined data sets from various sources while also attempting to unify their ground truths using Image Processing techniques (detailed in Sect. 3.2). This paper makes the following major contributions:

- Compilation of a dataset consisting of 9000+ images of cracks on diverse surfaces with annotations for semantic segmentation. The compiled data is also categorized into linear, branched, and webbed.
- Further refinement of the ground truth masks by denoising and unifying the annotations using Image Processing techniques per the dataset's requirements.
- An end-to-end pipeline has been devised for efficient localization and segmentation of cracks for images of arbitrary lighting, orientation, resolution and background.
- We explore various rule-based and data-driven techniques for crack segmentation to establish a benchmark on our dataset.
- Finally, we present an approach to fuse unsupervised self-attention mechanisms into CNNs for improvising over the SOTA models.

2   Related Works

2.1   Rule Based Image Processing Techniques

Rule-based Image Processing methods have been traditionally used for image-based crack segmentation. The techniques can broadly be classified into two categories, i.e., Edge detection and Thresholding [22].

Thresholding complies with the rule of classifying the pixels of a given image into groups of objects and background; the former includes those pixels with gray values more than a specific threshold, and the latter comprises those with gray values equal to and below the threshold. Choosing an appropriate threshold value is the main challenge in this method. Using a histogram to choose a threshold value gives satisfactory results for segmenting images of road surfaces [29].

The purpose of the edge-based algorithm is to segment an image based on the discontinuity detection technique [43]. The focus of edge-based algorithms is on the linear features that mainly correspond to crack boundaries and interesting object surface details. Based on the comparative study, the algorithms calculate the magnitude for each pixel and use double Thresholding to discover potential edges. Even though useful for edge detection, these algorithms can only detect edges and cannot extrapolate over the entire crack surface; this makes edge detection ineffective for binary segmentation.

2.2   Data Driven Methods

The lack of generalizability of rule-based methods to diverse environments with different backgrounds and lighting has promoted research into the data-driven detection of cracks. These techniques comprise of data-hungry Deep Learning models that are either fully convolutional, attention based or a mix of both, requiring the training of millions of parameters.

While traditional object detection algorithms re-purposed classifiers to perform detection, **YOLO** [35] framed object detection as a regression problem to spatially separated bounding boxes and associated class probabilities. A single neural network directly predicts bounding boxes and class probabilities from full images in one evaluation. We have trained a YOLO-v5 object detector to localize cracks in images of arbitrary size to crop them before pixel-wise segmentation.

The image-to-Image translation is the spatial transformation of images from one domain to another. This field has grown immensely since **Pix2Pix GAN** [18] was introduced. GANs [14] consist of two parts: a generative model G that captures the data distribution and a discriminative model D that estimates the probability that a sample came from the training data rather than G. The training procedure for G is to maximize the probability of D making a mistake. We pose semantic segmentation of cracks using Pix2Pix as an image-to-image translation problem, where the generator translates an input RGB image into a binary segmentation mask for cracks, and the discriminator tries to distinguish b/w real and fake samples conditioned on the input. The **U-Net** [37] architecture is used for semantic segmentation, which consists of a contracting path to capture context and a symmetric expanding path that enables precise localization. This has been utilized as a backbone for the Generator of GANs.

DeepLab [5] highlights convolution with upsampled filters, or 'atrous convolution' and ASPP 'Atrous Spatial Pyramidal Pooling', as a powerful tool in dense segmentation tasks. DMA-Net [44] proposes a multi-scale attention module in the decoder of DeepLabv3+ to generate an attention mask and dynamically assign weights between high-level and low-level feature maps for crack segmentation.

Swin-Unet [2] is an U-Net-like pure Transformer for image segmentation. The tokenized image patches are fed into the Transformer-based U-shaped Encoder-Decoder architecture with skip-connections for local-global semantic feature learning. Specifically, a hierarchical Swin Transformer with shifted windows is used as the encoder to extract context features. Moreover, a symmetric Swin Transformer-based decoder with a patch expanding layer is designed to perform the up-sampling operation to restore the spatial resolution of the feature maps.

DINO [3] highlights that self-supervised Vision Transformer features contain explicit information about the semantic segmentation of an image, which does not emerge as clearly with supervised Vision Transformers, nor with Convolutional Nets. Self-supervised ViT features explicitly contain the scene layout and, in particular, object boundaries. This information is directly accessible in the self-attention modules of the last block.

3   Dataset

The proposed dataset is the largest, most diverse and consistent crack segmentation dataset constructed so far. It contains a total of 9255 images, combining different smaller open source datasets. Therefore, the dataset has great diversity in the surface, background, lighting, exposure, crack width and type (linear, branched, webbed, non-crack). Smaller datasets like Eugen Muller [15] and Sylvie Chambon [4] were discarded due to poor quality and much fewer images. The DIC dataset [36] containing 400+ images was avoided because the images were augmented from a tiny set of 17 images and had very little variety. The Forest dataset [39] was excluded as it was found to be very similar to the CFD dataset.

This dataset is divided into training, validation and testing sets with 90%, 5% and 5% split. The final dataset is made available to public on Harvard Dataverse [38].

3.1 Dataset Details

Table 1 summarises the sub-datasets involved. This subsection will discuss the sub-datasets involved by providing an overview of data collection techniques, characteristics of image and dataset size.

Table 1. Sub dataset quantitative details and performance on benchmark models

Dataset	Size	Resolution	% of cracks	Pix2Pix U-Net		Deeplab ResNet *	
				mIoU	F1	mIoU	F1
Masonry	240	224 × 224	4.21	0.4685	0.0392	0.4986	0.0420
CFD	118	480 × 320	1.61	0.6100	0.3942	0.6232	0.4203
CrackTree200	175	800 × 600	0.31	0.5030	0.0799	0.5478	0.1132
Volker	427	512 × 512	4.05	0.6743	0.5423	0.8209	0.7955
DeepCrack	443	544 ×384	3.58	0.7207	0.6193	0.8311	0.8068
Ceramic	100	256 × 256	2.05	0.4783	0.0330	0.5095	0.0480
SDNET2018	1411	4032 × 3024	0	0.4865	N/A	0.5000	N/A
Rissbilder	2736	512 × 512	2.70	0.6050	0.3854	0.7512	0.6856
Crack500	3126	2000 × 1500	6.03	0.6495	0.4974	0.8230	0.8032
GAPS384	383	640 × 540	1.21	0.5716	0.2825	0.5965	0.3358

The **Masonry** [8] dataset contains images of masonry walls. They consist of images from the Internet and masonry buildings in the Groningen, Netherlands. The crack patches comprise of small (a couple of bricks) to larger (whole masonry walls) fields of view.

CFD [21] dataset is composed of images reflecting urban road surface conditions in Beijing, China. Each image has hand-labelled ground truth contours. **CrackTree200** [51], **Volker** [32] and **DeepCrack** [28] contain images with manually annotated cracks collected from pavements and buildings. They have challenges like shadows, occlusions, low contrast and noise.

Ceramic Cracks Database [19] was collected by students of the University of Pernambuco. The database has images of building facades with cracks on different colours and textures of ceramics.

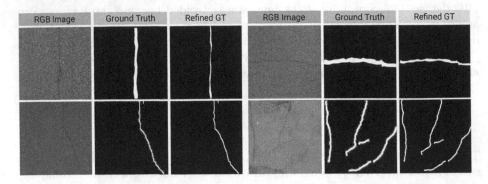

Fig. 2. Refined dataset. Hairline distortions have been tackled in the lower half of the images while the thickness has been reduced in the upper half of the images

SDNET2018 [9] contains various non-crack images from concrete bridge decks, walls, and pavements. The dataset includes cracks as narrow as 0.06 mm and as wide as 25 mm. The dataset also includes images with various obstructions, including shadows, surface roughness, scaling, edges, holes, and background debris.

Rissbilder [32] consists of varied types of architectural cracks, and **Crack500** [47] was collected at the Temple University campus using a smartphone as the data sensor. The **GAPS** [10] data acquisition occurred in the summer of 2015, so the measuring condition was dry and warm.

3.2   Dataset Refinement

The open source datasets had different resolutions and the ground truths had many distortions and discontinuities. They were also vastly inconsistent in their thickness(see Fig. 1). Many cracks had a hairlike boundary noise surrounding them. This led to scenarios where the model's metric score was less even for visually accurate predictions due to the distortions in the Ground Truths.

Firstly, we resize all the images to a standard size of 400×400. For images greater than this size, random crops of this size were extracted from images, and images with resolution less than it were linearly interpolated to the standard size. Once the uniformity in size was attained, we performed Gaussian blurring followed by morphological operations to remove noise from the image. A 3 × 3 kernel made up of ones was used to erode the mask, followed by dilation with the hairline distortions to remove it. Depending on the distortion (thickened or thinned), the images required either erosion or dilation. We used the contour area as the metric for finding the kind of distortion and operating on it, respectively.

We have summarised all the transformations performed on the sub-datasets in the Table 2 and sample refinements can be visualised in Fig. 2.

Table 2. Sub dataset refinement summary

Dataset	Distortion	Refinement
Masonry	None	None
CFD	None	None
CrackTree 200	Continuity errors; Hairline distortion	Dilation followed by Erosion with special kernels
Volker	Randomly thick cracks; Hairline distortion	Contour area based erosion; Morph opening
DeepCrack	Crack boundary irregular	None
Ceramic	None	None
SDNET2018	None	None
Rissbilder for Florian	Hairline distortion	Morph opening with increased kernel size
Crack500	Extra thick and hairline distortion	Erosion and morph opening
GAPS384	Random thickening of cracks	Erosion or Dilation on the basis of Contour size

3.3 Categorizing Dataset Based on Crack Types

The dataset has been divided into three subtypes, namely Linear, Branched and Webbed, based on the spatial patterns of cracks observed in the images. To achieve the classification for the entire dataset, we trained a small ResNet-18 pre-trained classifier on 150 images, with 50 belonging to each class. We manually labelled these 150 images to generate ground truths for classification. This

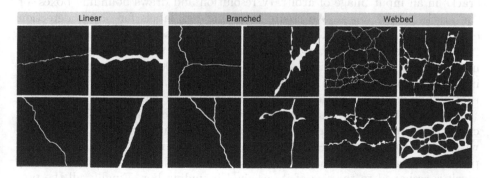

Fig. 3. Sample ground truths belonging to each category of crack type

methodology generalized well to our entire dataset. The cracks with no branches are classified as linear, Cracks with 2–5 branches are classified as branched, and cracks with more than five cracks are classified as webbed. This division of the dataset aims to benchmark and contrast the generalizing ability of benchmarked models across various crack modalities and complexity for detection. Figure 3 shows the cracks belonging to different classes, and Table 3 contains details of the crack types and their distribution in the dataset.

Table 3. Performance based on the type of crack

Dataset	Size	Percentage of cracks	Pix2Pix U-Net		Deeplab ResNet *	
			mIoU	F1	mIoU	F1
Linear	2369	4.94	0.6687	0.5455	0.8219	0.7991
Branched	4192	5.45	0.6184	0.4374	0.7940	0.7643
Webbed	1283	9.81	0.5963	0.4259	0.6488	0.5835
Non-crack	1411	0	0.4865	N/A	0.5000	N/A

4   Methodology and Experiments

This section will discuss the experiments conducted to benchmark and evaluate different models for crack segmentation. These experiments included an amalgamation of both supervised and unsupervised techniques. The experiments were performed on the refined dataset with a standard resolution of 400×400 for all images.

4.1   Pipeline

This subsection explains the details of our pipeline for end-to-end crack segmentation. The pipeline processes the input image through four steps (see Fig. 4). The first step consists of a YOLO-v5-based object detector that localizes the cracks in an input image of arbitrary resolution and draws bounding boxes for the cracks in the image. We create bounding box labels for object detection enclosing the crack using the extreme pixels of the segmented masks (ground truths). In the second step, cracks are cropped using the bounding box coordinates obtained in the previous step and resized to 400×400. The resized crack images are passed through the DINO model to extract unsupervised features that will be used as the prior for the segmentation model.

In the final step, the unsupervised features are concatenated with the resized crack image and fed into an FPN-based binary segmentation model to get pixelwise classification for the cracks in each image. The final post-processing step passes the segmentation maps through a denoiser; the denoised segmentation mask is reshaped to the size of its original bounding box. Finally, all the predictions for a given region are stitched together to produce the final output.

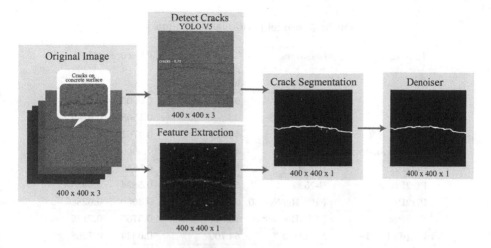

Fig. 4. General pipeline

The results on various techniques have been benchmarked in Table 4 and images with their predictions on the respective models are showcased in Fig. 5. Note: the purpose of the YOLO object detector is to concentrate the segmentation model's focus on a patch of cracks in case of high-resolution images.

4.2 Unsupervised Feature Extraction

In our first set of experiments, we attempt to generate the predicted masks without utilizing ground truth annotations.

Otsu Thresholding method failed to find the optimum threshold for images with noise. When the dimensions of the cracks were small relative to the background, the technique failed to identify the cracks. Otsu's thresholding method was also unsuccessful in effectively segmenting cracks when the images were taken under non-uniform lighting conditions.

DINO for ViT We used the ViT-Small model trained in a fully unsupervised fashion using fully self-supervised DINO [3] which generates self-attention maps for the input images. The generated maps were found to be very noisy, but the self-attention maps captured the branched cracks very successfully due to the global self-attention mechanism of Vision Transformers. Unlike Vision Transformers, CNN-based algorithms lack the global field of view and are dataset antagonistic due to weight sharing, due to which CNN-based algorithms underperform on branched and webbed cracks. The global attention maps generated by transformers via self-supervision will be further utilized in Sect. 4.4 to generate priors for CNN-based segmentation algorithms to mitigate the problems mentioned above.

Table 4. Model performance summary

Technique	Backbone	Trainable parameters in millions	F1 score	MIoU
Otsu	None	0	0.0609	0.3187
DINO	ViT-S	0	0.1568	0.5149
MaskRCNN	ResNet-50	44.178	0.4761	0.5213
SWIN transformer	U-Net	41.380	0.5009	0.6426
Pix2Pix	U-Net	54.421	0.5652	0.6666
Pix2Pix ✓	U-Net	54.422	0.5824	0.6953
Pix2Pix	FPN-ResNet-50	26.911	0.3950	0.6286
Pix2Pix ✓	FPN-ResNet-50	26.914	0.5765	0.7142
DeepLab V3+	Xception *	54.700	0.6344	0.7208
DeepLab V3+✓	Xception *	54.700	0.6608	0.7353
DeepLab V3+	ResNet-101 *	59.339	0.7060	0.7599
DeepLab V3+✓	ResNet-101 *	59.342	0.7238	0.7712

* - pre-training on ImageNet. ✓ - DINO features used as prior for training

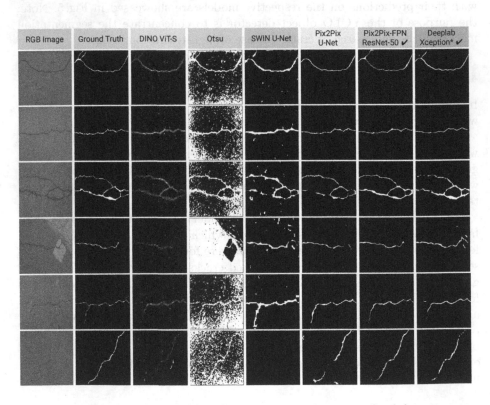

Fig. 5. Benchmarked model performance comparisons on refined dataset

4.3   Supervised Techniques

The unsupervised techniques gave noisy results and failed to capture the domain essence. Therefore, these techniques could not be used directly to infer results and merely act as priors for supervised techniques. In an attempt to improve the results, we conducted experiments using supervised techniques. Supervised techniques involve the utilization of segmented ground truths and they are more robust than unsupervised methods for this reason. We used four supervised Deep Learning algorithms for Image segmentation: MaskRCNN, DeepLab, Pix2Pix and Swin-Unet.

We trained **MaskRCNN** model using a ResNet-50 [16] backbone. The results obtained are significantly better than that of unsupervised techniques. However, we noticed two major issues with Mask-RCNN, the model does not generalize well to the thickness of the cracks despite the diversity in the dataset and is very susceptible to background noise such as debris and shadows, which led to a lot of false positives.

Swin-Unet [2] model was used on our dataset to measure the performance of transformer-based backbones trained end to end in a supervised fashion. The Swin Transformer blocks can model the long-range dependencies of different parts in a crack image from a local and global perspective, which leads to superior performance in the detection of branched and webbed crack.

We conducted two different experiments with **Pix2Pix** model, using Vanilla U-Net with skip connections and Feature Pyramidal Network with a resnet-50 backbone as the generators. The Patch GAN discriminator used in the Pix2Pix model improves the quality of predictions significantly. The models are able to detect finer cracks and the boundaries of cracks with higher precision. The experiments also highlight that Feature pyramidal networks are superior for segmentation tasks than U-Nets, as they achieve higher accuracy with less than half the number of parameters of U-Nets.

DeepLab V3+ using ResNet-101 and Xception [7] backbones pre-trained on imagenet as encoders. DeepLab is able to detect very fine-grained cracks even in adverse lighting conditions, the predicted masks generalize well to the thickness of the cracks, and in some cases, the model was able to detect cracks that were not annotated in the ground truth. The qualitative and quantitative results indicate DeepLab as the best model among our experiments.

4.4   Using Semi-supervised DINO Features as Prior for Segmentation

We conducted this experiment to verify our hypothesis that using semi-supervised DINO features prior to CNN-based image segmentation models leads to superior results on branched and webbed cracks. This is because global and local self-attention from transformers helps in establishing long-range dependencies in images and CNNs help in learning dataset-specific localized features. For this experiment, we concatenated the semi-supervised self-attention map generated from DINO as the fourth channel to our input image tensor. We conducted

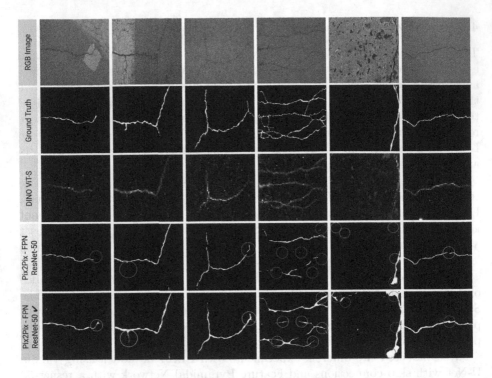

Fig. 6. Transformation in predictions when features are integrated

this experiment with all our CNN-based models and observed the following: FPN Pix2Pix(14% mIoU increase), U-Net Pix2Pix(4% mIoU increase), Resnet-101 DeepLab (2% mIoU increase) and Xception DeepLab (1% mIoU increase). To highlight the improvement in performance due to features, please refer Fig. 6. While integrating features definitely showed significant improvement in the predictions, it did have a couple of issues. Sometimes, the prediction was affected by the presence of noise in the background, resulting in an erroneous prediction, and at other times, the predicted crack turned out to be thicker than the actual, resulting in the loss of some fine details. These shortcomings have been highlighted in the last 2 columns of Fig. 6.

These experiments demonstrate that adding prior information to the input regarding the localization of cracks and scaling the network compute enough to process the additional information leads to a subsequent increase in model performance.

4.5   Comparative Analysis of Results

We performed a systematic quantitative (Table 5) and qualitative (Fig. 7) analysis of existing SOTA models for crack segmentation in the literature and compared their performances to the proposed models. All the models were evaluated

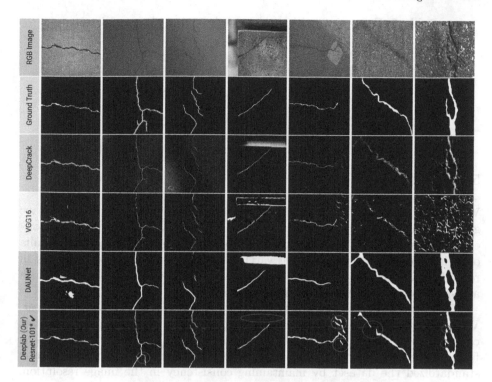

Fig. 7. Comparison with state-of-the-art models

on the test split of the refined dataset. We observed that DeepCrack scales well to different sizes of cracks due to the aggregation of multi-scale feature maps, but it does not accurately adapt to the varying thickness of cracks and is prone to a lot of noise in the output. DAU-Net and VGG16 do not perform well on untrained datasets; this highlights their inability to generalize across different crack surfaces and illumination conditions, which is largely attributed to the lack of diversity in their training data. However, our model is not perfect. As seen in the diagram, the bad predictions are marked with red circles. We observed that our model sometimes tended to have predictions with small discontinuities and gaps, resulting in the loss of crack information.

Analysis of results on different subsections of our dataset is presented in Table 1 and Table 3; we can observe that even though our dataset contains diverse crack images, the benchmarked models perform better for sub-datasets Volker, DeepCrack and Crack500 because they contain a larger number of images. Masonry and Ceramic datasets had very different background textures from the other images, which caused a downgrade in the model's performance. Analysis of the performance of benchmarked models based on the spatial types of cracks highlighted that, on average Linear cracks had the best predictions, followed by branched cracks, and the webbed cracks had the lowest quality of predictions.

Table 5. Comparison with previous works evaluated on our dataset

Technique	F1 score	MIoU
VGG16 [49]	0.3009	0.5444
DeepCrack [28]	0.3404	0.5813
DAU-Net [34]	0.5495	0.6857
DeepLab ✓ ResNet * (ours)	0.7238	0.7712

5   Conclusion

Within the past decade, Computer Vision techniques for crack detection have gained significant momentum and led to stimulating works and datasets. However, most works focus on a particular dataset and do not benchmark the results on other datasets, all while using varying metrics. This leads to problems in reproducibility and/or limitation in application to different types of cracks such as asphalt, concrete, masonry, ceramic and glass.

This paper tries to address the aforementioned vital issues that are pertinent. We do so by compiling various datasets and putting together a robust dataset consisting of 9000+ images with varying backgrounds, types of cracks and surfaces, and ground truth annotations. Further, we refined existing datasets and standardized the dataset by maintaining consistency in the image resolution. We further divide the dataset on the basis of the spatial arrangement of cracks into linear, branched and webbed. Finally, we benchmarked the state-of-the-art techniques on our final dataset.

Our results show that DeepLab with ResNet and XceptionNet as backbone perform the best. We further observe that the models perform best on images with linear cracks, and predictions' quality decreases for webbed and branched cracks. The paper highlights the advantages of using self-supervised attention in modelling long-range dependencies in vision tasks and presents a method to fuse semi-supervised attention feature maps with CNNs for enhanced crack detection. Adding the DINO features as prior results in significant improvements in the predictions with a marginal tradeoff in the number of trainable parameters.

Despite our best efforts in compiling a dataset containing cracks in different domains, the models trained by us have not been able to identify cracks well in the case of transparent backgrounds like glass. We believe in the need for more images belonging to such backgrounds so that models trained on them shall be more domain invariant. Further exploration into fusing semi-supervised and supervised training paradigms and better amalgamation of transformer and CNN architectures can be explored for better performance on images with branched and webbed cracks.

References

1. Akagic, A., Buza, E., Omanovic, S., Karabegovic, A.: Pavement crack detection using Otsu thresholding for image segmentation. In: 2018 41st International Convention on Information and Communication Technology, Electronics and Microelectronics (MIPRO), pp. 1092–1097. IEEE (2018)
2. Cao, H., et al.: Swin-Unet: Unet-like pure transformer for medical image segmentation (2021). https://doi.org/10.48550/ARXIV.2105.05537, https://arxiv.org/abs/2105.05537
3. Caron, M., et al.: Emerging properties in self-supervised vision transformers (2021). https://doi.org/10.48550/ARXIV.2104.14294, https://arxiv.org/abs/2104.14294
4. Chambo, S.: Aiglern. https://www.irit.fr/Sylvie.Chambon/Crack/Detection/Database.htm
5. Chen, L.C., Papandreou, G., Kokkinos, I., Murphy, K., Yuille, A.L.: DeepLab: semantic image segmentation with deep convolutional nets, atrous convolution, and fully connected CRFs (2016). https://doi.org/10.48550/ARXIV.1606.00915, https://arxiv.org/abs/1606.00915
6. Choi, W., Cha, Y.J.: SDDnet: real-time crack segmentation. IEEE Trans. Industr. Electron. **67**(9), 8016–8025 (2019)
7. Chollet, F.: Xception: deep learning with depthwise separable convolutions (2016). https://doi.org/10.48550/ARXIV.1610.02357, https://arxiv.org/abs/1610.02357
8. Dais, D., İhsan Engin Bal, Smyrou, E., Sarhosis, V.: Automatic crack classification and segmentation on masonry surfaces using convolutional neural networks and transfer learning. Autom. Constr. **125**, 103606 (2021). https://doi.org/10.1016/j.autcon.2021.103606, https://www.sciencedirect.com/science/article/pii/S0926580521000571
9. Dorafshan, S., Thomas, R., Maguire, M.: SDnet 2018: an annotated image dataset for non-contact concrete crack detection using deep convolutional neural networks. Data Brief **21** (2018). https://doi.org/10.1016/j.dib.2018.11.015
10. Eisenbach, M., et al.: How to get pavement distress detection ready for deep learning? A systematic approach. In: International Joint Conference on Neural Networks (IJCNN), pp. 2039–2047 (2017)
11. Fan, R., et al.: Road crack detection using deep convolutional neural network and adaptive thresholding (2019)
12. Flah, M., Suleiman, A.R., Nehdi, M.L.: Classification and quantification of cracks in concrete structures using deep learning image-based techniques. Cement Concrete Compos. **114**, 103781 (2020). https://doi.org/10.1016/j.cemconcomp.2020.103781, https://www.sciencedirect.com/science/article/pii/S0958946520302870
13. Gao, Z., Peng, B., Li, T., Gou, C.: Generative adversarial networks for road crack image segmentation. In: 2019 International Joint Conference on Neural Networks (IJCNN), pp. 1–8. IEEE (2019)
14. Goodfellow, I.J., et al.: Generative adversarial networks (2014). https://doi.org/10.48550/ARXIV.1406.2661, https://arxiv.org/abs/1406.2661
15. Ham, S., Bae, S., Kim, H., Lee, I., Lee, G.-P., Kim, D.: Training a semantic segmentation model for cracks in the concrete lining of tunnel. J. Korean Tunnel. Underground Space Assoc. **23**(6), 549–558 (2021)
16. He, K., Zhang, X., Ren, S., Sun, J.: Deep residual learning for image recognition. https://doi.org/10.48550/ARXIV.1512.03385, https://arxiv.org/abs/1512.03385
17. Hoang, N.D.: Detection of surface crack in building structure using image processing technique with an improved Otsu method for image thresholding. Adv. Civil Eng. **2018** (2018). https://doi.org/10.1155/2018/3924120

18. Isola, P., Zhu, J.Y., Zhou, T., Efros, A.A.: Image-to-image translation with conditional adversarial networks (2018)
19. Junior, G.S., Ferreira, J., Millán-Arias, C., Daniel, R., Junior, A.C., Fernandes, B.J.T.: Ceramic cracks segmentation with deep learning. Appl. Sci. **11**(13) (2021). https://doi.org/10.3390/app11136017, https://www.mdpi.com/2076-3417/11/13/6017
20. Kalfarisi, R., Wu, Z., Soh, K.: Crack detection and segmentation using deep learning with 3D reality mesh model for quantitative assessment and integrated visualization. J. Comput. Civ. Eng. **34**, 04020010 (2020)
21. Khalesi, S., Ahmadi, A.: Automatic road crack detection and classification using image processing techniques, machine learning and integrated models in urban areas: a novel image binarization technique (2020)
22. Kheradmandi, N., Mehranfar, V.: A critical review and comparative study on image segmentation-based techniques for pavement crack detection. Constr. Build. Mater. **321**, 126162 (2022). https://doi.org/10.1016/j.conbuildmat.2021.126162, https://www.sciencedirect.com/science/article/pii/S0950061821038940
23. Kim, B., Yuvaraj, N., Sri Preethaa, K., Arun Pandian, R.: Surface crack detection using deep learning with shallow CNN architecture for enhanced computation. Neural Comput. Appl. **33**(15), 9289–9305 (2021)
24. König, J., Jenkins, M.D., Mannion, M., Barrie, P., Morison, G.: Optimized deep encoder-decoder methods for crack segmentation. Digit. Sig. Process. **108**, 102907 (2021). https://doi.org/10.1016/j.dsp.2020.102907, https://doi.org/10.1016/j.dsp.2020.102907
25. Lau, S.L., Chong, E.K., Yang, X., Wang, X.: Automated pavement crack segmentation using U-net-based convolutional neural network. IEEE Access **8**, 114892–114899 (2020)
26. Lee, D., Kim, J., Lee, D.: Robust concrete crack detection using deep learning-based semantic segmentation. Int. J. Aeronaut. Space Sci. **20**(1), 287–299 (2019)
27. Li, S., et al.: Detection of concealed cracks from ground penetrating radar images based on deep learning algorithm. Constr. Build. Mater. **273**, 121949 (2021)
28. Liu, Y., Yao, J., Lu, X., Xie, R., Li, L.: DeepCrack: a deep hierarchical feature learning architecture for crack segmentation. Neurocomputing **338**, 139–153 (2019)
29. Mahler, D.S., Kharoufa, Z.B., Wong, E.K., Shaw, L.G.: Pavement distress analysis using image processing techniques. Comput.-Aided Civil Infrastr. Eng. **6**(1), 1–14 (1991)
30. Mandal, V., Uong, L., Adu-Gyamfi, Y.: Automated road crack detection using deep convolutional neural networks. In: 2018 IEEE International Conference on Big Data (Big Data), pp. 5212–5215. IEEE (2018)
31. Oliveira, H., Correia, P.L.: Road surface crack detection: improved segmentation with pixel-based refinement. In: 2017 25th European Signal Processing Conference (EUSIPCO), pp. 2026–2030. IEEE (2017)
32. Pak, M., Kim, S.: Crack detection using fully convolutional network in wall-climbing robot. In: Park, J.J., Fong, S.J., Pan, Y., Sung, Y. (eds.) Advances in Computer Science and Ubiquitous Computing. LNEE, vol. 715, pp. 267–272. Springer, Singapore (2021). https://doi.org/10.1007/978-981-15-9343-7_36
33. Park, S.E., Eem, S.H., Jeon, H.: Concrete crack detection and quantification using deep learning and structured light. Constr. Build. Mater. **252**, 119096 (2020)
34. Polovnikov, V., Alekseev, D., Vinogradov, I., Lashkia, G.V.: Daunet: deep augmented neural network for pavement crack segmentation. IEEE Access **9**, 125714–125723 (2021). https://doi.org/10.1109/ACCESS.2021.3111223

35. Redmon, J., Divvala, S., Girshick, R., Farhadi, A.: You only look once: unified, real-time object detection. In: 2016 IEEE Conference on Computer Vision and Pattern Recognition (CVPR), pp. 779–788 (2016). https://doi.org/10.1109/CVPR.2016.91
36. Rezaie, A., Achanta, R., Godio, M., Beyer, K.: Comparison of crack segmentation using digital image correlation measurements and deep learning. Constr. Build. Mater. **261**, 120474 (2020). https://doi.org/10.1016/j.conbuildmat.2020.120474, https://www.sciencedirect.com/science/article/pii/S095006182032479X
37. Ronneberger, O., Fischer, P., Brox, T.: U-net: convolutional networks for biomedical image segmentation (2015). https://doi.org/10.48550/ARXIV.1505.04597, https://arxiv.org/abs/1505.04597
38. Sharma, S., Balakrishnan, D., Kulkarni, S., Singh, S., Devunuri, S., Korlapati, S.C.R.: Crackseg9k: a collection of crack segmentation datasets (2022). https://doi.org/10.7910/DVN/EGIEBY
39. Shi, Y., Cui, L., Qi, Z., Meng, F., Chen, Z.: Automatic road crack detection using random structured forests. IEEE Trans. Intell. Transp. Syst. **17**(12), 3434–3445 (2016)
40. Shim, S., Kim, J., Cho, G.C., Lee, S.W.: Multiscale and adversarial learning-based semi-supervised semantic segmentation approach for crack detection in concrete structures. IEEE Access **8**, 170939–170950 (2020)
41. Silva, W.R.L.d., Lucena, D.S.d.: Concrete cracks detection based on deep learning image classification. In: Proceedings, vol. 2, p. 489. MDPI AG (2018)
42. Silva, W.R.L.d., Lucena, D.S.d.: Concrete cracks detection based on deep learning image classification. Proceedings **2**(8) (2018). https://doi.org/10.3390/ICEM18-05387, https://www.mdpi.com/2504-3900/2/8/489
43. Sonka, M., Hlavac, V., Boyle, R.: Segmentation, Ch. 5 of Image Processing. Analysis and Machine Vision, pp. 158–163. PWS Publishing (1999)
44. Sun, X., Xie, Y., Jiang, L., Cao, Y., Liu, B.: DMA-net: deeplab with multi-scale attention for pavement crack segmentation. IEEE Trans. Intell. Transp. Syste. 1–12 (2022). https://doi.org/10.1109/TITS.2022.3158670
45. Talab, A.M.A., Huang, Z., Xi, F., HaiMing, L.: Detection crack in image using Otsu method and multiple filtering in image processing techniques. Optik **127**(3), 1030–1033 (2016)
46. Yamane, T., Chun, P.J.: Crack detection from a concrete surface image based on semantic segmentation using deep learning. J. Adv. Concrete Technol. **18**(9), 493–504 (2020)
47. Yang, F., Zhang, L., Yu, S., Prokhorov, D.V., Mei, X., Ling, H.: Feature pyramid and hierarchical boosting network for pavement crack detection. IEEE Trans. Intell. Transp. Syst. **21**, 1525–1535 (2020)
48. Yang, F., Zhang, L., Yu, S., Prokhorov, D., Mei, X., Ling, H.: Feature pyramid and hierarchical boosting network for pavement crack detection (2019)
49. Zhang, L., Shen, J., Zhu, B.: A research on an improved Unet-based concrete crack detection algorithm. Struct. Health Monit. **20**(4), 1864–1879 (2021). https://doi.org/10.1177/1475921720940068, https://doi.org/10.1177/1475921720940068
50. Zhang, X., Rajan, D., Story, B.: Concrete crack detection using context-aware deep semantic segmentation network. Comput.-Aided Civil Infrastr. Eng. **34**(11), 951–971 (2019)
51. Zou, Q., Cao, Y., Li, Q., Mao, Q., Wang, S.: Cracktree: automatic crack detection from pavement images. Pattern Recogn. Lett. **33**(3), 227–238 (2012)
52. Özgenel, C.F., Sorguc, A.: Performance comparison of pretrained convolutional neural networks on crack detection in buildings (2018). https://doi.org/10.22260/ISARC2018/0094

PriSeg: IFC-Supported Primitive Instance Geometry Segmentation with Unsupervised Clustering

Zhiqi Hu$^{(\boxtimes)}$ and Ioannis Brilakis

Construction Information Technology Laboratory, Department of Engineering,
University of Cambridge, Cambridge, UK
{zh334,ib340}@cam.ac.uk

Abstract. One of the societal problems for current building construction projects is the lack of timely progress monitoring and quality control, causing over-budget costs, inefficient productivity, and poor performance. This paper addresses the challenge of high-accuracy primitive instance segmentation from point clouds with the support of IFC model as a core stage to facilitate maintaining a geometric digital twin during the construction stage. Keeping the geometry of a building digital twin dynamic and up-to-date will help monitor and control the progress and quality timely and efficiently. We propose a novel automatic method named *Priseg* to detect and segment the entire points corresponding to the as-designed instance by developing an IFC-based instance descriptor and unsupervised clustering algorithm. The proposed solution is robust in real complex environments, such as point clouds are noisy with high occlusions and clutter, the as-built status deviates from the as-designed model in terms of position, orientation, and scale.

Keywords: Digital Twin · Building Information Modelling · Point Cloud Data · Geometry · IFC Model · Design Intent · Instance Segmentation · Scan-vs-BIM · Maintenance of Geometric Digital Twin

1 Introduction

This research is about IFC-supported primitive object instance geometry segmentation with unsupervised clustering from Point Cloud Data (PCD). The output segmented data can be used to facilitate maintaining a building geometric Digital Twin (DT). By IFC, we refer here to Industry Foundation Classes that serve as a standardised digital description of the buildings. We use IFC as the supporting information in our research but do not entirely depend on it. By primitive, we refer here to three-dimensional (3D) basic constructive solid geometry including cuboid, cylinder, sphere, cone, pyramid, and torus [1]. By object instances, we refer here to objects from structural (e.g., walls, floors, ceilings, etc.), mechanical (e.g., piping segments, piping joint, ducts, etc.), and electrical (e.g., light fixtures, etc.) categories with different class labels and ID numbers

© The Author(s), under exclusive license to Springer Nature Switzerland AG 2023
L. Karlinsky et al. (Eds.): ECCV 2022 Workshops, LNCS 13807, pp. 196–211, 2023.
https://doi.org/10.1007/978-3-031-25082-8_13

[11]. By geometry, we refer here to 3D shapes of objects and their relationships. By segmentation from PCD, we refer here to detecting building object instances in the PCD and extracting all points corresponding to the instances at a given timestamp during the construction stage. By maintaining, we refer here to keeping DT's geometry dynamic and up-to-date based on the Design Intent (DI) (e.g., a final as-designed IFC model) to reflect the as-is status of a building. A DI serves as a benchmark for evaluating the construction outcomes. By a building DT, we refer here to a product and process information repository for storing and sharing physical and functional properties of a building over time with all Architectural, Engineering, Construction, and Operation (AECO) stakeholders throughout its lifecycle [20]. A DT is different from a Building Information Model (BIM) since a BIM only provides product information and can be updated at different timestamps of a DT's lifecycle.

One of the societal problems for current building construction projects is the poor timely progress monitoring and quality control. Construction companies often struggle with delivering large projects on budget and completing them with good performance on time [11]. The construction sector remains one of the least digitised sectors compared with media, finance, etc. It is significant to digitise and automate the process of design, construction, operation, and renovation of buildings to improve the projects' productivity and performance with the help of DTs. For example, a DT can help a specialist do clash detection between structural and mechanical components, simulate energy consumption in a building at the design stage, monitor the progress at the construction stage, and visualise the maintenance of a building during the operation stage. However, there are some limitations at present in using a DT.

The current state of practice in using a DT to facilitate construction progress monitoring and quality control still lacks automation, resulting in long-time feedback. Specifically, workers refer to a final client-approved IFC model as the DI to manually monitor the progress and quality of building components. Two commercial solutions named OpenSpace and Buildots provide automatic 2D image comparison methods to support progress monitoring. A worker needs to wear a helmet mounted by a 360° camera and walk through the construction site to capture a video. This image data stream is then compared with the DI to update the progress situation. However, these solutions only offer visual inspections but cannot update geometric information in 3D view. Also, it still requires manual effort to conduct quality control.

This research is designed for automatically maintaining a primitive geometric building DT on a high-accuracy level at different timestamps with complex environments during the construction stage. It is worth keeping the primitive geometry of DT up-to-date since the common building objects are designed and constructed with primitive shapes including cuboids and cylinders. The updated primitive geometry information of each object can support timely instance progress monitoring and quality control. In this paper, the authors propose a novel robust approach to automatically segment building object instances with primitive shapes (e.g., cuboid, cylinder, etc.) from noisy, occluded PCD with the reference of DI (e.g., as-designed IFC model) in a complex environment. High-accuracy object instance segmentation is a crucial step before generating and assigning meshes of as-built object instances to a DT to keep it updated. In particular, this paper presents the following contributions:

1. While most of the current methods only detect object instances in the Scan-vs-BIM system for progress monitoring, the proposed approach here can not only detect but also completely segment instances from PCD to provide the entire points corresponding to the instances. The extracted complete points can be used to generate solid model to support controlling construction quality.
2. Unlike the most current Scan-vs-BIM methods that entirely depend on the DI to detect or segment object instances from PCD, our proposed approach can be used in more complex and real environments where there are distinct deviations in terms of the position, orientation, and scale between the DI geometry and the as-built instances. Our approach takes advantage of the DI but does not fully depend on it, which makes the approach more robust and executable in the real environment to support progress monitoring and quality control.
3. Unlike the most current methods that are only reliable on the synthetic or simple dataset which is clean and complete, our proposed approach is robust on the real noisy PCD with high occlusions and clutter. For example, temporary building material or workers in front of the wall may be scanned into the PCD as noisy points to occlude that wall; using current methods may result in partial or irrelevant point extraction which cannot reflect the as-built geometry entirely and precisely.

The rest of this paper is organised as follows: background including the state of research is reviewed in Sect. 2; the proposed pipeline is introduced in Sect. 3; the implementation and experiments are shown in Sect. 4; discussion and conclusions are delivered in Sects. 5 and 6.

2   Background

Maintaining a building geometric DT requires three main stages [11]: 1) DI geometry to PCD registration ensures that the DI model (e.g., IFC model) is aligned with the as-built PCD into a same coordinate system. 2) PCD-vs-DI object instance detection and segmentation aims to detect and segment instances from the PCD with the help of the DI. 3) 3D representation from extracted PCD converts the points into information-rich 3D formats to finally support progress monitoring and quality control. We here focus on the second stage and discuss the detailed state of research below before summarising knowledge gaps and research objectives.

2.1   PCD-vs-DI Object Instance Detection and Segmentation

With regard to PCD-vs-DI object instance detection and segmentation, we summarised five method classes from recent papers: point-to-point comparison, point-to-surface comparison, feature-based method, Hough transform, and

RANSAC. The first three classes focus on instance detection while the last two classes deal with instance segmentation. The most salient object types detected and segmented are slabs, walls, columns, beams, cylindrical piping segments, and cuboid duct segments.

Point-to-point comparison was first proposed by [7] to detect points corresponding to DI instances in the as-built PCD. The authors calculated the ratio of the number of retrieved as-designed points to the total number of as-designed points to assess retrieval performance. This method has then been adopted to detect primary and temporary structural objects [23–25] and mechanical objects [6] for progress monitoring at the construction stage. The methods are robust in tracking structural instances but failed to detect mechanical instances with distinct spatial deviations against the DI. Point-to-surface comparison computes the ratio of the overlapping area between PCD and the DI [9,26]. In [22], the authors developed the surface coverage ratio calculation algorithm by using an alpha shape reconstructed from the orthogonal projection of points to make the method more robust. In [3,4,18], the authors used Euclidean distance to determine the nearest point to the DI geometry surface for instance detection. Feature-based methods use object features (e.g., position, size, normal, continuity) to detect instances. A three-feature (Lalonde, orientation, continuity) instance identification approach has been developed to measure building progress [14]. This strategy is robust in noisy environments but presupposes all object instances are DI-compliant. In [12], the authors use five features (length, size, colour, orientation, the number of connections) to detect clutter-free, prefabricated pipe settings. In [9], the authors computed the probability distribution of PCD and DI file geometric attributes to match objects. All suggested approaches for object instance detection cannot extract all instance points when PCD and DI geometry differ or when the PCD is cluttered.

Hough transform performs well in shape detection in a complex environment with noise. In [2], the authors projected 3D point slices along with the estimated object's normal orientation from the DI geometry before determining the 2D circle slices by Hough transform. However, this method requires the as-built position and dimension of the object instance are the same as the DI geometry. In [5], the authors combined the point-to-point comparison and Hough transform together to segment cylindrical MEP (Mechanical, Electrical and Plumbing) components. It overcomes the limitation from [6] but the computation is complex. It assumes that most cylindrical instances are built in the orthogonal direction, which leads to the methods with less robustness in complex environments. On the other hand, RANSAC is more resilient than all methods discussed above for segmenting instances from the as-built PCD. It was employed with the normal-based region growing method and K-Nearest Neighbours (KNN) to segment cylindrical pipe segments [17]. In [10,19], the authors used RANSAC and its variant (MLESAC) to segment cuboid-shaped instances, cylinder-shaped pipes, and cast-in-place footing. RANSAC-based algorithms can only recognise primitive-shape objects like cylinders, cuboids, and floors (planes).

2.2   Gaps in Knowledge and Research Objectives

We summarise the gaps in knowledge about instance detection and segmentation in the PCD-vs-DI system (also known as the Scan-vs-BIM system) as follows:

1. We do not yet know how to segment the entire valuable point cluster without missing valid points but discard any noisy points in the environment with high clutter and occlusions. For example, only a part of an object instance's surface is visible and captured in the PCD; some building materials (e.g., temporary objects in structural concrete work) and workers in front of the objects can also cause occlusions and clutter during scanning. The PCD captured in such complex environments cannot fully reflect the status of instances and thus may cause difficulties in instance segmentation and geometry reconstruction with high-accuracy.
2. We do not yet know how to segment the entire point cluster when there are distinct deviations in terms of the position, orientation, and scale between the DI geometry and the as-built object instances. Most current methods cannot guarantee to extract the complete and all relevant points corresponding to the object instance in the PCD-vs-DI environment.
3. We do not yet know how to detect and segment the entire point cluster for building object instances with non-primitive shapes, such as cross piping joints, sprinklers, terminals, and light fixtures. Most current methods are only workable for cuboid and cylindrical shapes.
4. We do not yet know how to detect object instances in the PCD when they are transparent, such as glass-made windows. To the best of our knowledge, the existing methods can only process opaque objects.

This paper proposes a novel automatic IFC-supported high-accuracy primitive instance geometry segmentation method named *PriSeg*. This method solves the research problems in the first and second gaps in knowledge and can be potentially adapted to deal with the third gap. The proposed solution will be introduced in the next section.

3   Proposed Solution

3.1   Scope and Overview

The geometry of a building DT needs to be updated regularly and iteratively during the construction stage to support progress monitoring and quality control. Figure 1 shows one loop for maintaining DT geometry. On the left side, object instances are detected and segmented from the PCD reflecting the as-built status with the help of DI (e.g., an IFC model). Consequently, the extracted PCD clusters are then labelled with their object types (e.g., columns, slabs). Finally, the labelled PCD clusters are tessellated and assigned to the IFC model as an updated timestamped geometry description. On the right side, the labelled PCD clusters can be used to facilitate progress monitoring and quality control.

By comparing with DI, the discrepancies between the actual and the designed instances are checked according to the design specifications. The result will be recorded in the IFC model for further decision making. An equivalent workflow is conducted at any follow-up timestamps to update the DT geometry dynamically with various timestamped tessellations. We propose a novel automatic method based on the instance descriptor and unsupervised clustering to automatically segment all relevant points corresponding to the designated instance fast and precisely.

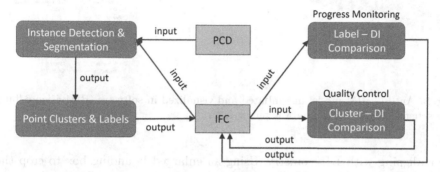

Fig. 1. DT geometry maintenance from the as-built PCD with the support of the DI IFC model to facilitate progress monitoring and quality control during the building construction stage: An overview

3.2   IFC-Based Instance Descriptor

IFC is an open and standardised schema intended to enable interoperability among BIM software applications in the AECO industry [15]. It has been widely used to generate as-designed BIM models for building construction and maintenance. Figure 2 instantiates and visualizes an IFC model [13] in the software BIMvision. Many applications including quantity take-off, model code compliance checking, and energy simulations can be done by IFC models. An IFC model follows a top-down hierarchy to express the properties of a building's structural, mechanical, and electrical objects. It contains an object's ID label (*identifier GUID*), dimension (*IfcBoundingBox*), location (*IfcObjectPlacement*), material properties (*IfcMaterial*), connection relationships (*IfcRelConnectsPathElements*), etc., which can be extracted and employed for PCD-vs-DI object instance detection and segmentation.

The current object instance descriptor only records the pose information and the dimension of the primitive shape "cuboid" [16, 19]. We propose a novel framework to generalise and standardise object instance descriptors based on the IFC schema. It contains pose, shape, and relation information to encode multiple object instance variables. The connection relation refers to walls connecting with slabs and other walls, or columns connecting with slabs. The whole as-built PCD needs to be cropped by an enlarged bounding box generated from the descriptor

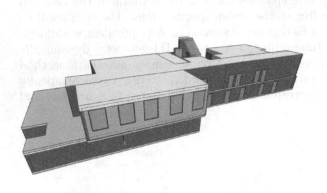

Fig. 2. An IFC BIM model instantiated and visualized in software (BIMvision 64bit). File schema: IFC2×3. The data is public from [13]

after aligning with DI geometry. Using an enlarged bounding box to crop the whole PCD can reduce the size of the input data into a small-scale cluster to decrease the computational complexity. Meanwhile, the bounding box is enlarged to make the method robust to deal with the situation when there are distinct deviations in terms of position, orientation, and scale between the DI and the as-built.

3.3   Unsupervised Clustering

Progressive Sample Consensus (PROSAC) was first proposed to exploit the linear ordering defined on the set of correspondences by a similarity function used in establishing tentative correspondences [8]. Random samples are initially added from the most certain matches to significantly speed up the outcome performance. PROSAC is adapted to extract the points representing primitive shapes such as planes and cylinders in the cropped PCD. The shape "plane" is used for cuboid detection (e.g., walls, slabs) while the shape "cylinder" is used for cylinder detection (e.g., columns). It should be noticed that all shapes defined here are infinite. For example, the definition of a plane is:

$$ax + by + cz + d = 0 \tag{1}$$

The plane defined here is infinite. Thus, the PROSAC-based primitive shape detection can only find the points representing that required shape in the enlarged bounding box but cannot distinguish which point belongs to the object instance and which point just belongs to the shape (e.g. noisy points). Therefore, we need to optimise the cluster to remove the noisy points.

Density-based Spatial Clustering of Application with Noise (DBSCAN) is an unsupervised data clustering algorithm [21]. It employs a simple minimum density estimation based on a threshold for the number of neighbours, $minPts$, within a radius ϵ to measure an arbitrary distance. Basically, each point in the PCD is classified as the core point, the border point, or the noisy point. Core points are objects that have more than $minPts$ neighbours within radius ϵ. If any of these neighbours becomes a core point again, their communities are included transitively. Non-core points in this set are known as border points (Fig. 3). DBSCAN does not partition the data based on the preset number of clusters, but extracts the dense clusters and remains sparse background as noise. The DBSCAN-based surface optimisation algorithm is therefore developed to distinguish the points belonging to the surfaces of an instance by removing outliers and minimising the noise. We adapt DBSCAN to cluster points into different clusters and employ the relationship descriptor to automatically rank and select clusters to extract final points corresponding to the object instance. Kd-tree is used here to accelerate the computation speed.

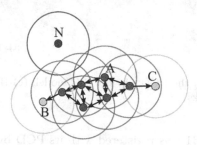

Fig. 3. Illustration of how DBSCAN works. Red dots (A): *core points*; yellow dots (B & C): *border points*; blue dot (N): *noise* [21] (Color figure online)

4 Implementation and Experiments

4.1 Data Acquisition and Pre-processing

ISPRS WG IV/5 Benchmark on Indoor Modelling is a public dataset [13]. It consists of six pairs of IFC files and PCDs captured by different sensors (i.e. Viametris iMS3D, Zeb-Revo, Leica C10, Zeb-1, Zeb Revo RT) in indoor environments of various complexities. We use the first dataset TUB1 to test the proposed solution for the proof of concept. The PCD and IFC model of TUB1 are shown in Fig. 4 (a) and (b). The PCD of TUB1 was captured by the Viametris iMS3D system (a mobile scanner with accuracy $\leq 3cm$). The indoor scene is comprised of 10 rooms on one floor which are enclosed by walls with different thicknesses, 23 open and closed doors, and 7 windows. The clutter in the PCD mostly corresponds to the presence of people and equipment during the survey.

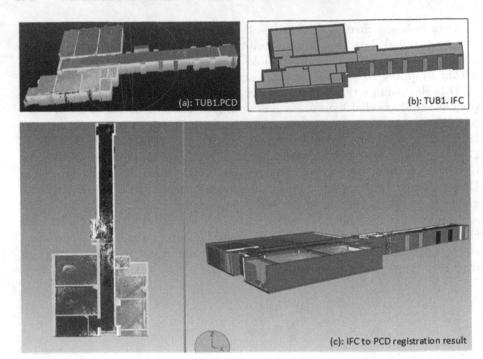

Fig. 4. TUB1 dataset: (a) the PCD; (b) the IFC model; (c) IFC to PCD registration, result shown in Navisworks

The IFC file in TUB1 was registered with its PCD by adjusting the origins and axes of two different coordinate systems into the same coordinate system via Revit and Recap (Processor: AMD Ryzen 5 5600X 6-Core Processor; RAM: 32 GB; GPU: AMD Radeon RX 6800) (Fig. 4 (c)). In the default setting of the standardised coordinate system, the origin is always set in the bottom left corner as (0,0,0) and the axes are aligned with the world truth north. A basic assumption is made here that the main walls of the building are vertical and the main floors are horizontal against the base plane of the ground.

4.2 Object Instance Segmentation

Four Types of Walls. The most common primitive object shapes in a typical building are cuboid and cylinder [11]. We, therefore, selected four types of walls based on the complexity in the TUB1 IFC file to conduct experiments to prove the feasibility and validity of the proposed solution (Fig. 5):

1. **Wall1**: This is a typical interior flat wall in the building without any furniture occlusions or other obstructions. The door built into the perpendicular wall was open during scanning so the PCD of one side of Wall 1 is partially occluded by the open door.

2. **Wall2**: This wall was built in a more complex environment than Wall 1 since it connects to another wall on the right side halfway along its main axis. The perpendicular wall also contains a door that was open during scanning. The PCD of Wall 2 has clutter and occlusions caused by the perpendicular wall and its door.
3. **Wall3**: This is a complex wall. The scale is larger, with many doors and perpendicular walls circled in yellow dots. All doors were open during scanning. The PCD of this wall has lots of occlusions and clutter.
4. **Wall4**: This is the most complicated wall among the four walls selected. The scale is the largest, with many doors and perpendicular walls. A Part of the wall is exterior. Some doors were open during scanning. The PCD of this wall has lots of occlusions and clutter.

Figure 6 illustrates an example of possible occlusions and clutter that occur in the PCD of walls, including tripods, small cabinets, and stuff on the floor.

Experimental Outcomes. The proposed pipeline was implemented in a software prototype written in C++ and Python. Figure 7 shows the results of the cropped PCD in the enlarged bounding box (a1 - a4), the final extracted points by our proposed method *Priseg* (b1 - b4), and the ground truth of instance segmentation (c1 - c4). The results of experiments were evaluated against the ground truth to verify the performance. We produced the ground truth by manually extracting the PCD corresponding to the four selected walls separately.

5   Discussion

Evaluation Metrics. We used point-to-point comparison to evaluate the experimental results of the proposed solution *Priseg* with the ground truth. Specifically, we first labelled each point in two datasets separately (the PCD from *Priseg* result and the PCD from ground truth) and then matched the corresponding points to calculate precision, recall, and IoU. This evaluation method is more accurate than surface-to-surface comparison since it computes the correspondence of points directly rather than transforming points to surfaces. The metrics are shown below:

Precision:

$$Precision = \frac{TP}{TP + FP} \tag{2}$$

where TP refers to true positive and FP refers to false positive.

Recall:

$$Recall = \frac{TP}{TP + FN} \tag{3}$$

where FN refers to false negative.

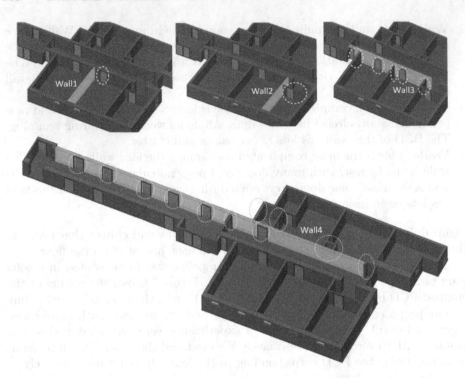

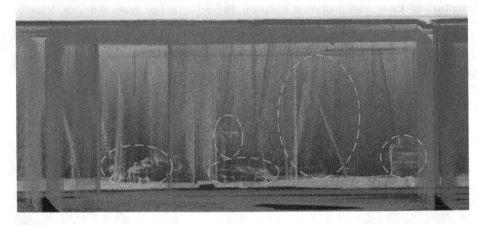

Fig. 5. The selected four types of walls (*in red*) to test the proposed solution. (*yellow dot circles*: highlight the instances that could cause clutter and occlusions during scanning) (Color figure online)

Fig. 6. An example of occlusions and clutter in the PCD of walls (circled in magenta) (Color figure online)

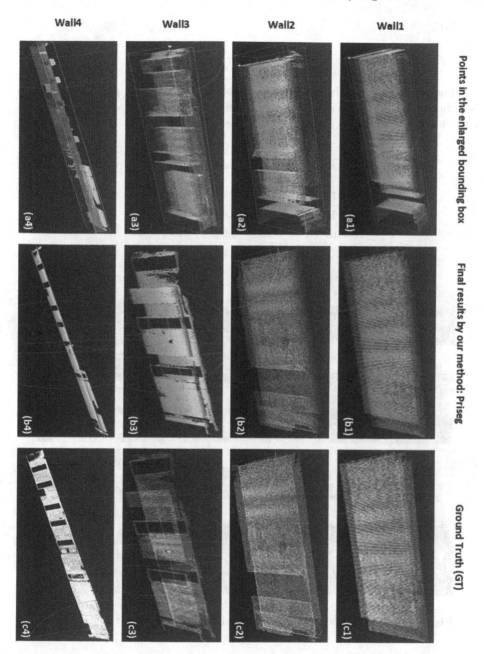

Fig. 7. Experimental results for four types of walls based on the complexity: (a1 – a4): points cropped in the enlarged bounding box; (b1 – b4): the results by our proposed algorithm *Priseg*; (c1 – c4): the ground truth

IoU:

$$IoU = \frac{TP}{TP + FP + FN} \tag{4}$$

The evaluation result is shown in Table 1. Overall, our proposed solution is robust with high precision and recall on primitive instance segmentation in complex environments. (precision >97%; recall >93%; IoU >91%).

Table 1. *Priseg* vs Ground Truth Evaluation: precision, recall, IoU

TUB1 Dataset	Wall1	Wall2	Wall3	Wall4
Precision	0.987	0.997	0.974	0.967
Recall	0.980	0.988	0.928	0.934
IoU	0.968	0.984	0.905	0.907

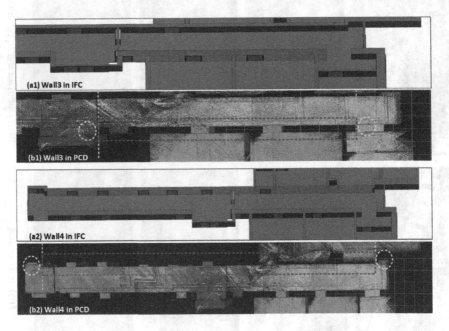

Fig. 8. Compare as-designed IFC model with as-built PCD for complex walls

Clutter, Occlusions, and Deviations. There are several factors that can influence the result of instance segmentation in the PCD-vs-DI environment. Figure 8 shows possible uncertainties resulting in discrepancies between the as-designed IFC model and as-built PCD (using Wall3 and Wall4 for example). Specifically, the clutter existing in yellow dot circles and magenta dot rectangles in the PCD may be difficult to fully remove and thus will cause false-positive segmentation results. The occlusions existing in magenta dot rectangles (e.g., doors, connected walls, windows. etc.) may result in less point retrieval nearby them, and thus will cause false-negative segmentation results. Lastly, the cyan dot lines show the edges of walls in the as-built PCD. Discrepancies in terms of position, orientation, and scale from the as-designed model may also result in false-negative or positive results of segmentation.

Applications. The proposed solution in this paper works fast, efficiently, and precisely for high-accuracy 3D primitive instance segmentation from PCD in a complex building environment. The segmented PCD can be used as the input to generate mesh for maintaining geometric DT. The timestamped meshes with the object's ID can also be employed to monitor the progress and control the quality during the building construction stage. For example, using a mesh to calculate the dimension of an as-built instance and assess it against as-designed values. Theoretically, our Algorithm *Priseg* is computationally simpler, more robust, and more accurate with higher average precision and recall than the current state of the art deep learning-based algorithms (e.g., MaskGroup with mPrevision 0.666, SoftGroup with mPrevision 0.753).

6   Conclusions

First, this paper discussed the current state of the art in detecting and segmenting point clusters from PCD in the PCD-vs-DI environment (also known as Scan-vs-BIM). The authors then proposed a novel solution named *"Priseg"* to automatically segment instances with high precision and recall from PCD with the support of DI. This IFC-supported instance geometry segmentation solution is robust when: 1) the input PCD has many occlusions and clutter; and 2) the as-built object instance has distinct deviations against the as-designed model in terms of position, orientation, and scale. The framework can also be theoretically extended to various object types with different shapes by applying Boolean operations.

In the future direction, we will focus on the generalisation of the object instance descriptor and the optimisation of the segmented point cluster to standardise the overall framework of geometric DT maintenance. We will compare our proposed solution with state-of-the-art deep learning algorithms on more datasets. Moreover, it is also worth improving the performance of registering complex PCD with its DI file, and optimising the gaps and truncation in the extracted point clusters to facilitate the geometry updating process.

Acknowledgments. This work is funded by European Commission's Horizon 2020 for CBIM (Cloud-based Building Information Modelling) European Training Network under agreement No.860555.

References

1. Requicha, A.A.G., Voelcker, H.: Constructive solid geometry. Production Automation Project, Univ. Rochester, Rochester, NY, Tech. Memo, vol. 25 (1977)

2. Ahmed, M.F., Haas, C.T., Haas, R.: Automatic detection of cylindrical objects in built facilities. J. Comput. Civ. Eng. **28**(3), 04014009 (2014)
3. Bariczová, G., Erdélyi, J., Honti, R., Tomek, L.: Wall structure geometry verification using TLS data and BIM model. Appl. Sci. **11**(24), 11804 (2021)
4. Bassier, M., Vermandere, J., De Winter, H.: Linked building data for construction site monitoring: a test case. In: ISPRS Annals of the Photogrammetry, Remote Sensing and Spatial Information Sciences vol. 2, pp. 159–165 (2022)
5. Bosché, F., Ahmed, M., Turkan, Y., Haas, C.T., Haas, R.: The value of integrating scan-to-BIM and scan-vs-BIM techniques for construction monitoring using laser scanning and BIM: The case of cylindrical MEP components. Autom. Constr. **49**, 201–213 (2015)
6. Bosché, F., Guillemet, A., Turkan, Y., Haas, C.T., Haas, R.: Tracking the built status of MEP works: Assessing the value of a scan-vs-BIM system. J. Comput. Civ. Eng. **28**(4), 05014004 (2014)
7. Bosche, F., Haas, C.T.: Automated retrieval of 3d cad model objects in construction range images. Autom. Constr. **17**(4), 499–512 (2008)
8. Chum, O., Matas, J.: Matching with PROSAC-progressive sample consensus. In: 2005 IEEE Computer Society Conference on Computer Vision and Pattern Recognition (CVPR'05), vol. 1, pp. 220–226. IEEE (2005)
9. Gao, T., Ergan, S., Akinci, B., Garrett, J.: Evaluation of different features for matching point clouds to building information models. J. Comput. Civ. Eng. **30**(1), 04014107 (2016)
10. Guo, J., Wang, Q., Park, J.H.: Geometric quality inspection of prefabricated MEP modules with 3d laser scanning. Autom. Constr. **111**, 103053 (2020)
11. Hu, Z., Fathy, Y., Brilakis, I.: Geometry updating for digital twins of buildings: a review to derive a new geometry-based object class hierarchy. In: Proceedings of the 2022 European Conference on Computing in Construction. Computing in Construction, vol. 3 (2022)
12. Kalasapudi, V.S., Turkan, Y., Tang, P.: Toward automated spatial change analysis of MEP components using 3d point clouds and as-designed BIM models. In: 2014 2nd International Conference on 3D Vision, vol. 2, pp. 145–152. IEEE (2014)
13. Khoshelham, K., Vilariño, L.D., Peter, M., Kang, Z., Acharya, D.: The ISPRS benchmark on indoor modelling. Int. Arch. Photogram. Remote Sens. Spat. Inf. Sci. **42**(2), W7 (2017)
14. Kim, C., Son, H., Kim, C.: Automated construction progress measurement using a 4d building information model and 3d data. Autom. Constr. **31**, 75–82 (2013)
15. Laakso, M., Kiviniemi, A., et al.: The IFC standard: a review of history, development, and standardization, information technology. ITcon **17**(9), 134–161 (2012)
16. Mansor, H., Shukor, S., Wong, R.: An overview of object detection from building point cloud data. In: Journal of Physics: Conference Series. vol. 1878, p. 012058. IOP Publishing (2021)
17. Nguyen, C.H.P., Choi, Y.: Comparison of point cloud data and 3d cad data for on-site dimensional inspection of industrial plant piping systems. Autom. Constr. **91**, 44–52 (2018)
18. Park, S., Ju, S., Yoon, S., Nguyen, M.H., Heo, J.: An efficient data structure approach for BIM-to-point-cloud change detection using modifiable nested octree. Autom. Constr. **132**, 103922 (2021)
19. Rausch, C., Haas, C.: Automated shape and pose updating of building information model elements from 3d point clouds. Autom. Constr. **124**, 103561 (2021)
20. Sacks, R., Brilakis, I., Pikas, E., Xie, H.S., Girolami, M.: Construction with digital twin information systems. Data-Centric Eng. **1** (2020)

21. Schubert, E., Sander, J., Ester, M., Kriegel, H.P., Xu, X.: DBSCAN revisited, revisited: why and how you should (still) use DBSCAN. ACM Trans. Database Syst. (TODS) **42**(3), 1–21 (2017)
22. Tran, H., Khoshelham, K.: Building change detection through comparison of a lidar scan with a building information model. In: International Archives of the Photogrammetry, Remote Sensing & Spatial Information Sciences (2019)
23. Turkan, Y., Bosche, F., Haas, C.T., Haas, R.: Automated progress tracking using 4d schedule and 3d sensing technologies. Autom. Constr. **22**, 414–421 (2012)
24. Turkan, Y., Bosché, F., Haas, C.T., Haas, R.: Toward automated earned value tracking using 3d imaging tools. J. Constr. Eng. Manag. **139**(4), 423–433 (2013)
25. Turkan, Y., Bosché, F., Haas, C.T., Haas, R.: Tracking of secondary and temporary objects in structural concrete work. Constr. Innov. **14**(2), 145–167 (2014)
26. Zhang, C., Arditi, D.: Automated progress control using laser scanning technology. Autom. Constr. **36**, 108–116 (2013)

Depth Contrast: Self-supervised Pretraining on 3DPM Images for Mining Material Classification

Prakash Chandra Chhipa[1]($\boxtimes$)(ID), Richa Upadhyay[1], Rajkumar Saini[1],
Lars Lindqvist[2], Richard Nordenskjold[2], Seiichi Uchida[3], and Marcus Liwicki[1]

[1] Machine Learning Group, EISLAB, Luleå Tekniska Universitet, Luleå, Sweden
{prakash.chandra.chhipa,richa.upadhyay,rajkumar.saini,
marcus.liwicki}@ltu.se
[2] Optimation Advanced Measurements AB, Luleå, Sweden
{lars.lindqvist,richard.nordenskjold}@optimation.se
[3] Human Interface Laboratory, Kyushu University, Fukuoka, Japan
uchida@ait.kyushu-u.ac.jp

Abstract. This work presents a novel self-supervised representation learning method to learn efficient representations without labels on images from a 3DPM sensor (3-Dimensional Particle Measurement; estimates the particle size distribution of material) utilizing RGB images and depth maps of mining material on the conveyor belt. Human annotations for material categories on sensor-generated data are scarce and cost-intensive. Currently, representation learning without human annotations remains unexplored for mining materials and does not leverage on utilization of sensor-generated data. The proposed method, Depth Contrast, enables self-supervised learning of representations without labels on the 3DPM dataset by exploiting depth maps and inductive transfer. The proposed method outperforms material classification over ImageNet transfer learning performance in fully supervised learning settings and achieves an F1 score of 0.73. Further, The proposed method yields an F1 score of 0.65 with an 11% improvement over ImageNet transfer learning performance in a semi-supervised setting when only 20% of labels are used in fine-tuning. Finally, the Proposed method showcases improved performance generalization on linear evaluation. The implementation of proposed method is available on GitHub (https://github.com/prakashchhipa/Depth-Contrast-Self-Supervised-Method).

Keywords: Computer vision · Material classification · Self-supervised learning · Contrastive learning

1 Introduction

Identifying and classifying materials on the conveyor belt is a fundamental process in the mining industry. It is essential to properly sort the materials before forwarding it for stockpiling or further processing. There is a need for real-time

L. Karlinsky et al. (Eds.): ECCV 2022 Workshops, LNCS 13807, pp. 212–227, 2023.
https://doi.org/10.1007/978-3-031-25082-8_14

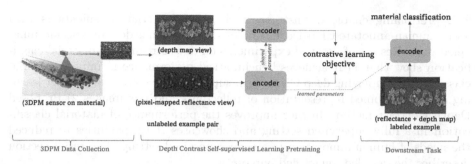

Fig. 1. The proposed approach comprises three steps: (1) Visual data collection of material on the conveyor belt through 3DPM sensor. It captures a depth map (height of material) and corresponding pixel-mapped reflectance data, the amount of light reflected from the imaged material registered by the camera sensor. (2) Self-supervised pre-training on unlabeled pair of the depth map and reflectance visual data using proposed method **Depth Contrast** exploiting depth of material as supervision signal from data, thus enabling self-supervised representation learning beyond RGB images. (3) Performing the downstream task of material classification using the pretrained encoder where input is formulated by combining depth map and reflectance visual data. This work describes pixel-mapped reflectance data as reflectance image and depth maps as raw image in further sections.

monitoring of the materials on the conveyor belt at mine sites for determining the destination for the material accurately and instantly. The traditional methods employed to identify materials or rock minerals are based on the physical and chemical properties of the material. Also, based on human vision (basically human experience). In the recent past, advancement in the field of Artificial Intelligence (AI) and Machine Learning (ML) has made it possible to successfully classify materials using data driven methods, more precisely intelligent computer vision methods [7,9,14]. This work aims to classify the materials using images from the 3DPM sensor [17] and transfer learning [23], Chap. 15 of [10], and employing Deep Learning (DL) architectures. The deep deep Convolutional Neural Networks (CNNs) like Efficient-net [19], DenseNet [13], ResNeXt [22] etc., are pretrained on large datasets such as ImageNet [8] and used to extract meaningful representations from new data. These high level representations are further used for classification.

This work proposes a self-supervised representation learning (SSL) method Depth Contrast, to learn rich representations without human annotations to showcase the capabilities to significantly reduce the need for human supervision for downstream tasks, e.g., material classification. The Depth Contrast self-supervised representation learning method employs the temperature scaled cross-entropy loss for contrastive learning based on the SimCLR method [4]. The proposed pretraining method Depth Contrast exploits supervision signal from data, e.g., depth information of material, thus enabling to learn self-supervised representations on pixel-mapped reflectance data and depth maps from 3DPM sensor, which is beyond the typical RGB visual. Learned representations through

SSL pretraining enable downstream tasks, e.g., material classifications with fewer human-annotated labeled examples, by leveraging domain-specific unlabeled data. Results of detailed experiments on downstream task material classification show that 1) It achieves benchmarked performance in mining material classification employing deep CNNs with supervised ImageNet [8] transfer learning on the combined representation of reflectance and raw images 2) Proposed Depth Contrast method further improves the performance of material classification in a fully supervised setting and showcases the capabilities to reduced the needs of human annotation in a semi-supervised setting. The result section describes the detailed, quantified outcomes.

1.1   Measurement System

3-Dimensional Particle Measurement (3DPM®) [17] is a system that estimates the particle size distribution of material passing on a conveyor belt. 3DPM can also determine particle shape, bulk volume flow and identify abnormalities such as the presence of boulders or rock bolts. Some benefits of online particle size distribution monitoring of bulk material on the conveyor are new control and optimization opportunities for grinding, crushing, and agglomerating. The system can be configured to individually estimate particle size distribution for a large variety of materials such as limestone, coke, agglomerated materials, metal ores, etc. In applications where different materials are transported on the same conveyor belt, the current material must be determined accurately for automated system configuration. First part of the Fig. 1 shows the schematic demonstration of the 3DPM system based on laser line triangulation, which produces 3D data with known real-world coordinates. The data used in this study comprises a depth map and corresponding pixel-mapped reflectance data. Reflectance data is the amount of light reflected from the imaged material registered by the camera sensor. The 3DPM data can be visualized as images but are different from RGB/gray images; however, they can be treated in similar ways. Table 1 shows the reflectance (left) and depth-map (right) images of cylindrical shaped material in first row, respectively.

1.2   Related Work

Many studies focused on similar material type or rock type identification using computer vision and machine learning techniques. In the past 5–6 years, the DL algorithms were used for material or rock classification using images, which yielded some remarkable results without the use of image pre-processing and feature extraction. The article [7] exploits deep CNNs for identification of rock images from Ordos basin and achieves an accuracy of 98.5%. In [14], a comparison of Support Vector Machine (SVM), a histogram of oriented gradient based random forest model, and comprehensive DL model (i.e., Inception-V3 + Color model), for mineral recognition in rocks indicates that the DL model has the best performance. A faster R-CNN model along with VGG16 for feature extraction is used in [15] for identifying nine different rock types and also hybrid images of multiple rocks. The designed model gave an excellent accuracy of 96% for independent rock types and around 80% for hybrid rock type images. Similar work in [9] employs transfer learning methods using two light-weighted CNNs SqueezeNet and MobileNets for classifying 28 types of rocks from images. These networks provide quick and accurate results on the test data; for example, SqueezeNet's accuracy is 94.55% with an average recognition time of 557 milliseconds for a single image. Since all these studies were carried out for different datasets and a different number of classes, it is not possible to compare them. Also, a few of the above studies focus on images of ores, some on rocks, mineral rocks, and many other related variants.

Most of the previous works for ore type or material classification have used 2-dimensional imaging systems that work with RGB images or grayscale images. The works which employ ML or DL are majorly carried out in supervised settings and do not exhibit much progress in utilizing the unlabeled data in either unsupervised or self-supervised settings. Self-supervised learning in computer vision attempts to learn representations from visuals without the need for annotations. There has been several approaches for self-supervised representation learning as (i) contrastive joint embedding methods (PIRL [16] SimCLR [4], SimCLRv2 [5], and MoCo [12]), (ii) quantization (SwAV [3] and DeepCluster [2]), (iii) knowledge distillation (BYOL [11] and SimSiam [6]), and (iv) redundancy reduction (BT [24] and VICReg [1]). This work specifically focuses on self-supervised methods based on contrastive learning (CL). The state-of-the-art contrastive learning methods SimCLR [4,5] and MoCo [12] learn by minimizing the distance between representations of different augmented views of the same visual example ('positive pairs') and maximizing the distance between the representations of augmented views from different visual examples('negative pairs'). However, applying self-supervised learning to leverage on utilization of unlabeled data is not explored beyond the natural images. Therefore, the current work proposes contrastive learning based novel self-supervised pretraining method Depth Contrast for learning representations on visual data (raw and reflectance images) from 3DPM sensor and further improving the performance of downstream task, online material type classification model from the 3DPM images while material passes on the conveyor belt. Many DL based models are investigated, and transfer learning is used to improve the task performance.

The rest of the paper is organized as follows. Section 2 discusses the dataset used in this study. Methodology and experiments are discussed in Sect. 3. Section 4 presents the results of the proposed work. Finally, we conclude in Sect. 5.

2   3DPM Dataset

For material classification, there are two modalities of data that are acquired using the 3DPM sensor. First are the reflectance images, which are 2D grayscale images of the material on the conveyor belt taken in red laser light (680 nm). Lighting conditions, color, and distance between the sensor and the conveyor belt i.e., base depth may affect the 2D imaging system [20]. The second modality is the 3D depth map of the bulk material, i.e., the distance between the material and the sensor. As the material is passed on the conveyor belt, data of multiple laser lines are collected at regular intervals, thereby creating a 3D depth image. 2D imaging systems have scale or perception bias, which is resolved in this 3D data as it is unaffected by the change in light and color of the material [21]. The 3DPM dataset is prepared for this work by collecting the data from sensors under similar lighting conditions. Further, the challenge of variable base depth, which is an issue for the depth map, is solved by subtracting the conveyor depth from all the data. Therefore, every reflectance image has a corresponding raw image (i.e., depth maps). Table 1 shows the sample 3DPM images of all the categories of material in the 3DPM dataset. In order to feed these images to the CNNs three arrangements were made, only raw images, only reflectance images, and respective reflectance and raw images are merged, forming a three-channel image. The experiments proved that combining the reflectance and raw images led to higher performance, refer to Table 3.

The 3DPM dataset has seven classes corresponding to the categories of mining material available on the conveyor belt. These classes are Cylindrical, Ore1, Ore2, Ore3, Mixed1, Mixed2, and Agglomerated (Aggl.). Some of the material categories belong to a specific type of material e.g., Cylindrical, Ore1, Ore2, Ore3, and Agglomerated, whereas the remaining categories, such as Mixed1 and Mixed2, represent the mixture of several materials present on the conveyor belt. Identification of Mixed categories remains challenging with manual visual inspections. The class distribution is shown in Table 2.

Table 1. Classwise raw and reflectance image pairs

3   Methodology and Experiments

This work aims to investigate the classification of materials using supervised and self-supervised learning methods. For faster and more efficient training with the limited (insufficient) data which is available at hand, transfer learning [9] is employed to derive essential features to achieve better classification performance.

Table 2. Types of materials and their distribution

Material name	Total images
Mixed 1	164
Mixed 2	122
Ore 1	860
Ore 2	698
Ore 3	503
Agglomerated (Aggl.)	616
Cylindrical	45

This section discusses transfer learning and the supervised and self-supervised techniques used in this work.

3.1 Transfer Learning

Transfer learning [18] is a method in ML that exploits the existing knowledge of a network to learn a distinct but relevant task efficiently. In transfer learning, the network is first well-trained on a very large dataset to solve a specific problem (maybe classification, regression, etc.). This mechanism is referred to as the source task. For another task i.e., target task, which has lesser data than the source task, the knowledge acquired by the source task is shared in network parameters (or weights) to accomplish the target task. The transfer of information can be carried out in two ways; the first is when the learned network parameters of the source task, act as a good starting point for the target task, and the complete network can further be fine-tuned while training on the target data, refereed as 'fine-tuning' in this article. The second is when the source network is used as it is to extract features for the target task, referred to as 'linear evaluation'. In this work, transfer learning is applied in both supervised and self-supervised settings. Already trained networks on ImageNet [8], which is a substantial multi-label image database, are utilized here for material type classification. These are further discussed in Sect. 3.2 and Sect. 3.3.

3.2 Supervised Downstream Task

Supervised learning [10] is a type of ML technique that uses human-annotated labels to predict the outcome of the learning problem. As the 3DPM dataset has labels along with the reflectance and raw images containing depth maps of the materials, thus applying supervised learning methods for downstream tasks of classification is reasonable to be applied in both supervised transfer learning Sect. 3.1 as well as self-supervised Depth Contrast learning Sect. 3.3. This work considers three pretrained networks i.e., Efficient-net b2 [19], DenseNet-121 [13] and ResNeXt-50 (32 × 4d) [22] for ImageNet [8] transfer learning method and EfficientNet-b2 [19] for self-supervised method for representation learning using

Depth Contrast method. Since the downstream task of material classification on the 3DPM dataset has seven classes corresponding to 7 categories of materials thus, the above-mentioned CNN networks are extended with a fully connected (FC) layer module having two FC layers, one hidden layer of 512, and one output layer of 7 units.

3.3 Depth Contrast - Self-supervised Learning

With a moderate amount of data availability in terms of reflectance and depth images as pair, we propose a method, "depth contrast", based on normalized temperature-scaled cross-entropy loss from SimCLR [4]. It models the two transformed views of input images from a pair of reflectance and depth images, respectively, and intends to maximize the agreement between the stated views of the same data example through contrastive loss. Derived from SimCLR [4], following are the components for Depth Contrast, mentioned in Fig. 2.

– A *domain specific prior* and *stochastic transformation* based module that transforms the synchronously yet independently generated pair of reflectance and depth views of the same example through uniform transformation of random cropping operation, denoted as $\tilde{x}_{ref}$ and $\tilde{x}_{dep}$ which are considers as positive pair. Both, reflectance and depth views are 32 bit images and depth view specifically captures the 3D depth of the material.
– A neural network *base encoder* $f(\cdot)$ which yields representations from transformed reflectance and depth views of data examples. Specifically, this work uses Efficient-net b2 [19] as base encoder based on empirical analysis of CNNs in transfer learning based supervised learning method and further obtains obtain $h_{ref} = f(\tilde{x}_{ref}) =$ Efficient-net$(\tilde{x}_{ref})$ and $h_{dep} = f(\tilde{x}_{dep}) =$ Efficient-net$(\tilde{x}_{dep})$ where $h_{ref}, h_{dep} \in$ & $\mathbb{R}^d$ are the output after the average pooling layer respectively.
– A small-scale multi-layer perceptron (MLP) network projection head $g(\cdot)$ that maps representations to the latent space where contrastive loss is applied. A MLP having three hidden layers of 2048, 2048, and 512 neurons and output layer with 128 neurons to obtain $z_{ref} = g(h_{ref}) = W^{(2)}\sigma(W^{(1)}h_{ref})$ and $z_{dep} = g(h_{dep}) = W^{(2)}\sigma(W^{(1)}h_{dep})$ where σ is a ReLU non-linearity.
– A contrastive loss function, normalized temperature-scaled cross entropy loss (NT-Xent) from SimCLR [4] is defined for a contrastive prediction task. For given a set $\tilde{x}_k$ including a positive pair of examples $\tilde{x}_{ref}$ and $\tilde{x}_{dep}$, the contrastive prediction task intend to identify $\tilde{x}_{dep}$ in $\{\tilde{x}_k\}_{k \neq ref}$ for a given $\tilde{x}_{ref}$.

The loss function for a positive pair of examples (ref, dep) is defined as

$$L_{ref,dep} = -log\frac{exp(sim(z_{ref}, z_{dep})/\tau)}{\sum_{k=1}^{2N} 1_{[k \neq ref]}exp(sim(z_{ref}, z_k)/\tau)} \tag{1}$$

In Eq. 1, where $1_{[k \neq ref]} \in 0,1$ is an indicator evaluating to 1 if $k \neq i$. Equations 2 and 3 define dot product between l_2 normalized respective quantities, and τ defines the temperature parameter. During pre-training, a randomly sampled

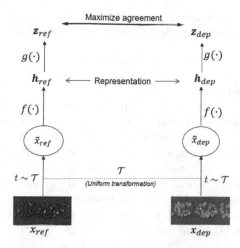

Fig. 2. Depth Contrast method. An uniform data augmentation operator is sampled from one random-crop transform $(t \sim T)$, applied to each pair of both views, reflectance and depth image from 3DPM sensor, considering correlated views. A base encoder network $f(\cdot)$ which is Efficient-net b2 and a projection head which is perceptron network $g(\cdot)$ are trained to maximize agreement using a normalized temperature-scaled cross entropy loss from SimCLR [4]. Upon completion of pre-training, encoder network $f(\cdot)$ and representation h used for downstream classification task

minibatch of N examples is randomly sampled. The contrastive prediction task is defined on pairs of reflectance and depth views derived from the minibatch, resulting in 2N data points. A sampling of negative examples is not performed explicitly. Instead, given a positive pair, the other $2(N - 1)$ augmented examples within a minibatch are considered negative examples. The loss value is computed across all positive pairs of reflectance and depth views, both (z_{ref}, z_{dep}) and (z_{dep}, z_{ref}), in each mini-batch.

3.4   Experiments

The experiments are designed for in-depth analysis of both types of input images, CNNs architectures, and comparative analysis of proposed self-supervised representation learning method Depth Contrast with ImageNet [8] pretrained models on a different level of labeled data availability. Broadly, training of material classification tasks is defined in two arrangements, fine-tuning and linear evaluation, respectively.

In fine-tuning training task, all the layers of Convolutional Neural Network (CNN) architecture and fully connected layers are trained on 3DPM dataset (refer to Table 3 whereas in the linear evaluation training task mentioned in Table 4, only fully connected layers are trainable and CNN architecture serve as a feature extractor.

Table 3. Performance comparison between ImageNet pretrained and Depth Contrast pretrained models **fine-tuned** on 3DPM dataset in semi and fully supervised setting

Exp. no.	Input	Model	Dataset split (in %)			Performance (in F1 score)	
			Train	Val.	Test	Train	Test
ImageNet pre-trained							
1	Raw images	EfficientNet	60	20	20	0.31 × 0.0142	0.30 × 0.0061
2	Reflectance images	EfficientNet	60	20	20	0.68 × 0.0081	0.69 × 0.0235
		3.1 EfficientNet	60	20	20	0.73 × 0.0077	**0.72 × 0.0294**
3	Raw + Reflectance	3.2 ResNeXt	60	20	20	0.75 × 0.0321	0.67 × 0.0841
		3.3 DenseNet	60	20	20	0.74 × 0.0297	0.72 × 0.0411
4	Raw + Reflectance	EfficientNet	10	10	20	0.65 × 0.0616	0.54 × 0.0231
Depth Contrast pre-trained (60% unlabelled data)							
5	Raw + Reflectance	EfficientNet	60	20	20	0.73 × 0.0152	**0.73 × 0.0135**
6	Raw + Reflectance	EfficientNet	10	10	20	0.71 × 0.0399	**0.65 × 0.0263**

Table 4. Performance comparison between ImageNet pretrained and Depth Contrast pretrained models **linear evaluation** on 3DPM dataset in semi and fully supervised setting

Exp. no.	Input	Model	Dataset split (in %)			Performance (in F1 score)	
			Train	Val.	Test	Train	Test
ImageNet pre-trained							
7	Raw + Reflectance	EfficientNet	60	20	20	0.74 × 0.0825	0.69 × 0.0431
8	Raw + Reflectance	EfficientNet	10	10	20	0.67 × 0.0364	0.52 × 0.0380
Depth Contrast pre-trained (60% unlabelled data)							
9	Raw + Reflectance	EfficientNet	60	20	20	0.74 × 0.0273	**0.72 × 0.0854**
10	Raw + Reflectance	EfficientNet	10	10	20	0.71 × 0.0624	**0.64 × 0.0147**

Comparative study of these two types of tasks allows for analyzing representation learning capabilities of the proposed method Depth Contrast in a more detailed manner. This work uses a 5-cross stratified validation technique in which each fold comprises 20% data ensuring class distributions to ensure robustness in performance. In general 60% data, equal to three folds used as the train set, 20%, one fold for the validation set, and the remaining 20%, one fold for the test set. This procedure is repeated five times to test each fold at once, and variations in results are reported in terms of standard deviation alongside the mean F1 score.

Fine-tuning experiments mentioned in Table 3, Experiment 1, 2, and 3.1 are focused on investigating the effectiveness of input image type for material classification and chooses Efficient-net b2 as base architecture based on preliminary analysis on several CNN architectures. Further, experiments 3.2 and 3.3 focus on exploring different CNN architectures in which input types remain a combination of raw and reflectance images, based on performance analysis of previous experiments. Based on preliminary empirical analysis, the combination of raw and reflectance images is composed of three-channel images and raw-reflectance-raw channels. Specifically, experiment 3.2 uses DenseNet 121 [13] and experiment 3.3 uses ResNeXt (31 × 4d) [22] architecture.

The purpose of experiment 4 in Table 3 is to benchmark the material classification performance in a limited labeled data setting while using the Efficient-net network over a combination of raw and reflectance images based on the performance of previous experiments. This experiment uses a single fold where 10% data for training and the remaining 10% data for validation. However, the test set remains the same for all folds. It is worth noticing that experiments 1 to 4 uses ImageNet [8] pretrained models.

Finally, Experiments 5 and 6 in Table 3 are focused on evaluating the effect of the proposed self-supervised representation learning method, Depth Contrast, in semi and fully supervised settings, respectively. Before these experiments, the Efficient-net b2 model is pretrained using the Depth Contrast method over training data without using labels. This pretraining allows the model to learn domain-specific representations in a self-supervised manner. Experiment 5 uses training data the same as previous experiments 1 to 3. Another side, experiment 6 evaluates the performance in a limited labeled data set to compare the effectiveness of Depth Contrast pretraining with experiment 4, while both the experiments use weights initialization from the Depth Contrast method.

Experiments dedicated to linear evaluation training tasks are mentioned in Table 4. Experiment 7 and 8 investigates the performance on ImageNet [8] pretrained model corresponds to fine-tuning experiments 3.1 and 4. Similarly, experiments 9 and 10 correspond to fine-tuning experiments 5 and 6 to investigate the performance of the Depth Contrast pretrained model. In a nutshell, a comparative analysis of the aforementioned experiments can analyze the effectiveness of different transfer learning mechanisms and the leverage of unlabelled sensor-generated data to improve performance in typical industrial settings.

All the fine-tuning experiments mentioned in Table 3 and linear evaluation experiments in Table 4 for material classification follow the *learning rate* of 0.00001, *Adam optimizer, batch size* of 16, *input image size* of 224 × 224, and *dropout regularization* of 0.3 for the hidden layer of FC layer module. Images are 32 bit, so current work only uses random crop as an augmentation technique. Further, Depth Contrast pretraining method uses the same input image size with uniform random crop to both input views, *batch size* of 256, a *learning rate* of 0.00005, *temperature parameter* value of 0.1, and the *MLP projector head* comprises three hidden layers with 2048-2048-512 neurons with ReLU and a batch normalization layer. The output layer of the projector head consists of 128 neurons during Depth Contrast pre-training. The architecture of the MLP projection head is chosen based on preliminary empirical analysis. Further, the F-score evaluation metric is used to accurately measure the performance of models while having class imbalance presents. The empirical analysis of network architectures and training parameters are confined by limited human & computing resources. Methodological development for Depth Contrast and its investigation remains the main focus.

4   Result and Discussion

This section analyzes the results on downstream tasks of material classifications while comparing the effect of the proposed pretraining method. It also extends the analysis of the learning capabilities of models concerning labeled data quantity and different training strategies, including fine-tuning and linear evaluation.

Table 5. Class-wise train and test performance for ImageNet pretrained and fine-tuned on 3DPM dataset in fully supervised setting

S. no.	Classes	Experiment no. 3.1					
		Train			Test		
		Precision	Recall	F1	Precision	Recall	F1
1	Mixed 1	0.31×0.0235	0.63×0.1522	0.41×0.0457	0.31×0.0589	0.61×0.1846	0.40×0.0643
2	Agglomerated	0.65×0.0755	0.84×0.0166	0.73×0.0539	0.64×0.0499	0.84×0.0425	0.72×0.0203
3	Mixed 2	0.70×0.2032	0.23×0.0405	0.34×0.0488	0.63×0.3924	0.22×0.1996	0.29×0.2164
4	Ore 1	0.89×0.0185	0.72×0.0527	0.79×0.0320	0.89×0.0113	0.72×0.0537	0.80×0.0294
5	Ore 2	0.89×0.0721	0.72×0.0433	0.79×0.0371	0.89×0.0841	0.71×0.0284	0.79×0.0300
6	Ore 3	0.69×0.0768	0.71×0.0450	0.70×0.0528	0.69×0.0744	0.70×0.0584	0.69×0.0610
7	Cylindrical	0.75×0.1912	0.58×0.0859	0.64×0.0908	0.84×0.1570	0.62×0.2558	0.67×0.2046

Table 6. Class-wise train and test performance for Depth Contrast pretrained and fine-tuned on 3DPM dataset in fully supervised setting

S. no.	Classes	Experiment no. 5					
		Train			Test		
		Precision	Recall	F1	Precision	Recall	F1
1	Mixed 1	0.30×0.0509	0.48×0.1288	0.36×0.0630	0.26×0.0393	0.56×0.2023	0.35×0.0675
2	Agglomerated	0.64×0.0295	0.87×0.0270	0.74×0.0172	0.66×0.0586	0.85×0.0552	0.74×0.0311
3	Mixed 2	0.70×0.1256	0.32×0.1026	0.43×0.1033	0.63×0.3049	0.20×0.1086	0.29×0.1546
4	Ore 1	0.93×0.0226	0.71×0.0198	0.80×0.0193	0.90×0.0330	0.73×0.0380	0.81×0.0179
5	Ore 2	0.88×0.0555	0.72×0.0391	0.79×0.0278	0.90×0.0361	0.74×0.0404	0.81×0.0250
6	Ore 3	0.68×0.0577	0.78×0.0307	0.73×0.0333	0.73×0.0365	0.74×0.0510	0.73×0.0193
7	Cylindrical	0.96×0.0359	0.45×0.1523	0.60×0.1421	0.89×0.1535	0.56×0.2940	0.62×0.2551

Results for experiments 1, 2, and 3.1 mentioned in Table 3 provides clear insights about input mode that combined representation of raw and reflectance image obtains higher performance on train and test set for fine-tuning on material classification. Commonly used Efficient-net b2 [19] model with the combined representation of raw and reflectance image input achieves an F1 score of 0.71 on the test set whereas the model with only reflectance image remain lower F1 score of 0.69 and model with only raw images performs significantly poor. These experiments direct the current work to use the combined representation of raw and reflectance images for further experiments to evaluate different CNN architectures and the proposed self-supervised method.

Further, the explorations of DenseNet 121 [13] and ResNeXt (31x4d) [22] CNN architectures obtains F1 score of 0.67 and 0.72 in experiments 3.2 and

3.3 respectively. It shows that Efficient-net b2 [19] in experiment 3.1 remains competitive concerning classification performance on the test set, lower standard deviation, and less prone to overfit on the train set. It is important to note that Efficient-net b2 [19] being a lightweight model with lesser learning parameters, is a good fit for the 3DPM dataset. Analysis from experiments 1 to 3.3 indicates that the combined representation of raw and reflectance image with Efficient-net b2 [19] architecture obtains benchmarked results and is a suitable candidate for further investigation on the proposed Depth Contrast method. The Depth Contrast pretrained model, which is fine-tuned in a fully supervised setting in experiment 5, outperforms other models. It not only achieves the highest F1 score of 0.73 with minimal standard deviation but also depicts no overfitting with the same F1 score on the train set.

More specifically, while comparing the performance of this model with ImageNet [8] pretrained model in the same fully supervised settings, it shows an improvement of 1% on the test set. This improvement in the downstream material classification task showcases the leverage of unlabeled data used by the proposed Depth Contrast pretraining method to learn efficient representations without human annotations.

Table 7. Class-wise train and test performance for Depth Contrast pretrained and fine-tuned on 3DPM dataset in semi supervised setting using only 20% of labelled data

S. no.	Classes	Experiment no. 6					
		Train			Test		
		Precision	Recall	F1	Precision	Recall	F1
1	Mixed 1	0.35 × 0.1273	0.51 × 0.0632	0.41 × 0.0763	0.30 × 0.0816	0.44 × 0.0785	0.34 × 0.0340
2	Agglomerated	0.56 × 0.0657	0.90 × 0.0470	0.69 × 0.0469	0.52 × 0.0757	0.85 × 0.0548	0.64 × 0.0575
3	Mixed 2	0.44 × 0.5110	0.09 × 0.0673	0.11 × 0.0694	0.17 × 0.3249	0.07 × 0.0758	0.07 × 0.0924
4	Ore 1	0.87 × 0.0761	0.76 × 0.0467	0.81 × 0.0221	0.83 × 0.0810	0.73 × 0.0481	0.77 × 0.0374
5	Ore 2	0.87 × 0.0645	0.65 × 0.1050	0.73 × 0.0587	0.87 × 0.0423	0.64 × 0.1256	0.73 × 0.0870
6	Ore 3	0.77 × 0.1103	0.56 × 0.1603	0.63 × 0.1199	0.76 × 0.1011	0.52 × 0.1292	0.61 × 0.1029
7	Cylindrical	0.74 × 0.4333	0.31 × 0.2556	0.41 × 0.2570	0.64 × 0.4165	0.22 × 0.2222	0.30 × 0.2547

Interestingly, The Depth Contrast pretrained model, which is fine-tuned in a semi-supervised setting using only 20% of labeled data in experiment 6, achieves a comparable F1 score of 0.65. The proposed method, Depth Contrast, clearly outperforms by 11% on ImageNet [8] pretrained model in the same semi-supervised settings mentioned in experiment 4. This comparative study show-cases the representation learning ability of the Depth Contrast method to yield performance generalization on unseen data even when trained on significantly reduced labeled data.

A similar performance is followed for linear evaluation tasks when pretrained CNN serves as a feature extractor, and only fully connected layers are trained. Table 4 showcases a comparative analysis of the Depth Contrast pretrained model with the ImageNet pretrained model in fully and semi-supervised settings. Depth Contrast pretrained and linear evaluated model in experiment 9 obtains

F1 score 0.72, which is 3% higher than ImageNet pretrained model in experiment 7 for same fully supervised setting. Similarly, Depth Contrast pretrained model from experiment 10 obtains F1 score of 0.64, which is 12% higher than the ImageNet model in the same semi-supervised setting. While linear evaluation clearly shows that ImageNet pretrained model suffers high overfitting in limited labeled data availability, whereas Depth Contrast leverages domain-specific unlabeled data and learns efficient representations for performance generalization.

Detailed class-wise results are mentioned in Table 5 and Table 6 for ImageNet and Depth Contrast pretrained model respectively in fully supervised setting. It is observed that material categories compounded by multiple materials, Mixed1 and Mixed2, perform poorly since their texture characteristics are not uniform, and the sample count is also low in the 3DPM dataset. A similar trend is followed in Depth Contrast pretrained model in the semi-supervised setting mentioned in Table 7. However, the recall and precision of other classes are better for the Depth Contrast pretrained model. Further, current work shares the following conceptual findings.

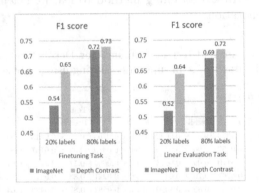

Fig. 3. Performance comparison of models pretrained on ImageNet and Depth Contrast Efficient-net in semi and fully supervised settings

4.1 Raw Image of Depth Maps Scales Representations and Improves Learning

Depth maps of the material on conveyor belts are directly associated with the shape and size of the material; thus, depth details are valuable information addition for learning representation. However, alone depth maps are not sufficient to define the material. It provides the reason that the combined representation of raw (depth details) and reflectance images yield optimal performance. This fact also forms the backbone of the Depth Contrast pretraining method for view pair modeling.

4.2 CNN with Reduced Parameters Performs Better

The architecture of Efficient-net b2 [19] offers adaptive properties for width, depth, and resolution of convolutional layers based on input image size. It seems that the model adapts conveniently based on dataset size and texture details, making it a suitable CNN architecture even with fewer learning parameters.

4.3 Self-supervised Method Depth Contrast Significantly Improves Downstream Task Performance and Reduces Human Annotations

As Depth Contrast constructs input pair view using reflectance image and depth map of the raw image, which allows learning representation comprehensively using both types of details textures from reflectance view and depth maps. During contrastive mechanism, it learns correlation between texture and depth of material, whereas negative pairs allow learning dissimilarity among texture and depth details across material categories. Performance on downstream tasks supports the fact about learning representations in the self-supervised manner as the method outperforms in fully and semi-supervised labeled data settings. Specifically, in limited labeled data scenarios, performance remains statistically significant ($p < 0.05$).

4.4 Self-supervised Method Depth Contrast Encourages Performance Generalization

Comparative analysis of fine-tuning and linear evaluation task in Fig. 3 shows that gap in classification performance between ImageNet pretrained and Depth Contrast pretrained model becomes significant when moving from fine-tuning to linear evaluation. It indicates the rich representation learning capabilities of the model due to Depth Contrast that when CNN model is used as only feature extractor, then also a performance on unseen data remain competent whereas the ImageNet pretrained model starts suffering in performance generalization. It also shows the potential of the self-supervised methods in the visual domain beyond the natural visual concepts that are challenging to understand and annotate by humans.

5 Conclusion and Future Scope

Current work explores the computer vision methods and deep CNNs on mining material classification on 3DPM sensor images. It proposes the self-supervised learning method Depth Contrast to improve the representation learning by leveraging sensor-generated data without human annotation. This work contributes the following:

– Obtains benchmark results of F1 score 0.72 on 3DPM dataset for mining material classification employing ImageNet [8] transfer learning in a fully supervised setting

- Contributes to a novel self-supervised pretraining method, Depth Contrast, which improves performance for mining material classification by obtaining an F1 score of 0.73 in a fully supervised setting. It obtains F1 score 0.65 semi-supervised setting which shows improvement of 11% over ImageNet [8] transfer learning. The proposed method depicts consistent performance improvement on fine-tuning and linear evaluation. It shows the applicability of self-supervised learning beyond the natural visual concept domain.
- The Depth Contrast method showcases the utilization and leverage of sensor-generated visual data to learn representations, significantly reducing the need for human supervision, thus making industrial AI automation more robust, cost-effective, and serving as lifelong learning. The proposed method's main idea is based on multiple input views from different sensors, which has potential possibilities to apply the method in different automation scenarios where views are in terms of video frames, visual captures from different angles, environment, etc.

The potential of self-supervised methods in the work indicates the well-versed possibilities to utilize the machine-generated unlabeled data to reduce the need for human supervision beyond natural visual concepts, reducing the cost and increasing efficiency. The proposed method is adaptable in all the domains where depth information is significant. Future work is subtle to investigate further the field of contrastive joint embedding and other self-supervised methods based on knowledge distillations and bootstrapping. Exploring more network architectures, including transformers, is also essential to elevate supervised performance with increased data.

References

1. Bardes, A., Ponce, J., LeCun, Y.: VICReg: variance-invariance-covariance regularization for self-supervised learning. arXiv preprint arXiv:2105.04906 (2021)
2. Caron, M., Bojanowski, P., Joulin, A., Douze, M.: Deep clustering for unsupervised learning of visual features. In: Proceedings of the European Conference on Computer Vision (ECCV), pp. 132–149 (2018)
3. Caron, M., Misra, I., Mairal, J., Goyal, P., Bojanowski, P., Joulin, A.: Unsupervised learning of visual features by contrasting cluster assignments. arXiv preprint arXiv:2006.09882 (2020)
4. Chen, T., Kornblith, S., Norouzi, M., Hinton, G.: A simple framework for contrastive learning of visual representations. In: International Conference on Machine Learning, pp. 1597–1607. PMLR (2020)
5. Chen, T., Kornblith, S., Swersky, K., Norouzi, M., Hinton, G.: Big self-supervised models are strong semi-supervised learners. arXiv preprint arXiv:2006.10029 (2020)
6. Chen, X., He, K.: Exploring simple Siamese representation learning. In: Proceedings of the IEEE/CVF Conference on Computer Vision and Pattern Recognition, pp. 15750–15758 (2021)
7. Cheng, G., Guo, W.: Rock images classification by using deep convolution neural network. In: Journal of Physics: Conference Series, vol. 887, p. 012089. IOP Publishing (2017)

8. Deng, J., Dong, W., Socher, R., Li, L.J., Li, K., Fei-Fei, L.: ImageNet: a large-scale hierarchical image database. In: CVPR 2009 (2009)
9. Fan, G., Chen, F., Chen, D., Dong, Y.: Recognizing multiple types of rocks quickly and accurately based on lightweight CNNs model. IEEE Access 8, 55269–55278 (2020). https://doi.org/10.1109/ACCESS.2020.2982017
10. Goodfellow, I.J., Bengio, Y., Courville, A.: Deep Learning. MIT Press, Cambridge (2016). http://www.deeplearningbook.org
11. Grill, J.B., et al.: Bootstrap your own latent: a new approach to self-supervised learning. arXiv preprint arXiv:2006.07733 (2020)
12. He, K., Fan, H., Wu, Y., Xie, S., Girshick, R.: Momentum contrast for unsupervised visual representation learning. In: Proceedings of the IEEE/CVF Conference on Computer Vision and Pattern Recognition, pp. 9729–9738 (2020)
13. Huang, G., Liu, Z., van der Maaten, L., Weinberger, K.Q.: Densely connected convolutional networks (2018)
14. Liu, C., Li, M., Zhang, Y., Han, S., Zhu, Y.: An enhanced rock mineral recognition method integrating a deep learning model and clustering algorithm. Minerals 9(9), 516 (2019)
15. Liu, X., Wang, H., Jing, H., Shao, A., Wang, L.: Research on intelligent identification of rock types based on faster R-CNN method. IEEE Access 8, 21804–21812 (2020). https://doi.org/10.1109/ACCESS.2020.2968515
16. Misra, I., van der Maaten, L.: Self-supervised learning of pretext-invariant representations. In: Proceedings of the IEEE/CVF Conference on Computer Vision and Pattern Recognition, pp. 6707–6717 (2020)
17. Optimation: 3DPM. https://optimation.se/en/3dpm/. Accessed 08 July 2021
18. Tan, C., Sun, F., Kong, T., Zhang, W., Yang, C., Liu, C.: A survey on deep transfer learning (2018)
19. Tan, M., Le, Q.V.: EfficientNet: rethinking model scaling for convolutional neural networks (2020)
20. Thurley, M.J.: Automated online measurement of particle size distribution using 3D range data. IFAC Proc. Vol. 42(23), 134–139 (2009)
21. Thurley, M.J.: Automated online measurement of limestone particle size distributions using 3D range data. J. Process Control 21(2), 254–262 (2011). https://doi.org/10.1016/j.jprocont.2010.11.011. Special Issue on Automation in Mining, Minerals and Metal Processing
22. Xie, S., Girshick, R., Dollár, P., Tu, Z., He, K.: Aggregated residual transformations for deep neural networks (2017)
23. Yosinski, J., Clune, J., Bengio, Y., Lipson, H.: How transferable are features in deep neural networks? (2014)
24. Zbontar, J., Jing, L., Misra, I., LeCun, Y., Deny, S.: Barlow twins: self-supervised learning via redundancy reduction. arXiv preprint arXiv:2103.03230 (2021)

Facilitating Construction Scene Understanding Knowledge Sharing and Reuse via Lifelong Site Object Detection

Ruoxin Xiong[1][iD], Yuansheng Zhu[2][iD], Yanyu Wang[1][iD], Pengkun Liu[1][iD], and Pingbo Tang[1(✉)][iD]

[1] Carnegie Mellon University, Pittsburgh, PA 15213, USA
{ruoxinx,yanyuwan,pengkunl,ptang}@andrew.cmu.edu
[2] Rochester Institute of Technology, Rochester, NY 14623, USA
yz7008@rit.edu

Abstract. Automatically recognizing diverse construction resources (*e.g.,* workers and equipment) from construction scenes supports efficient and intelligent workplace management. Previous studies have focused on identifying fixed object categories in specific contexts, but they have difficulties in accumulating existing knowledge while extending the model for handling additional classes in changing applications. This work proposes a novel lifelong construction resource detection framework for continuously learning from dynamic changing contexts without catastrophically forgetting previous knowledge. In particular, we contribute: (1) an Open-Construction Dataset with 31 unique object categories, integrating three large datasets for validating lifelong object detection algorithms; (2) an OpenConstruction Taxonomy, unifying heterogeneous label space from various scenarios; and (3) an informativeness-based lifelong object detector that leverages very limited examples from previous learning tasks and adds new data progressively. We train and evaluate the proposed method on the OpenConstruction Dataset in sequential data streams and show mAP improvements on the overall task. Code is available at https://github.com/YUZ128pitt/OpenConstruction.

Keywords: Construction site · Object detection · Common taxonomy · Object informativeness · Lifelong learning

1 Introduction

Construction scene images contain rich contextual information (*e.g.,* object location, categories, and relationships) for various construction resources, such as laborers, equipment, machines, tools, materials, and the environment, across different construction stages. Comprehensively monitoring and properly managing these construction resources across many different contexts contribute to the economy, quality, safety, and productivity of construction project performance [21,32]. Establishing computational methods that can continuously accumulate capabilities of recognizing new objects in updated contexts is critical for

L. Karlinsky et al. (Eds.): ECCV 2022 Workshops, LNCS 13807, pp. 228–243, 2023.
https://doi.org/10.1007/978-3-031-25082-8_15

supporting such cross-context management of various construction resources. Such methods should be able to keep learning from new data sets from new scenarios without losing the detection capability gained from previously used training data.

Over the last decade, the availability of ubiquitous on-site cameras and advanced deep neural networks (DNNs) have enabled the automatic scene understanding of construction activities from images and videos. Many researchers have explored deep learning-based methods for localizing and identifying construction site objects, including workers [14], building machinery and equipment [46], construction materials [24,40], and personal protective equipment [31, 47], from large-scale domain-specific datasets. However, these deep learning-based models are limited to a fixed number of object categories and construction activities in certain types of spatio-temporal contexts. They can hardly retain and utilize the previously learned knowledge for adapting to new tasks due to the "catastrophic forgetting" or "catastrophic inference" problems [29]. For example, if new object classes or instances are required for new tasks, the training process should be restarted from the beginning each time, and all previous and new data has to be collected during the model training.

Real-world applications in construction scenarios require the learning models' abilities to learn new objects in new scenes while keeping the concepts learned from previous scenes. However, the existing learning systems fail to satisfy such requirements. The followings discuss the two main reasons: (1) significant computational resources and storage space are required to re-train the model each time and fully access all previous and new training data. This constraint could also hinder the applications of automation and robotic technologies in construction scenarios, such as construction robotics, drones, and autonomous vehicles, which require consistent and timely interactions with the surrounding scenes with limited computational abilities [38]; (2) as the data are collected and stored from different organizations, previous training data may be unavailable due to data loss, privacy or cybersecurity concerns, and intellectual property rights [43]. Therefore, a universal construction scene understanding model demands the ability to sustain knowledge of objects and concepts learned from previously encountered scenarios while learning new tasks (e.g., new object instances and classes) - a lifelong/continuous learning model.

In this work, we develop a lifelong-based scheme that leverages the common concepts existing between different contexts and objects in those scenes to keep previously learned scene and object interpretation capabilities while learning new tasks. The developed approach enables a more efficient and scalable knowledge sharing and reuse of the concepts and objects in construction domains across different contexts. The main contributions of this work are:

– This study introduces a lifelong construction resource detection benchmark consisting of 31 object classes, 31,084 images, and 64,841 instances, namely the OpenConstruction Dataset, for identifying common concepts across scenes and validating lifelong object detection algorithms. We build such a dataset by integrating three existing large datasets, ACID [46], SODA [14], and MOCS [7], in continuous data stream settings.

- This study proposes a common taxonomy that captures common concepts critical for detecting similar objects in different contexts or tasks, namely OpenConstruction Taxonomy, providing hierarchical representations for unifying duplicate, conflicting, and new label spaces. This taxonomy can also support multi-scenario information exchanges and inference with other information sources based on the unified label space.

- This study develops a new informativeness-based lifelong learning algorithm to accumulate previous knowledge and learn new construction objects in continuous data streams that keep bringing new objects and contexts. The experimental results tested on the OpenConstruction Dataset show that our proposed method performed better than state-of-the-art methods for adapting to new scenes while keeping previously learned construction scene understanding capability.

2 Related Work

2.1 Object Detection

Automatic construction resource detection (e.g., worker, materials, machines, and tools) is fundamental for various operation-level applications, such as safety management [26,47], progress monitoring [13,44], productivity analysis [10,17,20], and material tracking [24,40]. The object detectors used for localizing and identifying these construction site objects can be generally classified into two groups: two-stage based methods, such as Fast/Faster R-CNN [16,36] and Cascade R-CNN [9], and one-stage based methods, such as Single Shot Detection (SSD) [27], You Only Look Once (YOLO) [34], and its variants [8,35]. In particular, two-stage object detectors first generate the region proposals by selective search [16] or a Region Proposal Network (RPN) [36], then classify the object classes for each region proposal. One-stage object detectors simultaneously predict the object bounding boxes and estimate class probabilities. Typically, the inference speed of one-stage detectors is faster than the two-stage detectors due to the single-network computation. However, the two-stage detectors can achieve better detection performance than the one-stage detectors.

2.2 Metadata Standards in Construction Domains

The semantic representations for construction scenes are subject to individual organizations and do not employ pre-defined and common taxonomy across various construction scenarios. The inconsistent and even conflicting label systems widely exist in the current datasets (e.g., mobile crane (ACID [46]) vs. vehicle crane (MOCS [7]), and worker (MOCS [7]) vs. person(SODA [14])), hindering the knowledge sharing, reuse, and exchange in the construction domains. Many existing building classification standards, such as National Building Specification [5], MasterFormat [4], and OmniClass [6], are developed for organizing and connecting construction information and specifications, while none of them can provide unified and consistent taxonomy for organizing construction site images in dynamic changing contexts. The classes listed in these standards have not covered various construction resources in the field. Similarly, existing construction data exchange standards, such as Industry Foundation Classes (IFC), also provide the schema of different building components [23]. However, they are not intended to classify available construction resources, such as the equipment and machines used in the workplace. Other industry standards and reports contain multiple construction resources but lack the classification taxonomy. For example, the Occupational Safety and Health Administration (OSHA) construction incidents investigation engineering reports [1] contains detailed illustrations of various incidents and accidents on the construction site and thus involves a wide range of construction resources. Since OSHA lists these reports separately, they do not have a classification taxonomy for the construction resources that appeared in these incidents. However, as the new instances and classes are increased continuously, an extensible taxonomy of construction resources is desired to organize massive and information-dense image data and support flexible data exchanges and reuse across various stakeholders [45].

2.3   Lifelong Learning

The parameters of trained machine learning models are usually fixed, limiting their abilities to handle new tasks or changing scenarios in real-world applications. In contrast, human beings can continuously learn and accumulate knowledge throughout their lives by forming and reusing concepts across scenes, enabling themselves to adapt to dynamic environments and new jobs. Inspired by the mechanism of human beings in concept formulation across different scenes, lifelong learning [41] aims to equip the machine learning models with concept reusing and adaptation across different scenes. Such algorithms can learn consecutive tasks without forgetting concepts encountered in previous contexts.

The bottleneck of lifelong learning is catastrophic forgetting, which refers to the phenomenon that the trained machine learning models tend to lose their previous knowledge when learning from new data due to the distribution shift between new and old training data. Current practices for mitigating catastrophic forgetting include parameter isolation, regularization-based techniques, replay techniques, or hybrid methods [12,29]. Particularly, as the knowledge is stored in the model parameters, parameter isolation methods prevent catastrophic forgetting by specifying networks' parameters for each task. For example, progressive neural networks [37] train columns of layers to execute a single task and fix them when learning other tasks. Instead of specifying the parameters, the regularization-based techniques, such as the Elastic Weights Consolidation (EWC) [22] prevent the parameters from changing too much via a regularizer such that the model could retain the previous knowledge. Replay techniques mix the previously seen samples with new samples and use the augmented data to retrain a model. In particular, the replay examples could either be a small subset of old data [28] or virtual samples derived from generative models [39].

3   Lifelong Construction Resource Detection Benchmark

3.1   Open-source Datasets for Detecting Construction Resources

Although many researchers have developed various datasets for detecting construction site objects, these individual datasets are designed for "static" evaluation protocols with limited object categories and scenarios. Typically, the data sources are acquired incrementally in real-world applications. This study proposed a new lifelong construction resource detection benchmark, namely the OpenConstruction Dataset, by integrating and unifying available datasets in sequential settings. Specifically, two main criteria are considered in selecting these data sources from the public datasets:

- *Data diversity*: The developed OpenConstruction Dataset aims to comprehensively integrate and cover various types of construction resources from public datasets in the community. For example, the three integrated large datasets have covered diverse categories of available open-source datasets.
- *Data quality*: We implemented the following three processes to ensure the data quality: (1) objects with tiny instance size ratios (less than 1.8%) are removed from the dataset. For example, some samples of workers and helmets in SODA [14] are very small and thus are deleted from the dataset. (2) Inaccurate annotations located out of the image areas were resized to the boundaries. (3) Class categories with unclear definitions were removed from the dataset, *e.g.,* "other vehicles" in the MOCS [7], to avoid ambiguous and conflicting detection results.

For constructing the OpenConstruction Dataset, we selected three large and diverse datasets, including ACID [46], SODA [14], and MOCS Dataset [7]. Descriptions and statistics of these open-source construction datasets are shown in Table 1. These diverse and heterogeneous datasets cover different construction resources from various scenarios. However, the label systems from these datasets are often duplicated, inconsistent, and even conflicting. Figure 1 shows some examples of images and their inconsistent annotations in these three datasets. The following section will develop a common taxonomy for transforming and unifying these heterogeneous label systems.

3.2   Label Space Transformation and Unification

Taxonomy Building. This study proposed a common and hierarchical taxonomy, namely OpenConstruction Taxonomy, for unifying duplicate, conflicting, and new object classes and building the mega-scale dataset of construction resources. Figure 2 shows the procedures for building and determining the object categories in the OpenConstruction Taxonomy. Referred to [15], we constructed the domain-specific taxonomy in an iterative process: (1) identify and extract common concepts by reviewing related international standards and industry reports, such as ISO/TR 12603 building construction machinery

Table 1. Data descriptions of three integrated construction resource datasets.

Datasets	Num. of categories	Object categories	# Images
ACID [46]	10	mobile crane, tower crane, cement truck, backhoe loader, wheel loader, compactor, dozer, dump truck, excavator, grader	10,071
SODA [14]	15	person, helmet, vest, board, wood, rebar, brick, scaffold, handcart, cutter, ebox, hopper, hook, fence, slogan	19,846
MOCS [7]	13	worker, static crane, hanging head, crane, roller, bulldozer, excavator, truck, loader, pump truck, concrete mixer, pile driving, other vehicle	41,668

Fig. 1. Examples of annotated images in three construction site object datasets. The inconsistent and conflicting labels across these overlapping examples include: *mobile crane* (ACID) vs. *vehicle crane* (MOCS), *worker* (MOCS) vs. *person* (SODA), and *cement truck* (ACID) vs. *truck* (MOCS).

and equipment [3] and OSHA Construction Incidents Investigation Engineering Reports [1], as well as existing domain-specific datasets. (2) Connect these concepts in a hierarchical structure combining top-down and bottom-up strategies. A top-down strategy first identifies major construction resources (*i.e.,* equipment, human, machine, material, and tool) and follows down to specific subclasses. This top-down strategy can help avoid integrating redundant, conflicting, and inconsistent concepts into the Taxonomy. On the other hand, a bottom-up strategy merges the specific labels into general groups based on their affiliated relationships. The bottom-up strategy enables the extensibility of the developed Taxonomy by continuously absorbing new classes from real-world applications.

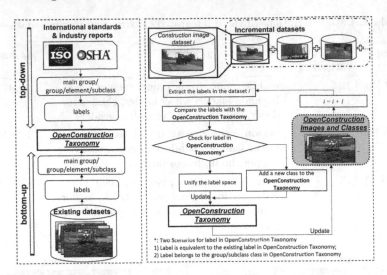

Fig. 2. Procedures for building the OpenConstruction Taxonomy.

Following the iterative top-down and bottom-up processes, we classify the diverse construction resources into four hierarchical levels: *main group* (level 1), *group* (level 2), *element* (level 3), and *subclass* (level 4), based on their specified purposes across various construction stages. The hierarchical structure of the OpenConstruction Taxonomy using the Protégé plugin [30], an ontology building and management system, is shown in Fig. 3.

Label Space Mapping. We manually transform and unify object labels assisted by the "string match" similarity using the spaCy tool [2]. However, as instances and classes steadily increase in new detection tasks, some new object categories may not be able to integrate into OpenConstruction Taxonomy in future applications. Therefore, we propose three extending steps that enable the continued growth of the OpenConstruction Taxonomy by analyzing their affiliated relationships with existing object categories:

- *Step 1: Link new subclasses.* The new concept is first examined as a new "subclass" using its affiliated relationship with the existing taxonomy. For example, for the new concept "wheeled backhoe loader", we add it as the new subclass of the "backhoe loader".
- *Step 2: Link new elements.* If a given concept cannot be linked through the "subclass", we query this element as new instances to its closest groups.
- *Step 3: Link new groups.* If no existing subclass and element are related to the new instance, we add it to the "group". For example, if a new concept is "personal protective equipment", we will integrate this class as the new group of "helmet" and "vest".

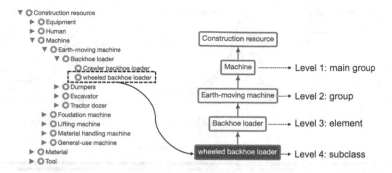

Fig. 3. Hierarchical structure of the OpenConstruction Taxonomy using the Protégé plugin [30]. The main groups consist of five categories: equipment, human, machine, material, and tool.

Dataset Statistics. After a label space transformation and unification process, this study introduces a lifelong construction resource detection benchmark, namely the OpenConstruction Dataset. This dataset comprises 31,084 images and 64,841 object instances, covering 31 diverse construction object categories in the workspace, including workers, tools, machines, materials, and equipment. As these three datasets are collected individually, the unified OpenCoinstruction Dataset can serve as a benchmark for validating and testing lifelong object detection algorithms. Figure 4 shows the data distributions and categories of the dataset. The results indicate that the developed dataset has a long tail distribution among object instances, categories, and instance sizes per image.

4 Informativeness-based Lifelong Learning for Construction Resource Detection

4.1 Preliminaries

Lifelong Object Detection Setup. Lifelong object detection setting requires the model's abilities to accommodate streaming data with new labels and avoid catastrophic forgetting. We define the continuous learning task at time timestamp i as T_i. The developed approach can continuously learn from a sequence of tasks $\mathcal{T} = \{T_1, T_2, T_3, \cdots\}$. In the developed comprehensive OpenConstruction benchmark, the streaming data come from three data sources that different organizations collect at various times (ACID, SODA, MOCS). The developed approach for constructing the OpenConstruction benchmark $\mathcal{T}_{Open}$ continuously learns from the different datasets. The mathematical representation of the developing process is $\mathcal{T}_{Open} = \{T_{ACID}, T_{SODA}, T_{MOCS}\}$. The developed lifelong object detection approach can label both known and unknown objects. For example, when learning the T_{MOCS} after learning T_{ACID}, the model will encounter both the *worker* (new instances) and the *concrete mixer* (new class). We use the informativeness-based approach to avoid catastrophic forgetting in lifelong object detection. The details are illustrated in Sect. 4.2.

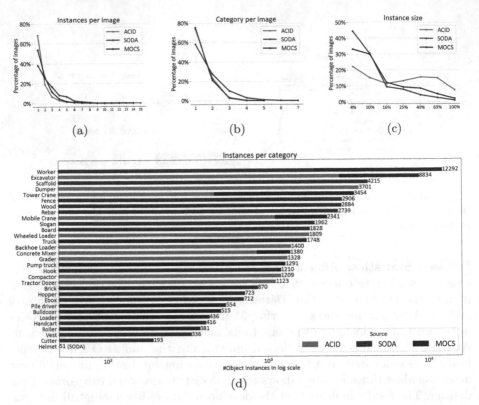

Fig. 4. Data distributions of the OpenConstruction Dataset. (a, b, c) distributions of the number of instances, categories, and instance size per image for ACID, SODA, and MOCS. (d) distributions of object instances in the OpenConstruction Dataset and their sources.

Model Family. Figure 5 shows the proposed informativeness-based lifelong learning framework for detecting construction resources incrementally. We use the Faster R-CNN [36], a two-stage object detector that consists of a backbone, an RPN, a region of interest (ROI) pooling, an ROI feature extractor, a bounding box regressor, and a classification head, as the basic detector. The box classifier and box regressor take the ROI features as the input and output the coordinates and class posterior probabilities of the bounding box.

4.2   Informativeness-based Lifelong Object Detector

We proposed a new training approach that enables lifelong object detection based on the Faster R-CNN [36] using streaming data. Previous approaches to training an object detection model require the availability of all training data simultaneously. The proposed training approach enables continuous learning based on Faster R-CNN through the following settings: 1) for the first task, we train the detector regularly; 2) for the subsequent tasks, we freeze the first three blocks

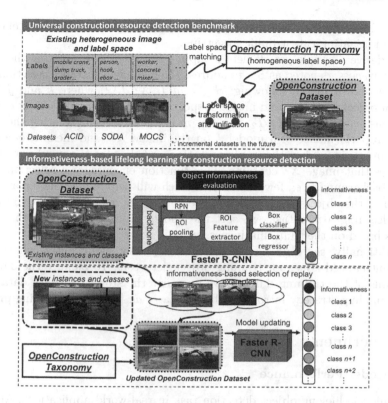

Fig. 5. Overview of the proposed informativeness-based lifelong learning framework for construction resource detection.

of the Encoder and train the fourth block, the ROI, RPN, classification head, and box regressor. Using such settings, the adapted Faster-RCNN can learn new tasks while retaining knowledge from previous tasks. Furthermore, we design a principle criteria to identify the desired relay examples to mitigate the catastrophic forgetting issue, which is a key challenge in lifelong learning tasks.

Replay Examples Selection. Fine-tuning the trained model in the new data inevitably impedes the model's ability to classify previously learned tasks, *i.e.,* catastrophic forgetting issues. Wang *et al.* [42]'s empirical results show that replay example-based methods outperform incremental fine-tuning and EWC methods in continuous object detection settings. The proposed informativeness-based lifelong learning approach uses the balanced memory bank [19,42] to keep instances from previous learning tasks and provides a principle to select the replaying instances. Previous studies select the old instances into the memory bank randomly [19,42]. The developed selection principle can better help the detector retains the knowledge of the corresponding task. Randomly selected replay examples may include non-representative instances (*i.e.,* objects that are partially hidden or objects in a complex background), resulting in less effectiveness for mitigating catastrophic forgetting.

We derive a simple yet efficient sampling function for evaluating the object informativeness, denoted as S^k, to address the inefficient instance selection challenge, as shown in the following: $S^k = m(f_i, X^j)/N^j$, where k and j is the index of object instances and image, respectively. f_j denotes the object detector which has learned task $\{T_1, T_2, \cdots, T_i\}$. m is a metric for evaluating the object detection performance of a model f on the image X^j, and mean average precision (mAP) is used in the implementation. N^j denotes the number of object instances that image X^j contains. This function encourages a selection of the simple background (*i.e.*, fewer objects), which will help reduce the replay examples where the objects not selected in the images are regarded as "background". When learning a new task, we sort the known objects and retain the top-30 objects based on their informativeness values.

Additionally, the object class distributions vary significantly, which could cause imbalanced issues for training object detectors. To balance the data distributions of the training dataset, we augment the stored and new data using the resampling strategy introduced by Gupta *et al.* [18] in the training process.

5   Experiments

5.1   Model Performance

To mimic the lifelong object detection task in real-world applications, we train and test all models on three individual datasets, *i.e.*, ACID [46], SODA [14], and MOCS [7], sequentially in three stages. Particularly, the new dataset is considered the new task, and the previous dataset is regarded as the old task for each training stage. Once a model is re-trained for learning a new task, the model weights stored in the previous task are used for subsequent training of the new task. This sequential training strategy can help evaluate the models' performance for continuously learning old and new tasks.

We use the standard mAP averaged for intersection over union (IoU)$\in$ [0.5: 0.05: 0.95], denoted as mAP@[.5, .95], to evaluate the proposed method. Four scenario settings are considered for model comparisons, all using the same object detector with different training strategies:

- *Joint training*: All instances and classes were added simultaneously for training an object detector in a typical training strategy, where all components of the objects are trainable.
- *Non-adaptation*: The model was trained in the same way as the *joint training* strategy, but the three datasets are sequentially exposed, and only the new data are used at each stage.
- *Fine-tuning*: Following the incremental fine-tuning in Wang *et al.* [42], we freeze the feature encoder after the first task and fine-tune the rest components of the detector (*i.e.*, RPN, box classifier, regressor).

- *iCaRL* [33]: *iCaRL* is proposed for incremental object classification. Following Wang *et al.* [42], we adapt *iCaRL* for lifelong object detection as a baseline for this study. On top of the *fine-tuning* strategy, this setting randomly selects the replay examples over each class and jointly trains the object detector with replay examples and new data.

Implementation Details. We use the same object detector for ours and all baselines methods, a Faster-RCNN model that is pre-trained on the MS COCO dataset [25]. We use the MMDetection toolbox [11] to build the test platform, running it on two NVIDIA RTX A6000 GPUs.

Lifelong Object Detection Results. Table 2 shows the mAP results on three individual tasks and the overall one, under off-line (*joint training*) and sequential training settings (*fine-tuning, non-adaptation*, and *iCaRL*). While the OpenConstruction dataset contains diverse construction scenes (31 object categories) and is highly imbalanced, the *joint training* strategy achieves more than 60% mAP, indicating that this unified dataset could also serve as a high-quality benchmark for detecting diverse construction resources in the civil engineering area.

Under the sequential learning settings, both *non-adaptation* and *fine-tuning* strategies get relatively low mAP on the first two tasks (T_{ACID} and T_{SODA}), particularly for the later one, meaning that they forget previously learned abilities to a large extend. On the other hand, because the T_{ACID} has more overlapped object categories with the last task than T_{SODA}, both methods retain some knowledge for learning this task. In terms of the final task, *non-adaptation* strategy achieves the highest mAP on the last task (T_{MOCS}), which even outperforms the *joint training* strategy by more than 10%. It could be explained that the *joint training* strategy learns the three tasks simultaneously, which is more complex than solely learning a single one. Both *non-adaptation* and *iCaRL* [33] strategies freeze the encoder after they learn the first task. Thus, they cannot learn new tasks well compared to trainable encoders (*i.e.,* Ours and *non-adaptation* strategy).

In particular, *iCaRL* [33] and ours store replay examples from previous tasks. The results of the first two tasks show that such a replaying technique can help mitigate the catastrophic problem. However, while the budget is set the same, ours achieves higher overall performance, especially for the T_{SODA}, indicating that our selected examples are more effective than randomly selected examples for retaining previous knowledge. Moreover, our method achieves the second best performance on the last task, owing to the trainable encoder. Finally, the overall performance shows the advantage of our approach, which could learn the new task well while retaining the ability to detect previous tasks. To summarize, our proposed method achieved a good balance of maintaining previous knowledge and the ability to learn a new task owing to the proposed selection strategy for replay examples and learnable encoder. Figure 6 visualizes three test examples using the joint training strategy, our method, and iCaRL.

Fig. 6. Prediction examples on ACID, SODA, and MOCS for Joint training (top), Ours (middle), and iCaRL (bottom).

Table 2. Lifelong object detection results on the OpenConstruction Dataset after sequential learning $\mathcal{T}_{Open} = \{\mathcal{T}_{ACID}, \mathcal{T}_{SODA}, \mathcal{T}_{MOCS}\}$ (Metric = mAP@[.5, .95]).

Methods	ACID	SODA	MOCS	Overall
Joint training	0.729	0.538	0.538	0.620
Non-adaptation	0.213	0.047	**0.650**	0.167
Fine-tuning	0.237	0.043	0.512	0.127
iCaRL [33]	**0.625**	0.204	0.502	0.358
Ours	0.555	**0.268**	0.566	**0.373**

Note: joint training is not trained in a continuous setting, thus only serving as a reference model here.

6   Conclusions

This paper proposed a novel scalable construction resource detection framework based on the lifelong learning scheme. Specifically, we construct a benchmark dataset with 31 unique object categories, namely the OpenConstruction Dataset, by integrating three large and heterogeneous datasets. To facilitate transforming and unifying heterogeneous label spaces, we develop a common taxonomy, namely the OpenConstruction Taxonomy. This taxonomy provides hierarchical annotations for linking inconsistent concepts actively. Finally, this study proposes a novel informativeness-based lifelong learning algorithm for continuous learning of new detection tasks by selecting the most informative examples in previous tasks. We train and evaluate the proposed method on three diverse datasets in continuous data streams, *i.e.*, ACID, SODA, and MOCS.

The experimental results show that our proposed method achieved 0.373 mAP on the overall task, outperforming the other three training strategies, *i.e.,* non-adaptation, fine-tuning, and iCaRL, for continuous construction scene understanding. The proposed lifelong construction resource detection framework can support efficient and scalable construction scene understanding knowledge sharing and reuse in real-world applications.

This study also has some limitations and will be improved in the future. First, we will integrate more datasets to increase the image samples of existing object categories such as helmets and vests and add new classes. Second, the taxonomy development and extensions are mostly completed by manual transformations. Future work will examine automatic methods to build such a taxonomy. Finally, this study did not fully handle the situations where object classes contained in the previous training dataset while not included in the subsequent learning tasks and thus considered as "background". We will model this issue as an open-world detection problem [19] to improve the model performance in the future.

Acknowledgments. We would like to thank the anonymous reviewers for their constructive comments. This material is based on work supported by Carnegie Mellon University's Manufacturing Futures Institute, the Nuclear Engineering University Program (NEUP) of the U.S. Department of Energy (DOE) under Award No. DE-NE0008864, and Bradford and Diane Smith Graduate Fellowship. The supports are gratefully acknowledged.

References

1. Construction incidents investigation engineering reports. https://www.osha.gov/construction/engineering. Accessed 12 Jul 2022
2. Industrial-strength natural language processing in Python. https://spacy.io/. Accessed 12 Jul 2022
3. ISO/TR 12603:2010(en), building construction machinery and equipment - classification. https://www.iso.org/standard/50886.html. Accessed 15 Jul 2022
4. Masterformat - construction specifications institute. https://www.csiresources.org/standards/masterformat. Accessed 15 Jul 2022
5. National building specification: connected construction information. https://www.thenbs.com/. Accessed 15 Jul 2022
6. Omniclass - construction specifications institute. https://www.csiresources.org/standards/omniclass. Accessed 15 Jul 2022
7. An, X., Zhou, L., Liu, Z., Wang, C., Li, P., Li, Z.: Dataset and benchmark for detecting moving objects in construction sites. Autom. Constr. **122**, 103482 (2021)
8. Bochkovskiy, A., Wang, C.Y., Liao, H.Y.M.: YOLOv4: optimal speed and accuracy of object detection. arXiv preprint arXiv:2004.10934 (2020)
9. Cai, Z., Vasconcelos, N.: Cascade R-CNN: high quality object detection and instance segmentation. IEEE Trans. Pattern Anal. Mach. Intell. **43**(5), 1483–1498 (2019)
10. Chen, C., Zhu, Z., Hammad, A.: Automated excavators activity recognition and productivity analysis from construction site surveillance videos. Autom. Constr. **110**, 103045 (2020)

11. Chen, K., et al.: MMDetection: Open MMLab detection toolbox and benchmark. arXiv preprint arXiv:1906.07155 (2019)
12. De Lange, M., et al.: A continual learning survey: defying forgetting in classification tasks. IEEE Trans. Pattern Anal. Mach. Intell. **44**(7), 3366–3385 (2022)
13. Dimitrov, A., Golparvar-Fard, M.: Vision-based material recognition for automated monitoring of construction progress and generating building information modeling from unordered site image collections. Adv. Eng. Informat. **28**(1), 37–49 (2014)
14. Duan, R., Deng, H., Tian, M., Deng, Y., Lin, J.: SODA: site object detection dataset for deep learning in construction. arXiv preprint arXiv:2202.09554 (2022)
15. El-Gohary, N.M., El-Diraby, T.E.: Domain ontology for processes in infrastructure and construction. J. Constr. Eng. Manag. **136**(7), 730–744 (2010)
16. Girshick, R.: Fast R-CNN. In: Proceedings of the IEEE International Conference on Computer Vision, pp. 1440–1448 (2015)
17. Gong, J., Caldas, C.H.: An object recognition, tracking, and contextual reasoning-based video interpretation method for rapid productivity analysis of construction operations. Autom. Constr. **20**(8), 1211–1226 (2011)
18. Gupta, A., Dollar, P., Girshick, R.: LVIS: A dataset for large vocabulary instance segmentation. In: Proceedings of the IEEE/CVF Conference on Computer Vision and Pattern Recognition, pp. 5356–5364 (2019)
19. Joseph, K., Khan, S., Khan, F.S., Balasubramanian, V.N.: Towards open world object detection. In: Proceedings of the IEEE/CVF Conference on Computer Vision and Pattern Recognition, pp. 5830–5840 (2021)
20. Kim, H., Bang, S., Jeong, H., Ham, Y., Kim, H.: Analyzing context and productivity of tunnel earthmoving processes using imaging and simulation. Autom. Constr. **92**, 188–198 (2018)
21. Kim, J.: Visual analytics for operation-level construction monitoring and documentation: state-of-the-art technologies, research challenges, and future directions. Front. Built Environ. **6**, 575738 (2020)
22. Kirkpatrick, J., et al.: Overcoming catastrophic forgetting in neural networks. Proc. National Acad. Sci. **114**(13), 3521–3526 (2017)
23. Laakso, M., Kiviniemi, A.: The IFC standard - a review of history, development, and standardization. J. Inf. Technol. Constr. **17**, 134–161 (2012)
24. Li, Y., Lu, Y., Chen, J.: A deep learning approach for real-time rebar counting on the construction site based on YOLOv3 detector. Autom. Constr. **124**, 103602 (2021)
25. Lin, T.Y., et al.: Microsoft COCO: Common objects in context. In: European Conference on Computer Vision, pp. 740–755 (2014)
26. Liu, J., Luo, H., Liu, H.: Deep learning-based data analytics for safety in construction. Autom. Constr. **140**, 104302 (2022)
27. Liu, W., et al.: SSD: Single shot multibox detector. In: European Conference on Computer Vision, pp. 21–37 (2016)
28. Lopez-Paz, D., Ranzato, M.A.: Gradient episodic memory for continual learning. In: Advances in Neural Information Processing Systems, vol. 30 (2017)
29. Menezes, A.G., de Moura, G., Alves, C., de Carvalho, A.C.: Continual object detection: a review of definitions, strategies, and challenges. arXiv preprint arXiv:2205.15445 (2022)
30. Musen, M.A.: The protégé project: a look back and a look forward. AI Matters **1**(4), 4–12 (2015)
31. Nath, N.D., Behzadan, A.H., Paal, S.G.: Deep learning for site safety: real-time detection of personal protective equipment. Autom. Constr. **112**, 103085 (2020)

32. Pham, H.T., Rafieizonooz, M., Han, S., Lee, D.E.: Current status and future directions of deep learning applications for safety management in construction. Sustainability **13**(24), 13579 (2021)

33. Rebuffi, S.A., Kolesnikov, A., Sperl, G., Lampert, C.H.: iCaRL: incremental classifier and representation learning. In: Proceedings of the IEEE Conference on Computer Vision and Pattern Recognition, pp. 2001–2010 (2017)

34. Redmon, J., Divvala, S., Girshick, R., Farhadi, A.: You only look once: unified, real-time object detection. In: Proceedings of the IEEE Conference on Computer Vision and Pattern Recognition, pp. 779–788 (2016)

35. Redmon, J., Farhadi, A.: YOLOv3: an incremental improvement. arXiv preprint arXiv:1804.02767 (2018)

36. Ren, S., He, K., Girshick, R., Sun, J.: Faster R-CNN: towards real-time object detection with region proposal networks. In: Advances in Neural Information Processing Systems, vol. 28 (2015)

37. Rusu, A.A., et al.: Progressive neural networks. arXiv preprint arXiv:1606.04671 (2016)

38. Shaheen, K., Hanif, M.A., Hasan, O., Shafique, M.: Continual learning for real-world autonomous systems: algorithms, challenges and frameworks. arXiv preprint arXiv: 2105.12374 (2021)

39. Shin, H., Lee, J.K., Kim, J., Kim, J.: Continual learning with deep generative replay. In: Advances in Neural Information Processing Systems, vol. 30 (2017)

40. Son, H., Kim, C., Hwang, N., Kim, C., Kang, Y.: Classification of major construction materials in construction environments using ensemble classifiers. Adv. Eng. Inf. **28**(1), 1–10 (2014)

41. Thrun, S.: Lifelong learning: a case study. Dept of Computer Science Carnegie Mellon University Pittsburgh PA, Tech. rep. (1995)

42. Wang, J., Wang, X., Shang-Guan, Y., Gupta, A.: Wanderlust: online continual object detection in the real world. In: Proceedings of the IEEE/CVF International Conference on Computer Vision, pp. 10829–10838 (2021)

43. Wang, Y., et al.: Characterizing perceived data sharing barriers and promotion strategies in civil engineering. In: Computing in Civil Engineering 2021, pp. 42–49 (2021)

44. Wang, Z., et al.: Vision-based framework for automatic progress monitoring of precast walls by using surveillance videos during the construction phase. J. Comput. Civil Eng. **35**(1), 04020056 (2021)

45. Wei, Y., Akinci, B.: Construction Scene Parsing (CSP): structured annotations of image segmentation for construction semantic understanding. In: Toledo Santos, E., Scheer, S. (eds.) ICCCBE 2020. LNCE, vol. 98, pp. 1152–1161. Springer, Cham (2021). https://doi.org/10.1007/978-3-030-51295-8_80

46. Xiao, B., Kang, S.C.: Development of an image data set of construction machines for deep learning object detection. J. Comput. Civil Eng. **35**(2), 05020005 (2021)

47. Xiong, R., Tang, P.: Pose guided anchoring for detecting proper use of personal protective equipment. Autom. Constr. **130**, 103828 (2021)

Model-Assisted Labeling
via Explainability for Visual Inspection
of Civil Infrastructures

Klara Janouskova[1], Mattia Rigotti[2], Ioana Giurgiu[2(✉)], and Cristiano Malossi[2]

[1] Visual Recognition Group, Faculty of Electrical Engineering,
Czech Technical University in Prague, Prague, Czechia
klara.janouskova@fel.cvut.cz
[2] IBM Research Zürich, Rüschlikon, Switzerland
{mrg,igi,acm}@zurich.ibm.com

Abstract. Labeling images for visual segmentation is a time-consuming task which can be costly, particularly in application domains where labels have to be provided by specialized expert annotators, such as civil engineering. In this paper, we propose to use attribution methods to harness the valuable interactions between expert annotators and the data to be annotated in the case of defect segmentation for visual inspection of civil infrastructures. Concretely, a classifier is trained to detect defects and coupled with an attribution-based method and adversarial climbing to generate and refine segmentation masks corresponding to the classification outputs. These are used within an assisted labeling framework where the annotators can interact with them as proposal segmentation masks by deciding to accept, reject or modify them, and interactions are logged as weak labels to further refine the classifier. Applied on a real-world dataset resulting from the automated visual inspection of bridges, our proposed method is able to save more than 50% of annotators' time when compared to manual annotation of defects.

Keywords: Civil infrastructure · Weakly supervised learning · Semantic segmentation · Model-assisted labeling

1 Introduction

Until recently visual inspection was exclusively a manual process conducted by reliability engineers. Not only is this dangerous due to the complexity of many civil engineering structures and the fact that some parts are hardly accessible. The main objective of the inspection is to assess the condition of an asset and determine whether repair or further maintenance operations are needed. Specifically, engineers make such decisions by analyzing the surfaces in search for defects, such as cracks, spalling, rust or algae, and assessing their severity, relative to their size and location in the structure.

K. Janouskova—Part of the work was done during the author's internship at IBM Research Zürich.

The advances in drone technology and its falling costs have recently pushed this laborious process of manual inspection progressively towards automation. Flying drones around a structure and using embedded high-resolution cameras to collect visual data from all angles not only speeds up the inspection process, but it also removes the human from potentially dangerous situations. In addition, thanks to the power of artificial intelligence capabilities, defects can be detected and localized with high precision automatically and presented to the reliability engineer for further analysis.

Typical approaches go beyond defect detection and generate fine-grained segmentation masks, which better characterize the defect. However, the drawback of these segmentation models is that they are fully supervised and therefore require a significant volume of high quality annotations at training time. Generating fine-grained segmentation masks is a manual task that involves a human expert deciding whether each pixel in the image belongs to a defect or not, which is time consuming and error-prone. Depending on the size of the images captured during inspection and the volume of defects present in a single image, annotating all defects per image can take hours, even with the aid of annotation tools like CVAT [26] or SuperAnnotate [30]. For example, it has been reported that single large (2048 × 1024) images depicting complex scenes require more than 90 min for pixel-level annotation [7].

The need for such expensive annotations can be alleviated by weakly supervised learning, in which a neural network is trained with cheaper annotations than explicit localization labels. In particular, weakly supervised segmentation methods can use image-level class labels [2,5,8,17], which require a single pixel annotation within the localized region of the target object. By using attribution maps obtained from a classifier, such as Grad-CAM [27], it is possible to identify the most important and discriminative regions of an image. However, these generated maps do not tend to cover the entire region of the target objects. Typical attempts to extend the maps manipulate either the image [19,29], or the feature map [11,32].

In this paper, we employ a different approach, based on adversarial-climbing, to extend the attributed regions of a target object [18]. This is opposed to an adversarial attack, which generates small perturbations of an image in order to change its classification output. As a result of applying adversarial climbing iteratively, the attribution map of the image gradually focuses on more extended regions of the target object, and can be used to generate fine-grained segmentation masks.

Specifically, we build a framework for model-assisted labeling of defects detected as a result of visual inspections of bridge structures from high resolution images. We train a classifier to recognize defect labels, apply Grad-CAM to generate segmentation masks and refine these masks with adversarial climbing. Once the masks have been generated, they are made available to the user through an interaction tool, where the expert is able to visualize, accept, reject or correct them, if need be (Fig. 1). We evaluate the approach on a real-world dataset and show that even after the first iteration, more than 50% of annotators' time is saved by refining the obtained masks instead of manually generating them. Moreover, the time saved is expected to increase in further iterations.

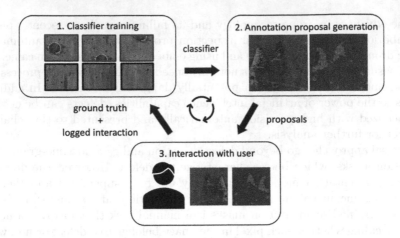

Fig. 1. Overview of the assisted labeling framework. First, a classifier is trained on weakly annotated images. Second, the classifier generates annotation proposals. Finally, the user interacts with the proposal (accept/modify/reject). The interaction is logged and used to extend the training set and improve the classifier, resulting in improved annotation proposals on future data.

2 Related Work

2.1 Weakly Supervised Segmentation and Localization

The vast majority of weakly supervised semantic segmentation and object localization methods depend on attribution maps obtained with approaches like Grad-CAM [27] from a trained classifier. While identifying the relevant regions of an image that have contributed to the classifier's decision is the goal, these regions tend to not be able to identify the whole region occupied by the target object.

Therefore, there have been many attempts to extend the attributed regions to ensure they cover more of the detected object. One popular approach is to manipulate the image [19,29]. For instance through erasure techniques [11,13, 22,31,32], already identified discriminative regions of the image are removed in an iterative manner, thus forcing the classifier to identify new regions of the object to be detected. However, the main drawback of the erasure approach is that there is a risk to generate wrong attribution maps when by erasing the discriminative regions of an image, the classifier's decision boundary changes. An alternative to image manipulation is feature map manipulation [6,16]. This produces a unified feature map by aggregating a variety of attribution maps from an image obtained by applying dropout to the network's feature maps.

Recently, adversarial climbing has been proposed to extend the attributed regions of the target object [18]. Applied iteratively to the attribution maps of a manipulated image results in a gradual identification of more relevant regions of the object. Regularization is additionally applied to avoid or reduce the activation of irrelevant regions, such as background or regions of other objects.

Unlike other approaches that require additional modules or different training techniques, applying adversarial climbing acts essentially as a post-processing step on top of the trained classifier. This makes it possible and easy to replace the underlying classifier's architecture or improve its performance without performing any changes to the backbone.

While adversarial climbing has been mainly applied for semantic segmentation, we employ it for instance segmentation, to generate precise and high-quality segmentation masks for fine-grained defects present in civil infrastructures. These masks go beyond providing localization cues for weakly supervised instance segmentation and defect localization. They significantly reduce the time required to manually annotate such defects at pixel-level, thus enabling downstream tasks such as supervised defect detection and segmentation at much lower costs.

2.2 Annotation Tools

Many annotation tools successfully deploy semi-supervised interactive annotation models, with different level of weak supervision at inference time. Traditional methods like GrabCut [25] which do not require fully-supervised pre-training exist, but are outperformed by learning-based strategies.

In DEXTR [23], at least four extreme points are required at inference time to infer segmentation while a bounding box and up to four correction points are used in [4]. In [14], a single click on an instance is enough to generate its segmentation mask. A crucial disadvantage of these approaches is that they do not have any localization ability and the detection of the defects fully relies on the human annotator. The performance of learning-based models also typically decreases with domain transfer. To obtain good performance on a new domain, full annotations are required. In [1], the problem of domain transfer is tackled by online fine-tuning.

3 Model-Assisted Labeling Framework

Instead of requiring segmentation masks from annotators, we propose to use weak labels consisting of classification labels. One label per image would make GPU training extremely challenging due to the large size of images in our dataset. Therefore, we ask the annotators to localize patches that contain defects. This approach is still substantially faster to input since they only require one click per defect. Similarly, negative samples require one click to indicate the absence of defects within a given image patch. These inputs are then used to generate a training dataset for a defect classifier by sampling crops around the annotated pixels.

Common approaches to weakly supervised learning with class-level supervision [3] use encoder architectures such as ResNet [10] to generate class activation maps (CAMs) [27]. Some work has gone into improving the resolution of the obtained CAMs using multi-scale inference [3,18] followed by post-processing steps like dense CRF [15], or aggregating activations from different levels of a ConvNet [9,12,28]. In a semi-supervised setup, [20] adopt a U-net architecture [24] and pre-train the encoder on a classification task, and then train the decoder to improve the mask starting with CAMs as a segmentation prior.

3.1 Proposal Generation and Refinement

Our weakly-supervised method to generate fine-grained segmentation masks consists of two steps. First, a deep neural network trained on a classification task is used to generate CAMs. Second, the CAMs go through a simple post-processing step to remove noise before connected component analysis, which gives the final annotation proposals. We generate the CAMs for all images as rejecting false positives only takes a negligible amount of time compared to polygon annotation and it forces the human annotators to check all the images for false positives/negatives. Optionally, most false negatives could easily be filtered out by applying a classifier to image patches. An example of the initial CAMs and the post-processed output is shown in Fig. 2.

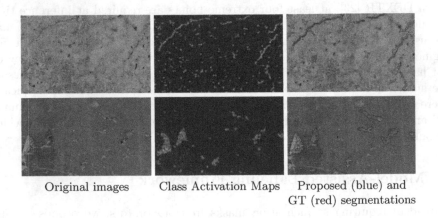

Original images   Class Activation Maps   Proposed (blue) and
                                              GT (red) segmentations

Fig. 2. Example images for the class 'crack'. First column: two examples of image of concrete surfaces containing cracks, and therefore labeled with the classification label 'crack'. Middle column: corresponding CAMs. Last column: corresponding proposed annotations obtained by filtering out noise in the CAMs by post-processing (blue), Ground Truth (GT) segmentation masks obtained by expert manual annotation from scratch (red). The proposed annotations generated by our method have large overlap with GT segmentation masks provided by expert annotators. (Color figure online)

Model Architecture. Similarly to [20], we adopt U-net [24], a segmentation architecture which aggregates features from different levels of the encoder at different resolutions. Instead of only using the pre-trained encoder for classification, we add the classification head directly on top of the decoder. This approach brings the advantage of having weights pre-trained on the target data as an initialization for subsequent fully supervised learning and increases the resolution of the final layer CAMs.

To further improve the resolution of the CAMs, we only build the U-net on top of the first two blocks of a Resnet34 encoder, avoiding resolution degradation from further downsampling. We also set the stride to 1 instead of the original 2 in the first convolutional layer, before the residual blocks. The model reduction does not lead to any classification performance degradation for the target application, however, Resnet34 produced better quality attribution masks then Resnet18. An overview of the architecture can be found in Fig. 3 and examples of CAMs with a different number of downsampling layers in Fig. 4.

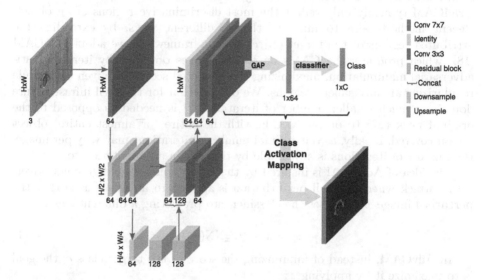

Fig. 3. Our U-net-based classifier for assisted labeling via explainability. Using a U-net architecture as the feature extractor of the classifier trained on (weak) classification labels allows us to generate CAMs with the same resolution as the input images using Grad-CAM, a standard gradient-based explainability method. The obtained high-resolution CAMs are then refined using anti-adversarial climbing (AdvCAM), post-processed, and used as proposal segmentation masks that can be further refined by annotators in a standard annotation tool.

Masks as Attribution Maps. GradCAM [27] is a gradient-based attribution method used to explain the predictions of a deep neural classifier by localizing class-discriminative regions of an image. It can be used for any layer but in the context of weakly supervised segmentation, the attribution maps of the final layer are typically used. When the last convolutional layer is followed by global average pooling (GAP) and a linear layer as classifier, these maps are commonly referred to as class activation maps (CAMs).

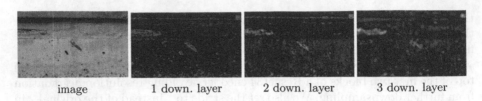

image   1 down. layer   2 down. layer   3 down. layer

Fig. 4. Sparsity vs. CAM resolution trade-off. 'Spalling' CAMs of networks with one, two and three downsampling layers are shown. The less downsampling layers, the sparser and with better resolution the maps are.

Refinement with Adversarial Climbing. The attribution maps obtained with GradCAM typically only reflect the most discriminative regions of an object. Recent methods aim to mitigate this in different ways, by extending the attributed regions of the target object. In our framework, we adopt AdvCAM [18], which produces the CAMs on top of images obtained by iterative anti-adversarial manipulation, maximizing the predicted score of a given class while regularizing already salient regions. We observe that for the civil infrastructure domain, a much smaller number of iterations (2) is needed as opposed to the original work (27) to output CAMs with the entire (or almost entire) object region covered. Ideally, as the optimal number of iterations may vary per image, the number of iterations is adjustable by the user in an annotation tool.

The idea of AdvCAM is inspired by that of an non-targeted gradient adversarial attack where a small perturbation is applied to an image x so that the perturbed image x' confuses the classifier into predicting a different class:

$$x' = x - \xi \nabla_x \text{NN}(x). \tag{1}$$

In AdvCAM, instead of minimizing the score of the target class c, the goal is to maximize it by applying:

$$x' = x + \xi \nabla_x y_c \tag{2}$$

where y_c is the logit of the target class.

This is referred to as anti-adversarial climbing and the procedure is iterative. Two forms of regularization are also introduced: i) the logits of the other classes are restricted to avoid increase in score for objects of classes close to the target class, and ii) attributions of already salient regions are restricted so that new regions are discovered.

Finally, the CAMs obtained from the adversarially-manipulated images are summed over all iterations and normalized. For more details, please refer to AdvCAM [18].

Post-processing. However, after the previous refinement step, the resulting CAMs contain noise, especially for images with highly structured background. Adversarial climbing typically further increases the amount of noise. Single threshold binarization either includes the noise for lower values, or defect parts are suppressed alongside the noise for higher values. Due to the increased resolution, the resulting activation maps are also sparser, sometimes leading to a defect split into multiple parts, especially for very thin cracks. We add two fast and simple post-processing steps after binarization with a low threshold value, $\theta = 0.1$, to address these issues. First, morphological closure is applied to counter the sparsity. Second, connected components are retrieved from the mask and all regions with an area below a threshold are filtered out. These steps effectively remove the majority of the noise while retaining the defect regions. The threshold value θ was selected according to the best performance on the validation set. The closure filter and minimal component area were selected based on observation of qualitative results. An example illustrating these steps and the post-processed result is shown in Fig. 5.

Original image   CAM   Thresholding   Closure   Filtering

Fig. 5. Postprocessing for noise removal from CAMs. The figure shows the essential post-processing step from the initial CAMs to the final proposed annotation. The CAMs are first binarized with a low threshold (in this case $\theta = 0.1$). We then apply morphological closure to the binary map, followed by filtering of connected components by area to remove small clusters of pixels.

3.2 Interactive Aspects

The quality of the automatically generated annotations varies considerably. We therefore treat them as proposals to be screened by human annotators in a selection phase, before training a segmentation model on them. The generated annotations are split into three groups:

- *Accept* – the proposed annotation covers the predicted defect and no modifications to the defect mask are needed;
- *Modify and accept* – the proposed annotation covers the predicted defect, but the mask needs refinement;
- *Reject* – the proposed annotation does not contain the predicted defect, as it is a false positive. Furthermore, the annotator checks if some defects have been missed by the model (false negatives), in which case a manual annotation process should take place.

Annotation modification. Depending on the proposed defect mask and the capabilities of the annotation tool used, the modification can take multiple forms. The most common one is the removal of false positive patches of "noise" from the neighbourhood of the defect, which can be done by simply clicking on them to erase them. Another commonly needed modification is removing or adding a

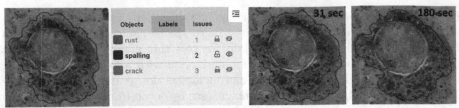

Proposed annotation   Refined annotation Manual annotation

Fig. 6. Difference in labeling time between manually refining a proposed annotation obtain with our method vs. manual annotation from scratch. The proposed annotation obtained through our method for a spalling defect is loaded in CVAT (left). It is manually refined by hand by a human annotator in 31 s (center). On the other hand, manual annotation from scratch of the same defect takes ∼6 times longer (180 s) than refining the proposed annotation (right).

small part of the defect, which requires moving/adding/removing polygon points, or with a brush tool. The interaction time can be further reduced by more sophisticated operations, for example, merging and splitting components, mask erosion and dilation, if the annotation tool in use allows for such operations.

Interaction logging. The interaction of the annotator with the proposals provides valuable information that can be exploited if logged. For instance, flagged false positive and false negative regions could be used to extend the weakly supervised training set, which consequently would improve the classifier and the subsequent generated proposals. The time spent on the modification of a proposal until it is accepted can also be used as a proxy for the sample difficulty, allowing for more efficient training strategies. This working modality of our proposal could be used in the future as part of a labeling pipeline combining active learning pipeline and weak supervision.

Demonstration. We show user interaction in CVAT, an open source annotation tool extensively used in various domains, including civil infrastructure. There, experts are able to visualize, accept, modify and reject the proposed annotations resulting from our framework. In Fig. 6, refining an annotation proposal for a spalling defect takes 31s, as opposed to manually generating the mask, which takes 180s. Additionally, the manual annotation is less fine-grained (i.e., fewer polygon points) than the proposal and reaching the same level of detail manually would extend well beyond 180s. The example shown here is pessimistic as the proposed and refined masks are very close and it is not clear edits were necessary. However, the refinement took only 16 % of the full annotation time.

4 Evaluation

4.1 Data Preparation

The defect dataset consists of 732 high resolution (5K × 5K pixels) images. The classification dataset is generated by sampling 5 crops of size 320 × 320 pixels around each positive (corresponding to a defect) user-annotated pixel. Given the extreme class-imbalance, where the least number of instances per class is 675 and the largest number of instances is 21,787 (see Table 1), we create a separate binary classification (defect/no defect) dataset for each defect type. Since there is significantly more negative pixel annotations, crops of negative class are sampled uniformly from each annotation so that there is the same number of positive and negative samples in each of the datasets. Performance of classifiers trained on multiclass datasets created by over/undersampling was inferior.

Table 1. Number of instances of each class in the dataset (highly imbalanced).

Defect	Crack	Spalling	Rust
Instances	21,787	675	1,078

4.2 Classifier Training

The models share the same architecture and were implemented in the PyTorch framework. Each model was trained for 6 h on 2 Nvidia A100 GPUs with the batch size of 32 and the AdamW [21] optimizer (learning rate 1e-4, weight decay 1e-2, all other parameters were kept at PyTorch default values). The best checkpoint was selected according to the highest f-measure on the validation set.

4.3 Estimation of Time Saved

To quickly estimate the annotation time reduction due to our weakly supervised model, we used the following procedure where users only estimate the percentage of time saved, as opposed to actually annotating the data. We first assumed that the user has at their disposal standard annotation tools such as brush and erasure, as well as the possibility to apply morphological operations such as dilation and erosion. We then split the test set instances (i.e. the connected components of the segmentation ground truth masks in the test patches, which are 265 in total) into 3 groups according to the estimated percentage of the time saved annotating the instances when using the output of our method as an initial annotation proposal within such a standard annotation tool. Specifically, group G_{95} are the instances for which, by visual inspection, we estimated modifying the CAM via the annotation tool provided a time saving above 95% over annotating the instance from scratch. Analogously, groups G_{75} and G_{50} are groups of instances where we estimated a time saving above 75% (but below 95%) and above 50% (but below 75%), respectively. Examples of instances from each group for all defects are shown in Fig. 7.

The ratio between the total time saved and the time needed to annotate was then simply estimated from the number of instances in each of the groups of connected components as:

$$\text{Relative time saving} = \frac{1}{N} \sum_{i \in \{95,75,50\}} \frac{i}{100} |G_i|, \tag{3}$$

where $|G_i|$ is the number of instances in the group.

This formula allowed us to estimate an average reduction of 52% in annotation time. This is broken down as follows for different types of defects: 57% for cracks, 58% for spalling and 40% for rust. Detailed results of this estimation procedure are reported in Table 2. The relatively low time saving on rust can be explained by the high sensitivity of the annotation proposal shape to the value of the threshold applied to the CAMs, meaning that each component would require fine manual tuning of the threshold within the annotation tool.

An important note on the limitations of this time estimation procedure is that on one hand, it does not take into account the instances that were missed by our method, meaning that in practice we assumed recall of the defects. On the other, the time estimation is very conservative. Less than 50 % time reduction is considered as 0 and a lower bound is used for the rest of the intervals. The final result of 52% annotation time saved should thus be considered as a very conservative lower bound and we plan to conduct a more detailed evaluation in the future.

Table 2. Estimated annotation time saved by our method in percentage over annotating the defect instances from scratch on a test set of 265 instances. Relative time saving is shown for each type of defect separately and in total averaged over all defects. The number of instances where the percentage of time saved $t \in [95, 100]$, $t \in [75, 95)$ and $t \in [50, 75)$ is also reported. Anything less than 50 is considered as no time saved on the instance.

	Instance count	95	75	50	Time saved (%)
Crack	111	19	40	30	57
Spalling	65	17	19	14	58
Rust	89	22	14	9	40
All defects	265	58	73	53	51

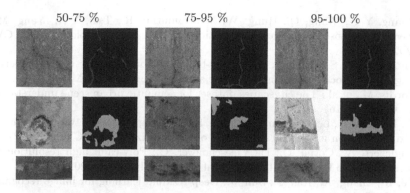

Fig. 7. Examples of instances according to the annotation time saved (in %). The first, second and last row show instances of cracks, spalling and rust, respectively.

5   Conclusions

We explored the use of weak labels from human annotators as a means to reduce the labeling time for a segmentation task in visual inspection, an application domain where the time of specialized annotators is particularly costly. The advantage of weak labels is that they are cheap to obtain because they require minimal interaction with the annotator. In our proposed approach, weak labels are used to train a classifier in order to generate proposals for segmentation masks by means of an explainability attribution method, followed by iterative adversarial climbing. Domain experts can then correct the proposed masks where needed by integrating this workflow in standard annotation tools like CVAT. Moreover, proposal segmentation masks can be used as pseudo-labels for unlabeled images, which can subsequently be employed to train supervised segmentation models, as well as to diagnose issues with ground truth labels from previous annotation campaigns.

Acknowledgement. This research was supported by Czech Technical University student grant SGS20/171/OHK3/3T/13. We would like to thank Finn Bormlund and Svend Gjerding from Sund&Bælt for the collection and annotation of the image data.

References

1. Acuna, D., Ling, H., Kar, A., Fidler, S.: Efficient interactive annotation of segmentation datasets with polygon-rnn++. In: CVPR, pp. 859–868 (2018)
2. Ahn, J., Kwak, S.: Learning pixel-level semantic affinity with image level supervision for weakly supervised semantic segmentation. In: CVPR (2018)
3. Ahn, J., Cho, S., Kwak, S.: Weakly supervised learning of instance segmentation with inter-pixel relations. In: CVPR, pp. 2209–2218 (2019)
4. Benenson, R., Popov, S., Ferrari, V.: Large-scale interactive object segmentation with human annotators. In: CVPR, pp. 11700–11709 (2019)

5. Chang, Y.T., Wang, Q., Hung, W.C., Piramuthu, R., Tsai, Y.H., Yang, M.H.: Weakly-supervised semantic segmentation via sub-category exploration. In: CVPR (2020)

6. Choe, J., Lee, S., Shim, H.: Attention-based dropout layer for weakly supervised single object localization and semantic segmentation. In: TPAMI (2020)

7. Cordts, M., et al.: The cityscapes dataset for semantic urban scene understanding. In: CVPR (2016)

8. E., K., Kim, S., Lee, J., Kim, H., Yoon, S.: Bridging the gap between classification and localization for weakly supervised object localization. In: CVPR (2022)

9. Englebert, A., Cornu, O., De Vleeschouwer, C.: Poly-cam: high resolution class activation map for convolutional neural networks. In: ICPR (2022)

10. He, K., Zhang, X., Ren, S., Sun, J.: Deep residual learning for image recognition. In: CVPR (2016)

11. Hou, Q., Jiang, P., Wei, Y., Cheng, M.M.: Self-erasing network for integral object attention. In: NeurIPS (2018)

12. Jiang, P.T., Zhang, C.B., Hou, Q., Cheng, M.M., Wei, Y.: Layercam: exploring hierarchical class activation maps for localization. IEEE Trans. Image Process. **30**, 5875–5888 (2021)

13. Ki, M., Uh, Y., Lee, W., Byun, H.: In-sample contrastive learning and consistent attention for weakly supervised object localization. In: ACCV (2020)

14. Koohbanani, N.A., Jahanifar, M., Tajadin, N.Z., Rajpoot, N.: NuClick: a deep learning framework for interactive segmentation of microscopic images. Med. Image Anal. **65**, 101771 (2020)

15. Krähenbühl, P., Koltun, V.: Efficient inference in fully connected crfs with gaussian edge potentials. NeurIPS **24**, 109–117 (2011)

16. Lee, J., Kim, E., Lee, S., Lee, J., Yoon, S.: Ficklenet: weakly and semi-supervised semantic image segmentation using stochastic inference. In: CVPR (2019)

17. Lee, J., Oh, S.J., Yun, S., Choe, J., Kim, E., Yoon, S.: Weakly supervised semantic segmentation using out-of-distribution data. In: CVPR (2022)

18. Lee, J., Kim, E., Yoon, S.: Anti-adversarially manipulated attributions for weakly and semi-supervised semantic segmentation. In: CVPR, pp. 4071–4080 (2021)

19. Li, K., Wu, Z., Peng, K.C., Ernst, J., Fu, Y.: Tell me where to look: Guided attention inference network. In: CVPR (2018)

20. Lin, D., Li, Y., Prasad, S., Nwe, T.L., Dong, S., Oo, Z.M.: CAM-UNET: class activation MAP guided UNET with feedback refinement for defect segmentation. In: ICIP, pp. 2131–2135 (2020)

21. Loshchilov, I., Hutter, F.: Decoupled weight decay regularization. In: ICLR (2019)

22. Mai, J., Yang, M., Luo, W.: Erasing integrated learning: a simple yet effective approach for weakly supervised object localization. In: CVPR (2020)

23. Maninis, K.K., Caelles, S., Pont-Tuset, J., Van Gool, L.: Deep extreme cut: from extreme points to object segmentation. In: CVPR, pp. 616–625 (2018)

24. Ronneberger, O., Fischer, P., Brox, T.: U-Net: convolutional networks for biomedical image segmentation. In: Navab, N., Hornegger, J., Wells, W.M., Frangi, A.F. (eds.) MICCAI 2015. LNCS, vol. 9351, pp. 234–241. Springer, Cham (2015). https://doi.org/10.1007/978-3-319-24574-4_28

25. Rother, C., Kolmogorov, V., Blake, A.: "GrabCut": interactive foreground extraction using iterated graph cuts. ACM Trans. Graph. (TOG) **23**(3), 309–314 (2004)

26. Sekachev, B., et al.: opencv/cvat: v1.1.0 (2020). https://doi.org/10.5281/zenodo.4009388

27. Selvaraju, R., Cogswell, M., Das, A., Vedantam, R., Parikh, D., Batra, D.: Grad-cam: visual explanations from deep networks via gradient-based localization. In: ICCV, pp. 618–626 (2017)
28. Shinde, S., Chougule, T., Saini, J., Ingalhalikar, M.: HR-CAM: precise localization of pathology using multi-level learning in CNNs. In: Shen, D., et al. (eds.) MICCAI 2019. LNCS, vol. 11767, pp. 298–306. Springer, Cham (2019). https://doi.org/10.1007/978-3-030-32251-9_33
29. Singh, K.K., Lee, Y.J.: Hide-and-seek: Forcing a network to be meticulous for weakly-supervised object and action localization. In: ICCV (2017)
30. SuperAnnotate: https://www.superannotate.com (2018)
31. Wei, Y., Feng, J., Liang, X., Cheng, M.M., Zhao, Y., Yan, S.: Object region mining with adversarial erasing: a simple classification to semantic segmentation approach. In: CVPR (2017)
32. Zhang, X., Wei, Y., Feng, J., Yang, Y., Huang, T.S.: Adversarial complementary learning for weakly supervised object localization. In: CVPR (2018)

A Hyperspectral and RGB Dataset for Building Façade Segmentation

Nariman Habili[1], Ernest Kwan[2], Weihao Li[1], Christfried Webers[1], Jeremy Oorloff[1], Mohammad Ali Armin[1(✉)], and Lars Petersson[1]

[1] CSIRO's Data61 (Imaging and Computer Vision Group), Canberra, Australia
nariman.habili@csiro.au, ali.armin@data61.csiro.au
[2] Australian National University, Canberra, Australia
https://data61.csiro.au/

Abstract. Hyperspectral Imaging (HSI) provides detailed spectral information and has been utilised in many real-world applications. This work introduces an HSI dataset of building facades in a light industry environment with the aim of classifying different building materials in a scene. The dataset is called the Light Industrial Building HSI (LIB-HSI) dataset. This dataset consists of nine categories and 44 classes. In this study, we investigated deep learning based semantic segmentation algorithms on RGB and hyperspectral images to classify various building materials, such as timber, brick and concrete. Our dataset is publicly available at CSIRO data access portal.

Keywords: Hyperspectral Imaging (HSI) · Building material · Imbalanced data

1 Introduction

Data on buildings and infrastructure are typically captured pre-disaster and post-disaster. In the case of pre-disaster inventory capture, the aim is to capture data on exposure, develop risk assessment models and inform governments and the private sector such as insurance companies about areas and infrastructure most vulnerable to disasters. On the other hand, a post-disaster survey is conducted to capture information about damage to buildings and other infrastructure. This information is used to assess the risk of future events, adapt and modify legacy structures and modify construction practices to minimise future losses. The information is also used to develop computational risk models or update existing models. Pre-disaster and post-disaster surveys are usually conducted by foot, with surveyors visiting every building in a built environment and recording information about the make-up of the building in a hand-held device. Information such as building type, building material (roof, walls etc.) and the level of damage sustained by a building (in case of a post-disaster survey) are recorded. This is a laborious task and the motivation to automate this task formed the basis to collect data for this research project. Classifying building

L. Karlinsky et al. (Eds.): ECCV 2022 Workshops, LNCS 13807, pp. 258–267, 2023.
https://doi.org/10.1007/978-3-031-25082-8_17

Fig. 1. Examples from our hyperspectral & RGB façade dataset, the first column shows an hyperspectral image with 204 bands, the second column shows the corresponding RGB image and the last one shows the ground truth labels.

materials in a single image scene is called facade segmentation, i.e., the process of labelling each pixel in the image to a class, e.g., concrete, metal, and vegetation. While RGB images with three bands are commonly used in existing facade datasets such as ECP [20], eTRIMS [11], Graz [16], and CMP [21], hyperspectral imaging (HSI) divides the spectrum into many more bands and is rarely used for this purpose [3]. In HSI, the recorded bands have fine wavelength resolutions and cover a wide range of wavelengths beyond the visible light. It provides significantly more information than RGB images and can be utilised in many real-world applications, such as agriculture, geosciences, astronomy and surveillance.

The dataset presented in this paper was collected as part of a broader project to classify building materials for pre-disaster inventory collection. Images were taken of building facades in a light industrial environment. The dataset consists of 513 hyperspectral images and their corresponding RGB images, and each hyperspectral image is composed of 204 bands with a spatial resolution of 512 × 512 pixels. 44 classes of material/context were labelled across the images. Table 1 compares our LIB-HSI dataset with previous facade segmentation datasets.

2   Related Work

Hyperspectral imaging provides detailed spectral information by sampling the reflective portion of the electromagnetic spectrum. Instead of using only the three bands red, green and blue (RGB), the light emitted and reflected by an object is captured by hyperspectral sensors as a spectrum of hundreds of contiguous narrow band channels. The intensity of these bands are registered in a hyperspectral image. The human eye can only see reflections in the visible light spectrum, but a hyperspectral image can contain reflections in ultraviolet and infrared as well. By obtaining the spectrum for each pixel in the image of a scene, machine learning models can be trained to find objects and identify materials. It has been widely used in many real-world applications such as remote sensing [15], food [5], agriculture [4], forestry [1] and in the medical field [14].

For each pixel in a hyperspectral image related to a material, there is a vector consisting of reflectance data. This vector is known as a spectral signature. These vectors form a hyperspectral data cube (w, h, b) for an image with width w, height h, and b bands. Whereas $b = 3$ in the case of RGB images, b can go up to hundreds in HSI. As different materials have their own spectral signatures, we can use this information to put pixels into groups. The hyperspectral image classification problem is the task of assigning a class label to each pixel vector based on its spectral or spectral-spatial properties. This is closely related to the semantic segmentation task in computer vision. Given an input image, the task is to output an image with meaningful and accurate segments.

Despite the increasing interest in applying deep learning to hyperspectral imaging, the maturity is still low in this field. Compared to RGB images, the number of datasets available for HSI are significantly fewer and they are mostly old images of low resolution. Public datasets [6] commonly used for benchmarking mostly consist of a single image that is captured by an Airborne Visible/Infrared Imaging Spectrometer (AVIRIS) sensor over some area. In addition, the boundary between segmentation and classification is unclear in HSI studies. Image classification usually refers to the task of giving one or more labels to the entire image, while image segmentation refers to assigning a class to each pixel in an image. A lot of research on HSI classification is actually doing semantic segmentation. For instance, state-of-the-art models in HSI classification [2,18,19] performs segmentation by applying classification models on a pixel by pixel basis. In this study we also investigate the semantic segmentation algorithms on the LIB-HSI dataset.

Table 1. Comparison of facade segmentation datasets

Datasets	Images	Classes	Modal
ECP [20]	104	7	RGB
eTRIMS [11]	60	8	RGB
Graz [16]	50	4	RGB
CMP [21]	606	12	RGB
LIB-HSI(ours)	513	44	RGB and hyperspectral

3   Dataset

The Specim IQ (Specim, Spectral Imaging Ltd.), a handheld hyperspectral camera, was used to acquire images for our LIB-HSI dataset. It is capable of capturing hyperspectral data, recovering the illumination spectrum of the scene (via a white reference panel or Spectralon) and deriving the reflectance cube as well as visualising the results. The wavelength range of the camera is 400 to 1000 nm nm. Each image consists of 204 bands and has a spatial resolution of 512×512 pixels. The hyperspectral images are saved in the ENVI format. In addition, the camera also converts the hyperspectral images to pseudo-RGB images, captures a separate RGB image (via its RGB sensor) and saves them all on a SD card.

Images for the LIB-HSI dataset were taken of building facades in a light industrial environment. For each building, multiple images from various locations and angles were captured. The characteristics of the illumination and sunlight are captured using a white reference panel. The derived reflectance image will contain errors if the lighting varies significantly across an image, e.g., due to shadows, as the Specim IQ camera will not be able to estimate the varying scene illumination with any accuracy. Therefore, all of the images were captured under shadow or overcast conditions. The LIB-HSI dataset consists of 513 images in total. The Scyven hyperspectral imaging software was used to visualize the images [7]. The dataset was annotated by an external data annotation company. Every pixel was assigned a class label, making the annotated dataset suitable for supervised semantic segmentation. There are 44 classes in total, which are shown in Table 2. Generally, they are grouped by their material class, some by their context. The goal is to correctly classify every pixel into one of the 44 classes and automate the identification of materials in scenes of buildings usually found in light industrial areas.

There are some issues with the dataset that may make it unfavourable for training a deep learning model. Firstly, a dataset of 513 images with 44 classes is unusually small for training a neural network for semantic segmentation. Popular public RGB datasets have tens of thousands, or often more, images. To increase the size of the dataset, we performed data augmentation. Each image is cropped into nine patches, each of size 256×256 pixels. These patches are then rotated by $0°$, $90°$, $180°$, and $270°$. Hence, $4 \times 9 = 36$ patches are created for every image. This resulted in over 15768 patches for training. For testing, the test set is kept in its original form.

In addition, the dataset suffers from a severe class imbalance. Many of the classes are only found in a few images and many only represent a small area of pixels. It will be very hard to recognize those classes compared to the majority classes for the network. How we choose to split the data set for training and testing needs to be carefully considered. Simply splitting it at random may lead to certain classes being completely omitted from training or testing. In order to have a roughly equal representation of classes in each set, we developed a new method to split the dataset. The proposed splitting method first preforms an analysis on the contents of the dataset (i.e., labels) and then makes the split according to the properties of the data, trying to retain the closest match between

Table 2. Classes contained in the dataset

Group	Superclass	Classes Contained
1	Miscellaneous	Whiteboard, Chessboard, Tripod, Metal-Frame, Miscellaneous, Metal-Vent, Metal-Knob/Handle, Plastic-Flyscreen, Plastic-Label, Metal-Label, Metal-Pipe, Metal-Smooth-Sheet, Plastic-Pipe, Metal-Pole, Timber-Vent, Plastic-Vent, Wood-Ground
2	Vegetation	Vegetation-Plant, Vegetation-Ground, Soil, Woodchip-Ground
3	Glass Window	Glass-Window, Glass-Door, Concrete-Window-Sill
4	Brick	Brick-Wall, Brick-Ground, Tiles-Ground
5	Concrete	Concrete-Ground, Concrete-Wall, Concrete-Footing, Concrete-Beam, Pebble-Concrete-Beam, Pebble-Concrete-Ground, Pebble-Concrete-Wall
6	Blocks	Block-Wall
7	Metal	Metal-Sheet, Metal-Profiled-Sheet
8	Door	Timber-Smooth-Door, Door-Plastic, Metal-Smooth-Door, Metal-Profiled-Door, Timber-Profiled-Door
9	Timber	Timber-Wall, Timber-Frame

the profile of the data in the test and training sets. This can be achieved by the following procedure: (i) load each label image and create a vector descriptor of the image that is simply the number of pixels that belong to each class contained within the image; (ii) create an initial random 80/20 split of the data; (iii) calculate the average vector descriptor of each set; (iv) calculate a score for how well the sets align (using L2 norm distance between average vectors); (v) randomly swap images between the two sets and maintain the swap if the score is lower after the swap (i.e., a closer match between two sets); and (vi) repeat the previous step until a minimum is obtained.

4   Experiments

In this section, first, we briefly explain the evaluation metrics and the semantic segmentation algorithms applied to LIB-HSI dataset; second, the implementation details are explained.

4.1   Metrics

We used the accuracy and the Intersection over Union (IoU) metric to evaluate the results of semantic segmentation on images. The accuracy is the fraction of pixels in the prediction having the same label as the corresponding pixel in the ground truth image. The IoU metric is defined for two sets, A and B. It is calculated by the formula

$$\mathrm{IoU}(A, B) = \frac{|A \cap B|}{|A \cup B|},$$

where $|X|$ denotes the size of a set X, and $\cap$ and $\cup$ are set intersection and union, respectively. Fixing an arbitrary order for all pixels in an image, the labels of all pixels in an image represent an ordered set. Thus two images can be compared using the IoU metric. Then the IoU is lower thresholded to 0.5 and the result linearly mapped to the interval $[0, 1]$. Let x be the IoU, the score y is then given by

$$y = 2 \max(x, 0.5) - 1.$$

For metrics per class, precision, recall and IoU are used. Precision measures the proportion of positive predictions being correct, while recall measures the proportion of actual positives being correctly predicted. First, we find the number of true positive (TP), false positive (FP) and false negative (FN) labels for each image. Then we calculate the metrics with

$$\mathrm{Precision} = \frac{TP}{TP + FP}$$
$$\mathrm{Recall} = \frac{TP}{TP + FN}$$
$$\mathrm{IoU} = \frac{TP}{TP + FP + FN}$$

During training, the accuracy and IoU based score of the model are computed on every epoch. After that, we select the best performing model on the validation set and test it on the test set. We used the metrics above to evaluate how well a model performs on the test set.

4.2   Segmentation Models

Among well established semantic segmentation algorithms applied on hyperspectral images, we chose Fully Convolutional Network (FCN) [13] and U-Net [17] with the ResNet backbone [8] in our experiments.

FCN is a deep neural network architecture used mainly for semantic segmentation [13]. It does not employ fully connected layers and solely uses locally connected ones, such as convolution, pooling, and upsampling. Since FCN avoids using dense layers, it means fewer parameters and makes the networks faster to train. It also suggests that an FCN can work for any image size because all connections are local. A downsampling path is used to extract and interpret the contextual information, and an upsampling path is used to help the localization.

U-Net is a CNN architecture developed for biomedical image segmentation [17]. The architecture is based on the FCN. U-Net consists of a contracting path and an expanding path. Hence it is a u-shaped architecture. The contracting path consists of a series of convolutional layers and activation functions, followed by max-pooling layers to capture features while reducing the spatial size. The expanding path is a symmetric path consisting of upsampling, convolutional layers, and concatenation with the corresponding feature map from the contracting path. An important aspect of U-Net is its ability to work with few training images, which is suited for HSI. In addition, we need an end-to-end network to output a segmentation map for an input image instead of CNNs used in HSI classification that output a single label for each input. U-Net is one of the most popular methods for image segmentation and is known to work for many different applications. In this work, we used a U-Net with the ResNet backbone [8], which was equipped with Squeeze-and-Excitation (SE) blocks [9].

4.3   Objective Function

To tackle the issues of the imbalanced dataset, we used focal loss [12]. Unlike cross-entropy loss, these losses are designed to penalise wrong classifications more and focus on training the hard samples.

4.4   Implementation

All weights were randomly initialised and trained using the back-propagation algorithm. The Adam optimizer [10] was used as it is popular and compares favorably to stochastic gradient descent (SGD) in practice. We used a learning rate of 0.001, but the ReduceLROnPlateau scheduler from PyTorch was used later to introduce a decay. The scheduler halves the learning rate when the validation loss stops improving. We used a batch size of 8 and trained the models for about 50 epochs.

5   Results and Discussion

The quantitative results of deep learning semantic segmentation algorithms on LIB-HSI dataset are presented in Table 3. The qualitative results are demonstrated in Fig. 2. Using only RGB images to classify building materials from an image showed lower performance in comparison to using hyperspectral images or using both hyperspectral images and RGB. This could be due to the rich information captured in different wavelengths in hyperspectral data. Adding RGB images to the input did marginally improve the results, which might be attributed to the fact that hyperspectral images already include some pixel information contained in RGB images. Inferred results from FCN as a patch-based method indicated a relatively higher performance in comparison to U-Net and therefore for the combination of RGB and hyperspectral images we only reported

Table 3. Semantic segmentation results using FCN and U-Net with ResNet50 backbone for classification of building materials from RGB, hyperspectral (HS).

Modality	Method	Accuracy	IoU score	Average class		
				Precision	Recall	IoU
RGB	FCN	0.829	0.387	0.619	0.636	0.443
RGB	U-Net	0.687	0.162	0.360	0.426	0.236
HS	FCN	0.868	0.544	0.704	0.691	0.547
HS	U-Net	0.750	0.180	0.465	0.510	0.318
RGB+HS	FCN	0.875	0.675	0.701	0.690	0.544

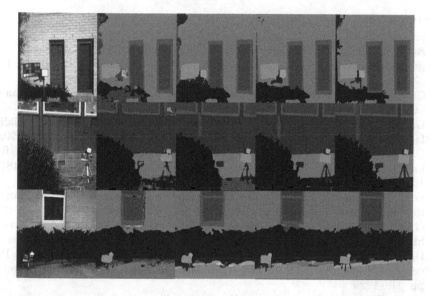

Fig. 2. Examples of semantic segmentation of the building materials using FCN and U-Net, from left to right panels; input RGB image, output of U-Net using RGB, output of FCN using only RGB, output of FCN using RGB and HSI, and the ground truth labels

the FCN results. More advanced deep learning models, including light-weight models are required to be investigated on LIB-HSI.

While using feature reduction algorithms, such as principal component analysis, is common in HSI segmentation [22,23], we decided not to use them as they could potentially increase the complexity while only offering a marginal improvement. In the future, we will investigate more advanced deep learning HSI dimension reduction methods.

6   Conclusion

In this paper we introduced a novel HSI and RGB building facade dataset, that can be accessed at CSIRO data access portal. Our LIB-HSI dataset was collected from a light industrial environment and covers a wide variety of building materials. We applied well established neural network algorithms on this dataset to segment the building materials. The results show that using both RGB and hyperspectral data can increase the performance of building material classification/segmentation. We envision that the LIB-HSI dataset can open up new research avenues for building material classification/segmentation.

References

1. Adão, T., et al.: Hyperspectral imaging: a review on UAV-based sensors, data processing and applications for agriculture and forestry. Remote Sens. **9**(11), 1110 (2017)
2. Chakraborty, T., Trehan, U.: Spectralnet: exploring spatial-spectral waveletcnn for hyperspectral image classification (2021)
3. Dai, M., Ward, W.O., Meyers, G., Tingley, D.D., Mayfield, M.: Residential building facade segmentation in the urban environment. Build. Environ. **199**, 107921 (2021)
4. Dale, L.M., et al.: Hyperspectral imaging applications in agriculture and AGRO-food product quality and safety control: a review. Appl. Spectrosc. Rev. **48**(2), 142–159 (2013)
5. Feng, Y.Z., Sun, D.W.: Application of hyperspectral imaging in food safety inspection and control: a review. Crit. Rev. Food Sci. Nutr. **52**(11), 1039–1058 (2012)
6. Graña, M., Veganzons, M., Ayerdi, B.: Hyperspectral remote sensing scenes. http://ehu.eus/ccwintco/index.php?title=Hyperspectral_Remote_Sensing_Scenes
7. Habili, N., Oorloff, J.: Scyllarus^{TM}: from research to commercial software. In: Proceedings of the ASWEC 2015 24th Australasian Software Engineering Conference, pp. 119–122 (2015)
8. He, K., Zhang, X., Ren, S., Sun, J.: Deep residual learning for image recognition. In: Proceedings of the IEEE Conference on Computer Vision and Pattern Recognition, pp. 770–778 (2016)
9. Hu, J., Shen, L., Sun, G.: Squeeze-and-excitation networks. In: Proceedings of the IEEE Conference on Computer Vision and Pattern Recognition, pp. 7132–7141 (2018)
10. Kingma, D.P., Ba, J.: Adam: a method for stochastic optimization (2017)
11. Korč, F., Förstner, W.: eTRIMS Image Database for interpreting images of man-made scenes (TR-IGG-P-2009-01) (2009). http://www.ipb.uni-bonn.de/projects/etrims_db/
12. Lin, T.Y., Goyal, P., Girshick, R., He, K., Dollár, P.: Focal loss for dense object detection. In: Proceedings of the IEEE International Conference on Computer Vision, pp. 2980–2988 (2017)
13. Long, J., Shelhamer, E., Darrell, T.: Fully convolutional networks for semantic segmentation. In: Proceedings of the IEEE Conference on Computer Vision and Pattern Recognition, pp. 3431–3440 (2015)
14. Lu, G., Fei, B.: Medical hyperspectral imaging: a review. J. Biomed. Opt. **19**(1), 010901 (2014)

15. Ma, L., Liu, Y., Zhang, X., Ye, Y., Yin, G., Johnson, B.A.: Deep learning in remote sensing applications: a meta-analysis and review. ISPRS J. Photogramm. Remote. Sens. **152**, 166–177 (2019)
16. Riemenschneider, H., Krispel, U., Thaller, W., Donoser, M., Havemann, S., Fellner, D., Bischof, H.: Irregular lattices for complex shape grammar facade parsing. In: 2012 IEEE Conference on Computer Vision and Pattern Recognition, pp. 1640–1647. IEEE (2012)
17. Ronneberger, O., Fischer, P., Brox, T.: U-Net: convolutional networks for biomedical image segmentation. In: Navab, N., Hornegger, J., Wells, W.M., Frangi, A.F. (eds.) MICCAI 2015. LNCS, vol. 9351, pp. 234–241. Springer, Cham (2015). https://doi.org/10.1007/978-3-319-24574-4_28
18. Roy, S.K., Krishna, G., Dubey, S.R., Chaudhuri, B.B.: Hybridsn: Exploring 3-d-2-d CNN feature hierarchy for hyperspectral image classification. IEEE Geosci. Remote Sens. Lett. **17**(2), 277–281 (2019)
19. Roy, S.K., Manna, S., Song, T., Bruzzone, L.: Attention-based adaptive spectral-spatial kernel ResNet for hyperspectral image classification. IEEE Trans. Geosci. Remote Sens. **59**(9), 7831–7843 (2020)
20. Teboul, O., Kokkinos, I., Simon, L., Koutsourakis, P., Paragios, N.: Shape grammar parsing via reinforcement learning. In: CVPR 2011, pp. 2273–2280. IEEE (2011)
21. Tyleček, R., Šára, R.: Spatial pattern templates for recognition of objects with regular structure. In: Weickert, J., Hein, M., Schiele, B. (eds.) GCPR 2013. LNCS, vol. 8142, pp. 364–374. Springer, Heidelberg (2013). https://doi.org/10.1007/978-3-642-40602-7_39
22. Zhang, S., Deng, Q., Ding, Z.: Hyperspectral image segmentation based on graph processing over multilayer networks. arXiv preprint. arXiv:2111.15018 (2021)
23. Zhao, J., Hu, L., Dong, Y., Huang, L., Weng, S., Zhang, D.: A combination method of stacked autoencoder and 3d deep residual network for hyperspectral image classification. Int. J. Appl. Earth Obs. Geoinf. **102**, 102459 (2021)

Improving Object Detection in VHR Aerial Orthomosaics

Tanguy Ophoff[✉][iD], Kristof Van Beeck[iD], and Toon Goedemé[iD]

EAVISE - PSI - ESAT, KU Leuven, Leuven, Belgium
{tanguy.ophoff,kristof.vanbeeck,toon.goedeme}@kuleuven.be
http://www.eavise.be/

Abstract. In this paper we investigate how to improve object detection on very high resolution orthomosaics. For this, we present a new detection model ResnetYolo, with a Resnet50 backbone and selectable detection heads. Furthermore, we propose two novel techniques to post-process the object detection results: a neighbour based patch NMS algorithm and an IoA based filtering technique. Finally, we fuse color and depth data in order to further increase the results of our deep learning model. We test these improvements on two distinct, challenging use cases: solar panel and swimming pool detection. The images are very high resolution color and elevation orthomosaics, taken from plane photography. Our final models reach an average precision of 78.5% and 44.4% respectively, outperforming the baseline models by over 15% AP.

Keywords: Aerial · Orthomosaic · Object detection · AI · Yolo · Sensor fusion

1 Introduction

Over the last years, geospatial data has become an essential source of information on which governments and industries base their decisions. Detailed and up-to-date geographic data has thus become of the utmost importance for many topics, including spatial planning, mobility, water resource management, energy, etc. For this reason, specialized companies are looking at methods to capture aerial images, in order to provide higher resolution imagery than satellites can offer. However, it is infeasible to manually work with this wealth of information and there is thus a need to automatically extract metadata from these orthomosaics. Artificial Intelligence (AI), being very good at pattern recognition, seems to be a suitable technology for this specific task.

In this paper, we investigate the potential of deep learning to extract such metadata. More specifically, we train and evaluate object detection models in order to detect solar panels and swimming pools from very high resolution (VHR) aerial orthomosaics with a ground resolution of 3 - 4 cm. In addition to the RGB data, the dataset also comprises of Light Detection and Ranging (LiDAR) data, providing detailed depth maps of the area. We first compare a general object

L. Karlinsky et al. (Eds.): ECCV 2022 Workshops, LNCS 13807, pp. 268–282, 2023.
https://doi.org/10.1007/978-3-031-25082-8_18

detection model, YoloV3 [16], with a model specifically designed for aerial object detection, D-Yolo [1]. Afterwards, we explore methods to improve the results of this pipeline:

- We create our own custom object detector, based on ResNet50 [10].
- We propose several optimizations to the post-processing pipeline, specifically targeted towards object detection on orthomosaics.
- We investigate the added value of fusing depth information with the color data into our deep learning model.

We validate our approach on two different use cases, namely solar panel and swimming pool detections. Solar panels are usually tightly clustered together and have an average size of around 50 pixels in length. This is in stark contrast with swimming pools, which are more dispersed throughout the images and have an average size of 200 pixels. These two use cases thus cover a wide range of possible variation for object detection in aerial imagery.

2 Related Work

Object detection has been a very active topic in the deep learning community for the past few decades. The various methodologies can be mainly divided in two different categories, being two-staged approaches and single-shot detectors. Pioneered by Girshick et al. [8,9], two-staged approaches first generate a selection of bounding boxes, called region proposals. These proposals are then classified and refined by a deep learning model in a second stage. While these techniques are orders of magnitude better than traditional computer vision approaches, they are computationally intensive and take a long time to process data. In this study, we need to cover vast areas of many km^2 at a very high resolution, resulting in an enormous search space. We therefore focus our attention to single-shot models. Firstly introduced by Redmon et al. [15] with the Yolo detector, single-shot object detectors reformulate the object detection task as a single regression problem, predicting both object location and classification label at the same time. This technique has the advantage of being much faster and also considers the full image data when performing classification, allowing single-shot detectors to use contextual cues to better detect and label objects. With YoloV2, they improve the model by starting from a fixed set of anchor boxes and regressing on the difference in size of the actual detection w.r.t. these anchors instead of regressing on the actual detection size [14]. Finally, YoloV3 performs detection at multiple scales with a feature pyramid network, making the model more robust at detecting objects of varying sizes [16].

When applying object detection on orthomosaics, the general approach is to cut the image into smaller patches, which are then individually processed by the deep learning pipeline [3]. Recently, several advancements have been made towards aerial object detection on orthomosaics [1,12,18]. Networks that perform less subsampling, or upsample their features with deconvolutions or feature pyramids tend to perform better, as the objects in aerial photography are usually smaller than their counterparts in natural images. Additionally, Van Etten [18] demonstrated that it is beneficial to first combine the detections of the different patches and perform the Non-Maximal Suppression (NMS) post-processing on the complete set of detections. This allows the different patches to overlap, providing a better coverage of the area, without creating duplicate detections. One problem with this approach is that it needs to compare all bounding boxes, taking a lot of processing power and memory. In this paper, we present a more efficient alternative.

Another factor to consider when training on these aerial images is the imbalance between non-empty patches with the targeted objects and empty patches without objects. Many datasets for remote sensing detection only contain image patches with the targeted objects present [5,11]. However, when considering the use cases of solar panel or swimming pool detection, there indeed seems to be a much higher number of empty patches in the orthographic imagery. As shown in [12], a higher accuracy can be reached by only training on non-empty patches. We investigate the potential of this technique in this paper as well.

Combining RGB and depth information in convolutional neural networks has already been successfully deployed for both classification [6,17] and object detection tasks [7,13,19]. Ophoff et al. [13] created a YoloV2-based RGBD detector, by running the RGB and depth channels in separate sub-networks, fusing the feature maps at various different stages in the network. They demonstrated that fusing the feature maps towards the end of the network yields the best results. In spite of that, to the best of our knowledge RGBD fusion has never been tried on aerial orthomosaics before.

3   Dataset

The data used in this study consists of RGB orthomosaics, taken by aerial plane photography over the region of Flanders, Belgium, with a ground sampling distance (GSD) of 3 cm. In addition to the RGB data, the setup also records LiDAR data, which is converted to a 2D digital elevation model (DEM) with a GSD of 25 cm. Both orthomosaics are stored as geolocalized images, using the Belgian Lambert 72 coordinate reference system (EPSG:31370).

The annotations are provided as horizontal bounding boxes, encoded as Geo-JSON files. They were provided by human annotators, based on visual inspection of the RGB orthomosaic data. By consequence, some inaccuracies can be expected w.r.t. both labeling and bounding box pixel accuracy. Nonetheless, our experiments establish that these annotations are of adequate quality for training and evaluation of our deep learning models.

In order to deploy detection models on orthomosaic imagery, we first split the data in distinct polygonal regions for training, validation and testing (See Fig. 1). As can be seen in Table 1, we approximately keep a 50-25-25% split between training, validation and testing areas.

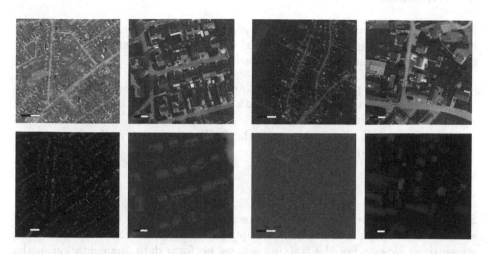

Fig. 1. Visual overview of our color and depth data (left solar panels; right swimming pools)

Table 1. Dataset statistics per split. Note that the validation and test sets have an overlap between patches, resulting in more patches per area. Non-empty patches denote image patches that contain at least one (part of an) object

	Train	Validation	Test
Solar panels			
Average object size	$47 \times 45 \pm 10$ px		
Patch size	416×416 px		
Region area	$5.0\,\text{km}^2$	$2.9\,\text{km}^2$	$2.2\,\text{km}^2$
Objects	16288	9504	7178
Non-empty patches	1753	1108	1045
Empty patches	24062	18187	18194
Swimming pools			
Average object size	$198 \times 198 \pm 101$ px		
Patch size	640×640 px		
Region area	$9.6\,\text{km}^2$	$3.2\,\text{km}^2$	$4.5\,\text{km}^2$
Objects	2017	359	624
Non-empty patches	2869	726	1221
Empty patches	20804	6156	8293

4   Methods

In this section, we first discuss the necessary steps we took in order to train any model on our orthomosaic dataset. Next, we look at various methods we propose to improve the accuracy of our deep learning pipeline:

- We improve the post-processing with a neighbour NMS and IoA filtering.
- We implement a custom model with a Resnet50 backbone and selectable multi-scale detection heads.
- By fusing depth information with the RGB color data, we further improve the accuracy of our model.

4.1   Baseline

Similarly to [12], we only train on non-empty patches, but perform validation and testing on both non-empty and empty patches, as this reflects the intended usage more closely. For the training set, we perform data augmentation in the form of HSV shifts, random horizontal and vertical flipping as well as XY-offset jittering. Positional augmentation is often implemented by taking minor random crops from the different sides of the image, or by adding a randomly sized border at each side. Since we start from a bigger orthomosaic, we implemented a random XY-offset jittering technique where we simply move the top-left point of our patch, resulting in patches with valid data everywhere. Additionally, we disable any overlap on the training dataset, as this is mostly redundant during training and the jitter augmentation already allows for different views of the same area. We train our models for a fixed number of epochs with a constant learning rate and perform a validation every 25 epochs. After 75% of the training is complete, we load the best previous weights and fine-tune the model further by lowering the learning rate with a factor of 10. For each use case and model, we perform an initial hyperparameter search with the Optuna library [2], maximizing the results on our validation dataset. The best parameters are then used for the remainder of the experiments.

4.2   Neighbour NMS

A major problem of processing patches of an image is that an object might be on the boundary between two patches. This makes it difficult for the model to correctly find and localize the object in question. We can solve this issue by allowing overlap between the different patches, but this introduces a new problem where we create duplicate detections on the overlapping regions of the patches. Van Etten [18] proposed to perform the NMS post-processsing on the combined detections from all the patches, thus filtering duplicate detections between patches as well. However, the memory requirements of this global NMS algorithm grow quadratically with the number of bounding boxes and is thus unsuited for processing large images, consisting of many patches. Moreover, only adjacent patches contain overlapping image data and should therefore be considered when performing NMS. Our proposed neighbour NMS approach loops through every patch of an image, selects all the detections of that patch and from adjacent patches in order to perform the NMS algorithm (See Fig. 2). This means that at most 9 patches will be considered together and thus the memory requirements will not grow w.r.t. the image size. Note that our algorithm is not completely equivalent with a globally applied NMS. When considering a certain patch, we also look at its adjacent patches, but not at the patches adjacent to those patches. This means that bounding boxes that are removed later - by a detection from another patch - are still allowed to filter other boxes in this first stage, similarly to the parallel fast-NMS implementation from Yolact [4].

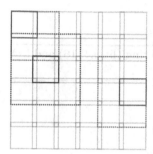  

Fig. 2. Left: Patch selection examples for our neighbour NMS algorithm. Middle: Example with local NMS. Right: Example with neighbour NMS. Note that there are less duplicate detections at patch boundaries with neighbour NMS

4.3 Intersection over Area Filtering

When detecting larger objects such as swimming pools, the object might be fully visible in one patch, but only partially visible in another. The model will then most likely detect the full object in the first patch, but only a partial object in the second patch. As the two different bounding boxes are not equal in size, they will not be filtered by the NMS algorithm (See Fig. 3).

Therefore, we developed an extra post-processing step, where we filter bounding boxes based on the intersection over area (IoA) instead of the intersection over union (IoU). When filtering detections, the traditional IoU-based NMS algorithm keeps the detection with the highest score, but our IoA filter keeps the detection with the biggest area. This can be problematic as it allows bounding boxes with a very low confidence to filter high confidence boxes. We thus rescale the IoA measurements, depending on the difference in confidence score between the detections (See Fig. 4). This allows detections to remove higher confidence detections as long as the difference in confidence is not too big. The complete neighbour NMS algorithm with IoA filtering is described in Algorithm 1.

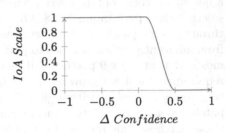

Fig. 3. A part of the pool is visible in the bottom patch. Without IoA filtering this patch does not get filtered, resulting in a false detection

Fig. 4. Confidence rescaling function that is used in the IoA filtering algorithm. $Y = 1 - smoothstep(x, 0.1, 0.5)$

Algorithm 1. Neighbour NMS algorithm with IoU filtering (lines 9-16) and IoA filtering (lines 17-26)

1: $D \leftarrow Detections\ sorted\ by\ descending\ confidence\ score$
2: $P \leftarrow Patches\ of\ an\ image$
3: $\lambda_{IoU} \leftarrow IoU\ Threshold$
4: $\lambda_{IoA} \leftarrow IoA\ Threshold$
5: **procedure** NEIGHBOURNMS($D, P, \lambda_{IoU}, \lambda_{IoA}$)
6:     **for** $p \in P$ **do**
7:         $P_{neighbours} \leftarrow selectNeighbouringPatches(p)$
8:         $D_{selected} \leftarrow selectDetections(D, P_{neighbours} \cup p)$

9:         **for** $d_i \in D_{selected}$ **do**
10:             **for** $d_j \in D_{selected}[i :]$ **do**
11:                 **if** $IoU(d_i, d_j) \geq \lambda_{IoU}$ **then**   ▷ Remove d_j from detections
12:                     $D \leftarrow D \setminus d_j$
13:                     $D_{selected} \leftarrow D_{selected} \setminus d_j$
14:                 **end if**
15:             **end for**
16:         **end for**

17:         **for** $d_i \in D_{selected}$ **do**
18:             **for** $d_j \in D_{selected}[i :]$ **do**
19:                 $ioa \leftarrow intersection(d_i, d_j)\ /\ area(d_i)$
20:                 $scale \leftarrow 1 - smoothstep(conf_j - conf_i, 0.1, 0.5)$
21:                 **if** $ioa * scale \geq \lambda_{IoA}$ **then**   ▷ Remove d_j from detections
22:                     $D \leftarrow D \setminus d_j$
23:                     $D_{selected} \leftarrow D_{selected} \setminus d_j$
24:                 **end if**
25:             **end for**
26:         **end for**
27:     **end for**
28: **end procedure**

4.4   ResnetYolo

Traditionally, deeper object detection models tend to perform better, as they are capable of modeling more complex features. In order to keep the computational and memory requirements of these models in check, we often subsample the data in the spatial dimension multiple times. However, this spatial subsampling has proven to be detrimental for aerial object detection, where we favor shallower models with spatially bigger feature maps [1,12,18]. Nonetheless, we argue that the used data is of a much higher resolution than traditional aerial or satellite imagery and thus explore the possibility of using deeper models with more subsampling. The initial results in Sect. 5 support this hypothesis, as the YoloV3 model outperforms D-Yolo for both our use cases.

We therefore create our own Yolo-based detector with a Resnet50 backbone [10]. On top of this backbone we add a feature pyramid network in order to combine shallower – but spatially bigger – feature maps with feature maps deeper in the network which are more subsampled. Finally, our model outputs four different detection heads at different spatial subsampling factors of 64, 32, 16 and 8 respectively (See Fig. 5).

For aerial object detection we are often interested in a single class of objects or in highly related classes of objects. This means that the targeted objects usually have a very similar size. There is thus little to no need for a multi-scale object detector. We thus made the four detection heads selectable, meaning you can customize whether to use a single detection scale or any combination of the four different scales. This ensures that out model is more generic and allows it to be used for detecting both small and large objects in varying scenarios.

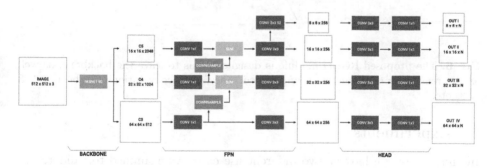

Fig. 5. The proposed ResnetYolo model architecture. Each head can be individually enabled or disabled

4.5   RGBD Fusion

As the dataset contains LiDAR based depth maps, we also look at fusing the color and depth data in the deep learning pipeline, in order to increase the accuracy of our model. A study of an RGBD fusion object detector on natural images showed promising results, as their RGBD models outperformed the RGB baseline on each experiment [13]. Additionally, they experimentally determined that fusing the color and depth data towards the end of the model gives the best results.

Consequently, we created a module that duplicates the backbone in order to evaluate it on both the color and depth data. By prepending a 1×1 convolution to the backbone that transforms the input to a fixed 3 channel tensor, we keep the backbone for both color and depth data identical. This allows us to use the same pretrained ImageNet weights for both branches of the backbone. Afterwards, we combine every individual output of the backbone by concatenating the tensors from each branch and running an extra convolution that reduces the number of channels back to the original (See Fig. 6). This allows the remainder of the network to be left unchanged and is roughly similar to the mid fusion of [13]. As reported in Sect. 5, we also experiment with different convolutional kernels for these last fusion convolutions.

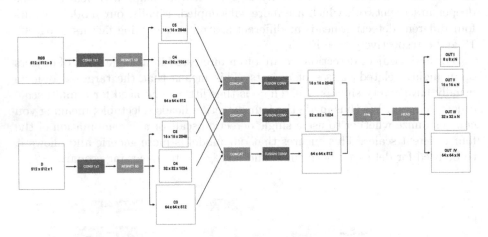

Fig. 6. The proposed RGBD module is designed so as to leave the backbone, as well as the remainder of the network unchanged. The fusion convolutions can be 1×1, 3×3, etc.

5   Experiments

In this section we look at two different use cases we examined to validate our techniques, solar panel and swimming pool detection. As explained in Sect. 3, these two cases cover a wide range of possible variation for object detection in aerial orthomosaics.

5.1   Solar Panel Detection

As solar panels are fairly small objects, we decide to use a patch size of 416×416 pixels, which is fairly standard practice. Before training, we compute 3 anchor boxes on the combined training and validation sets, with a K-means clustering algorithm. The resulting anchors are: 52.6×52.9, 51.6×38.5 and 37.7×40.2.

Post-processing. We start by training a standard YoloV3 [16] and D-Yolo [1] model as a baseline. Our training takes 250 epochs, and we perform a validation run every 25 epochs. At the end of training, we test the model that achieved the best validation run on our testing dataset, reaching an initial average precision (AP) of 62.96% for YoloV3 and 59.67% for D-Yolo.

We then validate our post-processing techniques on these two models. As a first test, we simply use a 50 pixel overlap between different patches. This however results in a significant drop in accuracy of 10% for both our models. By replacing the regular per-patch NMS with our neighbour NMS algorithm we manage to increase the results of our models by ~3% (See Fig. 7). Indeed, YoloV3 reaches an accuracy of 65.84% and D-Yolo reaches 63.04%, when using neighbour NMS with overlap. Finally, we also try filtering based on the IoA with different thresholds. However, this does not seem to increase the results by any significant margin for solar panels. Solar panels being relatively small, the chances of having partial solar panels in an image patch are relatively limited and as such it is expected that this technique only yields marginal effects. However, the fact that it does not decrease the results significantly allows one to enable this post-processing regardless of the targeted objects to detect. We nonetheless decide to disable the IoA filtering for the remainder of the solar panel experiments, and thus use our neighbour NMS algorithm with a 50 pixel overlap between patches.

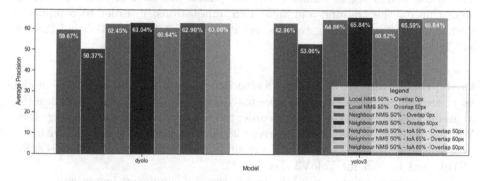

Fig. 7. Average Precision of D-Yolo and YoloV3 with various overlaps and post-processing methods on the solar panel data

ResnetYolo. Contrary to previous research [1,12,18], the deeper YoloV3 model outperforms D-Yolo on this data. We presume the reason for this is that the data is more detailed than previous work on orthomosaics, with its ground resolution of 3–4 cm.

Our even deeper ResnetYolo model further validates this hypothesis, as it outperforms YoloV3 by more than 10% AP. We experiment with our different selectable multi-scale heads and notice the first two, less subsampled detection heads perform best (See Table 2). By selecting both heads I&II together, we further improve the accuracy and achieve a final average precision of 76.11%.

Table 2. Results of ResnetYolo with different heads enabled on the solar panel data

ResnetYolo heads	I	II	III	IV	I&II
Average precision	75.73%	74.98%	60.25%	1.00%	**76.11%**

RGBD Fusion. Finally, we fuse color and depth data with our RGBD ResnetYolo model. As can be seen in Table 3, a 3 × 3 fusion kernel yields the best results, with an average precision of 78.46%.

Table 3. Results of ResnetYolo with fusion kernels on the solar panel data

Fusion kernels	1 × 1	3 × 3	5 × 5
Average precision	76.67%	**78.49%**	77.68%

5.2   Swimming Pool Detection

In contrast to solar panels, swimming pools are much larger objects (see Table 1). In order to limit the number of partial pools in a patch, we increase the patch size to 640 × 640 pixels. Similarly to solar panels, we use a K-means clustering algorithm on the training and validation annotations, in order to compute 3 anchor boxes. The resulting anchors are: 297.6 × 405.7, 268.3 × 231.9 and 123.2 × 124.3.

Post-processing. We again train a standard YoloV3 [16] and D-Yolo [1] model as a baseline. Seeing as there are more patches for swimming pools, we reduce the number of training epochs to 200, in order to keep the training times manageable. We still perform a validation run every 25 epochs, keeping the model with the best validation results. Our baseline results on swimming pools are 16.64% for D-Yolo and 13.96% for YoloV3 (See Fig. 8).

Even without neighbour NMS, we can reach a significant increase of almost 10% AP by adding a 100 pixel overlap between patches. These results get improved even further with our neighbour NMS algorithm by 2%. Because swimming pools are large objects, our IoA filtering algorithm has a better influence on the results. Indeed, when filtering the YoloV3 results with a IoA threshold of 65%, the model reaches an accuracy of 29.95%, which is almost 4% better than without. As the results with and without IoA filtering for D-Yolo are similar, we decide to test both methods on the ResnetYolo models.

ResnetYolo. Our deeper ResnetYolo model outperforms the baseline for swimming pools as well. In Table 4, we can clearly see that the more subsampled detection heads perform best in this scenario. This is to be expected, as swimming pools are larger objects.

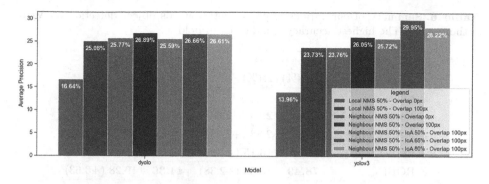

Fig. 8. Average Precision of D-Yolo and YoloV3 with various overlaps and post-processing methods on the swimming pool data

Additionally, the single detection head III outperforms the multi-scale detection with both heads III&IV. The sparse nature of swimming pools favors models that have a low density of potential detections per patch. A single-scale model thus trains best, as it outputs fewer detections.

Finally, we notice that IoA filtering greatly helps ResnetYolo, as it improves our best results by more than 5%. Our best model reaches an accuracy of 37.73%, outperforming the baseline by 7.8%. The remainder of the experiments for swimming pools is thus conducted with a single detection head 3 and an IoA threshold of 65%.

RGBD Fusion. Fusing color and depth data further increases our results. With an increase of 6.6%, our best fusion model uses 5×5 fusion kernels and reaches a total accuracy of 44.36% (See Table 5).

Table 4. Results of ResnetYolo with different heads enabled on the swimming pool data

ResnetYolo heads	I	II	III	IV	II&IV
No IoA	3.03	26.68	29.77	**32.04**	28.90
65% IoA	3.09	28.19	30.69	**37.73**	36.80

Table 5. Results of ResnetYolo with fusion kernels on the swimming pool data

Fusion Kernels	1×1	3×3	5×5
Average Precision	39.59	39.82	**44.36**

Table 6. Summary of our various improvements towards object detection on VHR orthomosaics. The highest accuracy is indicated in bold

	Solar panels		Swimming pools	
	$AP(\%)$	$\Delta(\%)$	$AP(\%)$	$\Delta(\%)$
Baseline	62.96		25.08	
✓ Neighbour NMS	65.84	2.88 (+2.88)	26.89	1.81 (+1.81)
✓ IoA Filtering	65.84	2.88 (+0.00)	29.95	4.87 (+3.06)
✓ ResnetYolo	76.11	13.15 (+10.27)	37.73	12.65 (+7.78)
✓ RGBD	**78.49**	15.53 (+2.38)	**44.36**	19.28 (+6.63)

6   Conclusions

In this paper we improved various facets of the deep learning pipeline for object detection on VHR aerial orthomosaics. We validated our techniques on two distinct use cases, showing the potential of each in challenging scenarios. The results of our experiments are summarized in Table 6.

By allowing overlap between adjacent patches in the orthomosaic and filtering the detection results with our neighbour NMS algorithm we increase the results of any model consistently by a few percentage points in AP.

Adding an extra filter in the NMS algorithm, based on the IoA between detections, gives an additional boost to the accuracy. The effect is most noticeable on larger objects, as is demonstrated with the swimming pool use case where we increased the results by more than 3%. Nonetheless, if properly thresholded, the extra filter has no negative effect on smaller objects such as solar panels and can thus be used as an intrinsic part of the object detection pipeline on orthomosaics.

Our proposed ResnetYolo architecture significantly outperforms the baseline on this high resolution dataset, increasing the results by 10.3% for the solar panel use case and 7.8% for swimming pools. By allowing to toggle the various multi-scale detection heads, the model is easily adaptable to many different scenarios, making it much more generally applicable than its alternatives.

Our last improvement consists of color and depth information fusion in our model. Different kinds of input data make for more robust models and gives a final boost in average precision.

Altogether our four contributions increase the results of our baseline by a significant margin. We demonstrated this by performing extensive experiments on two different use cases. Indeed, we increased the results by 15.5% AP for the use case of solar panels, reaching a final accuracy of 78.5%. For the use case of swimming pools, we reached an average precision of 44.36%, which equals an increase of 19.28% compared to our best baseline model.

Acknowledgements. This project was funded by VLAIO. We would like to thank Vansteelandt BV for preparing and providing the data used in this project.

References

1. Acatay, O., Sommer, L., Schumann, A., Beyerer, J.: Comprehensive evaluation of deep learning based detection methods for vehicle detection in aerial imagery. In: 2018 15th IEEE International Conference on Advanced Video and Signal Based Surveillance (AVSS), pp. 1–6. IEEE (2018)
2. Akiba, T., Sano, S., Yanase, T., Ohta, T., Koyama, M.: Optuna: a next-generation hyperparameter optimization framework. In: Proceedings of the 25rd ACM SIGKDD International Conference on Knowledge Discovery and Data Minding (2019)
3. Alganci, U., Soydas, M., Sertel, E.: Comparative research on deep learning approaches for airplane detection from very high-resolution satellite images. Remote Sens. **12**(3), 458 (2020)
4. Bolya, D., Zhou, C., Xiao, F., Lee, Y.J.: YOLACT: real-time instance segmentation. In: Proceedings of the IEEE/CVF International Conference on Computer Vision, pp. 9157–9166 (2019)
5. Ding, J., et al.: Object detection in aerial images: a large-scale benchmark and challenges. IEEE Trans. Pattern Anal. Mach. Intell. 1 (2021). https://doi.org/10.1109/TPAMI.2021.3117983
6. Eitel, A., Springenberg, J.T., Spinello, L., Riedmiller, M., Burgard, W.: Multimodal deep learning for robust RGB-D object recognition. In: 2015 IEEE/RSJ International Conference on Intelligent Robots and Systems (IROS), pp. 681–687. IEEE (2015)
7. Farahnakian, F., Heikkonen, J.: A comparative study of deep learning-based RGB-depth fusion methods for object detection. In: 2020 19th IEEE International Conference on Machine Learning and Applications (ICMLA), pp. 1475–1482 (2020). https://doi.org/10.1109/ICMLA51294.2020.00228
8. Girshick, R.: Fast R-CNN. In: ICCV, pp. 1440–1448 (2015). https://doi.org/10.1109/ICCV.2015.169
9. Girshick, R., Donahue, J., Darrell, T., Malik, J.: Rich feature hierarchies for accurate object detection and semantic segmentation. In: CVPR, pp. 580–587 (2014)
10. He, K., Zhang, X., Ren, S., Sun, J.: Deep residual learning for image recognition. In: Proceedings of the IEEE Conference on Computer Vision and Pattern Recognition, pp. 770–778 (2016)
11. Liu, Z., Wang, H., Weng, L., Yang, Y.: Ship rotated bounding box space for ship extraction from high-resolution optical satellite images with complex backgrounds. IEEE Geosci. Remote Sens. Lett. **13**(8), 1074–1078 (2016)
12. Ophoff, T., Puttemans, S., Kalogirou, V., Robin, J.P., Goedemé, T.: Vehicle and vessel detection on satellite imagery: A comparative study on single-shot detectors. Remote Sens. **12**(7), 1217 (2020)
13. Ophoff, T., Van Beeck, K., Goedemé, T.: Exploring RGB+depth fusion for real-time object detection. Sensors **19**(4) (2019). https://doi.org/10.3390/s19040866, https://www.mdpi.com/1424-8220/19/4/866
14. Redmon, J., Farhadi, A.: YOLO9000: better, faster, stronger. In: CVPR, pp. 6517–6525 (2017). https://doi.org/10.1109/CVPR.2017.690
15. Redmon, J., Divvala, S., Girshick, R., Farhadi, A.: You only look once: unified, real-time object detection. In: CVPR, pp. 779–788 (2016)
16. Redmon, J., Farhadi, A.: YOLOv3: an incremental improvement. Technical report (2018)

17. Schwarz, M., Schulz, H., Behnke, S.: RGB-D object recognition and pose estimation based on pre-trained convolutional neural network features. In: 2015 IEEE International Conference on Robotics and Automation (ICRA), pp. 1329–1335. IEEE (2015)

18. Van Etten, A.: You only look twice: rapid multi-scale object detection in satellite imagery. arXiv preprint arXiv:1805.09512 (2018)

19. Zhou, K., Paiement, A., Mirmehdi, M.: Detecting humans in RGB-D data with CNNs. In: 2017 Fifteenth IAPR International Conference on Machine Vision Applications (MVA), pp. 306–309 (2017). https://doi.org/10.23919/MVA.2017.7986862

Active Learning for Imbalanced Civil Infrastructure Data

Thomas Frick[1,2]([✉]), Diego Antognini[1], Mattia Rigotti[1], Ioana Giurgiu[1], Benjamin Grewe[2], and Cristiano Malossi[1]

[1] IBM Research, Zurich, Switzerland
diego.antognini@ibm.com, {fri,mrg,igi,acm}@zurich.ibm.com
[2] Institute of Neuroinformatics, UZH and ETH Zurich, Zurich, Switzerland
bgrewe@ethz.ch

Abstract. Aging civil infrastructures are closely monitored by engineers for damage and critical defects. As the manual inspection of such large structures is costly and time-consuming, we are working towards fully automating the visual inspections to support the prioritization of maintenance activities. To that end we combine recent advances in drone technology and deep learning. Unfortunately, annotation costs are incredibly high as our proprietary civil engineering dataset must be annotated by highly trained engineers. Active learning is, therefore, a valuable tool to optimize the trade-off between model performance and annotation costs. Our use-case differs from the classical active learning setting as our dataset suffers from heavy class imbalance and consists of a much larger already labeled data pool than other active learning research. We present a novel method capable of operating in this challenging setting by replacing the traditional active learning acquisition function with an auxiliary binary discriminator. We experimentally show that our novel method outperforms the best-performing traditional active learning method (BALD) by 5% and 38% accuracy on CIFAR-10 and our proprietary dataset respectively.

Keywords: Class imbalance · Active learning · Deep learning · Neural networks

1 Introduction

Civil infrastructures are constantly being monitored for critical defects, as failure to recognize deficiencies can end in disaster. Unfortunately, detailed inspections are time-consuming, costly, and sometimes dangerous for inspection personnel. We are working towards automating all aspects of visual inspections of civil infrastructures by recent advances in drone technology [6,18] that enable fast and remote visual inspection of inaccessible structures such as wind turbines, water dams, or bridges and simultaneous advances in instance segmentation using deep learning models [8,28,36] that facilitate accurate detection of surface defects on high resolution images. We are combining these novel technologies by utilizing drones equipped with high-resolution cameras to capture image material of the structures and deep neural networks to detect defects. Our goal is to decrease the duration of inspections, improve condition assessment frequency, and ultimately minimize harm to human health.

It is relatively simple to collect large amounts of data of civil infrastructures for our proprietary dataset. However, similar to many other projects applying deep learning, annotating the collected samples is incredibly costly and time-intensive as highly-trained engineers are required for the labeling process. To minimize costs, a popular approach is to annotate a small portion of the dataset first and then use Active Learning (AL) to select informative samples that should be labeled next. The goal is to optimize a trade-off between additional annotation cost (number of annotated samples) and an increase in model performance. AL has been successfully applied in medical imaging [24,33], astronomy [29], and surface defect detection [9]. For its application in industry, we observe two differences to the traditional AL setting:

First, previous research [3,11,20] has focused on starting the AL process with little to no data (100 samples or cold start). In contrast, industry projects often start the AL process with a larger pool of labeled training data: Despite the high annotation costs, a random set of initial data is labeled for a proof of concept. Only then data collection and labeling efforts are scaled up. In this phase, the efficient selection of samples is invaluable. Unfortunately, traditional AL strategies barely outperform random selection given a large initial dataset, as shown in our experiments.

Secondly, most academic datasets are class balanced, with each class having the same number of samples (e.g., CIFAR-10 [21] 5,000). In contrast, real-world industry datasets usually suffer from a long-tailed class distribution [25,39]. Moreover, the minority classes are often the most important ones. This is usually the case for civil structures: dangerous/critical defects rarely appear on drone scans of the structure as they are well maintained. Previous work on AL for imbalanced data [4,22,34] has shown that classical AL methods fail to select samples of the minority classes for heavily imbalanced datasets and therefore fail to improve model performance for the minority classes. These works focus on developing sample selection strategies that choose samples according to an uncertainty-based metric or a diversity-based approach.

We present a novel method that effectively and efficiently selects minority samples from a pool of unlabeled data for large datasets suffering from heavy class imbalance. Contrary to other AL methods that try to find informative samples from all classes, we limit ourselves to selecting samples for a single minority class, investing the total labeling budget for samples that improve model performance for only that minority class. To this end, our method replaces the AL acquisition function with a binary discriminator explicitly trained in a one-vs-all fashion (minority vs. majority classes) to distinguish between unlabeled minority and majority samples. In each cycle, the discriminator selects samples to be labeled next according to the highest prediction scores. We experimentally confirm that classical active learning methods fail to significantly improve model

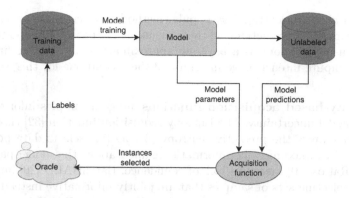

Fig. 1. Active learning cycle: a model is trained on a pool of labeled data. Given the model predictions and parameters, an acquisition function selects informative samples from an unlabeled pool of data which are sent to the oracle for labeling.

performance for CIFAR-10 [21] and our proprietary civil engineering dataset 3. Applying our method to our proprietary civil infrastructure dataset, we show a minority class recall improvement of 32% and an overall accuracy gain of 14% compared to the best-performing traditional AL method (BALD [17]).

2   Related Work

Active learning is a well studied problem [4,13,31] and especially advantageous for applications with high annotation costs because highly specialized annotators are needed. Active learning works by iterative selecting informative samples according to a query strategy from an unlabeled pool of data. The chosen samples are passed to the oracle (usually a human annotator) to be labeled. The goal is to improve model performance as much as possible while labeling as few samples as necessary. Settles et al. [31] identifies two different AL scenarios: In sample-based selective sampling, a stream of samples arrives one after another. Therefore, the algorithm decides whether to label or discard each sample without information about samples arriving in the future. In contrast, in pool-based active learning, the algorithm has access to the entire pool of unlabeled samples and needs to select samples to be labeled next. Our work is set in the second scenario.

Active Learning query strategies can be divided into geometric-based and uncertainty-based methods. Geometric-based approaches make use of the underlying feature space to select informative samples. Feature embeddings are necessary to have a meaningful feature space for an application to high-dimensional input data such as images. For our application, we focus on uncertainty-based methods as we assume that there is no pre-trained model available that can be used as an embedding. There is a wide variety of work on uncertainty estimation of deep neural networks [12]. Two popular methods that estimate uncertainty by generating multiple predictions on the same input are ensembling multiple models [23] or using Dropout as a bayesian approximation [11]. As the first option is highly compute-intensive, we make use of the second one for this work.

Uncertainty-based acquisition functions judge a samples informativeness based on model uncertainty. The Entropy acquisition function [32] chooses samples that maximise the predictive entropy, BALD [17] selects data points that are expected to maximise the information gained about the model parameters, Variation Ratios [10] measures lack of confidence. BatchBALD [20] improves on BALD by selecting sets of samples that are jointly informative instead of choosing data points that are informative individually. As BatchBALD is impractically compute-intensive for large batches and BALD has been shown to outperform other older methods, we focus only on Entropy and BALD for our experiments.

Imbalanced datasets have been widely studied (see Kaur et al. [19] for an extensive overview). Prior work frequently utilizes resampling techniques (e.g., oversampling and undersampling, or a combination of both) to balance the training data. While Hernandez et al. [16] show that simple resampling techniques can significantly improve model performance, Mohammed et al. [27] conclude that undersampling may discard informative majority class samples and therefore decrease majority class performance. There are synthetic sampling methods (SMOTE [7], ADSYN [14]) that generate new samples by interpolating in the underlying feature space. However, the underlying input feature space is too complex for image data to apply these methods successfully. Weighting samples according to inverse class frequency has also been successfully applied to prioritize underrepresented classes in training. For our work, we exclusively use oversampling as combining duplication of samples with image augmentation techniques results in a much more diverse set of minority samples than weighting the classes. One disadvantage of this method is the resulting much larger training pool and the longer training times compared to the class-weighted approach.

3   Background

As mentioned above, we work towards automating visual inspections of civil infrastructures. We have developed an automated inspection pipeline consisting of 4 consecutive stages: First, drones capture grids of high-resolution images, which are later stitched into full image scenes. Next, state-of-the-art deep learning methods analyze the image scenes and highlight defects such as cracks or rust. Finally, the size of detected defects is measured with a precision of 0.1 mm. The pipeline's output supports civil engineers in prioritizing further in-person inspections and maintenance activities.

Fig. 2. DJI Matrix 300 Hi-Res Grid: example of a 7 × 14-grid of 98 high-resolution images. Each individual image is 5184 × 3888 pixel in size resulting in a overall data size for a single pillar face of almost 1 GB.

3.1   Image Data

We have collected a civil infrastructure dataset consisting of high-resolution images of concrete bridge pillars. The data was collected in 2021 by certified drone pilots using DJI Matrix 300 [1] drones equipped with a Zenmuse H20 [2] lens attachment. Scenes were captured using DJI's Hi-Res Grid Photo mode as shown in Fig. 2. Drone pilots manually mark the area of the civil structure under inspection on an overview image. While hovering in place, the drone captures a grid of overlapping images by gimbaling the zoom lens. In total, 22 bridge pillars were scanned from each side, resulting in a dataset of over 22'000 raw images (5184 × 3888 pixels).

3.2   Instance Segmentation Annotations

In light of our focus on defect detection, we have invested considerable effort in creating a high-quality instance segmentation dataset where each defect is categorized and localized using mask annotations. In collaboration with our civil engineering experts, we developed extensive annotation guidelines focusing on the following six defect categories: rust, spalling, cracks, cracks with precipitation, net-cracks, and algae (see Fig. 3 for examples). Over six months, a team of annotators labeled 2500 images resulting in around 14'000 defect annotations. Given the well-maintained condition of the inspected structure, critical defects such as rust or cracks with precipitation appear infrequently. Consequently, our dataset suffers from a long-tailed class distribution with a maximum imbalance ratio of 1 to 130 (cracks with precipitation vs. cracks).

4   Method

This section first describes our novel method which replace the acquisition function with a binary discriminatory trained on the labeled pool to select minority class samples from the unlabeled pool. Secondly, it describes the dataset conversion algorithm which we use to convert our proprietary instance segmentation dataset to a classification dataset for the experiments.

4.1   Active Learning for Heavily Imbalanced Data

The goal of the classical AL setting is to improve the model performance by labeling additional informative samples. A model is trained on the initial pool of labeled data. Given the model's predictions and parameters, an acquisition function determines which samples from the unlabeled pool are sent to the oracle for labeling. The newly annotated samples are then added to the labeled pool, after which the cycle begins again by retraining the model. Traditional AL methods [3,11,17,20] strive to select samples for which the model performance for all classes improves most.

In contrast, we focus on improving model performance for a single pre-selected minority class. Given the high class imbalance of the initial AL dataset, we hypothesize that we do not need to find the most diverse or informative set of samples but that selecting and labeling any samples of the minority class will improve model performance. Therefore, our method replaces the traditional uncertainty-based AL acquisition function with an auxiliary binary classifier acting as a discriminator between the minority and majority classes. The discriminator is trained on the labeled AL pool in each cycle. Its predictions on the unlabeled pool are used to select samples for labeling. Selection is based on prediction scores, choosing the top-K samples.

For the training of the binary discriminator, binary labels are computed from the original multi-class training set in a one-vs-all fashion. As a result, the binary classification dataset is even more imbalanced than the original multi-class dataset. We improve discriminator performance by combating the bias of the class imbalance by applying the following modifications to the training procedure:

- oversample the positive class (the original minority class) until we reach balance with respect to the negative class (all majority classes);
- apply standard image augmentation techniques (flip, shift, scale, rotate, brightness, and contrast - see Albumentations [5]). As we apply the augmentations to the large number of minority samples generated from the oversampling in the first step, we end up with a highly diverse set of minority class samples;
- apply batch augmentations (MixUp [38], CutMix [37] - see timm [35] library) to further diversify the samples in each batch stabilizing the training procedure of the binary discriminator.

4.2   Instance Segmentation to Classification Dataset Conversion

As mentioned in Sect. 3, we are working towards automated detection of surface defects for civil infrastructures. We train instance segmentation models on the original dataset with instance mask annotations to detect the defects. Our annotations must be labeled with extraordinary precision due to the defects being of such small size (e.g., cracks in the order of millimeters). This additional time effort adds to the already high prize per annotation. Therefore, we have been working on a technique to use weakly supervised learning with class-level supervision to generate class activation maps (CAMs [30]) from which we extract fine segmentation masks. Unfortunately, there are still high annotation costs associated with class-level labels. We are trying to decrease these costs with active learning in this work. Consequently, we use class-level labels for our experiments which we extract from the original instance segmentation dataset using a patch-based dataset conversion algorithm.

The conversion algorithm works by extracting fixed-size patches from the original images. Intuitively, a patch can be assigned to a category if it depicts a piece of the original class; this can be either shape or texture. While the texture is informative enough for some objects to attribute the correct class, others are only correctly classifiable with information about their shape. As extracted patches only offer a small window into the original image, an object's shape information can be easily lost if a too small patch size is chosen.

The algorithm processes one image at a time: multiple patches are sampled for each instance and assigned the corresponding category. Additionally, patches are extracted from regions of the original image that do not contain any instances/defects and are assigned to an additional, newly introduced class "Background". We randomly sample instance patches such that the center point of the patch lies within the instance annotation and background patches such that the center point lies anywhere within the original image's borders. The algorithm rejects a sampled patch if it violates one of the following criteria: 1) it overlaps with an already chosen valid patch, 2) it intersects with an instance annotation belonging to a different category, 3) the patch breaches the boundaries of the original image. The algorithm extracts patches until the total number of required patches is reached or until the algorithm exceeds a total number of sampling attempts. Figure 3 shows the sampled patches for an example image as well as one example patch per category.

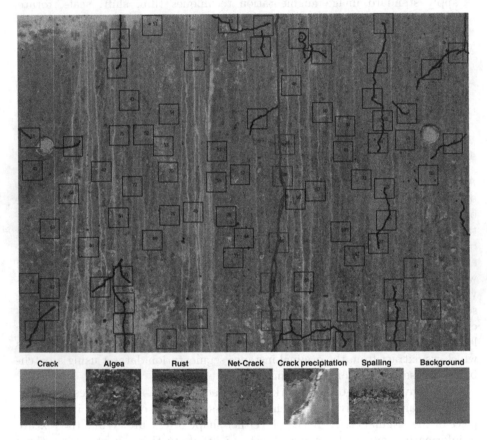

Fig. 3. Patch-based classification dataset conversion algorithm: we show instance segmentation polygons of the original image colored by class (red crack, yellow crack with precipitation, black background) and sampled patch boarders with the same color scheme. Below, we show one example patch per category of the original dataset plus one example patch for the additional background class. (Color figure online)

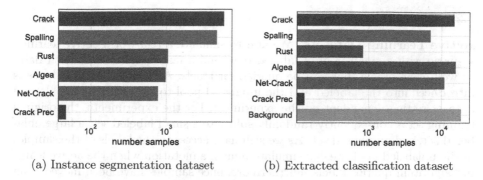

(a) Instance segmentation dataset   (b) Extracted classification dataset

Fig. 4. Class distribution of our civil infrastructure datasets: (a) shows statistics for our original instance segmentation dataset while (b) describes statistics for the classification dataset extracted with the path-based algorithm from the instance segmentation dataset (a). The additional background class in Figure (a) is a result of the background patch sampling of the algorithm.

5   Evaluation

In this section we first describe how we convert the image segmentation to a classification dataset with our patch extraction method to prepare our proprietary dataset. Then, we experimentally show how traditional active learning methods fail to improve minority class performance as the initial training data size increases. Finally, we run experiments showcasing model performance for traditional active learning methods, as well as our proposed method on CIFAR-10 and on the civil engineering classification dataset.

5.1   Civil Infrastructure Classification Dataset

We apply our instance segmentation to classification dataset conversion algorithm (see Sect. 4.2), extracting 160 × 160 pixel patches. Per image, the algorithm attempts to sample 100 class-patches and 10 background-patches with a maximum of 100 sampling attempts. The resulting classification dataset consists of a total of 67'162 samples. Figure 4 shows the class distribution for the original instance segmentation dataset and for the extracted classification dataset. Both suffer from heavy class imbalance.

We need a large enough dataset for our experiments to simulate a real-world active learning scenario with a small labeled and a large unlabeled pool. Accordingly, each category included in the final dataset has to contain enough samples such that it is possible to introduce artificial imbalance for each class during our experiments. Therefore, we remove all classes with less than 5'000 samples (crack with precipitation, rust). Additionally, we remove the net-crack class due to the visual similarity of its patches with the crack class. The final dataset consists of four classes: background, algae, crack, and spalling. It is split in a stratified fashion into 70% training set and 30% test set.

5.2   Experiment Setup

Active Learning Datasets. We aim to simulate a real-world active learning scenario with a small pool of labeled data and a large pool of unlabeled data, as well as an oracle that can be queried for labels. Artificial class imbalance is introduced into the labeled and the unlabeled pool to simulate the imbalanced dataset setting dependent on the experiment. For the experiments, the original training set is consequently randomly split into a small labeled set, a large unlabeled set, and an unused set. As we still have access to the labels of the samples of the unlabeled pool, we can simulate human annotation when the active learning algorithm queries labels. Furthermore, once samples have been moved from the unlabeled pool to the labeled pool (simulated labeling process), we can simulate a much larger unlabeled pool by moving the same number of samples per class from the unused pool to the unlabeled pool, restoring the original unlabeled data pool size and class balance. Finally, the test set is used as-is to evaluate the models after each time the AL algorithm queries new samples.

Model Training. We use the full ResNet18 [15,35] as a model backbone for our civil infrastructure dataset, but only use the first ResNet block for CIFAR-10 [21]. All models are trained for 50 epochs to convergence with the AdamW [26] optimizer with a learning rate of 0.0005 with the sample and batch augmentations highlighted above. For a fair comparison between the traditional AL

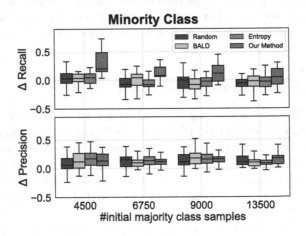

Fig. 5. Influence of initial AL training pool size: we report relative minority class performance gains for four dataset sizes with an increasing number of majority samples in the initial AL training pool. Performance differences are evaluated for each method, comparing model performance before the AL procedure with model performance after the procedure queried labels for 1000 samples. While performance gains of traditional AL methods decrease as the training dataset grows, our method retains more of the original performance increase.

methods and ours, we train all models with the improved training procedure of oversampling, image and batch augmentations.

Active Learning Process. We apply our method and traditional AL algorithms to each dataset under investigation. AL algorithms query labels five times for 200 samples per cycle. Each experiment is repeated three times selecting a class from the original dataset as the minority class for the artificially introduced imbalance procedure. All experiments use different randomly initialised model weights and a different random split of the data (labeled and unlabeled pool with artificial imbalance).

5.3   Results

Influence of Initial Training Pool Size. As a first experiment we evaluate all methods on CIFAR-10 [21] with increasing number of samples in the initial labeled pool. We create artificially imbalanced training datasets with a increasing number of samples for the majority class from 500 to 1500 while keeping the number of minority class samples constant at 50. To minimize the impact of the unlabeled set, it is kept constant with 120 minority samples and 3000 majority samples. Finally, we run the AL cycle for five rounds, selecting and labeling 200 samples per cycle. We measure the change in performance from the initial trained model to the model at the end of the AL procedure. As can be seen in Fig. 5, the delta in minority class performance drops off significantly for traditional active learning methods as the initial training dataset size increases. In contrast, our method retains more of the original performance gains as it is less sensitive to the initial data pool size. Specifically, the recall delta remains above zero as initial data size decreases while the classical AL method fall below zero, signaling a decrease in performance from the initial training set to the set with the additional 1000 labeled samples.

CIFAR-10. Next, we evaluate absolute model performance on CIFAR-10 [21], comparing traditional AL methods (BALD [17], Entropy [32]) with our novel method on model performance. Additionally, we also include a random acquisition function as a baseline. We create an initial AL dataset that is randomly split into unlabeled and labeled pool for each experiment with an artificial class imbalance of 50 minority samples to 1000 majority samples per class. The unlabeled pool consists of 300 minority samples and 3000 majority samples per class. Results in Fig. 6a show that traditional AL methods fail to significantly improve precision and recall for the minority class. Additionally, little performance improvement can be seen for the majority classes. We explain this with our experiments' much larger initial training pool compared to other publications. Our initial training set consists of few minority samples but a considerable amount of majority samples. The 200 additionally labeled samples per cycle do not yield much additional information compared to the existing larger training pool. Therefore, neither majority nor minority class performance improves. In

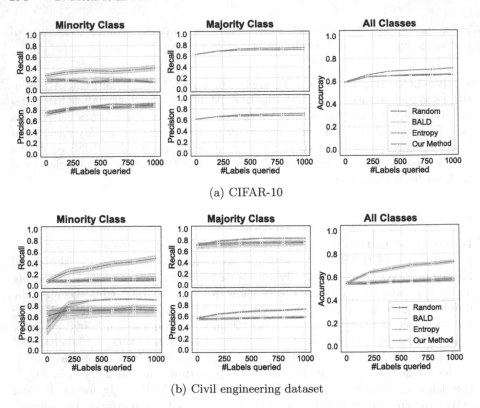

(a) CIFAR-10

(b) Civil engineering dataset

Fig. 6. Absolute model performance throughout the AL process: for each cycle, after labeling 200 additional samples, we report precision and recall for the minority class, macro average precision and recall for the majority classes, and overall accuracy for the CIFAR-10 dataset and our proprietary civil infrastructure dataset. Error bands show the standard error of the mean (SEM).

contrast, our method focuses only on the minority class, for which only a few samples are in the initial training pool. Therefore, even a few minority samples yield enough information to improve minority class performance considerably. Our method shows a clear performance improvement compared to traditional AL methods. Compared to the random baseline, our method improves on average by 25% recall of the minority class, while BALD only improves by 5% and the entropy method only improves by 0.3%. As a result, the overall accuracy of the model also increases significantly: 5% average improvement over the random baseline for our method, compared to only 0.6% for BALD and a 0.5% decrease for the entropy method.

Civil Engineering Dataset. We evaluate absolute model performance on our proprietary civil engineering dataset (see Sect. 5.1). We run experiments on an initial AL dataset with 50 minority and 1000 majority samples per class in the training set. The unlabeled pool consists of 300 minority and 2500 majority samples per class. As with CIFAR-10, the results in Fig. 6b show that traditional AL methods fail to improve precision and recall for the minority class. Meanwhile, our method shows clear improvement in minority class recall and precision as well as overall accuracy. Compared to the results on CIFAR-10, we see a larger performance improvement: compared to the random baseline our method improves by 38% minority recall, 18% minority precision, and 16% overall accuracy. BALD only improves by 6% minority recall and 2% overall accuracy while minority precision decreases by 4%. The entropy method improves only by 3% minority recall, 4% minority precision and 1% overall accuracy. We explain the higher overall accuracy gains compared to CIFAR-10 with the smaller number of classes in our dataset (4 compared to 10 in CIFAR-10). Given the overall accuracy is an average of class-specific performance, minority class improvement has a much larger impact on the majority classes.

6   Conclusion

We have presented a novel active learning method that replaces the traditional acquisition function with an auxiliary binary discriminator allowing the selection of minority samples even for initially large and imbalanced datasets. We have experimentally shown that our method outperforms classical AL algorithms on artificially imbalanced versions of CIFAR-10 and our proprietary civil engineering dataset when evaluated on minority class recall, precision, and overall classification accuracy. Consequently, our method facilitates the successful discovery and labeling of rare defects in the yet unlabeled pool of samples for our proprietary civil engineering dataset. Trained on the additional labeled data, our visual inspection defect detection models improve at supporting civil engineers' maintenance prioritization decisions for rare but critical defects.

Due to the large number of models trained per experiment, we are limited to dataset consisting of small images with many classes or large images with few classes. Additionally, our choices of datasets for experimentation were limited as we required many samples per class to simulate a large unlabeled pool. This excluded many popular large datasets as they often consist of many classes with a moderate amount of samples per class.

While we focus on classification datasets in this work, future research could extend our method to an application for the full instance segmentation dataset by creating sliding window patches and aggregating statistics over the full original image. Additionally, it would be interesting to develop our method further to work with multiple minority classes at a time. Finally, research should compare our method with feature space diversity-based AL methods when a pre-trained model (transfer learning or self-supervised learning) is available.

Acknowledgement. This work would not have been possible without Finn Bormlund and Svend Gjerding from Sund&Bælt. We would like to thank them for their collaboration, specifically for the collection of image data, for their expert annotations, and their tireless help with the annotation guidelines for the civil engineering dataset.

References

1. Matrice 300 RTK - Built Tough. Works Smart. https://www.dji.com/ch/photo
2. Zenmuse H20 Series - Unleash the Power of One. https://www.dji.com/ch/photo
3. Bayesian Active Learning (BaaL) (2022). https://github.com/baal-org/baal. Accessed 30 Sept 2019
4. Aggarwal, U., Popescu, A., Hudelot, C.: Active learning for imbalanced datasets. In: 2020 IEEE Winter Conference on Applications of Computer Vision (WACV), pp. 1417–1426. IEEE, Snowmass Village (2020). https://doi.org/10.1109/WACV45572.2020.9093475, https://ieeexplore.ieee.org/document/9093475/
5. Buslaev, A., Iglovikov, V.I., Khvedchenya, E., Parinov, A., Druzhinin, M., Kalinin, A.A.: Albumentations: fast and flexible image augmentations. Information **11**(2), 125 (2020). https://doi.org/10.3390/info11020125, https://www.mdpi.com/2078-2489/11/2/125
6. Chan, K.W., Nirmal, U., Cheaw, W.G.: Progress on drone technology and their applications: a comprehensive review. In: AIP Conference Proceedings, vol. 2030, no. 1, p. 020308 (2018). https://doi.org/10.1063/1.5066949, https://aip.scitation.org/doi/abs/10.1063/1.5066949
7. Chawla, N.V., Bowyer, K.W., Hall, L.O., Kegelmeyer, W.P.: SMOTE: synthetic minority over-sampling technique. J. Artif. Intell. Res. **16**, 321–357 (2002). https://doi.org/10.1613/jair.953, https://www.jair.org/index.php/jair/article/view/10302
8. Davtalab, O., Kazemian, A., Yuan, X., Khoshnevis, B.: Automated inspection in robotic additive manufacturing using deep learning for layer deformation detection. J. Intell. Manuf. **33**(3), 771–784 (2022). https://doi.org/10.1007/s10845-020-01684-w
9. Feng, C., Liu, M.Y., Kao, C.C., Lee, T.Y.: Deep active learning for civil infrastructure defect detection and classification. In: Computing in Civil Engineering 2017, pp. 298–306. American Society of Civil Engineers, Seattle (2017). https://doi.org/10.1061/9780784480823.036, http://ascelibrary.org/doi/10.1061/9780784480823.036
10. Freeman, L.C.: Elementary Applied Statistics: For Students in Behavioral Science. Wiley (1965). Google-Books-ID: r4VRAAAAMAAJ
11. Gal, Y., Ghahramani, Z.: Dropout as a Bayesian approximation: representing model uncertainty in deep learning (2016). arXiv:1506.02142 [cs, stat]
12. Gawlikowski, J., et al.: A survey of uncertainty in deep neural networks (2022). arXiv:2107.03342 [cs, stat]
13. Hanneke, S.: Theory of disagreement-based active learning. Found. Trends® Mach. Learn. **7**(2–3), 131–309 (2014). https://doi.org/10.1561/2200000037, https://www.nowpublishers.com/article/Details/MAL-037
14. He, H., Bai, Y., Garcia, E.A., Li, S.: ADASYN: adaptive synthetic sampling approach for imbalanced learning. In: 2008 IEEE International Joint Conference on Neural Networks (IEEE World Congress on Computational Intelligence), pp. 1322–1328 (2008). https://doi.org/10.1109/IJCNN.2008.4633969. iSSN 2161-4407
15. He, K., Zhang, X., Ren, S., Sun, J.: Deep residual learning for image recognition (2015). https://doi.org/10.48550/arXiv.1512.03385, arXiv:1512.03385 [cs]
16. Hernandez, J., Carrasco-Ochoa, J.A., Martínez-Trinidad, J.F.: An empirical study of oversampling and undersampling for instance selection methods on imbalance datasets. In: Ruiz-Shulcloper, J., Sanniti di Baja, G. (eds.) CIARP 2013. LNCS, vol. 8258, pp. 262–269. Springer, Heidelberg (2013). https://doi.org/10.1007/978-3-642-41822-8_33

17. Houlsby, N., Huszár, F., Ghahramani, Z., Lengyel, M.: Bayesian active learning for classification and preference learning (2011). https://doi.org/10.48550/arXiv.1112.5745, arXiv:1112.5745 [cs, stat]
18. Intelligence, I.: Drone technology uses and applications for commercial, industrial and military drones in 2021 and the future. https://www.businessinsider.com/drone-technology-uses-applications
19. Kaur, H., Pannu, H.S., Malhi, A.K.: A systematic review on imbalanced data challenges in machine learning: applications and solutions. ACM Comput. Surv. **52**(4), 1–36 (2020). https://doi.org/10.1145/3343440, https://dl.acm.org/doi/10.1145/3343440
20. Kirsch, A., van Amersfoort, J., Gal, Y.: BatchBALD: efficient and diverse batch acquisition for deep bayesian active learning (2019). arXiv:1906.08158 [cs, stat]
21. Krizhevsky, A.: Learning multiple layers of features from tiny images, p. 60, 8 April 2009
22. Kwolek, B., et al.: Breast cancer classification on histopathological images affected by data imbalance using active learning and deep convolutional neural network. In: Tetko, I.V., Kůrková, V., Karpov, P., Theis, F. (eds.) ICANN 2019. LNCS, vol. 11731, pp. 299–312. Springer, Cham (2019). https://doi.org/10.1007/978-3-030-30493-5_31
23. Lakshminarayanan, B., Pritzel, A., Blundell, C.: Simple and scalable predictive uncertainty estimation using deep ensembles (2017). https://doi.org/10.48550/arXiv.1612.01474, arXiv:1612.01474 [cs, stat]
24. Li, W., et al.: PathAL: an active learning framework for histopathology image analysis. IEEE Trans. Med. Imaging **41**(5), 1176–1187 (2022). https://doi.org/10.1109/TMI.2021.3135002
25. Liu, Z., Miao, Z., Zhan, X., Wang, J., Gong, B., Yu, S.X.: Large-scale long-tailed recognition in an open world, pp. 2537–2546 (2019). https://openaccess.thecvf.com/content_CVPR_2019/html/Liu_Large-Scale_Long-Tailed_Recognition_in_an_Open_World_CVPR_2019_paper.html
26. Loshchilov, I., Hutter, F.: Decoupled weight decay regularization (2019). https://doi.org/10.48550/arXiv.1711.05101, arXiv:1711.05101 [cs, math]
27. Mohammed, R., Rawashdeh, J., Abdullah, M.: Machine learning with oversampling and undersampling techniques: overview study and experimental results. In: 2020 11th International Conference on Information and Communication Systems (ICICS), pp. 243–248 (2020). https://doi.org/10.1109/ICICS49469.2020.239556. iSSN 2573-3346
28. Ren, R., Hung, T., Tan, K.C.: A generic deep-learning-based approach for automated surface inspection. IEEE Trans. Cybern. **48**(3), 929–940 (2018). https://doi.org/10.1109/TCYB.2017.2668395
29. Richards, J.W., et al.: Active learning to overcome sample selection bias: application to photometric variable star classification. Astrophys. J. **744**(2), 192 (2011). https://doi.org/10.1088/0004-637X/744/2/192
30. Selvaraju, R.R., Cogswell, M., Das, A., Vedantam, R., Parikh, D., Batra, D.: Grad-CAM: visual explanations from deep networks via gradient-based localization. Int. J. Comput. Vision **128**(2), 336–359 (2020). https://doi.org/10.1007/s11263-019-01228-7, arXiv:1610.02391 [cs]
31. Settles, B.: Active Learning Literature Survey. Technical Report, University of Wisconsin-Madison Department of Computer Sciences (2009), https://minds.wisconsin.edu/handle/1793/60660. Accessed 15–17 Mar 2012
32. Shannon, C.E.: A mathematical theory of communication p. 55 (1948)

33. Shi, X., Dou, Q., Xue, C., Qin, J., Chen, H., Heng, P.-A.: An active learning approach for reducing annotation cost in skin lesion analysis. In: Suk, H.-I., Liu, M., Yan, P., Lian, C. (eds.) MLMI 2019. LNCS, vol. 11861, pp. 628–636. Springer, Cham (2019). https://doi.org/10.1007/978-3-030-32692-0_72
34. Wang, X., Liu, B., Cao, S., Jing, L., Yu, J.: Important sampling based active learning for imbalance classification. Sci. China Inf. Sci. **63**(8), 182104 (2020). https://doi.org/10.1007/s11432-019-2771-0
35. Wightman, R., et al.: Comar: rwightman/pytorch-image-models: v0.6.5 Release (2022). https://doi.org/10.5281/ZENODO.4414861, https://zenodo.org/record/4414861
36. Yin, X., Chen, Y., Bouferguene, A., Zaman, H., Al-Hussein, M., Kurach, L.: A deep learning-based framework for an automated defect detection system for sewer pipes. Autom. Constr. **109**, 102967 (2020). https://doi.org/10.1016/j.autcon.2019.102967, https://www.sciencedirect.com/science/article/pii/S0926580519307411
37. Yun, S., Han, D., Oh, S.J., Chun, S., Choe, J., Yoo, Y.: CutMix: regularization strategy to train strong classifiers with localizable features (2019). https://arxiv.org/abs/1905.04899v2
38. Zhang, H., Cisse, M., Dauphin, Y.N., Lopez-Paz, D.: Mixup: beyond empirical risk minimization. arXiv:1710.09412 [cs, stat] (2018)
39. Zhang, Y., Kang, B., Hooi, B., Yan, S., Feng, J.: Deep long-tailed learning: a survey (2021). arXiv:2110.04596 [cs]

UAV-Based Visual Remote Sensing
for Automated Building Inspection

Kushagra Srivastava[1]($\boxtimes$), Dhruv Patel[1], Aditya Kumar Jha[2],
Mohhit Kumar Jha[2], Jaskirat Singh[3], Ravi Kiran Sarvadevabhatla[1],
Pradeep Kumar Ramancharla[1], Harikumar Kandath[1],
and K. Madhava Krishna[1]

[1] International Institute of Information Technology, Hyderabad, Hyderabad, India
kushagra2000@gmail.com,
{ravi.kiran,ramancharla,harikumar.k,mkrishna}@iiit.ac.in
[2] Indian Institute of Technology, Kharagpur, Kharagpur, India
{aditya.jha,mohhit.kumar.jha2002}@kgpian.iitkgp.ac.in
[3] University of Petroleum and Energy Studies, Dehradun, India

Abstract. Unmanned Aerial Vehicle (UAV) based remote sensing system incorporated with computer vision has demonstrated potential for assisting building construction and in disaster management like damage assessment during earthquakes. The vulnerability of a building to earthquake can be assessed through inspection that takes into account the expected damage progression of the associated component and the component's contribution to structural system performance. Most of these inspections are done manually, leading to high utilization of manpower, time, and cost. This paper proposes a methodology to automate these inspections through UAV-based image data collection and a software library for post-processing that helps in estimating the seismic structural parameters. The key parameters considered here are the distances between adjacent buildings, building plan-shape, building plan area, objects on the rooftop and rooftop layout. The accuracy of the proposed methodology in estimating the above-mentioned parameters is verified through field measurements taken using a distance measuring sensor and also from the data obtained through Google Earth. Additional details and code can be accessed from https://uvrsabi.github.io/.

Keywords: Building inspection · UAV-based remote sensing · Segmentation · Image stitching · 3D reconstruction

1 Introduction

Traditional techniques to analyze and assess the condition and geometric aspects of buildings and other civil structures involve physical inspection by civil experts

---

K. Srivastava and D. Patel—denotes equal contribution.

L. Karlinsky et al. (Eds.): ECCV 2022 Workshops, LNCS 13807, pp. 299–316, 2023.
https://doi.org/10.1007/978-3-031-25082-8_20

according to pre-defined procedures. Such inspections can be costly, risky, time-consuming, labour and resource intensive. A considerable amount of research has been dedicated to automating and improving civil inspection and monitoring through computer vision. This results in less human intervention and lower cost while ensuring effective data collection. Unmanned Aerial Vehicles (UAVs) mounted with cameras have the potential for contactless, rapid and automated inspection and monitoring of civil structures as well as remote data acquisition.

Computer vision-aided civil inspection has two prominent areas of application: damage detection and structural component recognition [1]. Studies focused on damage detection have used heuristic feature extraction methods to detect concrete cracks [2–4], concrete spalling [5,6], fatigue cracks [7,8] and corrosion in steel [9–11]. However, heuristic-based methods do not account for the information that is available in regions around the defect and have been replaced with deep learning-based methods. Image classification [12–14], object detection [15,16], semantic segmentation [17–19] based methods have been used to successfully detect and classify the damage type. On the contrary, structural component analysis involves detecting, classifying and studying the characteristics of a physical structure. Hand-crafted filters [20,21], point cloud-based [22–25], and deep learning-based [26–29] methods have been used to assess structural components like columns, planar walls, floor, bridges, beams and slabs. There also has been a emphasis on developing architectures for Building Information Modelling (BIM) [30–32] that involves analysis of physical features of a building using high resolution 3D reconstruction.

Apart from structural component recognition, it is also essential to assess the risk posed by earthquakes to buildings and other structural components. This is a crucial aspect of inspection in seismically active zones. Accurate seismic risk modeling requires knowledge of key structural characteristics of buildings. Learning-based models in conjunction with street imagery [33,34] have been used to perform building risk assessments. However, UAVs can also be used to obtain information in areas difficult to access by taking a large number of images and videos from several points and different angles of view. Thus, UAVs demonstrate huge potential when it comes to remote data acquisition for pre- and/or post-earthquake risk assessments [35].

The main contributions of this paper are given below.

1. Primarily, we automate the inspection of buildings through UAV-based image data collection and a post-processing module to infer and quantify the details. This in effect avoids manual inspection, reducing the time and cost.
2. We estimate the distance between adjacent buildings and structures. To the best of our knowledge, there has not been any work that has addressed this problem.
3. We develop an architecture that can be used to segment roof tops in case of both orthogonal and non-orthogonal view using a state-of-the-art semantic segmentation model.
4. The software library for post-processing collates different algorithms used in computer vision along with UAV state information to yield an accurate

estimation of the distances between adjacent buildings, building plan-shape, building plan area, objects on the rooftop, and rooftop layout. These parameters are key for the preparation of safety index assessment for buildings against earthquakes.

2 Related Works

2.1 Distance Between Adjacent Structures

The collision between adjacent buildings or among parts of the same building during strong earthquake vibrations is called pounding [36]. Pounding occurs due to insufficient physical separation between adjacent structures and their out-of-phase vibrations resulting in non-synchronized vibration amplitudes. Pounding can lead to the generation of a high-impact force that may cause either architectural or structural damage. Some reported cases of pounding include i) The earthquake of 1985 in Mexico City [37] that left more than 20% of buildings damaged, ii) Loma Prieta earthquake of 1989 [38] that affected over 200 structures, iii) Chi-Chi earthquake of 1999 [39] in central Taiwan, and iv) Sikkim earthquake (2006) [40]. Methods such as Rapid Visual Screening (RVS), seismic risk indexes, and vulnerability assessments have been developed to analyze the level of damage to a building [41]. In particular, RVS-based methods have been used for pre-and/or post-earthquake screening of buildings in earthquake-prone areas. The pounding effect is considered as a vulnerability factor by RVS methods like FEMA P-154, FEMA 310, EMS-98 Scale, NZSEE, OSAP, NRCC, IITK-GSDMA, EMPI and RBTE-2019 [42].

The authors in [43] present a UAV-based site survey using both Nadir and Oblique images for appropriate 3D modelling. The integration of nadir UAV images with oblique images ensures a better inclusion of facades and footprints of the buildings. Distances between the buildings in the site were manually measured from the generated dense point cloud. We use the 3D reconstruction of the structures from images in conjunction with conditional plane fitting for estimating the distance between adjacent structures.

2.2 Plan Shape and Roof Area Estimation

The relationship between the center of stiffness and gravity's eccentricity is influenced by shape irregularities, asymmetries, or concavities, as well as by building mass distributions. For any structure, if the centre of stiffness is moved away from the centre of gravity during ground motion, more torsion forces are produced [44]. When a building is shaken by seismic activity, this eccentricity causes structures to exhibit improper dynamic characteristics. Hence, the behavior of a building under seismic activity also depends on its 3D configuration, plan shape and mass distribution [45]. *Plan shape and Roof Area* is needed for calculating the Floor Space Index (FSI). FSI is the ratio of the total built-up area of all the floors to the plot area. FSI is a contributing factor in assessing the extent of the damage and is usually fixed by the expert committee.

Roof-top segmentation has been considered as a special case of 3D plane segmentation from point clouds and can be achieved through model fitting [46], region growing [47], feature clustering [48] and global energy optimization-based methods [49]. Studies focused on these methods have been tested on datasets where the roof was visible orthogonally through satellite imagery [50] and LiDAR point clouds [46–48]. The accuracy of these methods depends on how the roof is viewed. In case of a non-orthogonal view, these methods must be used in conjunction with some constraints. On the contrary, learning-based methods [51] have been developed that specifically segment out roofs. The neural networks employed in these methods have been trained on satellite imagery and do not perform well in non-orthogonal roof-view scenarios. Our approach is to segment out roofs when viewed both orthogonally and non-orthogonally by training a state-of-the-art semantic segmentation model on a custom roof-top dataset.

2.3   Roof Layout Estimation

Roof Layout Estimation refers to identifying and locating objects present on the roof such as air conditioner units, solar panels, etc. Such objects are usually non-structural elements (NSE). As the mass of the NSE increases, the earthquake response of the NSE starts affecting the whole building. Hence, they need to be taken into account for design calculations. Furthermore, the abundance of these hazardous objects may create instabilities on the roof making it prone to damage during earthquakes. Estimating the *Roof Layout* is not as trivial as in the case of satellite images, since the UAV has altitude limitations along with camera Field of View (FOV) constraints, thereby limiting us from obtaining a complete view of the roof in a single image. Moreover, we cannot rely on satellite images because it does not provide us with real time observation of our location of interest. Hence, we solve this problem by first stitching a large number of images with partially visible roofs to create a panoramic view of the roof and then we apply object detection and semantic segmentation to get the object and roof masks respectively.

Various techniques for image stitching can be roughly distinguished into three categories: direct technique [52–54], feature-based technique [55–57] and position-based technique [58]. The first category performs pixel-based image stitching by minimizing the sum of the absolute difference between overlapping pixels. These methods are scale and rotation variant and to tackle this problem, the second category focuses on extracting a set of images and matching them using feature based algorithms which includes SIFT, SURF, Harris Corner Detection. These methods are computationally expensive and fail in the absence of distinct features. The third category stitches images sampled from videos through their overlapping FOV. Due to the inability to obtain accurate camera poses, not much research has been conducted on this approach. In this paper, we present an efficient and reliable approach to make use of the camera poses and stitch a large set of images avoiding the problems of image drift and expensive computation associated with the first two categories.

3   Data Collection

This section discusses the methods for gathering data that were utilized to carry out the research experiments in this study. DJI Mavic Mini[1] UAV is used for gathering visual data because of its high-quality image sensory system with an adjustable gimbal.

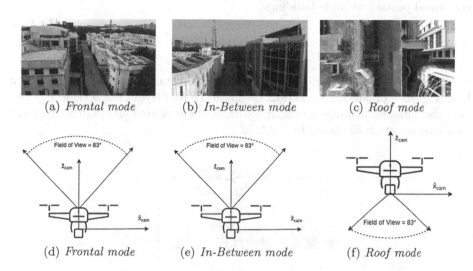

(a) Frontal mode   (b) In-Between mode   (c) Roof mode

(d) Frontal mode   (e) In-Between mode   (f) Roof mode

Fig. 1. Figures (a), (b), and (c) indicate the UAV's point of view while figures (d), (e), and (f) are representations of the respective coordinate system adopted.

For estimating the distance between adjacent structures, the images are collected in 3 different modes: *Frontal Mode*, *In-Between Mode* and *Roof Mode*. Figure 1(a) shows the frontal face of the two adjacent buildings for which data was collected. In this mode, we focus on estimating the distance between the two buildings by analyzing only their frontal faces through a forward-facing camera. This view is particularly helpful when there are impediments between the subject buildings and flying a UAV between them is challenging. In Fig. 1(b), the UAV was flown in-between the two buildings along a path parallel to the facade with a forward-facing camera. This mode enables the operators to calculate distances when buildings have irregular shapes. Lastly, for the *roof mode*, the UAV was flown at a fixed altitude with a downward-facing camera so as to capture the rooftops of the subject buildings. Figure 1(c) is a pictorial representation of the *roof mode*. The *roof mode* helps in tackling occlusions due to vegetation and other physical structures.

For *Rooftop Layout Estimation*, the UAV was flown at a constant height with a downward-facing camera, parallel to the plane of the roof. This helped in robust detection of NSE. To estimate the *Plan Shape and Roof Area*, a dataset

---

[1] UAV specification details can be found at the official DJI website: https://www. dji.com/mavic-mini.

comprising of around 350 images was prepared from the campus buildings and UrbanScene3D dataset [59]. The training set comprised of images scraped from the UrbanScene3D videos, *Buildings 4* and *6* and while the validation and test set comprised of the *Buildings 3, 5* and *7*. This was done to ensure that the model learns the characteristic features of a roof irrespective of the building plan shape. Out of these, 50 images had fully-visible buildings while the rest contained partially-visible buildings.

4   Methodology

We propose different methods to calculate the distance between the adjacent buildings using plane segmentation; estimate the roof layout using Object Detection and large scale image stitching; estimate the roof area and plan shape using roof segmentation as shown in Fig. 2.

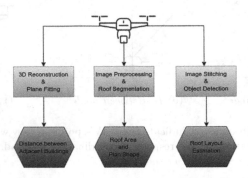

Fig. 2. Architecture of automated building inspection using the aerial images captured using UAV. The odometry information of UAV is also used for the quantification of different parameters involved in the inspection.

4.1   Distance Between Adjacent Buildings

We use plane segmentation to obtain the distance between the two adjacent buildings. We have divided our approach into three stages as presented in Fig. 3. In *Stage I*, images were sampled from the video captured by the UAV and panoptic segmentation was performed using a state-of-the-art network [60], to obtain vegetation-free masks. This removes trees and vegetation near the vicinity of the buildings and thus improves the accuracy of our module. Figure 4 shows the impact of panoptic segmentation for *frontal mode*. In *Stage II*, the masked images were generated from the binary masks and the corresponding images. The masked images are inputs to a state-of-the-art image-based 3D reconstruction library [61] which outputs a dense 3D point cloud and the camera poses through Structure-from-Motion. Our approach for all the three modes is same for the first two stages.

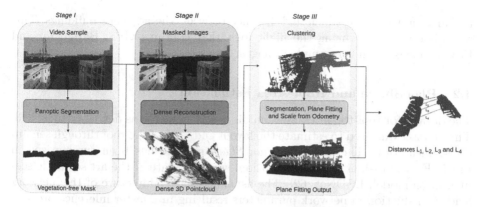

Fig. 3. Architecture for estimation of distance between adjacent structures.

(a) *Image sample*   (b) *Vegetation-free mask*

Fig. 4. Removal of vegetation from the sample images enhances the structural features of the buildings in the reconstructed 3D model leading to more accurate results.

In *Stage III*, we aim to extract planes from the given point cloud that are essential to identify structures such as roof and walls of the building. We employ the co-ordinate system depicted in Fig. 1. We can divide this task into two parts: i) Isolation of different building clusters and ii) Finding planes in each cluster. Isolation of the concerned buildings is done using euclidean clustering thereby creating two clusters. For instance, in *Roof Mode* the clusters are distributed on either side of the Y-axis. Similarly, for *In-Between* and *Frontal mode* the clustering happens about the Z-axis. In order to extract the planes of interest, we slice each cluster along a direction parallel to our plane of interest, into small segments of 3D points. For instance, in *Roof mode* we are interested in fitting a plane along the roof of a building; therefore, we slice the building perpendicular to ground normal, i.e., the Z-axis. Finally, Random Sample Consensus (RANSAC) algorithm is applied for each segment of 3D points to iteratively fit a plane and obtain a set of parallel planes as shown under the *Stage III* in Fig. 3.

Our approach selects a plane from the set of planes estimated in *Stage III* for each building based on the highest number of inliers. As stated above, for each mode, the selected planes for the adjacent buildings have the same normal unit vector. Further, we sample points on these planes to calculate the distance between the adjacent buildings at different locations. The scale estimation is done by using the odometry data received from the UAV and the estimated distance is scaled up to obtain the actual distance between the adjacent buildings. This was done by time-syncing the flight logs, that contains GPS, Barometer and IMU readings, with the sampled images.

4.2   Plan Shape and Roof Area Estimation

The dataset for roof-top of various buildings was collected as described in Sect. 3. This dataset was used to estimate the layout and area of the roof through semantic segmentation. The complete *Plan Shape* module has been summarized in Fig. 5. For the task of roof segmentation, we use a state-of-the-art semantic segmentation model, LEDNet [62]. The asymmetrical architecture of this network leads to reduction in network parameters resulting in a faster inference process. The split and shuffle operations in the residual layer enhances information sharing while the decoder's attention mechanism reduces complexity of the whole network. We subject the input images to a pre-processing module that removes distortion from the wide-angle images. Histogram equalization is also performed to improve the contrast of the image. Data augmentation techniques (4 rotations of 90° + horizontal flip + vertical flip) were used during inference to improve the network's performance and increase robustness. The single-channel grey-scale output is finally thresholded to obtain a binary mask. The roof area from the segmentation masks can be obtained by using Eq. 1 where C is the contour area (in $pixels^2$), obtained from the segmented mask, D is the depth of the roof from the camera (in m) and f is the focal length of the camera (in $pixels$) used.

$$Area\,(m^2) = C \times (D/f)^2 \tag{1}$$

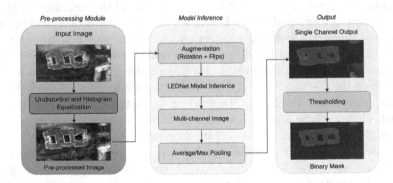

Fig. 5. Architecture of the *Plan Shape* module providing the segmented mask of the roof as output from the raw input image.

4.3 Roof Layout Estimation

The data for this module was collected as described in Sect. 3. Due to the camera FOV limitations and to maintain good resolution, it is not possible to capture the complete view of the roof in a single image, especially in the case of large sized buildings. Hence, we perform large scale stitching of partially visible roofs followed by NSE detection and roof segmentation. Figure 6 shows the approach adopted for *Roof Layout Estimation*.

Large Scale Aerial Image Stitching: We exploit the planarity of the roof and the fact that the UAV is flown at a constant height from the roof. Instead of opting for homography, that relates two geometric views in case of image stitching, we opt for affine transformations. Affine transformations are linear mapping methods that preserve points, straight lines, and planes.

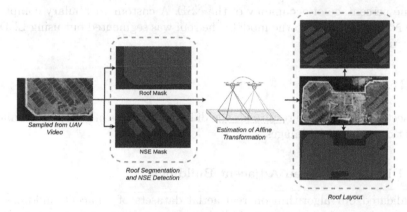

Fig. 6. We exploit the planarity of the roof and the fact that the distance between the UAV and the roof will be constant (the UAV is flown at a constant height). This enables us to relate two consecutive images through an affine transformation.

Let $I = \{i_1, i_2, i_3, ..., i_N\}$ represent an ordered set of images sampled from a video collected as per Sect. 3. The image stitching algorithm implemented has been summarized below:

1. Features were extracted in image i_1, using ORB feature detector and tracked in the next image, i_2 using optical flow. This helped in effective rejection of outliers.
2. The obtained set of feature matches across both the images were used to determine the affine transformation matrix using RANSAC.

3. Images i_1 and i_2 were then warped as per the transformation and stitched on a *canvas*.
4. Affine transformation was calculated between image i_3 and the previously warped image i_2 before it was stitched using steps 1 and 2. Image i_3 was then warped and stitched on the same *canvas*.
5. Step 4 was repeated for the next set of images, that is, affine transformation was calculated for image i_4 and the previously warped image i_3 before it was stitched.

Detecting Objects on Rooftop: For identification of NSE on the rooftop, we use a state-of-the art object detection model, Detic [63] because it is highly flexible and has been trained for large number of classes. In order to estimate the roof layout, it is essential to detect and locate the NSE as well as the roof from a query image. Note that we classify all the NSE as a single class. This information can then be represented as a semantic mask which will be to calculate the percentage of occupancy of the NSE. A custom vocabulary comprising of the NSE was passed to the model. The roof was segmented out using LEDNet as described in Sect. 4.2.

5   Results

This section presents the results for the different modules of automated building inspection using aerial images.

5.1   Distance Between Adjacent Buildings

We validated our algorithm on real aerial datasets of adjacent buildings and structures. In particular, we tested all the modes of this module on a set of adjacent buildings, *Buildings 1 and 2*, and also on *Building 3*, a U-shaped building. The resulting distances for all the modes can be visualized through Fig. 7. The corresponding distances visualized in Fig. 7 have been documented in Table 1 and 2. We obtain the ground truth from using a Time-of-Flight (ToF) based range measuring sensor[2]. This sensor has a maximum range of 60 m. We also compare the results with that from *Google Earth*. It must be noted that using *Google Earth*, it is not possible to measure some distances due to lack of 3D imagery.

---

[2] The ToF sensor can be found at: https://www.terabee.com/shop/lidar-tof-range-find ers/teraranger-evo-60m/.

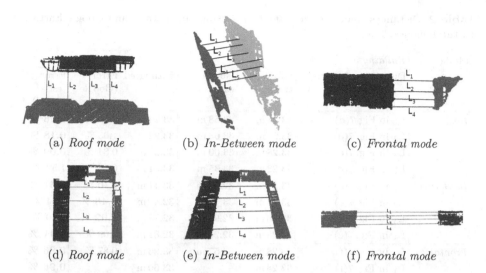

(a) *Roof mode*   (b) *In-Between mode*   (c) *Frontal mode*

(d) *Roof mode*   (e) *In-Between mode*   (f) *Frontal mode*

Fig. 7. (a), (b) and (c) and (d), (e) and (f) represent the implementation of plane fitting using piecewise-RANSAC in different views for *Buildings 1, 2* and *Building 3* respectively.

Table 1. Distances calculated for *Building 1 and 2* using our method and Google Earth for all three modes.

Mode	Buildings 1 and 2					
	Distance reference	Ground truth	Google earth	Estimated	Error (Google earth)	Error (estimated)
Roof	L_1 in Fig 7(a)	16.40 m	17.14 m	16.70 m	4.5%	**1.8%**
	L_2 in Fig. 7(a)	12.96 m	12.91 m	12.94 m	0.3%	**0.15%**
	L_3 in Fig. 7(a)	12.01 m	12.08 m	11.97 m	0.58%	**0.33%**
	L_4 in Fig. 7(a)	13.30 m	13.00 m	12.77 m	**2.31%**	3.98%
In-Between	L_1 in Fig. 7(b)	13.31 m	13.50 m	13.22 m	1.42%	**0.67%**
	L_2 in Fig. 7(b)	12.91 m	12.40 m	12.87 m	3.95%	**0.31%**
	L_3 in Fig. 7(b)	12.30 m	12.43 m	12.12 m	**1.00%**	1.39%
	L_4 in Fig. 7(b)	12.70 m	12.87 m	12.50 m	**1.33%**	1.57%
	L_5 in Fig. 7(b)	13.95 m	13.84 m	13.87 m	0.79%	**0.51%**
	L_6 in Fig. 7(b)	12.60 m	12.69 m	12.56 m	0.71%	**0.31%**
Frontal	L_1 in Fig. 7(c)	16.96 m	16.91 m	16.92 m	0.29%	**0.23%**
	L_2 in Fig. 7(c)	16.96 m	–	16.78 m	–	**1.06%**
	L_3 in Fig. 7(c)	16.96 m	–	17.13 m	–	**1.00%**
	L_4 in Fig. 7(c)	16.96 m	–	17.05 m	–	**0.53%**

Table 2. Distances calculated for *Building 3* using our method and Google Earth for all three modes.

Mode	Building 3					
	Distance reference	Ground truth	Google earth	Estimated	Error (Google earth)	Error (esti-mated)
Roof	L_1 in Fig 7(d)	33.28 m	33.58 m	33.26 m	0.90 %	**0.06 %**
	L_2 in Fig. 7(d)	33.28 m	33.63 m	33.22 m	1.05 %	**0.18 %**
	L_3 in Fig. 7(d)	33.28 m	33.00 m	33.28 m	0.84 %	**0.00 %**
	L_4 in Fig. 7(d)	33.28 m	33.35 m	33.81 m	**0.21 %**	1.59 %
In-Between	L_1 in Fig. 7(e)	33.28 m	32.58 m	33.11 m	2.10 %	**0.51 %**
	L_2 in Fig. 7(e)	33.28 m	33.12 m	32.40 m	**0.48 %**	2.64 %
	L_3 in Fig. 7(e)	33.28 m	32.94 m	32.78 m	**1.02 %**	1.50 %
	L_4 in Fig. 7(e)	33.28 m	32.25 m	32.81 m	3.09 %	**1.41 %**
Frontal	L_2 in Fig. 7(f)	33.28 m	33.57 m	33.26 m	0.87 %	**0.06 %**
	L_1 in Fig. 7(f)	33.28 m	–	33.60 m	-	**0.96 %**
	L_3 in Fig. 7(f)	33.28 m	–	33.59 m	-	**0.93 %**
	L_4 in Fig. 7(f)	33.28 m	–	33.99 m	-	**2.13 %**

5.2 Plan Shape and Roof Area Estimation

The roof area was estimated from images taken at different depths, that is, when the UAV was operated at different altitudes ranging from 50 m to 100 m. The module was tested on various campus buildings. The results in Table 3 were averaged out for all samples corresponding to the same building. The module estimates the roof area with an average difference of 4.7% with Google Earth data. Predicted roof masks of some buildings from LEDNet are shown in Fig. 8 and Fig. 9.

Building 3   Building 4

Building 5   Building 6

Fig. 8. Roof segmentation results for 4 buildings.

Fig. 9. We use LEDNet in both *Plan Shape and Roof Area Estimation* as well as *Roof Layout Estimation*. The trained model correctly segments out the roof in case of a non-orthogonal view, that is, when the downward-facing camera is not directly above the roof.

Table 3. Roof area estimation results

Building	Area measured using Google earth	Estimated area	Absolute difference	Percentage difference
Building 3	1859.77 m²	1939.84 m²	80.07 m²	4.3%
Building 4	350 m²	331.30 m²	18.70 m²	5.3%
Building 5	340 m²	329.55 m²	10.45 m²	3.1%
Building 6	3,127.60 m²	2936.82 m²	190.78 m²	6.1%

5.3   Roof Layout Estimation

Data for *Roof Layout Estimation* was collected as described in Sect. 3. Images were sampled at a frequency of 10 Hz. The video collected for *Roof Layout Estimation* was sampled at a frequency of 1 Hz generating 98 images. The results of image stitching can be visualized in Fig. 10(a) with the corresponding roof mask in Fig. 10(b) and the NSE mask in Fig. 10(c). The percentage occupancy was calculated by taking the ratio of object occupancy area ($pixels^2$) in Fig. 10(c) to total roof area ($pixels^2$) in Fig. 10(b). The final percentage occupancy obtained was 38.73%.

6   Discussion

We estimate the distance between adjacent structures using 3D reconstruction and conditional plane fitting and validate its performance on ground truth data from a ToF sensor. We also make a comparison of our proposed module with Google Earth and validate our superior performance. Moreover, it is not possible to employ Google Earth for this module universally due to the lack of 3D imagery for all the buildings. We validated our distance estimation algorithm and compared the results with the ground truth and Google Earth. We estimated the distance between adjacent structures with an average error of 0.94%, which is superior to Google Earth which performs with an average error of 1.36%. Our rooftop area estimation module performs with an average difference of 4.7% when compared to Google Earth. It is observed that the difference remains near-constant irrespective of the size of the rooftop area. Considering the irregular shape of the roofs, it is challenging to measure the ground truth of the roof area manually and it is also error-prone and resource-intensive. The roof layout

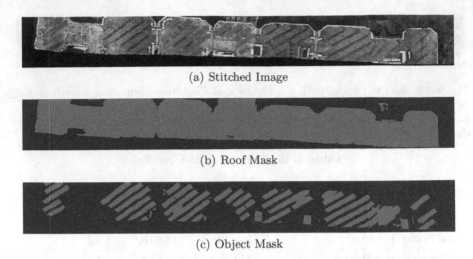

(a) Stitched Image

(b) Roof Mask

(c) Object Mask

Fig. 10. Results for roof layout estimation.

was estimated through semantic segmentation, object detection, and large-scale image stitching of 98 images. We also detected NSE on the rooftops and found their percentage occupancy to be 38.73%.

7   Conclusion

This paper presented an implementation of a considerable amount of approaches that have been developed, aiming at modeling the structure of buildings. Seismic risk assessment of buildings involves the estimation of several structural parameters. It is important to estimate the parameters which can describe the geometry of buildings, plan shape of the rooftops, and size of the buildings. In particular, we estimated the distances between adjacent buildings and structures, plan shape of a building, roof area, and percentage of area occupied by NSE. We plan to release these modules in the form of an open-source library that can be easily used by non-computer vision experts. Future work includes quantifying the flatness of ground, crack detection, and identification of water tanks and staircase exits that could help in taking preliminary precautions for earthquakes.

Acknowledgement. The authors acknowledge the financial support provided by IHUB, IIIT Hyderabad to carry out this research work under the project: IIIT-H/IHub/Project/Mobility/2021-22/M2-003.

References

1. Spencer, B.F., Hoskere, V., Narazaki, Y.: Advances in computer vision-based civil infrastructure inspection and monitoring. Engineering **5**(2), 199–222 (2019). ISSN 2095-8099. https://doi.org/10.1016/j.eng.2018.11.030, https://www.sciencedirect.com/science/article/pii/S2095809918308130

2. Abdel-Qader, I., Abudayyeh, O., Kelly, M.: Analysis of edge-detection techniques for crack identification in bridges. J. Comput. Civ. Eng. **17**, 10 (2003). https://doi.org/10.1061/(ASCE)0887-3801(2003)17:4(255)

3. Zhang, W., Zhang, Z., Qi, D., Liu, Y.: Automatic crack detection and classification method for subway tunnel safety monitoring. Sensors **14**(10), 19307–19328 (2014). ISSN 1424-8220. https://doi.org/10.3390/s141019307, https://www.mdpi.com/1424-8220/14/10/19307

4. Liu, Y.-F., Cho, S., Spencer, B., Fan, J.-S.: Concrete crack assessment using digital image processing and 3D scene reconstruction. J. Comput. Civ. Eng. **30**, 04014124 (2014). https://doi.org/10.1061/(ASCE)CP.1943-5487.0000446

5. Adhikari, R.S., Moselhi, O., Bagchi, A.: A study of image-based element condition index for bridge inspection. In: ISARC. Proceedings of the International Symposium on Automation and Robotics in Construction, vol. 30, p. 1. IAARC Publications (2013)

6. Paal, S., Jeon, J.-S., Brilakis, I., Desroches, R.: Automated damage index estimation of reinforced concrete columns for post-earthquake evaluations. J. Struct. Eng. **141**, 04014228 (2015). https://doi.org/10.1061/(ASCE)ST.1943-541X.0001200

7. Yeum, C.M., Dyke, S.J.: Vision-based automated crack detection for bridge inspection. Comput.-Aided Civ. Infrastruct. Eng. **30**(10), 759–770 (2015)

8. Jahanshahi, M.R., Chen, F.-C., Joffe, C., Masri, S.F.: Vision-based quantitative assessment of microcracks on reactor internal components of nuclear power plants. Struct. Infrastruct. Eng. **13**(8), 1013–1026 (2017)

9. Son, H., Hwang, N., Kim, C., Kim, C.: Rapid and automated determination of rusted surface areas of a steel bridge for robotic maintenance systems. Autom. Constr. **42**, 13–24 (2014)

10. Shen, H.-K., Chen, P.-H., Chang, L.-M.: Automated steel bridge coating rust defect recognition method based on color and texture feature. Autom. Constr. **31**, 338–356 (2013)

11. Medeiros, F.N.S., Ramalho, G.L.B., Bento, M.P., Medeiros, L.C.L.: On the evaluation of texture and color features for nondestructive corrosion detection. EURASIP J. Adv. Signal Process. **2010**, 1–7 (2010)

12. Cha, Y.-J., Choi, W., Büyüköztürk, O.: Deep learning-based crack damage detection using convolutional neural networks. Comput.-Aided Civ. Infrastruct. Eng. **32**(5), 361–378 (2017)

13. Zhang, L., Yang, F., Zhang, Y.D., Zhu, Y.J.: Road crack detection using deep convolutional neural network. In: 2016 IEEE International Conference on Image Processing (ICIP), pp. 3708–3712. IEEE (2016)

14. Atha, D.J., Jahanshahi, M.R.: Evaluation of deep learning approaches based on convolutional neural networks for corrosion detection. Struct. Health Monit. **17**(5), 1110–1128 (2018)

15. Yeum, C.M., Dyke, S.J., Ramirez, J.: Visual data classification in post-event building reconnaissance. Eng. Struct. **155**, 16–24 (2018)

16. Cha, Y.-J., Choi, W., Suh, G., Mahmoudkhani, S., Büyüköztürk, O.: Autonomous structural visual inspection using region-based deep learning for detecting multiple damage types. Comput.-Aided Civ. Infrastruct. Eng. **33**(9), 731–747 (2018)

17. Zhang, A., et al.: Automated pixel-level pavement crack detection on 3D asphalt surfaces using a deep-learning network. Comput.-Aided Civ. Infrastruct. Eng. **32**(10), 805–819 (2017)

18. Hoskere, V., Narazaki, Y., Hoang, T., Spencer Jr., B.F.: Vision-based structural inspection using multiscale deep convolutional neural networks. arXiv preprint arXiv:1805.01055 (2018)

19. Hoskere, V., Narazaki, Y., Hoang, T.A., Spencer Jr., B.F.: Towards automated post-earthquake inspections with deep learning-based condition-aware models. arXiv preprint arXiv:1809.09195 (2018)

20. Zhu, Z., Brilakis, I.: Concrete column recognition in images and videos. J. Comput. Civ. Eng. **24**(6), 478–487 (2010)

21. Koch, C., Paal, S.G., Rashidi, A., Zhu, Z., König, M., Brilakis, I.: Achievements and challenges in machine vision-based inspection of large concrete structures. Adv. Struct. Eng. **17**(3), 303–318 (2014)

22. Xiong, X., Adan, A., Akinci, B., Huber, D.: Automatic creation of semantically rich 3d building models from laser scanner data. Autom. Constr. **31**, 325–337 (2013)

23. Armeni, I., et al.: 3D semantic parsing of large-scale indoor spaces. In: Proceedings of the IEEE Conference on Computer Vision and Pattern Recognition, pp. 1534–1543 (2016)

24. Golparvar-Fard, M., Bohn, J., Teizer, J., Savarese, S., Peña-Mora, F.: Evaluation of image-based modeling and laser scanning accuracy for emerging automated performance monitoring techniques. Autom. Constr. **20**(8), 1143–1155 (2011)

25. Lu, R., Brilakis, I., Middleton, C.R.: Detection of structural components in point clouds of existing RC bridges. Comput.-Aided Civ. Infrastruct. Eng. **34**(3), 191–212 (2019)

26. Gao, Y., Mosalam, K.M.: Deep transfer learning for image-based structural damage recognition. Comput.-Aided Civ. Infrastruct. Eng. **33**(9), 748–768 (2018)

27. Liang, X.: Image-based post-disaster inspection of reinforced concrete bridge systems using deep learning with Bayesian optimization. Comput.-Aided Civ. Infrastruct. Eng. **34**(5), 415–430 (2019)

28. Yeum, C.M., Choi, J., Dyke, S.J.: Automated region-of-interest localization and classification for vision-based visual assessment of civil infrastructure. Struct. Health Monit. **18**(3), 675–689 (2019)

29. Narazaki, Y., Hoskere, V., Hoang, T.A., Fujino, Y., Sakurai, A., Spencer Jr., B.F.: Vision-based automated bridge component recognition with high-level scene consistency. Comput.-Aided Civ. Infrastruct. Eng. **35**(5), 465–482 (2020)

30. Dimitrov, A., Golparvar-Fard, M.: Vision-based material recognition for automated monitoring of construction progress and generating building information modeling from unordered site image collections. Adv. Eng. Inform. **28**(1), 37–49 (2014)

31. Golparvar-Fard, M., Pena-Mora, F., Savarese, S.: Automated progress monitoring using unordered daily construction photographs and IFC-based building information models. J. Comput. Civ. Eng. **29**(1), 04014025 (2015)

32. Hamledari, H., Davari, S., Azar, R., McCabe, B., Flager, F., Fischer, M.: UAV-enabled site-to-BIM automation: aerial robotic-and computer vision-based development of as-built/as-is BIMs and quality control. In: Construction research congress, pp. 336–346 (2017)

33. Pelizari, P.A., Geiß, C., Aguirre, P., María, H.S., Peña, Y.M., Taubenböck, H.: Automated building characterization for seismic risk assessment using street-level imagery and deep learning. ISPRS J. Photogram. Remote Sens. **180**, 370–386 (2021). ISSN 0924-2716. https://doi.org/10.1016/j.isprsjprs.2021.07.004, https://www.sciencedirect.com/science/article/pii/S0924271621001817

34. Gonzalez, D., et al.: Automatic detection of building typology using deep learning methods on street level images. Build. Environ. **177**, 106805 (2020). ISSN 0360-1323. https://doi.org/10.1016/j.buildenv.2020.106805, https://www.sciencedirect.com/science/article/pii/S0360132320301633

35. Hackl, J., Adey, B., Woźniak, M., Schümperlin, O.: Use of unmanned aerial vehicle photogrammetry to obtain topographical information to improve bridge risk assessment. J. Infrastruct. Syst. **24**, 04017041 (2018). https://doi.org/10.1061/(ASCE)IS.1943-555X.0000393

36. Miari, M., Choong, K.K., Jankowski, R.: Seismic pounding between adjacent buildings: identification of parameters, soil interaction issues and mitigation measures. Soil Dyn. Earthq. Eng. **121**, 135–150 (2019). ISSN 0267-7261. https://doi.org/10.1016/j.soildyn.2019.02.024, https://www.sciencedirect.com/science/article/pii/S0267726118313848

37. Carboney, J.A., García, H.J., Ortega, R., Iglesias, J.: The Mexico earthquake of September 19, 1985 - statistics of damage and of retrofitting techniques in reinforced concrete buildings affected by the 1985 earthquake. Earthq. Spectra **5** (1989). https://doi.org/10.1193/1.1585516

38. Kasai, K., Maison, B.F.: Building pounding damage during the 1989 Loma Prieta earthquake. Eng. Struct. **19**(3): 195–207 (1997). ISSN 0141-0296. https://doi.org/10.1016/S0141-0296(96)00082-X, https://www.sciencedirect.com/science/article/pii/S014102969600082X

39. Lin, J.-H., Weng, C.-C.: A study on seismic pounding probability of buildings in Taipei metropolitan area. J. Chin. Inst. Eng. **25**(2), 123–135 (2002). https://doi.org/10.1080/02533839.2002.9670687, https://doi.org/10.1080/02533839.2002.9670687

40. Kaushik, H.B., Da, K., Sahoo, D.R., Kharel, G.: Performance of structures during the Sikkim earthquake of 14 February 2006. Curr. Sci. **91**, 449–455 (2006)

41. Bektaş, N., Kegyes-Brassai, O.: Conventional RVS methods for seismic risk assessment for estimating the current situation of existing buildings: a state-of-the-art review. Sustainability **14**(5) (2022). ISSN 2071-1050. https://doi.org/10.3390/su14052583, https://www.mdpi.com/2071-1050/14/5/2583

42. Ramancharla, P., et al.: A primer on rapid visual screening (RVS) consolidating earthquake safety assessment efforts in India (2020)

43. Vacca, G., Dessì, A., Sacco., A.: The use of nadir and oblique UAV images for building knowledge. ISPRS Int. J. Geo-Inf. **6**, 393 (2017). https://doi.org/10.3390/ijgi6120393

44. Arnold, C., Reitherman, R.: Building Configuration and Seismic Design. Wiley, Hoboken (1982)

45. Sahar, L., Muthukumar, S., French, S.P.: Using aerial imagery and GIS in automated building footprint extraction and shape recognition for earthquake risk assessment of urban inventories. IEEE Trans. Geosci. Remote Sens. **48**(9), 3511–3520 (2010)

46. Chen, D., Zhang, L., Li, J., Liu, R.: Urban building roof segmentation from airborne lidar point clouds. Int. J. Remote Sens. **33**(20), 6497–6515 (2012). https://doi.org/10.1080/01431161.2012.690083

47. Vo, A.-V., Truong-Hong, L., Laefer, D.F., Bertolotto, M.: Octree-based region growing for point cloud segmentation. ISPRS J. Photogram. Remote Sens. **104**, 88–100 (2015). ISSN 0924-2716. https://doi.org/10.1016/j.isprsjprs.2015.01.011, https://www.sciencedirect.com/science/article/pii/S0924271615000283

48. Sampath, A., Shan, J.: Segmentation and reconstruction of polyhedral building roofs from aerial lidar point clouds. IEEE Trans. Geosci. Remote Sens. **48**(3), 1554–1567 (2010). https://doi.org/10.1109/TGRS.2009.2030180

49. Dong, Z., Yang, B., Hu, P., Scherer, S.: An efficient global energy optimization approach for robust 3D plane segmentation of point clouds. ISPRS J. Photogram. Remote Sens. **137**, 112–133 (2018). ISSN 0924-2716. https://doi.org/10.

1016/j.isprsjprs.2018.01.013, https://www.sciencedirect.com/science/article/pii/S0924271618300133

50. Li, W., He, C., Fang, J., Zheng, J., Fu, H., Yu, L.: Semantic segmentation-based building footprint extraction using very high-resolution satellite images and multi-source GIS data. Remote Sens. **11**(4) (2019). ISSN 2072-4292. https://doi.org/10.3390/rs11040403, https://www.mdpi.com/2072-4292/11/4/403

51. Qin, Y., Wu, Y., Li, B., Gao, S., Liu, M., Zhan, Y.: Semantic segmentation of building roof in dense urban environment with deep convolutional neural network: a case study using GF2 VHR imagery in China. Sensors **19**, 1164 (2019). https://doi.org/10.3390/s19051164

52. Bhat, A.S., Shivaprakash, A.V., Prasad, N.S., Nagaraj, C.: Template matching technique for panoramic image stitching. In: 2013 7th Asia Modelling Symposium, pp. 111–115 (2013). https://doi.org/10.1109/AMS.2013.22

53. Adwan, S., Alsaleh, I., Majed, R.: A new approach for image stitching technique using dynamic time warping (DTW) algorithm towards scoliosis X-ray diagnosis. Measurements **84**, 32–46 (2016). https://doi.org/10.1016/j.measurement.2016.01.039

54. Bonny, M., Uddin, M.: A technique for panorama-creation using multiple images. Int. J. Adv. Comput. Sci. Appl. **11** (2020). https://doi.org/10.14569/IJACSA.2020.0110293

55. Alomran, M., Chai, D.: Feature-based panoramic image stitching. In: 2016 14th International Conference on Control, Automation, Robotics and Vision (ICARCV), pp. 1–6 (2016). https://doi.org/10.1109/ICARCV.2016.7838721

56. Lowe, D.G.: Distinctive image features from scale-invariant keypoints. Int. J. Comput. Vision **60**(2), 91–110 (2004). ISSN 1573-1405. https://doi.org/10.1023/B:VISI.0000029664.99615.94

57. Zhang, Y., Yang, L., Wang, Z.: Research on video image stitching technology based on surf. In: 2012 Fifth International Symposium on Computational Intelligence and Design, vol. 2, pp. 335–338 (2012). https://doi.org/10.1109/ISCID.2012.235

58. Tsao, P., Ik, T.-U., Chen, G.-W., Peng, W.-C.: Stitching aerial images for vehicle positioning and tracking. In: 2018 IEEE International Conference on Data Mining Workshops (ICDMW), pp. 616–623. IEEE (2018)

59. Liu, Y., Xue, F., Huang, H.: UrbanScene3D: a large scale urban scene dataset and simulator (2021)

60. Wu, Y., Kirillov, A., Massa, F., Lo, W.-Y., Girshick, R.: Detectron2 (2019). https://github.com/facebookresearch/detectron2

61. Schönberger, J.L., Zheng, E., Frahm, J.-M., Pollefeys, M.: Pixelwise view selection for unstructured multi-view stereo. In: Leibe, B., Matas, J., Sebe, N., Welling, M. (eds.) ECCV 2016. LNCS, vol. 9907, pp. 501–518. Springer, Cham (2016). https://doi.org/10.1007/978-3-319-46487-9_31

62. Wang, Y., et al.: LEDNet: a lightweight encoder-decoder network for real-time semantic segmentation. In: 2019 IEEE International Conference on Image Processing (ICIP), pp. 1860–1864 (2019). https://doi.org/10.1109/ICIP.2019.8803154

63. Zhou, X., Girdhar, R., Joulin, A., Krähenbühl, P., Misra, I.: Detecting twenty-thousand classes using image-level supervision. arXiv preprint arXiv:2201.02605 (2021)

ConSLAM: Periodically Collected Real-World Construction Dataset for SLAM and Progress Monitoring

Maciej Trzeciak[1(✉)] [iD], Kacper Pluta[2] [iD], Yasmin Fathy[1] [iD], Lucio Alcalde[3],
Stanley Chee[3], Antony Bromley[3], Ioannis Brilakis[1] [iD], and Pierre Alliez[2] [iD]

[1] Department of Engineering, University of Cambridge, Cambridge, UK
{mpt35,yafa2,ib340}@cam.ac.uk
[2] Inria Sophia Antipolis-Méditerranée, Valbonne, France
{kacper.pluta,pierre.alliez}@inria.fr
[3] Laing O'Rourke, Dartford DA2 6SN, UK
{lalcalde,schee,abromley}@laingorourke.com

Abstract. Hand-held scanners are progressively adopted to workflows on construction sites. Yet, they suffer from accuracy problems, preventing them from deployment for demanding use cases. In this paper, we present a real-world dataset collected periodically on a construction site to measure the accuracy of SLAM algorithms that mobile scanners utilize. The dataset contains time-synchronised and spatially registered images and LiDAR scans, inertial data and professional ground-truth scans. To the best of our knowledge, this is the first publicly available dataset which reflects the periodic need of scanning construction sites with the aim of accurate progress monitoring using a hand-held scanner.

Keywords: Real-world dataset · SLAM · Construction progress monitoring · Point cloud

1 Introduction

The digitization of the geometry of existing infrastructure assets is a crucial step for creating an effective *Digital Twin* (DT) for many applications in *Architecture, Engineering and Construction* (AEC) industry. On the one hand, the growing adoption of mobile and hand-held scanning devices brings the hope of increased productivity with respect to capturing the geometric data. On the other, however, the underlying *Simultaneous Localization And Mapping (SLAM)* algorithms, which such scanners utilize, are not yet accurate enough to meet the requirements of demanding use cases such as engineering surveying. As a result, there is a mix of technologies used on construction sites.

Publicly available datasets serve as battlegrounds against which different methods compare their performance. Yet, there are very few of them that would enable the comparison of SLAM methods on construction sites. In fact, there is no publicly available dataset that would reflect the periodic need of scanning construction sites, with the view of accurate progress monitoring using a hand-held scanner. Our paper aims at addressing this problem.

L. Karlinsky et al. (Eds.): ECCV 2022 Workshops, LNCS 13807, pp. 317–331, 2023.
https://doi.org/10.1007/978-3-031-25082-8_21

Under the following link https://github.com/mac137/ConSLAM, we present a real-world dataset, the "**ConSLAM**", recorded by our prototypical hand-held scanner. The dataset consists of four sequences captured at the same floor of a construction site. We recorded one sequence approximately every month. Each sequence contains *Red-Green-Blue* (RGB) and *Near-InfraRed* (NIR) images of resolutions 2064 × 1544 and 2592 × 1944 pixels respectively, 16-beam Velodyne LiDAR scans and 9-axis *Inertial Measurement Unit* (IMU) data. The first three sensors were synchronised in time and recorded at 10 Hz while IMU was recorded at 400 Hz. The acquired sequences vary in their duration between five and nine minutes. For every sequence, we also include a *Ground-Truth* (GT) point cloud provided by a land surveying team. We used these point clouds to produce the ground-truth trajectories of our scanner, against which SLAM algorithms can measure their accuracy. Our hope is that all these modalities will enable further exploration of mobile mapping algorithms.

This paper is structured as follows. Section 2 provides a discussion on the existing datasets. Section 3 includes the description of our hand-held prototype, as well as other devices and methods we used to produce the complete dataset. In Sect. 4, we briefly describe the construction site and present the structure as well as availability of our dataset. We close the paper by discussing future steps in Sect. 5.

2   Existing Datasets

Mobile scanning systems are portable devices that integrate multiple sensors for obtaining detailed surveys of scanned scenes by creating 3D point cloud data. There are different sequential point cloud datasets, but very few, such as the Hilti SLAM challenge dataset [12] are both sequential as well as collected for construction sites.

The available sequential datasets can be categorized into two main types: synthetic and real-world. A synthetic dataset is artificially generated in a virtual world by simulating a real-world data acquisition system. A sequential dataset is collected as sequences of frames from a movable platform, e.g., vehicular or hand-held ones. Most studied sequential datasets are described in this section, and also summarized in Table 1.

KITTI[1] is a well-known benchmark collected mainly for 3D object detection scenarios [8,9]. The data includes six hours of traffic scenarios at 10–100 Hz using a system mounted on a moving vehicle with a driving speed up to 90 km/h. The system comprises data from high-resolution colour and greyscale stereo cameras, a LiDAR, a *Global Positioning System* (GPS) as well as IMU devices [8]. The set-up allows collecting data that are suitable for different tasks: stereography, optical flow, *Visual Odometry* (VO) and 3D object detection. For the visual odometry benchmark, the data contain 22 sequences of images, 11 of them being associated to ground-truth, and the remaining mainly contains raw sensor data. The ground-truth for VO is the output of GPS/IMU localization. The data is also provided along with the trajectories.

SemanticKITTI [1] is based on the KITTI dataset, mainly the sequences provided for the OV task. SemanticKITTI provides dense point-wise annotation for zero to ten sequences, while the other 11–21 sequences are used for testing, making the data suitable for various tasks. Three main tasks are proposed for SemanticKITTI; semantic

---

[1] http://www.cvlibs.net/datasets/kitti.

segmentation of a scene, semantic scene completion (i.e., predicting future semantic scenes), and semantic segmentation of multiple sequential scans.

SemanticPOSS [14] is a dataset that contains LiDAR scans with dynamic instances. It uses the same data format as SemanticKITT. Similar to KITTI, SemanticPOSS has been collected by a moving vehicle equipped with a Pandora module[2] and a GPS/IMU localization system to collect 3D point cloud data. The Pandora integrates cameras and LiDAR into the same module. The vehicle travelled a distance of around 1.5 kilometres on a road that includes many moving vehicles and walking and riding students. The collected data are annotated that each point contains unique instance labels for dynamic objects (car, people, rider). The data are suitable for predicting the accuracy of dynamic objects and people [7] and 3D semantic segmentation [14]. The data size of Semantic-POSS is limited compared to SemanticKITTI. Although there is a higher resolution on horizontal LiDAR scans, the spatial distribution of the LiDAR points is unbalanced [7].

SynthCity [10] is a synthetic labelled point cloud dataset generated from a synthetic full-colour mobile laser scanning with a predefined trajectory. Each point is labelled by one of nine categories: high vegetation, low vegetation, buildings, scanning artefacts, cars, hardscape, man-made terrain, and natural terrain. The synthetic point clouds are generated in urban/suburban environments modelled within Blender 3D graphics software[3]. The dataset has been released primarily for semantic per-point classification, where each point contains a local feature vector and a classification label. However, the dataset is unsuitable for instance segmentation as it does not include instances' identifiers.

The Grand Theft Auto V (GTA5) [15] is a synthetic sequential point cloud dataset that was generated based on the photo-realistic virtual world in the commercial video game "Grand Theft Auto V". The approach is based on creating large-scale pixel-level semantic segmentation by extracting a set of images from the game and then applying a pipeline to produce the corresponding label. The game includes different resource types, including texture maps and geometric meshes, combined to compose a scene, which facilitates establishing the associations between scene elements. GTA5 is three orders of magnitude larger than semantic annotations included in the KITTI dataset [8, 15]. The data contains 19 semantic classes, including road, building, sky, truck, person, traffic light and other objects on road scenes. The data was used for training semantic segmentation models and evaluated on two datasets, including KITTI [8] where the training phase included both real and synthetic data using minibatch stochastic gradient descent. The model trained with generated synthetic data within GTA5 outperforms the model trained without it by factor 2.6.

The nuTonomy scenes (nuScenes) [4] is a real-world dataset for collecting point cloud using six cameras, five radars and a LiDAR, each with a full 360-degree field of view. The data is fully annotated with 3D bounding boxes, mainly for autonomous driving scenarios, with available map information associated to the collected data. The data include trajectories as idealized paths that the movable platform should take; assuming there are no obstacles in the route. The data include 23 classes: road, pavement, ground, tree, building, pole-like, and others. Compared to the KITTI dataset, nuScenes has seven

---

[2] Please consult www.hesaitech.com.

[3] https://www.blender.org/.

times more object annotations and one hundred times more images. The dataset is currently suitable for 3D object detection and tracking, where tracking annotation is also available [5].

The Hilti SLAM 2021 challenge dataset [12] is a benchmark dataset that collects multiple sensor modalities of mixed indoor and outdoor environments with varying illumination conditions and along the trajectory. The indoor sequences portray labs, offices and construction environments, and the outdoor sequences were recorded in parking areas and on construction sites[4]. The data were collected by a handheld platform comprising multiple sensors: five AlphaSense cameras (stereo pair), two LiDARs (Ouster OS0-64 and Livox MID70), and three IMUs (ADIS16445) with accurate spatial and temporal calibration. The main aim of this dataset is to promote the development of new SLAM algorithms that attain both high accuracy and robustness for challenging real-world environments such as construction sites. In 2022, the same challenge was organized, but the data were collected mainly for construction sites and Sheldonian Theatre in Oxford, UK[5]. The data were collected by a sensor suite mounted on an aluminium platform for handheld operation. The suite consists of a Hesai PandarXT-32 and Sevensense Alphasense Core camera head with five 0.4MP global shutter cameras. The LiDAR and cameras are synchronised via *Precision Time Protocol* and all sensors are aligned within one millisecond.

To the best of our knowledge, the **ConSLAM** dataset is the first sequential dataset with a trajectory for a construction site that aims to capture the construction of a site over a few months. The data are collected by a hand-held mobile scanner comprising a LiDAR, RGB and NIR cameras and an IMU (shown in Fig. 1). All the modalities are synchronised in time and spatially registered. This aims to foster novel research and progress in evaluating SLAM approaches and trajectory tracking at construction sites and developing progress monitoring and quality control systems for the AEC community.

3   Methodology

In this section, we introduce the configuration of the hand-held prototype device, and our data acquisition and post-processing pipelines.

3.1   Sensors and Devices

The sensors used during data acquisition at the construction site are shown in Fig. 1. As shown in Fig. 1(a), our prototypical hand-held scanner consists of a LiDAR at the top (Velodyne VLP-16), an RGB camera (Alvium U-319c, 3.2 MP) located directly below the LiDAR, a NIR camera (Alvium 1800 U-501, 5.0 MP) located to the right of the RGB camera and an IMU (Xsens MTi-610), to the left of the RGB camera. They are rigidly attached to a custom-made aluminium frame with a handle at the bottom. All of them are connected to a laptop (MacBook Pro 2021) where data were recorded and

---

[4] https://hilti-challenge.com/dataset-2021.html.
[5] https://hilti-challenge.com/dataset-2022.html.

Table 1. Summary of existing sequential point cloud datasets

Name	Real/ Synthetic[a]	Indoor/ Outdoor[b]	Traject-ory[c]	Sector	Applications	Sensors
KITTI [9]	R	I	✓	Urban and transport	Autonomous vehicles [6, 13], 3D object detection and visual odometry [9]	Four colour and greyscale stereo cameras, a laser scanner (Velodyne), four Edmund optics lenses, GPS navigation systems
Semantic-KITTI [1]	R	O	✓	Urban/road	Semantic segmentation of a scene, semantic scene completion (i.e., predicting future semantic scenes), and semantic segmentation of multiple sequential scans [1]	Relying on the data collected by laser scanner (Velodyne) in the KITTI dataset
HILTI-OXFORD [12]	R	I &O	✓	Construction	Construction robotics, construction site environments	Five AlphaSense cameras (stereo pair), two LiDARs (Ouster OS0-64 and Livox MID70), and three IMUs (ADIS16445)
Semantic-POSS [14]	R	O	✗	Urban/road	Prediction accuracy of dynamic objects and people [7] and 3D semantic segmentation [14]	Pandora module (LiDAR, mono and colour cameras) and GPS/IMU
SynthCity [10]	S	O	✓	Urban/ suburban environments	Point cloud classification [10]	Mobile laser scanning
GTA5 [15]	S	O	✗	Urban/road	Semantic segmentation and scene understanding [15]	Frames extracted from "Grand Theft Auto V" video game; from a car perspective
nuScenes [4]	R	O	✓	Urban/road and autonomous driving	Object detection and tracking, segmentation [4]	Six cameras, five radars and one LiDAR, all with full 360° field of view
ConSLAM	R	O	✓	Construction	Progress monitoring & quality control, object detection and tracking	LiDAR (Velodyne VLP-16), RGB camera (Alvium U-319c, 3.2 MP), a NIR camera (Alvium 1800 U-501, 5.0 MP) and an IMU (Xsens MTi-610)(see Fig. 1)

[a] R: real-world, S: synthetic/artificial.
[b] I: indoor, O: outdoor.
[c] Indicates whether the data includes sensor's path, i.e., trajectory.

pre-processed. In addition, a Leica RTC 360[6] (see Fig. 1(b)) is used to collect precise scans which are later stitched together and geo-referenced by land surveyors. These scans serve as our ground-truth.

3.2 Intrinsic Calibrations of the Sensors

Both cameras, i.e., RGB and NIR, are intrinsically calibrated according to the Brown-Conardy model [3] with the camera intrinsic matrix and lens distortion coefficients stored along with the dataset. The resolution of distorted images is 2064×1544 and 2592×1944 pixels for the RGB and NIR cameras, respectively. Moreover, the LiDAR's intrinsic parameters have mostly default values as in the manufacturer's manual and the Velodyne driver[7] with a restricted range of 60 m and the parameter `lidar_timestamp_first_packet` is set to `true`.

---

[6] https://leica-geosystems.com/products/laser-scanners/scanners/leica-rtc360.

[7] https://github.com/ros-drivers/velodyne.

(a)   (b)

Fig. 1. Data acquisition: (a) our prototypical hand-held scanner, and (b) a static scanner used to collect ground-truth scans. Red, green and blue colours represent X-, Y- and Z-axes, respectively (Color figure online)

3.3 Extrinsic Calibrations of the Sensors

The LiDAR sensor is extrinsically calibrated in a pair with other sensors as follows: the LiDAR and the RGB camera, the LiDAR and the NIR camera, and the LiDAR and the IMU. The sensor frames are positioned against each other in our prototypical hand-held scanner, as shown in Fig. 1(a). It is worth-mentioning that when the scanner is held vertically (i.e., in its operational position), LiDAR's X-, RGB camera's Z- and NIR camera Z-axes face the front while IMU's X-axis faces backward. LiDAR's and IMU's Z-axes face upwards, while the RGB and NIR cameras' Y-axes face downwards. The remaining axes can be further deduced from the figure.

We used a method proposed by Beltrán et al. [2] to extrinsically calibrate the LiDAR with both the RGB camera and the NIR camera. We used the method used by VINS-Mono[8] for LiDAR-IMU calibration. Their respective matrices are stored along with the dataset.

3.4 Data Collection System of the Hand-Held Scanner

Our hand-held data collection system utilizes *Robot Operating System* (ROS)[9] as a backbone to process the data streams coming from all four sensors. Figure 2 presents the data processing pipeline in more detail.

We first launch the respective drivers of the four sensors, thus publishing the individual data messages to our ROS-based system as shown in the top layer in Fig. 2. Next, we synchronize the RGB camera, LiDAR and NIR camera in time using a standard ROS synchronization policy[10], based on matching messages whose difference in

---

[8] see https://github.com/chennuo0125-HIT/lidar_imu_calib, and https://blog.csdn.net/weixin_37835423/article/details/110672571.

[9] https://www.ros.org.

[10] https://wiki.ros.org/message_filters/ApproximateTime.

timestamps is smaller than 10 milliseconds. However, we decided to split this process into two because of problems encountered with our NIR camera. If the synchronization was matching timestamps from all three topics and any of the topics stopped working for a moment, the synchronization of all the three topics would stop too. Our NIR camera sporadically stops publishing images for a moment, which would effectively stop the synchronization of all three sensors. Instead, we decided to synchronise RGB images and LiDAR scans first and publish them on /pp_rgb/synced2points and /pp_points/synced2rgb topics respectively. NIR images are then synchronised with /pp_points/synced2rgb using the same synchronization policy and published on the /pp_nir/synced2points topic. This solution allows us to keep recording synchronized RGB images and lidar scans even when the NIR camera stops working for a moment.

During scanning, we also monitor the three synchronised topics and the IMU messages to make sure that our system actually receives data from the sensors. In the last step, we record the synchronised topics along with the IMU data (/imu/data) and store them as a standard bag file. We decided to record IMU messages 400 Hz because it is often the case that higher IMU rate improves the performance of SLAM algorithms [16].

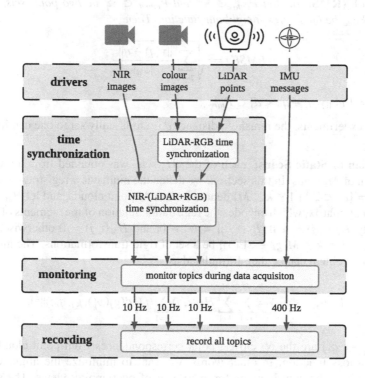

Fig. 2. Processing data streams on construction site

3.5 Ground-Truth Trajectories

This section discusses the ground-truth dataset and focuses on our post-processing pipeline used for its creation and refinement. We also discuss the issue of registering individual LiDAR scans to the ground-truth dataset to recover the ground-truth trajectory of our prototypical scanner. The ground-truth dataset contains four scans referred to as $GT_i, i = 1, \ldots, 4$, which were collected over a period of three months on an active construction site, as explained in Sect. 4.

Registration Error Metric: we faced two main issues during the registration process of datasets: varying overlap and geometric discrepancies. Such issues make a 3D point cloud registration difficult and require manual adjustments in order to ensure high precision of the data alignment. Such a registration can take up to a few dozens of minutes to be properly performed by an experienced person.

Nevertheless, we need to be able to provide a registration error metric for the datasets, hampered by the aforementioned issues. There exist several error metrics, which can be used for our purposes. We opted for the distance-constrained Root Mean Square Error (RMSE$_d$ for short), as provided in Definition 1.

Definition 1 (RMSE$_d$). *Let $P_{data} \subset \mathbb{R}^3$ and $P_{target} \subset \mathbb{R}^3$ be two point sets, and $\gamma : P_{target} \to P_{data}$ be the nearest-neighbour function. Then,*

$$RMSE_d = \sqrt{\sum_{q \in S_d} \frac{\|\gamma(q) - q\|^2}{|S_d|}}, \tag{1}$$

where $S_d = \{q \mid \|\gamma(q) - q\| < d\} \subset P_{target}$.

In our experiments, the threshold distance d is empirically set to one centimetre.

Registration of Static Scans: each of the GT_i sets, was obtained from a multi-view registration of M_i scans. In this section, we define the multi-view registration problem.

Let $\mathbb{P} = \{P_k \subset \mathbb{R}^3 \mid 1 \le k \le M\}$ denote a set of M point clouds, and let H_M denote a square binary matrix, which encodes the registration relation of the elements of $\mathbb{P}$. More specifically, $H_M(i, j) = 1$ if $|P_i \cap P_j| = N \gg 0$, and $H_M(i, j) = 0$ otherwise. Finally, let $\mathbb{G} = \{g_k \mid 1 \le k \le M, g_k \in \mathrm{SE}(3)\}$ be a set of rigid transformations. The multi-view registration problem can be then formulated as

$$E(g_1, \ldots, g_M) = \sum_{i=1}^{M} \sum_{j=1}^{M} H_M(i, j) \sum_{k=1}^{N_j} f_l(\|d(g_j(p_k^j), g_j(q_k^j))\|^2), \tag{2}$$

where $\{p_k^j \to q_k^j\}$ are the N_j closest point correspondences from point clouds P_i, P_j, and f_l is a loss function. In other words, we want to minimize the alignment error by summing up the contributions for every pair of overlapping views. The solutions $g_1, \ldots, g_M = argmin(E)$ are the rigid transformations that align the M clouds in the least squares sense. For more information, we refer the reader to a technical report authored by Adrian Haarbach [11].

Table 2. RMSE_d distance for ground-truths dataset. The measurements are recorded in centimetres

Dataset name	$\approx \min_{\mathrm{RMSE}_d}$	$\approx \max_{\mathrm{RMSE}_d}$	$\approx \mathrm{mean}_{\mathrm{RMSE}_d}$
GT_1	0.319	0.950	0.676
GT_2	0.262	0.980	0.605
GT_3	0.327	0.983	0.607
GT_4	0.360	0.902	0.637

Having registered the scans obtained from a terrestrial scanner, we have downsampled them using distance-based downsampling[11] with the threshold of five millimetres. Finally, we compute RMSE_d distances between overlapping point sets, see Table 2.

Registration of LiDAR Scans to Ground-Truth Scans: We create a ground-truth trajectory of the LiDAR by registering each LiDAR scan to the ground-truth scans. This means that we recover the true pose of each LiDAR scan with respect to the ground-truth scans received from land surveyors. To do that, we *play* each recorded bag file and save every LiDAR message to the PLY file format as an individual LiDAR scan. In addition, we run *Advanced Lidar Odometry and Mapping* (A-LOAM[12]) algorithm—an implementation of the LOAM algorithm proposed by Zhang et al. [17]—on each bag file and save odometric poses corresponding to the individual LiDAR scans as text files. The text files and the individual LiDAR scans are then matched based on their timestamps assigned during the data collection.

The ICP algorithm is then executed on each pose-scan pair. We extract edges from every LiDAR scan in the same way as it is done in A-LOAM and use the corresponding pose as an initial guess for the fine registration. The ICP is performed for a couple of iterations, starting with half a metre as an initial threshold for establishing correspondences to the closest points. With every iteration, the threshold decays by a factor of 0.85 and the algorithm converges when the RMSE/fitness is smaller than three centimetres.

The resulting six Degree of Freedom (6-DOF) transformations are stored as 4×4 matrices and are named after the corresponding LiDAR scans. In other words, these matrices represent the transformations that the LiDAR scans must undergo to be aligned with the ground-truth scans. The whole collection of transformations makes up the ground-truth trajectory. Figures 4 and 5 present photorealistic renderings of the ground-truth data and the position of the registered LiDAR scans.

We implemented a computer program to automate the registration process described above. However, there were still situations where our software was unsuccessful. This includes the following: (1) the drift by the LOAM algorithm run on the stream of LiDAR scans was high enough that its poses fed to our registration algorithm were too distant

---

[11] We used the distance-based downsampling implemented in CloudCompare 2.12.2. See https://www.cloudcompare.org.

[12] https://github.com/HKUST-Aerial-Robotics/A-LOAM.

to find correct correspondences between the extracted LiDAR features and the ground truth scans; (2) the exact trajectory our scanner followed is not fully covered by the ground-truth scans from the land surveying team, hence it was not possible to align the LiDAR scans from such places to the ground-truth scans. An example relating to the first issue can be seen in Fig. 3. We estimate that we have correctly registered around 80% of LiDAR scans to GT_1, about 80% of LiDAR scans to GT_2, approximately 60% of LiDAR scans to GT_3 and roughly 70% of LiDAR scans to GT_4.

4 Dataset: ConSLAM

We collected the four raw streams of data at a part of a story at Whiteley's in London which had been originally designed by John Belcher and John James Joass in 1911 as one of London's leading department stores. At the time of writing this paper, the building is undergoing redevelopment, which involved the demolition of the existing shopping centre behind a retained historic façade. The new development involved the creation of luxury retail, leisure and a residential scheme involving the construction of a new six to nine-story building.

4.1 Dataset Structure

Our dataset is structured as shown in Fig. 6. The directory includes five main files in the ZIP file format, four of them containing data from four individual scans carried once per month. The scans are numbered from 1 to 4, with the earliest scan marked with 1 and the oldest one marked with 4. In order to save space in Fig. 6, we encoded this fact with data_unpacked_x.zip where x = 1, ..., 4.

Each of the four data zipped files contains a recording.bag file recorded during scanning. This file can be *played* using rosbag[13] and contains four topics with the stream of RGB and NIR images, LiDAR points and IMU messages. groundtruth_scan.ply file is our ground-truth point cloud created by land surveyors as described in Sect. 3.5. Next, there are three folders rgb/, nir/ and lidar/ which contain messages unpacked

Fig. 3. Visualization of incorrectly registered LiDAR poses, which have been marked by the light-brown ellipse

---

[13] https://wiki.ros.org/rosbag.

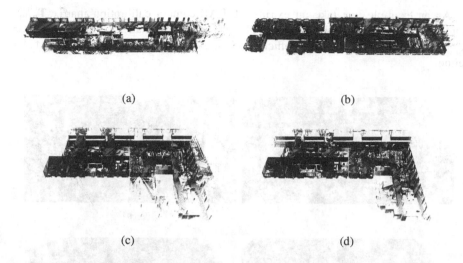

(a)   (b)

(c)   (d)

Fig. 4. Top-view visualization of the ground-truth datasets: GT_1 – (a), GT_2 – (b), GT_3 – (c), GT_4 – (d). Note that in some places we can see incorrectly registered LiDAR poses, i.e., (a) and (c)

from the `recording.bag` file. The corresponding LiDAR scans in `lidar/` and images in `rgb/`, `nir/` are named with the same timestamp coming from LiDAR scans recorded during scanning. The last folder `pose/` contains the ground-truth poses of the LiDAR sensor created as described in Sect. 3.5 and named also with the corresponding times-tamps. The collection of these poses makes up the ground-truth trajectory of the LiDAR sensor.

The file `data_calib.zip` includes all the calibration parameters including: RGB and NIR camera calibration matrices along with their distortion coeffi-cients in `calib_rgb.txt` and `calib_nir.txt` respectively. Moreover, there is a rigid-body transformation matrix between the LiDAR and the RGB camera in `calib_lidar2rgb.txt`, rigid-body transformation matrix between the LiDAR and the NIR camera in `calib_lidar2nir.txt`, and finally, a rotation matrix between the LiDAR and IMU in `calib_lidar2imu.txt`.

4.2 Practical Application: Projecting LiDAR Points onto Corresponding Images

Our **ConSLAM** dataset is available at https://github.com/mac137/ConSLAM, where a complete description and examples on how to use the data are provided.

As an example, we take an extrinsic LiDAR-camera calibration matrix $\mathbf{T}_{RGB}^{LiDAR}$ stored in `calib_lidar2rgb.txt` and RGB intrinsic camera matrix for distorted images $\mathbf{K}_{dist}^{RGB}$ along with five distortion coefficients $(k_1, k_2, k_3, k_4, k_5)$ from `calib_rgb.txt`. We define

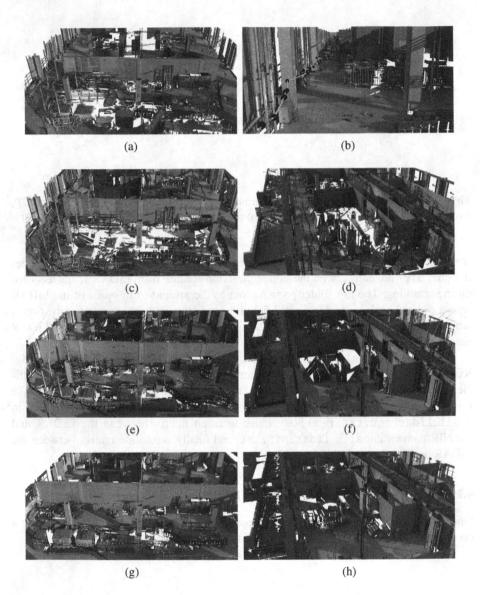

(a)

(b)

(c)

(d)

(e)

(f)

(g)

(h)

Fig. 5. Close-up visualization of the ground-truth datasets: GT_1 – (a-b), GT_2 – (c-d), GT_3 – (e-f), GT_4 – (g-h), together with the LiDAR positions depicted by red spheres. The LiDAR position have been connected to provide approximated paths

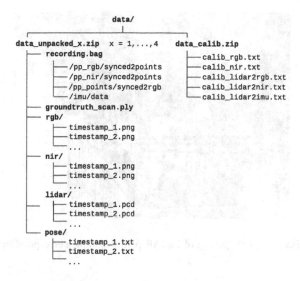

Fig. 6. Dataset folder structure

$$\mathbf{T}_{\text{RGB}}^{\text{LiDAR}} = \begin{bmatrix} \mathbf{R}_{\text{RGB}}^{\text{LiDAR}} & \mathbf{T}_{\text{RGB}}^{\text{LiDAR}} \\ \mathbf{0}_{1\times3} & 1 \end{bmatrix}, \tag{3}$$

where $\mathbf{R}_{\text{RGB}}^{\text{LiDAR}} \in SO(3)$ is a 3×3 rotation matrix from the LiDAR to the camera and $\mathbf{T}_{\text{RGB}}^{\text{LiDAR}}$ is a 3×1 translation vector also from the LiDAR to the camera.

Now, let us take an RGB image from the `rgb/` folder of any sequence from one to four, undistort it and compute the RGB camera intrinsic matrix for undistorted images $\mathbf{K}_{undist}^{\text{RGB}}$ using OpenCV[14] package, $\mathbf{K}_{dist}^{\text{RGB}}$ and $(k_1, k_2, k_3, k_4, k_5)$. We find the corresponding LiDAR scan in the `lidar/` folder using the file name of the image, and we iterate over its points. In order to project a single LiDAR point $\mathbf{x}_i = [x_i, y_i, z_i]^\top$ onto the undistorted images, we follow

$$\begin{bmatrix} u' \\ v' \\ w' \end{bmatrix} = \mathbf{K}_{undist}^{\text{RGB}} \begin{bmatrix} \mathbf{I}_{3\times3} \\ \mathbf{0}_{1\times3} \end{bmatrix}^\top \mathbf{T}_{\text{RGB}}^{\text{LiDAR}-1} \begin{bmatrix} x_i \\ y_i \\ z_i \\ 1 \end{bmatrix}, \tag{4}$$

and then the pixel coordinates $[u, v]^\top$ are recovered from the homogeneous coordinates as follows

$$\begin{bmatrix} u \\ v \end{bmatrix} = \begin{bmatrix} u'/w' \\ v'/w' \end{bmatrix}. \tag{5}$$

We refer the reader to Fig. 7, which shows an example of LiDAR points projected onto the corresponding image.

---

[14] https://opencv.org.

Fig. 7. Example of projecting of LiDAR points onto the corresponding image

5   Conclusion and Future Direction

We introduced a new real-world dataset, the "**ConSLAM**", recorded periodically by a hand-held scanner on a construction site. The dataset aims at facilitating the comparison of SLAM algorithms for periodic and accurate progress monitoring in construction. The dataset contains the ground-truth trajectories of the scanner, which allows for an accurate comparison of SLAM methods run on our recorded data streams.

In the future, we aim at registering of all LiDAR scans to the ground-truth scans, hence recovering the full ground-truth trajectory of our scanner. The objective is also to extend this dataset with ground-truth for semantic segmentation of images and point clouds. This will allow the development of progress monitoring systems based on the comparison of Design-Intent and As-Built, for example, using the volume of individual building elements. The semantic annotations will also allow prospective algorithms to measure the accuracy of inferred information in such popular tasks like object detection or instance segmentation.

Acknowledgment. The authors would like to thank Laing O'Rourke for allowing access to their construction site and for collecting the ground truth scans. We also acknowledge Romain Carriquiry-Borchiari of Ubisoft France for his help with rendering some of the figures. This work is supported by the EU Horizon 2020 BIM2TWIN: Optimal Construction Management & Production Control project under an agreement No. 958398. The first author would also like to thank BP, GeoSLAM, Laing O'Rourke, Topcon and Trimble for sponsoring his studentship funding.

References

1. Behley, J., et al.: SemanticKITTI: a dataset for semantic scene understanding of lidar sequences. In: Proceedings of the IEEE/CVF International Conference on Computer Vision, pp. 9297–9307 (2019)
2. Beltrán, J., Guindel, C., de la Escalera, A., García, F.: Automatic extrinsic calibration method for lidar and camera sensor setups. IEEE Trans. Intell. Transp. Syst. (2022). https://doi.org/10.1109/TITS.2022.3155228
3. Brown, D.: Decentering distortion of lenses. In: Photogrammetric Engineering, pp. 444–462 (1966)
4. Caesar, H., et al.: nuScenes: a multimodal dataset for autonomous driving. In: Proceedings of the IEEE/CVF Conference on Computer Vision and Pattern Recognition, pp. 11621–11631 (2020)
5. Chang, M.F., et al.: Argoverse: 3D tracking and forecasting with rich maps. In: Proceedings of the IEEE/CVF Conference on Computer Vision and Pattern Recognition, pp. 8748–8757 (2019)
6. Fritsch, J., Kuehnl, T., Geiger, A.: A new performance measure and evaluation benchmark for road detection algorithms. In: 16th International IEEE Conference on Intelligent Transportation Systems (ITSC 2013), pp. 1693–1700. IEEE (2013)
7. Gao, B., Pan, Y., Li, C., Geng, S., Zhao, H.: Are we hungry for 3D lidar data for semantic segmentation? A survey and experimental study. arXiv preprint arXiv:2006.04307 (2020)
8. Geiger, A., Lenz, P., Stiller, C., Urtasun, R.: Vision meets robotics: the KITTI dataset. Int. J. Robot. Res. 32(11), 1231–1237 (2013)
9. Geiger, A., Lenz, P., Urtasun, R.: Are we ready for autonomous driving? The KITTI vision benchmark suite. In: 2012 IEEE Conference on Computer Vision and Pattern Recognition, pp. 3354–3361. IEEE (2012)
10. Griffiths, D., Boehm, J.: SynthCity: a large scale synthetic point cloud. arXiv preprint arXiv:1907.04758 (2019)
11. Haarbach, A.: Multiview ICP. http://www.adrian-haarbach.de/mv-lm-icp/docs/mv-lm-icp.pdf
12. Helmberger, M., et al.: The Hilti SLAM challenge dataset. arXiv preprint arXiv:2109.11316 (2021)
13. Menze, M., Geiger, A.: Object scene flow for autonomous vehicles. In: Proceedings of the IEEE Conference on Computer Vision and Pattern Recognition, pp. 3061–3070 (2015)
14. Pan, Y., Gao, B., Mei, J., Geng, S., Li, C., Zhao, H.: SemanticPOSS: a point cloud dataset with large quantity of dynamic instances. In: 2020 IEEE Intelligent Vehicles Symposium (IV), pp. 687–693. IEEE (2020)
15. Richter, S.R., Vineet, V., Roth, S., Koltun, V.: Playing for data: ground truth from computer games. In: Leibe, B., Matas, J., Sebe, N., Welling, M. (eds.) ECCV 2016. LNCS, vol. 9906, pp. 102–118. Springer, Cham (2016). https://doi.org/10.1007/978-3-319-46475-6_7
16. Shan, T., Englot, B., Meyers, D., Wang, W., Ratti, C., Daniela, R.: LIO-SAM: tightly-coupled lidar inertial odometry via smoothing and mapping. In: IEEE/RSJ International Conference on Intelligent Robots and Systems (IROS), pp. 5135–5142. IEEE (2020)
17. Zhang, J., Singh, S.: LOAM: lidar odometry and mapping in real-time. In: Proceedings of Robotics: Science and Systems, Berkeley, USA (2014). https://doi.org/10.15607/RSS.2014.X.007

NeuralSI: Structural Parameter Identification in Nonlinear Dynamical Systems

Xuyang Li[✉][iD], Hamed Bolandi[iD], Talal Salem[iD], Nizar Lajnef[iD], and Vishnu Naresh Boddeti[iD]

Michigan State University, East Lansing, USA
{lixyan1,bolandih,salemtal,lajnefni,vishnu}@msu.edu

Abstract. Structural monitoring for complex built environments often suffers from mismatch between design, laboratory testing, and actual built parameters. Additionally, real-world structural identification problems encounter many challenges. For example, the lack of accurate baseline models, high dimensionality, and complex multivariate partial differential equations (PDEs) pose significant difficulties in training and learning conventional data-driven algorithms. This paper explores a new framework, dubbed NeuralSI, for structural identification by augmenting PDEs that govern structural dynamics with neural networks. Our approach seeks to estimate nonlinear parameters from governing equations. We consider the vibration of nonlinear beams with two unknown parameters, one that represents geometric and material variations, and another that captures energy losses in the system mainly through damping. The data for parameter estimation is obtained from a limited set of measurements, which is conducive to applications in structural health monitoring where the exact state of an existing structure is typically unknown and only a limited amount of data samples can be collected in the field. The trained model can also be extrapolated under both standard and extreme conditions using the identified structural parameters. We compare with pure data-driven Neural Networks and other classical Physics-Informed Neural Networks (PINNs). Our approach reduces both interpolation and extrapolation errors in displacement distribution by two to five orders of magnitude over the baselines. Code is available at https://github.com/human-analysis/neural-structural-identification.

Keywords: Neural differential equations · Structural system identification · Physics-informed machine learning · Structural health monitoring

1 Introduction

Structural-system identification (SI) [3,15,22,35,39,41] refers to methods for inverse calculation of structural systems using data to calibrate a mathematical or digital model. The calibrated models are then used to either estimate

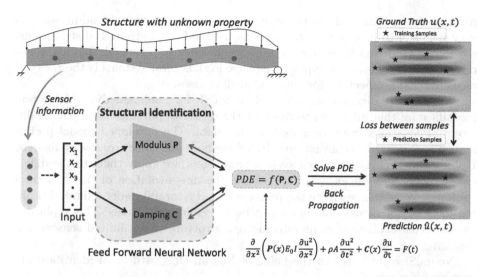

Fig. 1. Overview: We consider structures whose dynamics are governed by a known partial differential equation (PDE), but with unknown parameters that potentially vary in both space and time. These unknown parameters are modeled with neural networks, which are then embedded within the PDE. In this illustration, the unknown parameters, modulus P and damping C, vary spatially. The network weights are learned by solving the PDE to obtain the structural response (deflection in this case) and propagating the error between the predicted response and the measured ground truth response through the PDE solve and the neural networks.

or predict the future performance of structural systems and, eventually, their remaining useful life. Non-linear structural systems with spatial and temporal variations present a particular challenge for most inverse identification methods [4,14,21]. In dynamic analysis of civil structural systems, prior research efforts primarily focused on matching experimental data with either mechanistic models (i.e., known mechanical models) [31,38] or with black box models with only input/output information (i.e., purely data-driven approaches), [10,13,34]. Examples of these approaches include eigensystem identification algorithms [37], frequency domain decomposition [7], stochastic optimization techniques [30], and sparse identification [8]. A majority of these approaches, however, fail to capture highly non-linear behaviors.

In this paper, we consider the class of non-linear structural problems with unknown spatially distributed parameters (see Fig. 1 for an overview). The parameters correspond to geometric and material variations and energy dissipation mechanisms, which could be due to damping or other system imperfections that are not typically captured in designs. As an instance of this problem class, we consider forced vibration responses in beams with spatially varying parameters. The primary challenges in such problems arise from the spatially variable nature of the properties and the distributed energy dissipation. This is typical for built civil structures, where energy dissipation and other hard-to-model phenomena physically drive the dynamic response behavior. In addition, it is very common to have structural systems with unknown strength distributions, which can be driven by geometric non-linearities or indiscernible/hidden material weaknesses. Finally, a typical challenge in structural systems is the rarity of measured data, especially for extreme loading cases.

We propose a framework, dubbed NeuralSI, for nonlinear dynamic system identification that allows us to discover the unknown parameters of partial differential equations from measured sensing data. The developed model performance is compared to conventional PINN methods and direct regression models. Upon estimating the unknown system parameters, we apply them to the differential model and efficiently prognosticate the time evolution of the structural response. We also investigate the performance of NeuralSI under a limited training data regime across different input beam loading conditions. This replicates the expected challenges in monitoring real structures with limited sensors and sampling capabilities.

NeuralSI contributes to the fields of NeuralPDEs, structural identification, and health monitoring:

1. NeuralSI allows us to learn unknown parameters of fundamental governing dynamics of structural systems expressed in the form of PDEs.
2. We demonstrate the utility of NeuralSI by modeling the vibrations of nonlinear beams with unknown parameters. Experimental results demonstrate that NeuralSI achieves two-to-three orders of magnitude lower error in predicting displacement distributions in comparison to PINN-based baselines.
3. We also demonstrate the utility of NeuralSI in temporally extrapolating displacement distribution predictions well beyond the training data measurements. Experimental results demonstrate that NeuralSI achieves four-to-five orders of magnitude lower error compared to PINN-based baselines.

2   Related Work

Significant efforts have been directed toward physics-driven discovery or approximation of governing equations [15,21,26]. Such studies have further been amplified by the rapid development of advanced sensing techniques and machine learning methods [16,17,19,32]. Most of the work to date has mainly focused on ordinary differential equation systems [21,40]. Neural ODEs [9] have been widely adopted due to their capacity to learn and capture the governing dynamic behavior from directly collected measurements [2,28,40]. They represent a significant step above the direct fitting of a relation between input and output variables. In structural engineering applications, Neural ODEs generally approximate the time derivative of the main physical attribute through a neural network.

More recently, data-driven discovery algorithms for the estimation of parameters in differential equations are introduced. These methods typically referred to as physics-informed neural networks (PINNs) include differential equations, constitutive equations, and initial and boundary conditions in the loss function of the neural network and adopt automatic differentiation to compute derivatives of the network parameters [20, 29]. Variational Autoencoders were also learned to build baseline behavioral models, which were then used to detect and localize anomalies [23]. Many other applications have employed Neural ODE for dynamic structure parameter identification in both linear and nonlinear cases [2, 21, 40]. On the other hand, few studies have explored Neural PDEs in other fields such as message passing [6], weather and ocean wave data [11], and fluid dynamics [5]. In [18], a Graph Neural Network was used to solve flow phenomena, and a NeuralPDE solver package was developed in Julia [42] based on PINN.

3 Structural Problem – PDE Derivation

3.1 Problem Description

Many physical processes in engineering can be described as fourth-order time-dependent partial differential problems. Examples include the Cahn-Hilliard type equations in Chemical Engineering, the Boussinesq equation in geotechnical engineering, the biharmonic systems in continuum mechanics, the Kuramoto-Sivashinsky equation in diffusion systems [27] and the Euler-Bernoulli equation considered as an example case study in this paper. The Euler-Bernoulli beam equation is widely used in civil engineering to estimate the strength and deflection of beam structures. The dynamic beam response is defined by:

$$F(t) = \frac{\partial^2}{\partial x^2}\left(P(x)E_0 I \frac{\partial^2 u}{\partial x^2}\right) + \rho A \frac{\partial^2 u}{\partial t^2} + C(x)\frac{\partial u}{\partial t} \tag{1}$$

where $u(x,t)$ is the displacement as a function of space and time. $P(x)$ and E_0 are the modulus coefficient and the reference modulus value of the beam, I, ρ, and A are refereed to the beam geometry and density. F is the distributed force applied to the beam. $C(x)$ represents damping, which is related to energy dissipation in the structure. In this paper, we restrict ourselves only to spatial variation of the beam's properties and leave the most generalized case with variations in space and time of all variables for a future study.

The fourth-order derivative of the spatial variable and the second-order derivative of time describes the relation between the beam deflection and the load on the beam [1]. Figure 2 shows an illustration of the beam problem considered here, with the deflection $u(x,t)$ as the physical response of interest. The problem can also be formulated as a function of moments, stresses, or strains. The deflection formulation presents the highest order differentiation in the PDE. This was selected to allow for flexibility of the solution to be extended to other applications beyond structural engineering.

To accurately represent the behavior of a structural component, its properties need to be identified. Though the beam geometry is straightforward to measure, the material property and damping coefficient are hard to estimate. The beam reference modulus E_0 is expected to have an estimated range based on the choice of material (e.g., steel, aluminum, composites, etc.) but unforeseen weaknesses in the build conditions can introduce unexpected nonlinear behavior. One of the

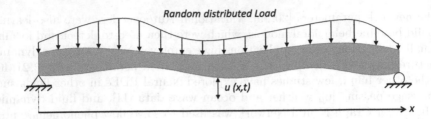

Fig. 2. Simply supported dynamic beam bending problem. Dynamic load can be applied to the structure with its values changing in time. The geometry, modulus, and other properties of the beam can also vary spatially with x. The deflection of the beam is defined as $u(x,t)$.

objectives of this work is to capture this indiscernible randomness from response measurements. In addition, as discussed above, the damping is unpredictable at the design stage and is usually calculated by experiments. For the simply supported beam problem, the boundary conditions are defined as:

$$\begin{cases} u(x=0,t) = 0; & u(x=L,t) = 0 \\ \frac{\partial^2 u(x=0,t)}{\partial x^2} = 0; & \frac{\partial^2 u(x=L,t)}{\partial x^2} = 0 \end{cases} \tag{2}$$

where L is the length of the beam. Initially, the beam is static and stable, so the initial conditions of the beam are:

$$\begin{cases} u(x,t=0) = 0 \\ \frac{\partial u(x,t=0)}{\partial t} = 0 \end{cases} \tag{3}$$

4   NeuralSI

4.1   Discretization of Space

To tackle this high-order PDE efficiently, a numerical approach based on the method of lines is employed to discretize the spatial dimensions of the PDE. Then the system is solved as a system of ordinary differential equations (ODEs). The implemented discretization for the spatial derivatives of different orders are expressed as:

$$A_4^* u / \Delta x^4 = \frac{\partial^4 u}{\partial x^4}; \quad A_3^* u / \Delta x^3 = \frac{\partial^3 u}{\partial x^3}; \quad A_2^* u / \Delta x^2 = \frac{\partial^2 u}{\partial x^2} \tag{4}$$

where in the fourth order discretization, A_4^* is a $N \times N$ modified band matrix (based on the boundary conditions), and the size depends on the number of elements used for the space discretization, and Δx is the distance between the adjacent elements discretized in the spatial domain. A similar principle is applied for other order derivatives.

4.2 The Proposed NeuralSI Schematic

A pictorial schematic of NeuralSI is shown in Fig. 1. The Julia differential equation package [28] allows for very efficient computation of the gradient from the ODE solver. This makes it feasible to be used for neural network backpropagation. Thus, the ODE solver can be considered as a neural network layer after defining the ODE problem with the required fields of initial conditions, time span, and any extra parameters. Inputs to this layer can either be output from the previous network layers or directly from the training data.

The network in NeuralSI for the beam problem takes as input the location of the deformation sensors installed on the structure for continuous monitoring of its response. A series of dense layers are implemented to produce the output, which are the parameters that represent the structural characteristics. The parameters are re-inserted into the pre-defined ODE to obtain the final output, i.e., the structure's dynamic response. The loss is determined by the difference between the dynamic responses predicted by NeuralSI and those measured by the sensors (ground truth).

4.3 Training Data Generation

For experimental considerations in future lab testing, we simulate in this case a beam with length, width, and thickness respectively of 40 cm, 5 cm, and 0.5 cm. The density ρ is 2700 kg/m^3 (aluminum as base material). The force $F(t)$ is defined as a nonlinear temporal function. Considering the possible cases of polynomial or harmonic material properties variations as an example [33], we integrate the beam with a nonlinear modulus $E(x)$ as a sinusoidal function. We use a range for the modulus from 70 GPa to 140 GPa (again using aluminum as a base reference). The damping coefficient $C(x)$ is modeled as a ramp function. The PDE can be rewritten and expressed as:

$$
\begin{aligned}
F(t) = E_0 I \left(\frac{\partial^2 P(x)}{\partial x^2} \frac{\partial^2 u}{\partial x^2} + 2 \frac{\partial P(x)}{\partial x} \frac{\partial^3 u}{\partial x^3} + P(x) \frac{\partial^4 u}{\partial x^4} \right) + \rho A \frac{\partial^2 u}{\partial t^2} + C(x) \frac{\partial u}{\partial t} \\
= E_0 I \left(A_2^* P(x) A_2^* u + 2 A_1^* u P(x) A_3^* u + P(x) A_4^* u \right) + \rho A \frac{\partial^2 u}{\partial t^2} + C(x) \frac{\partial u}{\partial t}
\end{aligned}
\tag{5}
$$

$$
F(t) = \begin{cases} 1000 & t \le 0.02\,\text{s} \\ 0 & t > 0.02\,\text{s} \end{cases}
\tag{6}
$$

where the estimated modulus reference E_0 is 70 GPa, and $P(x)$ and $C(x)$ are modulus coefficient and damping that can vary spatially with x. The pre-defined parameters $P_0(x)$ and $C_0(x)$ are shown in Fig. 3.

The PDE presented in (5) is solved via the differential equation package in Julia. The RK4 solver method is selected for this high-order PDE. The time span was set to 0.045 s to have 3 complete oscillations of the bending response. The number of spatial elements and time steps are chosen as 16 and 160 respectively for balancing the training time cost and response resolution (capture the peak deflections). The deflections $u(x, t)$ are presented as a displacement distribution of size 16×160, from which ground truth data is obtained for training.

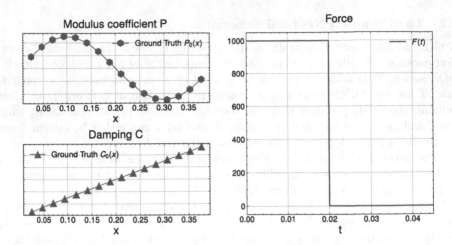

Fig. 3. Pre-defined structural properties and resultant dynamic response. Structural parameters P and C are defined as a sinusoidal and a ramp function. Force is applied as a step function of 1000 N and reduced to zero after 0.02 s.

4.4   Network Architecture and Training

The network architecture is presented as a combination of multiple dense layers and an PDE-solver layer. The input to the network is the spatial coordinates x for the measurements, and the network output is the prediction of the dynamic response $u(x, t)$. It is worth mentioning that the structural parameters P and C are produced from the multiple dense layers in separate networks, and the PDE layer takes those parameters to generate a response displacement distribution of size 16×160. The activation function for predicting the parameter P is a linear scale of the sigmoid function so that the output can be in a reasonable range. For the prediction of parameter C, the network of the same architecture is used, but the last layer does not take any activation function since the range of the damping value is unknown.

The modulus coefficient might be very high during training and lead to erroneous predictions with very high-frequency oscillations. So, we used minibatch training to escape local minima with a batch size of 16. The loss function is defined as the mean absolute error (MAE) between samples from the predicted and ground truth displacement distribution:

$$loss = \frac{1}{n} \sum_{i=1}^{n} |u - \hat{u}| \tag{7}$$

where n is the number of samples for training, u and $\hat{u}$ are the values from true and prediction dynamic responses at different training points in the same minibatch.

Furthermore, inspired by the effectiveness of positional embeddings for representing spatial coordinates in transformers [36], we adopt the same as well as

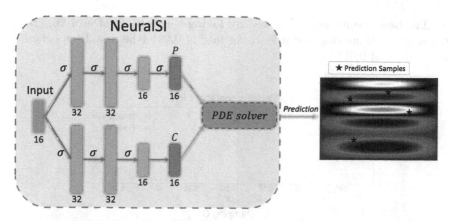

Fig. 4. NeuralSI network architecture and training. The network has several dense layers and the output is split into P and C. Those parameters are taken to the PDE solver for structural response prediction. Samples are taken randomly from the response for training the network.

for the spatial input to the network. It is worth noting that the temporal information in the measurements is only used as an aid for mapping and matching the predictions with the ground truth. We use ADAMW [24] as our optimizer, with a learning rate of 0.01 (Fig. 4).

5 Results and Performance

The evaluation of NeuralSI is divided into two parts. In the first part, we evaluate predictions of the parameters P and C from the trained neural network. We assume that each structure has a unique response. To determine how well the model is predicting the parameters, Fréchet distance [12] is employed to estimate the similarity between the ground truth and predicted functions. In this case, the predicted P and C are compared to the original P_0 and, C_0 respectively.

The second part of our evaluation is the prediction of the dynamic responses, which is achieved by solving the PDE using the predicted parameters. The metric to determine the performance of the prediction is the mean average error (MAE) between the predicted and ground truth displacement distribution. The prediction can be extrapolated by solving the PDE for a longer time span and compared with the extrapolated ground truth. The MAE is also calculated from the extrapolated data to examine the extrapolation ability of NeuralSI. Moreover, the dynamic response can be visualized on different elements separately (i.e., separate spatial locations x) for a more fine-grained comparison of the extrapolation results.

5.1 Results

We first trained and evaluated NeuralSI with different combinations of number and size of dense layers, percentage of data used for training, and minibatch size. The best results were achieved by taking a minibatch size of 16, training for a total of 20 epochs, and a learning rate of 0.001 (the first 10 epochs has a learning rate of 0.01).

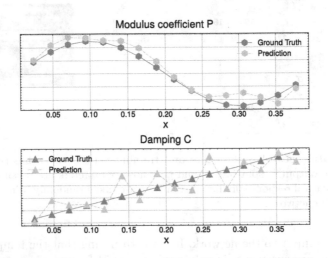

Fig. 5. Predicted beam parameters modulus coefficient (top) and damping (bottom). Observe that the modulus coefficient P matches well with the sinusoidal ground truth, since the modulus dominates the magnitude of the response. The damping C fluctuates as it is less sensitive than P, but the outputs still present a trend of increasing damping magnitude from the left end of the beam to the right end.

Figure 5 shows the output of modulus coefficient P and damping C from NeuralSI. For the most part, the predictions match well with the target modulus and damping, respectively. Compared to the modulus coefficient P, the predicted damping C has a larger error since it is less sensitive to the response. A small difference in damping magnitude will not affect the dynamic response as much as a change in the modulus parameter. However, the non-linearity of the modulus and damping are predicted accurately, and it is easy to identify whether the system is under-damped or over-damped based on the predicted damping parameters.

Figure 6 visualizes the ground truth and predicted dynamic displacement response, along with the error between the two. We observe that the maximum peak-peak value in the displacement error is only 0.3% of the ground truth. We also consider the ability of NeuralSI to extrapolate and display the dynamic response by doubling the prediction time span. It is worth mentioning that the peak error in temporal extrapolation does not increase much compared to the peak error in temporal interpolation. The extrapolation results are also examined at different elements from different locations. Figure 7 presents the response at the beam midspan and at quarter length. There are no observed discrepancies between the ground truth and the predicted response.

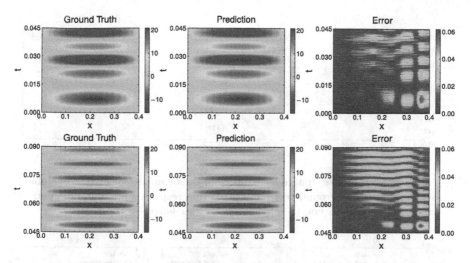

Fig. 6. NeuralSI predictions. The interpolation results (top row) are calculated from 0 to 0.045 s and temporal extrapolation results (bottom row) are from 0.045 s to 0.09 s. Peak error is only around 0.3% of the peak value from the ground truth, and the error magnitude remains the same for extrapolation.

5.2 Hyperparameter Investigation

Based on the parameters chosen above, we tested the effect of number of dense layers, training sample ratio and minibatch size on the parameter identification and prediction of dynamic responses.

Number of Layers: The number of layers is varied by consecutively adding an extra layer with 32 hidden units right after the input. From Fig. 8, the performance of the network is affected if the number of layers is below 4. This is explained by the fact that the network does not have sufficient capacity to precisely estimate the unknown structural parameters. It is noted that the size of the input and output are determined by the minibatch size and the number of elements used for discretization. A higher input or output size will automatically require a bigger network to improve prediction accuracy. Additionally, the Fréchet distance decreases as the size of the neural network increases, which demonstrates that the prediction of beam parameters is more accurate.

Sample Ratio: The number of training samples plays an important role in the model and in real in-field deployment scenarios. The number and the efficiency of sensor arrangements will be directly related to the number of samples required for accurately estimating the unknown parameters. It is expected that a reduced amount of data is sufficient to train the model given the strong domain knowledge (in the form of PDE) leveraged by NeuralSI. From Fig. 8, when 20% of the ground truth displacement samples are used for training, the loss drops noticeably. With an increased amount of training data, the network performance can

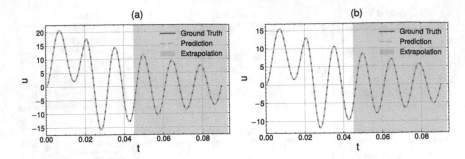

Fig. 7. Elemental response, spatial elements from the beam are selected to examine the temporal response. The ground truth and prediction responses are matching perfectly. (a) element at beam midspan; (b) element at quarter length of the beam.

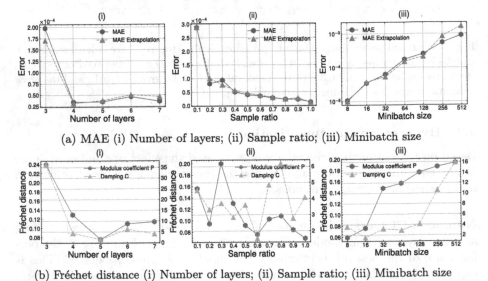

(a) MAE (i) Number of layers; (ii) Sample ratio; (iii) Minibatch size

(b) Fréchet distance (i) Number of layers; (ii) Sample ratio; (iii) Minibatch size

Fig. 8. Hyperparameter performance. A sufficient number of layers, more training samples, and small minibatch size will produce a good combination of hyperparameters and loss MAE (top row). The Fréchet distances (bottom row) are calculated for P and C respectively. The fluctuation of Fréchet distance for different sample ratio is because the values are relatively small.

still be improved. Furthermore, observe that there is a slight effect of data over-fitting when using the full amount of data for training. The Fréchet distance of damping is not stable since our loss function optimizes for accurately predicting the dynamic deflection response, instead of directly predicting the parameters. As such, the same error could be obtained through different combinations of those parameters.

Minibatch Size: The minibatch size plays an important role in the efficiency of the training process and the performance of the estimated parameters. It is worth mentioning that a smaller minibatch size helps escape local minima and reduces errors. However, this induces a higher number of iterations for a single epoch, which is computationally expensive. From Fig. 8 we observe that both the MAE error and the Fréchet distance are relatively low when the minibatch size is smaller than 32.

6   Comparison of NeuralSI with a Direct Response Mapping Deep Neural Network and a PINN

The NeuralSI framework is compared with traditional deep neural networks (DNN) and PINN methods. The tested DNN has 5 dense layers and a Tanh activation. The inputs are the spatial and temporal coordinates x and t, respectively, of the displacement response, and the output is the beam deflection $u(x, t)$ at that spatio-temporal position. The optimizer is LBFGS and the learning rate is 1.0. With a random choice of 20% samples, the loss stabilizes after 500 epochs.

The PINN method is defined with a similar strategy to existing solutions [20,29]. The Neural network consists of 5 dense layers with Tanh activation function. The loss is defined as a weighted aggregate of the boundary condition loss (second derivative of input x at the boundaries), governing equation loss (fourth-order derivative of x and second-order derivative of the t), and loss between the prediction and ground truth displacement response. We used LBFGS as the optimizer with a learning rate of 1.0. The training was executed for 3700 epochs.

The prediction of the dynamic deformation responses for the two baseline methods and NeuralSI and the corresponding displacement distribution errors are shown in Fig. 9. In NeuralSI, we used ImplicitEulerExtrapolation solver for a 4× faster inference. We further optimized the PDE function with Modeling-Toolkit [25], which provides another 10× speedup, for a total of 40× speedup over the RK4 solver used for training. Due to a limited amount of data for training, the DNN fails to predict the response. With extra information from the boundary conditions and equation, the PINN method results in an MAE loss of 0.344, and the prediction fits the true displacement distribution well. Most of the values in the displacement distribution error are small, except for some corners. But both methods fail to extrapolate the structural behavior temporally. The extrapolation of DNN predictions produces large discrepancies compared to the ground truth. Similarly, the PINN method fails to match the NeuralSI performance, while fairing much better than the predictions from the DNN, as expected due to the added domain knowledge. The MAE errors were computed and compared with the proposed method trained with 20% data as shown in Fig. 10.

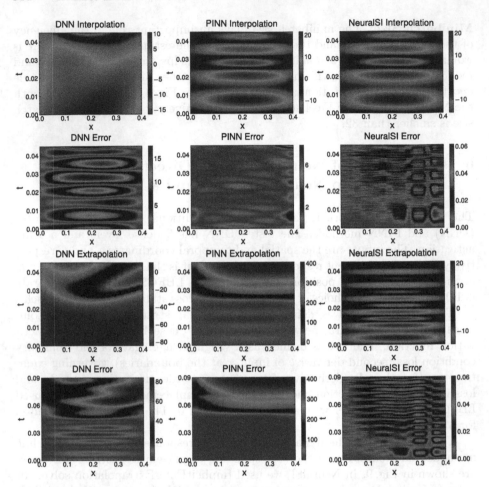

Fig. 9. Spatio-temporal displacement distribution predictions and comparisons between DNN, PINN and NeuralSI for both interpolation (top) and extrapolation (bottom). The DNN method fails to learn the interpolation response, while the PINN can predict most of the responses correctly, with only a few errors at the corners of the displacement response. Predictions from NeuralSI have two orders of magnitude lower error in comparison to PINN. With the learned structural parameters, NeuralSI maintains the same magnitude of error in extrapolation results. Both DNN and PINN completely fail at extrapolation and lead to considerable errors.

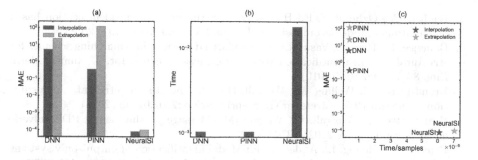

Fig. 10. Performance comparison between DNN, PINN, and NeuralSI for both interpolation and extrapolation (a) MAE, (b) Inference time, and (c) Trade-off between MAE and inference time. NeuralSI offers significantly lower error while being as expensive as solving the original PDE, thus offering a more accurate solution when the computational cost is affordable. NeuralSI obtains the extrapolation results by solving the whole time domain starting from $t = 0$, while DNN and PINN methods directly take the spatio-temporal information and solve for extrapolation.

7   Conclusion

In this paper, we proposed NeuralSI, a framework that can be employed for structural parameter identification in nonlinear dynamic systems. Our solution models the unknown parameters via a learnable neural network and embeds it within a partial differential equation. The network is trained by minimizing the errors between predicted dynamic responses and ground truth measurement data. A major advantage of the method is its versatility and flexibility; thus, it can be successfully extended to any PDEs with high-order derivatives and nonlinear characteristics. The trained model can be used to either explore structural behavior under different initial conditions and loading scenarios, which is vital for structural modeling or to determine high-accuracy extrapolation, also essential in systems' response prognosis. An example beam vibration study case was analyzed to demonstrate the capabilities of the framework. The estimated structural parameters and the dynamic response variations match well with the ground truth (MAE of 10^{-4}). The performance of NeuralSI is also shown to outperform direct regression significantly through deep neural networks and PINN methods by three to five orders of magnitude.

Acknowledgements. This research was funded in part by the National Science Foundation grant CNS 1645783.

References

1. Akinpelu, F.O.: The response of viscously damped Euler-Bernoulli beam to uniform partially distributed moving loads. Appl. Math. **3**(3), 199–204 (2012)
2. Aliee, H., Theis, F.J., Kilbertus, N.: Beyond predictions in neural odes: identification and interventions. arXiv preprint arXiv:2106.12430 (2021)

3. Bagheri, A., Ozbulut, O.E., Harris, D.K.: Structural system identification based on variational mode decomposition. J. Sound Vib. **417**, 182–197 (2018)

4. Banerjee, B., Roy, D., Vasu, R.: Self-regularized pseudo time-marching schemes for structural system identification with static measurements. Int. J. Numer. Meth. Eng. **82**(7), 896–916 (2010)

5. Brandstetter, J., Welling, M., Worrall, D.E.: Lie point symmetry data augmentation for neural PDE solvers. arXiv preprint arXiv:2202.07643 (2022)

6. Brandstetter, J., Worrall, D., Welling, M.: Message passing neural PDE solvers. arXiv preprint arXiv:2202.03376 (2022)

7. Brincker, R., Zhang, L., Andersen, P.: Modal identification of output-only systems using frequency domain decomposition. Smart Mater. Struct. **10**(3), 441 (2001)

8. Brunton, S.L., Proctor, J.L., Kutz, J.N.: Discovering governing equations from data by sparse identification of nonlinear dynamical systems. Proc. Natl. Acad. Sci. **113**(15), 3932–3937 (2016)

9. Chen, R.T., Rubanova, Y., Bettencourt, J., Duvenaud, D.K.: Neural ordinary differential equations. In: Advances in Neural Information Processing Systems, vol. 31 (2018)

10. Diez, A., Khoa, N.L.D., Makki Alamdari, M., Wang, Y., Chen, F., Runcie, P.: A clustering approach for structural health monitoring on bridges. J. Civil Struct. Health Monit. **6**(3), 429–445 (2016)

11. Dulny, A., Hotho, A., Krause, A.: NeuralPDE: modelling dynamical systems from data. arXiv preprint arXiv:2111.07671 (2021)

12. Eiter, T., Mannila, H.: Computing discrete fréchet distance (1994)

13. Entezami, A., Sarmadi, H., Behkamal, B., Mariani, S.: Big data analytics and structural health monitoring: a statistical pattern recognition-based approach. Sensors **20**(8), 2328 (2020)

14. Entezami, A., Shariatmadar, H., Sarmadi, H.: Structural damage detection by a new iterative regularization method and an improved sensitivity function. J. Sound Vib. **399**, 285–307 (2017)

15. Ghorbani, E., Buyukozturk, O., Cha, Y.J.: Hybrid output-only structural system identification using random decrement and Kalman filter. Mech. Syst. Signal Process. **144**, 106977 (2020)

16. Hasni, H., et al.: A new approach for damage detection in asphalt concrete pavements using battery-free wireless sensors with non-constant injection rates. Measurement **110**, 217–229 (2017)

17. Hasni, H., Alavi, A.H., Lajnef, N., Abdelbarr, M., Masri, S.F., Chakrabartty, S.: Self-powered piezo-floating-gate sensors for health monitoring of steel plates. Eng. Struct. **148**, 584–601 (2017)

18. Horie, M., Mitsume, N.: Physics-embedded neural networks: E(n)-equivariant graph neural PDE solvers. arXiv preprint arXiv:2205.11912 (2022)

19. Konkanov, M., Salem, T., Jiao, P., Niyazbekova, R., Lajnef, N.: Environment-friendly, self-sensing concrete blended with byproduct wastes. Sensors **20**(7), 1925 (2020)

20. Krishnapriyan, A., Gholami, A., Zhe, S., Kirby, R., Mahoney, M.W.: Characterizing possible failure modes in physics-informed neural networks. In: Advances in Neural Information Processing Systems, vol. 34, pp. 26548–26560 (2021)

21. Lai, Z., Mylonas, C., Nagarajaiah, S., Chatzi, E.: Structural identification with physics-informed neural ordinary differential equations. J. Sound Vib. **508**, 116196 (2021)

22. Lai, Z., Nagarajaiah, S.: Sparse structural system identification method for nonlinear dynamic systems with hysteresis/inelastic behavior. Mech. Syst. Signal Process. **117**, 813–842 (2019)
23. Li, X., Salem, T., Bolandi, H., Boddeti, V., Lajnef, N.: Methods for the rapid detection of boundary condition variations in structural systems. American Society of Mechanical Engineers (2022)
24. Loshchilov, I., Hutter, F.: Fixing weight decay regularization in adam (2018)
25. Ma, Y., Gowda, S., Anantharaman, R., Laughman, C., Shah, V., Rackauckas, C.: ModelingToolkit: a composable graph transformation system for equation-based modeling. arXiv preprint arXiv:2103.05244 (2021)
26. Maurya, D., Chinta, S., Sivaram, A., Rengaswamy, R.: Incorporating prior knowledge about structural constraints in model identification. arXiv preprint arXiv:2007.04030 (2020)
27. Modebei, M., Adeniyi, R., Jator, S.: Numerical approximations of fourth-order PDEs using block unification method. J. Nigerian Math. Soc. **39**(1), 47–68 (2020)
28. Rackauckas, C., et al.: Universal differential equations for scientific machine learning. arXiv preprint arXiv:2001.04385 (2020)
29. Raissi, M., Perdikaris, P., Karniadakis, G.E.: Physics-informed neural networks: a deep learning framework for solving forward and inverse problems involving nonlinear partial differential equations. J. Comput. Phys. **378**, 686–707 (2019)
30. Reynders, E., De Roeck, G.: Reference-based combined deterministic-stochastic subspace identification for experimental and operational modal analysis. Mech. Syst. Signal Process. **22**(3), 617–637 (2008)
31. Rezaiee-Pajand, M., Entezami, A., Sarmadi, H.: A sensitivity-based finite element model updating based on unconstrained optimization problem and regularized solution methods. Struct. Control. Health Monit. **27**(5), e2481 (2020)
32. Salehi, H., Burgueño, R., Chakrabartty, S., Lajnef, N., Alavi, A.H.: A comprehensive review of self-powered sensors in civil infrastructure: state-of-the-art and future research trends. Eng. Struct. **234**, 111963 (2021)
33. Salem, T., Jiao, P., Zaabar, I., Li, X., Zhu, R., Lajnef, N.: Functionally graded materials beams subjected to bilateral constraints: structural instability and material topology. Int. J. Mech. Sci. **194**, 106218 (2021)
34. Sarmadi, H., Entezami, A., Daneshvar Khorram, M.: Energy-based damage localization under ambient vibration and non-stationary signals by ensemble empirical mode decomposition and Mahalanobis-squared distance. J. Vib. Control **26**(11–12), 1012–1027 (2020)
35. Tuhta, S., Günday, F.: Multi input multi output system identification of concrete pavement using N4SID. Int. J. Interdisc. Innov. Res. Dev. **4**(1), 41–47 (2019)
36. Vaswani, A., et al.: Attention is all you need. In: Advances in Neural Information Processing Systems, vol. 30 (2017)
37. Yang, X.M., Yi, T.H., Qu, C.X., Li, H.N., Liu, H.: Automated eigensystem realization algorithm for operational modal identification of bridge structures. J. Aerosp. Eng. **32**(2), 04018148 (2019)
38. Yin, T., Jiang, Q.H., Yuen, K.V.: Vibration-based damage detection for structural connections using incomplete modal data by Bayesian approach and model reduction technique. Eng. Struct. **132**, 260–277 (2017)
39. Yuen, K.V., Au, S.K., Beck, J.L.: Two-stage structural health monitoring approach for phase I benchmark studies. J. Eng. Mech. **130**(1), 16–33 (2004)
40. Zhang, T., et al.: ANODEV2: a coupled neural ODE framework. In: Advances in Neural Information Processing Systems, vol. 32 (2019)

41. Zhou, X., He, W., Zeng, Y., Zhang, Y.: A semi-analytical method for moving force identification of bridge structures based on the discrete cosine transform and FEM. Mech. Syst. Signal Process. **180**, 109444 (2022)
42. Zubov, K., et al.: NeuralPDE: automating physics-informed neural networks (PINNs) with error approximations. arXiv preprint arXiv:2107.09443 (2021)

A Geometric-Relational Deep Learning Framework for BIM Object Classification

Hairong Luo[1,2]([✉]), Ge Gao[1,2], Han Huang[1,2], Ziyi Ke[1], Cheng Peng[1], and Ming Gu[1,2]

[1] School of Software, Tsinghua University, Beijing, China
{luohr22,h-huang20}@mails.tsinghua.edu.cn, gaoge@tsinghua.edu.cn,
{ziyike,guming}@mail.tsinghua.edu.cn
[2] Beijing National Research Center for Information Science and Technology(BNRist),
Tsinghua University, Beijing, China

Abstract. Interoperability issue is a significant problem in Building Information Modeling (BIM). Object type, as a kind of critical semantic information needed in multiple BIM applications like scan-to-BIM and code compliance checking, also suffers when exchanging BIM data or creating models using software of other domains. It can be supplemented using deep learning. Current deep learning methods mainly learn from the shape information of BIM objects for classification, leaving relational information inherent in the BIM context unused. To address this issue, we introduce a two-branch geometric-relational deep learning framework. It boosts previous geometric classification methods with relational information. We also present a BIM object dataset—IFCNet++, which contains both geometric and relational information about the objects. Experiments show that our framework can be flexibly adapted to different geometric methods and relational features do act as a bonus to general geometric learning methods, obviously improving their classification performance, thus reducing the manual labor of checking models and improving the practical value of enriched BIM models.

Keywords: BIM · Object classification · Semantic enrichment · Relational feature · Deep learning

1 Introduction

Interoperability issues, as an essential problem in the application of BIM, still affect the practical value of BIM models. Proper use of the BIM technique without interoperability problems requires BIM data to be shared and exchanged conveniently and undamaged between software of different professions. Now most BIM software support IFC as a standard data exchange schema, which plays a crucial role in enabling interoperability [20].

This work was supported by the National Key Research and Development Program of China (2021YFB1600303).

However, because IFC contains entity and relationship definitions of various AEC subdomains, the schema defined by IFC is complex and redundant [9], which causes unacceptable mismatch and reliability problems [9,28]. Lack of formal logic rigidness in IFC [9] also makes mapping of BIM elements to IFC types arbitrary and susceptible to misclassifications [20]. These phenomena are a cause of interoperability problems and hamper the advance of BIM. Fixing erroneous, misrepresented, contradictory, or missing data that appears during model data exchange remains to be laborious and frustrating [1]. This poses a challenge to the reuse of BIM models in downstream tasks. Semantic enrichment techniques [1,2,7,20,23,25,28–30,32,37] solve the interoperability problem by exploiting existing numeric, geometric, or relational information in the model to infer new semantic information.

Object classification integrity is a fundamental yet critical requirement that needs to be satisfied using semantic enrichment. Object type provides hints about an object's function, location, size, etc. However, IFC does not ensure correct mapping between BIM objects and their corresponding IFC types [23]. Missing or incorrect object type usually occurs due to the inconsistent definition of an object's role in different AEC subdomains. Supplementing the object type information can improve the usability and practical value of BIM models.

Deep learning applications have been explored in various fields in recent years, including BIM object classification. By inputting the objects extracted from an IFC file to a trained deep learning model, the model is able to check the integrity of BIM element to IFC class mappings and identify discrepancies [20]. They first represent BIM objects as pure geometric representations, such as voxels, meshes, 2D views, or point clouds, then classify the objects using 3D geometric learning models like MVCNN [34]. This approach neglects the relational information between objects in the BIM context, which might also provide guidance.

Starting from this intuition, we put forward a geometric-relational deep learning framework that learns the geometric and relational features on different branches and fuses them as a unified object descriptor. Particularly, we propose a relational feature extractor and a feature fusion module in the framework. The two modules serve to extract high-level relational features of BIM objects and fuse them with geometric features extracted by the geometric feature extractor, respectively. This framework can be applied to most existing models and robustly boost their performance, because almost all mainstream geometric deep learning models can serve as the geometric feature extractor. We select MVCNN, DGCNN [38] and MVViT (a 3D deep learning model adapted from Vision Transformer [8]) as geometric feature extractors in the framework and propose three corresponding models, namely Relational MVCNN (RMVCNN), Relational DGCNN (RDGCNN) and Relational MVViT (RMVViT). Experiments show that with the addition of relational features, the BIM object classification abilities of these models are noticeably improved to varying degrees. This proves the efficacy and flexibility of our framework.

As for the data, there is still a lack of BIM object datasets that contain objects' relations in the BIM context. We propose the IFCNet++ dataset to fill this vacancy. We attach selected representative relational features to each BIM object in the dataset, along with their geometric shapes. We use this dataset in all the experiments for training and testing our models.

To sum up, the contributions of this paper are as follows:

- Proposing a geometric-relational deep learning framework to utilize both geometric and relational information of BIM objects simultaneously for BIM object classification;
- Putting forward three BIM object classification models based on the geometric-relational framework, and achieving better classification results than their baseline models using additional relational information.
- Proposing IFCNet++, a BIM object dataset containing geometric and relational information for BIM object classification task.
- The efficacy and flexibility of our framework to fully exploit the relational information are demonstrated by comprehensive experiments.

2 Related Work

2.1 3D Object Recognition

3D object recognition is a longstanding problem in computer vision and computer graphics. BIM objects can be seen as 3D objects with semantics, so we can utilize existing object recognition methods to classify them. Early works concentrate on designing 3D local feature descriptors to solve a variety of 3D problems [3,5,12,14,15,21,22,24,26,27,33,36], including 3D object recognition. These handcrafted descriptors are required to be descriptive and robust [13], but they have difficulty in deciding a priori what constitutes meaningful information and what is the result of noise and external factors [18]. In recent years, deep learning based methods attract great attention. Wu et al. [39] classify objects represented in voxel form using a 3D convolutional network. Su et al. [34] propose an architecture that synthesizes a single compact 3D shape descriptor of an object using image features extracted from the object's multiple views. The synthesis is done by a view-pooling layer. Qi et al. [31] design a deep learning model that directly takes the point cloud as input to carry out classification or segmentation tasks. This architecture is invariant to input permutation. Wang et al. [38] proposed to dynamically generate graph structures based on the input point cloud, and use EdgeConv modules to perform convolution operations on the graphs. Our geometric-relational framework can be applied to these geometric learning methods to boost their performance on BIM object classification.

2.2 BIM Object Classification

BIM object classification is a fundamental task of BIM semantic enrichment. Relative methods can be divided into deductive methods and inductive methods

[4]. Deductive methods require the design of explicit and unique rules for the particular object. They usually ensure certainty in conclusions, but the labor of devising rules for all possible pairs of types makes these methods practically intractable [20]. As an example, Ma *et al.* [23] propose a procedure for establishing a knowledge base that associates objects with their features and relationships, and a matching algorithm based on a similarity measurement between the knowledge base and facts. Machine learning approaches are representative of inductive methods. These methods generate their own rules inductively by optimizing the weights of features in a model [20]. Koo *et al.* [20] use support vector machines to check the semantic integrity of mappings between BIM elements and IFC classes. Kim *et al.* [17] use 2D CNN to classify furniture entities according to their images. Koo *et al.* [18] compare the classification results of MVCNN and PointNet on wall subtypes and door subtypes. Koo *et al.* [19] also use the same two models to classify BIM objects in road infrastructure. Collins *et al.* [6] encode BIM objects using two kinds of graph encodings and utilize a graph convolutional network to create meaningful local features for subsequent classification. Emunds *et al.* [11] propose an efficient neural network based on sparse convolutions to learn from point cloud representation of BIM objects. This study is also dedicated to solving the BIM object classification problem using deep learning methods.

2.3 BIM Object Datasets

Sufficiently large and comprehensive datasets are the key to the training of deep learning models. Currently, there are two relevant datasets, IFCNet [10] and BIMGEOM [6]. IFCNet is a dataset of single-entity IFC files spanning a broad range of IFC classes containing both geometric and single-object semantic information. BIMGEOM is assembled of building models from both industry and academia. It consists of structural elements, equipment and interior furniture types. Both datasets don't contain any relational information of objects, thus not suitable for our framework. We propose IFCNet++ which involves certain relationships between objects and use it to train our models.

3   Geometric-Relational Deep Learning Framework

The target of our research is to create a reasonable deep learning framework to simultaneously learn the geometric and relational information for BIM object classification. Object shape data and relational data are usually presented in different forms, which makes learning these features using a single network branch difficult. So we adopt a two-branch method intuitively. The two branches can use different network architectures to process the two kinds of information and extract a geometric descriptor and a relational descriptor of the object. These descriptors are fused to get a unified object descriptor.

We show the overview of our geometric-relational deep learning framework in Fig. 1. It consists of three main modules, i.e. geometric feature extractor,

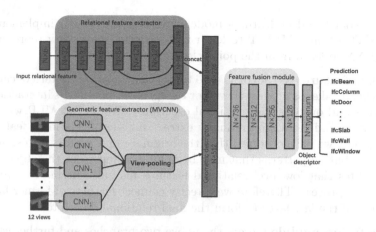

Fig. 1. Geometric-relational framework overview. The geometric feature extractor represents a common geometric learning backbone. In this figure we use the MVCNN backbone. It extracts the geometric feature of the BIM object from its shape representation. The relational feature extractor extracts relational features from raw relational data. It also connects low-level features to high-level features to form the complete relational feature. The feature fusion module fuses these two kinds of features and outputs the object descriptor, which can be sent to a classifier to get the corresponding type label of the object

relational feature extractor and feature fusion module. The geometric feature extractor learns to represent a BIM object's shape as a geometric descriptor. The relational feature extractor extracts relational features of different levels and connects them to form the relational descriptor. The feature fusion module mixes the two descriptors and outputs the final object descriptor. This descriptor is a more complete abstraction of the BIM object than a pure geometric descriptor, and can be used by a classifier to perform classification more accurately. Next, we will introduce the module designs and implementation details of our framework.

3.1   Module Designs

In the following, we make a detailed explanation of the design intuitions and detailed structures of the three modules in our framework.

Geometric feature extractor serve to learn from objects' shapes. It takes the raw geometric data of BIM objects as input and outputs their high-level geometric descriptors. A BIM object's shape can be represented in various forms, like multiple views, point clouds, voxels, etc. To properly extract geometric features from these representations, we do not fix the geometric feature extractor to a certain design. Instead, any geometric deep learning method that can extract a geometric descriptor from an object can be the geometric extractor. This provides flexibility to the geometric input form and our framework design. It also empowers our framework to boost geometric methods' of any kind with relational information. The classification result of our framework would also be improved as better geometric deep learning models are proposed. In our implementation, we use MVCNN and MVViT to learn from objects' multi-view representation and DGCNN to learn from the point cloud representation.

Relational feature extractor is used to learn the relation pattern of each BIM object type. It takes as input a 1D vector to represent an object's interaction with the context and outputs the relational descriptor. We use an MLP with batch normalization and ReLU to gradually extract high-level relational features. In the meantime, we have observed that the original input vectors also show some simple distribution patterns that may be helpful to represent the object's type. This indicates that low-level relational features may also be instructive to the classification process. Therefore we directly connect features of former layers to the feature of the last layer to form the final relational descriptor.

Feature fusion module merges the above two branches and further studies a unified BIM object descriptor using the extracted geometric feature and relational feature. It is designed as an MLP with batch normalization and ReLU, too, but without connections between different layers. It first concatenates the two descriptors, then uses the MLP to extract the object descriptor, which can be sent to a classifier.

3.2 Implementation Details

According to the observation that the geometric aspect of a BIM object contains more useful information and can provide more clues about the object's type label than its relational aspect, we design the geometric branch as the main branch and the relational branch as a supplement. Specifically, the geometric feature extractor is more complicated than the relational counterpart. And the geometric descriptor size is larger than the relational descriptor size. This viewpoint can be observed in the following implementation details.

Because we directly utilize existing geometric learning methods as the geometric feature extractor, we skip this part and begin the introduction with the relational feature extractor. As shown in Fig. 1, this branch consists of 6 linear layers. It gradually embeds the input 6-dimensional relation vector into a high-level feature space of 128 dimensions. We also concatenate the feature of the second and the fourth layer to the feature of the last layer to form the final relational descriptor. The total length of the descriptor is 224. The simple design of the relational branch guarantees that it takes up less computational resources, verifying the viewpoint in the previous paragraph.

The feature fusion module is composed of four linear layers. It maps the concatenation of the geometric descriptor and the relational descriptor down to a 128-dimension feature space. The feature length of the first module layer is the same as the length of the input, so it may vary as the geometric feature extractor changes. The second to the last layers output features of fixed length. Detailed length information is shown in Fig. 1. We use a linear layer with a softmax operation as the classifier, which outputs the predicted probability of each BIM object type.

4   Implementation

4.1   Relational Models

We select MVCNN and DGCNN as our baseline for multi-view based method and point cloud based method, respectively. To test our framework on a Transformer based model to explore its generalization, we modify Vision Transformer to a multi-view based geometric learning method. We use its backbone to extract a classification token for each view, and max-pool the tokens to obtain the geometric feature descriptor of the object. The model is called Multi-view Vision Transformer or MVViT. We transform these three models into corresponding relational models using our framework, which are called RMVCNN, RDGCNN and RMVViT, respectively.

4.2   Training Configurations

All the models are implemented based on PyTorch and trained using NVIDIA RTX 3090. The batch size is 64. Adam optimizer is used with $\beta_1 = 0.9$, $\beta_2 = 0.999$ and $\epsilon = $1e-8. We use cross-entropy loss as the loss function.

In the implementation, we adopt pre-trained ResNet34 [16] as the backbone network of MVCNN and RMVCNN. For MVViT and RMVViT, we use pre-trained ViT-Base as the backbone. Each baseline and its relational model share the same learning rate and weight decay. The training epochs are tailored for each model to get relatively good performance.

5   IFCNet++ Dataset

5.1   Dataset Overview

Fig. 2. Dataset samples: (from left to right in reading order) IfcBeam, IfcColumn, IfcDoor, IfcFlowFitting, IfcFlowSegment, IfcFlowTerminal, IfcPlate, IfcRailing, IfcSlab, IfcWall, IfcWindow

We focus on BIM object classification based on supervised learning. Supervised learning methods use labeled datasets to train models to learn data distribution.

There are two representative BIM object datasets, which are IFCNet and BIM-GEOM. These two datasets present BIM objects using file formats such as ifc, obj and png. Except for the ifc format, which contains varying degrees of single-object semantics like size and material, other formats only contain geometric information about the objects. However, BIM models are rich in semantics. That means BIM objects are not pure geometric shapes. They exist in the building context and have various spatial and topological relations with other objects. So these relationships should be recorded to preserve the BIM nature when making BIM object datasets. Therefore, we propose IFCNet++ as an enhanced BIM object dataset. It contains BIM objects' geometric shapes and interactions with other objects. We collect 9228 objects belonging to 11 most common types in the dataset. The 11 types are selected based on their appearance frequency and importance in BIM models. They can cover most objects of interest in a BIM model, and cover object types in both architectural models and MEP models. An overview of the dataset is shown in Fig. 2 and Table 1.

Table 1. Data distribution overview of IFCNet++ dataset

BIM object type	Training	Testing	Total
IfcBeam	66	27	93
IfcColumn	64	27	91
IfcDoor	939	402	1341
IfcFlowFitting	103	43	146
IfcFlowSegment	115	49	164
IfcFlowTerminal	273	116	389
IfcPlate	1400	600	2000
IfcRailing	210	90	300
IfcSlab	929	397	1326
IfcWall	1400	600	2000
IfcWindow	965	413	1378
Total	6464	2764	9228

5.2   Relational Feature Design

IFCNet++ focuses on recording relational information of objects. We extract relational information from relational items in IFC files. IFC schema has defined abundant relationships among objects. But in most cases, few relationships have been implemented in an IFC file. Some relationships may also get lost because of interoperability problems. Therefore, we select four subtypes of IfcRelationship that appear in most IFC files and are easy to extract, i.e. IfcRelConnectsElement, IfcRelFillsElement, IfcRelAggregates and IfcRelVoidsElement.

Another problem is how to represent these relationships in the dataset. A direct idea is to build a connected relation graph of a BIM model. That is, each object represents a node and each relational item represents a connecting edge. However, the selected relationships usually exist in local regions. They can not connect all the objects in a whole graph, so we discard the graph based solution. Instead, we adopt a counting method and attach a simple 1D vector to each BIM object. Each vector consists of six numbers. Each number represents how many times an object is quoted in a certain relationship attribute. The six numbers correspond to the following six attributes:

- IfcRelConnectsElement.RelatingElement
- IfcRelConnectsElement.RelatedElement
- IfcRelAggregates.RelatingObject
- IfcRelAggregates.RelatedObjects
- IfcRelVoidsElement.RelatingBuildingElement
- IfcRelFillsElement.RelatedBuildingElement

Despite being simple, this vectorized form has a strong representative ability for the relational information, as the simple local relational structure can be well represented by this counting approach. Besides, this form is convenient for later processes using MLP. It also maintains a good trade-off between data size and the ability to represent relational information.

5.3   Data Collection and Processing

We collected the BIM objects in IFCNet++ from more than ten IFC files. We first split the model into individual objects. Then we extract all the objects of interest in obj format to get their geometric representation. Deduplication is performed on the collected objects. We consider an object as a duplication if it can overlap with another object after translation and rotation. Next, we extract the four selected relationship items and count how many times an object is quoted by a certain relationship attribute. Finally, we attach the counted vectors to the corresponding objects. The collected object distribution is very unbalanced across object types. For example, wall objects tend to appear in large amounts for a BIM model. So we randomly select at most 2000 objects for a certain object type. We split the training set and the testing set by a ratio of 7:3.

Besides, we need to further process the obj files to get the proper input format of multi-view based and point cloud based methods. We render each object to get a 12-view representation using the rendering method of [35]. We also convert objects to point cloud form using the code in [10]. We show an example of our 12-view representation and point cloud representation in Fig. 3.

6   Experiments

In this section, we show the experimental results of our framework. Experiments on the three pairs of baseline and relational models show the performance boost gained by taking relations into the learning process.

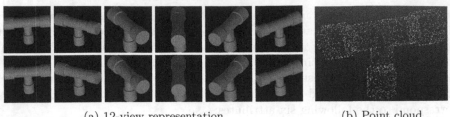

(a) 12-view representation   (b) Point cloud

Fig. 3. Two representations of an IfcFlowFitting object in IFCNet++

6.1   Testing Metrics

We test the models on the test set of IFCNet++ and show their classification results in Table 2.

Table 2. Classification results of our trained models on the IFCNet++ test set

Model	Accuracy	Balanced accuracy	Precision	Recall	F1 score
MVCNN	0.9732	0.9549	0.9742	0.9732	0.9734
DGCNN	0.9801	0.9536	0.9809	0.9801	0.9802
MVViT	0.9797	0.9527	0.9812	0.9797	0.9799
RMVCNN	**0.9917**	**0.9750**	**0.9918**	**0.9917**	**0.9916**
RDGCNN	0.9902	0.9624	0.9906	0.9902	0.9903
RMVViT	0.9841	0.9581	0.9857	0.9841	0.9842

The baselines can already reach a high precision of 97% and a balanced accuracy of 95% merely utilizing the geometric information. This illustrates that most BIM objects can be correctly classified by their shapes. However, with the addition of relational features, the three relational models can get better results on all the metrics than their corresponding baselines. This intuitively shows that the composition of selected relationships can effectively represent the local relation patterns of each object type. And our framework can learn these patterns and fuse them with geometric features to refine the object descriptors.

Noticeably, even though MVCNN doesn't perform very well relative to the other two baselines, RMVCNN not only performs best on all the metrics, but also gains the most improvement with each metric improved by about 2%. RDGCNN and RMVViT have been improved by about 1% and 0.5% on each metric, respectively. This shows even with the same input relational features, certain geometric models can gain better improvement using our framework. The relational feature space fuses best with the geometric feature space learned by MVCNN to gain the most improvement. So the key to better classification results is to find a proper geometric method that fits our framework well.

Table 3. Classification accuracy of RMVCNN organized by object types

Object type	Total	Correctly classified	Accuracy (%)
IfcBeam	27	27	100.0
IfcColumn	27	24	88.9
IfcDoor	402	399	99.3
IfcFlowFitting	43	43	100.0
IfcFlowSegment	49	43	87.8
IfcFlowTerminal	116	116	100.0
IfcPlate	600	598	99.7
IfcRailing	90	89	98.9
IfcSlab	397	390	98.2
IfcWall	600	599	99.8
IfcWindow	413	413	100.0
Total	2764	2741	99.2

We list the classification accuracy of RMVCNN organized by object types in Table 3 to further explore its classification ability. RMVCNN performs well on 9 of the 11 types, reaching an accuracy higher than 98%. This means RMVCNN can effectively learn geometric and relational features of most types. Accuracy on the other two types is slightly lower than 90%. This may be partially explained by the small quantities of training samples of these types.

6.2 Confusion Rate

The geometric learning baselines are prone to be confused by object types that contain geometrically similar objects. To quantitatively analyze this trend, we propose the notion of confusion rate. Suppose A and B are two object types, we define a model's confusion rate between a pair of types A and B as:

$$c = \frac{m_{AB} + m_{BA}}{n_A + n_B}. \tag{1}$$

Here m_{AB} is the amount of BIM objects of type A misclassified as type B. m_{BA} is the amount of BIM objects of type B misclassified as type A. n_A and n_B represent the amounts of objects of type A and type B in the test set.

We compute the confusion rates of MVCNN and RMVCNN on the test set. We then sort the type pairs in descending order of the confusion rates of MVCNN and list the results of the first ten pairs in Table 4.

In the ten type pairs that MVCNN is prone to confuse, RMVCNN's confusion rates have obviously reduced on nine of them. The confusion rates of five type pairs have even been reduced to zero. It shows that for most type pairs that share similar object shapes, relational information can help the model find the difference between them according to the union of their shapes and context

Table 4. Confusion rates of MVCNN and RMVCNN. The first ten type pairs sorted in descending order of MVCNN's confusion rates are selected

Type 1	Type 2	MVCNN (%)	RMVCNN (%)
IfcColumn	IfcFlowSegment	9.2	3.9
IfcBeam	IfcFlowSegment	5.3	5.3
IfcDoor	IfcWall	1.2	0
IfcPlate	IfcSlab	1.1	0.2
IfcSlab	IfcWall	0.9	0.4
IfcBeam	IfcWall	0.6	0
IfcFlowFitting	IfcFlowTerminal	0.6	0
IfcFlowSegment	IfcFlowTerminal	0.6	0
IfcFlowTerminal	IfcPlate	0.6	0
IfcColumn	IfcWall	0.5	0.3

information. What's more, the highest confusion rate has come down from 9.2% of MVCNN to 5.3% of RMVCNN. This implies that relational information plays an important part in lowering both the average level and the upper limit of confusion rates. Our framework performs well in alleviating the model's confusion situations.

6.3 Corrected Classification Results

We display some of the BIM objects that are misclassified by MVCNN but correctly classified by RMVCNN in Fig. 4. Figure 4a shows an IfcFlowTerminal. Because its shape looks like a joint of two pipelines, it is misclassified as an IfcFlowFitting by MVCNN. However, with the assistance of its relational features, RMVCNN can judge its type correctly. A similar problem happens when MVCNN tries to classify the IfcWindow in Fig. 4b. According to the thin columns on its surface, this window looks much like a straight railing, so MVCNN classifies it as an IfcRailing wrongly. This mistake is also avoided by RMVCNN. The IfcDoor in Fig. 4c is misclassified as IfcWindow by MVCNN for its window-like frame and also gets corrected by RMVCNN. These examples clearly demonstrate our framework's advantage in distinguishing BIM objects with misleading geometric shapes.

6.4 Computational Cost

To figure out the trade-off between performance boost and computational cost introduced by our framework, we list the number of parameters and calculations of MVCNN and RMVCNN in Table 5.

By applying our framework, we only introduce about 1.1 M params (5%) and 1.1 M MACs (0.0025%) to RMVCNN relative to MVCNN. Considering the

(a) IfcFlowTerminal mis-   (b) IfcWindow misclassi-   (c) IfcDoor misclassified
classified as IfcFlowFit-   fied as IfcRailing   as IfcWindow
ting

Fig. 4. Demonstration of three BIM objects misclassified by MVCNN but correctly classified by RMVCNN

Table 5. Number of parameters and calculations of MVCNN and RMVCNN

Model	Params (M)	MACs (M)
MVCNN	21.290	44049.020
RMVCNN	22.407	44050.135
Cost introduced	1.117	1.115

fact that MVCNN can already reach a very high classification accuracy, further improvement is hard to achieve. However, our framework uses a relatively low price to push its performance boundary by an obvious margin. It avoids adding a huge amount of computational cost like some large-scale models. This result shows the superiority and efficiency of our method.

6.5 Ablation Study

We conduct ablation studies on RMVCNN and remove each of the three component modules to look into their contributions. When removing the geometric or relational feature extractor, we also abandon the corresponding input information. The results are shown in Table 6. When removing the geometric feature extractor, the fundamental part of the framework, the results are very poor and could not be trusted. Because the relational feature extractor acts as an auxiliary part, the results of the framework without it look good but slightly suffer. The experiment without the feature fusion module shows better results, but could not reach the results of the full RMVCNN. This illustrates that the fusion process does help in extracting a more compact and representative object descriptor.

Table 6. Ablation studies. We remove each of the three modules in RMVCNN to validate their contribution in recognizing BIM objects

Model	Accuracy	Balanced accuracy	Precision	Recall	F1 score
w/o geometric	0.5380	0.2839	0.4748	0.5380	0.4030
w/o relational	0.9805	0.9667	0.9813	0.9805	0.9807
w/o fusion	0.9881	0.9695	0.9883	0.9881	0.9881
full RMVCNN	0.9917	0.9750	0.9918	0.9917	0.9916

7   Conclusion

We focus on the BIM object classification problem to ease the interoperability problem of BIM software. We first propose a two-branch geometric-relational deep learning framework. It introduces relational information of BIM objects to assist pure geometric deep learning methods which neglect the relational information inherent in BIM models. Geometric descriptors and relational descriptors are extracted by the two branches, respectively. They are mixed by the feature fusion module to generate the final object descriptors. Our design of the geometric feature extractor makes the framework applicable to most existing geometric learning methods, including CNN based and Transformer based methods. And the framework can always boost their classification performance to different degrees. It shows the efficacy and flexibility of our framework.

Then, to fill the vacancy in BIM object datasets with relationships, we collect the IFCNet++ dataset. It contains BIM objects' geometric representation and certain local relationships. The relational information is stored in a vectorized form, easy to be processed. Though simple, the relationships are representative enough to help the geometric methods achieve a performance gain.

We follow our framework to put forward three relational models based on different geometric learning baselines and carry out experiments on them. We found that only with little additional cost introduced, our relational models can utilize objects' relations to better distinguish between BIM types with similar looks. They compensate for the weakness of pure geometric-based methods.

The limitation of our research lies in three aspects:

- The relationships of interest are explicitly presented in IFC files. They may also be wrongly labeled or lost, affecting the performance of our method.
- IFCNet++ dataset only covers major BIM object types. It still needs to be enriched with more object types, so that it can be used to train a more generalized deep learning model.
- The BIM object types are coarsely defined. They do not include subtype information of objects to provide fine-grained type information and domain-specific knowledge required in some AEC subdomains.

References

1. Bazjanac, V., Kiviniemi, A.: Reduction, simplification, translation and interpretation in the exchange of model data. In: Cib w, vol. 78, pp. 163–168 (2007)
2. Belsky, M., Sacks, R., Brilakis, I.: Semantic enrichment for building information modeling. Comput.-Aided Civ. Infrastruct. Eng. **31**(4), 261–274 (2016)
3. Bennamoun, M., Guo, Y., Sohel, F.: Feature selection for 2 d and 3 d face recognition. Wiley Encycl. Electr. Electr. Eng., 1–28 (1999)
4. Bloch, T., Sacks, R.: Comparing machine learning and rule-based inferencing for semantic enrichment of BIM models. Autom. Constr. **91**, 256–272 (2018)
5. Bronstein, A.M., Bronstein, M.M., Guibas, L.J., Ovsjanikov, M.: Shape google: geometric words and expressions for invariant shape retrieval. ACM Trans. Graph. (TOG) **30**(1), 1–20 (2011)
6. Collins, F.C., Braun, A., Ringsquandl, M., Hall, D.M., Borrmann, A.: Assessing IFC classes with means of geometric deep learning on different graph encodings. In: Proceedings of the 2021 European Conference on Computing in Construction (2021)
7. Daum, S., Borrmann, A.: Processing of topological BIM queries using boundary representation based methods. Adv. Eng. Inf. **28**(4), 272–286 (2014)
8. Dosovitskiy, A., et al.: An image is worth 16x16 words: transformers for image recognition at scale. arXiv preprint arXiv:2010.11929 (2020)
9. Eastman, C.M., Jeong, Y.S., Sacks, R., Kaner, I.: Exchange model and exchange object concepts for implementation of national BIM standards. J. Comput. Civ. Eng. **24**(1), 25–34 (2010)
10. Emunds, C., Pauen, N., Richter, V., Frisch, J., van Treeck, C.: IFCNet: a benchmark dataset for IFC entity classification. In: EG-ICE 2021 Workshop on Intelligent Computing in Engineering, p. 166. Universit atsverlag der TU Berlin (2021)
11. Emunds, C., Pauen, N., Richter, V., Frisch, J., van Treeck, C.: Sparse-BIM: classification of IFC-based geometry via sparse convolutional neural networks. Adv. Eng. Inf. **53**, 101641 (2022)
12. Gao, Y., Dai, Q.: View-based 3d object retrieval: challenges and approaches. IEEE Multimed. **21**(3), 52–57 (2014)
13. Guo, Y., Bennamoun, M., Sohel, F., Lu, M., Wan, J., Kwok, N.M.: A comprehensive performance evaluation of 3d local feature descriptors. Int. J. Comput. Vis. **116**(1), 66–89 (2016)
14. Guo, Y., Sohel, F., Bennamoun, M., Lu, M., Wan, J.: Rotational projection statistics for 3d local surface description and object recognition. Int. J. Comput. Vis. **105**(1), 63–86 (2013)
15. Guo, Y., Sohel, F., Bennamoun, M., Wan, J., Lu, M.: An accurate and robust range image registration algorithm for 3d object modeling. IEEE Trans. Multimed. **16**(5), 1377–1390 (2014)
16. He, K., Zhang, X., Ren, S., Sun, J.: Deep residual learning for image recognition. In: Proceedings of the IEEE Conference on Computer Vision and Pattern Recognition, pp. 770–778 (2016)
17. Kim, J., Song, J., Lee, J.-K.: Recognizing and classifying unknown object in BIM using 2D CNN. In: Lee, J.-H. (ed.) CAAD Futures 2019. CCIS, vol. 1028, pp. 47–57. Springer, Singapore (2019). https://doi.org/10.1007/978-981-13-8410-3_4
18. Koo, B., Jung, R., Yu, Y.: Automatic classification of wall and door BIM element subtypes using 3D geometric deep neural networks. Adv. Eng. Inform. **47**, 101200 (2021)

19. Koo, B., Jung, R., Yu, Y., Kim, I.: A geometric deep learning approach for checking element-to-entity mappings in infrastructure building information models. J. Comput. Des. Eng. **8**(1), 239–250 (2021)
20. Koo, B., La, S., Cho, N.W., Yu, Y.: Using support vector machines to classify building elements for checking the semantic integrity of building information models. Autom. Constr. **98**, 183–194 (2019)
21. Lai, K., Bo, L., Ren, X., Fox, D.: A scalable tree-based approach for joint object and pose recognition. In: Twenty-fifth AAAI Conference on Artificial Intelligence (2011)
22. Lei, Y., Bennamoun, M., Hayat, M., Guo, Y.: An efficient 3D face recognition approach using local geometrical signatures. Pattern Recogn. **47**(2), 509–524 (2014)
23. Ma, L., Sacks, R., Kattell, U.: Building model object classification for semantic enrichment using geometric features and pairwise spatial relations (2017)
24. Matei, B., et al.: Rapid object indexing using locality sensitive hashing and joint 3D-signature space estimation. IEEE Trans. Pattern Anal. Mach. Intell. **28**(7), 1111–1126 (2006)
25. Mazairac, W., Beetz, J.: BIMQL-an open query language for building information models. Adv. Eng. Inform. **27**(4), 444–456 (2013)
26. Mian, A.S., Bennamoun, M., Owens, R.: Three-dimensional model-based object recognition and segmentation in cluttered scenes. IEEE Trans. Pattern Anal. Mach. Intell. **28**(10), 1584–1601 (2006)
27. Mian, A.S., Bennamoun, M., Owens, R.A.: A novel representation and feature matching algorithm for automatic pairwise registration of range images. Int. J. Comput. Vis. **66**(1), 19–40 (2006)
28. Olofsson, T., Lee, G., Eastman, C.: Case studies of BIM in use. Electron. J. Inf. Technol. Constr. **13**, 244–245 (2008)
29. Pauwels, P., Terkaj, W.: EXPRESS to OWL for construction industry: towards a recommendable and usable ifcOWL ontology. Autom. Constr. **63**, 100–133 (2016)
30. Pazlar, T., Turk, Ž: Interoperability in practice: geometric data exchange using the IFC standard. J. Inf. Technol. Constr. (ITcon) **13**(24), 362–380 (2008)
31. Qi, C.R., Su, H., Mo, K., Guibas, L.J.: Pointnet: deep learning on point sets for 3d classification and segmentation. In: Proceedings of the IEEE Conference on Computer Vision and Pattern Recognition, pp. 652–660 (2017)
32. Qin, F.W., Li, L.Y., Gao, S.M., Yang, X.L., Chen, X.: A deep learning approach to the classification of 3D CAD models. J. Zhejiang Univ. SCI. C **15**(2), 91–106 (2014)
33. Shang, L., Greenspan, M.: Real-time object recognition in sparse range images using error surface embedding. Int. J. Comput. Vis. **89**(2), 211–228 (2010)
34. Su, H., Maji, S., Kalogerakis, E., Learned-Miller, E.: Multi-view convolutional neural networks for 3D shape recognition. In: Proceedings of the IEEE International Conference on Computer Vision, pp. 945–953 (2015)
35. Su, J.C., Gadelha, M., Wang, R., Maji, S.: A deeper look at 3D shape classifiers. In: Proceedings of the European Conference on Computer Vision (ECCV) Workshops (2018)
36. Tombari, F., Salti, S., Di Stefano, L.: Performance evaluation of 3d keypoint detectors. Int. J. Comput. Vis. **102**(1), 198–220 (2013)
37. Venugopal, M., Eastman, C.M., Sacks, R., Teizer, J.: Semantics of model views for information exchanges using the industry foundation class schema. Adv. Eng. Inform. **26**(2), 411–428 (2012)

38. Wang, Y., Sun, Y., Liu, Z., Sarma, S.E., Bronstein, M.M., Solomon, J.M.: Dynamic graph CNN for learning on point clouds. ACM Trans. Graph. (TOG) **38**(5), 1–12 (2019)
39. Wu, Z., et al.: 3d shapenets: a deep representation for volumetric shapes. In: Proceedings of the IEEE Conference on Computer Vision and Pattern Recognition, pp. 1912–1920 (2015)

Generating Construction Safety Observations via CLIP-Based Image-Language Embedding

Wei Lun Tsai[(✉)], Jacob J. Lin, and Shang-Hsien Hsieh

National Taiwan University, Taipei, Taiwan
william041107@gmail.com

Abstract. Safety inspections are standard practices to prevent accidents from happening on construction sites. Traditional workflows require an inspector to document the violations through photos and textual descriptions explaining the specific incident with the objects, actions, and context. However, the documentation process is time-consuming, and the content is inconsistent. The same violation could be captioned in various ways, making the safety analysis tricky. Research has investigated means to improve the documentation process efficiency through applications with standardized forms and develop language understanding models to analyze the safety reports. Nevertheless, it is still challenging to streamline the entire documentation process and accurately compile the reports into meaningful information. We propose an image-language embedding model that automatically generates textual safety observations through the Contrastive Language-Image Pre-trained (CLIP) fine-tuning and CLIP prefix captioning designed based on the construction safety context. CLIP can obtain the contrastive features to classify the safety attribute types for images, and CLIP prefix captioning generates the caption from the given safety attributes, images, and captions. The framework is evaluated through a construction safety report dataset and could create reasonable textual information for safety inspectors.

Keywords: Image captioning · Safety inspection · Construction safety · CLIP

1 Introduction

Construction safety is one of the most critical issues for project management. Current practices utilize various tools to document safety violations, including paper and pen, smartphone applications, cameras, and text messages. Data collected from traditional tools are not easy to analyze and require extra digitalization effort [4]. In recent years, companies have moved from traditional tools to smartphone applications that try to streamline the documentation process and eliminate human error possibilities while converting data [17,26]. Although the tools' workflow might differ, they all have one expected final deliverable - a safety report with all the violations presented with photos and textual descriptions.

© The Author(s), under exclusive license to Springer Nature Switzerland AG 2023
L. Karlinsky et al. (Eds.): ECCV 2022 Workshops, LNCS 13807, pp. 366–381, 2023.
https://doi.org/10.1007/978-3-031-25082-8_24

The textual description explains the safety violation usually through specifying the objects, actions, and context directly related to companies' or official agencies' safety regulations [13,26]. While there are particular formats to follow, the description differs from person to person. Recent studies have developed language models to understand the description and create a knowledge base to map safety regulation automatically [19]. Although preliminary results show excellent performance, it is still challenging due to the variety of expressions in the textual description. On the other hand, photos provide visual references and proof of the violation that could be used for better language understanding where multiple research shows promising results [27]. This paper aims to generate construction safety observation through image-language embedding automatically.

Currently, studies have developed vision language models to understand the features between images and text. Image captioning is one of the major implementations of the vision language model, in which the model will predict the relative text according to the input image [10]. This technique can identify the relationship between images and text, and generate a sentence describing them. However, most of these methods focus on describing general construction photos [2,14] and would need to be adopted for construction safety-specific scenarios. Furthermore, various pre-trained vision language models based on different frameworks and modalities have been developed to address image encoding and text generation tasks [8]. Contrastive Language-Image Pre-training (CLIP) [20] is one of the robust vision-language pre-trained models with a dual encoder for both text and image data. The objective of contrastive learning corresponds to determining the types of captions and violations for safety violation images. In this research, we use CLIP prefix tuning and add image features encoded by CLIP as a prefix [16] for the language model to generate text descriptions, violations, and related regulations.

In summary, we develop four modules, including the data set development, CLIP fine-tuning, CLIP prefix captioning, and user interface implementation. We created captioning data sets related to the safety inspection procedure and added additional labeled attributes for contrastive learning strategies. The CLIP model is fine-tuned to obtain the contrastive feature with different combinations of attributes from caption types and violation types. The CLIP prefix can translate the prediction embeddings from the previous CLIP model to the captions or violation/regulations lists. Finally, a smartphone application was designed to implement the captioning model and provide an annotation function as an end-to-end application for safety inspectors conducting a safety inspection.

2   Related Work

Several Natural Language Processing (NLP) techniques and algorithms have been proposed to solve a wide variety of textual problems, including knowledge extraction from documents and information for construction or safety management. The following discussion about NLP will focus on Vision-based Natural Language Generation in Construction. The Vision Language Model (VLM) is the main approach of Natural Language Generation (NLG) to retrieving textual information from images. VLM will obtain features by encoding images and generating textual descriptions to explain contextual information, which is the main concept of image captioning. We will introduce different aspects of VLM and image captioning applications.

Liu et al. [14] addressed a structured linguistic description of the construction activity scene and manifested the construction scene via the image captioning technique. The linguistic description divides a scene's sentences by main objects, main actions, and main attributes for five different construction activities: cart transportation, masonry work, rebar work, plastering, and tiling. Each construction scene should contain at least one construction activity and five descriptions following the instruction of the MS COCO caption format [3]. The captioning model is constructed with a visual encoder and sequential decoder. The basic CNN framework, VGG-16, and ResNet-50 are applied as the encoder for images and LSTM is selected as the decoder RNN. Finally, generated testing results evaluated by humans present the feasibility of applying image captioning into practice. This paper attempts to bridge the gap between visual information and natural language sentences. Whereas the construction activities discussed in this research only contain five common activities. There are still various and diverse scenarios for natural language describing construction activities that lack real-world application scenes to implement this captioning method.

Xiao et al. [25] performed a feasibility study on the potential of image captioning in construction scenes by developing a vision language model from the computer vision community. The construction dataset following linguistic schema was developed and focused on construction equipment images. More advanced vision language models, combining ResNet101 [9] and Transformer [24] rather than the basic CNN-RNN method were implemented to build the state-of-art image captioning model providing a detailed attention mechanism. This research demonstrated the feasibility of image captioning in the construction domain. The captioning results reached 86.6% and 41.2% of F1 scores for recognizing scene objects and activities.

On the other hand, CLIP is a vision language model proposed by OpenAI [20]. CLIP connects images and texts in an automatic way that does not rely on labeling data. By encoding a series of image and text features to the same dimension space, the cosine similarity between each image and text features will be calculated to compare the similarity of each image and text pair. The idea of contrastive was indicated in the comparison between the similarity across the same image or text [23]. While the prediction is the images and texts pair with the highest similarity, it implies the contrastive relationship between image and text candidates or text and image candidates. Therefore, contrastive learning can be applied to classify [6] and recognize the attributes in an image.

Consequently, NLG has various implementations in different scenarios to extract information from documents and images. For example, question answering retrieving from building regulations, construction scene understanding via UAV-acquired images, and textual and visual encoder-decoder-based image captioning models generating descriptions explaining activities and scene objects. In recent years, the language model evolving with Transformer, like BERT, GPT-2, Text-To-Text Transfer Transformer (T5) [21], has dominated the mainstream solutions for NLP and computer vision (CV) [8] which is well pre-trained and developed to improve downstream tasks like question answering and image captioning. However, these primary techniques have not been attempted and validated in construction scenarios. There is no research discussing the application for safety inspection reports which contain rich text and image pairs. Moreover, the correlations between safety inspection images and the violation lists or regulation lists from companies or OHSA have not been discussed yet.

Therefore, the NLG combined with the VLM technique performs comprehensive knowledge retrieval and restores knowledge to human-understandable language which can be the novel solution for safety inspectors obtaining related violations and regulations from taken images. Namely, the knowledge base can bridge the gap between violation images and regulation and be embedded into the proposed image captioning model based on the state-of-art vision language model.

3 Methodology

In this paper, we propose a construction safety inspection workflow with Construction CLIP and CLIP prefix captioning techniques to learn the domain knowledge from the safety reports. Construction CLIP classifies the safety violation and caption type via a contrastive learning approach. We have also developed a user interface to implement the image captioning model in practice. Figure 1 illustrated the overview of this research which contains four main modules which are data set development, Construction CLIP fine-tuning, CLIP prefix captioning, and user interface implementation. The overview shows the comprehensive process of an end-to-end workflow from contrastive features learning with images taken from construction sites and attribute annotations to model implementation onto a user interface that can be applied by safety inspectors in a real-world scenario. The following section will introduce the methods and procedures for each module in depth.

3.1 Dataset Development

Labelling Strategy for Safety Violation Captioning Data Sets. A novel dataset capturing safety observations text descriptions, safety attributes, and images are presented in this paper. In order to build the additional attribute information for contrastive learning, we label four more keys for caption annotations regard to each image, which are *caption_type*, *violation_type*, *violation_list*, and *objects*, according to the categories applied for classifying observation results in different hierarchy. *caption_type*, including *status* and *violation*, stands for whether this observation containing safety violations. *violation_type* represents the types of the violations indicated to the violation categories refer to Regulations of Occupational Safety and Health Act (OSHA) from Ministry of Labor in Taiwan [12]. *violation_list* is a combination of the description describing safety

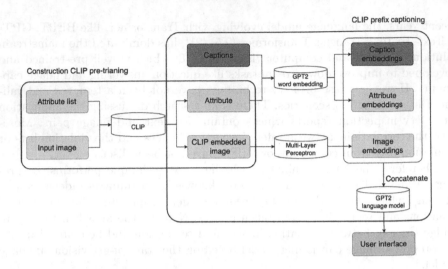

Fig. 1. The overview of proposed framework

violations according to the internal regulation list in the company or the safety regulations from OHSA. Specifically, *violation_list* can offer a consistent and unified narrative for the same violation scene. *objects* shows the object mentioned in the *violation_list*. Table 1 and 2 shows the image amounts of each types for *caption_type* and *violation_type*.

Table 1. The images amounts with different caption types

	Status	Violation	Total
Caption type	194	612	806

Table 2. The images amounts with different violation types

	Falling	PPE	Electric	Workspace	Material
Violation type	437	129	49	49	31
	Explosion	Puncture	Mechanical	Transport	Total
Violation type	30	26	24	12	787

Status and Safety Violations Image Collection. We collect the images taken by three professional and experienced safety inspectors belonging to different companies with a distinct format. Overall, 806 text and image pairs describing observation results for safety inspection in practical construction sites were labeled which will be split into the training set and testing set with an 80 percent of training ratio. Last, Table 3 shows the example for caption data with additional attributes.

Table 3. The example for caption data with additional attribute

Sample images	Ground truth caption data
	– caption_type: violation – violation_type: falling – violation_list: Openings are without guardrail – caption: – file_name: data1.jpg – objects: openings, guardrail
	– caption_type: violation – violation_type: falling – violation_list: Openings are without guardrail – caption: No fall prevention for openings – file_name: data2.jpg – objects: openings, guardrail

3.2 Construction CLIP Fine-Tuning

Model Selection. Transformers replace commonly used recurrent layers in encoder-decoder architectures and achieve high performance in linguistic-based tasks. Moreover, Vision Transformer (ViT) was introduced to replace conventional CNN-based image encoder with Transformer blocks [7]. By splitting an image into patches and obtaining sequential positional embeddings to record the position for each patch, these patches are input to a Transformer block in the same way as tokenized words in NLP applications. With the fewest modification, ViT performs relatively high performance for many image classification data sets showing the robust capability to extract image features for additional objectives. In conjunction with a multi-head self-attention Transformer and Vision Transformer as text and vision encoder, CLIP attains powerful vision language modeling for a variety of tasks during a large-scale data set fine-tuning [20]. The next subsection will clarify the architecture and fine-tuning process of Construction CLIP.

Fine-Tuning Strategy. Figure 2 illustrates the proposed architecture for Construction CLIP fine-tuning. Additional information mentioned in the labelling strategy section, *caption_type* and *violation_type*, will be the reference attribute list for dual-stream CLIP text embedding. The images in the caption data sets are split into subsets during fine-tuning according to the label of *caption_type* and *violation_type* for each image. The elements in the collection list of *caption_type* and *violation_type* pair with all possible combinations for every two elements to learn the ability to distinguish pair-wise images and types contrastingly. However, owing to lacking amount of captioning data, the amounts for each *caption_type* and *violation_type* are imbalanced. If the amounts of two types are not the same, the type with fewer amounts will be accessed repeatedly to ensure all data in more amounts are paired and fine-tuned. After fine-tuning, both attribute lists will encode with dual-stream CLIP and calculate cosine similarity with the input image. Finally, the attributes with highest similarity in *caption_type* and *violation_type* are corresponding to the caption type and violation type for the input image.

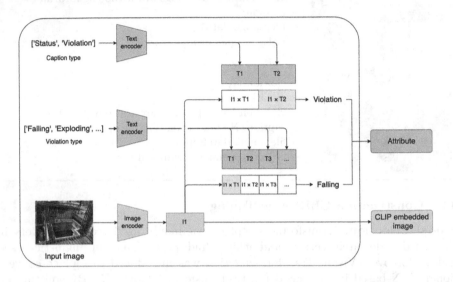

Fig. 2. The architecture of construction CLIP fine-tuning

3.3   CLIP Prefix Captioning

Prefix Tuning with Attributes. In our work, on account of additional attributes classified by Construction CLIP, we can engage more information to the language model providing extra instruction and conditions for text generation. In terms of the concept of conditional generation according to the CLIP prefix framework proposed by Ron et al. [16], the language model tends to generate sentences regarding the input words which is the condition for text generation [5,21,22]. Formula 1 represents the objective to predict next word from given contents $y_{1:i-1}$ and other input x. After each token is produced, that token is added to the original input sequence and becomes a new input sequence to the model in the next iteration which makes RNNs effective, called *auto-regression* [1]. For adding supplementary attributes with embedded captions, the language model should be able to generate captions more corresponding to images and original captions. Eventually, the predicted attributes from CLIP will be embedded by GPT-2 tokenizer and concatenated with image and captions embeddings which embedded attributes and images are regarded as the prefix for concatenated embeddings and input to the language model.

$$P(y_i|y_{1:i-1}, x) \tag{1}$$

Figure 3 reveals the architecture of CLIP prefix captioning. Based on the architecture of CLIP prefix captioning, the additional information, *caption_type* and *violation_type*, predicted automatically through fine-tuned Construction CLIP model are adjoined as concatenated embeddings, including captions and attributes embeds by GPT-2 tokenizer based on BPE and CLIP embedded image translated by MLP.

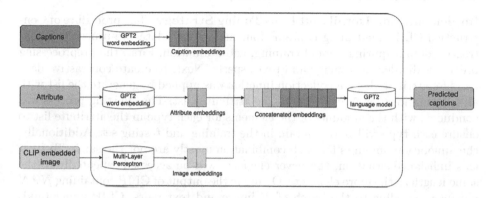

Fig. 3. The architecture of prefix captioning with attributes

Decoding Strategies. With a trained conditional language model, decoding methods decide how to generate texts from the language model also playing an important role to generate the final outputs of the language model. Due to the support of the auto-regressive language generation, GPT-2 can generate word sequences based on the prediction from every iteration applying the original sequence and padding with a new prediction. Equation 2 explains that auto-regressive language generation is based on the product of the probability for each conditional distribution decomposed from the probability distribution [18].

$$P(y_{1:N}|x) = \prod_{i=1}^{N} P(y_i|y_{1:i-1}, x), \text{ with } y_{1:0} \neq 0 \qquad (2)$$

x is the initial context from input embeddings. The length N of the output sequence is determined dynamically by when the time step i generated the *End Of Sentence* (EOS) token from $P(y_i|y_{1:i-1}, x)$ in real-time. There are several prominent decoding strategies have been established, for example, Greedy search, Beam search, Top-K sampling, Top-p sampling, and Temperature. We applied beam search for N = 3, and temperature for 0.5 as decoding strategies.

To sum up, with the additional attributes describing the types of caption and violation, concatenated embeddings, including encoded captions, encoded attributes, and translated CLIP embeddings image features, are inputted to the transformer-based GPT-2 language model to obtain a more comprehensive and precise description for the captioning. The caption data can be descriptions in natural language, the safety violation list, and the regulations list from internal companies corresponding to different application scenarios for the CLIP prefix captioning model.

4 Experiment

4.1 Construction CLIP Fine-Tuning

Implementation Detail and Fine-Tuning Strategy. The procedure of Construction CLIP fine-tuning contains four main steps: data pre-processing, contrastive data acquiring, model training, and evaluation. First, the preprocessing process is divided into image and text aspects. Next, to create contrastive data pairs for attribute lists, a collection function was applied to generate the list with all combinations for N elements in the attributes list. The splitting process was conducted with the amounts of the elements for each type in the attribute list to ensure each type will both contain in the training and testing set. Additionally, the annotation amounts for each combination mostly are not the same. To treat this imbalance situation, the fewer elements will be accessed repeatedly to the same length as the more elements. Owing to the output of CLIP containing $N \times N$ pairings according to the batch of N image and text pairs, CLIP understands a multi-modal embedding space through training both text and image encoder cooperatively to the objective of maximizing the cosine similarity. We applied AdamW optimizer [11] with decoupled weight decay regularization [15] and the linear scheduler with a warmup. To evaluate the performance for the classification results of caption and violation types after Construction CLIP fine-tuning, the basic accuracy for multi-class classification was applied.

The procedure of Construction CLIP fine-tuning is conducted in two different settings according to the violation type with 9 elements. The number of elements N for combinations is set into 2 and 9. The combinations with every 2 elements can make the model understand the contrastive feature between every two types of data and distinguish them more correctly. The combinations with 9 elements can imitate the scenario during inference which applies the list with all violation types. Moreover, to obtain both advantages with a different number of elements N for combinations, there are two more settings to chain the fine-tuned weights jointly. Fine-tuning with $N = 9$ set after fine-tuning $N = 2$ set and doing conversely in violation type can gain the ability to learn the robust ability to distinguish contrastive feature for $N = 2$ set and the ability to determine the type from $N = 9$ set standing for all violation type. The fine-tuning process was monitored by the loss and the accuracy of the training set and the accuracy of the testing set checking at the end of each epoch. In both fine-tuning steps with different N, the model for each epoch with the highest accuracy in the testing set will be chosen as the final weights. After fine-tuning in two different settings for violation types, we pick a final model fine-tuned with $N = 9$ set after fine-tuning $N = 2$ set to obtain the ability to distinguish all 9 violations types which match the final application scenario. Table 4 shows the highest accuracy for training and testing set in two settings and our final decision model.

Table 4. The accuracy with different numbers of elements N for combinations

Number of elements N for combinations	Training accuracy	Testing accuracy
$N = 9$	0.99	0.49
$N = 2$ and 9	0.99	0.49

Evaluation and Discussion. According to the figure, without many training epochs, the loss and the accuracy decreased and increased dramatically which means the model has a powerful ability to fit the training data in Fig. 4. However, the trend of the testing accuracy shown in Fig. 5 is not proportional to the training accuracy. Due to learning the detail and noise in the training data and limiting the ability to identify unseen data not included in the training set, the model is overfitting during the fine-tuning process which contains low bias and high variance. Also, owing to the imbalance of data, the type with fewer images was accessed repeatedly which is another cause for the overfitting situation. Overfitting can limit by several solutions. First, to enhance the training data, there are only 806 pairs of image captioning data compared to the size of the CLIP pre-trained data is largely relatively small. On the other hand, reducing the layer of the transformer block could be one of the solutions. About fine-tuned jointly, we found that after chaining different settings of N elements of the combination, the accuracy and loss from training data will increase to 0.99 and decrease to 0, respectively. However, there is a limited increase in testing accuracy. It demonstrates these strategies are not well set to learn with fewer data in the different given labels.

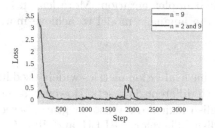 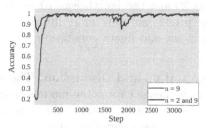

Fig. 4. Training loss and accuracy curve ending with $N = 9$ elements of combinations

4.2   CLIP Prefix Captioning

Implementation Detail. The main procedure of CLIP prefix captioning contains three main steps: data pre-processing, model training, and evaluation. Regarding data pre-processing, textual data and visual data are processed separately as well. Textual data, for example, caption and attribute, was tokenized by a tokenizer from GPT2 which is also based on BPE. Both caption and attribute tokens were padded to the same length and masked to identify whether tokens

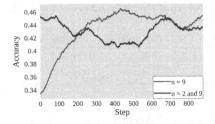

Fig. 5. Testing accuracy ending with $N = 9$ elements of combinations

should be attended to or not. A period also was added to the end of each caption to imply the end of the sentence. On the other hand, visual data is encoded by the image encoder from fine-tuned Construction CLIP model. Next, GPT-2 is the chosen language model to perform the text generation task. The input caption and attribute tokens were concatenated together and embedded by word embeddings from GPT-2. The input encoded image was converted into GPT-2 space by different mapping types, MLP and Transformer. Then both textual visual embeddings were concatenated again as the input for the language to generate captions. The loss function was applied cross-entropy loss comparing the output logits and labeled captions. The optimizer and scheduler were the same as the ones applied in Construction CLIP fine-tuning which is AdamW and the linear scheduler with the warmup. Overall, the image embeddings and the attribute embeddings are regarded as a prefix for the input concatenated embeddings as the controller to steer what generates in the language model in Fig. 6. We increased the length of the image and attribute prefix embeddings was 20 and the length was 10 in the original implementation. More length for the prefix can store more information for the language model to achieve more accuracy and fluent generation.

Evaluation and Discussion. There are some evaluation metrics widely used in various image captioning models trained with different data sets. However, some of the evaluation metrics will consider English lexical database, or graph-based semantic representation also only in English. Therefore, BLEU and ROUGE are chosen as the evaluation metric. Despite applying automatic metrics can acquire the evaluation results fast and objective, there is still a large gap between automatic evaluation metrics and human judgment. Thus, we will consider the human evaluation due to the complex scenario in our target images.

The training loss was decreasing as we expect which means the prefix captioning model with the additional attribute can generate similar captions compared to the ground truth in Fig. 7. There is no figure of the testing loss curve because the loss curve only exists during backpropagation. However, the prediction results, in the end, still do not reach feasible conditions for various scenarios in construction sites. Some prediction results are shown in Table 5 and 6 with different completeness for prediction and ground truth captions which were eval-

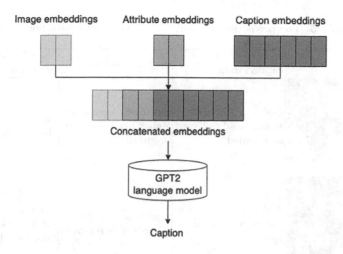

Fig. 6. The process for concatenating embeddings

uatcd by a human. Also, some of the captions tend to generate exactly captions with different generation conditions. It stands for the situation of model collapse, which means only a few patterns the model will generate after prefix tuning. The results of the automatic evaluation function are also shown in Table 7 comparing with the baseline model applying CNN and LSTM model trained with ACID caption data set [25]. The ACID caption data set was established in a linguistic schema which implies a higher score for the image captioning task. Compared with different types, the method using a transformer was expected to achieve better performance due to the attention mechanism. However, the MLP method attained higher BLEU and ROUGE scores which implies the model with more complexity will overfit data caused by insufficient data amount.

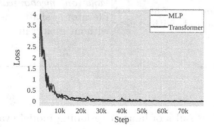

Fig. 7. Loss curves for training data applying different mapping types

Table 5. The prediction of attributes and captions with few errors

Sample images	Ground truth caption data	MLP prediction	Transformer prediction
	violation puncture Rebar head not bent or guarded	*violation puncture* Rebar head not bent or guarded	*violation puncture* Rebar not bent
	violation falling Guardrail is damaged	*violation falling* Guardrail is removed	*violation mechanical* Grinder is without guard shield

Table 6. The prediction of attributes and captions with errors

Sample images	Ground truth caption data	MLP prediction	Transformer prediction
	violation falling Opening is without safety guardrail	*status falling* Openings are not covered with safety net	*violation falling* Long safety net is not restored after the mud work
	violation falling Openings are not covered with safety net	*violation mechanical* Elevator shaft is without guardrail	*violation mechanical* Formwork winches shall not hang scaffolding

Table 7. Results of image captioning model evaluation

Mapping type	BLEU-1	BLEU-2	BLEU-3	BLEU-4	ROUGE-1	ROUGE-L
Baseline*	0.606	0.506	0.398	0.320	–	0.560
Transformer	0.249	0.242	0.333	0.249	0.058	0.057
MLP	0.454	0.447	0.525	0.454	0.119	0.118

*The baseline score was performed by Xiao et al. [25].

5   Conclusion

In this paper, we propose a construction safety inspection framework assisted with image captioning. The system uses a novel transformer-based vision language model to classify the caption type and violation types from images. By obtaining additional attributes, the vision language model can generate photo descriptions with multiple embeddings for safety inspection purposes. The proposed framework consists of four main modules: dataset development, Construction CLIP fine-tuning, CLIP prefix captioning, and user interface implementation. We created an image captioning dataset from previous inspection reports. A construction safety knowledge base is created based on the dataset and can be embedded into related language models for a safety inspection. We use CLIP prefix tuning and add image features encoded by CLIP as a prefix for the language model to generate text descriptions, violations, and related regulations. Finally, the captioning model was implemented into a smartphone application to conduct caption generating automatically to assist safety inspectors.

Acknowledgement. We thank Dr. Shuai Tang for the discussions and suggestions. The project is supported in part by MOST, Taiwan 110-2622-E-002-039, 110-2222-E-002-002-MY3.

References

1. Alammar, J.: The illustrated gpt-2 (visualizing transformer language models) [blog post] (2019). https://jalammar.github.io/illustrated-gpt2/
2. Bang, S., Kim, H.: Context-based information generation for managing UAV-acquired data using image captioning. Autom. Constr. **112**, 103116 (2020). https://doi.org/10.1016/j.autcon.2020.103116. https://www.sciencedirect. com/science/article/pii/S0926580519308519
3. Chen, X., et al: Microsoft coco captions: data collection and evaluation server (2015). https://doi.org/10.48550/ARXIV.1504.00325. https://arxiv.org/abs/1504. 00325
4. Cheng, M.Y., Kusoemo, D., Gosno, R.A.: Text mining-based construction site accident classification using hybrid supervised machine learning. Autom. Constr. **118**, 103265 (2020). https://doi.org/10.1016/j.autcon.2020.103265. https://www. sciencedirect.com/science/article/pii/S092658051931341X
5. Chiu, S., Li, M., Lin, Y.T., Chen, Y.N.: Salesbot: Transitioning from chit-chat to task-oriented dialogues (2022). https://doi.org/10.48550/ARXIV.2204.10591. https://arxiv.org/abs/2204.10591
6. Conde, M.V., Turgutlu, K.: Clip-art: contrastive pre-training for fine-grained art classification. In: 2021 IEEE/CVF Conference on Computer Vision and Pattern Recognition Workshops (CVPRW), pp. 3951–3955 (2021). https://doi.org/ 10.1109/CVPRW53098.2021.00444
7. Dosovitskiy, A., et al.: An image is worth 16x16 words: Transformers for image recognition at scale (2020). https://doi.org/10.48550/ARXIV.2010.11929. https:// arxiv.org/abs/2010.11929

8. Du, Y., Liu, Z., Li, J., Zhao, W.X.: A survey of vision-language pre-trained models (2022). https://doi.org/10.48550/ARXIV.2202.10936. https://arxiv.org/abs/2202.10936

9. He, K., Zhang, X., Ren, S., Sun, J.: Deep residual learning for image recognition (2015). https://doi.org/10.48550/ARXIV.1512.03385. https://arxiv.org/abs/1512.03385

10. Hossain, M.Z., Sohel, F., Shiratuddin, M.F., Laga, H.: A comprehensive survey of deep learning for image captioning. ACM Comput. Surv. **51**(6), 1–36 (2019). https://doi.org/10.1145/3295748

11. Kingma, D.P., Ba, J.: Adam: A method for stochastic optimization (2014). https://doi.org/10.48550/ARXIV.1412.6980. https://arxiv.org/abs/1412.6980

12. Ministry of Labor, T.: Regulations of occupational safety and health act (2022). https://law.moj.gov.tw/LawClass/LawAllPara.aspx?pcode=N0060009

13. Lin, J.R., Hu, Z.Z., Li, J.L., Chen, L.M.: Understanding on-site inspection of construction projects based on keyword extraction and topic modeling. IEEE Access **8**, 198503–198517 (2020). https://doi.org/10.1109/ACCESS.2020.3035214

14. Liu, H., Wang, G., Huang, T., He, P., Skitmore, M., Luo, X.: Manifesting construction activity scenes via image captioning. Autom. Constr. **119**, 103334 (2020). https://doi.org/10.1016/j.autcon.2020.103334. https://www.sciencedirect.com/science/article/pii/S0926580520309146

15. Loshchilov, I., Hutter, F.: Decoupled weight decay regularization (2017). https://doi.org/10.48550/ARXIV.1711.05101. https://arxiv.org/abs/1711.05101

16. Mokady, R., Hertz, A., Bermano, A.H.: Clipcap: Clip prefix for image captioning. arXiv preprint arXiv:2111.09734 (2021)

17. Pal, A., Hsieh, S.H.: Deep-learning-based visual data analytics for smart construction management. Autom. Constr. **131**, 103892 (2021). https://doi.org/10.1016/j.autcon.2021.103892. https://www.sciencedirect.com/science/article/pii/S0926580521003435

18. von Platen, P.: How to generate text: using different decoding methods for language generation with transformers[blog post] (2020). https://huggingface.co/blog/how-to-generate

19. Qady, M.A., Kandil, A.: Concept relation extraction from construction documents using natural language processing. J. Constr. Eng. Manage. **136**(3), 294–302 (2010). https://doi.org/10.1061/(ASCE)CO.1943-7862.0000131

20. Radford, A., et al.: Learning transferable visual models from natural language supervision (2021). https://doi.org/10.48550/ARXIV.2103.00020. https://arxiv.org/abs/2103.00020

21. Raffel, C., et al.: Exploring the limits of transfer learning with a unified text-to-text transformer. J. Mach. Learn. Res. **21**(140), 1–67 (2020). http://jmlr.org/papers/v21/20-074.html

22. Tang, J., Zhao, T., Xiong, C., Liang, X., Xing, E., Hu, Z.: Target-guided open-domain conversation. In: Proceedings of the 57th Annual Meeting of the Association for Computational Linguistics, pp. 5624–5634. Association for Computational Linguistics, Florence, Italy (2019). https://doi.org/10.18653/v1/P19-1565. https://aclanthology.org/P19-1565

23. Tian, Y., Krishnan, D., Isola, P.: Contrastive multiview coding. arXiv preprint arXiv:1906.05849 (2019)

24. Vaswani, A., et al.: Attention is all you need (2017). https://doi.org/10.48550/ARXIV.1706.03762. https://arxiv.org/abs/1706.03762

25. Xiao, B., Wang, Y., Kang, S.C.: Deep learning image captioning in construction management: a feasibility study. J. Constr. Eng. Manage. **148**(7), 04022049 (2022). https://doi.org/10.1061/(ASCE)CO.1943-7862.0002297
26. Zhang, H., Chi, S., Yang, J., Nepal, M., Moon, S.: Development of a safety inspection framework on construction sites using mobile computing. J. Manage. Eng. **33**(3), 04016048 (2017). https://doi.org/10.1061/(ASCE)ME.1943-5479.0000495
27. Zhong, B., He, W., Huang, Z., Love, P.E., Tang, J., Luo, H.: A building regulation question answering system: a deep learning methodology. Adv. Eng. Inf. **46**, 101195 (2020). https://doi.org/10.1016/j.aei.2020.101195. https://www.sciencedirect.com/science/article/pii/S1474034620301658

W30 - AI-Enabled Medical Image Analysis: Digital Pathology and Radiology/COVID-19

W30 - AI-Enabled Medical Image Analysis: Digital Pathology and Radiology/COVID-19

Deep Learning has made rapid advances in the performance of medical image analysis, challenging physicians in their traditional fields. In the pathology and radiology fields, in particular, automated procedures can help to reduce the workload of pathologists and radiologists and increase the accuracy and precision of medical image assessment, which is often considered subjective and not optimally reproducible. In addition, Deep Learning and Computer Vision demonstrate the ability/potential to extract more clinically relevant information from medical images than what is possible in current routine clinical practice by human assessors. Nevertheless, considerable development and validation work lie ahead before AI-based methods can be fully integrated into medical departments.

The workshop on AI-enabled Medical Image Analysis (AIMIA) at ECCV 2022 aimed to foster discussion and presentation of ideas to tackle the challenges of whole slide image and CT/MRI/X-ray analysis/processing and identify research opportunities in the context of digital pathology and radiology/COVID-19.

The workshop invited high-quality original contributions targeted in several contexts, such as using self-supervised and unsupervised methods to enforce shared patterns emerging directly from data, developing strategies to leverage few (or partial) annotations, promoting interpretability in both model development and/or the results obtained, or ensuring generalizability to support medical staff in their analysis of data coming from multi-centres, multi-modalities, or multi-diseases.

October 2022

Jaime S. Cardoso
Stefanos Kollias
Sara P. Oliveira
Mattias Rantalainen
Jeroen van der Laak
Cameron Po-Hsuan Chen
Diana Felizardo
Ana Monteiro
Isabel M. Pinto
Pedro C. Neto
Xujiong Ye
Luc Bidaut
Francesco Rundo
Dimitrios Kollias
Giuseppe Banna

Harmonization of Diffusion MRI Data Obtained with Multiple Head Coils Using Hybrid CNNs

Leon Weninger[1]([✉])(iD), Sandro Romanzetti[2,3](iD), Julia Ebert[2,3](iD), Kathrin Reetz[2,3](iD), and Dorit Merhof[1](iD)

[1] Institute of Imaging and Computer Vision, RWTH Aachen University, Aachen, Germany
`leon.weninger@lfb.rwth-aachen.de`
[2] Department of Neurology, RWTH Aachen University, Aachen, Germany
[3] JARA-BRAIN Institute Molecular Neuroscience and Neuroimaging, Forschungszentrum Jülich and RWTH Aachen University, Aachen, Germany

Abstract. In multisite diffusion MRI studies, different acquisition settings can introduce a bias that may overshadow neurological differentiations of the study population. A variety of both classical harmonization methods such as histogram warping as well as deep learning methods have been recently proposed to solve this problem and to enable unbiased multisite studies. However, on our novel dataset—it consists of acquisitions on the same scanner with the same acquisition parameters, but different head coils—available methods were not sufficient for harmonization. Nonetheless, solving the challenge of harmonizing this difficult dataset is relevant for clinical scenarios. For example, a successful harmonization algorithm would allow the continuation of a clinical study even when the employed head coil changes during the course of the study. Even though the differences induced by the change of the head coil are small, they may lead to missed or false associations in clinical studies.

We propose a harmonization method based on known operator hybrid 3D convolutional neural networks. The employed neural network utilizes a customized loss that includes diffusion tensor metrics directly into the harmonization of raw diffusion MRI data. It succeeds a preliminary histogram warping harmonization step. The harmonization performance is evaluated with diffusion tensor metrics, multi-shell microstructural estimates, and perception metrics. We further compare the proposed method to a previously published deep learning algorithm and standalone intensity warping. We show that our approach successfully harmonizes the novel dataset, and that it performs significantly better than the previously published algorithms.

Keywords: Harmonization · Diffusion MRI · Known operator learning

Large-scale group studies such as the Human Connectome Project or the Alzheimer Disease Neuroimaging Initiative showcase the benefits of acquiring data from a large number of individuals. However, a transfer of results on these

© The Author(s), under exclusive license to Springer Nature Switzerland AG 2023
L. Karlinsky et al. (Eds.): ECCV 2022 Workshops, LNCS 13807, pp. 385–396, 2023.
https://doi.org/10.1007/978-3-031-25082-8_25

datasets to local clinical data, where a different diffusion MRI (dMRI) scanner is used, is difficult, as different scanners induce biases. For this reason, various works have already addressed harmonization of inter-site biases, with the purpose of aligning the diffusion signals between two scanner systems [17].

Recently, it has been shown that dMRI images, while being theoretically a quantitative measure, are also biased if subjects are scanned on the same scanner, but with different head coils and acquisition parameters [15]. In our study, we have acquired diffusion data of 24 subjects on one scanner with two different head coils. The two acquisitions were conducted with the same acquisition parameters and in direct succession to reduce acquisition variety and to enable a high and uniform image quality. The same scanner hard- and software and reconstruction algorithms were used to narrow the acquisition bias down to the different head coils. The resulting image differences are, thus, lower compared to what has been found in previous harmonization studies, making the harmonization task even more challenging. Even though, these differences cannot be ignored, as they can still lead to misinterpretation of study results [15]. Further, a two head coil scenario is indeed realistic in clinical studies. For example, a head coil may break halfway through a study, or the typically used head coil may be to small for individuals with a large skull. To successfully harmonize this dataset, we present a novel method, a known operator hybrid 3D convolutional neural network, and compare it to previous methods.

While this proposed method harmonizes raw diffusion-attenuated images, it is of utmost importance to asses harmonization performance on derived parametric maps such as fractional anisotropy (FA) and mean diffusivity (MD)—metrics typically used in dMRI based studies—, as well as on differences in fiber direction, which influences derived structural connectivity metrics. Next to these metrics, the harmonization quality is assessed with multi-shell microstructural biomarkers and with deep learning based perceptual image similarity measurement. The neurite orientation dispersion and density imaging (NODDI) technique [22] is used to estimate microstructural complexity and fiber direction. For perceptual differences, the learned perceptual image patch similarity (LPIPS) metric [23] is applied.

Related Work. The most straightforward way to match the distribution of diffusion-attenuated signals across scan settings is the method of moments, i.e., matching the spherical mean and spherical variance of the two settings [8]. While this method is very practical, it is also limited, as it cannot correct nonlinear signal intensity distortions or locally varying signal drift. More degrees of freedom in harmonization are possible when using rotation invariant spherical harmonic features for determining a voxel-wise matching of signals [5,11,12]. This approach further ensures that the shape of the diffusion signal is changed, but not its orientation. Other recent approaches to harmonization rely on deep learning algorithms [13] or sparse dictionaries [19]. A variety of these algorithms have been evaluated in the multi-shell dMRI harmonization challenge [20]. In this challenge, several algorithms showed promising results [14]. One example is the

SHResNet [9], a spherical harmonics based deep residual network for harmonization. This SHResNet is compared to the proposed method in this work. We further use a standalone intensity warping technique termed MICA (Multisite Image Harmonization by cumulative distribution function Alignment) [21] as benchmark.

Contributions. We introduce an approach that merges histogram warping with a 3D convolutional neural network (CNN) and known operator learning [10]. As global differences are removed by histogram warping in a first step, the CNN used for the second step thus only needs to correct location- or diffusion gradient-dependent effects. Further, while the harmonization procedure operates on the raw diffusion-attenuated images, the inclusion of a fully differentiable diffusion tensor fitting layer in the deep learning framework ensures a correct harmonization of derived parametric maps. In contrast, previous approaches either harmonize raw diffusion-attenuated images without taking account of the effect on derived parametric maps, or harmonize only the derived parametric maps. Lastly, a neural network discriminator is further employed to ensure that the harmonization approach itself does not introduce any unwanted effect in the data, e.g., that the harmonization performance is not achieved through denoising. We provide comparisons of this method with a previous top-performing deep learning method and with a standalone histogram warping approach.

1 Materials and Methods

1.1 Image Data and Preprocessing

24 healthy subjects (male, 21–30 years) underwent MRI examinations on a 3-Tesla Siemens Prisma MRI, both with a 20-channel and a 64-channel head coil. The study was approved by the local ethics committee of the Medical Faculty of the University of the RWTH Aachen (EK029/19) and conducted in accordance with the standards of Good Clinical Practice and the Declaration of Helsinki. The scanning protocol included a 3D T1-weighted image with the 64-channel head coil as well as the same dMRI sequence for both head coils. The dMRI sequences consisted of 289 single acquisitions, with 90 diffusion directions each for the b-values 1000, 2000 and 3000 $\frac{s}{mm^2}$ as well as 19 b0 images. Further pulse sequence settings were as follows for the T1-scan: repetition time (TR) = 2,400 ms, echo time (TE) = 2.22 ms, 176 slices with a slice thickness of 0.8 mm, flip angle = 8°, field of view (FoV) = 282 mm, voxel size = 0.8 mm isotropic, and 170 × 268 mm matrix. The dMRI data was acquired using an echo planar imaging sequence with TR = 4000 ms, TE = 103.6 ms, FoV = 213 mm, 100 slices with a slice thickness of 0.8 mm, and voxel resolution = 0.8 mm isotropic. The sequence was acquired two times, once with anterior-posterior phase encoding, and once with posterior-anterior phase encoding. Ensuring to reduce the acquisition variety as much as possible, the same hardware setup was used and the scanner software remained the same throughout the study.

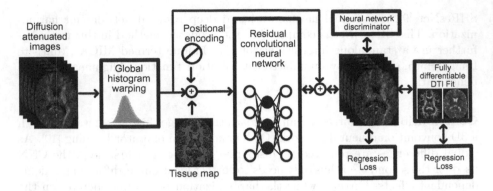

Fig. 1. The harmonization pipeline. The histogram warping removes global biases, while the neural network removes more fine-grained differences. The diffusion-attenuated images, the tissue map and positional encoding are used as input for the neural network. The loss is three-fold, with a standard regression loss, a adversarial discriminator as well as a microstructure loss using a fully differentiable diffusion tensor fitting layer. Image shows single-shell harmonization. Multi-shell harmonization is performed through individual harmonization of each shell.

The dMRI images were preprocessed using FSL. Using TOPUP [1], EPI distortions were corrected by use of the reverse phase-encoded data. Eddy currents were corrected with FSL Eddy [2] after BET brain-extraction on the b0 images of the EPI distortions-corrected data. Using the ANTs CorticalThickness procedure, the brain was extracted from the T1 image and white matter, deep gray matter and cortical gray matter masks were generated.

From the preprocessed dMRI data, FA images were created and used for an affine registration of the T1 image, and transformation of the brain segmentation to the dMRI space. The raw 20-channel head coil dMRI acquisitions were transformed with nearest neighbor interpolation to the 64-channel head coil acquisition by affine registration of the FA images. All registrations were carried out using ANTs [3]. Finally, the diffusion orientations were matched across scanners and subjects through the use of 8th order spherical harmonics, and the dMRI images were divided by the mean b0 image to obtain diffusion-attenuated data. The dataset is publicly available[1].

1.2   Methods

We propose a two-step dMRI harmonization method: First, a histogram warping step removes global bias. Second, a patch-based 3D residual CNN, enhanced by positional encoding and microstructure information, removes locally varying and gradient-dependent bias. This harmonization pipeline is visualized in Fig. 1.

For the histogram warping, the probability density function of intensities of the input diffusion-attenuated data is adapted to the intensities of the target

---

[1] https://www.lfb.rwth-aachen.de/download/hdd_dataset.

data using a non-linear transformation. Specifically, by mapping of the sampled intensities x_i of all diffusion-attenuated images of all training subjects, the cumulative distribution function (CDF) $F_I(x)$ of the input space is transformed to the CDF $F_T(x)$ of the target space. As the sampled intensities may not be the same in both distribution, $F_T(x)$ is linearly interpolated to match the discrete data points in $F_I(x)$. The transformation of values of an unseen acquisition in input space is then obtained according to this mapping. Sampled intensities are again interpolated to match the discrete data points in $F_I(x)$. A similar intensity warping technique termed MICA has been proposed for standalone harmonization of structural MRI data [21].

The second step consists of a CNN that is optimized for the harmonization task with a standard regression loss, as well as through diffusion tensor imaging (DTI) and adversarial loss functions. The network structure is based on the U-Net [18]. However, the bottleneck of this U-Net structure is composed of four residual blocks as found in the ResNet [7]. The path between the input and these residual blocks consists of two convolutional layers with subsequent instance normalization and LeakyReLU activation functions (negative slope of $s = 0.2$). The path between the residual blocks and the outputs is built of the same blocks, with a further $1 \times 1 \times 1$ convolution as a final layer. As the target data is closely related to the input data, the neural network predicts only the difference between the two. In other words, the original input image is added to the output of the network (Fig. 1).

Further, in order to include differences in FA and MD directly into the loss function, we implemented the DTI fitting process [4] using singular value decomposition, as described in [6], in PyTorch to benefit from the PyTorch reverse automatic differentiation system [16]. It is hence possible to back-propagate the difference in harmonized FA and MD and the target FA and MD back to the CNN.

The discriminator network is built by four convolutional layers, each with instance normalization and LeakyReLU activation functions. The architecture is based on the discriminator presented by Zhu et al. [24]. In the four convolutional layers, the number of feature maps is increased in two steps from the initial number of diffusion directions to 128 and then to 256 directions, before using a last convolutional layer with only one feature map as output. A sigmoid activation function maps the output to the value range $(0, 1)$. From this output, a three-dimensional pooling layer is used in order to obtain a single output, which is compared against the target (real or fake) during training.

In total, the generator is trained with a combination of four different loss functions: raw regression loss, FA loss, MD loss, and discriminator loss. At the beginning of the training process, they are scaled to an equal weighting. The initial scaling is kept constant throughout the training.

Other neural network settings are as follows: An Adam optimizer with a learning rate of 0.0001 is used. The patch size is $40 \times 40 \times 40$, with patches randomly cropped. Training is carried out on an Nvidia GeForce 2080 Ti GPU in 600 epochs.

1.3   Comparison Methods

We compare the proposed approach against the adapted MICA method as a standalone harmonization tool, and against the deep learning method SHResNet [9]. SHResNet is a spherical harmonics and spherical convolution based algorithm for multi-shell diffusion harmonization that showed good performance in the MUSHAC challenge [14]. While using a patch-based input, it outputs a single voxel and not a 3D structure, and does not incorporate a fully differentiable DTI fit or an adversarial discriminator for the loss function.

2   Results

To enable an assessment of the performance of our approach, we compared the proposed approach to the SHResNet [9], the standalone MICA method and the baseline difference between the two images. For evaluation, the 24 subjects were randomly split into groups of four subjects for a six-fold cross validation. Cross-validation splits were performed before application of the whole pipeline and were kept consistent for each single step. For MICA and the proposed approach, separate but identical harmonization pipelines were used for each of the three diffusion shells (b = 1000 s/mm², b = 2000 s/mm², b = 3000 s/mm²). The three harmonized shells were merged for the final result.

Three different evaluations were carried out: (1) comparison of the single-shell metrics used for training of the proposed neural network, raw diffusion-attenuation and the DTI-metrics FA and MD, (2) differences in multi-shell microstructure modeling with NODDI, (3) perception-based similarity using the LPIPS metric. For all evaluations, significance of superior results was tested with the Wilcoxon signed-rank test on a per-subject basis against the best result of either baseline or the second best approach, i.e., it was tested if a significant number of subjects showed better harmonization results than with methods previously available.

First, raw diffusion-attenuated harmonization performance as well as DTI-derived metrics were compared on a per-shell basis. The mean squared error (MSE) of the diffusion-attenuated signal was evaluated individually for the three different b-value shells. FA and MD maps were also generated individually for each diffusion shell. These results can be seen in Table 1.

Our proposed method outperformed all other approaches on the three derived metrics, and is the only evaluated method that outperformed the baseline on all measurements. However, for the raw diffusion-attenuated errors, the SHResNet approach achieved the best results. Exemplary harmonization results are visualized in Fig. 2.

Second, we fitted NODDI [22] to the complete multi-shell data, and used it to quantify the intra head coil effects and the harmonization performance on microstructural estimates. The MSE of the orientation dispersion index (ODI) of neurites and the neurite density index (NDI) were derived for the different approaches. Further, the effects on the fiber direction was compared through the mean orientation of the Watson distribution fitted to the neurite compartment.

Table 1. Difference between the two head coil acquisitions before and after harmonization for the three b-value shells. Diffusion attenuation (raw data), FA, and MD differences were measured in mean squared error (MSE). The best results are marked in **bold**. Significantly ($p < 0.005$) better results than the second best approach are further labeled with an asterisk*.

	Before	SHResNet	MICA	Proposed
Raw data b1000 (in 10^{-3})	9.39 ± 0.79	$\mathbf{5.82 \pm 0.49}^*$	8.85 ± 0.75	8.58 ± 7.07
Raw data b2000 (in 10^{-3})	7.27 ± 0.55	$\mathbf{4.33 \pm 0.32}^*$	6.82 ± 0.52	6.62 ± 0.50
Raw data b3000 (in 10^{-3})	6.18 ± 0.49	$\mathbf{3.71 \pm 0.32}^*$	5.71 ± 0.44	5.48 ± 0.41
MD b1000 (in 10^{-9})	29.6 ± 6.54	33.9 ± 7.42	29.5 ± 6.60	$\mathbf{28.2 \pm 6.36}^*$
MD b2000 (in 10^{-9})	9.01 ± 1.75	9.83 ± 1.78	8.71 ± 1.62	$\mathbf{8.40 \pm 1.53}^*$
MD b3000 (in 10^{-9})	4.50 ± 0.75	4.75 ± 7.00	3.92 ± 0.58	$\mathbf{3.80 \pm 0.60}^*$
FA b1000 (in 10^{-3})	8.09 ± 0.88	9.34 ± 1.06	7.87 ± 0.84	$\mathbf{7.62 \pm 0.82}^*$
FA b2000 (in 10^{-3})	6.95 ± 0.62	8.45 ± 0.84	6.99 ± 0.60	$\mathbf{6.74 \pm 0.59}^*$
FA b3000 (in 10^{-3})	6.79 ± 0.51	8.12 ± 0.76	6.84 ± 0.51	$\mathbf{6.53 \pm 0.50}^*$

Table 2. Harmonization effects on multi-shell NODDI microstructural estimates. ODI: orientation dispersion index, NDI: neurite density index. The differences between target image and original or harmonized image are given in MSE. The best results are marked in **bold**. Significantly ($p < 0.005$) better results than the second best approach are further labeled with an asterisk*.

	Before	SHResNet	MICA	Proposed
ODI (in 10^{-3})	0.852 ± 0.17	1.008 ± 0.18	0.874 ± 0.17	$\mathbf{0.845 \pm 0.17}^*$
NDI (in 10^{-3})	0.954 ± 0.17	0.897 ± 0.24	0.847 ± 0.16	$\mathbf{0.792 \pm 0.14}^*$
Fiber direction	0.235 ± 0.031	0.237 ± 0.025	0.235 ± 0.031	$\mathbf{0.231 \pm 0.30}$

These NODDI results are shown in Table 2. Our algorithm significantly improves image similarities on both NDI and ODI ($p < 0.005$), with effects especially pronounced for the NDI. The similarity of the fiber orientations between the two acquisitions is also improved, but the improvements remain non-significant.

Third, as simple signal smoothing already leads to improvements in most signal comparison evaluations [14], we evaluated the raw diffusion-attenuated image differences with a perception-based metric. The LPIPS metric was chosen, as it closely matches human perception [23]. It was calculated individually over all diffusion-attenuated images, and averaged over images and subjects. Our proposed approach performs significantly better than the baseline and the two comparison algorithms for all three b-value shells (Table 3). It should be noted, that, due to lower signal intensities, the MSE errors on raw data, FA, and MD decrease with increasing b-value (Table 1). Meanwhile, the LPIPS scores increase with the b-value, indicating higher dissimilarities between the images for stronger diffusion weightings.

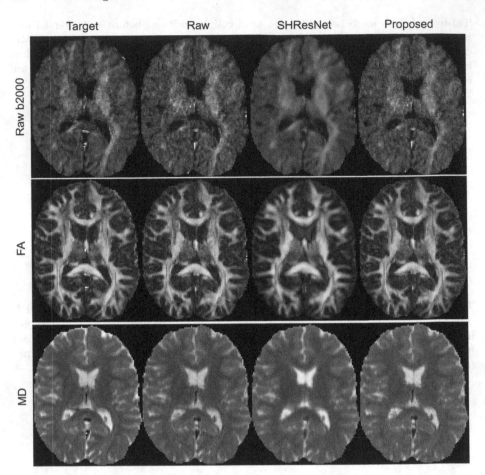

Fig. 2. Exemplary target and input image, together with the harmonization results using the SHResNet method and our proposed method. The upper row shows a slice of a raw diffusion-attenuated image (b2000), the middle row the derived FA values, and the lower row the derived MD values.

In Fig. 3, we visualize exemplary effects of the three different harmonization approaches. In general, MICA and the proposed method generally distort the signal much less than the SHResNet. By design, the MICA method does not include location-dependent effects, and cannot smooth the signal. In contrast, our proposed method includes a location-dependent effect. An example of this effect is marked with green circles in the figure.

Table 3. Perception metrics on raw diffusion-attenuated data. LPIPS: Learned Perceptual Image Patch Similarity. Lower scores denote more similar image patches. The best results are marked in **bold**. Significantly (p < 0.005) better results than the second best approach are further labeled with an asterisk*.

	Before	SHResNet	MICA	Proposed
LPIPS b1000 (in 10^{-2})	4.91 ± 0.44	10.46 ± 0.92	4.68 ± 0.42	$\mathbf{4.52 \pm 0.40}^*$
LPIPS b2000 (in 10^{-2})	6.87 ± 0.58	15.41 ± 1.29	6.70 ± 0.57	$\mathbf{6.46 \pm 0.54}^*$
LPIPS b3000 (in 10^{-2})	8.32 ± 0.75	18.87 ± 1.61	8.15 ± 0.73	$\mathbf{7.78 \pm 0.71}^*$

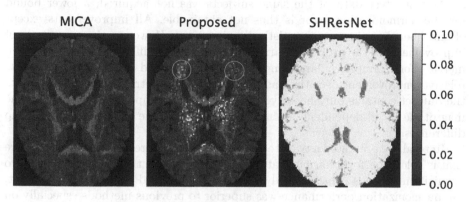

Fig. 3. Absolute values of changes made by the harmonization algorithm, maximum intensity projection for one subject over all $b = 2000 \frac{s}{mm^2}$ acquisitions. Left: MICA. Middle: The proposed method. Right: SHResNet approach. The green circles showcase location-dependent effects: the area in the left circle shows much stronger harmonization effects than the contralateral side. (Color figure online)

3   Discussion and Conclusion

The proposed harmonization method shows superior results compared to the other methods for the DTI metrics FA and MD, for the NODDI microstructural estimates, and for the perceptual similarity. Moreover, it is the only one of the evaluated methods that consistently obtains better harmonization results than the baseline. The DTI results are especially important as FA and MD maps are the most used markers in dMRI based clinical studies. Furthermore, our model enables an improvement of harmonized multi-shell microstructural estimates and fiber direction, which reduces biases in tractography between raw and harmonized data. The present improvements are, thus, crucial for an inclusion of harmonization methods in practical applications.

However, compared to the SHResNet, our method under-performs on the raw diffusion-attenuated data. This difference in performance on raw- and derived metrics can be explained by smoothing effects of the different approaches. The SHResNet strongly smoothes the output, which leads to superior MSE results on the raw values. In contrast, by design, the MICA method does not allow any smoothing of values. Similarly, the adversarial discriminator used in the training of the CNN discourages smoothing effects of the neural network, as the output images need to be indistinguishable from the real target images. Thus, if only the MSE on the raw data is used as evaluation metric, an algorithm with strong smoothing will lead to superficially superior performance. Nonetheless, such superficially superior performance actually introduces further bias, as harmonized images exhibit different properties (i.e., smoothing) than the target dataset, with adverse consequences for diffusivity-derived metrics.

In the end, it remains unclear how optimal the final harmonization result is. As test-retest data of the same subjects was not acquired, a lower bound for the harmonization error is thus not obtainable. All improvements except of the fiber direction were statistically significant. However, the magnitude of improvements of the proposed harmonization method differs with the evaluation metric. For example, a large improvement is achieved for NDI, whereas the relative improvement over the baseline for the ODI metric is small. Further, as the dataset contains only healthy young adults, it is impossible to compare the magnitude of coil-dependent effects against the effect size of pathologies or group differences.

In total, we presented a two-step hybrid known operator 3D CNN architecture that is able to perform harmonization in a setting where the difference in the two acquisition settings consisted only of mismatching head coils. On this dataset, the harmonization performance was superior to previous methods especially on derived metrics.

Acknowledgements. This work was funded by the German Research Foundation (Deutsche Forschungsgemeinschaft, DFG) under project number 269953372 (IRTG 2150) and project number 417063796, and supported by the Brain Imaging Facility of the Interdisciplinary Center for Clinical Research (IZKF) Aachen within the Faculty of Medicine at RWTH Aachen University.

References

1. Andersson, J.L., Skare, S., Ashburner, J.: How to correct susceptibility distortions in spin-echo echo-planar images: application to diffusion tensor imaging. Neuroimage **20**, 870–888 (2003). https://doi.org/10.1016/S1053-8119(03)00336-7
2. Andersson, J.L., Sotiropoulos, S.N.: An integrated approach to correction for off-resonance effects and subject movement in diffusion MR imaging. Neuroimage **125**, 1063–1078 (2016). https://doi.org/10.1016/j.neuroimage.2015.10.019
3. Avants, B., Tustison, N., Song, G.: Advanced normalization tools (ANTS). Insight J. 1–35 (2008). https://doi.org/10.54294/uvnhin
4. Basser, P., Mattiello, J., LeBihan, D.: MR diffusion tensor spectroscopy and imaging. Biophys. J . **66**(1), 259–267 (1994). https://doi.org/10.1016/S0006-3495(94)80775-1
5. Cetin Karayumak, S., Kubicki, M., Rathi, Y.: Harmonizing diffusion MRI data across magnetic field strengths. In: Frangi, A.F., Schnabel, J.A., Davatzikos, C., Alberola-López, C., Fichtinger, G. (eds.) MICCAI 2018. LNCS, vol. 11072, pp. 116–124. Springer, Cham (2018). https://doi.org/10.1007/978-3-030-00931-1_14

6. Hasan, K.M., Parker, D.L., Alexander, A.L.: Comparison of gradient encoding schemes for diffusion-tensor MRI. J. Magn. Reson. Imaging **13**(5), 769–780 (2001). https://doi.org/10.1002/jmri.1107
7. He, K., Zhang, X., Ren, S., Sun, J.: Deep residual learning for image recognition. In: IEEE Conference on Computer Vision and Pattern Recognition (CVPR), pp. 770–778 (2016)
8. Huynh, K.M., Chen, G., Wu, Y., Shen, D., Yap, P.T.: Multi-site harmonization of diffusion MRI data via method of moments. IEEE Trans. Med. Imaging **38**(7), 1599–1609 (2019). https://doi.org/10.1109/TMI.2019.2895020
9. Koppers, S., Bloy, L., Berman, J.I., Tax, C.M.W., Edgar, J.C., Merhof, D.: Spherical harmonic residual network for diffusion signal harmonization. In: Bonet-Carne, E., Grussu, F., Ning, L., Sepehrband, F., Tax, C.M.W. (eds.) MICCAI 2019. MV, pp. 173–182. Springer, Cham (2019). https://doi.org/10.1007/978-3-030-05831-9_14
10. Maier, A., Köstler, H., Heisig, M., Krauss, P., Yang, S.H.: Known operator learning and hybrid machine learning in medical imaging - a review of the past, the present, and the future. CoRR abs/2108.04543 (2021)
11. Mirzaalian, H., et al.: Multi-site harmonization of diffusion MRI data in a registration framework. Brain Imaging Behav. **12**(1), 284–295 (2017). https://doi.org/10.1007/s11682-016-9670-y
12. Mirzaalian, H., et al.: Harmonizing diffusion MRI data across multiple sites and scanners. In: Navab, N., Hornegger, J., Wells, W.M., Frangi, A.F. (eds.) MICCAI 2015. LNCS, vol. 9349, pp. 12–19. Springer, Cham (2015). https://doi.org/10.1007/978-3-319-24553-9_2
13. Moyer, D., Ver Steeg, G., Tax, C.M.W., Thompson, P.M.: Scanner invariant representations for diffusion MRI harmonization. Magn. Reson. Med. **84**(4), 2174–2189 (2020). https://doi.org/10.1002/mrm.28243
14. Ning, L., et al.: Cross-scanner and cross-protocol multi-shell diffusion MRI data harmonization: algorithms and results. Neuroimage **221**, 117128 (2020). https://doi.org/10.1016/j.neuroimage.2020.117128
15. Panman, J.L., et al.: Bias introduced by multiple head coils in MRI research: an 8 channel and 32 channel coil comparison. Front. Neurosci. **13**, 729 (2019). https://doi.org/10.3389/fnins.2019.00729
16. Paszke, A., et al.: PyTorch: an imperative style, high-performance deep learning library. In: Wallach, H., Larochelle, H., Beygelzimer, A., d'Alché-Buc, F., Fox, E., Garnett, R. (eds.) Advances in Neural Information Processing Systems, vol. 32, pp. 8024–8035. Curran Associates, Inc. (2019)
17. Pinto, M.S., et al.: Harmonization of brain diffusion MRI: concepts and methods. Front. Neurosci. **14**, 396 (2020). https://doi.org/10.3389/fnins.2020.00396
18. Ronneberger, O., Fischer, P., Brox, T.: U-Net: convolutional networks for biomedical image segmentation. In: Navab, N., Hornegger, J., Wells, W.M., Frangi, A.F. (eds.) MICCAI 2015. LNCS, vol. 9351, pp. 234–241. Springer, Cham (2015). https://doi.org/10.1007/978-3-319-24574-4_28
19. St-Jean, S., Viergever, M.A., Leemans, A.: Harmonization of diffusion MRI data sets with adaptive dictionary learning. Hum. Brain Mapp. **41**(16), 4478–4499 (2020). https://doi.org/10.1002/hbm.25117
20. Tax, C.M., et al.: Cross-scanner and cross-protocol diffusion MRI data Harmonisation: a benchmark database and evaluation of algorithms. Neuroimage **195**, 285–299 (2019). https://doi.org/10.1016/j.neuroimage.2019.01.077
21. Wrobel, J., et al.: Intensity warping for multisite MRI harmonization. Neuroimage **223**, 117242 (2020). https://doi.org/10.1016/j.neuroimage.2020.117242

22. Zhang, H., Schneider, T., Wheeler-Kingshott, C.A., Alexander, D.C.: NODDI: practical in vivo neurite orientation dispersion and density imaging of the human brain. Neuroimage **61**(4), 1000–1016 (2012)
23. Zhang, R., Isola, P., Efros, A.A., Shechtman, E., Wang, O.: The unreasonable effectiveness of deep features as a perceptual metric. In: CVPR (2018)
24. Zhu, J.Y., Park, T., Isola, P., Efros, A.A.: Unpaired image-to-image translation using cycle-consistent adversarial networks. In: IEEE International Conference on Computer Vision (ICCV) (2017)

CCRL: Contrastive Cell Representation Learning

Ramin Nakhli[1]⬤, Amirali Darbandsari[2(✉)], Hossein Farahani[1],
and Ali Bashashati[1,3]⬤

[1] School of Biomedical Engineering, University of British Columbia,
Vancouver, Canada
{ramin.nakhli,h.farahani,ali.bashashati}@ubc.ca
[2] Department of Electrical and Computer Engineering,
University of British Columbia, Vancouver, Canada
a.darbandsari@ubc.ca
[3] Department of Pathology and Laboratory Medicine,
University of British Columbia, Vancouver, Canada

Abstract. Cell identification within the H&E slides is an essential pre-requisite that can pave the way towards further pathology analyses including tissue classification, cancer grading, and phenotype prediction. However, performing such a task using deep learning techniques requires a large cell-level annotated dataset. Although previous studies have investigated the performance of contrastive self-supervised methods in tissue classification, the utility of this class of algorithms in cell identification and clustering is still unknown. In this work, we investigated the utility of Self-Supervised Learning (SSL) in cell clustering by proposing the Contrastive Cell Representation Learning (CCRL) model. Through comprehensive comparisons, we show that this model can outperform all currently available cell clustering models by a large margin across two datasets from different tissue types. More interestingly, the results show that our proposed model worked well with a few number of cell categories while the utility of SSL models has been mainly shown in the context of natural image datasets with large numbers of classes (e.g., ImageNet). The unsupervised representation learning approach proposed in this research eliminates the time-consuming step of data annotation in cell classification tasks, which enables us to train our model on a much larger dataset compared to previous methods. Therefore, considering the promising outcome, this approach can open a new avenue to automatic cell representation learning.

Keywords: Self-supervised learning · Contrastive learning · Cell representation learning · Cell clustering

1 Introduction

Cells are the main components that determine the characteristics of tissues and, through mutual interactions, can play an important role in many aspects including tumor progression and response to therapy [19,23,30]. Therefore, cell identification can be considered as the first and essential building block for many

L. Karlinsky et al. (Eds.): ECCV 2022 Workshops, LNCS 13807, pp. 397–407, 2023.
https://doi.org/10.1007/978-3-031-25082-8_26

tasks, including but not limited to tissue identification, slide classification, T-cell infiltrating lymphocytes analysis, cancer grade prediction, and clinical phenotype prediction [21,25]. In clinical practice, manual examination of the Whole Slide Images (WSI), multi-gigapixel microscopic scans of tissues stained with Hematoxylin & Eosin (H&E), is the standard and widely available approach for cell type identification [1]. However, due to the large number of cells and their variability in texture, not only is the manual examination time-consuming and expensive, but it also introduces intra-observer variability [5]. Although techniques such as immunohistochemistry (IHC) staining can be used to identify various cell types in a tissue, they are expensive, are not routinely performed for clinical samples, and require a deep biological knowledge for biomarker selection [26].

With the exponential growth of machine learning techniques in recent years, especially deep learning, computational models have been proposed to accelerate the cell identification process [15,29]. However, these models need large cell-level annotated datasets to be trained on [2,15]. Collecting such datasets is time-consuming and costly as the pathologists have to annotate tens of thousands of cells present in H&E slides, and this procedure has to be carried out for any new tissue type. To mitigate this, several research groups have moved to crowdsourcing [2]. Nevertheless, providing such pipelines is still difficult in practice. Therefore, all the aforementioned problems set a strong barrier to the important cell identification task which is a gateway to more complex analyses.

In this paper, using state-of-the-art deep learning methods, we investigate the utility of contrastive self-supervised learning to obtain representations of cells without any kind of supervision. Moreover, we show that clusters of these representations are associated with specific types of cells enabling us to apply the proposed model for cell identification on routine H&E slides in a massive scale. Furthermore, our results demonstrate that our trained model can outperform all the existing counterparts.

Therefore, the contributions of this work can be summarized as: 1) the first work to study the utility of self-supervised learning in cell representation learning in H&E images; 2) introducing a novel framework for this purpose; 3) outperforming all the existing unsupervised baselines with a large margin.

2   Previous Works

2.1   Self-supervised Learning

Self-Supervised Learning (SSL) is a technique to train a model without any human supervision in a way that the generated representations capture the semantics of the image. Learning from the pseudo-labels generated by applying different types of transformations has been a popular approach in the early ages of this technique. Learning local position of image patches [28], rotation angle prediction [22], and color channel prediction [35] are some of the common transformations used to this end.

Recent studies have shown that using similar transformations in a contrastive setting can significantly enhance the quality of the representations to an extent that makes it possible to even outperform supervised methods after fine-tuning [10]. In this setting, models pull the embeddings of two augmentations of the same image (one called query and the other called positive sample) together while they push the embedding of other images (negative samples) as far as possible. SimCLR [7,8] proposes using the same encoder network to encode the query, positive, and negative samples while MoCo [10,17] introduces using the momentum encoder to encode the positive and negative samples, which is updated by the weights of the query image encoder. On the other hand, BYOL [16] and DINO [6] remove the need for negative samples.

In addition to the applications of the self-supervised learning in image classification [6,8,10] and object detection [32], several papers have shown benefits of this approach in medical imaging. Azizi et al. [3] show their model can outperform supervised models across multiple medical imaging datasets by pulling the embeddings of two different views from the same patient under different conditions, and [33] applies self-supervised learning for cell detection. while [33] applies self-supervised learning for cell segmentation, [11] uses SimCLR to learn representations on patches of H&E slides. Furthermore, they show that increasing the dataset size and variety improves the performance of the model on the patch classification task. In contrast, [34] shows that using another type of contrastive learning architecture reduces the final performance of patch classification compared to the pre-trained models on the ImageNet dataset [14]. Although some of the aforementioned studies investigate the influence of contrastive self-supervised learning in the context of histopathology patch classification, to the best of our knowledge, the applications of self-supervised techniques for cell labelling (as opposed to patch classification) are largely ignored. Furthermore, the contradictory results of the aforementioned studies (one showing the superiority of SSL over the ImageNet pre-trained model, and the other showing the opposite) warrant further investigation of these models in the context of cell clustering and labelling in histopathology.

2.2 Cell Classification in Histopathology

The utility of machine learning algorithms in cell classification is an active and important area of research. Earlier works have been mainly focused on extracting hand-crafted features from cell images and applying machine learning classification models to perform this task. For example, using the H&E images, [27] extracts cytological features from cells and applies Support Vector Machines (SVM) to separate the cell types. [13] uses the size, color, and texture of the cells to assign a score to each cell based on which they can classify the cell. Later works combined the deep and hand-crafted features to improve the accuracy of cell classification [12]. However, recent studies (for example, [15,29]) show that cell classification accuracy can significantly improve, solely based on deep learning-based features.

Although these studies have shown promising performance, they require a large annotated dataset which is difficult to collect. A few recent studies have focused on unsupervised cell classification to address this problem. For example,

Hue et al. [20] take advantage of the InfoGAN [9] design to provide a categorical embedding for images, based on which they can differentiate between cell types and Vununu et al. [31] propose using Deep Convolution Auto-encoder (DCAE) which learns feature embeddings by performing image reconstruction and clustering at the same time. However, all of these works are focused on only one tissue type. In this study, we present the first contrastive self-supervised cell representation learning framework using H&E images and show that this design can consistently outperform all the currently available baselines across two tissue types.

3   Method

Given an image of a cell, our objective is to learn a robust representation for the cell which can be used for down-stream tasks such as cell clustering. Figure 1 depicts an overview of our proposed self-supervised method, called CCRL (Contrastive Cell Representation Learning). The main goal of our framework is to provide the same representation for different views of a cell. More specifically, this framework consists of two branches, query and key, which work on different augmentations of the same image.

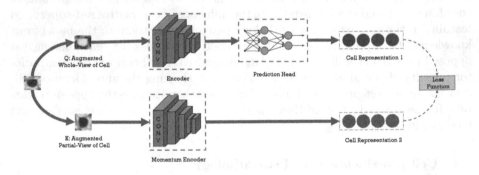

Fig. 1. Overview of the framework

In this design, cell embeddings are learned by pulling the embeddings of two augmentations of the same image together, while the representations of other images are pushed away. Consider the input image batch of $X = x_1, x_2, ..., x_N$ where x_i is a small crop of the H&E image around a cell in a way that it only includes that specific cell. Two different sets of augmentations are applied to X to generate $Q = \{q_i | i = 1, ..., N\}$ and $K = \{k_i | i = 1, ..., N\}$ where N is the batch size. These sets are called query and key, respectively, and q_i and k_j are the augmentations of the same image if and only if $i = j$. The query batch is encoded using a backbone model, a neural network of choice, while the keys are encoded using a momentum encoder, which has the same architecture as the backbone. Using a momentum encoder can be viewed as keeping an ensembling of the query model throughout its training, providing more robust representations. This momentum encoder is updated using the Eq. 1 in which θ_k^t is the parameter of momentum encoder at time t, m is the momentum factor, and θ_q^t is the parameter of the backbone at time t

$$\theta_k^t = m\theta_k^{t-1} + (1 - m)\theta_q^t. \tag{1}$$

Consequently, the obtained query and key representations are passed through separate Multi-Layer Perceptron (MLP) layers called projector heads. Although the query projector is trainable, the key projector is updated with momentum using the weight of the query projector head. Similar to the momentum encoder, the key projector can be considered as an ensembling of that of the query branch. We have restricted these models to be 2-layer MLPs with the input size of 512, hidden size of 128, and output size of 64. In addition to the projector head, we use an extra MLP on the query side of the framework, called prediction head. This extra network enables us to provide asymmetricity in the design of our model (as apposed to using a deeper projector head which keeps the design symmetric), providing more flexible representations on the query branch for competing with the ensembled representations coming from the key branch. This network is a 2-layer MLP with input, hidden, and output sizes of 64, 32, and 64, respectively. Similar to the last fully-connected layers of a conventional classification network, the projection and prediction heads provide more representation power to the model.

Finally, the models are trained using the Eq. 2, pulling the positive and pushing the negative embeddings

$$L_{q_i}^{cell} = -\log \frac{\exp \frac{\|f_q(q_i)\|^2 \cdot \|f_k(k_i)\|^2}{\tau}}{\sum_{j=0}^{N+Q} \exp \frac{\|f_q(q_i)\|^2 \cdot \|f_k(k_j)\|^2}{\tau}}. \tag{2}$$

In this equation, τ is the temperature which controls the sharpness of the similarity distribution, Q is the number of items stored in the queue from the key branch, $\|x\|^2$ is the second-order normalization of x, f_q is the equal function for the combination of the backbone, query projection head, and query prediction head, and f_k shows the equal function for the momentum encoder and the key projection head.

As mentioned above, we use an external memory bank to store the processed key representations. The stored representation will be used in the contrastive setting as negative samples, and they are limited to $65,536$ samples throughout the training.

Additionally, we incorporated a local-global connection technique to ensure that the model is always focusing on the whole view representation of the cell throughout the training process. To this end, only one of the two augmentation pipelines includes cropping operations. This pipeline generates local regions of the cell image while the images generated by the other augmentation pipeline are global, containing the whole-cell view. The rest of the operations are the same in both pipelines, and they include color jitter (brightness of 0.4, contrast of 0.4, saturation of 0.4, and hue of 0.1), gray-scale conversion, Gaussian blur (with a random sigma between 0.1 and 2.0), horizontal and vertical flip, and rotation (randomly selected between 0 to 180°).

At inference time, cell embeddings were generated from the trained momentum encoder and were clustered by applying the K-means algorithm. It is worth mentioning that one can use either the encoder or momentum encoder for embedding generation; however, the momentum encoder provides more robust representations since it aggregates the learned weights of the encoder network from all of the training steps (an ensembling version of the encoder throughout training). We refer to this technique as ensembling in the rest of this article.

4   Experiments

4.1   Evaluation Metrics

As the main goal of this work was to provide a framework for clustering of cell types, we evaluated the performance of the models using Adjusted Mutual Information (AMI), Adjusted Random Index (ARI), and Purity of the identified cell clusters by the model and the ground truth labels.

AMI captures the agreement between two sets of assignments using the amount of mutual information that exists between these sets. However, it is adjusted to mitigate the effect of chance in the score.

ARI is the chance adjusted form of the Rand Index, which calculates the quality of the clustering based on the number of instance pairs.

Purity measures how the samples within each cluster are similar to each other. In other words, it demonstrates if each cluster is a mixture of different classes.

4.2   Datasets

To demonstrate the utility and performance of our proposed model, we used two publicly available datasets (CoNSeP [15] and NuCLS [2]) representing two different tissue types with cell-level annotations. Although the annotations were not used in the training step, we leveraged them to evaluate the performance of different models on the test set.

The CoNSeP dataset consisted of 41 H&E tiles from colorectal tissues extracted from 16 whole slide images of a single patient. All tiles were in 40× magnification scale with the size of 1,000 × 1,000 pixels. Cell types included 7 different categories of normal epithelial, malignant epithelial, inflammatory, endothelial, muscle, fibroblast, and miscellaneous. However, as suggested by the original paper, normal and malignant epithelial were grouped into the epithelial category, and the muscle, fibroblast, and endothelial cells were grouped into the spindle-shaped category. Therefore, the final 4 groups included epithelial, oval-shaped, inflammatory, and miscellaneous.

The NuCLS dataset included 1744 H H&E tiles of breast cancer images from the TCGA dataset collected from 18 institutions. The tiles had different sizes, but they were roughly 300×300 pixels. There were 12 different cell types available in this dataset: tumor, fibroblast, lymphocyte, plasma, macrophage, mitotic, vascular endothelium, myoepithelium, apoptotic body, neutrophil, ductal epithelium, and eosinophil. However, as suggested by the paper, these subtypes were grouped together into 5 superclasses including tumor (containing tumor and mitotic cells), stromal (containing fibroblast, vascular endothelium, and macrophage), sTILs (containing lymphocyte and plasma cells), apoptotic cells, and others.

Details of each dataset can be found in Table 1.

Table 1. Dataset details

	CoNSeP	NuCLS
Type count	4	5
Cell count	24,319	51,986
Tile count	41	1,744
Tile size	$1,000 \times 1,000$	300×300
Tissue type	Colorectal	Breast

4.3 Data Preparation

The aforementioned datasets included patch-level images, while we required cell-level ones for the training of the model. To generate such data, we used the instance segmentation provided in each of the datasets to find cells and crop a small box around them. We adopted an adaptive window size to extract these images, whose size was equal to twice the size of the cell in CoNSeP and equal to the size of the cell in the NuCLS dataset. The images were resized to 32×32 pixels before being fed into our proposed framework.

4.4 Implementation Details

The code was implemented in Pytorch, and the model was run on a V100 GPU. The batch size was set to 1024, the queue size to 65536, and pre-activated ResNet18 [18] was used for the backbone. The model was trained using Adam optimizer for 500 epochs with a starting learning rate of 0.001, a cosine learning rate scheduler, and a weight decay of 0.0001. We also adopted a 10-epoch warm-up step. The momentum factor in the momentum encoders was set to 0.999, and the temperature was set to 0.07.

4.5 Results

The results of unsupervised clustering of CCRL can be found in Table 2 as well as that of the baselines. We compared the performance of our model with five different baseline and state-of-the-art models. The pre-trained ImageNet used the weights trained on the ImageNet dataset to generate the cell embeddings. The second baseline model used morphological features to produce a 30-dimensional feature vector, consisting of geometrical and shape attributes [4]. The third baseline method utilized the Manual Features (MF), a combination of Scale-Invariant Feature Transform (SIFT) and Local Binary Patterns (LBP) features, proposed by [20]. We also compared the results of our model with two state-of-the-art unsupervised cell clustering methods. The DCAE [31] model adopted a deep convolution auto-encoder model alongside a clustering layer to learn cell embeddings by preforming an image reconstruction task. Also, the authors of [20] developed a generative adversarial model for cell clustering by increasing the mutual information between the cell representation and a categorical noise vector.

Table 2. Unsupervised clustering performance comparison.

Model	CoNSeP			NuCLS		
	AMI	ARI	Purity	AMI	ARI	Purity
Pre-trained ImageNet	7.3%	7%	42.7%	9.3%	7.8%	56.7%
Morphological [4]	12.7%	1.3%	48.8%	21.1%	18.8%	66.1%
Manual Features [20]	9.5%	6.4%	45.5%	11.25%	7.8%	56.2%
Auto-Encoder [31]	10.1%	7.3%	50.5%	8.3%	7.2%	56.8%
InfoGAN [20]	14.8%	15.7%	**58.4%**	14%	12.6%	62%
CCRL (Ours)	**24.2%**	**21.7%**	51.8%	**22.8%**	**24%**	**68.3%**

As can be seen in Table 2, our model can outperform its counterparts with a large margin in terms of different clustering metrics across all datasets.

4.6 Ablation Study

Ablation studies were performed on three most important components of our framework: 1) local-global connection technique; 2) inference with the ensemble model; 3) query prediction head.

Tables 3 demonstrates the effect of ablation of each component. Based on these experiments, all components are essential for learning effective cell representations regardless of the tissue type. In this part, we used NuCLS dataset for ablation, as it has been collected from multiple patients and diverse locations.

Table 3. Ablation study. First and second performances are highlighted and underlined, respectively.

	AMI	ARI	Purity
w/o Local-Global	21.9%	20.4%	**69.7%**
w/o Ensembling	21.1%	19.2%	67.2%
w/o Prediction Head	22%	20.9%	68.1%
w/ All	**22.8%**	**24%**	68.3%

5 Discussion and Conclusion

Cell identification is a gateway to many complex tissue analysis applications. However, due to the large number of cells in H&E slides, manual execution of such a task is very time-consuming and resource-intensive. Although several research studies have provided machine learning models to classify cells in an automatic manner, they still require a large dataset that is manually annotated. In this paper, we investigated the utility of self-supervised learning in the context of cell representation learning by designing a self-supervised model with designated architecture for the task of cell representation learning. The quality of the representations was measured based on the clustering performance, by applying the K-means algorithm on top of these representations and measuring the cluster enrichment in specific cell types. Our experiments confirm that the SSL training improves the clustering metrics compared to currently available unsupervised methods.

It is worthwhile to mention that SSL frameworks are mostly evaluated on natural images (e.g., ImageNet dataset), which include a large number of categories. However, in our case, the number of classes (i.e., cell types) is extremely small (4 and 5 classes for the CoNSeP and NuCLS datasets, respectively, versus 1000 classes for ImageNet). This means that, on average, 25% and 20% of the negative samples for the CoNSeP and NuCLS dataset are false-negatives, respectively, while this ratio is only 0.1% for the ImageNet dataset. Therefore, our findings show that the proposed SSL framework can operate well when a small number of classes or categories exists.

This paper is the first attempt to apply contrastive self-supervised learning to cell identification in H&E images. The proposed model enables us to achieve robust cell representation using an enormous amount of unlabeled data which can simply be generated by scanning routine H&E stained slides in the clinical setting. Furthermore, due to the unsupervised learning nature of the framework, the proposed model has the potential to identify novel cell types that may have been overlooked by pathologists. In addition to the above-mentioned benefits that an SSL framework could provide in the context of cell classification, these models are also robust to long-tail distributions in the data [24]; hence addressing the common issue of rare cell populations in pathology (e.g., tumor budding, mitotic figures). Therefore, we hope that this work motivates researchers and serves as a step towards more utilization of unsupervised learning in pathology applications, especially in the context of cell-level information representation.

References

1. Alturkistani, H.A., Tashkandi, F.M., Mohammedsaleh, Z.M.: Histological stains: a literature review and case study. Glob. J. Health Sci. **8**(3), 72 (2016)
2. Amgad, M., et al.: NuCLS: a scalable crowdsourcing, deep learning approach and dataset for nucleus classification, localization and segmentation. arXiv preprint arXiv:2102.09099 (2021)
3. Azizi, S., et al.: Big self-supervised models advance medical image classification. arXiv preprint arXiv:2101.05224 (2021)
4. Bhaskar, D., et al.: A methodology for morphological feature extraction and unsupervised cell classification. bioRxiv, p. 623793 (2019)
5. Boyle, P., Langman, M.J.: ABC of colorectal cancer: epidemiology. BMJ **321**(Suppl. S6) (2000)
6. Caron, M., et al.: Emerging properties in self-supervised vision transformers. arXiv preprint arXiv:2104.14294 (2021)
7. Chen, T., Kornblith, S., Norouzi, M., Hinton, G.: A simple framework for contrastive learning of visual representations. In: International Conference on Machine Learning, pp. 1597–1607. PMLR (2020)
8. Chen, T., Kornblith, S., Swersky, K., Norouzi, M., Hinton, G.: Big self-supervised models are strong semi-supervised learners. arXiv preprint arXiv:2006.10029 (2020)
9. Chen, X., Duan, Y., Houthooft, R., Schulman, J., Sutskever, I., Abbeel, P.: Info-GAN: interpretable representation learning by information maximizing generative adversarial nets. In: Proceedings of the 30th International Conference on Neural Information Processing Systems, pp. 2180–2188 (2016)
10. Chen, X., Fan, H., Girshick, R., He, K.: Improved baselines with momentum contrastive learning. arXiv preprint arXiv:2003.04297 (2020)
11. Ciga, O., Xu, T., Martel, A.L.: Self supervised contrastive learning for digital histopathology. Mach. Learn. Appl. **7**, 100198 (2022)
12. Cruz-Roa, A.A., Arevalo Ovalle, J.E., Madabhushi, A., González Osorio, F.A.: A deep learning architecture for image representation, visual interpretability and automated basal-cell carcinoma cancer detection. In: Mori, K., Sakuma, I., Sato, Y., Barillot, C., Navab, N. (eds.) MICCAI 2013. LNCS, vol. 8150, pp. 403–410. Springer, Heidelberg (2013). https://doi.org/10.1007/978-3-642-40763-5_50
13. Dalle, J.R., Li, H., Huang, C.H., Leow, W.K., Racoceanu, D., Putti, T.C.: Nuclear pleomorphism scoring by selective cell nuclei detection. In: WACV (2009)
14. Deng, J., Dong, W., Socher, R., Li, L.J., Li, K., Fei-Fei, L.: ImageNet: a large-scale hierarchical image database. In: 2009 IEEE Conference on Computer Vision and Pattern Recognition, pp. 248–255. IEEE (2009)
15. Graham, S., et al.: Hover-net: simultaneous segmentation and classification of nuclei in multi-tissue histology images. Med. Image Anal. **58**, 101563 (2019)
16. Grill, J.B., et al.: Bootstrap your own latent: a new approach to self-supervised learning. arXiv preprint arXiv:2006.07733 (2020)
17. He, K., Fan, H., Wu, Y., Xie, S., Girshick, R.: Momentum contrast for unsupervised visual representation learning. In: Proceedings of the IEEE/CVF Conference on Computer Vision and Pattern Recognition, pp. 9729–9738 (2020)
18. He, K., Zhang, X., Ren, S., Sun, J.: Identity mappings in deep residual networks. In: Leibe, B., Matas, J., Sebe, N., Welling, M. (eds.) ECCV 2016. LNCS, vol. 9908, pp. 630–645. Springer, Cham (2016). https://doi.org/10.1007/978-3-319-46493-0_38
19. Heindl, A., Nawaz, S., Yuan, Y.: Mapping spatial heterogeneity in the tumor microenvironment: a new era for digital pathology. Lab. Invest. **95**(4), 377–384 (2015)

20. Hu, B., et al.: Unsupervised learning for cell-level visual representation in histopathology images with generative adversarial networks. IEEE J. Biomed. Health Inform. **23**(3), 1316–1328 (2018)
21. Javed, S., et al.: Cellular community detection for tissue phenotyping in colorectal cancer histology images. Med. Image Anal. **63**, 101696 (2020)
22. Komodakis, N., Gidaris, S.: Unsupervised representation learning by predicting image rotations. In: International Conference on Learning Representations (ICLR) (2018)
23. Levy-Jurgenson, A., Tekpli, X., Kristensen, V.N., Yakhini, Z.: Spatial transcriptomics inferred from pathology whole-slide images links tumor heterogeneity to survival in breast and lung cancer. Sci. Rep. **10**(1), 1–11 (2020)
24. Liu, H., HaoChen, J.Z., Gaidon, A., Ma, T.: Self-supervised learning is more robust to dataset imbalance. arXiv preprint arXiv:2110.05025 (2021)
25. Martin-Gonzalez, P., Crispin-Ortuzar, M., Markowetz, F.: Predictive modelling of highly multiplexed tumour tissue images by graph neural networks. In: Reyes, M., et al. (eds.) IMIMIC/TDA4MedicalData -2021. LNCS, vol. 12929, pp. 98–107. Springer, Cham (2021). https://doi.org/10.1007/978-3-030-87444-5_10
26. van Muijen, G.N., et al.: Cell type heterogeneity of cytokeratin expression in complex epithelia and carcinomas as demonstrated by monoclonal antibodies specific for cytokeratins nos. 4 and 13. Exp. Cell Res. **162**(1), 97–113 (1986)
27. Nguyen, K., Jain, A.K., Sabata, B.: Prostate cancer detection: fusion of cytological and textural features. J. Pathol. Inform. **2** (2011)
28. Noroozi, M., Favaro, P.: Unsupervised learning of visual representations by solving jigsaw puzzles. In: Leibe, B., Matas, J., Sebe, N., Welling, M. (eds.) ECCV 2016. LNCS, vol. 9910, pp. 69–84. Springer, Cham (2016). https://doi.org/10.1007/978-3-319-46466-4_5
29. Sirinukunwattana, K., Raza, S.E.A., Tsang, Y.W., Snead, D.R., Cree, I.A., Rajpoot, N.M.: Locality sensitive deep learning for detection and classification of nuclei in routine colon cancer histology images. IEEE Trans. Med. Imaging **35**(5), 1196–1206 (2016)
30. Son, B., Lee, S., Youn, H., Kim, E., Kim, W., Youn, B.: The role of tumor microenvironment in therapeutic resistance. Oncotarget **8**(3), 3933 (2017)
31. Vununu, C., Lee, S.H., Kwon, K.R.: A strictly unsupervised deep learning method for hep-2 cell image classification. Sensors **20**(9), 2717 (2020)
32. Xie, E., et al.: DetCo: unsupervised contrastive learning for object detection. In: Proceedings of the IEEE/CVF International Conference on Computer Vision, pp. 8392–8401 (2021)
33. Xie, X., Chen, J., Li, Y., Shen, L., Ma, K., Zheng, Y.: Instance-aware self-supervised learning for nuclei segmentation. In: Martel, A.L., et al. (eds.) MICCAI 2020. LNCS, vol. 12265, pp. 341–350. Springer, Cham (2020). https://doi.org/10.1007/978-3-030-59722-1_33
34. Zhang, L., Amgad, M., Cooper, L.A.: A histopathology study comparing contrastive semi-supervised and fully supervised learning. arXiv preprint arXiv:2111.05882 (2021)
35. Zhang, R., Isola, P., Efros, A.A.: Colorful image colorization. In: Leibe, B., Matas, J., Sebe, N., Welling, M. (eds.) ECCV 2016. LNCS, vol. 9907, pp. 649–666. Springer, Cham (2016). https://doi.org/10.1007/978-3-319-46487-9_40

Automatic Grading of Cervical Biopsies by Combining Full and Self-supervision

Mélanie Lubrano[1,2], Tristan Lazard[2,3,4], Guillaume Balezo[1],
Yaëlle Bellahsen-Harrar[5], Cécile Badoual[5], Sylvain Berlemont[1],
and Thomas Walter[2,3,4]($\boxtimes$)

[1] KEEN EYE, 75012 Paris, France
[2] Centre for Computational Biology (CBIO), Mines Paris, PSL University,
75006 Paris, France
thomas.walter@mines-paristech.fr
[3] Institut Curie, PSL University, 75005 Paris, France
[4] INSERM, U900, 75005 Paris, France
[5] Department of Pathology, Hôpital Européen Georges-Pompidou,
APHP, Paris, France

Abstract. In computational pathology, predictive models from Whole Slide Images (WSI) mostly rely on Multiple Instance Learning (MIL), where the WSI are represented as a bag of tiles, each of which is encoded by a Neural Network (NN). Slide-level predictions are then achieved by building models on the agglomeration of these tile encodings. The tile encoding strategy thus plays a key role for such models. Current approaches include the use of encodings trained on unrelated data sources, full supervision or self-supervision. While self-supervised learning (SSL) exploits unlabeled data, it often requires large computational resources to train. On the other end of the spectrum, fully-supervised methods make use of valuable prior knowledge about the data but involve a costly amount of expert time. This paper proposes a framework to reconcile SSL and full supervision, showing that a combination of both provides efficient encodings, both in terms of performance and in terms of biological interpretability. On a recently organized challenge on grading Cervical Biopsies, we show that our mixed supervision scheme reaches high performance (weighted accuracy (WA): 0.945), outperforming both SSL (WA: 0.927) and transfer learning from ImageNet (WA: 0.877). We further shed light upon the internal representations that trigger classification results, providing a method to reveal relevant phenotypic patterns for grading cervical biopsies. We expect that the combination of full and self-supervision is an interesting strategy for many tasks in computational pathology and will be widely adopted by the field.

Keywords: Mixed supervision · Histopathology · Whole-slide classification · Self-supervised learning

---

Supplementary Information The online version contains supplementary material available at https://doi.org/10.1007/978-3-031-25082-8_27.

1   Introduction

Computational Pathology is concerned with the application of Artificial Intelligence (AI) to the automatic analysis of Whole Slide Images (WSI). Examples include cancer subtyping [5], prediction of gene mutations [15,26] or genetic signatures [8,15,16]. Predictive models operating on WSI as inputs, faces two main challenges: first, WSIs are extremely large and cannot be fed directly into traditional neural networks due to memory constrains. Second, expert annotations are laborious to attain, costly and prone to subjectivity. The most popular methods today rely on Multiple Instance Learning (MIL), which frames the problem as a bag classification task. The strategy is thus to split WSIs into small workable images (tiles), to encode the tiles by a NN and then to build a predictive model on the agglomeration of these tile encodings. Tile representations are crucial to the downstream WSI classification task. One common approach consists of initializing the feature extractor by pre-training on natural images, such as ImageNet. While such encodings are powerful and generic, they do not lie within the histopathological domain. Different strategies have thus been developed to reach more appropriate tile encodings using different levels of supervision.

A first strategy aims to learn tile features with full supervision [1,10]. For this, one or several experts manually review a large number of tiles and sort them into meaningful classes. While the model may thus benefit from medically relevant prior knowledge, the process is time consuming and costly. A second strategy consists of learning tile representations through self-supervision, leveraging the unannotated data. It has proven its efficacy [19,23] and even its superiority to the fully supervised scheme [7]. However, this approach has a non-negligible computational cost, as training necessitates around 1000 h of computation on a standard GPU [7]. Moreover, it is not guaranteed that the obtained encodings are optimal for the prediction task we are trying to solve. While both techniques come have undeniable advantages, we hypothesized that combining them could allow us to benefit from the best of both worlds.

Motivated by these observations we propose a joint-optimization process mixing self, full and weak supervision (Fig. 1). Our contributions are:

- We propose a method for mixed supervision that combines the power of generic encodings obtained from self-supervision with the biomedical meaningfulness obtained from full supervision.
- We measure the trade-off in performance between the number of annotations and the computational cost of training a self-supervised model, thus providing guidelines to train a clinically impactful classifier with a limited budget in expert and/or computational workload.
- With activation optimization (AM), we further show that the learned feature encodings extract biologically meaningful information, recognized by pathologists as true histopathological criteria for the grading of cervical biopsies. Of note, we demonstrate for the first time that AM can point to key tissue phenotypes. We thus provide a complementary method for network introspection in Computational Pathology.
- We used a cost-sensitive loss to boost the performance of grading by deep learning.

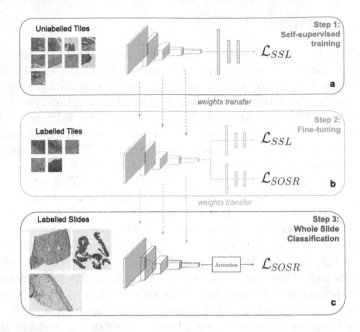

Fig. 1. Mixed Supervision Process: a) A self-supervised model (SimCLR) is trained on unlabeled tiles extracted from the slides. Feature extractor and contrastive layer weights are transferred to the joint-optimization architecture **b)** Joint-optimization model is trained on the labeled tiles of the dataset. The feature extractor weights are transferred to the WS classification model. **c)** WSI classification model is trained on the 1015 whole slide images.

2   Related Work

Medical data is often limited. For this reason, one might want to take advantage of all the available data even if annotations might not be homogeneous and even though they might be difficult to exploit because multiple levels of supervision are available. AI applications have usually been dichotomized between supervised and unsupervised methods, spoiling the potential of combining several types of annotations. For this reason, mixing supervision for medical images analysis has gained interest in past years [13,17,18].

For instance, in [20] the author showed that combining global labels and local annotations by training in a multi-task setting, the capacities of the model to segment brain tumors on Magnetic Resonance Images were improved.

In [24], the author introduced a mixed supervision framework for metastasis detection building on the CLAM [19] architecture. CLAM is a variant of the popular attention based MIL [14] with 2 extensions: first, in order to make the method applicable in a multi-class setting, class-specific attention scores are learned and applied. Second, the last layer of the tile encoding network is trained to also predict the top and bottom attention scores, thus mimicking tile-level annotations. In [24], the authors highlight the limitations of this instance-classification approach and propose to leverage a low number of fully annotated slides to train the attention mechanism. In a second step, they propose to turn to a standard MIL training (using only slide-level annotations). Even with few annotated slides, this approach allows to boost classification performance and is thus the first demonstration of the benefits of mixed supervision in computational pathology. However, there are also some limitations. First, the method relies on exhaustive annotation of selected slides: for the annotated slides, all the key regions are annotated pixel-wise. Second, due to the CLAM architecture, the approach only fine-tunes a single dense layer downstream the pre-trained feature extractor. Third, the algorithm has been designed for an application case in which the slide and tile labels coincide (tumour presence). This however is not always the case: when predicting genetic signatures, grades or treatment responses, it is unclear how tile and slide level annotations relate to each other. In this article, we propose to overcome these limitations. We propose to combine self-supervised learning with supervision prior to training the MIL network. We thus start from more powerful encodings, that are not only capable of solving the pretext task of self-supervised learning, but also the medical classification task that comes with the annotated tiles. Consequently, this method does not require full-slide annotations, optimizes the full tile encoding network and does not come with any constraint regarding the relationship between tile and slide level annotations.

3   Dataset and Problem Setting

The Tissue Net Challenge [9] organized in 2020, the *Société Française de Pathologie (SFP)* and the *Health Data Hub* aimed at developing methods to automatically grade lesions of the uterine cervix in four classes according to their severity.

Fully Supervised Dataset. 5926 annotated Regions of Interest (ROIs) of fixed size 300×300 micrometers were provided. Each ROI had roughly the same size as a tile at 10x magnification and were labeled by the severity of the lesion it contained: "normal" (0) if tissue was normal, (1) low grade dysplasia or (2) high grade dysplasia if it presented precancerous lesions that could have malignant potential and (3) invasive squamous carcinoma.

Weakly Supervised Dataset. The dataset was composed of 1015 WSIs acquired from 20 different centers in France at an average resolution of 0.234 $\pm$ 0.0086 mpp (40X). The class of the WSI corresponded to the class of the most severe lesions it contained (grade from 0 to 3). Slide labels were balanced across the dataset.

Misclassification Costs. Misclassification errors do not lead to equally serious consequences. Accordingly, a panel of pathologists established a cost matrix that assigns to each combination of true class and prediction $(i, j) \in \{0, 1, 2, 3\}^2$ a severity score $0 \leqslant C_{i,j} \leqslant 1$ (Table 1). The metric used in the challenge to evaluate and rank the submissions was computed from the average of these misclassification costs. More precisely, if $P(S)$ was the prediction of a slide S labelled $l(S)$, the challenge metric M_{WA} was:

$$M_{WA} = \frac{1}{N} \sum_S (1 - C_{l(S),P(S)}) \tag{1}$$

with N the number of samples.

Table 1. Weighted Accuracy Error Table - Error table to weight misclassification according to their gap with the ground truth

Ground truth	Benign (pred)	Low-grade (pred)	High-grade (pred)	Carcinoma (pred)
Benign	0.0	0.1	0.7	1.0
Low-grade	0.1	0.0	0.3	0.7
High-grade	0.7	0.3	0.0	0.3
Carcinoma	1.0	0.7	0.3	0.0

4   Proposed Architecture

4.1   Multiple Instance Learning and Attention

In Multiple Instance Learning, we are given sets of samples $B_k = \{x_i | i = 1 \dots N_k\}$, also called bags. The annotation y_k we are given refers only to the bags and not the individual samples. In our case, the bags correspond to the slides, the samples to the tiles. We assume, that such tile-level labels exist in principle, but that we just do not have access to them. The strategy is to first map each tile x_i to its encoding z_i, which is then mapped to a scalar value a_i, often referred to as attention score. The tile representations z_i and attention scores a_i are then agglomerated to build the slide representation s_k which is then further processed by a neural network. The agglomeration can be based on tile selection [2,6], or on an attention mechanism [14], which is today the most widely used strategy.

4.2   Self-supervised Learning

Self-supervised learning provides a framework to train neural networks without human supervision. The main goal of self-supervised learning is to learn to extract efficient features with inputs and labels derived from the data itself using a pretext task. Many self-supervised approaches are based on contrastive learning in the feature space. SimCLR, a simple framework relying on data augmentation was introduced in [3]. Powerful feature representations are learned by maximizing agreement between differently augmented views of the same data point via a contrastive loss $\mathcal{L}_{SSL}$ applied in the feature space. Details can be found in Supplementary Materials.

4.3   Cost-Sensitive Training

Instead of the traditional cross-entropy loss we used a cost-aware classification loss, the Smooth-One-Sided Regression Loss $\mathcal{L}_{SOSR}$. First introduced to train SVMs in [25], this objective function was smoothed and adapted for backpropagation in deep networks in [4]. When using this loss, the network is trained to predict the class-specific risk rather than a posterior probability; the decision function chooses the class minimizing this risk. The SOSR loss is defined as follows:

$$\mathcal{L}_{SOSR} = \sum_i \sum_j ln(1 + exp(2_{i,j} \cdot (\hat{c}_i - \mathcal{C}_{i,j}))) \qquad (2)$$

with $2_{i,j} = -1_{i \neq j} + 1_{i=j}$, $\hat{c}_i$ the i-th coordinate of the network output and $\mathcal{C}$ the error table.

4.4   Mixed Supervision

To be tractable, training of attention-MIL architectures requires freezing the feature extractor weights. While SSL allows the feature extractor to build meaningful representations [7,23], they are not specialized to the actual classification problems we try to solve. Several studies have shown that such SSL models benefit from specific fine-tuning to the downstream task [3]. We therefore added a training step to leverage the tile-level annotation and fine-tune the self-supervised model. However, as the final WSI classification task is not identical to the tile classification task, we suspect that fine-tuning solely on the tile classification task may over-specialize the feature extractor and thus sacrifice the generalizability of SSL (and for this reason ultimately also degrading the WSI classification performances). To avoid this, we developed a training process that optimizes the self-supervised and tile-classification objectives jointly. Two different heads, plugged before the final classification layer, are used to compute both loss functions $\mathcal{L}_{SSL}$ and $\mathcal{L}_{SOSR}$ The final objective $\mathcal{L}$ is then:

$$\mathcal{L} = \beta \mathcal{L}_{SSL} + (1 - \beta)\mathcal{L}_{SOSR} \qquad (3)$$

where β is a tuned hyperparameter. We found $\beta = 0.3$ (see Supplementary).

5   Understanding the Feature Extractor with Activation Maximization

To further understand the features learned by the different pre-training policies, we used Activation Maximization (AM) to visualize extracted features and provide an explicit illustration of the specificity learned. Methods to generate pseudo-images maximizing a feature activation have been introduced in [11]. This technique consists in synthesizing the images that will maximize one feature activation and can be summarized as follows [21]: if we consider a trained classifier with set of parameters θ that map an input image $x \in \mathbb{R}^{h \times w \times c}$, $(h$

and w are the height and width and c the number of channels) to a probability distribution over the classes, we can formulate the following optimization problem:

$$x^* = \arg\max_x(\sigma_i^l(\theta, x)) \tag{4}$$

where $\sigma_i^l(\theta, x)$ is the activation of the neuron i in a given layer l of the classifier. This formulation being a non-convex problem, local maximum can be found by gradient ascent, using the following update step:

$$x_{t+1} = x_t + \epsilon \frac{\partial \sigma_i^l(\theta, x)}{\partial x_t} \tag{5}$$

The optimization process starts with a randomly initialized image. After a few steps, it generates an image which can help to understand what information is being captured by the feature. As we try to visualize meaningful representations of the features, some regularization steps are applied to the random noise input (random crop and rotations to generate more stable visualization, details can be found in Supplementary Materials). To generate filter visualization within the HE space, we transformed the RGB random image to HE input by color deconvolution [22]. This preprocessing allowed to generate images with histology-like colors when converted back to the RGB space. To select the most meaningful features for each class, we trained a Lasso classifier without bias to classify the extracted feature vectors into the four classes of the dataset for the four pre-training policies. The feature vectors for each tile were first normalized and divided element-wise by the vector of features' standard deviation across all the tiles. The L1 regularization factor λ was set to 0.01. Details about Lasso training can be found in Supplementary Materials. Contribution scores for each feature were therefore derived from the weights of the Lasso linear classifier: negative weights were removed and remaining positive weights were divided by their sum to obtain contribution scores $\in [0, 1]$. By filtering out the negative weights, the contribution score corresponds to the proportion of attribution among the features positively correlated to a class, and allows to select feature capturing semantic information related to the class, leaving out those containing information for other classes.

6 Experimental Setting

6.1 WSI Preprocessing

Preprocessing on a downsampled version of the WSIs was applied to select only tissue area and non-overlapping tiles of 224×224 pixels were extracted at a resolution of 1 mpp. (Details in Supplementary Materials).

6.2 Data Splits for Cross-Validation

To measure the performances of our models we performed 3-fold cross-validation for all our training settings. Because the annotated tiles used in our joint-optimization step were directly extracted from the slides themselves, we carefully split the tiles such that tiles in different folds were guaranteed to originate from different slides. The split divided the slides and tiles into a training set (70%), a validation set (10%) and a test set (20%). All subsequent performance results are then reported as the average and standard deviation of the performance results on each of these 3 test folds.

6.3 Feature Extractor Pre-training

The feature extractor is initialized with pre-trained weights obtained with three distinct supervision policies: fully supervised, self-supervised or a mix of supervision. These three policies rely on the fine-tuning of a DenseNet121 [12], pre-trained on ImageNet. The fully-supervised architecture is fine-tuned solely on the tile classification task. The SSL architecture is derived from SimCLR framework and is trained on an unlabeled dataset of 1 million tiles extracted from the slides. Finally for the mixed-supervised architecture, a supervised branch is added to the previous SSL network and trained using the mixed objective function (see Fig. 1 and Eq. 3) on the fully supervised dataset. Technical details of these three training settings are available in the supplementary material.

6.4 Whole Slide Classification

After tiling the slides, the frozen feature extractor (DenseNet121) was applied to extract meaningful representations from the tiles. This feature extractor was initialized sequentially with the pre-trained weights mentioned above and generated as many sets of features. These bags of features were then used to train the Attention-MIL model with SOSR loss applied slide-wise. (Supplementary Materials).

6.5 Feature Visualization

To select the most relevant features, we trained an unbiased linear model on the feature vectors extracted from the annotated tiles. The feature vectors were standardised. The weights of the linear model were used to determine which features were the most impactful for each class. Feature visualizations were generated for the selected features and for each set of pre-trained weights (best training from cross-validation). We extracted the tiles expressing the most of these features by selecting the feature vectors with the higher activation for the concerned feature (Supplementary Materials).

7 Results

7.1 Self-supervised Fine-Tuning

We saved the checkpoints of the self-supervised feature extraction model at each epoch of training, allowing us to investigate the amount of training needed to reach good WSI classification performances. We computed the embeddings of the whole dataset with each of the checkpoints and trained a WSI classifier from them. Figure 2 reports the performances of WSI classification models for each of these checkpoints. SSL training led to a higher Weighted Accuracy than using ImageNet weights after 3 epochs and resulted in a gain of +4.8% after 100 epochs. Interestingly, as little as 6 epochs of training are enough to gain 4% of Weighted Accuracy: a significant boost in performance is possible with 50 GPU-hours of training. We then observe a small increase in performance until the 100th epoch.

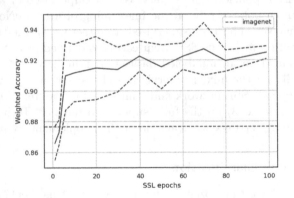

Fig. 2. Weighted Accuracy (WA) for SSL and ImageNet pretraining - WA (obtained by 3-fold cross validation) as a function of the number of epochs for SSL training. Solid line: average WA, lines above and below: ± standard deviation, respectively. Horizontal line: WA obtained by pretraining on ImageNet.

7.2 Pre-training Policy Comparison

To compare the weights obtained with the various supervision levels, we ran a 3-fold cross-validation on the WS classification task and summarized the results in

Table 2. The results indicate that the SSL pre-training substantially improves the WSI classification performance. In contrast, we see that initializing the feature extractor with fully-supervised weights gives an equivalent or poorer performance than any other initialization. SSL pre-training allows us to extract rich features that are generic, yet still relevant to the dataset (unlike ImageNet). On the other hand, fully supervised features are probably too specific and seem not to represent the full diversity of the image data. The joint-optimization process manages to balance out generic and specialized features without neutralizing them: mixing the supervision levels brings significant improvements (+2%) to the performance, leading to a Weighted Accuracy of 0.945.

Next, we compared the cost-sensitive loss (Eq. 2) with the cross-entropy loss. Our results show that with ImageNet weights the SOSR loss improves the WA by 1% and the accuracy by 3%.

In conclusion, the combination of the SSL pre-trained model, its fully supervised fine-tuning, and the cost-sensitive loss leads to a notable improvement of 8 WA points over the baseline MIL model with ImageNet pretraining.

Table 2. Comparison of pre-training policies: performance depending on the pre-training regime (first column) and the classification loss (second column).

	Downstream loss	Accuracy	Weighted accuracy
ImageNet	Cross entropy	0.758 ± 0.034	0.865 ± 0.023
ImageNet	$\mathcal{L}_{SOSR}$	0.787 ± 0.032	0.877 ± 0.029
Supervised	$\mathcal{L}_{SOSR}$	0.772 ± 0.055	0.874 ± 0.027
SSL	$\mathcal{L}_{SOSR}$	0.803 ± 0.016	0.925 ± 0.006
Mixed	$\mathcal{L}_{SOSR}$	$\mathbf{0.845 \pm 0.028}$	$\mathbf{0.945 \pm 0.005}$

7.3 Number of Annotations vs Number of Epochs

We have seen that the combination of SSL and supervised pre-training lead to improved WSI classification. To further investigate the relationship between these two supervision regimes, we trained models with only some of the fully supervised annotations (15, 65, 100%) on top of intermediate SSL checkpoints. Results are reported in Table 3. It appears that without SSL pre-training (or with too few epochs of training), the supervised fine-tuning does not bring additional improvement for WSI classification. This is in line with the work of Chen et al. [3] that showed that an SSL model is up to 10x more label efficient than a supervised one. However, while fine-tuning the models by mixed supervision with too few annotations (15%) leads to a slight drop in WSI classification performances, we observe an improvement of 2 points of WA when using 100% of the tile annotations. Finally, we see a diminution of the standard deviations across splits for the different pre-training policies, showing better stability for longer SSL training and more annotations. We draw different conclusions from these observations:

- SSL is always preferable to only investing in annotations.
- The supervised fine-tuning needs enough annotations to bring an improvement to the WSI classification task. We can note however that even when considering the 100% annotation settings, the supervised dataset (approx. 5000 images) is still rather small in comparison to traditional image datasets.
- A full SSL training is mandatory to leverage this small amount of supervised data.

Table 3. Performance (WA) depending on SSL training time and number of annotations

	0 Annot.	~1 Annot./slide (1015 tiles)	~4 Annot./slide (3901 tiles)	~6 Annot./slide (5926 tiles)
ImageNet (no SSL)	0,877 ± 0.029	0.872 ± 0.024	0.872 ± 0.023	0,874 ±0.027
SSL-epoch10	0,912 ± 0.019	0,907± 0.024	0,903 ± 0.029	0,916 ± 0.019
SSL-epoch50	0,915 ± 0.014	0,913 ± 0.024	0,916 ± 0.014	0,914 ± 0.022
SLL-epoch100	0,925 ± 0.006	0,916 ± 0.010	0,921 ± 0.010	**0,945 ± 0.005**

7.4   Feature Visualization

We generated the pseudo-images of the most important features for each class and each pre-training policy and extracted the related tiles. Figure 3 displays the most important features along with their most activating tiles for the class "Normal" (0). Although interpretation of such pseudo images must be treated carefully, we notice that the features obtained with SSL, supervised and mixed training are indubitably more specialized to histological data than those obtained with ImageNet. Some histological patterns, such as nuclei, squamous cells or basal layers are clearly identifiable in the generated images. The extracted tiles are strongly correlated with class-specific biomarkers. Feature **e** represents a normal squamous maturation, i.e. a layer of uniform and rounded basal cells, with slightly larger and bluer nuclei than mature cells. Features **c** and **d** highlight clouds of small regular and rounded nuclei (benign cytological signs). Feature **g** and **h** are characteristic of squamous cells (polygonal shapes, stratified organization lying on a straight basal layer). Interestingly, features extracted with the supervised method (**g, h**) manage to sketch a normal epithelium with high resemblance, the features are more precise. On the other hand, features extracted with SSL (**c, d**) highlight true benign criteria but do not entirely summarize a normal epithelium (no basal maturation). The mixed model displays both, suggesting that mixed supervision highlights pathologically relevant patterns to a larger extent than the other regimes [27].

In Fig. 4 we can further identify class-related biomarkers for dysplasia and carcinoma grade. Tiles with visible koilocytes (cells with a white halo around the nucleus) have been extracted from the top features for Low Grade class. Koilocytes are symptomatic of infection by Human Papillomavirus and are a key

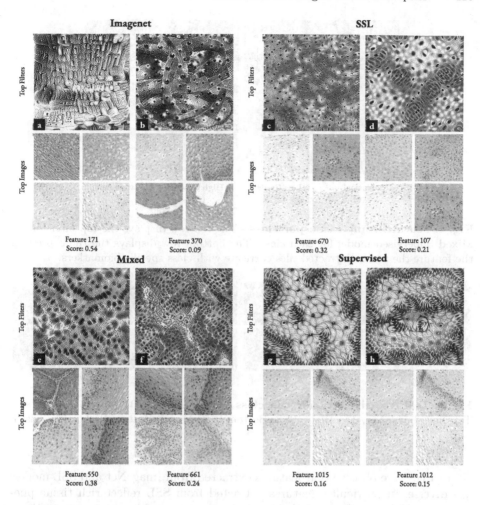

Fig. 3. Feature Visualization - Top Features for class "Normal" (0) and tiles expressing the most the feature. Features obtained with SSL and Mixed Supervision are clearly related to histopathological patterns. Nuclei, squamous cells, basal layers and other histological morphologies are identifiable. Similar visualization for other classes are available in the Supplementary Materials.

element for this diagnosis (almost always responsible for precancerous lesions in the cervix, [27]). High Grade (2) generated image represents disorganised cells with a high nuclear-to-cytoplasmic ratio, marked variations in size and shape and loss of polarity. For the class "Carcinoma" (3), we observe irregular clusters of cohesive cells with very atypical nuclei, separated by a fibrous texture that can be identified as stroma reaction. All these criteria have been identified in [27] as key elements for diagnosis of dysplasia and invasive carcinoma.

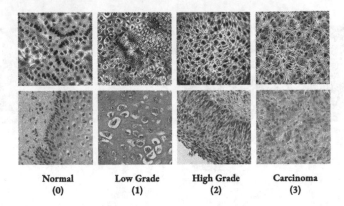

Normal   **Low Grade**   **High Grade**   **Carcinoma**
(0)   (1)   (2)   (3)

Fig. 4. Feature comparison per class - The top row displays the top filter for the Mixed Supervised model for each class. The bottom row displays the tile expressing the feature the most. Extracted tiles correlate with class-specific biomarkers.

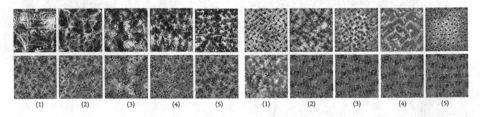

(1)   (2)   (3)   (4)   (5)   (1)   (2)   (3)   (4)   (5)

Fig. 5. Feature Diversity for the class "Carcinoma" (3) (top 5 features) - Top Left: ImageNet, Top Right: SSL. Bottom Left: Mixed, Bottom Right: Supervised.

In Fig. 5, we observe that features extracted from ImageNet and SSL models are diverse, in particular, features extracted from SSL reflect rich tissue phenotypes which correlates to their generic capacities of image representations. On the other hand, features extracted with supervised and mixed methods are more redundant. Visualizations for other classes are available in Supplementary Materials.

8 Discussion

In pathology, expert annotations are usually hard to obtain. However, we are often in a situation where a small proportion of labeled annotation exists but not in sufficient quantities to support fully supervised techniques. Yet, even in small quantities, expert annotations carry meaningful information that one could use to enforce biological context to deep learning models and make sure that networks learn appropriate patterns. On the other hand, self-supervised methods have proven their efficacy to extract generic features in the histopathological domain and their usefulness for downstream supervision tasks, even in the absence of massive ground truth data. Methods capable of reconciling self-supervision with strong supervision can therefore be useful and open the door to better performance.

In this paper, we presented a way to inject the fine-grained tile level information by fine-tuning the feature extractor with a joint optimization process. This process thus allows to combine self- with full supervision during the encoder training. We applied our method to the TissueNet Challenge, a challenge for the automatic grading of cervix cancer, that provided annotations at the slide and tile level, thus representing an appropriate use case to validate our method of mixed supervision. We also propose in this study insights and guidelines for the training of a WSI classifier in the presence of tile annotations. First, we showed that SSL is always beneficial to our downstream WSI classification tasks. Fine-tuning pre-trained weights with SSL for only 50 h brings a 4% improvement over WSI classification weighted accuracy, and near to 5% when fine-tuning for longer (100 epochs). Second, a small set of annotated tiles can bring benefit to the WSI classification task, up to 2% of weighted accuracy for a supervised dataset of around 5000 images. Such a set of tiles can be obtained easily by asking the pathologist to select a few ROIs that guided their decision while labeling the WSIs, which can be achieved without a strong time commitment. However this boost in performance can be reached only if the feature extractor is pre-trained with SSL, and for sufficiently long: SSL unlocks the supervised fine-tuning benefits.

To further understand the differences between the range of supervision used to extract tile features, we conducted qualitative analysis on features visualizations by activation maximization and observed that features obtained from SSL, supervised or mixed trainings were more relevant for histological tasks and that class-discriminative patterns were indeed identified by the model.

The scope of this study contains by design three limitations. First, SSL models were trained by fine-tuning already pre-trained weights on imagenet. This may explain the rapid convergence and boost in performance observed; however it may also underestimate this boost if the SSL models were trained from scratch. We did not compare SSL trained from scratch and fine-tuned SSL, and left it to future work. Second, all the conclusions reached are conditioned by the fact that we do not fine-tune the feature extractor network during the WSI classification training. Keeping these weights frozen, and even pre-computing the tile representations brings a large computational benefit (both in memory and speed of computations), but prevents the feature extractor from specializing during the WSI classification training. Third, the tendency observed in Table 3 of better performances correlated with larger numbers of annotations is modest and would require more annotations to validate it. Application to different task such as rare diseases dataset with few data could show better benefit.

To conclude, we present a method that provides an interesting alternative to using full supervision, pre-training on unrelated data sets or self-supervision. We convincingly show that the learned feature representations are both leading to higher performance and providing intermediate features that are more adapted to the problem and point to relevant cell and tissue phenotypes. We expect that the mixed supervision will be adopted by the field and lead to better models.

Acknowledgments. The authors thank Etienne Decencière for the thoughful discussions that help the project. ML was supported by a CIFRE PhD fellowship founded by KEEN EYE and ANRT (CIFRE 2019/1905). TL was supported by a Q-Life PhD fellowship (Q-life ANR-17-CONV-0005). This work was supported by the French government under management of ANR as part of the "Investissements d'avenir" program, reference ANR-19-P3IA-0001 (PRAIRIE 3IA Institute).

References

1. Bera, K., Schalper, K.A., Rimm, D.L., Velcheti, V., Madabhushi, A.: Artificial intelligence in digital pathology-new tools for diagnosis and precision oncology. Nat. Rev. Clin. Oncol. **16**(11), 703–715 (2019)
2. Campanella, G., et al.: Clinical-grade computational pathology using weakly supervised deep learning on whole slide images. Nat. Med. **25**(8), 1301–1309 (2019). https://doi.org/10.1038/s41591-019-0508-1. https://www.nature.com/articles/s41591-019-0508-1
3. Chen, T., Kornblith, S., Norouzi, M., Hinton, G.: A simple framework for contrastive learning of visual representations (2020). https://arxiv.org/abs/2002.05709v3
4. Chung, Y.A., Lin, H.T., Yang, S.W.: Cost-aware pre-training for multiclass cost-sensitive deep learning. IJCAI (2016)
5. Coudray, N., et al.: Classification and mutation prediction from non-small cell lung cancer histopathology images using deep learning. Nat. Med. **24**(10), 1559–1567 (2018). https://doi.org/10.1038/s41591-018-0177-5. https://www.nature.com/articles/s41591-018-0177-5
6. Courtiol, P., Tramel, E.W., Sanselme, M., Wainrib, G.: Classification and disease localization in histopathology using only global labels: a weakly-supervised approach. arXiv:1802.02212 [cs, stat] (2020)
7. Dehaene, O., Camara, A., Moindrot, O., de Lavergne, A., Courtiol, P.: Self-supervision closes the gap between weak and strong supervision in histology (2020). https://arxiv.org/abs/2012.03583v1
8. Diao, J.A., et al.: Human-interpretable image features derived from densely mapped cancer pathology slides predict diverse molecular phenotypes. Nat. Commun. **12**(1), 1613 (2021). https://doi.org/10.1038/s41467-021-21896-9. https://www.nature.com/articles/s41467-021-21896-9
9. DrivenData: TissueNet: Detect Lesions in Cervical Biopsies. https://www.drivendata.org/competitions/67/competition-cervical-biopsy/page/254/
10. Ehteshami Bejnordi, B., et al.: Consortium: diagnostic assessment of deep learning algorithms for detection of lymph node metastases in women with breast cancer. JAMA **318**(22), 2199–2210 (2017). https://doi.org/10.1001/jama.2017.14585
11. Erhan, D., Bengio, Y., Courville, A., Vincent, P.: Visualizing higher-layer features of a deep network. Technical report, Univeristé de Montréal (2009)
12. Huang, G., Liu, Z., van der Maaten, L., Weinberger, K.Q.: Densely connected convolutional networks. Technical report (2016). https://ui.adsabs.harvard.edu/abs/2016arXiv160806993H

13. Huang, Y.J., et al.: Rectifying supporting regions with mixed and active supervision for rib fracture recognition. IEEE Trans. Med. Imaging **39**(12), 3843–3854 (2020). https://doi.org/10.1109/TMI.2020.3006138

14. Ilse, M., Tomczak, J.M., Welling, M.: Attention-based deep multiple instance learning (2018). https://arxiv.org/abs/1802.04712v4

15. Kather, J.N., et al.: Pan-cancer image-based detection of clinically actionable genetic alterations. Nat. Cancer **1**(8), 789–799 (2020). https://doi.org/10.1038/s43018-020-0087-6. https://www.nature.com/articles/s43018-020-0087-6

16. Lazard, T., et al.: Deep learning identifies morphological patterns of homologous recombination deficiency in luminal breast cancers from whole slide images. Cell Rep. Med. **3**(12), 100872 (2022). https://doi.org/10.1016/j.xcrm.2022.100872

17. Li, J., et al.: Hybrid supervision learning for pathology whole slide image classification. In: de Bruijne, M., et al. (eds.) MICCAI 2021. LNCS, vol. 12908, pp. 309–318. Springer, Cham (2021). https://doi.org/10.1007/978-3-030-87237-3_30

18. Li, Z., et al.: Thoracic disease identification and localization with limited supervision. In: Proceedings of the IEEE Conference on Computer Vision and Pattern Recognition (2018)

19. Lu, M.Y., Williamson, D.F.K., Chen, T.Y., Chen, R.J., Barbieri, M., Mahmood, F.: Data-efficient and weakly supervised computational pathology on whole-slide images. Nat. Biomed. Eng. **5**(6), 555–570 (2021). https://doi.org/10.1038/s41551-020-00682-w. https://www.nature.com/articles/s41551-020-00682-w

20. Mlynarski, P., Delingette, H., Criminisi, A., Ayache, N.: Deep learning with mixed supervision for brain tumor segmentation. J. Med. Imaging **6**(3), 034002 (2019). https://doi.org/10.1117/1.JMI.6.3.034002. https://www.spiedigitallibrary.org/journals/journal-of-medical-imaging/volume-6/issue-3/034002/Deep-learning-with-mixed-supervision-for-brain-tumor-segmentation/10.1117/1.JMI.6.3.034002.full

21. Nguyen, A., Yosinski, J., Clune, J.: Understanding neural networks via feature visualization: a survey. arXiv:1904.08939 [cs, stat] (2019)

22. Ruifrok, A.C., Johnston, D.A.: Quantification of histochemical staining by color deconvolution. Anal. Quant. Cytol. Histol. **23**(4), 291–299 (2001)

23. Saillard, C., et al.: Identification of pancreatic adenocarcinoma molecular subtypes on histology slides using deep learning models. J. Clin. Oncol. **39**(15_Suppl.), 4141 (2021). https://doi.org/10.1200/JCO.2021.39.15suppl.4141. https://ascopubs.org/doi/abs/10.1200/JCO.2021.39.15suppl.4141

24. Tourniaire, P., Ilie, M., Hofman, P., Ayache, N., Delingette, H.: Attention-based multiple instance learning with mixed supervision on the camelyon16 dataset. In: Proceedings of the MICCAI Workshop on Computational Pathology, pp. 216–226. PMLR (2021). https://proceedings.mlr.press/v156/tourniaire21a.html

25. Tu, H.H., Lin, H.T.: One-sided support vector regression for multiclass cost-sensitive classification, p. 8 (2010)

26. Weitz, P., et al.: Transcriptome-wide prediction of prostate cancer gene expression from histopathology images using co-expression based convolutional neural networks. arXiv preprint arXiv:2104.09310 (2021)

27. WHO: Colposcopy and treatment of cervical intraepithelial neoplasia: a beginners' manual (2020). https://screening.iarc.fr/colpochap.php?chap=2

When CNN Meet with ViT: Towards Semi-supervised Learning for Multi-class Medical Image Semantic Segmentation

Ziyang Wang[1]([⊠]), Tianze Li[2], Jian-Qing Zheng[3], and Baoru Huang[4]

[1] Department of Computer Science, University of Oxford, Oxford, UK
ziyang.wang@cs.ox.ac.uk
[2] Canford School, Wimborne, UK
[3] The Kennedy Institute of Rheumatology, University of Oxford, Oxford, UK
[4] Department of Surgery and Cancer, Imperial College London, London, UK

Abstract. Due to the lack of quality annotation in medical imaging community, semi-supervised learning methods are highly valued in image semantic segmentation tasks. In this paper, an advanced consistency-aware pseudo-label-based self-ensembling approach is presented to fully utilize the power of Vision Transformer (ViT) and Convolutional Neural Network (CNN) in semi-supervised learning. Our proposed framework consists of a feature-learning module which is enhanced by ViT and CNN mutually, and a guidance module which is robust for consistency-aware purposes. The pseudo labels are inferred and utilized recurrently and separately by views of CNN and ViT in the feature-learning module to expand the data set and are beneficial to each other. Meanwhile, a perturbation scheme is designed for the feature-learning module, and averaging network weight is utilized to develop the guidance module. By doing so, the framework combines the feature-learning strength of CNN and ViT, strengthens the performance via dual-view co-training, and enables consistency-aware supervision in a semi-supervised manner. A topological exploration of all alternative supervision modes with CNN and ViT are detailed validated, demonstrating the most promising performance and specific setting of our method on semi-supervised medical image segmentation tasks. Experimental results show that the proposed method achieves state-of-the-art performance on a public benchmark data set with a variety of metrics. The code is publicly available (https://github.com/ziyangwang007/CV-SSL-MIS).

1 Introduction

Medical image segmentation is an essential task in computer vision and medical image analysis community where deep learning methods have shown dominated position recently. The promising results of current deep learning study

Supplementary Information The online version contains supplementary material available at https://doi.org/10.1007/978-3-031-25082-8_28.

not only relies on architecture engineering of CNN [9,28,34,43], but also on sufficient high-quality annotation of data set [2,11,13,48]. The most common situation of clinical medical image data, however, is with a small amount of labelled data and a large number of raw images such as CT, ultrasound, MRI, and videos from laparoscopic surgery [17,30,44,47]. In recent studies of neural network architecture engineering, the performance of purely self-attention-based, Transformer [40], outperforms CNN and RNN because of the ability of modeling long-range dependencies [13,27]. Following the above concern of data situation, and the recent success in network architecture engineering, we hereby proposed a **Semi-S**upervised medical image **S**emantic **S**egmentation framework aiming to fully utilize the power of CNN and ViT simultaneously, called **S4CVnet**. The framework of S4CVnet consists of a feature-learning module, and a guidance module, which is briefly sketched in Fig. 1. This setting is inspired by the Student-Teacher style framework [15,39,48], that the perturbation is applied to the student network, and the parameters of the teacher network are updated through Exponential Moving Average (EMA) [26] which makes the teacher network much robust to guide the learning of the student network with pseudo label under consistency-aware concern. To utilize the feature-learning power of CNN and ViT simultaneously and avoid the barrier caused by the different architecture of two networks, we hereby come up with a dual-view co-training approach in the feature-learning module [10,47]. Two different views of networks infer pseudo labels simultaneously to expand the size of the data set with raw data, complementing and beneficial to each other during the training process. One feature-learning network is also considered a bridge to be applied network perturbation and transfer of learning knowledge via the Student-Teacher style scheme.

The contributions of S4CVnet is fourfold and discussed as follows:

- an enhanced dual-view co-training module aiming to fully utilize the feature-learning power of CNN and ViT mutually is proposed. Both CNN and ViT are with the same U-shape Encoder-Decoder style segmentation network for fair comparison and exploration,
- a robust guidance module based on computational efficient U-shape ViT is proposed, and a consistency-aware Student-Teacher style approach via EMA is properly designed,
- an advanced semi-supervised multi-class medical image semantic segmentation framework is proposed, evaluated on a public benchmark data set with a variety of evaluation measures, and keeps state-of-the-art against other semi-supervised methods under the same setting and feature information distribution to our best of knowledge [30,32,39,41,42,45,47,48,51,52],
- a topological exploration study of all alternative supervision modes with CNN and ViT, as well as an ablation study, is validated to present a whole picture of utilizing CNN and ViT in a semi-supervised manner, and demonstrates the most proper setting and promising performance of S4CVnet.

2   Related Work

Semantic Segmentation. The convolutional neural network (CNN) for image semantic segmentation, as a dense prediction task, has been widely studied since 2015, i.e. FCN [28]. It is the first CNN-based network trained with a supervised fashion for pixels-to-pixels prediction tasks. Then, the subsequent study of segmentation was dominated by CNN with three aspects of contribution: backbone network, network blocks, and training strategy. For example, one of the most promising backbone networks is UNet [34], which is an Encoder-Decoder style network with skip connections to efficiently transfer multi-scale semantic information. A variety of advanced network blocks to further improve CNN performance such as attention mechanism [24,35], residual learning [16], densely connected [18], dilated CNN [8] have been applied to the backbone network, UNet, which results in a family of UNet [21,23,44,46]. The CNN for dense prediction tasks, however, is lack of ability of modelling long-range dependencies in recent studies, and is defeated by Transformer, a purely self-attention-based network, that originated from natural language processing [40]. The Transformer was widely explored in computer vision tasks, i.e. Vision Transformer (ViT) [13], around classification, detection, and segmentation tasks [4,6,27,38]. In this paper on the backbone aspect, we focus on exploring the feature-learning power of CNN and ViT simultaneously, enabling both of them beneficial to each other, and specifically tackling a semi-supervised dense prediction task based on a multi-view co-training self-ensembling approach.

Semi-supervised Semantic Segmentation. Besides the study of backbone networks and network blocks, the training strategy is also an essential study depending on the different scenarios of data set, such as weakly-supervised learning to tackle low-quality annotations [5,36,53], noisy annotations [44], multi-rater annotations [25], and mixed-supervised learning for multi-quality annotations [33]. The most common situation of medical imaging data is with a small amount of labelled data and a large amount of raw data due to the high labelling cost, so semi-supervised learning is significantly valuable to be explored. Co-training, and self-training are two widely studied approaches in semi-supervised learning. Self-training, also known as self-labelling, is to initialize a segmentation network with labelled data at first. Then the pseudo segmentation masks on unlabelled data are generated by the segmentation network [7,20,29,31,50,54]. A condition is set for the selection of pseudo segmentation masks, and the segmentation network is retrained by expanding training data several times. GAN-based approaches mainly studied how to set the condition using discriminator learning for distinguishing the predictions and the ground-truth segmentation [19,37]. The other approach is Co-training, which is usually to train two separate networks as two views. These two networks thus expand the size of training data and complement each other. Deep Co-training was firstly proposed in [32] pointing out the challenge of utilizing co-training with a single data set, i.e. 'collapsed neural networks'. Training two networks on the same data set cannot enable multi-view feature learning because two networks will necessarily end up similar.

Disagreement-based Tri-net was proposed with three views which improved with diversity augmentation for pseudo label editing to solve 'collapsed neural networks' by 'View Differences' [12,45]. Uncertainty estimation is also an approach to enable reliable pseudo labels to be utilized to train other views [10,48,49]. Current key studies of Co-training mainly on: (a) enabling the diversity of two views, and (b) properly/confidently generating pseudo labels for retraining networks. In this paper of the training strategy aspect, we adopt two completely different segmentation networks to encourage the difference between two views in the feature-learning module. Furthermore, inspired by the Student-Teacher style approach [26,39], a ViT-based guidance module is developed which is much more robust with the help of the feature-learning module via perturbation, and average model weights [14,26]. The guidance module is able to confidently and properly supervise the two networks of the feature-learning module in the whole semi-supervision process via pseudo label.

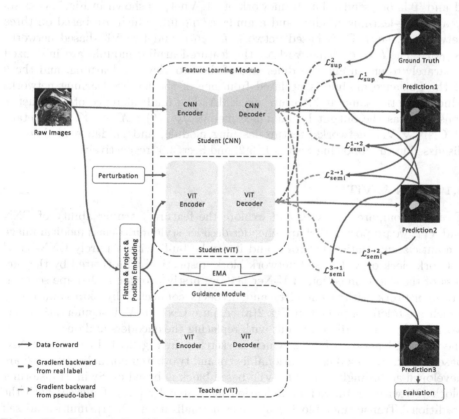

Fig. 1. The Framework of S4CVnet. It is a semi-supervised medical image semantic segmentation framework with the power of CNN and ViT, which consists of a feature-learning module (CNN & ViT), and a guidance module (ViT). The supervision mechanism is illustrated by minimizing the difference (also known as *Loss*) between prediction and (pseudo) labels.

3   Methodology

In generic semi-supervised learning for image segmentation tasks, $\mathbf{L}$, $\mathbf{U}$ and $\mathbf{T}$ normally denote a small number of labelled data, a large amount of unlabeled data, and a testing data set. We denote a batch of labeled data as $(\boldsymbol{X}_l, \boldsymbol{Y}_{gt}) \in \mathbf{L}$, $(\boldsymbol{X}_t, \boldsymbol{Y}_{gt}) \in \mathbf{T}$ for labeled training and testing data with its corresponding ground truth, and a batch of only raw data as $(\boldsymbol{X}_u) \in \mathbf{U}$ in the unlabeled data set, where $\boldsymbol{X} \in \mathbb{R}^{h \times w}$ representing a 2D gray-scale image. $\boldsymbol{Y}_p$ is the dense map predicted by a segmentation network $f(\theta) : \boldsymbol{X} \mapsto \boldsymbol{Y}_p$ with the θ as the parameters of the network f. $\boldsymbol{Y}_p$ can be considered as a batch of pseudo label for unlabeled data $(\boldsymbol{X}_u, \boldsymbol{Y}_p) \in \mathbf{U}$ for retraining networks. Final evaluation results are calculated based on the differences between Y_p and Y_{gt} of $\mathbf{T}$. The training of S4CVnet framework is to minimize the sum of supervision loss $Loss_{sup}$ and the semi-supervision loss $Loss_{semi}$ which are based on the difference of inference of each network with Y_{gt}, and Y_p, respectively. There is no overlap between $\mathbf{L}$, $\mathbf{U}$ and $\mathbf{T}$ in our study. The framework of S4CVnet, as shown in Fig. 1, consists of a feature-learning module and a guidance module which are based on three networks f, i.e. a CNN-based network $f_{CNN}(\theta)$, and two ViT-based networks $f_{ViT}(\theta)$. The θ of each network of the feature-learning module are initialized separately to encourage the difference of the two views of learning, and the θ of the guidance module is updated from one of the feature-learning networks which have the same architecture via EMA. The final inference of S4CVnet is considered as the output by guidance module $f_{ViT}(\bar{\theta}) : \boldsymbol{X} \mapsto \boldsymbol{Y}$. The details of CNN & ViT networks, feature-learning module, and guidance module are discussed in the following Sect. 3.1, 3.2, and Sect. 3.3, respectively.

3.1   CNN & ViT

To fairly compare, analyse, and explore the feature learning ability of CNN and ViT, we propose a U-shape encoder-decoder style multi-class medical image semantic segmentation network, and it can be built with a purely CNN-based network block or ViT-based network block, respectively. Motivated by the success of the skip connection of U-Net [34], we firstly propose a U-shape segmentation network with 4 encoders and decoders connected by skip connections which are briefly sketched in Fig. 2(a). A pure CNN or ViT segmentation network hereby can be directly built with replacing the encoders and decoders with the proposed network blocks which is sketched in Fig. 2(b). In each of CNN-based block, two 3×3 convolutional layers and two batch normalization [22] are developed accordingly [34]. The ViT-based block is based on Swin-Transformer block [27] with no further modification motivated by [4,6]. Different with the traditional Transformer block [13], layer normalization LN [1], multi-head self attention, residual connection [16], MLP with GELU are developed with shift-window which results in window-based multi-head self attention (WMSA) and shifted window-based multi-head self attention (SWMSA). Both of WMSA and SWMSA are applied in the two successive transformer blocks respectively shown

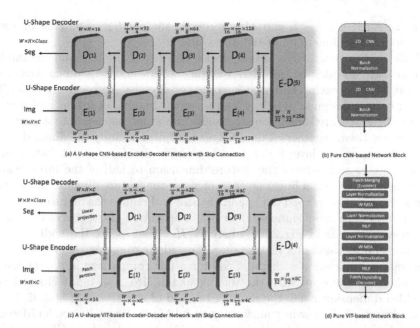

Fig. 2. The Backbone Segmentation Network. (a, c)a U-shape CNN-based or ViT-based encoder-decoder style segmentation network, (b, d)a pure CNN-based or ViT-based network block. These two network blocks can be directly applied to the U-shape encoder-decoder network resulting in a purely CNN- or ViT-based segmentation network.

on the upside of Fig. 2(b). The details of data pipeline through self-attention-based WMSA, SWMSA, MLP for feature learning of ViT are summarised in Eqs. 1, 2, 3, 4, and 5, where $i \in 1 \cdots L$, and L is the number of blocks. The self-attention mechanism comprises three point-wise linear layers mapping tokens to intermediate representations: quires Q, keys K, and values V, introduced in Eq. 5. In this way, the transformer block maps input sequence $Z_0 = [z_{0,1} \cdots z_{0,N}]$ positions to $Z_L = [z_{L,1}, ..., z_{L,N}]$, and the much richer sufficient semantic feature information (global dependencies) is fully extracted and collected through the ViT-based block.

$$Z_{i-1} = \text{WMSA}(\text{LN}(Z_{i-1})) + Z_{i-1} \tag{1}$$

$$Z_i = \text{MLP}(\text{LN}(Z_i)) + Z_i \tag{2}$$

$$Z_{i+1} = \text{SWMSA}(\text{LN}(Z_i)) + Z_i \tag{3}$$

$$Z_{i+1} = \text{MLP}(\text{LN}(Z_{i+1})) + Z_{i+1} \tag{4}$$

$$\text{MSA}(Z') = \text{softmax}(\frac{QK}{\sqrt{D}})V \tag{5}$$

where $Q, K, V \in \mathbb{R}^{M^2 \times d}$, and M^2 represents the number of patches in a window, and d is the dimension of the query and key.

Unlike conventional CNN-based blocks with downsampling and upsampling between each encoder or decoder, merging layers and expanding layers are designed between each ViT-based encoder, or decoder, respectively [4,6]. The merging layer is designed to reduce 2 times of the number of tokens and increase 2 times of the feature dimension. It divides the input patches into 4 parts and concatenates them together. A linear layer is applied to unify the dimension to 2 times. The expanding layer is designed to reshape the size of input feature maps 2 times bigger, and reduces the feature dimension to half of the input feature map dimension. It uses a linear layer to increase the feature dimension, and then rearranges operation is used to expand the size and reduce the feature dimension to a quarter of the input dimension. A brief illustration of the size of the feature map in each step is in Fig. 2(a), where W, H, C represents the width, height, and channel dimension of a feature map. Considering making the ViT the same computational efficiency with the CNN for a fair comparison and complement each other, we come up with the setting: patch size is 4, input channel is 3, embedded dimension is 96, the number of head of self-attention is 3, 6, 12, 24, the window size is 7, 2 swin-transformer-based blocks for each encoder/decoder, and the ViT is pre-trained with ImageNet [11]. More details of CNN and ViT backbone with setting is available in Appendix.

3.2   Feature-Learning Module

The semi-supervised learning, especially in the pseudo-label-based approach, has been studied in image segmentation [3,12,32]. It incorporates segmentation inference on unlabeled data from one network as the pseudo label to retrain the other network, i.e. multi-view co-training approach [15,30]. Motivated by the recent success of cross pseudo label supervision [10], which has two networks $f(\theta_1)$, $f(\theta_2)$ with same architecture but initialized separately to encourage the difference of dual views, we further propose a feature-learning module aiming to explore the power of ViT and CNN mutually. Besides parameters of two networks are of course initialized separately, two completely different architectures of networks $f_{\text{CNN}}(\theta_1)$, $f_{\text{ViT}}(\theta_2)$ are designed to benefit each other via multiview learning thus boost the performance of dual-view learning. The proposed feature-learning module to generate pseudo label can be illustrated as:

$$P_1 = f_{\text{CNN}}(X; \theta_1), P_2 = f_{\text{ViT}}(X; \theta_2). \tag{6}$$

where θ_1, θ_2 demonstrate network are initialized separately, P_1, P_2 represent the segmentation inference with $f_{\text{CNN}}(\theta_1)$, $f_{\text{ViT}}(\theta_2)$, respectively. The pseudo label based on P_1, P_2 then is utilized to supervise and complement each other. The CNN is mainly based on the local convolution operation, but the ViT is to model the global dependencies of feature through self-attention [13], so two segmentation inferences P_1, P_2 have different properties of prediction, and no explicit constraints to enforce two inferences similar. The supervision detail of simultaneously complementing each other (update parameters of ViT and CNN) is discussed in Sect. 3.4.

3.3   Guidance Module

Except for the feature-learning module to enable two networks to learn from the data, a robust guidance module is designed under the consistency-aware concern to boost the performance and also act as the final module for evaluation of S4CVnet. Inspired by temporal ensembling [26], and self-ensembling [39], the guidance network is to further supervise the perturbed networks and minimize the inconsistency. In a training process, the perturbation is firstly applied to the one of a network in the feature-learning module. Secondly, the parameter of the network is updated iteratively with back prorogation. Then the network of guidance module is updated via exponential moving average (EMA) from the feature-learning module. Finally, a much more robust guidance module which is more likely to be correct than the feature learning network is then to supervise two feature-learning networks with the consistency concern. In S4CVnet, the guidance module is based on ViT which has the same architecture as the ViT in the feature-learning network, so that guidance ViT can be constantly updated through EMA of the parameter of the ViT network learning from the data [26]. The proposed guidance module to generate pseudo label can be illustrated as:

$$P_3 = f_{\mathrm{ViT}}(X; \overline{\theta}). \tag{7}$$

where $\overline{\theta}$ demonstrates the network ViT is based on averaging network weights rather than directly trained by the data. $\overline{\theta}$ is updated based on the parameter of feature-learning ViT model θ_t on past training step t, which can be illustrated as $\overline{\theta} = \alpha\theta_{t-1} + (1 - \alpha)\theta_t$. α is a weight factor which is calculated as the $\alpha = 1 - \frac{1}{t+1}$. P_3 represents the segmentation inference with $f_{\mathrm{ViT}}(\overline{\theta})$, which is used to supervise the feature-learning module following the consistency-aware concern. The supervision details of feature-learning ViT and CNN by guidance module are discussed in Sect. 3.4.

3.4   Objective

The training objective is to minimize the sum of the supervision loss $\mathcal{L}_{\mathrm{sup}}$ and the semi-supervision $\mathcal{L}_{\mathrm{semi}}$ among the three networks $f_{\mathrm{CNN}}(\theta_1), f_{\mathrm{ViT}}(\theta_2)$, and $f_{\mathrm{ViT}}(\overline{\theta})$, so the overall loss of S4CVnet being optimized during training is detailed in Eq. 8:

$$\mathcal{L} = \mathcal{L}_{\mathrm{sup1}} + \mathcal{L}_{\mathrm{sup2}} + \lambda_1(\mathcal{L}_{\mathrm{semi1}} + \mathcal{L}_{\mathrm{semi2}}) + \lambda_2(\mathcal{L}_{\mathrm{semi3}} + \mathcal{L}_{\mathrm{semi4}}) \tag{8}$$

where λ_1, λ_2 are the weight factor of cross-supervision dual-view loss and consistency-aware loss, and it is updated every 150 iterations [26]. It is a trade-off weight that keeps increasing during the training process to make S4CVnet focus on labelled data when initialize, and then move focus to unlabeled data with our proposed semi-supervision approach. This is made under the assumption of the S4CVnet can gradually infer much reliable pseudo label confidently. The weight factor is briefly indicated in Eq. 9.

$$\lambda = e^{-5\times(1-t_{\text{iteration}}/t_{\text{maxiteration}})^2} \tag{9}$$

where t indicates the current iteration number in a complete training process. Each of $\mathcal{L}_{\text{sup}}$ and $\mathcal{L}_{\text{semi}}$ are discussed as follows:

The semi-supervision loss among each network $\mathcal{L}_{\text{semi}}$ are calculated based on Cross-Entropy CE as shown in Eq. 10:

$$\mathcal{L}_{\text{semi}} = \text{CE}\big(\text{argmax}(f_1(\boldsymbol{X};\theta), f_2(\boldsymbol{X};\theta))\big) \tag{10}$$

here we simply four $\mathcal{L}_{\text{semi}}$ losses with a pair of $(f_1(\boldsymbol{X};\theta), f_2(\boldsymbol{X};\theta))$, where the pair can be $(f_{\text{CNN}}(\boldsymbol{X};\theta_1), f_{\text{ViT}}(\boldsymbol{X};\theta_2))$, $(f_{\text{ViT}}(\boldsymbol{X};\theta_2), f_{\text{CNN}}\boldsymbol{X};\theta_1))$, $(f_{\text{ViT}}(\boldsymbol{X};\overline{\theta}), f_{\text{ViT}}(\boldsymbol{X};\theta_2))$, and $(f_{\text{ViT}}(\boldsymbol{X};\overline{\theta}), f_{\text{CNN}}(\boldsymbol{X};\theta_1))$.

The supervision loss $\mathcal{L}_{\text{sup}}$ for each network is calculated based on both CE and the Dice Coefficient Dice as shown in Eq. 11:

$$\mathcal{L}_{\text{sup}} = \frac{1}{2} \times \big(\text{CE}(Y_{\text{gt}}, f(\boldsymbol{X};\theta)) + \text{Dice}(Y_{\text{gt}}, f(\boldsymbol{X};\theta))\big) \tag{11}$$

Here we simply two $\mathcal{L}_{\text{sup}}$ supervision losses with a network $f(\boldsymbol{X};\theta)$, which can be considered as $f_{\text{CNN}}(\theta_1)$, and $f_{\text{ViT}}(\theta_2)$, because each network trained with labeled data Y_{gt} set is directly with the same way. The S4CVnet and all other baseline methods reported in Sect. 4 are with the same loss design including CE, and Dice for $\mathcal{L}_{\text{sup}}$ and $\mathcal{L}_{\text{semi}}$ in order to conduct a fair comparison.

4　Experiments and Results

Data Set. Our experiments validate the S4CVnet and all other baseline methods on the MRI ventricle segmentation data set from the automated cardiac diagnosis MICCAI Challenge 2017 [2]. The data is from 100 patients (nearly 6 000 images) covering different distributions of feature information, across five evenly distributed subgroups: normal, myocardial infarction, dilated cardiomyopathy, hypertrophic cardiomyopathy, and abnormal right ventricle. All images are resized to 224×224. 20% of images are selected as the testing set, and the rest of the data set is for training (including validation).

Implementation Details. Our code has been developed under Ubuntu 20.04 in Python 3.8.8 using Pytorch 1.10 and CUDA 11.3 using four Nvidia GeForce RTX 3090 GPU, and Intel(R) Intel Core i9-10900K. The runtimes averaged around 5 h, including the data transfer, training, inference and evaluation. The data set is processed for 2D image segmentation purposes. S4CVnet is trained for 30,000 iterations, the batch size is set to 24, the optimizer is SGD, and the learning rate is initially set to 0.01, momentum is 0.9, and weight decay is 0.0001. The network weight is saved and evaluated on the validation set every 200 iterations, and the network of guidance module with the best validation performance is used for final testing. The setting is also applied to other baseline methods directly without any modification.

Backbone. The S4CVnet consists of two types of networks as shown in Fig. 1. One is CNN-based segmentation network with skip connection, UNet [34], and the other one is ViT-based segmentation network with shift window [27] and skip connection, Swin-UNet [4]. For a fair comparison, two networks are both with U-shaped architecture with purely CNN- or ViT-based blocks as encoders and decoders. The tiny version of ViT is selected in this study to make the computational cost and training efficiency similar to CNN.

Baseline Methods. All methods including S4CVnet and other baseline methods are trained with the same hyper-parameter setting, and the same distribution of features. The randomly selection of test set, labelled train set and unlabeled train set are only conducted once and then tested with all baseline methods together as well as S4CVnet. The baseline methods reported includes: MT [39], DAN [52], ICT [41], ADVENT [42], UAMT [51], DCN [32], CTCT [30] with CNN as the backbone segmentation network.

Evaluation Measures. The direct comparison experiments between S4CVnet and other baseline methods are conducted with a variety of evaluation metrics including similarity measures: Dice, IOU, Accuracy, Precision, Sensitivity, and Specificity, which are the higher the better. The difference measures are also investigated: Hausdorff Distance (HD), and Average Surface Distance (ASD), which are the lower the better. The mean value of these metrics is reported, because the data set is a multi-class segmentation data set. The full evaluation measures are reported when comparing S4CVnet against other baseline methods, and the topological exploration of all alternative frameworks. IOU as the most common metric is also selected to report the performance of all baseline methods and S4CVnet under the assumption of different ratios of labelled data/total data. IOU, Sensitivity, and Specificity are selected to report the ablation study of different networks with different combinations of our proposed contribution.

Qualitative Results. Figure 3 illustrates eight randomly selected sample raw images with related predicted images against the published ground truth, where Yellow, Red, Green and Black represent as True Positive (TP), False Positive (FP), False Negative (FN) and True Negative (TN) inferences at pixel level, respectively. This illustrates how S4CVnet can give rise to fewer FP pixels and lower ASD compared to other methods.

Quantitative Results. Table 1 reports the direct comparison of S4CVnet against other semi-supervised methods including similarity measures and difference measures when the ratio of assumed labelled data/total data is 10%. The best result of different measures on the table is in **Bold**. A line chart in logarithmic scale is briefly sketched in Fig. 4(a), where the X-axis is the ratio of labelled data/total data, and Y-axis is the IOU performance, illustrating the valuable performance of S4CVnet against other baseline methods, especially in a

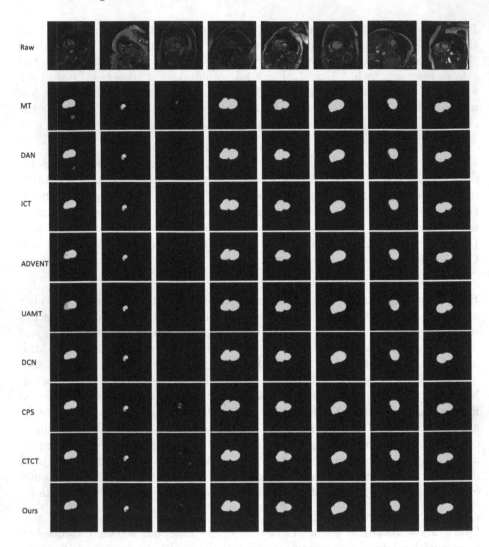

Fig. 3. Sample Qualitative Results on MRI Cardiac Test Set. Yellow, Red, Green, and Black Indicate True Positive, False Positive, False Negative, and True Positive of Each Pixel. (Color figure online)

low ratio of labelled data/total data. Details of quantitative results of S4CVnet and baseline methods under different assumption of ratio of labelled data/total data is in Appendix. A histogram indicating the cumulative distribution of IOU performance of prediction image is briefly sketched in Fig. 4(b), where the X-axis is the IOU threshold and the Y-axis is the number of predicted images on the test set, demonstrating S4CVnet is more likely to predict images with high IOU against other methods.

Table 1. Direct comparison of semi-supervised frameworks on MRI cardiac test set

Framework	mDice↑	mIOU↑	Acc↑	Pre↑	Sen↑	Spe↑	HD↓	ASD↓
MT [39]	0.8860	0.8034	0.9952	0.8898	0.8829	0.9720	9.3659	2.5960
DAN [52]	0.8773	0.7906	0.9947	0.8721	0.8832	0.9743	9.3203	3.0326
ICT [41]	0.8902	0.8096	0.9954	0.8916	0.8897	0.9745	11.6224	3.0885
ADVENT [42]	0.8728	0.7836	0.9947	0.8985	0.8517	0.9601	9.3203	3.5026
UAMT [51]	0.8683	0.7770	0.9946	0.8988	0.8416	0.9582	8.3944	2.2659
DCN [32]	0.8809	0.7953	0.9951	0.8915	0.8714	0.9690	8.9155	2.7179
tri [10]	0.8918	0.7906	0.9947	0.8721	0.8832	0.9743	**7.2026**	2.2816
CTCT [30]	0.8998	0.8245	0.9959	0.8920	0.9083	0.9825	9.6960	2.7293
S4CVnet	**0.9146**	**0.8478**	**0.9966**	**0.9036**	**0.9283**	**0.9881**	12.5359	**0.6934**

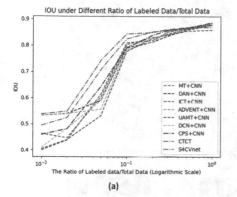
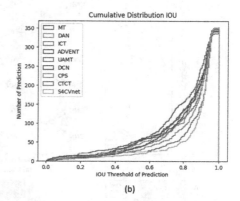

(a)   (b)

Fig. 4. The Performance of S4CVnet Against Other Baseline Methods. (a) The line chart of mIOU results on the test set with different assumptions of the ratio of label/total data for training. (b) The histogram chart indicates the cumulative distribution of IOU performance of the predicted image on the test set.

Ablation Study. In order to analyze the effects of each of the proposed contributions and combinations including the setting of the network, the mechanism of the feature-learning module and guidance module, and the robustness of each network of S4CVnet, extensive ablation experiments have been conducted and reported in Table 2. ✗ indicates either a network of feature-learning module or a guidance module is removed, and all alternative network settings (CNN or ViT) are explored. In different combinations of the proposed contribution, all the available networks are also tested separately, and the ablation study demonstrates that the proposed S4CVnet is with the most proper setting to fully utilize the power of CNN and ViT via student-teacher guidance scheme and dual-view co-training feature-learning approach in semi-supervised image semantic segmentation.

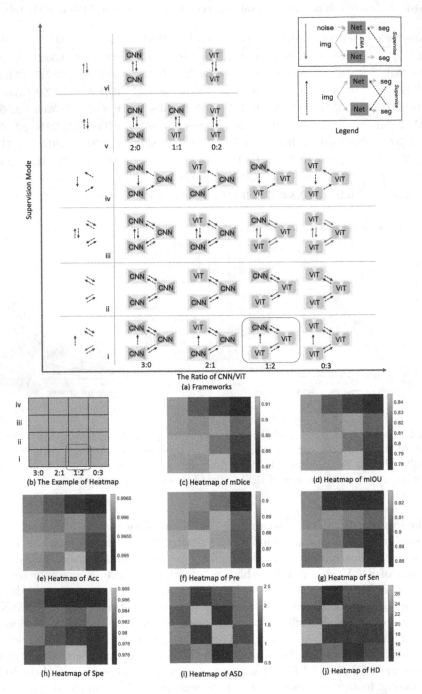

Fig. 5. The Topological Exploration of the Network (CNN&ViT), and Semi-Supervised Supervision Mode (Student-Teacher Style & Pseudo-Label). (Color figure online)

Table 2. Ablation studies on contributions of architecture and modules

Learning module		Guidance module	Test network	IOU↑	Sen↑	Spe↑
Network A	Network B	Network C				
ViT	ViT	✗	A	0.8034	0.8829	0.9720
ViT	ViT	✗	B	0.8135	0.9036	0.9821
CNN	CNN	✗	A	0.7906	0.8832	0.9743
CNN	CNN	✗	B	0.8231	0.8967	0.9761
✗	CNN	CNN	B	0.7345	0.8094	0.9586
✗	CNN	CNN	C	0.7660	0.8481	0.9585
✗	ViT	ViT	B	0.8159	0.9032	0.9822
✗	ViT	ViT	C	0.7359	0.8415	0.9716
ViT	ViT	ViT	A	0.8096	0.8995	0.9817
ViT	ViT	ViT	B	0.8194	0.9078	0.9833
ViT	ViT	ViT	C	0.8183	0.9037	0.9822
CNN	CNN	CNN	A	0.8399	0.9225	0.9848
CNN	CNN	CNN	B	0.8432	0.9189	0.9848
CNN	CNN	CNN	C	0.8345	0.9168	0.9828
CNN	ViT	ViT	A	0.8341	0.9135	0.9825
CNN	ViT	ViT	B	0.8354	0.9177	0.9839
CNN	ViT	ViT	C	**0.8478**	**0.9283**	**0.9881**

Supervision Mode Exploration. Besides the ablation study to explore the different settings and combination of networks, feature-learning module, and guidance module, we fully explore the semi-supervised learning in medical image semantic segmentation through topological exploration of all alternative supervision modes of CNN and ViT. The full list of alternative frameworks is illustrated in Fig. 5, where two supervision mode is briefly sketched in the legend of the figure. ⟶ indicates the Student-Teacher style supervision mode, and --→ indicate cross pseudo-label-based supervision mode. Figure 5(a) briefly illustrates all alternative frameworks with two axes, the Y-axis with different supervision modes from three networks to two networks, and the X-axis with the ratio of the number of CNN/ViT networks, and the proposed S4CVnet is in a red bounding box. All frameworks shown in the Fig. 5(a) have been tested and reported with heatmap format directly. Figure 5(b) is an example heatmap to indicate the supervision mode and ratio of CNN/ViT information depending on the position of heatmap with a red bounding box to illustrate where is the S4CVnet as well. Figure 5(c, d, e, f, g, h, i, j) represent the heatmap with mDice, mIOU, accuracy, precision, sensitivity, specificity, ASD, and HD validation performance, which demonstrate a whole picture of semi-supervised learning for medical semantic segmentation with CNN and ViT, and the denominating position of our proposed S4CVnet. The details of the quantitative results of topological exploration is in Appendix.

5　Conclusions

In this paper, we introduce an advanced semi-supervised learning framework in medical image semantic segmentation, S4CVnet, aiming to fully utilize the power of CNN and ViT simultaneously. S4CVnet consists of a feature-learning module and a guidance module. The feature-learning module, a dual-view feature learning approach, is proposed to enable two networks to complement each other via pseudo-label supervision. The guidance module is based on averaging network weights to supervise the learning modules under the consistency concern. Our proposed methods is evaluated with a variety of evaluation metrics and different assumption of the ratio of labelled data/total data against other semi-supervised learning baselines with the same hyperparameters settings and keeps the state-of-the-art position on a public benchmark data set. Besides a comprehensive ablation study, a topological exploration with CNN and ViT illustrates a whole picture of utilizing CNN and ViT in semi-supervised learning.

References

1. Ba, J.L., Kiros, J.R., Hinton, G.E.: Layer normalization. arXiv preprint arXiv:1607.06450 (2016)
2. Bernard, O., et al.: Deep learning techniques for automatic MRI cardiac multi-structures segmentation and diagnosis: is the problem solved? IEEE Trans. Med. Imaging 37(11), 2514–2525 (2018)
3. Blum, A., Mitchell, T.: Combining labeled and unlabeled data with co-training. In: Proceedings of the Eleventh Annual Conference on Computational Learning Theory, pp. 92–100 (1998)
4. Cao, H., et al.: Swin-UNet: UNet-like pure transformer for medical image segmentation. arXiv preprint arXiv:2105.05537 (2021)
5. Chang, Y.T., et al.: Weakly-supervised semantic segmentation via sub-category exploration. In: Proceedings of the IEEE/CVF Conference on Computer Vision and Pattern Recognition, pp. 8991–9000 (2020)
6. Chen, J., et al.: TransUNet: transformers make strong encoders for medical image segmentation. arXiv preprint arXiv:2102.04306 (2021)
7. Chen, L.-C., et al.: Naive-student: leveraging semi-supervised learning in video sequences for urban scene segmentation. In: Vedaldi, A., Bischof, H., Brox, T., Frahm, J.-M. (eds.) ECCV 2020. LNCS, vol. 12354, pp. 695–714. Springer, Cham (2020). https://doi.org/10.1007/978-3-030-58545-7_40
8. Chen, L.C., Papandreou, G., Kokkinos, I., Murphy, K., Yuille, A.L.: DeepLab: semantic image segmentation with deep convolutional nets, atrous convolution, and fully connected CRFs. IEEE Trans. Pattern Anal. Mach. Intell. 40(4), 834–848 (2017)
9. Chen, L.C., Zhu, Y., Papandreou, G., Schroff, F., Adam, H.: Encoder-decoder with atrous separable convolution for semantic image segmentation. In: Proceedings of the European Conference on Computer Vision, pp. 801–818 (2018)

10. Chen, X., et al.: Semi-supervised semantic segmentation with cross pseudo supervision. In: CVPR (2021)
11. Deng, J., et al.: ImageNet: a large-scale hierarchical image database. In: 2009 IEEE Conference on Computer Vision and Pattern Recognition, pp. 248–255. IEEE (2009)
12. Dong-DongChen, W., WeiGao, Z.H.: Tri-net for semi-supervised deep learning. In: Proceedings of Twenty-Seventh International Joint Conference on Artificial Intelligence, pp. 2014–2020 (2018)
13. Dosovitskiy, A., et al.: An image is worth 16×16 words: transformers for image recognition at scale. arXiv preprint arXiv:2010.11929 (2020)
14. Gal, Y., Ghahramani, Z.: Dropout as a Bayesian approximation: representing model uncertainty in deep learning. In: International Conference on Machine Learning, pp. 1050–1059. PMLR (2016)
15. Han, B., et al.: Co-teaching: robust training of deep neural networks with extremely noisy labels. In: Advances in Neural Information Processing Systems, vol. 31 (2018)
16. He, K., et al.: Deep residual learning for image recognition. In: Proceedings of the IEEE Conference on Computer Vision and Pattern Recognition (2016)
17. Huang, B., et al.: Simultaneous depth estimation and surgical tool segmentation in laparoscopic images. IEEE Trans. Med. Robot. Bionics 4(2), 335–338 (2022)
18. Huang, G., Liu, Z., Van Der Maaten, L., Weinberger, K.Q.: Densely connected convolutional networks. In: Proceedings of the IEEE Conference on Computer Vision and Pattern Recognition, pp. 4700–4708 (2017)
19. Hung, W.C., et al.: Adversarial learning for semi-supervised semantic segmentation. In: 29th British Machine Vision Conference, BMVC 2018 (2018)
20. Ibrahim, M.S., et al.: Semi-supervised semantic image segmentation with self-correcting networks. In: Proceedings of the IEEE/CVF Conference on Computer Vision and Pattern Recognition, pp. 12715–12725 (2020)
21. Ibtehaz, N., Rahman, M.S.: MultiResUNet: rethinking the U-Net architecture for multimodal biomedical image segmentation. Neural Netw. **121**, 74–87 (2020)
22. Ioffe, S., Szegedy, C.: Batch normalization: accelerating deep network training by reducing internal covariate shift. In: International Conference on Machine Learning, pp. 448–456. PMLR (2015)
23. Isensee, F., et al.: nnU-Net: self-adapting framework for U-Net-based medical image segmentation. In: Handels, H., Deserno, T., Maier, A., Maier-Hein, K., Palm, C., Tolxdorff, T. (eds.) Bildverarbeitung für die Medizin 2019. I, p. 22. Springer, Wiesbaden (2019). https://doi.org/10.1007/978-3-658-25326-4_7
24. Jaderberg, M., Simonyan, K., Zisserman, A., et al.: Spatial transformer networks. In: Advances in Neural Information Processing Systems, vol. 28 (2015)
25. Ji, W., et al.: Learning calibrated medical image segmentation via multi-rater agreement modeling. In: Proceedings of the IEEE/CVF Conference on Computer Vision and Pattern Recognition, pp. 12341–12351 (2021)
26. Laine, S., Aila, T.: Temporal ensembling for semi-supervised learning. arXiv preprint arXiv:1610.02242 (2016)
27. Liu, Z., et al.: Swin transformer: hierarchical vision transformer using shifted windows. In: Proceedings of the IEEE/CVF International Conference on Computer Vision, pp. 10012–10022 (2021)
28. Long, J., Shelhamer, E., Darrell, T.: Fully convolutional networks for semantic segmentation. In: Proceedings of the IEEE Conference on Computer Vision and Pattern Recognition, pp. 3431–3440 (2015)

29. Luo, X., et al.: Efficient semi-supervised gross target volume of nasopharyngeal carcinoma segmentation via uncertainty rectified pyramid consistency. In: de Bruijne, M., et al. (eds.) MICCAI 2021. LNCS, vol. 12902, pp. 318–329. Springer, Cham (2021). https://doi.org/10.1007/978-3-030-87196-3_30

30. Luo, X., et al.: Semi-supervised medical image segmentation via cross teaching between CNN and transformer. arXiv preprint arXiv:2112.04894 (2021)

31. Mendel, R., de Souza, L.A., Rauber, D., Papa, J.P., Palm, C.: Semi-supervised segmentation based on error-correcting supervision. In: Vedaldi, A., Bischof, H., Brox, T., Frahm, J.-M. (eds.) ECCV 2020. LNCS, vol. 12374, pp. 141–157. Springer, Cham (2020). https://doi.org/10.1007/978-3-030-58526-6_9

32. Qiao, S., Shen, W., Zhang, Z., Wang, B., Yuille, A.: Deep co-training for semi-supervised image recognition. In: Proceedings of the European Conference on Computer Vision, pp. 135–152 (2018)

33. Reiß, S., Seibold, C., Freytag, A., Rodner, E., Stiefelhagen, R.: Every annotation counts: multi-label deep supervision for medical image segmentation. In: Proceedings of the IEEE/CVF Conference on Computer Vision and Pattern Recognition, pp. 9532–9542 (2021)

34. Ronneberger, O., Fischer, P., Brox, T.: U-Net: convolutional networks for biomedical image segmentation. In: Navab, N., Hornegger, J., Wells, W.M., Frangi, A.F. (eds.) MICCAI 2015. LNCS, vol. 9351, pp. 234–241. Springer, Cham (2015). https://doi.org/10.1007/978-3-319-24574-4_28

35. Woo, S., et al.: CBAM: convolutional block attention module. In: Proceedings of the European Conference on Computer Vision (ECCV), pp. 3–19 (2018)

36. Song, C., et al.: Box-driven class-wise region masking and filling rate guided loss for weakly supervised semantic segmentation. In: Proceedings of the IEEE/CVF Conference on Computer Vision and Pattern Recognition, pp. 3136–3145 (2019)

37. Souly, N., Spampinato, C., Shah, M.: Semi supervised semantic segmentation using generative adversarial network. In: Proceedings of the IEEE International Conference on Computer Vision, pp. 5688–5696 (2017)

38. Strudel, R., Garcia, R., Laptev, I., Schmid, C.: Segmenter: transformer for semantic segmentation. In: Proceedings of the IEEE/CVF International Conference on Computer Vision, pp. 7262–7272 (2021)

39. Tarvainen, A., et al.: Mean teachers are better role models: weight-averaged consistency targets improve semi-supervised deep learning results. In: Advances in Neural Information Processing Systems (2017)

40. Vaswani, A., et al.: Attention is all you need. In: Advances in Neural Information Processing Systems (2017)

41. Verma, V., et al.: Interpolation consistency training for semi-supervised learning. In: International Joint Conference on Artificial Intelligence (2019)

42. Vu, T.H., et al.: Advent: adversarial entropy minimization for domain adaptation in semantic segmentation. In: Proceedings of the IEEE/CVF Conference on Computer Vision and Pattern Recognition, pp. 2517–2526 (2019)

43. Wang, Z.: Deep learning in medical ultrasound image segmentation: a review. arXiv preprint arXiv:2002.07703 (2020)

44. Wang, Z., et al.: RAR-U-Net: a residual encoder to attention decoder by residual connections framework for spine segmentation under noisy labels. In: 2021 IEEE International Conference on Image Processing (ICIP). IEEE (2021)

45. Wang, Z., Voiculescu, I.: Triple-view feature learning for medical image segmentation. In: Xu, X., Li, X., Mahapatra, D., Cheng, L., Petitjean, C., Fu, H. (eds.) REMIA 2022. LNCS, vol. 13543, pp. 42–54. Springer, Cham (2022). https://doi.org/10.1007/978-3-031-16876-5_5

46. Wang, Z., Voiculescu, I.: Quadruple augmented pyramid network for multi-class Covid-19 segmentation via CT. In: 2021 43rd Annual International Conference of the IEEE Engineering in Medicine Biology Society (EMBC) (2021)
47. Wang, Z., et al.: Computationally-efficient vision transformer for medical image semantic segmentation via dual pseudo-label supervision. In: IEEE International Conference on Image Processing (ICIP) (2022)
48. Wang, Z., Zheng, J.Q., Voiculescu, I.: An uncertainty-aware transformer for MRI cardiac semantic segmentation via mean teachers. In: Yang, G., Aviles-Rivero, A., Roberts, M., Schönlieb, C.B. (eds.) MIUA 2022. LNCS, vol. 13413, pp. 497–507. Springer, Cham (2022). https://doi.org/10.1007/978-3-031-12053-4_37
49. Xia, Y., et al.: 3D semi-supervised learning with uncertainty-aware multi-view co-training. In: Proceedings of the IEEE/CVF Winter Conference on Applications of Computer Vision, pp. 3646–3655 (2020)
50. You, X., et al.: Segmentation of retinal blood vessels using the radial projection and semi-supervised approach. Pattern Recogn. 44(10–11), 2314–2324 (2011)
51. Yu, L., Wang, S., Li, X., Fu, C.-W., Heng, P.-A.: Uncertainty-aware self-ensembling model for semi-supervised 3D left atrium segmentation. In: Shen, D., et al. (eds.) MICCAI 2019. LNCS, vol. 11765, pp. 605–613. Springer, Cham (2019). https://doi.org/10.1007/978-3-030-32245-8_67
52. Zhang, Y., Yang, L., Chen, J., Fredericksen, M., Hughes, D.P., Chen, D.Z.: Deep adversarial networks for biomedical image segmentation utilizing unannotated images. In: Descoteaux, M., Maier-Hein, L., Franz, A., Jannin, P., Collins, D.L., Duchesne, S. (eds.) MICCAI 2017. LNCS, vol. 10435, pp. 408–416. Springer, Cham (2017). https://doi.org/10.1007/978-3-319-66179-7_47
53. Zhou, B., et al.: Learning deep features for discriminative localization. In: Proceedings of the IEEE Conference on Computer Vision and Pattern Recognition, pp. 2921–2929 (2016)
54. Zoph, B., et al.: Rethinking pre-training and self-training. In: Advances in Neural Information Processing Systems, vol. 33, pp. 3833–3845 (2020)

Using Whole Slide Image Representations from Self-supervised Contrastive Learning for Melanoma Concordance Regression

Sean Grullon[1]($\boxtimes$), Vaughn Spurrier[1], Jiayi Zhao[1], Corey Chivers[1], Yang Jiang[1], Kiran Motaparthi[2], Jason Lee[3], Michael Bonham[1], and Julianna Ianni[1]

[1] Proscia, Inc., Philadelphia, USA
sean.grullon@proscia.com
[2] Department of Dermatology, University of Florida College of Medicine, Gainesville, USA
[3] Department of Dermatology, Sidney Kimmel Medical College at Thomas Jefferson University, Philadelphia, USA

Abstract. Although melanoma occurs more rarely than several other skin cancers, patients' long term survival rate is extremely low if the diagnosis is missed. Diagnosis is complicated by a high discordance rate among pathologists when distinguishing between melanoma and benign melanocytic lesions. A tool that provides potential concordance information to healthcare providers could help inform diagnostic, prognostic, and therapeutic decision-making for challenging melanoma cases. We present a melanoma concordance regression deep learning model capable of predicting the concordance rate of invasive melanoma or melanoma in-situ from digitized Whole Slide Images (WSIs). The salient features corresponding to melanoma concordance were learned in a self-supervised manner with the contrastive learning method, SimCLR. We trained a SimCLR feature extractor with 83,356 WSI tiles randomly sampled from 10,895 specimens originating from four distinct pathology labs. We trained a separate melanoma concordance regression model on 990 specimens with available concordance ground truth annotations from three pathology labs and tested the model on 211 specimens. We achieved a Root Mean Squared Error (RMSE) of 0.28 ± 0.01 on the test set. We also investigated the performance of using the predicted concordance rate as a malignancy classifier, and achieved a precision and recall of 0.85 ± 0.05 and 0.61 ± 0.06, respectively, on the test set. These results are an important first step for building an artificial intelligence (AI) system capable of predicting the results of consulting a panel of experts and delivering a score based on the degree to which the experts would agree on a particular diagnosis. Such a system could be used to suggest additional testing or other action such as ordering additional stains or genetic tests.

Keywords: Self supervised learning · Contrastive learning · Melanoma · Weak supervision · Multiple instance learning · Digital pathology

© The Author(s), under exclusive license to Springer Nature Switzerland AG 2023
L. Karlinsky et al. (Eds.): ECCV 2022 Workshops, LNCS 13807, pp. 442–456, 2023.
https://doi.org/10.1007/978-3-031-25082-8_29

1   Introduction

More than 5 million diagnoses of skin cancer are made each year in the United States, about 106,000 of which are melanoma of the skin [1]. Diagnosis requires microscopic examination of hematoxylin and eosin (H&E) stained, paraffin wax embedded biopsies of skin lesion specimens on glass slides. These slides can be manually observed under a microscope, or digitally on a Whole Slide Image (WSI) scanned on specialty hardware. The 5-year survival rate of patients with metastatic malignant melanoma is less than 20% [15]. Melanoma occurs more rarely than several other types of skin cancer, and its diagnosis is challenging, as evidenced by a high discordance rate among pathologists when distinguishing between melanoma and benign melanocytic lesions ($\sim$40% discordance rate; e.g. [7,10]). The high discordance rate highlights that greater scrutiny is likely needed to arrive at an accurate melanoma diagnosis, however patients receive diagnoses only from a single dermatopathologist in many instances. This tends to increase the probability of misdiagnosis, where frequent over-diagnosis of melanocytic lesions results in severe costs to a clinical practice and additional costs and distress to patients. [26]. In this scenario, the decision-making of the single expert would be further informed by knowledge of a likely concordance level among a group of multiple experts in a given case under consideration. Additional methods of providing concordance information to healthcare providers could help further inform diagnostic, prognostic, and therapeutic decision-making for challenging melanoma cases. A method capable of predicting the results of consulting a panel of experts and delivering a score based on the degree to which the experts would agree on a particular diagnosis would help reduce melanoma misdiagnosis and subsequently improve patient care.

The advent of digital pathology has brought the revolution in machine learning and artificial intelligence to bear on a variety of tasks common to pathology labs. Campanella et al. [2] trained a model in a weakly-supervised framework that did not require pixel-level annotations to classify prostate cancer and validated on $\sim$10,000 WSIs sourced from multiple countries. This represented a considerable advancement towards a system capable of use in clinical practice for prostate cancer. However, some degree of human-in-the-loop curation was performed on their data set, including manual quality control such as post-hoc removal of slides with pen ink from the study. Pantanowitz et al. [17] used pixel-wise annotations to develop a model trained on $\sim$550 WSIs that distinguishes high-grade from low-grade prostate cancer. In dermatopathology, the model developed in [22] classified skin lesion specimens between six morphology-based groups (including melanoma), was tested on $\sim$5099 WSIs, provided automated quality control to remove WSI patches with pen ink or blur, and also demonstrated that use of confidence thresholding could provide a high accuracy.

The recent application of deep learning to digital pathology has predominately leveraged the use of pre-trained WSI tile representations, usually obtained by using feature extractors pre-trained on the ImageNet [6] data set. The features learned by such pre-training is dominated by features present in natural-scene images, which are not guaranteed to generalize to histopathology images. Such representations can limit the reported performance metrics and affect model robustness. It has been shown in [13] that self-supervised pre-training on WSIs improved the downstream performance in identifying metastastic breast cancer.

In this work, we present a deep learning regression model capable of predicting from WSIs the concordance rate of consulting a panel of experts on rendering a case diagnosis of invasive melanoma or melanoma in-situ. The deep learning model learns meaningful feature representations from WSIs through self-supervised pre-training, which are used to learn the concordance rate through weakly-supervised training.

2   Methods

2.1   Data Collection and Characteristics

The melanoma concordance regression model was trained and evaluated on 1,412 specimens (consisting of 1,722 WSIs) from three distinct pathology labs. The first lab consists for 611 suspected melanoma specimens from a leading dermatopathology lab in a top academic medical center (Department of Dermatology at University of Florida College of Medicine), denoted as *University of Florida*. The second lab consisted of 605 suspected melanoma specimens distributed across North America, but re-scanned at The Department of Dermatology at University of Florida College of Medicine, denoted as *Florida - External*. The third lab consisted of 319 suspected specimens from Jefferson Dermatopathology Center, Department of Dermatology & Cutaneous Biology, Thomas Jefferson University denoted as *Jefferson*. The WSIs consisted exclusively of H&E-stained, formalin-fixed, paraffin-embedded dermatopathology tissue and were all scanned using a 3DHistech P250 High Capacity Slide Scanner at an objective power of 20X, corresponding to $0.24\,\mu m$/pixel. The diagnostic categories present in our data set are summarized in Table 1.

The annotations for our data set were provided by at least three board-certified pathologists who reviewed each melanocytic specimen. The first review was the original specimen diagnosis made via glass slide examination under a microscope. At least two and up to four additional dermatopathologists independently reviewed and rendered a diagnosis digitally for each melanocytic specimen. The patient's year of birth and gender were provided with each specimen upon review. Two dermatopathologists from the United States reviewed all 1,412 specimens in our data set and up to two additional dermatopathologists reviewed a subset of our data set. A summary of the number of concordant reviews in this study is given in Table 2.

The concordance reviews are converted to a concordance rate by calculating the fraction of dermatopathologists who rendered a diagnosis of melanoma in-situ or invasive melanoma. The concordance rate runs from 0.0 (No dermatopathologist rendered a melanoma in-situ/invasive melanoma diagnosis) to 1.0 (all dermatopathologists rendered a melanoma in-situ/invasive melanoma diagnosis). It has been previously noted [20] that the concordance rate itself is correlated with the likelihood that a specimen is malignant. The concordant labels present in

Table 1. Specimen counts of each of the pathologies in the data set, broken-out into specific diagnostic categories.

Diagnostic morphology	Number of specimens
Melanoma In Situ	607
Invasive Melanoma	306
Mild-to-moderate Dysplastic Nevus	209
Conventional Melanocytic Nevus	123
Severe Dysplastic Nevus	49
Spitz Nevus with Spindle Cell Morphology	43
Spitz Nevus	36
Junctional Nevus	27
Dermal Nevus	23
Blue Nevus	20
Halo Nevus	11
Total	1412

our data set include 0.0, 0.25, 0.33, 0.5, 0.67, 0.75, and 1.0. For training, validating, and testing, we divided this data set into three partitions by sampling at random without replacement with 70% of specimens used for training, and 15% used for each of validation and testing. 990 specimens were used for training, 211 specimens were used for validation, and 211 specimens were used for testing.

Table 2. Number of concordant reviews in our data set.

Number of dermatopathologists	Number of specimens
Three	687
Four	216
Five	509
Total	1412

2.2   Melanoma Concordance Regression Deep Learning Architecture

The Melanoma Concordance Regression deep learning pipeline consists of three main components: quality control, feature extraction and concordance regression. A diagram of the pipeline is shown in Fig. 1. Each specimen was first segmented into tissue-containing regions through Otsu's method [16], subdivided into 128×128 pixel tiles, and extracted at an objective power of 10X. Each tile was passed through the quality control and feature extraction components of the pipeline.

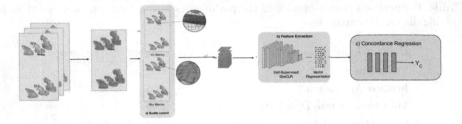

Fig. 1. The stages of the Melanoma Concordance Regression pipeline are: Quality Control, Feature Extraction, and Regression. All single specimen WSIs were first passed through the tiling stage, then the quality control stage consisting of ink and blur filtering. The filtered tiles were passed through the feature extraction stage consisting of a self-supervised SimCLR network pre-trained on WSIs with a ResNet50 backbone to obtain embedded vectors. Finally, the vectors were propagated through the regression stage consisting of fully connected layers to obtain a concordance prediction.

Quality Control. Quality control consisted of ink and blur filtering. Pen ink is common in labs migrating their workload from glass slides to WSIs where the location of possible malignancy was marked. This pen ink represented a biased distractor signal in training that is highly correlated with melanoma. Tiles containing pen ink were identified by a weakly supervised model trained to detect inked slides. These tiles were removed from the training and validation data and before inference on the test set. We also sought to remove areas of the image that were out of focus due to scanning errors by setting a threshold on the variance of the Laplacian over each tile [18,19].

Self-supervised Feature Extraction. The next component of the Melanoma Concordance Regression pipeline extracted informative features from the quality controlled tiles. To capture higher-level features in these tiles, we trained a self-supervised feature extractor based on the contrastive learning method proposed in [3,4] known as SimCLR. SimCLR relies on maximizing agreement in the latent space between representations of two augmented views of the same image. In particular, we maximized agreement between two augmented views of 128×128 pixel tiles in our data set.

In order to capture as much variety in real-world WSIs as possible, we trained a dermatopathology feature extractor neural network with the Sim-CLR contrastive learning strategy using skin specimens from four labs and three distinct scanners. The skin specimens originated from both from a sequentially-accessioned workflow from these labs as well as a curated data set to capture the morphological diversity of different skin pathologies. The curated data set included various types of basal and squamous cell carcinomas, benign to moderately atypical melanocytic nevi, atypical melanocytic nevi, melanoma in-situ, and invasive melanoma. We note that the feature extractor training set consisted of wider variety of skin pathologies than the concordance regression model, which was trained and tested only on the skin pathologies outlined in Table 1. We included WSIs from the University of Florida and Jefferson that were scanned with 3D Histech P250 scanners. We also included WSIs from another top medical center, the Department of Pathology and Laboratory Medicine at Cedars-Sinai Medical Center, which were scanned with a Ventana DP 200 scanner. We finally included WSIs from an undisclosed partner lab in western Europe that were scanned with a Hamamatsu NanoZoomer XR scanner. We note that WSIs from Cedars-Sinai and the undisclosed partner lab were only used for training the feature extractor and not the concordance regression model, as concordance review annotations were not available for these labs.

We randomly sampled 26,209 tiles from Florida and 57,147 tiles from the remaining three labs for a total of 83,356 tiles randomly sampled from 10,895 specimens for use during training of the feature extraction network. Each tile was sampled from WSIs extracted at an objective power of 10x. We set the temperature hyperparameter, (τ) to $\tau = 0.1$, batch size $= 128$, and the learning rate $= 0.001$ during training. (We note that the feature extractor is trained separately from the Melanoma Concordance Regression model described in Sect. 2.2.) We randomly divided 80% of the tiles for training, and 20% for validation. We used the ResNet50 [11] backbone for training, and the Normalized Temperature-scaled Cross Entropy (NT-Xent) loss function as in [3]. We followed the same augmentation strategies in [4] applied to the tiles. We used a temperature hyperparemter value of 0.1. We achieved a minimum NT-Xent loss value on the validation set of 2.6. As a point of comparison, we note that the minimum validatation NT-Xent loss when training with the ImageNet data set with a ResNet50 backbone is 4.4 [8]. However, given that tiles from WSIs are vastly different from landscape images in ImageNet, this may not be meaningful. Once the feature extractor was trained, it was deployed to the Melanoma Concordance Regression pipeline in order to embed each tile from the melanoma consensus regression data set (consisting of 1,722 WSIs) into a latent space of 2048-channel vectors.

Melanoma Concordance Regression. The melanoma concordance regression model predicted a value representing the fraction of dermatopathologists that is concordant with a diagnosis of Melanoma In-Situ or Invasive Melanoma. The model consisted of four fully-connected layers (two layers of 1024 channels each, followed by two of 512 channels each). Each neuron in these four layers was ReLU activated. The model was trained under a weakly-supervised multiple-instance learning (MIL) paradigm. Each embedded tile propagated through the feature extractor described in Sect. 2.2 was treated as an instance of a bag containing all quality-assured tiles of a specimen. Embedded tiles were aggregated using sigmoid-activated attention heads [12]. The final layer after the attention head was a linear layer that takes 512 channels as input and outputs a concordance rate prediction. To help prevent over-fitting, the training data set consisted of augmented versions (inspired by Tellez *et al.* [25]) of the tiles. Augmentations were generated with the following augmentation strategies: random variations in brightness, hue, contrast, saturation, (up to a maximum of 15%), Gaussian noise with 0.001 variance, and random 90° image rotations. We trained the melanoma concordance regression model with the Root mean squared error (RMSE) loss function. We regularized model training by using the dropout method [24] with 20% probability.

3   Results

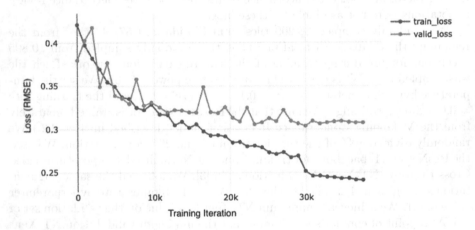

Fig. 2. Melanoma Concordance Regression Training and Validation Loss Curves. The minimum validation RMSE loss was found to be 0.30. The RMSE loss on the test set was found to be 0.28 ± 0.01.

To demonstrate the performance of the Melanoma Concordance Regression model, we first show the training and validation loss curves in Fig. 2. The minimum validation loss was found to be 0.30. We then calculated the RMSE and R^2 on the test set both across laboratory sites and for individual laboratory sites with 90% confidence intervals derived from bootstrapping with replacement. The RMSE across laboratory sites on the test set was calculated to be 0.28 ± 0.01 and the R^2 was found to be 0.51 ± 0.05. The lab-specific regression performance of the model on the test set is summarized in Table 3. We note that the RMSE performance is consistent across sites and are within the error bars derived from bootstrapping within replacement. The correlation between

Table 3. Regression metrics across the individual laboratory sites in our test data set. Errors are 90% confidence intervals derived through bootstrapping.

Metric	All sites	University of Florida	Florida - External	Jefferson
RMSE	0.28 ± 0.01	0.27 ± 0.02	0.26 ± 0.02	0.3 ± 0.03
R^2	0.51 ± 0.05	0.49 ± 0.09	0.52 ± 0.08	0.36 ± 0.13

the melanoma concordance predictions and the ground truth labels is shown in Fig. 3.

We next assessed the goodness-of-fit of the regression model by calculating the standardized residuals:

$$e_i = \frac{y_{pred,i} - y_{true,i}}{\sigma} \tag{1}$$

where i is the ith data point, and σ is the standard deviation of the residuals. The P-P plot comparing the cumulative distribution functions of the standardized residuals to a Normal distribution is shown in Fig. 4. The standardized residuals were found to be consistent with a Normal distribution by performing the Shaprio-Wilk test [23], where the p-value was found to be 0.97.

3.1   Malignant Classification

Table 4. Classification metrics when using the predicted concordance value as a malignant classifier across the individual laboratory sites in our test data set. Errors are 90% confidence intervals derived through bootstrapping with replacement.

Metric	All sites	University of Florida	Florida - External	Jefferson
AUC	0.89 ± 0.02	0.86 ± 0.04	0.9 ± 0.04	0.85 ± 0.06
Precision	0.86 ± 0.05	0.89 ± 0.08	1.0 ± 0.0	0.7 ± 0.15
Recall	0.61 ± 0.06	0.65 ± 0.09	0.5 ± 0.1	0.74 ± 0.12
Specificity	0.97 ± 0.02	0.94 ± 0.04	1.0 ± 0.0	0.9 ± 0.05

As mentioned in Sect. 2.1, increased inter-pathologist agreement on melanoma correlates with malignancy. We therefore investigated the performance of using the predicted concordance rate of the melanoma concordance model as a binary classifier to classify malignancy. We derived malignancy binary ground truth labels by defining a threshold value to binarize the ground truth concordance label, where a specimen with an observed concordance rate above this threshold value received a label of malignant, and below this threshold value received a label of not malignant. We performed a grid search on possible ground truth thresholds, and chose the threshold that maximized the Area Underneath

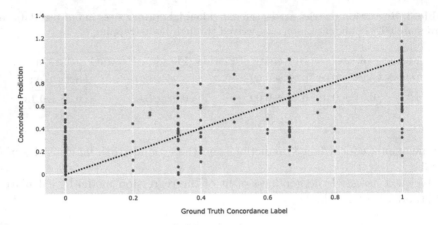

Fig. 3. The correlation of melanoma concordance predictions against the ground truth concordance label.

the Receiver Operating Characteristic (ROC) Curve (AUC). We also found the same ground truth threshold maximized the average precision of the precision-recall curve. We found that a threshold of 0.85 on the ground truth concordance rate to label malignancy maximizes both AUC and average precision, and yielded an AUC value of 0.89 ± 0.02 and an average precision of 0.81 ± 0.04. The classification metrics with this threshold are shown both across laboratory sites and individual laboratory sites in Table 4. The ROC and precision-recall curves of using the melanoma concordance prediction as a malignant classifier is shown in Fig. 5 both across laboratory sites as well as individual laboratory sites. We note that the classification performance are within the error bars derived from bootstrapping with replacement across sites, with the exception of the low false positive rate at Florida - External, which exhibits very high precision.

3.2   Ablation Studies

Binary Classifier Ablation Study. We compared the performance of using the concordance rate as a malignancy classifier with a model trained to perform a binary classification task. In particular, we configured the final linear layer of the deep learning model to predict one of two classes: malignant or not malignant. We subsequently trained a binary classification model with the cross entropy loss function. We used a threshold of 0.85 on the ground truth to annotate malignancy derived from Sect. 3.1. We performed this ablation study specifically on data from the University of Florida, and the results are shown in Table 5. It can be seen that using the concordance rate to classify malignancy yields better performance than training a dedicated binary classifier.

Feature Extractor Ablation Study. We mentioned in Sect. 1 that features learned by pre-training on ImageNet are not guaranteed to generalize to

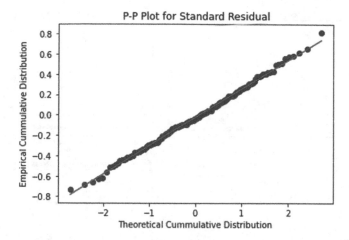

Fig. 4. A Probability-Probability plot is used as a goodness-of-fit in order to compare the cumulative distribution function of the standardized residuals of the melanoma concordance regression model to a Gaussian distribution. The P-P plot demonstrates the standardized residuals of melanoma concordance regression are normally distributed. The residuals were found to be consistent with a Normal distribution by performing the Shapiro-Wilk test on the standardized residuals. The p-value of the test was found to be 0.97.

histopathology image and could limit performance metrics and model robustness. To test this claim, we first visually examined SimCLR feature vectors to investigate visual coherence with morphological features. In particular, we projected the SimCLR feature vectors of 1,440 selected tiles to two dimensions with the Uniform Manifold Approximation and Projection (UMAP [14]) algorithm. We then arranged the tiles on a grid where the two-dimensional UMAP coordinates are mapped to the nearest grid coordinate, which arranged the tiles onto an evenly spaced plane. The arranged tiles are shown in Fig. 6. It can be seen that the learned feature vectors displays a coherence with respect to morphological features.

Table 5. Results of the binary classifier ablation experiment. A binary malignancy classifier was trained with the cross entropy loss function on data from the University of Florida, and the results are compared to using the predicted concordance rate from the regression model to classify malignancy.

Metric	Melanoma concordance regression	Binary classifier
AUC	0.86 ± 0.04	0.67 ± 0.03
Precision	0.89 ± 0.08	0.57 ± 0.07
Recall	0.62 ± 0.08	0.51 ± 0.07
Specificity	0.97 ± 0.02	0.75 ± 0.04

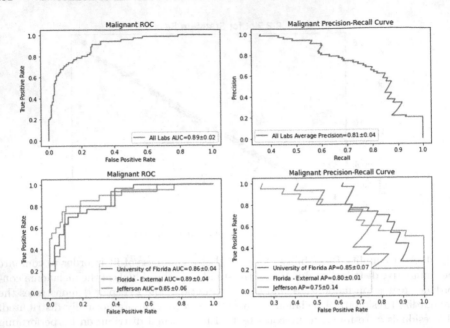

Fig. 5. Receiver Operating Characteristic (ROC) and Precision-Recall curves for using the melanoma concordance prediction as a malignant classifier on the test data set. The top row shows the curves across laboratory sites, and the bottom row shows the curves for individual sites. The AUC value for malignant classification was found to be 0.89 ± 0.02 and the average precision (AP) was found to be 0.81 ± 0.04. The bottom row demonstrates the curves for individual laboratory sites.

We next performed an ablation experiment to quantify the impact that our self-supervised feature extractor has on model performance. In particular, we ablated the SimCLR embedder by propagating the WSI tiles specifically from the University of Florida through a ResNet50 [11] pre-trained on the ImageNet [6] data set to embed each input tile into 2048 channel vectors. The embedded vectors were used to train a subsequent melanoma concordance regression model. The resulting model performance for this feature ablation experiment is summarized in Table 6 for both the raw regression metrics as well as the malignant classification task. (The same 0.85 threshold on the ground truth concordance rate defined in Sect. 3.1 was used to define malignancy) It can be seen that SimCLR results in higher performance in both the concordance regression and malignant classification tasks. In particular, there is a 14% improvement in RMSE in using the SimCLR embedder over imagenet for the regression task, and a 27.5% improvement in recall for the malignant classification task.

Fig. 6. Grid of SimCLR feature vectors arranged by their embedded vector values. First, SimCLR feature vectors of 1,440 selected tiles were projected onto two dimensions with the UMAP algorithm. We next arranged the WSI tiles on a grid where the two-dimensional UMAP coordinates are mapped to the nearest grid coordinate, which arranged the tiles onto an evenly spaced plane.

Table 6. Feature Extractor ablation study results from the University of Florida. feature vectors derived from SimCLR with a ResNet50 model backbone trained on Whole Slide Images were compared to feature vectors derived from a pre-trained ResNet50 model on the ImageNet data set. Concordance regression metrics and classification metrics are shown. Classification metrics were derived by utilizing the predicted concordance value as a malignant classifier.

Metric	SimCLR trained with WSIs	ResNet50 trained with ImageNet
RMSE	0.27 ± 0.02	0.31 ± 0.02
R^2	0.49 ± 0.09	0.15 ± 0.15
AUC	0.86 ± 0.04	0.75 ± 0.06
Precision	0.89 ± 0.08	0.76 ± 0.07
Recall	0.62 ± 0.08	0.45 ± 0.08
Specificity	0.97 ± 0.02	0.71 ± 0.08

4 Conclusions

We presented in this paper a melanoma concordance regression model that was trained across three laboratory sites that demonstrates regression performance of 0.28 ± 0.01 on the test set across the three laboratory sites in our data set.

The performance of the melanoma concordance regression model was limited by the number of dermatopathologists performing concordant reviews on our data set. From Table 2, the average number of dermatopathologists reviewing a case is 3.7 pathologists, resulting in a concordance rate resolution of 0.27. The melanoma concordance regression model that we built is at the limit of this resolution. The regression performance could be further improved by collecting more concordance reviews from dermatopathologists. We maximized the melanoma concordance regression performance by learning informative features through self-supervised training with SimCLR. We also demonstrated that the predicted melanoma concordance rate can be used as a malignant classifier, as the concordance rate is correlated with the likelihood of malignancy. We adjusted the threshold of malignancy to maximize both AUC and average precision, where we found the AUC value to be 0.89 ± 0.02 and the average precision to be 0.81 ± 0.04.

These results are an important first step for building an AI system capable of predicting the results of consulting a panel of experts and delivering a score based on the degree to which the experts would agree on a particular diagnosis or opinion. Upon further improvement, a concordance score reliably representing a panel of experts can additionally be used as a diagnostic assist to suggest additional testing or other action, for example the ordering of additional stains or a genetic test. Additionally, dermatopathologists utilizing a concordance score as a diagnostic assist can empower them to speed up malignancy diagnosis, increase confidence in case sign out, and lower the misdiagnosis rate, thereby increasing sensitivity and improving patient care. The possibility also exists that a concordance score could be fed into a combined-test that incorporates the results of a multi-gene assay (e.g., Castle MelanomaDx [9]) in order to enhance the performance of the test. Finally, such a concordance AI system can be extended to any pathology that exhibits a high discordance rate, such as breast cancer staging [21] and Gleason grading of prostate cancer [5].

Acknowledgments. The authors thank the support of Jeff Baatz, Ramachandra V. Chamarthi, Nathan Langlois, and Liren Zhu at Proscia for their engineering support; Theresa Feeser, Pratik Patel, and Aysegul Ergin Sutcu at Proscia for their data acquisition and Q&A support; and Dr. Curtis Thompson at CTA and Dr. David Terrano at Bethesda Dermatology Laboratory for their consensus annotation support.

References

1. American Cancer Society. Cancer facts and figures 2021. American Cancer Society, Inc. (2021)
2. Campanella, G., et al.: Clinical-grade computational pathology using weakly supervised deep learning on whole slide images. Nat. Med. **25**(8), 1301–1309 (2019)
3. Chen, T., Kornblith, S., Norouzi, M., Hinton, G.: A simple framework for contrastive learning of visual representations. arXiv preprint arXiv:2002.05709 (2020)
4. Chen, T., Kornblith, S., Swersky, K., Norouzi, M., Hinton, G.: Big self-supervised models are strong semi-supervised learners. arXiv preprint arXiv:2006.10029 (2020)

5. Coard, K.C., Freeman, V.L.: Gleason grading of prostate cancer: level of concordance between pathologists at the University Hospital of the West Indies. Am. J. Clin. Pathol. **122**(3), 373–376 (2004)

6. Deng, J., Dong, W., Socher, R., Li, L.J., Li, K., Fei-Fei, L.: Imagenet: a large-scale hierarchical image database. In: IEEE Conference on Computer Vision and Pattern Recognition, pp. 248–255 (2009)

7. Elmore, J.G., et al.: Pathologists' diagnosis of invasive melanoma and melanocytic proliferations: observer accuracy and reproducibility study. BMJ **357** (2017)

8. Falcon, W., Borovec, J., Brundyn, A., Harsh Jha, A., Koker, T., Nitta, A.: Pytorch lightning bolts self supervised pre-trained models documentation (2022). https:// pytorch-lightning-bolts.readthedocs.io/en/latest/self_supervised_models.html

9. Fried, L., Tan, A., Bajaj, S., Liebman, T.N., Polsky, D., Stein, J.A.: Technological advances for the detection of melanoma: advances in molecular techniques. J. Am. Acad. Dermatol. **83**(4), 996–1004 (2020)

10. Gerami, P., et al.: Histomorphologic assessment and interobserver diagnostic reproducibility of atypical spitzoid melanocytic neoplasms with long-term follow-up. Am. J. Surg. Pathol. **38**(7), 934–940 (2014)

11. He, K., Xiangyu, Z., Ren, S., Sun, J.: Deep residual learning for image recognition. arXiv preprint arXiv:1512.03385 (2015)

12. Ilse, M., Tomczak, J., Welling, M.: Attention-based deep multiple instance learning. In: Proceedings of the 35th International Conference on Machine Learning, pp. 2127–2136 (2018)

13. Li, B., Li, Y., Eliceiri, K.W.: Dual-stream multiple instance learning network for whole slide image classification with self-supervised contrastive learning. In: Proceedings of the IEEE/CVF Conference on Computer Vision and Pattern Recognition, pp. 14318–14328 (2021)

14. McInnes, L., Healy, J., Saul, N., Großberger, L.: UMAP: uniform manifold approximation and projection. J. Open Source Softw. **3**(29), 861 (2018). https://doi.org/ 10.21105/joss.00861

15. Noone, A.M., et al.: Cancer incidence and survival trends by subtype using data from the surveillance epidemiology and end results program, 1992–2013. Cancer Epidemiol. Prev. Biomark. **26**(4), 632–641 (2017)

16. Otsu, N.: A threshold selection method from gray-level histograms. IEEE Trans. Syst. Man Cybern. **9**(1), 62–66 (1979). https://doi.org/10.1109/TSMC.1979. 4310076

17. Pantanowitz, L., et al.: An artificial intelligence algorithm for prostate cancer diagnosis in whole slide images of core needle biopsies: a blinded clinical validation and deployment study. Lancet Digit. Health **2**(8), e407–e416 (2020). https:// doi.org/10.1016/S2589-7500(20)30159-X. https://www.sciencedirect.com/science/ article/pii/S258975002030159X

18. Pech-Pacheco, J.L., Cristóbal, G., Chamorro-Martinez, J., Fernández-Valdivia, J.: Diatom autofocusing in brightfield microscopy: a comparative study. In: Proceedings 15th International Conference on Pattern Recognition, ICPR-2000, vol. 3, pp. 314–317. IEEE (2000)

19. Pertuz, S., Puig, D., García, M.Á.: Analysis of focus measure operators for shape-from-focus. Pattern Recogn. **46**, 1415–1432 (2013)

20. Piepkorn, M.W., et al.: The MPATH-Dx reporting schema for melanocytic proliferations and melanoma. J. Am. Acad. Dermatol. **70**(1), 131–141 (2014)

21. Plichta, J.K., et al.: Clinical and pathological stage discordance among 433,514 breast cancer patients. Am. J. Surg. **218**(4), 669–676 (2019)

22. Sankarapandian, S., et al.: A pathology deep learning system capable of triage of melanoma specimens utilizing dermatopathologist consensus as ground truth. In: Proceedings of the ICCV 2021 CDpath Workshop (2021)
23. Shapiro, S.S., Wilk, M.B.: An analysis of variance test for normality (complete samples). Biometrika **52**(3–4), 591–611 (1965). https://doi.org/10.1093/biomet/52.3-4.591
24. Srivastava, N., Hinton, G., Krizhevsky, A., Sutskever, I., Salakhutdinov, R.: Dropout: a simple way to prevent neural networks from overfitting. J. Mach. Learn. Res. **15**(56), 1929–1958 (2014). http://jmlr.org/papers/v15/srivastava14a.html
25. Tellez, D., et al.: Quantifying the effects of data augmentation and stain color normalization in convolutional neural networks for computational pathology. Med. Image Anal. **58**, 101544 (2019)
26. Welch, H.G., Mazer, B.L., Adamson, A.S.: The rapid rise in cutaneous melanoma diagnoses. N. Engl. J. Med. **384**(1), 72–79 (2021)

Explainable Model for Localization of Spiculation in Lung Nodules

Mirtha Lucas[1]([⊠])(iD), Miguel Lerma[2](iD), Jacob Furst[1](iD), and Daniela Raicu[1](iD)

[1] DePaul University, Chicago, IL 60604, USA
mlucas3@depaul.edu, {jfurst,draicu}@cdm.depaul.edu
[2] Northwestern University, Evanston, IL 60208, USA
mlerma@math.northwestern.edu

Abstract. When determining a lung nodule malignancy one must consider the spiculation represented by spike-like structures in the nodule's boundary. In this paper, we develop a deep learning model based on a VGG16 architecture to locate the presence of spiculation in lung nodules from Computed Tomography images. In order to increase the expert's confidence in the model output, we apply our novel Riemann-Stieltjes Integrated Gradient-weighted Class Activation Mapping attribution method to visualize areas of the image (spicules). Therefore, the attribution method is applied to the layer of the model that is responsible for the detection of the spiculation features. We show that the first layers of the network are specialized in detecting low-level features such as edges, the last convolutional layer detects the general area occupied by the nodule, and finally, we identify that spiculation structures are detected at an intermediate layer. We use three different metrics to support our findings.

Keywords: Artificial intelligence · XAI · Computer-aided detection · CAD · Imaging informatics

1 Introduction

Spiculation is one of the features used by medical experts to determine if a lung nodule is malignant [18,30]. It is defined as the degree to which the nodule exhibits spicules, spike-like structures, along its border. This feature can be observed using imaging detection, which is easy to perform and causes less discomfort than alternative diagnosis methods such as a biopsy.

Automatic detection of features such as nodule spiculation by computer-aided detection (CAD) systems can help medical experts in the diagnosis process. For these systems to be adopted in clinical practice, their output has to be not only accurate, but also explainable in order to increase the trust between the human and the technology.

In this work, our main goal is to introduce a new approach to identify highly spiculated lung nodules using a convolutional network, and provide a visual

L. Karlinsky et al. (Eds.): ECCV 2022 Workshops, LNCS 13807, pp. 457–471, 2023.
https://doi.org/10.1007/978-3-031-25082-8_30

explanation by highlighting the locations of the spicules. Furthermore, we want to determine what part of the network is responsible for the detection of spiculation. Convolutional neural networks (CNN) perform a bottom-up process of feature detection, beginning with low level features (such as edge detection) at the layers that are closer to the input, to high level features (such as the general location of an object) at layers closer to the output [12]. This is so because the kernels processing the outputs of each layer cover only a small area of the layer, and they can pay attention only to that small area. As the information flows from input to output it gets integrated into complex combinations of lower level features that can be interpreted as higher level features. In the particular problem studied here, i.e., detection of spiculation in lung nodules, we are interested in locating the defining elements of spiculation (the spicules) in the input image, and also what part of the network (which layer) plays the main role in detecting this particular feature. This is important because common attribution methods used to explain classification of images (like the ones discussed in the next section) are applied to a pre-selected layer of the network, so we need to determine which layer provides the strongest response to the presence of the feature that we are trying to detect. Our hypothesis is that spicule detection happens at some intermediate layer of the network, not necessarily the last one. In the process of testing this hypothesis we make the following contributions:

- We use transfer learning on a network pretrained on a large dataset of images (ImageNet) to be used on CT scans of lung nodules to detect high/marked spiculation.
- We apply a novel attribution method to locate the spicules in nodules classified with high/marked spiculation.
- We identify the layer of the network that captures the "spiculation" feature.

The approach and methods used here are easy to generalize to other problems and network architectures, hence they can be seen as examples of a general approach by which not only the network output is explained, but also the hidden parts of the network are made more transparent by revealing their precise role in the process of feature detection.

2   Previous Work

Spiculation has been used for lung cancer screening [1,17,19,20], and its detection plays a role in computer-aided diagnosis (CAD) [11]. New tools for cancer diagnosis have been made available with the development of deep learning, reaching an unprecedented level of accuracy which is even higher than that of a general statistical expert [8]. Following this success, the need of developing tools to explain the predictions of artificial systems used in CAD quickly arose. They take the form of attribution methods that quantify the impact that various elements of a system have in providing a prediction.

In the field of attribution methods for networks processing images there are a variety of approaches. One frequently used is Gradient-weighted Class Activation Mapping (Grad-CAM) [25], which produces heatmaps by highlighting areas of the image that contribute most to the network output. It uses the gradients of a target class output with respect to the activations of a selected layer. The method is easy to implement, and it has been used in computer-aided detection [15], but it does not work well when outputs are close to saturation levels because gradients tend to vanish. There are derivatives of Grad-CAM, such as Grad-CAM++ [3], that use more complex ways to combine gradients to obtain heatmaps, but they are still potentially affected by problems when those gradients become zero or near zero.

An attribution method that not only is immune to the problem of vanishing gradients, but can also be applied to any model regardless of its internal structure, is Integrated Gradients (IG) [27]. This method deals with the given model as a (differentiable) multivariate function for which we do not need to know how exactly its outputs are obtained from its inputs. IG is a technique to attribute the predictions of the model to each of its input variables by integrating the gradients of the output with respect to each input along a path in the input space interpolating from a baseline input to the desire input. One problem with this approach is that it ignores the internal structure of an explainable system and makes no attempt to understand the roles of its internal parts. While ignoring the internal structure of the model makes the attribution method more general, it deprives it from potentially useful information that could be used to explain the outputs of the model.

The attribution method used here, our novel Riemann-Stieltjes Integrated Gradient-weighted Class Activation Mapping (RSI-Grad-CAM), has the advantage of being practically immune to the vanishing gradients problem, and having the capability to use information from inner layers of the model [13].

3   Methodology

To detect spiculated nodules and localize the boundaries that present characteristics specific to spiculation, we employ the following steps:

1. Train a neural network to classify lung nodules by spiculation level.
2. Use an attribution method capable to locate the elements (spicules) that contribute to make the nodule having high spiculation.
3. Provide objective quantitative metrics showing that the attribution method was in fact able to locate the spicules.

Additionally, we are interested in determining what part of the network is responsible for the detection of spiculation versus other characteristics of the nodule, such as the location of its contour (edge detection) and the general area occupied by the nodule. This is important because it tells us where to look in the network for the necessary information concerning the detection of the feature of interest (spiculation).

In the next subsections we introduce the dataset, deep learning model, and attribution method used. Then, we explain the metrics used to determine the impact of each layer of the network in the detection of spiculation.

3.1 Dataset

This work uses images taken from the Lung Nodule Image Database Consortium collection (LIDC-IDRI), consisting of diagnostic and lung cancer screening thoracic computed tomography (CT) scans with marked-up annotated lesions by up to 4 radiologists [9]. The images have been selected from CT scans containing the maximum area section of 2687 distinct nodules from 1010 patients. Nodules of three millimeters or larger have been manually identified, delineated, and semantically characterized by up to four different radiologists across nine semantic characteristics, including spiculation and malignancy. Given that ratings provided by the radiologists do not always coincide, we took the mode of the ratings as reference truth. Ties were resolved by using the maximum mode.

The size of each CT scan is 512×512 pixels, but most nodules fit in a 64×64 pixel window. Pixel intensities represent radio densities measured in Hounsfield units (HU), and they can vary within a very large range depending on the area of the body. Following guidelines from [7] we used the $[-1200, 800]$ HU window recommended for thoracic CT scans, and mapped it to the 0–255 pixel intensity range commonly used to represent images.

The radiologists rated the spiculation level of each nodule in an ordinal scale from 1 (low/no spiculation) through 5 (high/marked spiculation) [5]. The number of nodules in each level is shown in Table 1. spiculation labels changed according to [5].

Table 1. Distribution of nodules by spiculation level.

Spiculation level	Number of nodules
1 No spiculation	1850
2	415
3	180
4	122
5 Marked spiculation	120
Total	2687

We aim to detect spike-like structures, similar to the binary split in [19], we combine the nodules in two classes to denote low-level spiculation (Class 1 - level 1) and high-level spiculation (Class 2 - levels 4 and 5). After eliminating nodules with sizes in pixels less than 6×6 (too small to significantly encode content information) and larger than 64×64 (there was only one nodule with that size in our dataset) we obtained a total of 1714 nodules in Class 1 and 234 nodules in Class 2. Each class is divided in training and testing sets in a proportion of 80/20. The final number of samples in each class is shown in Table 2. Images of 64×64 pixels with the nodule in the center are produced by clipping the original images. Figure 1 shows a few examples of images of nodules after clipping.

Table 2. Distribution of nodules by spiculation class.

Spiculation class	Total nodules	Training set	Testing set
1 (low spiculation)	1714	1371	343
2 (high spiculation)	234	187	47
Totals	1948	1558	390

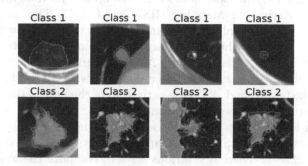

Fig. 1. Sample nodules. The window is 64 by 64 pixels, and the red line is the contour drawn by a radiologist. The images in the top row belong to class 1 (low spiculation), the ones in the bottom row belong to class 2 (high spiculation). (Color figure online)

To achieve class balancing in the training set we add images obtained by random rotations and flips from the images of nodules with high spiculation (Class 2). In order to avoid losing the corners of the images during the rotations we work with clippings of size 128×128 pixels, and reclip to the final size 64×64 after rotation. We do not perform class balancing in the testing set.

3.2 Classifier Network

We built a classifier using transfer learning on a VGG16 network pretrained on ImageNet [2,23,26]. We used the base section of the VGG16 excluding its top fully connected layers, and added a global average pooling layer at the end, plus a fully connected layer with 512 outputs and ReLU activation function, followed by a fully connected layer with 1 output and sigmoid activation function. An n-class classifier network typically has n output units, but for a binary classifier one output unit suffices. The outputs are numerical, with target 0 representing Class 1, and target 1 representing Class 2.

We performed transfer learning in two steps:

1. Model training: We froze all its layers except our two last fully connected layers and the last convolutional layer of each of its five blocks. The reason to retrain deep hidden layers is to help the network learn low level features of images that belong to a domain different from the original ImageNet on which it was pretrained.

2. Model parameter tuning: We kept training the network with only the last (fully connected) layers unfrozen.

The loss function used in both trainings was the Mean Squared Error.

3.3 Attribution Method

We apply our novel attribution method, Riemann-Stieltjes Integrated Gradient-weighted Class Activation Mapping (RSI-Grad-CAM). This method can be applied to any convolutional network, and works as follows. First we must pick a convolutional layer A, which is composed of a number of feature maps, also called channels, $A^1, A^2, \ldots, A^N$ (where N is the number of feature maps in the picked layer), all of them with the same dimensions. If A^k is the k-th feature map of the picked layer, and A_{ij}^k is the activation of the unit in the position (i, j) of the k-th feature map, then, a localization map or "heatmap" is obtained by combining the feature maps of the chosen layer using weights w_k^c that capture the contribution of the k-th feature map to the output y^c of the network corresponding to class c. There are various ways to compute the weights w_k^c. For example Grad-CAM, introduced in [25], uses the gradient of the selected output y^c with respect to the activations A_{ij}^k averaged over each feature map, as shown in Eq. (1). Here Z is the size (number of units) of the feature map.

$$w_k^c = \overbrace{\frac{1}{Z} \sum_i \sum_j}^{\substack{\text{global} \\ \text{average pooling}}} \underbrace{\frac{\partial y^c}{\partial A_{ij}^k}}_{\substack{\text{gradients} \\ \text{via backprop}}} \tag{1}$$

Given that Grad-CAM is vulnerable to the vanishing gradients problem that occurs when the gradients are zero or near-zero [6], we propose to use our novel RSI-Grad-CAM method which handles the vanishing gradient problem. RSI-Grad-CAM computes the weights w_k^c using integrated gradients in the following way. First we need to pick a baseline input I_0 (when working with images I_0 is typically a black image). Then, given an input I, we consider the path given in parametric form $I(\alpha) = I_0 + \alpha(I - I_0)$, where α varies between 0 and 1, so that $I(0) = I_0$ (baseline) and $I(1) = I$ (the given input). When feeding the network with input $I(\alpha)$, the output corresponding to class c will be $y^c(\alpha)$, and the activations of the feature map k of layer A will be $A_{ij}^k(\alpha)$. Then, we compute the weights by averaging the integral of gradients over the feature map, as shown in Eq. (2).

$$w_k^c = \frac{1}{Z} \sum_{i,j} \int_{\alpha=0}^{\alpha=1} \frac{\partial y^c(\alpha)}{\partial A_{ij}^k} \, dA_{ij}^k(\alpha) \tag{2}$$

The integral occurring in Eq. (2) is the Riemann-Stieltjes integral of function $\partial y^c(\alpha)/\partial A_{ij}^k$ with respect to function $A_{ij}^k(\alpha)$ (see [16]). For computational purposes this integral can be approximated with a Riemann-Stieltjes sum:

$$w_k^c = \frac{1}{Z} \sum_{i,j} \left(\sum_{\ell=1}^{m} \left\{ \frac{\partial y^c(\alpha_\ell)}{\partial A_{ij}^k} \times \Delta A(\alpha_\ell) \right\} \right) \tag{3}$$

where $\Delta A(\alpha_\ell) = A_{ij}^k(\alpha_\ell) - A_{ij}^k(\alpha_{\ell-1})$, $\alpha_\ell = \ell/m$, and m is the number of interpolation steps.

The next step consists in combining the feature maps A^k with the weights computed above, as shown in Eq. (4). Note that the combination is also followed by a Rectified Linear function $\text{ReLU}(x) = \max(x, 0)$, because we are interested only in the features that have a positive influence on the class of interest.

$$L_{\text{Grad-CAM}}^c = \text{ReLU} \underbrace{\left(\sum_k w_k^c A^k \right)}_{\text{linear combination}} \tag{4}$$

After the heatmap has been produced, it can be normalized and upsampled via bilinear interpolation to the size of the original image, and overlapped with it to highlight the areas of the input image that contribute to the network output corresponding to the chosen class.

We also generated heatmaps using other attribution methods, namely Grad-CAM [25], Grad-CAM++ [3], Integrated Gradients [27], and Integrated Grad-CAM [24], but our RSI-Grad-CAM produced the neatest heatmaps, as shown in Fig. 3.

3.4 Metrics

In order to test the quality of our attribution method, we need to ensure that the metrics are not affected by limitations in the classification power of the neural network. Consequently, during the evaluation we use only sample images that have been correctly classified by the network (Table 4).

Since the spicules occur at the boundary of the nodule, we measure the localization power of our attribution method by determining to what extent the heatmaps tend to concentrate on the contour of the nodule (annotated by one of the radiologists) compared to other areas of the image. To that end, we compute the average intensity value of the heatmap along the contour $(Avg(contour))$, and compare it to the distribution of intensities of the heatmap on the whole image. Using the mean μ and standard deviation σ of the intensities of the heatmap on the image, we assign a z-score to the average value of the heatmap along the contour using the formula

$$z_{Avg(contour)} = \frac{Avg(contour) - \mu}{\sigma} \tag{5}$$

This provides a standardized value that we can compare across different images. We expect our attribution method will produce a larger $z_{Avg(contour)}$ for high spiculation nodules compared to low spiculation nodules. We also study how the difference between high and low spiculation $z_{Avg(contour)}$ varies depending on which layer we pick to apply our attribution method.

The first approach consisted in comparing boxplots of $z_{Avg(contour)}$ for low and high spiculation (see Fig. 4). That provided a first (graphical) evidence of the power of our attribution method to recognize spiculation as a feature that depends on characteristics of the contour of a nodule. An alternative approach consists in comparing the cumulative distribution functions (cdfs) of $z_{Avg(contour)}$ for high and low spiculation nodules (see Fig. 5).

Next, we measured the difference between the distributions of $z_{Avg(contour)}$ for low and high spiculation nodules using three different approaches, two of them measuring distances between distributions, and the third one based on an hypothesis testing in order to determine which distribution tends to have larger values.

In the first approach we used the *Energy distance* between two real-valued random variables with cumulative distributions functions (cdfs) F and G respectively, given by the following formula [22, 28]:

$$D(F, G) = \left(2 \int_{-\infty}^{\infty} (F(x) - G(x))^2 \, dx \right)^{1/2} \tag{6}$$

Recall that the cdf F of a random variable X is defined $F(x) = P(X \leq x) =$ probability that the random variable X is less than or equal x.

Our second approach uses the 1-dimensional *Wasserstein distance* [10, 29], which for (1-dimensional) probability distributions with cdfs F, G respectively is given by the following formula [21]:

$$W(F, G) = \int_{-\infty}^{\infty} |F(x) - G(x)| \, dx \tag{7}$$

There are other equivalent expressions for Energy and 1-dimensional Wasserstein distances, here we use the ones based on cdfs for simplicity (they are L^p-distances between cdfs). The distributions of $z_{Avg(contour)}$ are discrete, and cdfs can be used in this distribution, therefore formulas (6) and (7) are still valid [4].

The Energy distance (D) and Wasserstein distance (W) tell us how different two distributions are, but they don't tell which one tends to take larger values. In our tests we gave a sign to the metrics equal to that of the difference of the means of the distributions. More specifically, assume the mean of distributions with cdfs F and G are respectively μ_F and μ_G. Then, we define the signed metrics as follows:

$$D_{signed}(F, G) = \text{sign}(\mu_G - \mu_F) \, D(F, G)$$
$$W_{signed}(F, G) = \text{sign}(\mu_G - \mu_F) \, W(F, G) \tag{8}$$

where $\text{sign}(x)$ is the sign function, i.e., $\text{sign}(x) = x/|x|$ if $x \neq 0$, and $\text{sign}(0) = 0$.

The third approach consists in using the one-sided *Mann-Whitney U rank test* between distributions [14]. For our purposes this test amounts to the following. Let C_1 be the set of samples with low spiculation, and let C_2 be set of samples with high spiculation. For each sample s_1 with low spiculation let u_1 be the z-score of the average of the heatmap over the contour of s_1. Define analogously u_2 for each sample s_2 with high spiculation. Let U_1 = number of sample pairs (s_1, s_2) such that $u_1 > u_2$, and U_2 = number of sample pairs (s_1, s_2) such that $u_1 < u_2$ (if there are ties $u_1 = u_2$ then U_1 and U_2 are increased by half the number of ties each). Then, we compute $(U_2 - U_1)/(U_1 + U_2)$, which ranges from -1 to 1. If the result is positive that will indicate that u_2 tends to take larger values than u_1, while a negative value will indicate that u_1 tends to take larger values than u_2. The one-sided Mann-Whitney U rank test also assigns a p-value to the null hypothesis $H_0 \equiv P(u_1 \geq u_2) \geq 1/2$ (where P means probability) versus the alternate hypothesis $H_1 \equiv P(u_1 < u_2) > 1/2$.

4   Results

We used the VGG16 classifier network described in Sect. 3.2 trained using the following parameters (for training and fine tuning): learning rate = 0.00001, batch size = 32, number of epochs = 10. The performance on the testing set is shown in the classification report, Table 3. The accuracy obtained was 91%.

Table 3. Classification report on the test set.

Class	Precision	Recall	F1-score	Support
Class 1 (low spic.)	0.94	0.96	0.95	343
Class 2 (high spic.)	0.64	0.57	0.61	47
accuracy			0.91	390
macro avg	0.79	0.77	0.78	390
weighted avg	0.91	0.91	0.91	390

Table 4. Distribution of nodules by spiculation class.

Class	Total nodules	Correctly classified
Low spiculation	1714	1606
High spiculation	234	137
Total	1948	1743

Next, we provide evaluation results of our attribution method. In order to focus the evaluation on the attribution method rather than the network performance, we used only images of nodules that were correctly classified. The last column of Table 4 indicates the number of correctly classified nodules from each class.

We computed heatmaps using RSI-Grad-CAM at the final convolutional layer of each of the five blocks of the VGG16 network, using a black image as baseline. We used index 0 through 4 to refer to those layers, where 0 represents the last layer of the first block and 4 refers to the last layer of the last block, as shown in Table 5.

Table 5. Layers used for evaluating the attribution method.

Layer index	Layer name
0	block1_conv2
1	block2_conv2
2	block3_conv3
3	block4_conv3
4	block5_conv3

Figure 2 shows heatmaps produced by our RSI-Grad-CAM method at the final layer of the final convolutions block for a highly spiculated nodule. The contour of the nodule appears in red.

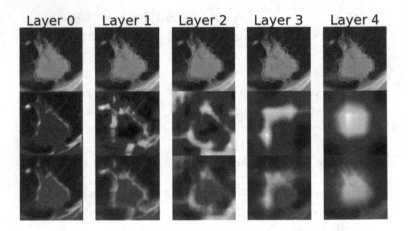

Fig. 2. RSI-Grad-CAM heatmaps for a highly spiculated nodule. Each column corresponds to a layer. The original images are at the top, heatmaps in the middle, and overlays at the bottom. The read line represents the contour annotated by the radiologist.

For comparison, Fig. 3 shows heatmaps produced by various attribution methods for a highly spiculated nodule at layer 3 (except for Integrated Gradients that applies to the layer input only).

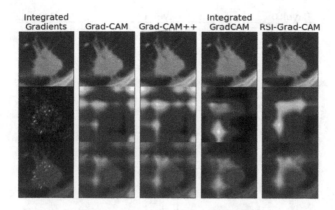

Fig. 3. Heatmaps generated by various attribution methods.

We notice that in all the layers heatmaps tend to highlight more the contour of the highly spiculated nodule, but this is not yet evidence of detection of spicules in the contour. We expect the first layers of the network to be specialized in low level features such as edge detection, while the last layer may detect the general area occupied by the nodule. As stated in our hypothesis, we expect spicule detection to happen at some intermediate layer. The boxplots of $z_{Avg(contour)}$ in (Fig. 4) provide evidence in favor of this hypothesis.

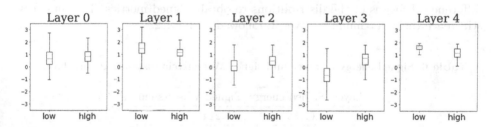

Fig. 4. Boxplots of z-scores of average values of heatmaps on contours for correctly classified samples.

The comparison of the cdfs of the z-scores of average values of heatmaps on contours (Fig. 5) shows that the values tend to be larger for high spiculation nodules with respect to low spiculation nodules at layers 2 and 3, with the maximum difference at layer 3.

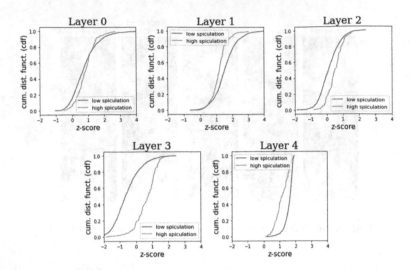

Fig. 5. Cdfs of z-scores of average values of heatmaps on contours for correctly classified samples.

The Energy and Wasserstein distances between the distributions of z-scores of average values of heatmaps on contours for low and high spiculation nodules at each layer are shown in Table 6. In order to include information about what distribution takes higher values we multiplied the computed distance by the difference of means of the distributions to obtain signed metrics. The numbers still reach their maximum at layer 3, followed by layer 2.

Table 6. Signed energy distance and signed Wasserstein distance at each layer.

Layer	Signed energy	Signed Wasserstein
0	0.184	0.214
1	−0.384	−0.374
2	0.402	0.374
3	**1.038**	**1.182**
4	−0.558	−0.428

The results for the 1-sided Mann-Whitney U rank test are shown in Table 7. The barplot shows that the maximum of $(U_2 - U_1)/(U_1 + U_2)$ happens at layer 3, which is consistent with the results obtained using Energy and Wasserstein metrics. The p-values obtained strongly favor the hypothesis that the z-scores of average values of heatmaps on contours are larger for high spiculation nodules at layers 2 and 3, with the maximum at layer 3.

Table 7. One-sided Mann-Whitney U rank test.

Layer	$\frac{U_2 - U_1}{U_1 + U_2}$	p-value
0	0.109	0.017
1	−0.342	1.0
2	0.373	2.01×10^{-13}
3	**0.730**	$\mathbf{4.09 \times 10^{-43}}$
4	−0.602	1.0

5 Conclusions

We have used a modified VGG16 network retrained with a transfer learning technique to classify low and high spiculation nodules from images from the LIDC-IDRI database. Then, we applied the RSI-Grad-CAM attribution method to locate the elements of the images that contribute to the spiculation, i.e., spicules located at the boundary of the nodules. Furthermore, we were interested in determining what part of the network detects the spiculation feature. Common attribution methods are applied to a pre-selected layer of the network, so we need to determine which layer provides the strongest response to the presence of the feature that we are aiming to detect. Also, it is important to highlight that some features may be hard to locate directly in the model input, so methods based on heatmaps highlighting areas of the input may be less useful than the ones able to identify what internal parts of the model perform the detection of a given feature.

The metrics used to compare the distributions of average values of heatmaps on contours corresponding to low and high spiculation nodules were Energy distance, Wasserstein distance, and the 1-sided Mann-Whitney U rank test. All three test favor the hypothesis that the spiculation feature is detected at the last layer of the fourth convolutional block of the network (layer index 3, an intermediate hidden layer rather than the last one). In practice this means that, for our network, an explanation for the detection of spiculation can be provided by the heatmap produced at the last layer of the fourth convolutional block, since that heatmap tends to highlight the spicules occurring at the contour of the nodule.

The work performed here has been restricted to one network architecture (VGG16) performing binary classification, one image domain (images of lung nodules from the LICD-IDRI dataset), and one semantic feature (spiculation). Further work can be made to adapt the methods used here to other network architectures (e.g. ResNet, Siamese networks, etc.), multiclass classification (e.g. by adding the middle spiculation levels), data domains (e.g. natural language), and features (e.g. sample similarity).

References

1. Andrejeva, L., Geisel, J.L., Harigopa, M.: Spiculated masses, breast imaging. Oxford Medicine Online (2018). https://doi.org/10.1093/med/9780190270261.003. 0025
2. Bengio, Y.: Deep learning of representations for unsupervised and transfer learning. In: Proceedings of the 2011 International Conference on Unsupervised and Transfer Learning Workshop, vol. 27, pp. 17–37 (2011)
3. Chattopadhyay, A., Sarkar, A., Howlader, P., Balasubramanian, V.N.: Grad-CAM++: generalized gradient-based visual explanations for deep convolutional networks. In: 2018 IEEE Winter Conference on Applications of Computer Vision (WACV) (2018). https://doi.org/10.1109/wacv.2018.00097
4. Devore, J.L., Berk, K.N.: Discrete random variables and probability distributions. In: Devore, J.L., Berk, K.N. (eds.) Modern Mathematical Statistics with Applications. STS, pp. 96–157. Springer, New York (2012). https://doi.org/10.1007/978-1-4614-0391-3_3
5. Hancock, M.C., Magnan, J.F.: Lung nodule malignancy classification using only radiologist-quantified image features as inputs to statistical learning algorithms: probing the lung image database consortium dataset with two statistical learning methods. J. Med. Imaging 3(4), 044504 (2016)
6. Hochreite, S.: The vanishing gradient problem during learning recurrent neural nets and problem solutions. Int. J. Uncertain. Fuzziness Knowl.-Based Syst. 06(02), 107–116 (1998)
7. Hofer, M.: CT Teaching Manual, A Systematic Approach to CT Reading. Thieme Publishing Group (2007)
8. Huanga, S., Yang, J., Fong, S., Zhao, Q.: Artificial intelligence in cancer diagnosis and prognosis: opportunities and challenges. Cancer Lett. 471(28), 61–71 (2020)
9. Armato III, S.G., et al.: Lung image database consortium: developing a resource for the medical imaging research community. Radiology 232(3), 739–748 (2004). https://doi.org/10.1148/radiol.2323032035
10. Kantorovich, L.V.: Mathematical methods of organizing and planning production. Manage. Sci. 6(4), 366–422 (1939)
11. Lao, Z., Zheng, X.: Multiscale quantification of tissue spiculation and distortion for detection of architectural distortion and spiculated mass in mammography. In: M.D., R.M.S., van Ginneken, B. (eds.) Medical Imaging 2011: Computer-Aided Diagnosis, vol. 7963, pp. 468–475. International Society for Optics and Photonics, SPIE (2011). https://doi.org/10.1117/12.877330
12. LeCun, Y., Bengio, Y., Hinton, G.: Deep learning. Nature 521, 436–444 (2015)
13. Lucas, M., Lerma, M., Furst, J., Raicu, D.: Visual explanations from deep networks via Riemann-Stieltjes integrated gradient-based localization (2022). https://arxiv.org/abs/2205.10900
14. Mann, H., Whitney, D.: On a test of whether one of two random variables is stochastically larger than the other. Ann. Math. Stat. 18(1), 50–60 (1947)
15. Mendez, D.M.M., Bermúdez, A., Tyrrell, P.N.: Visualization of layers within a convolutional neural network using gradient activation maps. J. Undergraduate Life Sci. 14(1), 6 (2020)
16. Protter, M.H., Morrey, C.B.: The Riemann-Stieltjes integral and functions of bounded variation. In: Protter, M.H., Morrey, C.B. (eds.) A First Course in Real Analysis. Undergraduate Texts in Mathematics. Springer, New York (1991). https://doi.org/10.1007/978-1-4419-8744-0_12

17. Nadeem, W.C.S., Alam, S.R., Deasy, J.O., Tannenbaum, A., Lu, W.: Reproducible and interpretable spiculation quantification for lung cancer screening. Comput. Methods Programs Biomed. (2020). https://doi.org/10.1016/j.cmpb.2020.105839

18. Paci, E., et al.: Ma01.09 mortality, survival and incidence rates in the italung randomised lung cancer screening trial (Italy). J. Thorac. Oncol. **12**(1), S346–S347 (2017)

19. Qiu, B., Furst, J., Rasin, A., Tchoua, R., Raicu, D.: Learning latent spiculated features for lung nodule characterization. In: Annual International Conference of the IEEE Engineering in Medicine and Biology Society, pp. 1254–1257 (2020). https://doi.org/10.1109/EMBC44109.2020.9175720

20. Qiu, S., Sun, J., Zhou, T., Gao, G., He, Z., Liang, T.: Spiculation sign recognition in a pulmonary nodule based on spiking neural p systems. BioMed Res. Int. **2020** (2020). https://doi.org/10.1155/2020/6619076

21. Ramdas, A., Garcia, N., Cuturi, M.: On Wasserstein two sample testing and related families of nonparametric tests. Entropy **19**(2), 47 (2017). https://doi.org/10.3390/e19020047

22. Rizzo, M.L., Székely, G.J.: Energy distance. Wiley Interdiscip. Rev. Comput. Stat. **8**(1), 27–38 (2015)

23. Russakovsky, O., et al.: ImageNet large scale visual recognition challenge. Int. J. Comput. Vision **115**(3), 211–252 (2015). https://doi.org/10.1007/s11263-015-0816-y

24. Sattarzadeh, S., Sudhakar, M., Plataniotis, K.N., Jang, J., Jeong, Y., Kim, H.: Integrated Grad-CAM: sensitivity-aware visual explanation of deep convolutional networks via integrated gradient-based scoring (2021). https://arxiv.org/abs/2102.07805

25. Selvaraju, R.R., Cogswell, M., Das, A., Vedantam, R., Parikh, D., Batra, D.: Grad-CAM: visual explanations from deep networks via gradient-based localization. Int. J. Comput. Vision **128**(2), 336–359 (2019). https://doi.org/10.1007/s11263-019-01228-7

26. Simonyan, K., Zisserman, A.: Very deep convolutional networks for large-scale image recognition (2015). https://arxiv.org/abs/1409.1556

27. Sundararajan, M., Taly, A., Yan, Q.: Axiomatic attribution for deep networks. In: Precup, D., Teh, Y.W. (eds.) Proceedings of the 34th International Conference on Machine Learning. Proceedings of Machine Learning Research, vol. 70, pp. 3319–3328. PMLR (2017). https://proceedings.mlr.press/v70/sundararajan17a.html

28. Szekely, G.J.: E-statistics: the energy of statistical samples. Technical report, Bowling Green State University, Department of Mathematics and Statistics (2002)

29. Waserstein, L.N.: Markov processes over denumerable products of spaces, describing large systems of automata. Problemy Peredači Informacii **5**(3), 6–72 (1969)

30. Winkels, M., Cohena, T.S.: Pulmonary nodule detection in CT scans with equivariant CNNs. Med. Image Anal. **55**, 15–26 (2019)

Self-supervised Pretraining for 2D Medical Image Segmentation

András Kalapos[(✉)] and Bálint Gyires-Tóth

Department of Telecommunications and Media Informatics,
Faculty of Electrical Engineering and Informatics, Budapest University of Technology
and Economics, Műegyetem rkp. 3., Budapest 1111, Hungary
{kalapos.andras,toth.b}@tmit.bme.hu

Abstract. Supervised machine learning provides state-of-the-art solutions to a wide range of computer vision problems. However, the need for copious labelled training data limits the capabilities of these algorithms in scenarios where such input is scarce or expensive. Self-supervised learning offers a way to lower the need for manually annotated data by pretraining models for a specific domain on unlabelled data. In this approach, labelled data are solely required to fine-tune models for downstream tasks. Medical image segmentation is a field where labelling data requires expert knowledge and collecting large labelled datasets is challenging; therefore, self-supervised learning algorithms promise substantial improvements in this field. Despite this, self-supervised learning algorithms are used rarely to pretrain medical image segmentation networks. In this paper, we elaborate and analyse the effectiveness of supervised and self-supervised pretraining approaches on downstream medical image segmentation, focusing on convergence and data efficiency. We find that self-supervised pretraining on natural images and target-domain-specific images leads to the fastest and most stable downstream convergence. In our experiments on the ACDC cardiac segmentation dataset, this pretraining approach achieves 4–5 times faster fine-tuning convergence compared to an ImageNet pretrained model. We also show that this approach requires less than five epochs of pretraining on domain-specific data to achieve such improvement in the downstream convergence time. Finally, we find that, in low-data scenarios, supervised ImageNet pretraining achieves the best accuracy, requiring less than 100 annotated samples to realise close to minimal error.

Keywords: Self-supervised learning · Medical image segmentation · Pretraining · Data-efficient learning · cardiac MRI segmentation

1 Introduction

The success and popularity of machine learning in the last decade can be traced largely to the progress in supervised learning research. However, these methods require copious amounts of annotated training data. Therefore, their capabilities are limited in scenarios where annotated data is scarce or expensive. One

L. Karlinsky et al. (Eds.): ECCV 2022 Workshops, LNCS 13807, pp. 472–484, 2023.
https://doi.org/10.1007/978-3-031-25082-8_31

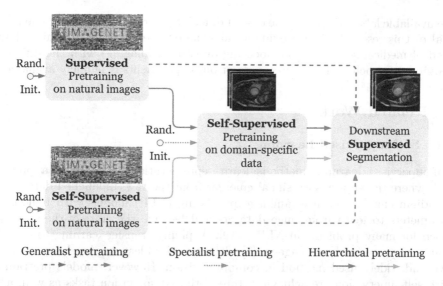

Fig. 1. Pretraining pipelines that we investigate in this paper.

such domain is biomedical data analysis where labelling often requires expert knowledge. In most cases, a limited number of medical professionals are available to perform certain annotation tasks and their availability for such activities is a strong constraint. Even collecting consistent and comparable unlabelled datasets at large scale is challenging due to different data acquisition equipment and practices at different medical centres. For rare diseases, the challenges of data acquisition are even greater. The amount of available data in the medical imaging field is increasing in public datasets or national data banks (e.g., UK Biobank [23]), however, even in these data sources, more unlabelled data could be available than with annotations that fit certain research interests.

Self-supervised learning [3,4,9,11,12,16,18,24,27,28] (SSL) is an approach to unsupervised representation learning that attracted great research interest in recent years. Its goal is to use unlabelled data to learn representations that can be fine-tuned for a wide variety of downstream supervised learning tasks. The learned representation should be general, task agnostic and high-level to be useful for many different downstream tasks. Being able to pretrain a model on unlabelled data could significantly reduce the need for annotated data and opens the possibility to train larger and potentially more accurate models than what is feasible with purely supervised learning.

In this paper, we elaborate on different supervised and self-supervised pretraining approaches and analyse their transfer learning capability on a downstream medical image segmentation task. Investigated pretraining approaches include training steps on natural images and target-task-specific medical images. Our primary research interest is to assess the ability of self-supervised pretraining algorithms to learn representations that improve the data efficiency or accuracy on the target segmentation task. We study how downstream accuracy scales with the available labelled data in the case of different pretraining methods. The main goal of this research is to provide a guideline on the annotated training data-need of medical segmentation (focusing on cardiac imaging) with the different training approaches. The source code or our experiments is publicly available.[1]

2 Related Work

2.1 Self-supervised Learning

Self-supervised learning methods learn representations by solving a pretext task where the supervisory signal emerges from the raw, unlabelled data. This paradigm enabled natural language processing (NLP) models with billions of parameters to learn from huge datasets and became the state-of-the-art approach for many problems in NLP [8,19]. Applying transfer learning to models that were pretrained on Imagenet [7] with supervised learning has been an effective and widely used method in computer vision. However, models pretrained with self-supervision are achieving state of the art for vision tasks as well, and the possibility of learning useful visual representations from uncurated and unlabelled data has been demonstrated [10].

However, constructing good pretext tasks for image data is not as intuitive as for text due to the high dimensionality and continuous nature of images. Early approaches used heuristic pretext tasks such as solving a jigsaw puzzle [18], predicting the relative position of image patches [9], colourization [28] and denoising [24]. Recent methods ([3,4,11,12,16,27] etc.) rely on the instance discrimination task, which treats augmented versions of every instance from a dataset as a separate class. A set of random augmentations are applied to each image, resulting in various *views* of it. The learning objective is then to assign similar latent vectors to different views of the same image, but different ones to views of different images. Methods that use such an instance discrimination task primarily differ in the set of augmentations and the approach to enforce this similarity of the latent vectors.

These methods aim to train a model that acts as an encoder, mapping images to an embedded vector space, also called latent vectors. These encoders could then be incorporated into downstream tasks as pretrained models.

The most common **transfer learning** approach in computer vision is fine-tuning a model that was pretrained for supervised image classification on ImageNet [7]. Another approach receiving notable attention recently is self-supervised pretraining either on ImageNet or on domain-specific data (i.e. on data that is similar or related to the problem tackled by transfer learning). Reed et al. [20] investigate self-supervised pretraining pipelines including ImageNet pretraining (generalist pretraining), domain-specific (or specialist) pretraining and the sequential combination of the previous two, which they refer to as "hierarchical pretraining".

---

[1] https://github.com/kaland313/SSL-MedSeg.

2.2   Self-supervised Learning for Medical Image Processing

Medical image recognition problems are commonly solved using transfer learning from natural images, even though the distribution, frequency pattern, and important features of medical images are not the same as those of natural images [17]. Self-supervised pretraining on domain-specific data offers an alternative approach to this task, eliminating the mismatch in source and target dataset characteristics, and allowing a methodologically more precise pretraining. Moreover, domain-specific pretraining has the potential to reduce the number of labelled data, which has high costs in the medical domain.

Recently, a handful of papers have been published that adapt self-supervised learning methods to pretrain models on medical image analysis problems such as histopathology [5], chest X-ray and dermatology classification [1,22]. These works show that MoCo [12] and SimCLR [4] are effective pretraining methods on medical images without major modifications to the algorithms. They highlight the capability of these algorithms to outperform supervised ImageNet pretraining by 1–5% classification accuracy on various medical image recognition datasets.

3   Methods and Data

3.1   Pretraining Approaches

Training pipelines consist of one or two pretraining stages followed by the downstream supervised segmentation training (see Fig. 1). All reported metrics and learning curves are from the final segmentation training step. Between each training stage, we transfer the weights of the encoder network to the next stage, but every stage uses additional layers, which we initialize randomly. As explained in Sect. 3.2, the applied self-supervised learning algorithm trains an encoder through two subnetworks (called the projector and predictor). The downstream segmentation network also has an additional module after the encoder: a decoder. All these subnetworks are randomly initialized at the beginning of each training phase using Kaiming-uniform initialization [13], while the weights of the encoder are always loaded from the previous stage. In the first stage of every pipeline, the encoder is random-initialised as well.

Generalist Pretraining. In this paper, we refer to supervised and self-supervised pretraining on natural images as generalist pretraining (see Fig. 1). For both learning modes, this approach assumes that the representations learned on natural images are general and useful for other problems and domains. A benefit of transfer learning from natural images is the ample availability of common models pretrained on ImageNet. In our experiments, we use such pretrained models instead of running supervised and self-supervised training on ImageNet.

Specialist Pretraining. Opposing the previous assumption, we conduct experiments on self-supervised pretraining on images that come from the same domain and dataset as the labelled data for the target segmentation problem.

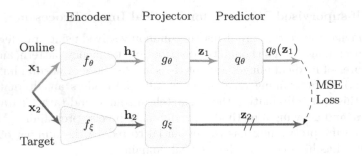

Fig. 2. Architecture of BYOL for a pair of views: x_1, x_2. Modified from [11]

Hierarchical Pretraining. Hierarchical pretraining combines the previous two approaches, aiming to keep general representations learned on natural images but also fine-tune these to better match the target domain. In practice, this includes fully training a model on natural images, then continue its training on images from the target domain. We extend the notion of hierarchical pretraining as opposed to that of Reed et al. [20], who only refer to self-supervised pretraining on natural images as the first step of hierarchical pretraining. We elaborate on both supervised and self-supervised learning for the training step on natural images.

3.2   Applied Self-supervised Learning Method

Grill et al. [11] proposed the Bootstrap Your Own Latent (BYOL) self-supervised learning method, which relies on the instance-discrimination task described in Sect. 2.1 and a special asymmetric architecture to avoid representation collapse. BYOL trains an encoder (f_θ on Fig. 2) to learn the representations of another fixed encoder (f_ξ on Fig. 2) for x_1, x_2 different views of the same input image. In other words, the f_ξ fixed network produces some h_2 representation of the x_2 augmented version of the image, which the trained network should reproduce, even though its input is a different x_1 augmentation of the image. Empirical evidence shows, that the trained f_θ "online" network can learn better representations than that of the fixed f_ξ "target" network. This allows for an iterative update procedure where the fixed network's weights are also slowly updated, based on the improved weights of the online network. The latter one is trained by gradient-based learning, while the target network is updated using the exponential moving average of the online network's weights. Gradients are not propagated through the target network, however, as training converges the weights of the two networks become similar and can be used for downstream tasks.

To achieve stable training, BYOL requires a q_θ predictor network in the online branch of the network. BYOL trains the online branch using the mean squared error between the $q_\theta(z_1)$ output of the predictor and the z_2 output of the g_ξ target projector. See Fig. 2 for an illustration of BYOL's asymmetric architecture.

3.3 Cardiac Segmentation Dataset

The "Automated Cardiac Diagnosis Challenge" dataset [2] is a publicly available dataset aiming to improve deep learning-based automated cardiac diagnosis. The dataset contains cardiac MRI records from 100 patients, distributed evenly in five groups, one healthy and four with different cardiac conditions. For each patient, the complete cardiac cycle[2] is covered by the recordings, where each recording consists of 3D MRI scans, which include 6–18 high-resolution 2D slices. This gives approximately 25 thousand slices in total for the dataset. The resolution of the slices varies from 150×150 to 500×500 pixels.

Segmentation labels are provided for the end-diastolic (ED) and end-systolic (ES) frames of the cardiac cycle, which gives 1900 slices with segmentation masks. For the rest of the frames, accurate segmentation labels are not available, however, self-supervised learning algorithms can leverage this data as well. We refer to the slices of annotated frames as the labelled 2D segmentation dataset and include all 25k slices (labelled and unlabelled) in the unlabelled dataset. One can consider this a semi-supervised learning dataset, with labels only available for a subset of all slices.

Segmentation masks are provided with four classes, left and right ventricle (LV, RV), myocardium (MYO), and background (see Fig. 5 as an example).

3.4 Model Architecture

We selected a U-Net [21] architecture for the segmentation training due to the successful application of this architecture for medical image segmentation. The architecture and implementation of the encoder match with the ones we pretrain, which is necessary for transferring the pretrained weights to the segmentation model. Encoders pretrained on ImageNet expect three-channel, RGB images as input, however, MRI scans are monochromatic. Therefore, the first layer of these encoders needs to be adjusted for single-channel input, which we achieve by depth-wise summation of the weights of the first layer (after [25]).

4 Experiments

Our primary research interest is to assess the scaling of downstream accuracy with the amount of available labelled training data for different pretraining strategies. We hypothesize that efficient pretraining approaches improve the accuracy of downstream training for limited data.

4.1 Training Phases

Natural Images. In all our experiments, we pretrained ResNet-50 [14] encoders, which we also used as the encoder of a U-Net segmentation network.

---

[2] One cardiac cycle is the period between two heartbeats.

We selected this architecture for its frequent use in self-supervised learning literature. We used weights from Wightman [25] instead of training an encoder from scratch in our experiments that utilized supervised ImageNet pretraining. For BYOL pretraining on ImageNet, we obtained weights from a third-party GitHub repository[3]. This model was trained for 300 epochs on 32 NVIDIA V100 GPUs.

Domain-Specific. For self-supervised pretraining on the ACDC dataset, we apply simple augmentations that are reasonable on monochromatic medical images: random resizing and cropping, horizontal flipping and brightness, and contrast perturbations. For this training phase we adapt the augmentations to the ACDC dataset, but keep most hyperparameters as published in [6].

In experiments with ImageNet pretraining preceding domain-specific SSL training, we adapt the weights of the input layer from three-channel to single-channel as explained in Sect. 3.4. In these experiments, we run SSL pretraining for 25 epochs, which takes 45 min on a single NVIDIA V100 GPU. For specialist pretraining, we randomly initialize the encoder at the beginning of the SSL training and run it for 400 epochs (which takes almost 10 h).

Downstream Segmentation. We apply the same augmentations as in the domain-specific SSL stage and train for 150 epochs to reach convergence in all experiments, which takes 20 min on average with one V100 GPU. Furthermore, we use Jaccard loss and periodic cosine annealing learning rate schedule. The U-Net's decoder is relatively small with 9 million trainable parameters, compared to the encoder's 23 million. With such a small decoder we expect that segmentation performance relies more on the encoder than in the case of a symmetric architecture with a large encoder. The decoder uses nearest neighbour upsampling and two convolutional layers at each upsampling stage. We implement downstream segmentation training based on an open-source repository for segmentation models and methods [26].

4.2 Data-Efficient Learning

We aim to test the hypothesis that efficient pretraining improves the downstream accuracy in low-annotated-data scenarios. To test this hypothesis, we run segmentation training on subsets of the labelled dataset, ranging from 1 sample to the full dataset. We also vary the pretraining strategy preceding the segmentation training. Each training runs for the same number of steps which is equivalent to 150 epochs on the full dataset and each run was repeated with 10 different randomly set initial parameter states (i.e. random seeds).

---

[3] https://github.com/yaox12/BYOL-PyTorch; Accessed on 4 May 2022.

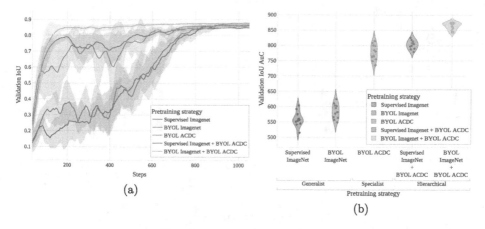

Fig. 3. (a) Learning curves for segmentation training on the full labelled dataset, preceded by different pretraining pipelines. Solid lines show means, shaded area represents ± standard deviation of 10 runs. (b) Violin plot of the area under the learning curves on the left. Ordered according to medians. Colours for different pretraining pipelines are consistent across all figures.

5 Results

5.1 Convergence and Stability

Figures 3a shows learning curves for downstream segmentation training with different pretraining approaches. The figure shows that hierarchical pretraining methods converge faster than generalist or specialist pretraining on this semantic segmentation task. This indicates that representations learned during both stages retain information useful for the target task. However, two notable differences separate these hierarchical methods as well. First, pretraining involving only self-supervised steps leads to faster and more stable convergence compared to supervised ImageNet pretraining followed by domain-specific SSL pretraining. Second, the final segmentation Jaccard index (Intersection over Union, IoU score) for the latter approach is significantly lower than for any other method. Figure 3a also reveals that self-supervised only hierarchical pretraining (BYOL ImageNet + BYOL ACDC) leads to downstream convergence in ∼200 steps, while other methods reach close to maximal IoU in 4–5 times more steps. The area under the learning curves (see Fig. 3b) highlight the stability of self-supervised only hierarchical pretraining even more clearly (BYOL ImageNet + BYOL ACDC on the figure). This figure also indicates that specialist self-supervised pretraining (BYOL ACDC) leads to similar results as mixing supervised and self-supervised pretraining (Supervised ImageNet + BYOL ACDC). Our findings show that among the investigated approaches, self-supervised only pretraining (i.e. BYOL ImageNet + BYOL ACDC pretraining) produces representations that achieve the fastest and most stable downstream convergence.

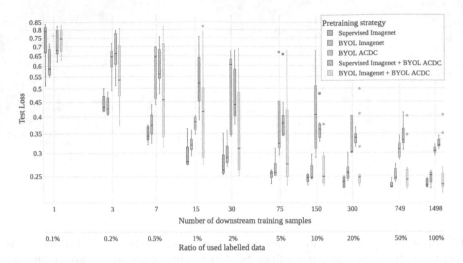

Fig. 4. Downstream segmentation test loss (Jaccard loss), for different annotated data quantities and pretraining pipelines. We select individual MRI slices randomly from slices of all patients to acquire training samples. Each box represents statistics of 10 runs with different random seeds. Boxes are grouped for each data setting, resulting in offsets from the true x-axis location, which must be taken into consideration when interpreting the figure.

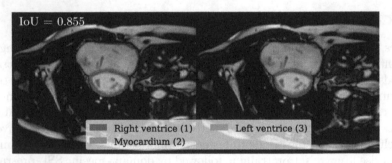

Fig. 5. Left: An example image from the ACDC dataset [2] with ground truth segmentation masks overlaid. Right: The same input image with predictions of a model that was pretrained using specialist pretraining (BYOL [11]).

5.2   Data-Efficiency

Figure 4 shows how downstream error scales with the available annotated data, in the case of different pretraining methods. Contradicting our hypothesis, the figure reveals that with low data quantities generalist pretraining approaches have lower test error compared to hierarchical pretraining.

Hestness et al. [15] investigates similar scaling behaviour as visible on Fig. 4. They show the existence of a power-law region, where the reduction of test error follows a power-law scaling with increasing data quantities. Hestness et al. [15]

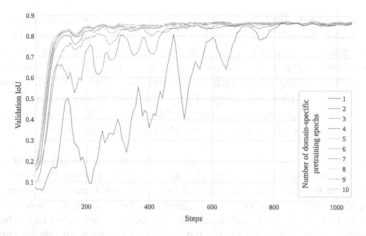

Fig. 6. Downstream learning curves after different number of domain-specific pretraining epochs in a hierarchical pretraining pipeline. Starting with a model pretrained using BYOL on ImageNet, we run BYOL pretraining on the ACDC dataset, save the encoder after every epoch, and execute downstream training based on each of these. The figure shows learning curves for these downstream segmentation trainings.

also point out that by increasing the amount of training data above a certain threshold, the improvement of test error starts to decrease slower than a power-function of the data quantity. We can observe both scaling regions on Fig. 4. Below 2% of the total available training data, that is 30 MRI slices, generalist pretraining approaches seem to follow power-law scaling (linear scaling on a log-log plot), while above 5% of available training data, the reduction of error is notably slower, indicating the transition to the so-called irreducible error regime. The amount of training data that corresponds to this transition is important to machine-learning development and decision-making, because investing more in annotated data above this region has a much smaller return in error reduction, than with fewer data. In simple terms, labelling a given number of additional samples is increasingly less profitable above this transition region. In our experiments this transition region is around 30–75 randomly selected slices of MRI records, annotating more slices still improves downstream performance, however, the expected improvement after adding an annotated sample reduces.

5.3 Domain-Specific Pretraining Epochs

We experiment with running the domain-specific step for a different number of epochs. As shown on Fig. 6, we find that 3–4 epochs are sufficient to achieve the fast-converging, stable learning curves that we presented in Sect. 5.1.

6 Conclusion

In this work we investigate supervised and self-supervised approaches to pretraining convolutional encoders for medical image segmentation. In addition, we study how these affect the data efficiency of the downstream training. We find that combining self-supervised learning on natural images and target-task-specific images (in other words hierarchical pretraining) leads to fast and stable downstream convergence, 4–5 times faster compared to the widely used ImageNet pretraining. We also find that, with supervised ImageNet pretraining, only a few dozen annotated samples achieve good segmentation results on cardiac MRI segmentation. In our experiments on the ACDC dataset, 30–75 randomly selected annotated slices are sufficient to achieve close to minimal error.

Acknowledgment. The research presented in this paper, carried out by Budapest University of Technology and Economics has been supported by the Hungarian National Laboratory of Artificial Intelligence funded by the NRDIO under the auspices of the Hungarian Ministry for Innovation and Technology. We thank for the usage of the BME Joker and ELKH Cloud GPU infrastructure (https://science-cloud.hu/) that helped us achieve the results published in this paper. We gratefully acknowledge the support of NVIDIA Corporation with the donation of the NVIDIA GPU also used for this research.

References

1. Azizi, S., et al.: Big self-supervised models advance medical image classification. In: Proceedings of the IEEE/CVF International Conference on Computer Vision (ICCV), pp. 3478–3488 (2021). https://openaccess.thecvf.com/content/ICCV2021/papers/Azizi_Big_Self-Supervised_Models_Advance_Medical_Image_Classification_ICCV_2021_paper.pdf

2. Bernard, O., et al.: Deep learning techniques for automatic MRI cardiac multi-structures segmentation and diagnosis: is the problem solved? IEEE Trans. Med. Imaging **37**(11), 2514–2525 (2018). https://doi.org/10.1109/TMI.2018.2837502

3. Caron, M., Misra, I., Mairal, J., Goyal, P., Bojanowski, P., Joulin, A.: Unsupervised learning of visual features by contrasting cluster assignments. In: Larochelle, H., Ranzato, M., Hadsell, R., Balcan, M.F., Lin, H. (eds.) Advances in Neural Information Processing Systems, vol. 33, pp. 9912–9924. Curran Associates, Inc. (2020). https://proceedings.neurips.cc/paper/2020/file/70feb62b69f16e0238f741fab228fec2-Paper.pdf

4. Chen, T., Kornblith, S., Norouzi, M., Hinton, G.: A simple framework for contrastive learning of visual representations. In: Daumé III, H., Singh, A. (eds.) Proceedings of the 37th International Conference on Machine Learning. Proceedings of Machine Learning Research, vol. 119, pp. 1597–1607. PMLR (2020). https://proceedings.mlr.press/v119/chen20j.html

5. Ciga, O., Xu, T., Martel, A.L.: Self supervised contrastive learning for digital histopathology. Mach. Learn. Appl. **7**, 100198 (2021)

6. da Costa, V.G.T., Fini, E., Nabi, M., Sebe, N., Ricci, E.: solo-learn: a library of self-supervised methods for visual representation learning. J. Mach. Learn. Res. **23**(56), 1–6 (2022). https://jmlr.org/papers/v23/21-1155.html

7. Deng, J., Dong, W., Socher, R., Li, L.J., Li, K., Fei-Fei, L.: Imagenet: a large-scale hierarchical image database. In: 2009 IEEE Conference on Computer Vision and Pattern Recognition, pp. 248–255 (2009). https://doi.org/10.1109/CVPR.2009.5206848

8. Devlin, J., Chang, M.W., Lee, K., Toutanova, K.: BERT: pre-training of deep bidirectional transformers for language understanding. In: Proceedings of the 2019 Conference of the North American Chapter of the Association for Computational Linguistics: Human Language Technologies, Volume 1 (Long and Short Papers), Minneapolis, Minnesota, pp. 4171–4186. Association for Computational Linguistics (2019). https://doi.org/10.18653/v1/N19-1423. https://www.aclweb.org/anthology/N19-1423

9. Doersch, C., Gupta, A., Efros, A.A.: Unsupervised visual representation learning by context prediction. In: 2015 IEEE International Conference on Computer Vision (ICCV), pp. 1422–1430 (2015). https://doi.org/10.1109/ICCV.2015.167

10. Goyal, P., et al.: Self-supervised Pretraining of Visual Features in the Wild (2021). https://arxiv.org/abs/2103.01988

11. Grill, J.B., et al.: Bootstrap your own latent - a new approach to self-supervised learning. In: Larochelle, H., Ranzato, M., Hadsell, R., Balcan, M.F., Lin, H. (eds.) Advances in Neural Information Processing Systems, vol. 33, pp. 21271–21284. Curran Associates, Inc. (2020). https://proceedings.neurips.cc/paper/2020/file/f3ada80d5c4ee70142b17b8192b2958e-Paper.pdf

12. He, K., Fan, H., Wu, Y., Xie, S., Girshick, R.: Momentum contrast for unsupervised visual representation learning. In: Proceedings of the IEEE/CVF Conference on Computer Vision and Pattern Recognition (CVPR) (2020). https://openaccess.thecvf.com/content_CVPR_2020/html/He_Momentum_Contrast_for_Unsupervised_Visual_Representation_Learning_CVPR_2020_paper.html

13. He, K., Zhang, X., Ren, S., Sun, J.: Delving deep into rectifiers: surpassing human-level performance on imagenet classification. In: Proceedings of the IEEE International Conference on Computer Vision (ICCV) (2015)

14. He, K., Zhang, X., Ren, S., Sun, J.: Deep residual learning for image recognition. In: Proceedings of the IEEE Conference on Computer Vision and Pattern Recognition (CVPR) (2016)

15. Hestness, J., et al.: Deep Learning Scaling is Predictable, Empirically (2017). https://doi.org/10.48550/ARXIV.1712.00409. https://arxiv.org/abs/1712.00409

16. Hénaff, O.J., et al.: Data-efficient image recognition with contrastive predictive coding. In: Daumé III, H., Singh, A. (eds.) Proceedings of the 37th International Conference on Machine Learning. Proceedings of Machine Learning Research, vol. 119, pp. 4182–4192. PMLR (2020). https://proceedings.mlr.press/v119/henaff20a.html

17. Morid, M.A., Borjali, A., Del Fiol, G.: A scoping review of transfer learning research on medical image analysis using ImageNet. Comput. Biol. Med. **128**, 104115 (2021). https://doi.org/10.1016/j.compbiomed.2020.104115. https://www.sciencedirect.com/science/article/pii/S0010482520304467

18. Noroozi, M., Favaro, P.: Unsupervised learning of visual representations by solving jigsaw puzzles. In: Leibe, B., Matas, J., Sebe, N., Welling, M. (eds.) ECCV 2016. LNCS, vol. 9910, pp. 69–84. Springer, Cham (2016). https://doi.org/10.1007/978-3-319-46466-4_5

19. Radford, A., Narasimhan, K., Salimans, T., Sutskever, I.: Improving language understanding by generative pre-training (2018). https://openai-assets.s3.amazonaws.com/research-covers/language-unsupervised/language_understanding_paper.pdf

20. Reed, C.J., et al.: Self-supervised pretraining improves self-supervised pretraining. In: Proceedings of the IEEE/CVF Winter Conference on Applications of Computer Vision (WACV), pp. 2584–2594 (2022). https://openaccess.thecvf.

com/content/WACV2022/html/Reed_Self-Supervised_Pretraining_Improves_Self-Supervised_Pretraining_WACV_2022_paper.html

21. Ronneberger, O., Fischer, P., Brox, T.: U-Net: convolutional networks for biomedical image segmentation. In: Navab, N., Hornegger, J., Wells, W.M., Frangi, A.F. (eds.) MICCAI 2015. LNCS, vol. 9351, pp. 234–241. Springer, Cham (2015). https://doi.org/10.1007/978-3-319-24574-4_28

22. Sowrirajan, H., Yang, J., Ng, A.Y., Rajpurkar, P.: MoCo pretraining improves representation and transferability of chest X-ray models. In: Medical Imaging with Deep Learning, pp. 728–744. PMLR (2021)

23. Sudlow, C., et al.: UK biobank: an open access resource for identifying the causes of a wide range of complex diseases of middle and old age. PLOS Med. **12**(3), 1–10 (2015). https://doi.org/10.1371/journal.pmed.1001779. https://doi.org/10.1371/journal.pmed.1001779

24. Vincent, P., Larochelle, H., Bengio, Y., Manzagol, P.A.: Extracting and composing robust features with denoising autoencoders. In: Proceedings of the 25th International Conference on Machine Learning, ICML 2008, pp. 1096–1103. Association for Computing Machinery, New York (2008). https://doi.org/10.1145/1390156.1390294

25. Wightman, R.: PyTorch Image Models (2019). https://github.com/rwightman/pytorch-image-models. https://doi.org/10.5281/zenodo.4414861

26. Yakubovskiy, P.: Segmentation Models Pytorch (2020). https://github.com/qubvel/segmentation_models.pytorch

27. Zbontar, J., Jing, L., Misra, I., LeCun, Y., Deny, S.: Barlow twins: self-supervised learning via redundancy reduction. In: Meila, M., Zhang, T. (eds.) Proceedings of the 38th International Conference on Machine Learning. Proceedings of Machine Learning Research, vol. 139, pp. 12310–12320. PMLR (2021). https://proceedings.mlr.press/v139/zbontar21a.html

28. Zhang, R., Isola, P., Efros, A.A.: Colorful image colorization. In: Leibe, B., Matas, J., Sebe, N., Welling, M. (eds.) ECCV 2016. LNCS, vol. 9907, pp. 649–666. Springer, Cham (2016). https://doi.org/10.1007/978-3-319-46487-9_40

CMC_v2: Towards More Accurate COVID-19 Detection with Discriminative Video Priors

Junlin Hou[1], Jilan Xu[1,3], Nan Zhang[2], Yi Wang[3], Yuejie Zhang[1(✉)], Xiaobo Zhang[4(✉)], and Rui Feng[1,2,4(✉)]

[1] School of Computer Science, Shanghai Key Laboratory of Intelligent Information Processing, Fudan University, Shanghai, China
{jlhou18,jilanxu18,20210860062,yjzhang}@fudan.edu.cn
[2] Academy for Engineering and Technology, Fudan University, Shanghai, China
[3] Shanghai AI Laboratory, Shanghai, China
[4] National Children's Medical Center, Children's Hospital of Fudan University, Shanghai, China
zhangxiaobo0307@163.com, fengrui@fudan.edu.cn

Abstract. This paper presents our solution for the 2nd COVID-19 Competition, occurring in the framework of the AIMIA Workshop at the European Conference on Computer Vision (ECCV 2022). In our approach, we employ the winning solution last year which uses a strong 3D Contrastive Mixup Classification network (CMC_v1) as the baseline method, composed of contrastive representation learning and mixup classification. In this paper, we propose CMC_v2 by introducing natural video priors to COVID-19 diagnosis. Specifically, we adapt a pre-trained (on video dataset) video transformer backbone to COVID-19 detection. Moreover, advanced training strategies, including hybrid mixup and cutmix, slice-level augmentation, and small resolution training are also utilized to boost the robustness and the generalization ability of the model. Among 14 participating teams, CMC_v2 ranked 1st in the 2nd COVID-19 Competition with an average Macro F1 Score of 89.11%.

Keywords: COVID-19 detection · Hybrid CNN-transformer · Contrastive learning · Hybrid mixup and cutmix

1 Introduction

The Coronavirus Disease 2019 SARS-CoV-2 (COVID-19), identified at the end of 2019, is a highly infectious disease, leading to an everlasting worldwide pandemic and collateral economic damage [29]. Early detection of COVID-19 is crucial to the timely treatment of patients, and beneficial to slowdown or even break viral transmission. COVID-19 detection aims to identify COVID from non-COVID cases. Among several COVID-19 detection means, chest computed tomography (CT) has been recognized as a key component in the diagnostic procedure for COVID-19. In CT, we resort to typical radiological findings to confirm

L. Karlinsky et al. (Eds.): ECCV 2022 Workshops, LNCS 13807, pp. 485–499, 2023.
https://doi.org/10.1007/978-3-031-25082-8_32

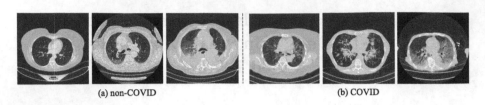

(a) non-COVID   (b) COVID

Fig. 1. Some examples of (a) non-COVID and (b) COVID cases from the COV19-CT-DB dataset. The non-COVID category includes no pneumonia and other pneumonia cases. The COVID category contains COVID-19 cases of different severity levels.

COVID-19, including ground glass opacities, opacities with rounded morphology, crazy-paving pattern, and consolidations [3]. As a CT volume contains hundreds of slices, delivering a convincing diagnosis from these data demands a heavy workload on radiologists. Relying on manual analysis is barely scalable considering the surging increasing number of infection cases. Regarding this, there is an urgent need for accurate automated COVID-19 diagnosis approaches.

Recently, deep learning approaches have achieved promising performance in fighting against COVID-19. They have been widely applied to various medical practices, including the lung and infection region segmentation [2,10,18,27] as well as the clinical diagnosis and assessment [9,22,25,28]. Though a line of works [9,10,28] has been employed for COVID-19 detection via CT analysis and yielded effective results, it is still worth pushing its detection performance to a new level in a faster and more accurate manner for a better medical assistant experience. Improving this performance is non-trivial, since the inner variances between CT scans of COVID are huge and its differences with some non-COVID like pneumonia are easily overlooked. Specifically, CT scans vary greatly in imaging across different devices and hospitals (Fig. 1), and they share several similar visual manifestations with other types of pneumonia. Further, the scarcity of CT scans of COVID-19 due to regulations in the medical area makes these challenges even harder, as we cannot simply turn to a deep model to learn these mentioned characteristics with a big number of annotated scans from scratch.

To tackle these challenges, we exploit video priors along with the given limited number of CT scans to learn an effective feature space for COVID-19 detection, along with contrastive training and some hybrid data augmentation means for further data-efficient learning. Specifically, we employ the advanced 3D contrastive mixup classification network (CMC-COV19D, abbr. CMC_v1) [8], the winner in the ICCV 2021 COVID-19 Diagnosis Competition of AI-enabled Medical Image Analysis Workshop [13], as a baseline. CMC_v1 introduces contrastive representation learning to discover discriminative representations of COVID-19 cases. Besides, a joint training loss is devised by combining the classification loss, mixup loss, and contrastive loss. In this work, we propose CMC_v2 by introducing the following mechanisms customized for 3D models. (1) To capture the long-range lesion span across the slices in the CT scans, we adopt a hybrid CNN-transformer model, i.e. Uniformer [17] as the backbone network.

The combination of convolution and self-attention reduces the network parameters and computational costs. It relieves the potential overfitting when deploying 3D models to small-scale medical datasets. Besides, we empirically show that initializing the model with 3D weights pre-trained on video datasets is promising as modeling the relationship among slices is critical for COVID-19 detection. (2) We develop a hybrid mixup and cutmix augmentation strategy to enhance the models' generalization ability. Due to the limited memory, a gather-and-dispatch mechanism is also customized for the modern Distributed DataParallel (DDP) scheme in Multi-GPU training. (3) We showcase both the 2D slice-level augmentation and the small resolution training bring improvements. By applying the intra-and-inter model ensemble [8], CMC_v2 won the first prize in the 2nd COVID-19 detection challenge of the Workshop "AI-enabled Medical Image Analysis - Digital Pathology & Radiology/COVID19 (AIMIA)". CMC_v2 significantly outperforms the baseline model provided by the organizers by 16% Macro F1 Score.

The remainder of this paper is organized as follows. Section 2 reviews related works. In Sect. 3, we first recap the CMC_v1 network, the basis of CMC_v2, and then introduce the newly proposed modules in CMC_v2. Section 4 describes the COV19-CT-DB dataset used in this paper. Section 5 provides the experimental settings and results. Section 6 concludes our work.

2 Related Work

2.1 COVID-19 Detection

Numerous deep learning approaches have made great efforts to separate COVID patients from non-COVID subjects. Despite the binary classification, the task is challenging as the non-COVID cases include both common pneumonia subjects and non-pneumonia subjects.

The majority of deep learning approaches are based on Convolutional Neural Networks (CNN). [25] was a pioneering work that designed a CNN model to classify COVID-19 and typical viral pneumonia. Song et al. [22] proposed a deep learning-based CT diagnosis system (Deep Pneumonia) to detect patients with COVID-19 from patients with bacteria pneumonia and healthy people. Li et al. [18] developed a 3D COVNet based on ResNet50, aiming to extract both 2D local and 3D global features to classify COVID-19, CAP, and non-pneumonia. Xu et al. [31] introduced a location-attention model to categorize COVID-19, Influenza-A viral pneumonia, and healthy cases. It took the relative distance-from-edge of segmented lesion candidates as extra weight in a fully connected layer to offer distance information.

Recently, Vision Transformer (ViT) has demonstrated its potentials by achieving competitive results on a variety of computer vision tasks. Relevant studies have also been conducted on the COVID-19 diagnosis. Gao et al. [6] used a ViT based on the attention models to classify COVID and non-COVID CT images. To integrate the advantages of convolution and transformer for COVID-19 detection, Park et al. [20] presented a novel architecture that utilized

CNN as a feature extractor for low-level Chest X-ray feature corpus, upon which Transformer was trained for downstream diagnosis tasks with the self-attention mechanism.

2.2 Advanced Network Architecture

In our approach, we adopt two representative deep learning architectures as the backbones, namely ResNeSt-50 and Uniformer-S. Here, we briefly introduce the closely related ResNet and Transformer architectures and their variants.

In the family of ResNets, ResNet [7] introduced a deep residual learning framework to address the network degradation problem. ResNeXt [30] established a simple architecture by adopting group convolution in the ResNet bottleneck block. ResNeSt [33] presented a modular split-attention block within the individual network blocks to enable attention across feature-map groups.

Although CNN models have shown promising results, the limited receptive field makes it hard to capture global dependency. To solve this problem, Vision Transformer (ViT) [4] was applied to the sequences of image patches for an image classification task. Later on, Swin Transformer [19] proposed to use shifted windows between consecutive self-attention layers, which had the flexibility to model at various scales and had linear computational complexity with respect to the image size. Multi-scale Vision Transformer [5] connected the seminal idea of multi-scale feature hierarchies with transformer models for video and image recognition. Pyramid Vision Transformer [26] used a progressive shrinking pyramid to reduce the computations of large feature maps, which overcame the difficulties of porting Transformer models to various dense prediction tasks and inherits the advantages of both CNN and Transformer. Unified transformer (Uni-Former) [17] sought to integrate the merits of convolution and self-attention in a concise transformer format, which can tackle both local redundancy and global dependency. To achieve the balance between accuracy and efficiency, we adopt Uniformer as the default backbone network.

3 Methodology

The overall framework of our model is shown in Fig. 2. In this section, we review the baseline method CMC_v1 [8] firstly and then introduce several simple and effective mechanisms to boost the detection performance.

3.1 Recap of CMC_v1

CMC_v1 employs the contrastive representation learning (CRL) as an auxiliary task to learn discriminative representations of COVID-19. CRL is comprised of the following components. 1) A stochastic data augmentation module $A(\cdot)$, which transforms an input CT x_i into a randomly augmented sample $\tilde{x}_i$. Two augmented volumes are generated from each input CT scan. 2) A base encoder $E(\cdot)$, mapping the augmented CT sample $\tilde{x}_i$ to its feature representation $r_i =$

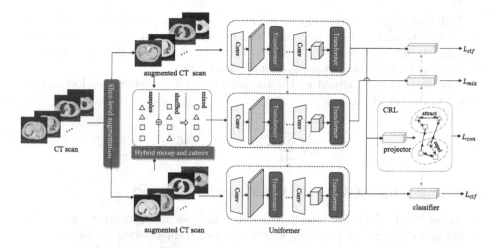

Fig. 2. Overview of our CMC_v2 network for COVID-19 detection.

$E(\tilde{x}_i) \in \mathbb{R}^{d_e}$. 3) A projection network $P(\cdot)$, used to map the representation vector r_i to a relative low-dimension vector $z_i = P(r_i) \in \mathbb{R}^{d_p}$. 4) A classifier $C(\cdot)$, classifying the vector $r_i \in \mathbb{R}^{d_c}$ to the final prediction.

Contrastive Representation Learning. Given a minibatch of N CT volumes and their labels $\{(x_i, y_i)\}$, we can generate a minibatch of $2N$ samples $\{(\tilde{x}_i, \tilde{y}_i)\}$ after data augmentations. Inspired by the supervised contrastive loss [11], we define the positives as any augmented CT samples from the same category, whereas the CT samples from different classes are considered as negative pairs. Let $i \in \{1, \ldots, 2N\}$ be the index of an arbitrary augmented sample, the contrastive loss function is defined as:

$$\mathcal{L}_{con}^i = \frac{-1}{2N_{\tilde{y}_i} - 1} \sum_{j=1}^{2N} \mathbb{1}_{i \neq j} \cdot \mathbb{1}_{\tilde{y}_i = \tilde{y}_j} \cdot \log \frac{\exp(z_i^T \cdot z_j / \tau)}{\sum_{k=1}^{2N} \mathbb{1}_{i \neq k} \cdot \exp(z_i^T \cdot z_k / \tau)}, \quad (1)$$

where $\mathbb{1} \in \{0, 1\}$ is an indicator function, and $\tau > 0$ denotes a scalar temperature hyper-parameter. $N_{\tilde{y}_i}$ is the total number of samples in a minibatch that have the same label $\tilde{y}_i$.

Mixup Classification. CMC_v1 adopts the mixup [34] strategy during training to further boost the generalization ability of the model. For each augmented CT sample $\tilde{x}_i$, the mixup sample and its label are generated as:

$$\tilde{x}_i^{mix} = \lambda \tilde{x}_i + (1 - \lambda)\tilde{x}_p, \quad \tilde{y}_i^{mix} = \lambda \tilde{y}_i + (1 - \lambda)\tilde{y}_p, \quad (2)$$

where p is randomly selected indice; λ is the balancing coefficient. The mixup loss is defined as the cross-entropy loss of mixup samples:

$$\mathcal{L}_{mix}^i = \text{CrossEntropy}(\tilde{x}_i^{mix}, \tilde{y}_i^{mix}). \quad (3)$$

Different from the original design [34] where they replaced the classification loss with the mixup loss, we merge the mixup loss with the standard cross-entropy classification loss $\mathcal{L}_{clf}^i = \mathrm{CrossEntropy}(\tilde{x}_i, \tilde{y}_i)$ to enhance the classification ability on both mixup samples and raw samples.

The total loss is defined as the combination of the contrastive loss, mixup loss, and classification loss:

$$\mathcal{L} = \frac{1}{2N} \sum_{i=1}^{2N} (\mathcal{L}_{con}^i + \mathcal{L}_{mix}^i + \mathcal{L}_{clf}^i). \tag{4}$$

3.2   Improving COVID-19 Detection with CMC_v2

To boost the COVID-19 detection performance, we incorporate natural video priors into CMC_v1 by adapting an efficient pre-trained video backbone to our task, and develop a hybrid data augmentation strategy to increase data efficiency.

Transfer Learning with a Stronger Backbone and Pre-training. In CMC_v1, a 3D ResNeSt-50 model [33] is employed as the backbone network for feature extraction. Although 3D convnets capture local volume semantics efficiently, they are incapable of modeling long-range dependencies between spatial/temporal features explicitly. For simplicity, we refer 'temporal' to the relationship among different CT slices in this paper. Recent works on Vision Transformer [4] managed to encode long-range information using self-attention. However, global self-attention is computationally inefficient and transformer models only demonstrate superior results when huge data is available. Compared with natural image datasets, COVID-19 image datasets have a smaller scale and the model is prone to overfitting. To alleviate this issue, we adopt a video transformer named Uniformer [17], a novel hybrid CNN-transformer model which integrates the advantages of convolution and self-attention in spatial-temporal feature learning while achieving the balance between accuracy and efficiency. In particular, Uniformer replaces the naive transformer block with a Uniformer block, which is comprised of a Dynamic Position Embedding (DPE) layer, a Multi-Head Relation Aggregator (MHRA) layer, and a Feed-Forward Network (FFN).

Furthermore, we experimentally find that training the model from scratch leads to poor results. In transfer learning, it is a common practice to initialize the model on downstream tasks with weights pre-trained on a large-scale ImageNet dataset. To initialize the 3D model, CMC_v1 inflated the ImageNet pre-trained 2D weights to the 3D model. This is achieved by either copying the 2D weights to the center of the 3D weights or repeating the 2D weights along the third dimension. However, these inflated 3D weights may not excel at modeling the temporal relationship between different slices. To address this issue, we directly initialize the model with 3D weights pre-trained on video action recognition datasets, i.e. k400 [1]. We empirically prove that k400 pre-training yields better results than inflated weight initialization in this task.

Hybrid Mixup and Cutmix Strategy. In CMC_v1, the mixup strategy is introduced to generate diverse CT samples. These pseudo samples are beneficial for improving the model's generalization ability. Similar to a mixup, cutmix [32] replaces a local region in the target image with the corresponding local region sampled in the source image. To combine the merits of both, we develop a hybrid mixup and cutmix strategy. In each iteration, we select one strategy with equal probability. This hybrid strategy works well on the traditional Data Parallel (DP) mechanism [21] in multi-GPU training. However, it's challenging to scale to the modern Distributed Data Parallel (DDP) mechanism. The original batch size on each GPU is set to 1 in our case because the effective batch size is 4 after two-view augmentation and hybrid mixup and cutmix strategy, reaching the memory limit on each GPU (The shape of the mini-batch tensors on each GPU is $4 \times T \times 3 \times H \times W$). As the DDP mechanism starts an individual process on each GPU, the hybrid strategy is directly employed on each GPU individually. Performing the hybrid mixup and cutmix strategy on the augmented views of the same image does not align with the original effect. To make it work, we gather all the samples from the GPUs, conduct the hybrid mixup and cutmix over all the samples, and dispatch the generated samples back to each GPU. It guarantees that the hybrid strategy is performed across different CT scans in the current mini-batch. The hybrid mixup and cutmix strategy boost the model's generalization ability.

Slice-Level Augmentation. The data augmentation strategies used in CMC_v1 are 3D rescaling, 3D rotation, and color jittering on all the slices. To further increase the data diversity, we follow the common practice in video data processing and perform different 2D augmentations on each slice, termed as SliceAug. SliceAug achieves slightly better performance than 3D augmentation while having a comparable pre-processing time.

Small Resolution Training. Prior works [4,23] have demonstrated the effectiveness of using small image resolution during training and large resolution during validation/testing. This mechanism bridges the gap between the image size mismatch caused by the random resized cropping during training and center cropping during testing [24]. Besides, the small resolution makes training more efficient. In the experiments, we use the resolution of 192×192 and 224×224 for training and testing, respectively.

4 Dataset

We evaluate our proposed approach on the COV19-CT-Database (COV19-CT-DB) [12]. The COV19-CT-DB contains chest CT scans marking the existence of COVID-19. It consists of about 1,650 COVID and 6,100 non-COVID chest CT scan series from over 1,150 patients and 2,600 subjects. In total, 724,273 slices correspond to the CT scans of the COVID category and 1,775,727 slices correspond to the non-COVID category. Data collection was conducted in the period from September 1, 2020 to November 30, 2021. Annotation of each CT scan was obtained by 4 experienced medical experts and showed a high degree of agreement (around 98%). Each 3D CT scan includes a different number of slices, ranging from 50 to 700. This variation in the number of slices is due to the context of CT scanning. The database is split into training, validation, and testing sets. The training set contains 1,992 3D CT scans (1,110 non-COVID cases and 882 COVID cases). The validation set consists of 504 3D CT scans (289 non-COVID cases and 215 COVID cases). The testing set includes 5,281 scans and the labels are not available during the challenge.

5   Experiments

5.1   Implementation Details

All CT volumes are resized from $(T, 512, 512)$ to $(128, 224, 224)$, where T denotes the number of slices. For training, data augmentations include random resized cropping on the transverse plane, random cropping on the vertical section to 64, rotation, and color jittering. We employ the 3D ResNeSt-50 and Uniformer-S as the backbones in our experiments. The value of parameter d_e is 2,048/512 for ResNeSt-50/Uniformer-S, and d_p is set to 128. All networks are optimized using the Adam algorithm with a weight decay of 1e-5. The initial learning rate is set to 1e-4 and then divided by 10 at 30% and 80% of the total number of training epochs. The networks are trained for 100 epochs. Our methods are implemented in PyTorch and run on eight NVIDIA Tesla A100 GPUs.

5.2   Evaluation Metrics

To evaluate the performance of the proposed method, we adopt the same official protocol of 2nd COVID-19 Competition as the evaluation metric. We report F1 Scores for non-COVID and COVID categories as well as the Macro F1 Score for overall comparison. The Macro F1 Score is defined as the unweighted average of the class-wise/label-wise F1 Scores. We also present ROC curves and Area Under Curve (AUC) for each category.

5.3   Ablation Studies on COVID-19 Detection Challenge

We conduct ablation studies on the validation set of COVID-19 detection challenge to show the impact of each component of our proposed methods. We first analyze the effects of different backbones, and then we discuss the effectiveness of the CMC_v1 framework and the choice of various pre-training methods. Finally, we investigate the impact of the new components in our CMC_v2, i.e. slice-level augmentation (SliceAug), hybrid mixup and cutmix strategy (Hybrid), and small resolution training (SmallRes).

Table 1. The results on the validation set of COVID-19 detection challenge.

ID	Method	Param	FLOPs	Pre-train	Macro F1	F1	
						Non-COVID	COVID
1	ResNet50-GRU [12]	-	-	-	77.00	-	-
2	ResNeSt-50	52.8M	371.9G	ImageNet	89.89	91.27	88.52
3	Uniformer-S	21.2M	230.1G	ImageNet	90.98	92.08	89.88
4	CMC_v1 (R)	57.3M	371.9G	ImageNet	91.98	93.14	90.82
5	CMC_v1 (U)	21.5M	230.1G	ImageNet	92.26	93.11	91.42
6	CMC_v1 (U)	21.5M	230.1G	k400_16 × 4	92.48	93.28	91.67
7	CMC_v1 (U)	21.5M	230.1G	k400_16 × 8	92.70	93.41	91.99
8	CMC_v2 (U, SliceAug)	21.5M	230.1G	k400_16 × 8	93.07	93.94	92.20
9	CMC_v2 (U, Hybrid)	21.5M	230.1G	k400_16 × 8	93.29	94.12	92.45
10	CMC_v2 (U, Hybrid+SmallRes)	21.5M	169.1G	k400_16 × 8	93.30	94.07	92.52

Backbone Network. To analyze the effects of architectures, we compare different backbone models, and the results are shown in the first three rows of Table 1. The reported result of the baseline approach 'ResNet50-GRU' [12] is 77.00% Macro F1 Score. This model is based on CNN-RNN architecture [14–16], where the CNN part performs local analysis on each 2D slice, and the RNN part combines the CNN features of the whole 3D CT scan. Compared to the baseline, our 3D ResNeSt-50 and Uniformer-S backbones achieve more than 12% improvements on the Macro F1 Scores. Specifically, the Uniformer-S achieves better performance on all the metrics, surpassing ResNeSt-50 by 1.09% Macro F1 Score, 0.81% and 1.36% F1 Scores for non-COVID and COVID classes. Besides, the Uniformer-S greatly reduces the network parameters and computational costs. The results demonstrate the long-range dependencies modeling ability of Uniformer-S, which is important to capture the relationships between different CT slices.

Analysis of CMC_v1. We evaluate the effectiveness of the previous CMC_v1 network. The 4th and 5th rows in Table 1 show the results of CMC_v1 (R) and CMC_v1 (U), where the R and U denote ResNeSt-50 and Uniformer-S backbones, respectively. CMC_v1 on both backbones can achieve significant performance improvements. In particular, CMC_v1 (U) obtains 92.26% on Macro F1 Score, 93.11% and 91.42% on F1 Scores for non-COVID and COVID categories. The results demonstrate the generality of the CMC_v1, which can consistently improve the COVID-19 detection performance with different backbones.

Pre-training Schemes. We compare three pre-training methods, namely ImageNet, k400_16 × 4, and k400_16 × 8. ImageNet pre-training inflates the 2D pretrained weights to our 3D models. K400 pre-training denotes 3D weights pretrained on the video action recognition dataset k400, where 16 × 4 and 16 × 8 indicate the sampling 16 frames with frame stride 4 and 8, respectively. It can

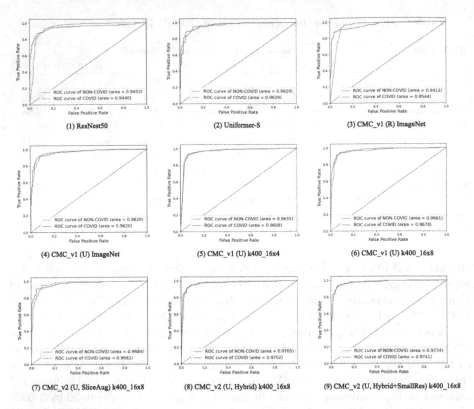

Fig. 3. The ROC curves and AUC scores of different networks.

be seen from the 5th to 7th rows in Table 1, CMC_v1 (U) with k400_16 × 8 pre-training weights outperforms the other two methods on all metrics. Based on the above results, we choose the Uniformer-S with k400_16 × 8 pre-training weights as the default backbone for our proposed CMC_v2.

Analysis of CMC_v2. In this part, we investigate the impact of our newly proposed components in CMC_v2, including slice-level augmentation (SliceAug), hybrid mixup and cutmix strategy (Hybrid), and small resolution training (SmallRes). The experimental results in the 8th row of Table 1 indicate that CMC_v2 (U, SliceAug) can improve the performance on all metrics compared with the CMC_v1 (U). The slice-level augmentation can further increase the data diversity and benefit COVID-19 detection performance. As for the hybrid mixup and cutmix strategy, the CMC_v2 (U, Hybrid) achieves further improvement by 0.59% Macro F1 Score, 0.71% COVID F1 Score, and 0.46% non-COVID F1 Score, compared with the CMC_v1 (U) that only employs the single mixup strategy. Our hybrid mixup and cutmix strategy generates diversified data for improving the model's generalization ability in COVID-19 detection. When we adopt the small resolution training mechanism, the CMC_v2 (U, Hybrid+SmallRes)

Table 2. The leaderboard on the 2nd COVID-19 detection challenge.

Rank	Teams	Macro F1	F1	
			Non-COVID	COVID
1	FDVTS (Ours)	89.11	97.31	80.92
1	ACVLab	89.11	97.45	80.78
3	MDAP	87.87	96.95	78.80
4	Code 1055	86.18	96.37	76.00
5	CNR-IEMN	84.37	95.98	72.76
6	Dslab	83.78	96.22	71.33
7	Jovision-Deepcam	80.82	94.56	67.07
8	ICL	79.55	93.77	65.34
9	etro	78.72	93.48	63.95
10	ResNet50-GRU [12]	69.00	83.62	54.38

achieves the best performance with minimal computational costs among all models. It obtains 93.30% on Macro F1 Score, 94.07% on non-COVID F1 Score, and 92.52% on COVID F1 Score. In particular, this model shows the superior recognition ability for the COVID-19 category among all other approaches.

In addition, we present the ROC curves and AUC of our models in Fig. 3. The AUC results of all the models reach more than 0.94 for both non-COVID and COVID classes. Especially, the full version of CMC_v2 (U, Hybrid+SmallRes) obtains the highest AUC Scores (0.9734 and 0.9741 for non-COVID and COVID, respectively) among all settings.

5.4 Results on COVID-19 Detection Challenge Leaderboard

Table 2 shows the results of our method and other participants on the testing set of 2nd COVID-19 detection challenge. Our method ensembles all the CMC_v2, including CMC_v2 (U, SliceAug), CMC_v2 (U, Hybrid), and CMC_v2 (U, Hybrid+SmallRes) following the strategy in [8]. The final prediction of each CT scan is obtained by averaging the predictions from individual models. We also adopt a test time augmentation (TTA) operation to boost the generalization ability of our models on the testing set. It can be seen from Table 2 that our proposed method ranks first in the challenge with 89.11% Macro F1 Score. Compared to other methods, our model achieves significant improvement on the F1 Score for the COVID category (80.92%), indicating the ability to distinguish COVID cases from non-pneumonia and other types of pneumonia correctly.

5.5 Visualization Results

To verify the interpretability of our model, we visualize the results using Class Activation Mapping (CAM) [35]. As illustrated in Fig. 4, we select four COVID-19 CT scans from the validation set of COV19-CT-DB dataset. In each group,

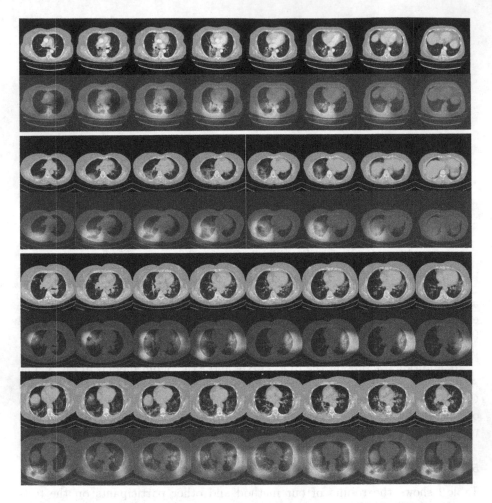

Fig. 4. The visualization results on the COVID-19 CT scans.

the upper row shows the series of CT slices, and the lower row presents the corresponding CAM results. In the first group, it can be seen that the attention maps focus on the local infection regions accurately. In the second group, the wide range of infection regions can also be covered. In the third and fourth groups, the infections in bilateral lungs can also be located precisely. These attention maps provide convincing interpretability for the COVID-19 detection results, which is helpful for real-world clinical diagnosis.

6   Conclusions

In this paper, we propose a novel and practical solution winning COVID-19 detection at the 2nd COVID-19 Competition. Based on the CMC_v1 network, we further develop the CMC_v2 network with substantial improvements, including the CNN-transformer video backbone, hybrid mixup and cutmix strategy, slice-level augmentation, and small resolution training mechanism. The experimental results demonstrate that the new components boost the COVID-19 detection performance and the generalization ability of the model. On the testing set, our method ranked 1st in the 2nd COVID-19 Competition with 89.11% Macro F1 Score among 14 participating teams.

Acknowledgement. This work was supported by the Scientific & Technological Innovation 2030 - "New Generation AI" Key Project (No. 2021ZD0114001; No. 2021ZD0114000), and the Science and Technology Commission of Shanghai Municipality (No. 21511104502; No. 21511100500; No. 20DZ1100205).

References

1. Carreira, J., Zisserman, A.: Quo vadis, action recognition? A new model and the kinetics dataset. In: Proceedings of the IEEE Conference on Computer Vision and Pattern Recognition, pp. 6299–6308 (2017)
2. Chen, J., Wu, L., Zhang, J., Zhang, L., Gong, D., Zhao, Y., et al.: Deep learning-based model for detecting 2019 novel coronavirus pneumonia on high-resolution computed tomography. Sci. Rep. **10**(1), 1–11 (2020)
3. Chung, M., et al.: CT imaging features of 2019 novel coronavirus (2019-nCoV). Radiology (2020)
4. Dosovitskiy, A., et al.: An image is worth 16x16 words: transformers for image recognition at scale. arXiv preprint arXiv:2010.11929 (2020)
5. Fan, H., et al.: Multiscale vision transformers. In: Proceedings of the IEEE/CVF International Conference on Computer Vision, pp. 6824–6835 (2021)
6. Gao, X., Qian, Y., Gao, A.: COVID-VIT: classification of COVID-19 from CT chest images based on vision transformer models. arXiv preprint arXiv:2107.01682 (2021)
7. He, K., Zhang, X., Ren, S., Sun, J.: Deep residual learning for image recognition. In: 2016 IEEE Conference on Computer Vision and Pattern Recognition, pp. 770–778 (2016)
8. Hou, J., Xu, J., Feng, R., Zhang, Y., Shan, F., Shi, W.: CMC-COV19D: contrastive mixup classification for COVID-19 diagnosis. In: Proceedings of the IEEE/CVF International Conference on Computer Vision, pp. 454–461 (2021)
9. Hou, J., et al.: Periphery-aware COVID-19 diagnosis with contrastive representation enhancement. Pattern Recogn. **118**, 108005 (2021)
10. Jin, S., et al.: AI-assisted CT imaging analysis for COVID-19 screening: building and deploying a medical AI system in four weeks. MedRxiv (2020)
11. Khosla, P., et al.: Supervised contrastive learning. In: Annual Conference on Neural Information Processing Systems 2020 (2020)
12. Kollias, D., Arsenos, A., Kollias, S.: AI-MIA: COVID-19 detection & severity analysis through medical imaging. arXiv preprint arXiv:2206.04732 (2022)
13. Kollias, D., Arsenos, A., Soukissian, L., Kollias, S.: MIA-COV19D: COVID-19 detection through 3-D chest CT image analysis. arXiv preprint arXiv:2106.07524 (2021)
14. Kollias, D., et al.: Deep transparent prediction through latent representation analysis. arXiv preprint arXiv:2009.07044 (2020)

15. Kollias, D., Tagaris, A., Stafylopatis, A., Kollias, S., Tagaris, G.: Deep neural architectures for prediction in healthcare. Complex Intell. Syst. **4**(2), 119–131 (2018)

16. Kollias, D., et al.: Transparent adaptation in deep medical image diagnosis. In: TAILOR, pp. 251–267 (2020)

17. Li, K., et al.: Uniformer: unifying convolution and self-attention for visual recognition. arXiv preprint arXiv:2201.09450 (2022)

18. Li, L., Qin, L., Xu, Z., Yin, Y., Wang, X., Kong, B., et al.: Artificial intelligence distinguishes COVID-19 from community acquired pneumonia on chest CT. Radiology **296**, 200905 (2020)

19. Liu, Z., et al.: Swin transformer: hierarchical vision transformer using shifted windows. In: Proceedings of the IEEE/CVF International Conference on Computer Vision, pp. 10012–10022 (2021)

20. Park, S., et al.: Vision transformer using low-level chest X-ray feature corpus for COVID-19 diagnosis and severity quantification. arXiv preprint arXiv:2104.07235 (2021)

21. Paszke, A., et al.: Pytorch: an imperative style, high-performance deep learning library. In: Advances in Neural Information Processing Systems, vol. 32 (2019)

22. Song, Y., Zheng, S., Li, L., Zhang, X., Zhang, X., Huang, Z., et al.: Deep learning enables accurate diagnosis of novel coronavirus (COVID-19) with CT images. MedRxiv (2020)

23. Touvron, H., Cord, M., Douze, M., Massa, F., Sablayrolles, A., Jégou, H.: Training data-efficient image transformers & distillation through attention. In: International Conference on Machine Learning, pp. 10347–10357. PMLR (2021)

24. Touvron, H., Vedaldi, A., Douze, M., Jégou, H.: Fixing the train-test resolution discrepancy. In: Advances in Neural Information Processing Systems, vol. 32 (2019)

25. Wang, S., Kang, B., Ma, J., Zeng, X., Xiao, M., Guo, J., et al.: A deep learning algorithm using CT images to screen for corona virus disease (COVID-19). Eur. Radiol. **31**, 1–9 (2021)

26. Wang, W., et al.: Pyramid vision transformer: a versatile backbone for dense prediction without convolutions. In: Proceedings of the IEEE/CVF International Conference on Computer Vision, pp. 568–578 (2021)

27. Wang, X., Deng, X., Fu, Q., Zhou, Q., Feng, J., Ma, H., et al.: A weakly-supervised framework for COVID-19 classification and lesion localization from chest CT. IEEE Trans. Med. Imaging **39**(8), 2615–2625 (2020)

28. Wang, Z., Xiao, Y., Li, Y., Zhang, J., Lu, F., Hou, M., et al.: Automatically discriminating and localizing COVID-19 from community-acquired pneumonia on chest X-rays. Pattern Recogn. **110**, 107613 (2020)

29. WHO: Coronavirus disease (COVID-19) pandemic (2022). https://www.who.int/emergencies/diseases/novel-coronavirus-2019

30. Xie, S., Girshick, R., Dollár, P., Tu, Z., He, K.: Aggregated residual transformations for deep neural networks. In: Proceedings of the IEEE Conference on Computer Vision and Pattern Recognition, pp. 1492–1500 (2017)

31. Xu, X., Jiang, X., Ma, C., Du, P., Li, X., Lv, S., et al.: A deep learning system to screen novel coronavirus disease 2019 pneumonia. Engineering **6**(10), 1122–1129 (2020)

32. Yun, S., Han, D., Oh, S.J., Chun, S., Choe, J., Yoo, Y.: Cutmix: regularization strategy to train strong classifiers with localizable features. In: Proceedings of the IEEE/CVF International Conference on Computer Vision, pp. 6023–6032 (2019)

33. Zhang, H., et al.: Resnest: split-attention networks. arXiv preprint arXiv:2004.08955 (2020)

34. Zhang, H., Cisse, M., Dauphin, Y.N., Lopez-Paz, D.: mixup: beyond empirical risk minimization. arXiv preprint arXiv:1710.09412 (2017)
35. Zhou, B., Khosla, A., Lapedriza, A., Oliva, A., Torralba, A.: Learning deep features for discriminative localization. In: Proceedings of the IEEE Conference on Computer Vision and Pattern Recognition, pp. 2921–2929 (2016)

COVID Detection and Severity Prediction with 3D-ConvNeXt and Custom Pretrainings

Daniel Kienzle[✉], Julian Lorenz, Robin Schön, Katja Ludwig, and Rainer Lienhart

Augsburg University, 86159 Augsburg, Germany
{daniel.kienzle,julian.lorenz,robin.schoen,katja.ludwig, rainer.lienhart}@uni-a.de

Abstract. Since COVID strongly affects the respiratory system, lung CT-scans can be used for the analysis of a patients health. We introduce a neural network for the prediction of the severity of lung damage and the detection of a COVID-infection using three-dimensional CT-data. Therefore, we adapt the recent ConvNeXt model to process three-dimensional data. Furthermore, we design and analyze different pretraining methods specifically designed to improve the models ability to handle three-dimensional CT-data. We rank 2nd in the *1st COVID19 Severity Detection Challenge* and 3rd in the *2nd COVID19 Detection Challenge*.

Keywords: Machine learning · COVID detection · Severity prediction · Medical image analysis · CT scans · 3D data

1 Introduction

The last few years have been strongly shaped by the COVID-19 pandemic, with a considerable amount of cases ending deadly. For the treatment of patients it is crucial to predict the severity of lung damage caused by a SARS-CoV-2 infection accurately. The lung damage is visually detectable by visible groundglass opacities and mucoid impactions on the slices of a patients CT-scan [34]. Thus, it might be beneficial to automatically process CT-scans for the diagnosis of the patients.

In this paper, we introduce a neural network to automatically analyze CT-scans. We train our model to classify the severity of lung damage caused by SARS-CoV-2 into four different categories. The model is trained and evaluated using the COV19-CT-DB database [19]. Additionally, we transfer our architecture and training pipeline to the detection of SARS-CoV-2 infections in CT-scans and train a separate model for this task. Consequently, we show that our method can easily be transferred to multiple COVID-related analyses of CT-scans. We rank 2nd in the *1st COVID19 Severity Detection Challenge* and 3rd in the *2nd*

D. Kienzle, J. Lorenz and R. Schön—Authors contributed equally.

L. Karlinsky et al. (Eds.): ECCV 2022 Workshops, LNCS 13807, pp. 500–516, 2023.
https://doi.org/10.1007/978-3-031-25082-8_33

COVID19 Detection Challenge. Moreover, our model is especially good at identifying the most severe cases that are most important to detect in a clinical setting.

As medical datasets are small in comparison to common computer-vision datasets, the application of large computer-vision architectures is not straight forward as they tend to overfit very quickly. As a result, the development of a good pretraining pipeline as well as the utilization of additional data is essential in order to get adequate results.

Since medical datasets are comparably small, the validation split is as well in most cases very small. However, evaluating the models performance on a single small validation set leads to non-representative results as the validation set is not representative for the overall data distribution. Furthermore, the evaluation on a single small dataset could cause overfitting of the hyperparameters to the validation set characteristics and, therefore, reduces the models test-set performance. As a result, it is very important to use strategies like cross-validation in order to get a better estimate of the models performance.

Goal of this paper is to develop a neural network that is capable of automatically predicting four degrees of severity of lung damage from a patents lung CT-scan. In addition, we also adapt our architecture to predict infections with the SARS-CoV-2 virus using CT-scans. In order to improve the performance on these two tasks, our main contributions are:

1. We adapt the recent ConvNeXt architecture [27] to process three dimensional input-data.
2. We introduce multiple techniques for pretraining of our architecture in order to increase the ability of our network to handle three-dimensional CT-scans.

2   Related Work

The idea of using neural networks for the prediction of certain properties visible in medical data has developed to increasing levels of importance in the last few years (examples can be found in [1,40,43]). The authors of [4,20,21] and [22] have used CNNs and Recurrent Neural Networks (RNNs) for the prediction of Parkinson's disease on brain MRI and DaT scans. In [29,33,39] NN-based methods for the detection of lung cancer are developed.

Since its occurrence in late 2019, a considerable number of articles have concerned themselves with using neural networks for the purpose of predicting a potential SARS-CoV-2 infection from visual data. The authors of [31] propose neural networks for the usage of CT scans as well as chest x-rays, whereas [32] puts a lot of focus on computational efficiency and design a lightweight network, in order to be able to also run on CPU hardware. In the wake of this development, there have also been methods which combine neural and non-neural components. In [5] the authors first extract the features by means of neural network backbone, and the utilize an optimization algorithm in combination with a local search method before feeding the resulting features into a classifier. The method of [24] applies a fusion based ranking model, based on reparameterized Gompertz function, after the neural network has already produced its output probabilities. In [22] and [20] the authors also make use of clustering in order to carry out a further analysis of the produced features vectors, and classify the CT scans according to their proximity to the cluster centers.

In the context of last years ICCV there has been a challenge with the aim of detecting COVID-19 from CT images [19]. The winners of this contest [16] were using contrastive learning techniques in order to improve their networks performance. [30] and [25] use 3D-CNNs for the detection of the disease, whereas the teams in [45] and [41] happen to use transformer-based architectures. In addition to that, [3] proposes the usage of AutoGluon [12] as an AutoML based approach.

Our architectures is, similar to other previously existing approaches for the processing of 3D-data, based on the idea, that 2D architectures can be directly extended to 3D architectures [7,23,36]. In our case we use 3D modification of the ConvNeXt architecture [27]. This particular approach is characterized by architectural similarities to MobileNets [17,37] and Vision Transformers (ViT, [11,26]).

In the medical field, due to privacy restrictions that protect the patients' data, the potential amount of training data is rather sparse. This especially holds, when it comes to datasets that accompany particular benchmarks. However, datasets for similar tasks may be exploited for pretraining and multitask learning, if one can assume that insights from one datasets might benefit the main objective. This inspired the authors of [42] and [13] to pretrain the model on pretext tasks, which are carried out on medical data. The authors of [28] show that pretraining on ImageNet is useful, by the virtue of the sheer amount of data. In some particular cases, we might have access to a larger amount of data while lacking labels. The publications [38] and [8] use semi-supervised learning techniques to overcome this particular situation.

3 Methods

Goal of this work is to develop a neural network architecture capable of predicting the severity of a SARS-CoV-2 infection and to transfer the method to the task of infection detection. We apply our models to the COV19-CT-DB database [19]. The train and test set of this database consist of 2476 scans for the task of infection detection and 319 images for severity prediction. Each scan is composed of multiple two-dimensional image slices (166 slices on average). We concatenate these slices into a three-dimensional tensor and apply cubic spline interpolation to get tensors of the desired spatial dimension. In this section we introduce the key methods for improving the automatic analysis of this data.

3.1 ConvNeXt 3D

The architecture we utilize is a three-dimensional version of the recent ConvNeXt architecture [27]. This architecture type is especially characterized by multiple alterations, which have already proven themselves to be useful in the context of Vision Transformers, and were applied to the standard ResNet [14].

For example, ConvNeXt has a network stem that patchifies the image using non-overlapping convolutions followed by a number of blocks with a compute ratio of (3 : 3 : 9 : 3) that make up the stages of the network. The influence which MobileNetV2 [37] had on this type of architecture is expressed by the introduction of inverted bottleneck blocks and the usage of depthwise convolutions, which, due to their computational efficiency, allow for an unusually large kernel size of 7 × 7. Additional distinguishing properties of this architecture are the replacement of Batch Normalization by Layer Normalization, the usage of less activation functions and the replacement of the ReLU activation function by the GELU activation [15].

The standard ConvNeXt architecture in its initial form was conceptualized for the purpose of processing 2D images with 3 color channels, whereas we want to process 3D computational tomography scans that only have one color channel (initially expressed in Hounsfield Units [6]). We adapted the ConvNeXt architecture to our objective by using 3D instead of 2D convolutions. In order to be able to make use of potentially pre-existing network weights, we apply kernel weight inflation techniques to the 2D networks parameters as described in Sect. 3.2.

3.2 Pretraining

In contrast to ordinary computer vision datasets like e.g. ImageNet [10], medical datasets are usually considerably smaller. In order to still be able to train large neural networks with these datasets we utilize various pretraining techniques. As our data consists of three-dimensional gray-scale tensors as input data instead of two-dimensional RGB-images and we use 3D-convolutions instead of 2D-convolutions, it is not possible to directly use the publicly-available pretrained ConvNeXt weights. In this section we present various possibilities for the initialization of our network with pretrained weights.

For 2D models, it is common to pretrain a model on the task of ImageNet classification. As our data consists of gray-scale tensors, we implemented a ConvNeXt pretraining with gray-scale ImageNet images to obtain weights for a two-dimensional ConvNeXt model. To use those weights for our 3D model, we propose three different inflation techniques for the two-dimensional weights of the pretrained 2D model. We will refer to these as *full inflation*, *1G inflation* and *2G inflation*.

Let $K \in \mathbb{R}^{I \times O \times H \times W}$ be the 2D kernel weight tensor, and $K^\uparrow \in \mathbb{R}^{I \times O \times H \times W \times D}$ be the 3D kernel weights after inflation. For these kernel weights we denote by I the input channels, O the output channels, H the height, W the width and D the additional dimension of the 3D kernel. Also, let i, o, h, w, d denote all possible positions along the aforementioned dimensions. γ is a normalization factor that normalizes the inflated tensor $K^\uparrow$ to have the L2 norm of the 2D kernel K.

The first way, called *full inflation*, is the commonly used option of simply copying the weights along the new tensor axis [7]. This can be described as an equation of the form

$$\forall i, o, h, w, d \ : \ K^\uparrow_{i,o,h,w,d} = K_{i,o,h,w} \cdot \gamma. \tag{1}$$

In *1G inflation*, we use a Gaussian weight $\mathcal{N}(\cdot, \mu, \sigma)$ in order to create different weights, that are the largest in the kernel center:

$$\forall i, o, h, w, d : K^{\uparrow}_{i,o,h,w,d} = \left(K_{i,o,h,w} \cdot \mathcal{N}\left(d, \frac{D}{2}, \frac{D}{8}\right) \right) \cdot \gamma \quad (2)$$

The third, that is referred to *2G inflation* is based on multiplying the 2D weights along 2 axes:

$$\forall i, o, h, w, d :$$

$$K^{\uparrow}_{i,o,h,w,d} = \left(K_{i,o,h,w} \cdot \mathcal{N}\left(d, \frac{D}{2}, \frac{D}{8}\right) + K_{i,o,h,w} \cdot \mathcal{N}\left(w, \frac{W}{2}, \frac{W}{8}\right) \right) \cdot \gamma \quad (3)$$

The different inflation approaches to create 3D kernels are visualized in Fig. 1.

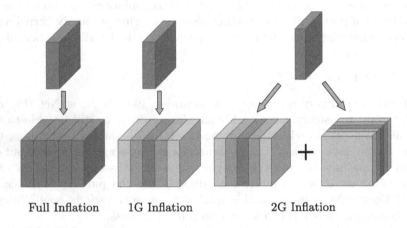

Fig. 1. A visualization of the different inflation approaches to generate 3D kernels. *Full inflation* simply copies the weights along the new axis. *1G inflation* also copies the weights along the new axis but multiplies them with Gaussian weights. *2G inflation* acts similar to 1G inflation but add the weights after going over two dimensions.

Since the images in the ImageNet database are very distinct from CT-images as used in this paper, we introduce various further ways to adjust the model to three-dimensional CT-scans. For instance, we use an additional dataset designed for lung-lesion segmentation in CT scans [2,9,35] and the STOIC dataset created for SARS-CoV-2 severity prediction [34]. As those datasets consist of CT-scans of SARS-CoV-19 infected patients similar to the COV19-CT-DB database, we assume that pretraining with these additional datasets will be beneficial for our model performance and will increase in robustness as it is able to deal with a greater variety of data.

Since the lung damage caused by SARS-CoV-2 is visually detectable in lung CT-scans, we use the segmentation dataset to pretrain our model to segment lung lesions. By directly showing the damaged lung regions to the network we hope to provide a reasonable bias for learning to predict the severity. Furthermore, a segmentation pretraining is beneficial as the segmentation task is more robust to overfitting in contrast to a classification task. This is important as it enables us to apply large-scale architectures to the small medical datasets.

The STOIC dataset provides two categories of severity for each patients CT-scans. Even though the categories in the STOIC dataset are different to the categories in the COV19-CT-DB database we assume, nevertheless, that pretraining with the STOIC dataset teaches the network a general understanding of severity.

In order to be able to generate segmentation masks with our model additionally to severity classification outputs, we extend our architecture similarly to the Upernet architecture [44]. Our architecture is explained in Fig. 2

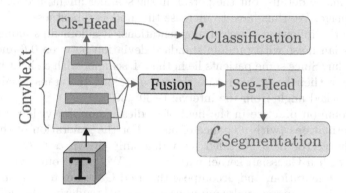

Fig. 2. Visualization of the pretraining architecture. An *input-tensor* T is processed by the *ConvNeXt* model. To generate a classification label (e.g. for severity prediction or infection detection) the output-features of the last block are processed by a task-specific *classification head* in order to generate the class probabilities. In order to compute a segmentation mask, the output-features of every block of the ConvNeXt architecture are upsampled and concatenated similar to [44]. This is further processed by a *segmentation-head* in order to produce the segmentation output.

In this work we compare 4 different pretraining methods:

1. We directly use our inflated grayscale ImageNet pretraining weights. This approach is referred to as *ImageNet model.*
2. We train our network for the task of segmentation on the segmentation dataset starting from inflated ImageNet weights (1.). This approach is referred to as *segmentation model.*
3. Pseudo-labels are generated with the segmentation model (2.) in order to get segmentation masks for the COV19-CT-DB database. We train a new model for the task of segmentation on the COV19-CT-DB database using the pseudo-labels as ground-truth. The model is initialized with our inflated ImageNet weights (1.). This is referred to as *segmia model.*

4. Pseudo-labels are generated with the segmentation model (2.) to get segmentation masks for the STOIC dataset. We train a new model to jointly optimize severity classification using real labels and segmentation using pseudo-labels for the STOIC dataset. This model is initialized with our inflated ImageNet weights (1.). This is referred to as *multitask model*

After each pretraining method, we finetune our model for either severity prediction or infection detection on the COV19-CT-DB database.

3.3   Approaches for Increased Robustness

When automatically analyzing CT images, we have to account for multiple potential forms of irregularities. Not only might different CT scanners result in varying image details, but the person in the scanner might also lie differently on every image. We thus have to increase the models robustness.

We use the following classical augmentations: Random flips along all three axes, Gaussian noise with random standard deviation between 0.6 and 0.8, and Gaussian blur. Since some patients lie in the CT scanner with a slight inclination to the side of their body, we rotate the tensors along the transversal axis by a randomly picked angle from the interval $(-30°, 30°)$.

As is common practice in the field of medical computer-vision, we also use elastic deformations (with a chance of 50%). For the generation of those deformations the vector field is scaled by a randomly drawn $\alpha \in (1, 7)$ and then smoothed with a Gaussian kernel with $\sigma = 35$. We create our own GPU compatible implementation, and decompose the used Gaussian filter along its axes. This results in an augmented computational speed, rendering the deformations viable for fast online computation.

During manual inspection, we came across some tensors that were oriented in a different way (for example vertically instead of horizontally) or with patients facing a different direction. In order to make our network robust to these variations, we add an augmentation that simulates different orientations: For each of the three axes (x-, y- and z-axis) we randomly pick a multiple of 90°, and rotate the tensor by this angle. Since the misoriented tensors still constitute a considerable minority of all cases we do not apply this augmentation to every tensor during training, but only with a probability of 25%.

In addition, we apply random crops with a probability of 50%. Therefore, we first rescale the scan to a resolution of $(256 \times 256 \times 256)$ and take a random crop of $(224 \times 224 \times 224)$. When no random crop is applied, we directly rescale to the latter resolution. Rescaling of the tensors is performed with cubic spline interpolation and the rescaled tensors are precomputed prior to the training to decrease computation time. When inspecting the data, we discovered some tensors where the slice resolution happened to be internally inconsistent between different slices of the same CT. In those cases, we discarded the inconsistent slices. In order to stabilize our performance, we kept a second copy of our model whose weights were not learned directly, but which is an exponential moving average (EMA) of the trained models weights. This copy of the network is used for the evaluations and final predictions.

Besides data augmentation, we also use 5-fold cross validation to improve model robustness. We split the public training set into five folds with almost equal size and make sure that each class is evenly distributed across all folds. For example, each fold for severity prediction contains 12 or 13 moderate cases. Each fold forms the validation set for one model, the remaining folds serve as the training set. This way we get 5 models that are trained and evaluated on different datasets. To get the final predictions for the official test set, we predict every case from the test set using all of the 5 models and take the mean of the model outputs. Before averaging the outputs, we apply Softmax to bring all values to the same scale.

4   Experiments

Most experiments were performed for the task of severity prediction because of two reasons. First, it is supposedly the more challenging task due to the classification into four classes and, second, it is computationally less expensive due to a smaller dataset allowing for a greater number of experiments.

We evaluate our models using the COV19-CT-DB database. As the validation set is comparatively small, evaluating the models' performance on its validation set leads to non-representative results as the validation set is not representative for the data. Furthermore, evaluation on a single validation set easily causes overfitting of the hyperparameters. As a result, we do not use the validation set to analyse the models performance and instead perform 5-fold cross-validation only on the training set. The average performance of the 5 training runs is used as an estimate for the performance. In order to generate predictions for the validation set and test set we use an ensemble consisting of the 5 models from the cross-validation runs. The predictions of the models are averaged after application of the softmax to produce the final prediction. In addition, we also add the results obtained only with the 5th model.

4.1   Preliminary Experiments

As the frequency of the classes in the COV19-CT-DB database is not evenly distributed, we recognized that our neural network trained with ordinary cross-entropy as loss function has problems detecting the less frequent classes. Especially patients with critical severity are not correctly classified. As it is especially important to recognize the critical cases, we introduced a weight in order to balance the cross entropy. Thus, the loss for every class is multiplied with the normalized class-frequency. The severity-prediction results for the ordinary cross entropy (*CE*) and balanced cross entropy (*balanced CE*) are given in Table 1. The cross-validation performance of both loss functions is very similar. Even though the ordinary cross-entropy performs a little bit better, we chose to use the balanced cross entropy for our further experiment since we think that the balanced cross entropy improves the performance for the underrepresented critical cases. We assume that the good performance of our challenge submission (see Table 4) for critical cases is partly because of the balanced cross-entropy. Consequently, we would advise to use balanced cross-entropy instead of ordinary cross-entropy in a clinical setting.

Table 1. Comparison of balanced cross entropy with ordinary cross entropy. The models are initialized with full ImageNet initialization. The cross validation results and the results for the official validation set are reported. The ensemble predictions are marked with a †. F1 scores are macro F1 scores

Loss	F1 cross val	F1 val	F1 val mild	F1 val moderate	F1 val severe	F1 val critical
balanced CE	64.99	$62.82^\dagger$	$82.93^\dagger$	$57.14^\dagger$	$57.89^\dagger$	$53.33^\dagger$
CE	**65.37**	$60.10^\dagger$	$82.05^\dagger$	$50.00^\dagger$	$55.00^\dagger$	$53.33^\dagger$

4.2 Comparison of Pretrainings

One main goal of this paper is to enhance the severity-prediction performance by introducing several pretrainings as explained in Sect. 3.2. Results for the various ImageNet inflation methods are added in Table 2.

Table 2. Comparison of ImageNet initialization. The cross validation results and the results for the official validation set are reported. The ensemble predictions are marked with a †. F1 scores are macro F1 scores

Initialization	F1 cross validation	F1 validation
Full	65.67	$\mathbf{61.28}^\dagger$
1G	65.81	$60.71\ 59.93^\dagger$
2G	**69.61**	$56.92^\dagger$

According to these results, the performance is best for 2G inflation in terms of the cross-validation metrics and, consequently, we assume that using multiple geometrical-oriented planes is beneficial as it better utilizes all three dimensions.

Results for the various pretraining methods can be seen in Table 3. For some of those experiments, the scores for the official test set of the COV19-CT-DB database are available. As this test set is substantially larger than the official validation set, the performance estimate is considered as more accurate. Thus, we use this metric in addition to our cross-validation results to interpret our performance. The results clearly indicate that the cross-validation metrics is a much better estimate of the models performance than the validation-set metrics since the *segmentation*, *segmia* and *multitask model* perform better than the

Table 3. Comparison of models initialized with different pretrainings. Performance is evaluated for the severity-prediction task. Random-initialization is denoted as *Random*. The ensemble predictions are marked with a †. F1 scores are macro F1 scores.

Pretraining	F1 cross validation	F1 test	F1 validation
Random	62.71	–	56.46†
ImageNet (full)	65.67	45.73†	61.28†
Segmentation	67.25	46.21†	61.28†
Segmia	66.48	48.85†	63.05†
Multitask	**68.18**	**48.95†**	58.77†

ImageNet model on both cross-validation metrics and test-set metrics. Moreover, the best model in terms of cross-validation metrics performs also best on the test-set. As a result, we strongly advise to evaluate models intended for clinical usage based on cross-validation metrics.

Even though cross validation gives well reasoned clues about the models performance, a large gap between the cross-validation score and the test-set metrics can be observed. Because the test-set performance is worse by a large margin, we suppose that the test-set statistics do not fully match the train-set characteristics and, thus, there could be a small domain shift in the test-set data. As a result, good test-set results can only be achieved with a robust model and it seems that the utilization of additional datasets and the use of pseudo-labels both increase the robustness of the model significantly.

In Table 3 it is clearly visible that all pretrainings yield significantly better results than a randomly initialized model in terms of the cross-validation metrics and the *segmentation, segmia* and *multitask* models score is higher than the score of the *ImageNet model*. Consequently, it can be concluded that a pretraining utilizing segmentation labels is highly favorable. As the *multitask model* outperforms the other variants on the test-set as well as on the cross-validation metrics, we think that a pretraining with a task similar to the final task as well as the utilization of segmentation pseudo-labels is very beneficial. We suppose that the models gain a greater robustness due to the usage of the additional datasets in the pretraining pipeline. As the STOIC dataset used for the multitask pretraining is comparably large, the *multitask model* seems to be especially robust, thus performing best on the test-set. Subsequently, we recommend combining a task similar to the final task with a segmentation task for superior pretraining results.

4.3 Challenge Submission Results

We participate in two challenges hosted in the context of the Medical Image Analysis (MIA) workshop at ECCV 2022 [18]. The two tasks were the detection of COVID infections and the prediction of the severity the patient is experiencing. We rank 2nd in the *1st COVID19 Severity Detection Challenge* and 3rd in the *2nd COVID19 Detection Challenge*. In this section, we present our submissions and further discuss the results. Our submission code is published at https://github.com/KieDani/Submission_2nd_Covid19_Competition.

Since we apply 5-fold cross-validation, we suggest to use an ensemble of the 5 trained models to generate the validation-set as well as test-set predictions. However, due to a coding mistake, we only use the fifth model instead of the full ensemble during the challenge. As a result, this causes deviations from the cross-validation estimate as our model is only trained with 80% of the training data. Nevertheless, it is even more impressive that we still achieved such good results and, thus, our architecture and pretraining pipeline are very well suited for COVID-related tasks.

In contrast to using the fifth model for predictions, we advise to use either an ensemble of the 5 models or to train a single model with all available data based on the settings found with cross validation.

1st COVID19 Severity Detection Challenge: The ranking for the winning teams is shown in Table 4. It can be seen that our method is by a large margin the best in predicting *Severe* and *Critical* cases. We suppose this is achieved through the utilization of the balanced cross entropy and we emphasize that this property is exceptionally important for clinical use cases.

Table 4. Comparison of best submissions of the winning teams in the 1st COVID19 Severity Detection Challenge. Performance is evaluated for the severity-prediction task. Our prediction is calculated only with the 5th model of the 5-fold cross-validation. F1 scores are macro F1 scores.

Team	F1 test	F1 test mild	F1 test moderate	F1 test severe	F1 test critical
1st: FDVTS	**51.76**	58.97	**44.53**	58.89	44.64
2nd: Ours	51.48	**61.14**	34.06	**61.91**	**48.83**
3rd: CNR-IEMN	47.11	55.67	37.88	55.46	39.46

Since it was possible to submit up to 5 different solutions to the challenge, we list our submissions in Table 5. The *segmia model* performs best. Furthermore, the comparison of submission 2 and 5 indicates that the random-orientation augmentation increases the performance. However, as the test-set scores are very similar the augmentation does not have a great effect on the test set and, thus, is negligible for this challenge. We suppose that this is due to fewer CT-scans with deviating orientation in the test set compared to the train and validation sets.

Table 5. Our submissions to the 1st COVID19 Severity Detection Challenge. The ensemble predictions are marked with a †. Predictions marked with $\overline{\underline{\text{五}}}$ are calculated only with the 5th model of the 5-fold cross-validation. Usage of the random-orientation augmentation is denoted with *ROr*. F1 scores are macro F1 scores.

Submission #	Pretraining	F1 cross validation	F1 test	F1 validation
1	ImageNet (Full)	65.67	$46.67^{\overline{\underline{五}}}$	$67.21^{\overline{\underline{五}}}$ $61.28^{\dagger}$
2	Segmentation	67.25	$49.36^{\overline{\underline{五}}}$	$63.43^{\overline{\underline{五}}}$ $61.28^{\dagger}$
3	Segmia	66.48	$\mathbf{51.48}^{\overline{\underline{五}}}$	$60.89^{\overline{\underline{五}}}$ $63.05^{\dagger}$
4	Multitask	68.18	$46.01^{\overline{\underline{五}}}$	$55.51^{\overline{\underline{五}}}$ $58.77^{\dagger}$
5	Segmentation ROr	71.74	$49.90^{\overline{\underline{五}}}$	$60.02^{\overline{\underline{五}}}$ $62.68^{\dagger}$

We added an example for a correctly classified and an incorrectly classified CT-scan in Fig. 3.

2nd COVID19 Detection Challenge: In addition to severity-detection, we used our architecture and training-pipeline also to train our model for the task of infection detection and participated in the *2nd COVID19 Detection challenge*. The ranking for the winning teams is depicted in Table 6. Because it was also possible to submit up to 5 solutions, our submissions can be seen in Table 7. We achieve the best results without cross validation using the *segmentation model*. This is probably due to the coding mistake mentioned above as this model is trained with 100% of the training data in contrast to 80%. Nearly the same performance is achieved using the (fifth) *multitask model*, thus indicating that the multitask pretraining is a good choice for infection detection as well. As submission 2 to 5 are considerably better than submission 1, we conclude that our custom pretrainings improve the results in contrast to the ImageNet model for the infection-detection task, too.

Table 6. Comparison of the best submissions of the winning teams in the 2nd COVID19 Detection Challenge. Our prediction is calculated only with the 5th model of the 5-fold cross-validation. F1 scores are macro F1 scores.

Team	F1 test	F1 test non-COVID	F1 test COVID
1st: ACVLab	89.11	97.45	80.78
1st: FDVTS	89.11	97.31	80.92
2nd: MDAP	87.87	96.95	78.80
3rd: Ours	86.18	96.37	76.00

Correctly Classified   Underestimated

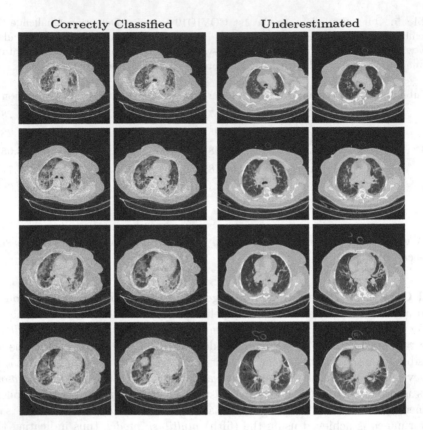

Fig. 3. Qualitative example slices for the severity prediction. The CT scan on the *left* has been correctly identified to display a patient that is in a critical state. The CT scan on the *right* has been predicted to be in a moderate state although the correct prediction would have been a severe state.

Furthermore, by analyzing submission 1 to 3, we deduce that the cross-validation results are good estimates for the models performance as the order of the scores matches the test-set scores. Moreover, since the gap between cross-validation metrics and test-set metrics is considerably smaller than for the severity prediction task, we reason that the dataset statistics of the train-set and the test-set are much more similar for the infection-detection task. We guess that the statistics are more similar in this challenge because the dataset size is substantially larger and, consequently, we emphasize the need to use larger datasets in order to get valid performance estimates for clinical usage.

Table 7. Our submissions to the 2nd COVID19 Detection Challenge. F1 scores are macro F1 scores. The * denotes that no cross validation was used. The ensemble predictions are marked with a †. Predictions marked with $\bar{\pi}$ are calculated only with the 5th model of the 5-fold cross-validation. Usage of the random-orientation augmentation is denoted with *ROr*.

Submission #	Pretraining	F1 cross validation	F1 test	F1 validation
1	ImageNet (full)	91.73	$82.13^{\bar{\pi}}$	$86.60^{\bar{\pi}}$ $89.71^{\dagger}$
2	Multitask	93.53	$86.02^{\bar{\pi}}$	$87.80^{\bar{\pi}}$ $88.79^{\dagger}$
3	Segmia	93.33	$83.63^{\bar{\pi}}$	$88.31^{\bar{\pi}}$ $89.22^{\dagger}$
4	Segmia ROr*	–	83.93	92.03
5	Segmentation*	–	**86.18**	93.48

5   Conclusion

In this paper, we analyzed various pretraining techniques designed to enhance SARS-CoV-2 severity-prediction performance of our neural network and show that the performance can be significantly increased utilizing segmentation labels and additional datasets. Additionally, we show that our architecture and pretraining pipeline can easily be transferred to the task of infection detection and, thus, our method can be regarded as a general method to enhance COVID-related CT-scan analysis.

The pretraining methods were applied to a three-dimensional ConvNeXt architecture and a finetuning for the COV19-CT-DB dataset was performed. We achieved 2nd rank in the *1st COVID19 Severity Detection Challenge* and 3rd rank in the *2nd COVID19 Detection Challenge*, consequently proving that our method yields competitive results.

In addition to that, we introduced the balanced cross-entropy and argued that this loss-function is important for clinical use cases. We emphasize that our model achieved best results in detecting the most-severe cases.

Altogether, we presented a framework for severity prediction as well as infection detection and achieved good performance by applying this framework to the ConvNeXt architecture. We encourage further research based upon our framework to enhance the diagnosis options in clinical use cases.

References

1. Abdou, M.A.: Literature review: efficient deep neural networks techniques for medical image analysis. Neural Comput. Appl. 1–22 (2022)
2. An, P., et al.: CT images in Covid-19 [data set]. The Cancer Imaging Archive (2020)
3. Anwar, T.: Covid19 diagnosis using autoML from 3D CT scans. In: Proceedings of the IEEE/CVF International Conference on Computer Vision (ICCV) Workshops, pp. 503–507, October 2021

4. Arsenos, A., Kollias, D., Kollias, S.: A large imaging database and novel deep neural architecture for Covid-19 diagnosis. In: 2022 IEEE 14th Image, Video, and Multidimensional Signal Processing Workshop (IVMSP), pp. 1–5. IEEE (2022)

5. Basu, A., Sheikh, K.H., Cuevas, E., Sarkar, R.: Covid-19 detection from CT scans using a two-stage framework. Expert Syst. Appl. **193**, 116377 (2022)

6. Buzug, T.M.: Einführung in die Computertomographie: mathematisch-physikalische Grundlagen der Bildrekonstruktion. Springer, Heidelberg (2011). https://doi.org/10.1007/978-3-642-18593-9

7. Carreira, J., Zisserman, A.: Quo Vadis, action recognition? A new model and the kinetics dataset. In: 2017 IEEE Conference on Computer Vision and Pattern Recognition (CVPR), pp. 4724–4733 (2017)

8. Chaitanya, K., Erdil, E., Karani, N., Konukoglu, E.: Local contrastive loss with pseudo-label based self-training for semi-supervised medical image segmentation, December 2021

9. Clark, K., et al.: The cancer imaging archive (TCIA): maintaining and operating a public information repository. J. Digit. Imaging **26**(6), 1045–1057 (2013)

10. Deng, J., Dong, W., Socher, R., Li, L.J., Li, K., Fei-Fei, L.: Imagenet: a large-scale hierarchical image database. In: 2009 IEEE Conference on Computer Vision and Pattern Recognition, pp. 248–255 (2009)

11. Dosovitskiy, A., et al.: An image is worth 16x16 words: transformers for image recognition at scale. In: International Conference on Learning Representations (2021)

12. Guo, J., et al.: Gluoncv and gluonnlp: deep learning in computer vision and natural language processing. J. Mach. Learn. Res. **21**, 23:1–23:7 (2020)

13. Hatamizadeh, A., Nath, V., Tang, Y., Yang, D., Roth, H., Xu, D.: Swin unetr: swin transformers for semantic segmentation of brain tumors in MRI images. arXiv preprint arXiv:2201.01266 (2022)

14. He, K., Zhang, X., Ren, S., Sun, J.: Deep residual learning for image recognition. In: Proceedings of the IEEE Conference on Computer Vision and Pattern Recognition (CVPR), June 2016

15. Hendrycks, D., Gimpel, K.: Bridging nonlinearities and stochastic regularizers with gaussian error linear units. CoRR abs/1606.08415 (2016). https://arxiv.org/abs/1606.08415

16. Hou, J., Xu, J., Feng, R., Zhang, Y., Shan, F., Shi, W.: CMC-Cov19d: contrastive mixup classification for Covid-19 diagnosis. In: Proceedings of the IEEE/CVF International Conference on Computer Vision (ICCV) Workshops, pp. 454–461, October 2021

17. Howard, A.G., et al.: Mobilenets: efficient convolutional neural networks for mobile vision applications. arXiv preprint arXiv:1704.04861 (2017)

18. Kollias, D., Arsenos, A., Kollias, S.: AI-MIA: Covid-19 detection & severity analysis through medical imaging. arXiv preprint arXiv:2206.04732 (2022)

19. Kollias, D., Arsenos, A., Soukissian, L., Kollias, S.: MIA-Cov19d: Covid-19 detection through 3-D chest CT image analysis. arXiv preprint arXiv:2106.07524 (2021)

20. Kollias, D., et al.: Deep transparent prediction through latent representation analysis. arXiv preprint arXiv:2009.07044 (2020)

21. Kollias, D., Tagaris, A., Stafylopatis, A., Kollias, S., Tagaris, G.: Deep neural architectures for prediction in healthcare. Complex Intell. Syst. **4**(2), 119–131 (2018)

22. Kollias, D., et al.: Transparent adaptation in deep medical image diagnosis. In: TAILOR, pp. 251–267 (2020)

23. Kopuklu, O., Kose, N., Gunduz, A., Rigoll, G.: Resource efficient 3D convolutional neural networks. In: Proceedings of the IEEE/CVF International Conference on Computer Vision Workshops (2019)

24. Kundu, R., Basak, H., Singh, P.K., Ahmadian, A., Ferrara, M., Sarkar, R.: Fuzzy rank-based fusion of CNN models using Gompertz function for screening Covid-19 CT-scans. Sci. Rep. **11**(1), 1–12 (2021)

25. Liang, S., Zhang, W., Gu, Y.: A hybrid and fast deep learning framework for Covid-19 detection via 3D chest CT images. In: Proceedings of the IEEE/CVF International Conference on Computer Vision (ICCV) Workshops, pp. 508–512, October 2021

26. Liu, Z., et al.: Swin transformer: hierarchical vision transformer using shifted windows. In: Proceedings of the IEEE/CVF International Conference on Computer Vision (ICCV) (2021)

27. Liu, Z., Mao, H., Wu, C.Y., Feichtenhofer, C., Darrell, T., Xie, S.: A convnet for the 2020s. In: Proceedings of the IEEE/CVF Conference on Computer Vision and Pattern Recognition (CVPR) (2022)

28. Loh, A., et al.: Supervised transfer learning at scale for medical imaging. arXiv preprint arXiv:2101.05913 (2021)

29. Mhaske, D., Rajeswari, K., Tekade, R.: Deep learning algorithm for classification and prediction of lung cancer using CT scan images. In: 2019 5th International Conference On Computing, Communication, Control and Automation (ICCUBEA), pp. 1–5 (2019)

30. Miron, R., Moisii, C., Dinu, S., Breaban, M.E.: Evaluating volumetric and slice-based approaches for Covid-19 detection in chest CTs. In: Proceedings of the IEEE/CVF International Conference on Computer Vision (ICCV) Workshops, pp. 529–536, October 2021

31. Mukherjee, H., Ghosh, S., Dhar, A., Obaidullah, S.M., Santosh, K., Roy, K.: Deep neural network to detect Covid-19: one architecture for both CT scans and chest x-rays. Appl. Intell. **51**(5), 2777–2789 (2021)

32. Polsinelli, M., Cinque, L., Placidi, G.: A light CNN for detecting Covid-19 from CT scans of the chest. Pattern Recogn. Lett. **140**, 95–100 (2020)

33. Rao, P., Pereira, N.A., Srinivasan, R.: Convolutional neural networks for lung cancer screening in computed tomography (CT) scans. In: 2016 2nd International Conference on Contemporary Computing and Informatics (IC3I), pp. 489–493 (2016)

34. Revel, M.P., et al.: Study of thoracic CT in Covid-19: the stoic project. Radiology (2021)

35. Roth, H.R., et al.: Rapid artificial intelligence solutions in a pandemic-the Covid-19-20 lung CT lesion segmentation challenge. Res. Square (2021)

36. Ruiz, J., Mahmud, M., Modasshir, Md., Shamim Kaiser, M., Alzheimer's Disease Neuroimaging Initiative: 3D DenseNet ensemble in 4-way classification of Alzheimer's disease. In: Mahmud, M., Vassanelli, S., Kaiser, M.S., Zhong, N. (eds.) BI 2020. LNCS (LNAI), vol. 12241, pp. 85–96. Springer, Cham (2020). https://doi.org/10.1007/978-3-030-59277-6_8

37. Sandler, M., Howard, A., Zhu, M., Zhmoginov, A., Chen, L.C.: Mobilenetv 2: inverted residuals and linear bottlenecks. In: Proceedings of the IEEE Conference on Computer Vision and Pattern Recognition (CVPR), June 2018

38. Seibold, C.M., Reiß, S., Kleesiek, J., Stiefelhagen, R.: Reference-guided pseudo-label generation for medical semantic segmentation. In: Proceedings of the AAAI Conference on Artificial Intelligence, vol. 36, pp. 2171–2179 (2022)

39. Shakeel, P.M., Burhanuddin, M., Desa, M.I.: Automatic lung cancer detection from CT image using improved deep neural network and ensemble classifier. Neural Comput. Appl. 1–14 (2020)

40. Suganyadevi, S., Seethalakshmi, V., Balasamy, K.: A review on deep learning in medical image analysis. Int. J. Multimedia Inf. Retrieval **11**(1), 19–38 (2022)

41. Tan, W., Liu, J.: A 3D CNN network with BERT for automatic Covid-19 diagnosis from CT-scan images. In: Proceedings of the IEEE/CVF International Conference on Computer Vision (ICCV) Workshops, pp. 439–445, October 2021

42. Tang, Y., et al.: Self-supervised pre-training of swin transformers for 3D medical image analysis. In: Proceedings of the IEEE/CVF Conference on Computer Vision and Pattern Recognition, pp. 20730–20740 (2022)

43. Wang, J., Zhu, H., Wang, S.H., Zhang, Y.D.: A review of deep learning on medical image analysis. Mob. Netw. Appl. **26**(1), 351–380 (2021)

44. Xiao, T., Liu, Y., Zhou, B., Jiang, Y., Sun, J.: Unified perceptual parsing for scene understanding. In: Ferrari, V., Hebert, M., Sminchisescu, C., Weiss, Y. (eds.) ECCV 2018. LNCS, vol. 11209, pp. 432–448. Springer, Cham (2018). https://doi.org/10.1007/978-3-030-01228-1_26

45. Zhang, L., Wen, Y.: A transformer-based framework for automatic Covid19 diagnosis in chest CTs. In: Proceedings of the IEEE/CVF International Conference on Computer Vision (ICCV) Workshops, pp. 513–518, October 2021

Two-Stage COVID19 Classification Using BERT Features

Weijun Tan[1,2](✉) (iD), Qi Yao[1], and Jingfeng Liu[1]

[1] Jovision-Deepcam Research Institute, Shenzhen, China
{sz.twj,sz.yaoqi,sz.ljf}@jovision.com,
{weijun.tan,qi.yao,jingfeng.liu}@deepcam.com
[2] LinkSprite Technology, Longmont, CO 80503, USA
weijun.tan@linksprite.com

Abstract. We propose an automatic COVID1-19 diagnosis framework from lung CT-scan slice images using double BERT feature extraction. In the first BERT feature extraction, A 3D-CNN is first used to extract CNN internal feature maps. Instead of using the global average pooling, a late BERT temporal pooing is used to aggregate the temporal information in these feature maps, followed by a classification layer. This 3D-CNN-BERT classification network is first trained on sampled fixed number of slice images from every original CT scan volume. In the second stage, the 3D-CNN-BERT embedding features are extracted for every 32 slice images sequentially, and these features are divided into fixed number of segments. Then another BERT network is used to aggregate these features into a single feature followed by another classification layer. The classification results of both stages are combined to generate final outputs. On the validation dataset, we achieve macro F1 score 92.05%; and on the testing dataset, we achieve macro F1 84.43%.

Keywords: 3D CNN · BERT · COVID-19 · Classification · Diagnosis · Feature extraction

1 Introduction

There are a lot of research on automatic diagnosis of COVID-19 since the breakout of this terrible pandemic. Among the techniques to diagnosis COVID-19, X-ray and CT-scan images are studied extensively. In this paper, we present an automatic diagnosis framework from chest CT-scan slice images using BERT feature extraction for its extraordinary representing capability with both spatial and temporal attention. The goal is to classify COVID-19, and non-COVID-19 from a volume of CT-scan slice images of a patient. We use the dataset provided in the ECCV2022 MIA-COV19D challenge [11], which is a follow-up challenge of the ICCV2021 MIA-COV19D challenge [10]. Since there is no slice annotation in this dataset, 2D CNN classification network on single slice image is not considered. Instead, 3D CNN network based methods are explored.

L. Karlinsky et al. (Eds.): ECCV 2022 Workshops, LNCS 13807, pp. 517–525, 2023.
https://doi.org/10.1007/978-3-031-25082-8_34

A 3D network has been successfully used in many tasks including video under-standing (recognition, detection, captioning). A 3D CNN network is much more powerful then 2D CNN feature extraction followed by RNN feature aggregation [10,11]. 3D network is also used in COVID-19 diagnosis, where the slice images at different spacing form a 3D series of images. The correlation between slice images is just analogous to the temporal information in videos.

Transformer is a one of break-through technologies in machine learning in recent years [21]. It is first used in language model, then extended to a lot other areas in machine learning. Standard transformer is one directional, and the BERT is a bidirectional extension of it [2]. It is also first used language model, and later extended to other areas including video action recognition [8]. The work [19] is the first to use a 3D CNN network with BERT (CNN-BERT) for video action recognition [8] in classification of COVID-19 from CT-scan images.

We follow the approaches and framework used in [19], including the prepro-cessing, 3D-CNN-BERT network. Instead of using the MLP for all slice image feature aggregation, we propose to use a second BERT network in the second stage which is more powerful than a simple MLP. The classification results of both stages are combined to generate the final decision for every patient.

In the first stage, 3D CNN-BERT is used to extract and aggregate a CNN feature for a fixed number of slice images. A classification layer is used on the feature to classify the input to be COVID or non-COVID. In the second stage, we extract the embedding feature vector of all available sets of images for every CT-scan volume. Since the number of images are different, we divide the features into a fixed number of segments (typically 16) using linear interpolation. Then a second BERT is used to aggregate these features into a single feature followed by a few full-connection classification layers. The classification results from both stages are combined to generate the final outputs.

We evaluate the first stage, the second stage individually and combined. Experiment results show that the combined decision gives best performance. On the validation dataset, we achieve macro F1 score 0.9163.

2   Related Work

Deep learning has been successful in a lot of medical imaging tasks. Some examples are [12–14]. Since the outbreak of the COVID-19 pandemic, a lot of researches have been done to automatically diagnose on CT scan images or X-ray images. For a latest review, please refer to [15].

Based on the feature extraction and classification approaches, there are basi-cally three types of methods, the 2D CNN method, 2D+1D method, and 3D CNN method. When the slice annotation is not available, either the 2D+1D method or the 3D CNN method can be used. In the 2D+1D method, feature is first extracted on 2D slice image, then multiple features are aggregated used 1D RNN or other network [6,10,11]. Since the slice annotation is not available, some weak supervision using the patient annotation is used. In the 3D CNN, 3D CNN is directly used on selected slice images for the patient classification. There are many new developments of 3D CNN for COVID-19 diagnosis recently. We list some of them published in 2022 [1,5,7,17,18]. The last one [5] uses 3D CNN and a contrastive learning.

3   Data Preprocessing and Preparation

In this section we review the preprocessing and preparation method used in [19], which we use in this work as well.

3.1   Slice Image Filtering and Lung Segmentation

The first task of preprocessing is to select good slice images for training and validation. This is because at the beginning and end of a patient's CT scan images, many images are useless. Some work simply drop a fixed percentage of images at both ends [22], but we think that method is too coarse. Another motivation for this task is to filter out closed lung images [16], because when the lung is closed, the image is also useless. We first use traditional morphological transforms to find the bounding box of the lung, and the mask of the lung contour. We find the max area of lungs out of all slice images of a patient. Then a percentage threshold is set. Slice images whose lung area is less than this thresholds are discarded.

The second task is to find more accurate lung mask using an UNet. We use an UNet because the morphological lung segmentation is too coarse and miss many important details of the true lung mask, particularly near the edges, and on images with infection lesions. We reuse trained checkpoint in [19] to process all the images in the new ECCV22-MIA dataset [11]. After that, a refinement is used to fill the holes and discontinued edges. Shown in Fig. 1 are an example of the preprocessing steps.

3.2   Preparing Input for 3D CNN

The 3D CNN we use requires to use a fixed number of images as input. In our work, we use 32 slice images. However, since the number of available slice images is varying, we need to use resampling strategy to generate the input slice images. There are two cases down-sampling and up-sampling. We use the same resampling method as in [19], which gets the idea from [4]. On the training dataset, random sampling is used, while on the validation and test datasets, a symmetrical and uniform resampling is used.

In the training, validation and testing time of the first stage 3D-CNN-BERT classification network, only one set of images is selected in every epoch. However, in the second stage BERT classification network, multiple sets of images are selected as much as possible sequentially from the beginning to the end of all slice images of a patient.

Another big factor of the input is the modality of the input data. An naive idea is to simply use the RGB image. However in [22], the authors find that

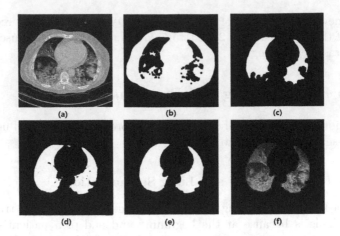

Fig. 1. An example of morphological segmentation and UNet segmentation: (a) raw image, (b) after binarization, (c) morphological segmented mask, (d) UNet segmented mask, (e) refined UNet segmented mask, (f) masked lung image. The figure is from [19].

adding the segmented mask to the RGB image can help the performance. In [19] the authors go further to study the RGB image, the mask and the masked lung image. They find that the combination of these three modalities give the best performance. So in this work we keep using this modality.

4   Classification Networks

In this paper, we explore two stages of classification networks. In the first level, a 3D CNN-BERT network [8] is used. In the second level, feature vectors of all available set of slices images are generated before the classification layer in the first stage. After equally divided to a given number of segments, a second BERT is used to aggregate these features to a single feature vector for every CT-scan volume. This feature is sent to 3 FC layers to do classification. The diagram is shown in Fig. 2.

4.1   First Stage 3D CNN-BERT Network

We reuse the 3D CNN-BERT network in [8]. This architecture utilizes BERT feature aggregation to replace widely uses global pooling. In this work, we borrow the idea and apply it to COVID-19 classification. In this architecture, the selected 32 slice images are propagated through a 3D CNN architecture and then a BERT feature pooling. In order to perform classification with BERT, an classification token (xcls) is appended at the input and a classification embedding vector (ycls) is generated at the output. This embedding vector is sent the FC layer, producing the classification output.

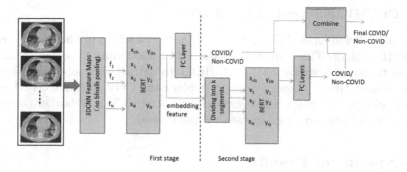

Fig. 2. Network architecture of new two-stage classification networks using BERT feature.

In [8] many different 3D backbone networks are studied, including the R(2+1)D [20]. On two main video action recognition benchmark datasets, this architecture achieves best performance making new SOTA records. For more detail of this architecture, please refer to [8].

In this paper, we use the R(2+1)D backbone [20] with a Resnet34 [3]. We use 32 slice images as input. The input image size is set to 224×224, while 112×112 is used in [8]. The embedding vector (ycls) is generated and saved for use in the second stage classification network.

4.2   Second Stage BERT Classification Network

The 3D CNN-BERT network produces a classification result for a single set of input slice images. For CT-volumes where there are more than one set of slice images, we want to process all available slice images in order not to miss some useful information. Therefore, we propose to use a second stage BERT feature aggregation and classification network.

First, the embedding features are generated for every 32 sequential slice images. If the number of images is less than 32, or the number of images in a sequential set if less than 32, then the same resampling method as in Sect. 3.2 is used. After that, these features are equally divided into a given number, e.g. 16, segments by taking the mean of features among the available features in that segment. This can be done using simple linear interpolation. Depending on the available number of features and number of segments, upsampling or downsampling may be used.

These given number of segment features are sent to the second BERT network. A classification token is generated inside the BERT, and an output classification vector is sent to the FC layers to do classification. Other outputs are not used.

4.3   COVID-19 Severity Classification

In the second challenge of the ECCV22-MIA, the task is to predict the severity of a COVID-patient. There are four classes: Mild, Moderate, Severe, and Critical. We use the same first stage 3D-CNN-BERT to extract the embedding features. In the second BERT stage, we simply replace the COVID/non-COVID binary classification to the four class severity classification. The standard cross entropy loss instead of the binary cross entropy loss is used.

5   Experiment Results

5.1   Dataset

In the ECCV22-MIA-COV19D dataset called COVID19-CT-DB [10], there are 1992 training CT-scan volumes, 504 validation volumes, and 5544 test volumes. In each volume, there are various number of slice images, ranging from 1 to more than a thousand. There are no slice annotations, so 2D CNN classification is not possible if not using extra datasets.

Most of the sizes of the images are 512×512. However, there are quite a lot images whose sizes are not so. So we first do a quick check, if the size of an image is not 512×512, it is resized to 512×512 before any other preprocessing is applied.

For the Unet segmentation, we use the annotated dataset we find on Kaggle and from CNBC [23]. In the CNBC annotation, three types of annotations - lung field, Ground-glass opacity, and Consolidation. We merge all three types to the lung field.

No other datasets are used in training or validation of the 3D CNN-BERT network or the MLP network.

5.2   Implementation Details

Both stages of the BERT networks are implemented in one Pytorch framework. In the first stage, the input image size is set to 224×224. We choose to use 32 slice images as input and use the R(2+1)D backbone. In the second stage, the input feature size is 512. Features are equally divided to 4, 8 or 16 segments to find the best performance. We use the Adam optimizer with a initial learning rate 1E-5. A reduced learning rate on plateau with a factor 0.1 and 5 epoch patience is used. The training runs at most 200 epochs with early stopping. The validation accuracy is used to select the best model.

Input Data Modality: In [19], they study the modality of the slice images. They compare the RGB image only, RGB image and mask [22], RGM image, mask, and masked image. Their ablation study shows that the last one performs the best. In this work we keep using this modality as input to the first stage 3D-CNN-BERT network.

5.3   Detection Challenge Results

After an ablation study to select parameters, we choose a few top performers to bench mark our algorithm. On the COVID19-CT-DB [10] validation dataset, the benchmark results are listed in Table 1. In this table, the macro F1 scores are presented.

On the COVID19-CT-DB validation dataset [10], the best macro F1 scores from the first stage 3D CNN-BERT and the second stage BERT classification networks are 90.73% and 90.72% respectively. The combined macro F1 score from both stages is 92.05%. On the COVID19-CT-DB test dataset, the best macro F1 score we achieve is 84.43%. The winning teams [6], [5] of this detection challenge have macro F1 score 0.8911.

Table 1. Macro F1 scores of the detection challenge on the ECCV22 COVID19-CT-DB datasets.

Dataset	Method	1st stage	2nd stage	Combined
Test	Baseline	-	-	0.6900
Test	[6]	–	–	0.8911
Test	[5]	–	–	0.8911
Validation	Ours	0.9073	0.9072	0.9205
Test	Ours	–	–	0.8443

5.4   Severity Challenge Results

On the COVID19-CT-DB [10] validation dataset, the severity classification results are listed in Table 2, where the macro F1 scores are presented. Please note that for this task, only the second stage BERT network gives results. We achieve an macro F1 score 0.685. On the test dataset, the macro F1 score is

Table 2. Macro F1 scores of the severity challenge on the ECCV22-MIA datasets.

Dataset	Method	Macro F1 score
Test	Baseline [11]	0.4030
Test	[5]	0.5176
Test	[9]	0.5148
Validation	Ours	0.6855
Test	Ours	0.4149

6   Conclusions

In this paper we present a two stage classification network where in the first stage a 3D CNN-BERT network is used, and in the 2nd stage, a second BERT network is used to aggregate features from all slice images into a single feature for classification. For the detection challenge, on the validation dataset, our best F1 score is 92.05%; and on the test dataset, our best F1 score is 84.33%.

References

1. Bao, G., et al.: Covid-mtl: multitask learning with shift3d and random-weighted loss for covid-19 diagnosis and severity assessment. Pattern Recogn. **124**, 108499 (2022)
2. Devlin, J., Chang, M.W., Lee, K., Toutanova, K.: Bert: pre-training of deep bidirectional transformers for language understanding. arXiv (2018)
3. He, K., Zhang, X., Ren, S., Sun, J.: Deep residual learning for image recognition. CVPR (2015)
4. He, X., et al.: Automated model design and benchmarking of 3d deep learning models for covid-19 detection with chest ct scans. Proceedings of the AAAI Conference on Artificial Intelligence (2021)
5. Hou, J., Xu, J., Feng, R., Zhang, Y.: Fdvts's solution for 2nd cov19d competition on covid-19 detection and severity analysis (2022)
6. Hsu, C.C., Tsai, C.H., Chen, G.L., Ma, S.D., Tai, S.C.: Spatiotemporal feature learning based on two-step lstm and transformer for ct scans. arXiv 2207.01579 (2022)
7. Huang, Z., et al.: Multi-center sparse learning and decision fusion for automatic covid-19 diagnosis. Appl. Soft Comput. **115**, 108088 (2022)
8. Kalfaoglu, M.E., Kalkan, S., Alatan, A.A.: Late temporal modeling in 3D CNN architectures with BERT for action recognition. In: Bartoli, A., Fusiello, A. (eds.) ECCV 2020. LNCS, vol. 12539, pp. 731–747. Springer, Cham (2020). https://doi.org/10.1007/978-3-030-68238-5_48
9. Kienzle, D., Lorenz, J., Schön, R., Ludwig, K., Lienhart, R.: Custom pretrainings and adapted 3d-convnext architecture for covid detection and severity prediction. arXiv 2206.15073 (2022)
10. Kollias, D., Arsenos, A., Soukissian, L., Kollias, S.: Mia-cov19d: Covid-19 detection through 3-d chest ct image analysis. arXiv preprint arXiv:2106.07524 (2021)
11. Kollias, D., Arsenos, A., Soukissian, L., Kollias, S.: Ai-mia: Covid-19 detection and severity analysis through medical imaging. arXiv preprint 2206.04732 (2022)
12. Kollias, D., et al.: Deep transparent prediction through latent representation analysis. arXiv preprint arXiv:2009.07044 (2020)
13. Kollias, D., Tagaris, A., Stafylopatis, A., Kollias, S., Tagaris, G.: Deep neural architectures for prediction in healthcare. Complex Intell. Syst. **4**(2), 119–131 (2018)
14. Kollias, D., et al.: Transparent adaptation in deep medical image diagnosis. In: TAILOR, pp. 251–267 (2020)
15. Liu, F., Chen, D., Zhou, X., Dai, W., Xu, F.: Let AI perform better next time: a systematic review of medical imaging-based automated diagnosis of covid-19: 2020–2022. Appl. Sci. **12**(8) (2022)

16. Rahimzadeh, M., Attar, A., Sakhaei, S.M.: A fully automated deep learning-based network for detecting covid-19 from a new and large lung ct scan dataset. medRxiv (2020)

17. Riahi, A., Elharrouss, O., Al-Maadeed, S.: Bemd-3dcnn-based method for covid-19 detection. Comput. Biol. Med. **142**, 105188 (2022)

18. Sobahi, N., Sengur, A., Tan, R.S., Acharya, U.R.: Attention-based 3d CNN with residual connections for efficient ECG-based covid-19 detection. Comput. Biol. Med. **143**, 105335 (2022)

19. Tan, W., Liu, J.: A 3d CNN network with bert for automatic covid-19 diagnosis from CT-scan images. In: ICCV Workshops (2021)

20. Tran, D., Wang, H., Torresani, L., Ray, J., Lecun, Y., Paluri, M.: A closer look at spatiotemporal convolutions for action recognition. In: CVPR (2018)

21. Vaswani, A., et al.: Attention is all you need (2017)

22. Wang, X., et al.: A weakly-supervised framework for covid-19 classification and lesion localization from chest CT. IEEE Trans. Med. Imaging **39**, 2615–2625 (2020)

23. Zhang, K., Liu, X., Shen, J., et al.: Clinically applicable ai system for accurate diagnosis, quantitative measurements and prognosis of covid-19 pneumonia using computed tomography. Cell (April 2020)

PVT-COV19D: COVID-19 Detection Through Medical Image Classification Based on Pyramid Vision Transformer

Lilang Zheng[1], Jiaxuan Fang[1], Xiaorun Tang[1], Hanzhang Li[2], Jiaxin Fan[2], Tianyi Wang[1], Rui Zhou[1(✉)], and Zhaoyan Yan[1(✉)]

[1] School of Information Science and Engineering, Lanzhou University, Lanzhou, China
zr@lzu.edu.cn, yanchy16@lzu.edu.cn
[2] School of Mathematics and Statistics, Lanzhou University, Lanzhou, China

Abstract. With the outbreak of COVID-19, a large number of relevant studies have emerged in recent years. We propose an automatic COVID-19 diagnosis model based on PVTv2 and the multiple voting mechanism. To accommodate the different dimensions of the image input, we classified the images using the Transformer model, sampled the images in the dataset according to the normal distribution, and fed the sampling results into the PVTv2 model for training. A large number of experiments on the COV19-CT-DB dataset demonstrate the effectiveness of the proposed method. Our method won the sixth place in the (2nd) COVID19 Detection Challenge of ECCV 2022 Workshop: AI-enabled Medical Image Analysis - Digital Pathology & Radiology/COVID19. Our code is publicly available at https://github.com/MenSan233/Team-Dslab-Solution.

1 Introduction

The Coronavirus Disease 2019 (COVID-19) is caused by the severe acute respiratory syndrome coronavirus 2 (SARS-CoV-2) and has become a public health emergency worldwide [10]. In order to prevent the further spread of COVID-19 and promptly treat infected patients, early detection and isolation are important for a successful response to the COVID-19 pandemic.

Infection caused by COVID-19 can severely affect the respiratory tract and create a layer of lesions in the lungs, which can interfere with the normal function of the lungs [1]. The medical imaging characteristics of chest can help to detect infected or not with COVID-19 quickly. Imaging features of the chest can be obtained by CT (computed tomography) scans and X-rays. Compared to X-rays, CT scans perform better, for example, X-rays give a two-dimensional view, while CT scans are made up of 3D views of the lungs, which help to check for disease and its location. In addition, machines are used for CT scans in almost every country, and the use of chest CT scans to detect COVID-19 has attracted the attention of many researchers [7].

L. Zheng and J. Fang—Equal contribution.

L. Karlinsky et al. (Eds.): ECCV 2022 Workshops, LNCS 13807, pp. 526–536, 2023.
https://doi.org/10.1007/978-3-031-25082-8_35

However, rapid and accurate detection of COVID-19 using chest CT scans requires the guidance of specific experts, and the limited number of radiologists in different regions makes it challenging to provide specialized clinicians in each hospital [24]. As more people need to be tested for COVID-19, medical staff are overwhelmed, reducing their focus on correctly diagnosing COVID-19 cases and confirming results. Therefore, it is necessary to distinguish COVID-19 infection cases in time. With the help of deep learning-based algorithms, medical staff can quickly exclude uninfected cases at first and allocate more medical resources to the infected ones. Therefore, to reduce the human involvement in using chest CT images to detect COVID-19, a more effective model for detecting COVID-19 through medical images is urgently needed.

In the past research, due to the breakthrough success of deep learning in the field of image recognition, many researchers have applied deep learning methods to the medical field [19]. Combined with the latest research progress in big data analysis of medical images, Lei Cai et al. introduced the application of intelligent imaging and deep learning in the field of big data analysis and early diagnosis of diseases, especially the classification and segmentation of medical images [4]. Amjad Rehman et al. introduced deep learning-based detection of COVID-19 using CT and X-ray images and data analysis of global spread [27]. Rajit Nair et al. [23] developed a fully automated model to predict COVID-19 using chest CT scans. The performance of the proposed method was evaluated by classifying CT scans of community-acquired pneumonia (CAP) and other non-pneumonic cases. The proposed deep learning model is based on ResNet 50, named CORNet, for the detection of COVID-19, and a retrospective and multicenter analysis was also performed to extract visual features from volumetric chest CT scans during COVID-19 detection [23].

For a long time, various computer vision tasks have been developed by using CNN (Convolutional Neural Network) as the backbone. This paper proposes a simple CNN-free backbone for dense prediction tasks, the PVT-COV19D. PVT (Pyramid Vision Transformer) [31] introduces a pyramid structure to the Transformer model, which enables various dense prediction tasks such as detection, segmentation, etc. In this paper, the method is applied to COVID-19 detection.

2   Related Work

After the outbreak of the COVID-19 pandemic, there have been studies showing that the deep learning method may be a potential method to detect COVID-19 [29]. A lot of researchers have used deep learning methods for diagnosis, mainly CNN on CT scan images or X-ray images. Unlike using CNN for classification, in this paper, we use PVT [31], which is a simple backbone without CNN, for intensive prediction tasks.

2.1   CNN

In the past decade, compared with traditional machine learning and computer vision technology, deep learning has achieved the most advanced image recognition tasks. Similarly, the deep learning correlation scheme is widely used in the field of medical signal processing [19]. In [15] CNN+RNN network was used to input CT scanning images to distinguish between COVID-19 and non-COVID cases. In [22], the 3-D ResNet models were used to detect COVID-19, and the authors used volumetric 3-D CT scanning to distinguish it from other common pneumonia (CP) and normal cases. Wang et al. [34] used 3D DeCoVNet to detect COVID-19, which took the CT volume of its lung mask as the input. In [20] D. Kollias et al. classified medical images using transparent adaptation (TA) method. In [18] D. Kollias et al. proposed a method that extracts the potential information from the trained deep neural network (DNN), derives the concise representation, and analyzes it in an effective and unified way to achieve the purpose of prediction.

Among these classification methods, some use pure 2-D networks to predict slice images. Furthermore, in order to judge lung CT, some methods use 2-D networks to generate embedded feature vectors for each image, then merge all feature vectors into global feature vectors and classify them using some fully connected (FC) layers. This is known as a 2-D plus 1-D network [9,21]. Besides, voting methods are usually used [6,13,25]. The third method is the pure 3-D CNN network, which does not require slice annotation. It uses one or all available slice images as input, and the 3-D network processes all of these input images at once in the 3D channel space [12,34].

2.2   The Vision Transformer

The weight of the convolution kernel is fixed after training, so it is difficult to adapt to the change of input. Therefore, some methods based on self-attention have been proposed to alleviate this problem. For example, the classic non-local [33], which attempts to model the long-distance dependence in the time domain and space domain, improves the accuracy of video classification. CCNet [14] proposed cross-attention to reduce the amount of computation in non-local. Stand-alone self-attention [26] tried to replace the convolution layer with a local self-attention unit. AANet [3] combines self-attention and convolution. DETR [5] used a Transformer decoder to model target detection as an end-to-end dictionary query problem, and successfully removed post-processing such as NMS. Based on DETR, deformable DETR [36] further introduces a deformable attention layer to focus on the sparse set of context elements, which makes convergence faster and performance better.

Different from the mature CNN, the backbone of the Vision Transformer(ViT) is still in its early stage of development. Recently, ViT [8] established an image classification model using the pure Transformer, which takes a group of patches as image input. DeiT [30] further extended ViT through a new distillation method. T2T ViT [35] gradually connects the marks in the overlapping sliding window into one mark. TNT [11] uses internal and external Transformer blocks to generate pixel embedding and patch embedding respectively. Because the characteristic graph of ViT output is single-scale, even the normal input size will consume a lot of computing overhead and memory overhead for ViT.

2.3   Pyramid Vision Transformer (PVT)

To reduce the cost of computing overhead and memory, Wang et al. proposed Pyramid Vision Transformer (PVTv1) [31]. PVTv1 introduced the pyramid structure into the Transformer and designed a pure Transformer backbone for intensive prediction tasks. So multi-scale feature maps can be generated for intensive prediction tasks. Similar to CNN's backbone, PVTv1 also has four stages, and each stage generates feature maps of different scales. Figure 1 shows the overall architecture diagram of the PVTv1. As with ViT, PVTv1 encodes the image with a 4*4 patch, and processes the image into a set of nonoverlapping patch sequences, ignoring some local continuity of the image. In addition, both ViT and PVTv1 adopt fixed size position coding, which is not friendly to any size image.

Wang et al. proposed PVTv2 [32] to improve the PVTv1. It can get more local continuous images and feature maps that have the same linear complexity as CNN and allows more flexible input.

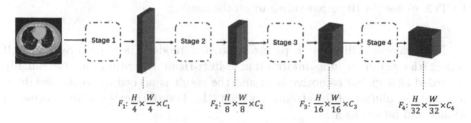

$$F_1: \frac{H}{4} \times \frac{W}{4} \times C_1 \qquad F_2: \frac{H}{8} \times \frac{W}{8} \times C_2 \qquad F_3: \frac{H}{16} \times \frac{W}{16} \times C_3 \qquad F_4: \frac{H}{32} \times \frac{W}{32} \times C_4$$

Fig. 1. Overall architecture of Pyramid Vision Transformer (PVT) [31].

3   Methodology

To adapt the different dimensional inputs of the image, we can use a PVTv2 model for feature extraction by treating an image as a sequence of patches. The pyramid Transformer framework is to be used to generate multi-scale feature maps for the classification tasks. Our method uses four stags to generate different scales of feature maps, which share a similar architecture, the stage consists of a patch embedding layer and a Transformer layer.

PVTv2 can obtain more local continuity of images and feature maps, it can handle variable resolution inputs more flexibly while having the same linear complexity as CNN. Simply stack multiple independent Transformer encoders and gradually reduce the input resolution through Patch Embedding in each stage. Our model (Fig. 2) samples the pictures of the dataset according to the normal distribution, 10 groups of one person's pictures are sampled, each has 8 pictures.

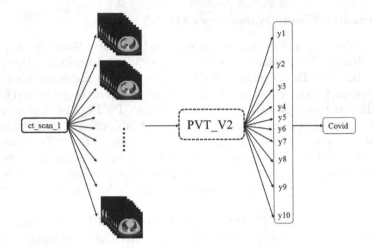

Fig. 2. The architecture of our model: **a)** sample 10 groups of pictures from the dataset according to the normal distribution as input image. **b)** use each group as the input of PVTv2. **c)** use the 10 output voting to get the result.

Each sample will input 8 pictures into the model, these pictures can well reflect the patient's lung information. Multi-Head Attention can be roughly regarded as a global receptive field and the result is according to the weighted average convolution of the attention weight, the Transformer's feature expression ability will be stronger.

4   Dataset

The COVID19-CT-Database (COV19-CT-DB) [2] consists of chest CT scans that are annotated for the existence of COVID-19. Data were aggregated from many hospitals, containing anonymized human lung CT scans with signs of COVID-19 and without signs of COVID-19. The COV19-CT-DB database consists of about 7750 chest CT scan series. It consists of 1,650 COVID and 6,100 non-COVID chest CT scan series. Annotation of each CT slice has been performed by 4 very experienced (each with over 20 years of experience) medical experts; two radiologists and two pulmonologists. Labels provided by the 4 experts showed a high degree of agreement (around 98%).

One difference of COV19-CT-DB from other existing datasets is its annotation by medical experts (labels have not been created as a result of just positive RT-PCR testing).

Each of the scans includes a different number of slices, ranging from 50 to 700. The database has been split into training, validation, and testing sets. The training set contains, in total, 1993 cases. These include 882 COVID-19 cases and 1110 non-COVID-19 cases. The validation set consists of 484 cases, 215 of them represent COVID-19 cases and 269 of them represent non-COVID-19 cases. Both include different numbers of CT slices per CT scan, ranging from 50 to 700 [17].

5   Experiments

5.1   Data Pre-processing

At first, CT images were extracted from DICOM files. Due to some slices of CT scan might be useless for recognizing COVID-19 (e.g., top/bottom slices might not contain chest information), the slices selection of the CT scan in the training phase is essential, as well as in the evaluation phase. During the training phase, CT scan sections were sampled according to the normal distribution, a small number of CT scan sections were collected from the top and bottom, and more CT scan sections were collected from the central part with a higher possibility of the lesion. And then these selected sections were enhanced and normalized. In image analysis, the quality of the image directly affects the design of the recognition algorithm and the accuracy of the effect, so before image analysis (feature extraction, segmentation, matching, recognition, etc.), it needs to be preprocessed. The main purpose of image preprocessing is to eliminate irrelevant information in images, recover useful real information, enhance the detectability of relevant information, simplify data to the maximum extent, and improve the reliability of classification. First, image enhancement is performed on the data [28]. The main purpose of image enhancement is to improve the image quality and recognizability, so that the image is more conducive to observation or further analysis and processing. Image enhancement technology usually highlights or enhances some features of the image, such as edge information, contour information, and contrast, so as to better display the useful information of the image and improve the use value of the image. Image enhancement technology is under a certain standard, the processed image is better than the original image. Then, normalized to the range of [0, 1].

5.2   Implementation Details

A lung CT scan may contain dozens to hundreds of pictures, most of the useful information is concentrated in the middle of these pictures. Thus, we use normal sampling each time, and randomly select a batch of 8 pictures in the CT scan of the same case as the input of the neural network, those pictures in the middle of a case are more likely to be input to the model for training, which makes the model to converge faster.

We set the target of each positive CT scan to 1, and set the target of each negative CT scan to -1, using MSEloss as the loss function, and AdamW as the optimizer with an initial learning rate of 1×10^{-4}. In the evaluation stage, we introduced multiple voting mechanisms, inputting a batch into the model will get several outputs, and then the outputs will be averaged, and the plus or minus sign of this average will be used as one evaluation standard. The above steps will be performed n times. If more than half of the averages are plus, the case is considered positive, otherwise, it is negative.

From the experimental results, in the multiple voting mechanisms, most of the single votes of positive cases are all positive, and most of the single votes of negative cases are all negative. But sometimes a single vote for a positive case will result in a negative number, and a single vote for a negative case will result in a positive number. How to distinguish these ambiguous votes is a key point of our classification. In these cases, the proportion of single votes that are not true to the total votes affects the classification accuracy of our model. Experiments show that in the case of non-uniform single voting results, many results can eventually lead to correct classification results in the multiple voting mechanisms. More importantly, it makes our classification results more stable.

Table 1. Comparison of the proposed method with the baseline on the validation set.

Method	Macro F1 Score
Baseline	77.00%
PVT-COV19D (ours)	**88.17%**

Table 2. The results on the test set of Baseline and PVT-COV19D.(s: single prediction, m: multiple voting mechanism.)

Method	Macro F1 Score	F1 (NON-COVID)	F1 (COVID)
Baseline	69.00%	83.62%	54.38%
PVT-COV19D (s)	81.69%	95.74%	67.64%
PVT-COV19D (s)	81.04%	95.47%	66.61%
PVT-COV19D (s)	82.38%	95.85%	68.92%
PVT-COV19D (m)	**83.78%**	**96.22%**	**71.33%**
PVT-COV19D (m)	83.53%	96.19%	70.87%

Our models are trained with a batch size of 8 on one 2080Ti GPU, the training and validation set are partitioned by [17], where the number of training and validation CT scans are 1, 992 and 484. After training for 60 epochs on the validation set, the positive accuracy of our model reached 84.11%, the negative accuracy reached 91.54%, and the validation set prediction accuracy reached 88.19%, the macro F1 score reached 0.8817 for this binary classification. The performance comparison between the proposed and baseline [16] is showed in Table 1. Compared with baseline, the results of the model on the test set are significantly improved, the F1 score reaches 83.78%, and we show the five results in Table 2.

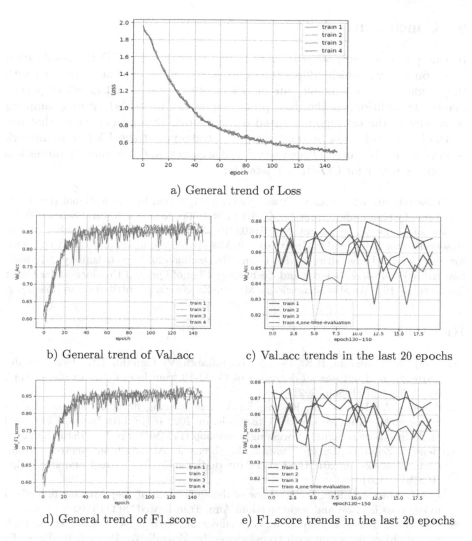

a) General trend of Loss

b) General trend of Val_acc

c) Val_acc trends in the last 20 epochs

d) General trend of F1_score

e) F1_score trends in the last 20 epochs

Fig. 3. We selected the results of four training sessions, each with 150 epochs. Train 1/2/3 are experiments using multiple voting mechanism. Train 4 is a single prediction experiment. Acc trends and F1 score trends in the last 20 epochs show the advantages of our multiple voting mechanism over single prediction.

The results of our experiments in Fig. 3 show that compared with the single prediction method, our multiple voting mechanisms are more stable in the process of predicting whether a case is negative or positive, which helps to effectively avoid misjudgment. In addition, as shown in Table 2 and Fig. 3, the prediction accuracy of the multiple voting mechanisms have an average 2% to 3% improvement in results compared to the single prediction.

6   Conclusions

In this paper, we propose a PVT-COV19D model for COVID-19 classification based on PVTv2, which adopts the multiple voting mechanism. Compared with the single prediction model, our model has more stable and excellent performance. In addition, in the data processing stage, we added normal sampling to accelerate the convergence speed of the model. Experiments show that our method is feasible. The normal sampling strategy and the PVTv2 framework achieve a good balance between speed and accuracy, and our model requires less computing power for COVID-19 detection.

Acknowledgements. This work was partially supported by the National Key R&D Program of China under Grant No. 2020YFC0832500, National Natural Science Foundation of China under Grant No. 61402210, Ministry of Education-China Mobile Research Foundation under Grant No. MCM20170206, Fundamental Research Funds for the Central Universities under Grant No. lzujbky-2021-sp43, lzujbky-2019-kb51 and lzujbky-2018-k12, Science and Technology Plan of Qinghai Province under Grant No.2020-GX-164. We also acknowledge Mr. Rui Zhao for his contribution to this paper.

References

1. Ahuja, S., Panigrahi, B.K., Dey, N., Rajinikanth, V., Gandhi, T.K.: Deep transfer learning-based automated detection of Covid-19 from lung CT scan slices. Appl. Intell. **51**(1), 571–585 (2021)
2. Arsenos, A., Kollias, D., Kollias, S.: A large imaging database and novel deep neural architecture for Covid-19 diagnosis. In: 2022 IEEE 14th Image, Video, and Multidimensional Signal Processing Workshop (IVMSP), pp. 1–5. IEEE (2022)
3. Bello, I., Zoph, B., Le, Q., Vaswani, A., Shlens, J.: Attention augmented convolutional networks. In: 2019 IEEE/CVF International Conference on Computer Vision (ICCV) (2020)
4. Cai, L., Gao, J., Zhao, D.: A review of the application of deep learning in medical image classification and segmentation. Ann. Transl. Med. **8**(11) (2020)
5. Carion, N., Massa, F., Synnaeve, G., Usunier, N., Kirillov, A., Zagoruyko, S.: End-to-end object detection with transformers. In: Vedaldi, A., Bischof, H., Brox, T., Frahm, J.-M. (eds.) ECCV 2020. LNCS, vol. 12346, pp. 213–229. Springer, Cham (2020). https://doi.org/10.1007/978-3-030-58452-8_13
6. Chaudhary, S., Sadbhawna, S., Jakhetiya, V., Subudhi, B.N., Baid, U., Guntuku, S.C.: Detecting Covid-19 and community acquired pneumonia using chest CT scan images with deep learning. In: 2021 IEEE International Conference on Acoustics, Speech and Signal Processing (ICASSP), ICASSP 2021 (2021)
7. Chung, M., et al.: CT imaging features of 2019 novel coronavirus (2019-nCoV). Radiology (2020)
8. Dosovitskiy, A., et al.: An image is worth 16 × 16 words: transformers for image recognition at scale. arXiv preprint arXiv:2010.11929 (2020)
9. Garg, P., Ranjan, R., Upadhyay, K., Agrawal, M., Deepak, D.: Multi-scale residual network for Covid-19 diagnosis using CT-scans. In: 2021 IEEE International Conference on Acoustics, Speech and Signal Processing (ICASSP), ICASSP 2021 (2021)

10. Giri, B., Pandey, S., Shrestha, R., Pokharel, K., Ligler, F.S., Neupane, B.B.: Review of analytical performance of Covid-19 detection methods. Anal. Bioanal. Chem. **413**(1), 35–48 (2021)

11. Han, K., Xiao, A., Wu, E., Guo, J., Xu, C., Wang, Y.: Transformer in transformer. In: Advances in Neural Information Processing Systems, vol. 34, pp. 15908–15919 (2021)

12. He, X., et al.: Automated model design and benchmarking of 3D deep learning models for Covid-19 detection with chest CT scans. arXiv preprint arXiv:2101.05442 (2021)

13. Heidarian, S., et al.: Covid-fact: a fully-automated capsule network-based framework for identification of Covid-19 cases from chest CT scans. Front. Artif. Intell. **4**, 598932 (2021)

14. Huang, Z., Wang, X., Huang, L., Huang, C., Wei, Y., Liu, W.: CCNet: criss-cross attention for semantic segmentation. In: Proceedings of the IEEE/CVF International Conference on Computer Vision, pp. 603–612 (2019)

15. Khadidos, A., Khadidos, A.O., Kannan, S., Natarajan, Y., Mohanty, S.N., Tsaramirsis, G.: Analysis of Covid-19 infections on a CT image using deepsense model. Front. Public Health **8**, 599550 (2020)

16. Kollias, D., Arsenos, A., Kollias, S.: AI-MIA: Covid-19 detection & severity analysis through medical imaging. arXiv preprint arXiv:2206.04732 (2022)

17. Kollias, D., Arsenos, A., Soukissian, L., Kollias, S.: MIA-COV19D: Covid-19 detection through 3-D chest CT image analysis. In: Proceedings of the IEEE/CVF International Conference on Computer Vision, pp. 537–544 (2021)

18. Kollias, D., et al.: Deep transparent prediction through latent representation analysis. arXiv preprint arXiv:2009.07044 (2020)

19. Kollias, D., Tagaris, A., Stafylopatis, A., Kollias, S., Tagaris, G.: Deep neural architectures for prediction in healthcare. Complex Intell. Syst. **4**(2), 119–131 (2018)

20. Kollias, D., et al.: Transparent adaptation in deep medical image diagnosis. In: Heintz, F., Milano, M., O'Sullivan, B. (eds.) TAILOR 2020. LNCS (LNAI), vol. 12641, pp. 251–267. Springer, Cham (2021). https://doi.org/10.1007/978-3-030-73959-1_22

21. Li, L., et al.: Artificial intelligence distinguishes Covid-19 from community acquired pneumonia on chest CT. Radiology (2020)

22. Li, Y., Xia, L.: Coronavirus disease 2019 (Covid-19): role of chest CT in diagnosis and management. Am. J. Roentgenol. **214**(6), 1280–1286 (2020)

23. Nair, R., Alhudhaif, A., Koundal, D., Doewes, R.I., Sharma, P.: Deep learning-based Covid-19 detection system using pulmonary CT scans. Turk. J. Electr. Eng. Comput. Sci. **29**(8), 2716–2727 (2021)

24. Panahi, A.H., Rafiei, A., Rezaee, A.: FCOD: fast Covid-19 detector based on deep learning techniques. Inform. Med. Unlocked **22**, 100506 (2021)

25. Rahimzadeh, M., Attar, A., Sakhaei, S.M.: A fully automated deep learning-based network for detecting Covid-19 from a new and large lung CT scan dataset. Biomed. Signal Process. Control **68**, 102588 (2021)

26. Ramachandran, P., Parmar, N., Vaswani, A., Bello, I., Levskaya, A., Shlens, J.: Stand-alone self-attention in vision models. In: Advances in Neural Information Processing Systems, vol. 32 (2019)

27. Rehman, A., Saba, T., Tariq, U., Ayesha, N.: Deep learning-based Covid-19 detection using CT and X-ray images: current analytics and comparisons. IT Prof. **23**(3), 63–68 (2021)

28. Son, T., Kang, J., Kim, N., Cho, S., Kwak, S.: URIE: universal image enhancement for visual recognition in the wild. In: Vedaldi, A., Bischof, H., Brox, T., Frahm, J.-M. (eds.) ECCV 2020. LNCS, vol. 12354, pp. 749–765. Springer, Cham (2020). https://doi.org/10.1007/978-3-030-58545-7_43

29. Song, Y., et al.: Deep learning enables accurate diagnosis of novel coronavirus (Covid-19) with CT images. IEEE/ACM Trans. Comput. Biol. Bioinf. **18**(6), 2775–2780 (2021)

30. Touvron, H., Cord, M., Douze, M., Massa, F., Sablayrolles, A., Jégou, H.: Training data-efficient image transformers & distillation through attention. In: International Conference on Machine Learning, pp. 10347–10357. PMLR (2021)

31. Wang, W., et al.: Pyramid vision transformer: a versatile backbone for dense prediction without convolutions. In: Proceedings of the IEEE/CVF International Conference on Computer Vision, pp. 568–578 (2021)

32. Wang, W., et al.: PVT v2: improved baselines with pyramid vision transformer. Comput. Vis. Media **8**(3), 415–424 (2022)

33. Wang, X., Girshick, R., Gupta, A., He, K.: Non-local neural networks. In: Proceedings of the IEEE Conference on Computer Vision and Pattern Recognition, pp. 7794–7803 (2018)

34. Wang, X., et al.: A weakly-supervised framework for Covid-19 classification and lesion localization from chest CT. IEEE Trans. Med. Imaging **39**(8), 2615–2625 (2020)

35. Yuan, L., et al.: Tokens-to-token ViT: training vision transformers from scratch on imagenet. In: Proceedings of the IEEE/CVF International Conference on Computer Vision, pp. 558–567 (2021)

36. Zhu, X., Su, W., Lu, L., Li, B., Wang, X., Dai, J.: Deformable DETR: deformable transformers for end-to-end object detection. arXiv preprint arXiv:2010.04159 (2020)

Boosting COVID-19 Severity Detection with Infection-Aware Contrastive Mixup Classification

Junlin Hou[1], Jilan Xu[1], Nan Zhang[2], Yuejie Zhang[1(✉)], Xiaobo Zhang[3(✉)], and Rui Feng[1,2,3(✉)]

[1] School of Computer Science, Shanghai Key Laboratory of Intelligent Information Processing, Fudan University, Shanghai, China
{jlhou18,jilanxu18,20210860062,yjzhang,fengrui}@fudan.edu.cn
[2] Academy for Engineering and Technology, Fudan University, Shanghai, China
[3] National Children's Medical Center, Children's Hospital of Fudan University, Shanghai, China
zhangxiaobo0307@163.com

Abstract. This paper presents our solution for the 2nd COVID-19 Severity Detection Competition. This task aims to distinguish the Mild, Moderate, Severe, and Critical grades in COVID-19 chest CT images. In our approach, we devise a novel infection-aware 3D Contrastive Mixup Classification network for severity grading. Specifically, we train two segmentation networks to first extract the lung region and then the inner lesion region. The lesion segmentation mask serves as complementary information for the original CT slices. To relieve the issue of imbalanced data distribution, we further improve the advanced Contrastive Mixup Classification network by weighted cross-entropy loss. On the COVID-19 severity detection leaderboard, our approach won the first place with a Macro F1 Score of 51.76%. It significantly outperforms the baseline method by over 11.46%.

Keywords: COVID-19 severity detection · Chest CT images · Infection-aware contrastive mixup classification

1 Introduction

The Coronavirus Disease 2019 SARS-CoV-2 (COVID-19) is an ongoing pandemic, which spreads fast worldwide since the end of 2019 [35]. To date (July, 2022), the number of infected people is still increasing rapidly. In clinical practice, thoracic computed tomography (CT) has been recognized to be a reliable complement to RT-PCR assay [7]. From CT scans, a detailed overview of COVID-19 radiological patterns can be clearly presented, including ground glass opacities, rounded opacities, enlarged intra-infiltrate vessels, and later more consolidations [5]. In fighting against COVID-19, two important clinical tasks can be conducted

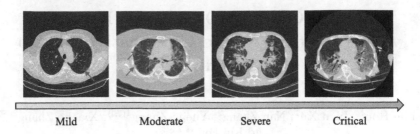

Fig. 1. Examples of COVID-19 CT images with different severity levels from the COV19-CT-DB dataset. The red arrows indicate the locations of the main lesion regions. In the mild case, the lesions only account for small regions in the lung. In the critical case, the lesions spread over the periphery of the lung and are indistinguishable from the outer tissue. (Color figure online)

in chest CT scans, namely identification and severity assessment. Early identification is vital to the slowdown of viral transmission, whereas severity assessment is significant for clinical treatment planning.

Recently, deep learning approaches have achieved excellent performance in discriminating COVID-19 from non-pneumonia or other types of pneumonia [3,12,16,34]. For instance, Chen et al. [3] detected the suspicious lesions for COVID-19 diagnosis using a UNet++ model. Javaheri et al. [16] designed the CovidCTNet to differentiate COVID-19 from other lung diseases. Xu et al. [36] developed a deep learning system with a location-attention mechanism to categorize COVID-19, Influenza-A viral pneumonia, and healthy subjects. Despite the great success in COVID-19 identification, the automatic assessment of COVID-19 severity in CT images is still a very challenging task.

In this paper, we present a deep-learning based solution for the 2nd COV19D Competition of the Workshop "AI-enabled Medical Image Analysis - Digital Pathology & Radiology/COVID19 (AIMIA)", which occurs in conjunction with the European Conference on Computer Vision (ECCV) 2022. This paper mainly focuses on the COVID-19 Severity Detection Challenge. As illustrated in Fig. 1, the goal of this task is to categorize the severity of COVID-19 into four stages, including Mild, Moderate, Severe, and Critical. Each severity category is determined by the presence of ground glass opacities and the pulmonary parenchymal involvement. This task is of great significance in real-world scenarios. Especially for patients with severe or critical grades, extensive care and attention are necessary in the clinical practice. The main challenge of the severity detection task is three-fold: (1) Compared with the binary classification task (COVID vs. non-COVID), the severity detection requires more fine-grained lesion features to determine the final grade. (2) The accurate annotation of grades requires domain knowledge and it is labor-expensive. The limited amount of data makes the task difficult for deep-learning based methods. (3) In the original CT volume, it is hard to identify the lesions of interest from substantial non-lesion voxels.

In order to address the aforementioned challenges, we propose a novel infection-aware Contrastive Mixup Classification network. This framework is constructed in a segmentation-classification manner. In particular, we initially extract the lung regions and segment the inner lesion regions using two individual segmentation networks. The obtained lesion segmentation mask serves as complementary information for fine-grained classification. The location, as well as the number of lesions, are both crucial in determining the severity grade. After lesion segmentation, an advanced 3D Contrastive Mixup Classification network (CMC-COV19D) in the previous work [11] is applied. It won the first price in the ICCV 2021 COVID-19 Diagnosis Competition of AI-enabled Medical Image Analysis Workshop [21]. The CMC-COV19D framework introduces contrastive representation learning to discover more discriminative representations of COVID-19 cases. It includes a joint training loss that combines the classification loss, mixup loss, and contrastive loss. We further improve the classification loss with weight balancing strategy to enhance the attention on minor classes (i.e. moderate and critical). Experimental results on the COVID-19 severity detection challenge shows that our approach achieves a Marco F1 Score of 51.76%, surpassing all the participants.

The remainder of this paper is organized as follows. In Sect. 2, we summarize related works on segmentation of lung regions and lesions, and COVID-19 severity assessment. Section 3 introduces our proposed infection-aware Contrastive Mixup Classification framework in detail. Section 4 describes the two datasets. Extensive experiments are conducted in Sect. 5 to evaluate the performance of our network. The conclusion is given in Sect. 6.

2   Related Work

2.1   Segmentation of Lung Regions and Lesions

Segmentation is an important step in AI-based COVID-19 image processing and analysis [32]. It extracts the regions of interest (ROI), including lung, lobes, and infected regions in CT scans for further diagnosis. A large number of researches [1,14,39] utilized U-Net [30] to segment lung fields and lesions. For example, Qi et al. [29] used U-Net to delineate the lesions in the lung and extract radiometric features of COVID-19 patients with the initial seeds given by a radiologist for predicting hospital stay. Moreover, Jin et al. [17] proposed an COVID-19 screening system, in which the lung regions are first detected by UNet++ [41]. The Inf-Net [6] aimed to segment the COVID-19 infection regions for quantifying and evaluating the disease progression. Shan et al. [31] proposed a VB-Net for segmentation of lung, lung lobes and lung infection, which provided accurate quantification data for medical studies. Motivated by the above approaches, we train segmentation networks to extract infection-aware clues for severity grading.

2.2   COVID-19 Severity Assessment

The study of COVID-19 severity assessment plays an essential role in clinical treatment planning. Tang et al. [33] established a random forest model based on

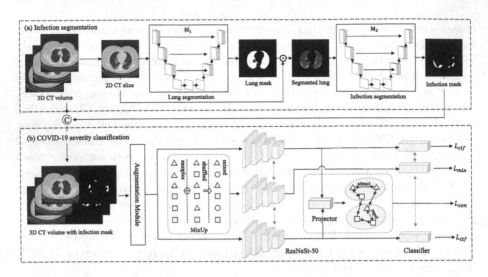

Fig. 2. Overview of our infection-aware CMC network, comprised of the infection segmentation stage and the COVID-19 severity classification stage.

quantitative features for COVID-19 severity assessment (non-severe or severe). Chaganti et al. [2] presented a method that automatically segmented and quantified abnormal CT patterns commonly presented in COVID-19. Huang et al. [14] assessed a quantitative CT image parameter defined as the percentage of lung opacification (QCT-PLO) and classified the severity into four classes (mild, moderate, severe and critical). Pu et al. [28] developed an automated system to detect COVID-19, quantify the extent of disease, and assess the progression of the disease. Feng et al. [8] designed a novel Lesion Encoder framework to detect lesions in chest CT scans and encode lesion features for automatic severity assessment. He et al. [10] proposed a synergistic learning framework for automated severity assessment of COVID-19 in 3D CT images. In our work, we first identify the infected regions, and then utilize segmented lesion masks as complementary information for the original CT slices.

3   Methodology

Figure 2 illustrates the overview of our infection-aware Contrastive Mixup Classification (CMC) framework in a combined segmentation-classification manner. The framework is divided into two stages. The first stage is the infection segmentation that generates lesion masks of a CT volume slice by slice (Fig. 2(a)). The lesion masks of an input CT volume help the classification model focus on important regions and better classify COVID-19 severity. The second stage is to take a CT volume and its 3D lesion masks as the input and output the COVID-19 severity prediction (Fig. 2(b)). We adopt our previous work, the CMC network, as the classification model, which has shown excellent COVID-19 detection performance.

In the following subsections, we will elaborate the infection segmentation procedure in detail, and then describe the CMC network for COVID-19 severity classification.

3.1 Infection Segmentation

As shown in Fig. 2(a), given an input CT volume $x = \{\mathbf{x}_t\}_{t=1}^T$, we perform slice-level infection segmentation by the following two steps.

Step 1: lung region extraction. For each slice $\mathbf{x}_t \in \mathbb{R}^{H \times W}$ in the CT volume, we train a lung segmentation model M_1 to predict the lung region mask $M_1(\mathbf{x}_t; \theta) \in \mathbb{R}^{H \times W}$, where θ are the network parameters. Extracting the lung regions of the CT volume removes the redundant background region as the background varies across different patients and screening devices. Moreover, lung extraction eases the subsequent lesion segmentation model as the lesions usually appear at the periphery of the lung. We empirically show that with simple training recipes, the lung segmentation model yields high segmentation accuracy. After training, the segmented CT slice $\hat{\mathbf{x}}_t \in \mathbb{R}^{H \times W}$ is obtained by multiplying the lung segmentation mask with the original CT slice:

$$\hat{\mathbf{x}}_t = M_1(\mathbf{x}_t; \theta) \odot \mathbf{x}_t \tag{1}$$

where $\odot$ denotes the Hadamard product.

Step 2: lesion segmentation. Given each CT slice $\hat{\mathbf{x}}_t$, we obtain the infection mask $I_t = M_2(\hat{\mathbf{x}}_t; \phi)$ by training another segmentation model M_2 parameterized by ϕ. We refer the infection mask to the binary map where 1 indicates the pixel with COVID-19 infection and 0 represents the non-infected region. Notably, M_2 is trained on the CT slices with the segmented lung region, and thus the noisy background and artifacts are avoided. This reduces the model's potential false positive predictions over the background. The input to the classification model is defined as the concatenation of the infection mask and the original CT image:

$$\bar{x}_t = [I_t, \mathbf{x}_t] = [M_2(M_1(\mathbf{x}_t; \theta) \odot \mathbf{x}_t; \phi), \mathbf{x}_t]. \tag{2}$$

Segmentation Models. We employ several state-of-the-art models for lung and infection segmentation, including DeepLabV3 [4], FPN [25], U-Net [30], and U-Net++ [41]. Different backbones such as ResNet [9] and DenseNet [13] with ImageNet pre-trained weights are adopted as the encoder layers of segmentation models. All segmentation models are trained and validated on the COVID19-CT-Seg dataset [26]. The models that perform best on the validation set are then transferred to the COV19-CT-DB dataset to generate lung and infection masks for each CT volume.

Loss Function. We use the Dice loss [27] as the objective function to optimize lung and infection segmentation networks. For each category c, the Dice loss is defined as follows:

$$\mathcal{L}_{Dice}^c = 1 - 2 \times \frac{\sum_{n=1}^{N} y_n^c \hat{y}_n^c}{\sum_{n=1}^{N}(y_n^c + \hat{y}_n^c) + \epsilon}, \tag{3}$$

where y_n and $\hat{y}_n$ denote the label and the prediction of each pixel n, respectively. ϵ is a small constant to avoid zero division.

3.2 Infection-Aware COVID-19 Severity Classification

As illustrated in Fig. 2(b), the CT volume $X \in \mathbb{R}^{T \times H \times W}$ and its infection mask $I \in \mathbb{R}^{T \times H \times W}$ are concatenated per channel to produce $\bar{X} \in \mathbb{R}^{2 \times T \times H \times W}$ as the input. We adopt the Contrastive Mixup Classification (CMC) network for COVID-19 severity detection.

The CMC network is mainly comprised of the following four components.

- **A data augmentation module** $A(\cdot)$. This module transforms an input CT volume x into an augmented volume $\tilde{x}$ randomly. In our work, two augmented volumes are generated from each input CT scan.
- **A feature extractor** $E(\cdot)$. It maps the augmented CT sample $\tilde{x}$ to a representation vector $r = E(\tilde{x}) \in \mathbb{R}^{d_e}$ in d_e-dimensional latent space. We adopt an inflated 3D ResNeSt-50 to extract features of each CT scan.
- **A small projection network** $P(\cdot)$. This network is used to map the high-dimension vector r to a relative low-dimension vector $z = P(r) \in \mathbb{R}^{d_p}$. A MLP (multi-layer perception) can be employed as the projection network.
- **A classifier network** $C(\cdot)$. It classifies the vector $r \in \mathbb{R}^{d_e}$ to the COVID-19 severity prediction using fully connected layers.

We train the CMC network using a well-designed adaptive loss, composed of contrastive loss, mixup loss, and weighted classification loss.

Contrastive Loss. The CMC network employs the contrastive learning as an auxiliary task to learn more discriminative representations of each COVID-19 severity categories. Formally, given a minibatch of $2N$ augmented CT volumes and the labels $\{(\tilde{x}_i, \tilde{y}_i)\}_{i=1,...,2N}$, we define the positives as any augmented CT samples from the same category, while those from different classes are considered as negative pairs [19]. Let $i \in \{1, \ldots, 2N\}$ be the index of an arbitrary augmented sample, the contrastive loss function is defined as:

$$\mathcal{L}_{con}^i = \frac{-1}{2N_{\tilde{y}_i} - 1} \sum_{j=1}^{2N} \mathbb{1}_{i \neq j} \cdot \mathbb{1}_{\tilde{y}_i = \tilde{y}_j} \cdot \log \frac{\exp(z_i^T \cdot z_j / \tau)}{\sum_{k=1}^{2N} \mathbb{1}_{i \neq k} \cdot \exp(z_i^T \cdot z_k / \tau)}, \tag{4}$$

where $N_{\tilde{y}_i}$ is the number of samples in a minibatch that share the same label $\tilde{y}_i$, $\mathbb{1} \in \{0, 1\}$ is an indicator function, and τ denotes a temperature parameter.

Mixup Loss. To boost the generalization ability of the model, the mixup strategy [38] is also adopted in the training procedure. For each augmented CT volume $\tilde{x}_i$, we generate a mixed sample $\tilde{x}_i^{mix}$ and its mixed label $\tilde{y}_i^{mix}$ as:

$$\tilde{x}_i^{mix} = \lambda \tilde{x}_i + (1 - \lambda)\tilde{x}_p, \ \tilde{y}_i^{mix} = \lambda \tilde{y}_i + (1 - \lambda)\tilde{y}_p, \tag{5}$$

where p is a random index. The mixup loss is defined as the cross-entropy loss of mixed samples:

$$\mathcal{L}_{mix}^i = \text{CrossEntropy}(\tilde{x}_i^{mix}, \tilde{y}_i^{mix}). \tag{6}$$

Weighted Classification Loss. In addition to the mixup loss on the mixed samples, we also employ the cross-entropy loss on raw samples for severity detection. It is observed from Table 1 that there are imbalance data distributions on the COV19-CT-DB dataset. To alleviate this problem, we adopt the weighted cross entropy loss. Each sample is scaled by the weight proportional to the inverse of the percentage of the sample for each class c in the training set, denoted as $\alpha = [\alpha^1, \ldots, \alpha^c]$. Therefore, the weighted classification loss is defined as:

$$\mathcal{L}_{clf}^i = -\alpha \tilde{y}_i^T \log \hat{y}_i, \tag{7}$$

where $\tilde{y}_i$ denotes the one-hot vector of ground truth label, and $\hat{y}_i$ is predicted probability of the sample x_i $(i = 1, \ldots, 2N)$.

Adaptive Joint Loss. Finally, we merge the contrastive loss, mixup loss, and weighted classification loss into a combined objective function with learnable weights [18]:

$$\mathcal{L} = \frac{1}{2N\sigma_1^2} \sum_{i=1}^{2N} \mathcal{L}_{con}^i + \frac{1}{2N\sigma_2^2} \sum_{i=1}^{2N} (\mathcal{L}_{mix}^i + \mathcal{L}_{clf}^i) + \log \sigma_1 + \log \sigma_2, \tag{8}$$

where σ_1, σ_2 are utilized to learn the adaptive weights of the three losses.

4   Dataset

4.1   COV19-CT-DB Database

The COV19-CT-Database (COV19-CT-DB) [20] provides chest CT scans that are marked for the existence of COVID-19. It contains about 1,650 COVID and 6,100 non-COVID chest CT scan series from over 1,150 patients and 2,600 subjects. The CT scans were collected in the period from September 1, 2020 to November 30, 2021. Each CT scan was annotated by 4 experienced medical experts, which showed a high degree of agreement (around 98%). The number of slices of each 3D CT scan ranging from 50 to 700. The variation in the number of slices is due to the context of CT scanning.

In particular, the COV19-CT-DB further provides the severity annotations on a small portion of the COVID-19 cases. The descriptions and number of

Table 1. The number of samples and descriptions of four severity categories on the COV19-CT-DB database.

Category	Severity	Training	Validation	Description
1	Mild	85	22	Few or no ground glass opacities. Pulmonary parenchymal involvement $\leq 25\%$ or absence
2	Moderate	62	10	Ground glass opacities. Pulmonary parenchymal involvement 25–50%
3	Severe	85	22	Ground glass opacities. Pulmonary parenchymal involvement 50–75%
4	Critical	26	5	Ground glass opacities. Pulmonary parenchymal involvement $\geq 75\%$

samples of the four severity categories are shown in detail in Table 1. The severity categories include Mild, Moderate, Severe, and Critical, in the range from 1 to 4. The training set contains, in total, 258 3D CT scans (85 Mild cases, 62 Moderate cases, 85 Severe cases, and 26 Critical cases). The validation set consists of 61 3D CT scans (22 Mild cases, 10 Moderate cases, 22 Severe cases, and 5 Critical cases). The testing set includes 265 scans and the labels are not available during the challenge.

4.2 COVID-19-CT-Seg Dataset

The COVID-19-CT-Seg dataset [26] provides 20 public COVID-19 CT scans from the Coronacases Initiative and Radiopaedia. All the cases contain COVID-19 infections with the proportion ranging from 0.01% to 59%. The left lung, right lung, and infections (i.e., all visibly affected regions of the lungs) were annotated, refined and verified by junior annotators, radiologists, and a senior radiologist, progressively. The whole lung mask includes both normal and pathological regions. In total, there are more than 300 infections with over 1,800 slices.

5 Experimental Results

5.1 Implementation Details

Lung and Infection Segmentation. We employ multiple segmentation models [4,25,30,41] with the ImageNet pre-trained encoder [9,13] to perform lung and infection segmentation on 2D CT slices. The input resolution is resized to 256×256. Random crop, random horizontal flip, vertical flip, and random rotation are applied as forms of data augmentation to reduce overfitting. The initial learning rate is set to 1e-3 and then divided by 10 at 30% and 80% of the total number of training epochs. The segmentation models are trained for 100 epochs with the Stochastic Gradient Descent (SGD) optimizer. The batch size is 8. We select two models that perform best on the validation set as the final lung and infection segmentation models.

Table 2. The segmentation results on the COVID-19-CT-Seg dataset (unit: %).

Model	Encoder	Lung segmentation		Infection segmentation	
		mIoU	lung IoU	mIoU	Infection IoU
FPN	ResNet-34	96.81	94.50	77.88	56.46
Deeplabv3	ResNet-34	95.01	91.44	78.67	58.10
UNet	ResNet-34	96.96	94.75	79.67	59.93
UNet	DenseNet-121	97.54	95.75	77.52	55.73
UNet++	ResNet-34	97.37	95.46	**80.36**	**61.36**
UNet++	DenseNet-121	**97.56**	**95.79**	80.04	60.71

COVID-19 Severity Detection. We employ an inflated 3D ResNest50 as the backbone for COVID-19 severity detection. Each CT volume is resampled from $T \times 512 \times 512$ to $64 \times 256 \times 256$, where T denotes the number of slices. Data augmentations include random resized crop and random contrast changes. Other augmentations such as flip and rotation are also tried, but no significant improvement is yielded. We optimize the networks using the Adam algorithm with a weight decay of 1e-5. The initial learning rate is set to 1e-4 and then divided by 10 at 30% and 80% of the total number of training epochs. Our models are trained for 100 epochs with the batch size 4. Our methods are implemented in PyTorch and run on four NVIDIA Tesla V100 GPUs.

5.2 Evaluation Metrics

For the lung and infection segmentation, we report mean Intersection over Union (mIoU), and IoU of the foreground. For the COVID-19 severity detection, we adopt the Macro F1 Score as the evaluation metric for overall comparison. The score is defined as the unweighted average of the class-wise/label-wise F1 Scores. We also show the F1 Score for each severity category. Furthermore, we utilize a model explanation technique, namely Class Activation Mapping (CAM) [40], to conduct visual interpretability analysis [15]. The CAM can identify the important regions that are most relevant to the predictions.

5.3 Results of Lung and Infection Segmentation

The lung and infection segmentation results on the COVID-19-CT-Seg dataset are shown in Table 2. Among all segmentation models, the UNet++ models achieve better performance on both lung and infection segmentation tasks. The UNet++ with ResNet-34 encoder reaches the highest infection segmentation results with 80.36% mean IoU and 61.36% IoU for infections. The Unet++ with DenseNet-121 encoder performs best on the lung segmentation task. It obtains 97.56% mIoU and 95.79% IoU for lung regions.

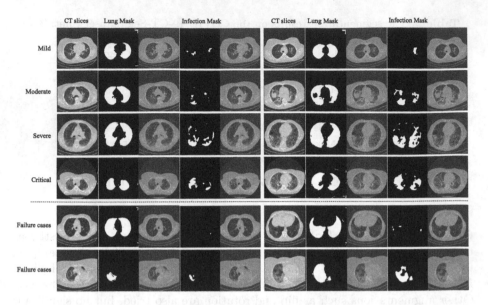

Fig. 3. An illustration of the segmentation results on the COV19-CT-DB dataset.

After the segmentation on the COVID-19-CT-Seg dataset, we utilize the two trained models to inference the lung and infection regions on the CT scans from COV19-CT-DB dataset. As the COV19-CT-DB dataset lacks the segmentation annotations, we illustrate several representative visual results in Fig. 3. For each severity category, the segmentation results of two slices from a 3D CT scan are presented in the first four rows. As can be apparently seen, our lung segmentation model can predict the whole lung regions accurately, which shows its strong generalization ability. In addition, our infection segmentation model can inference most infection areas from mild to critical precisely.

We also analyze the failure cases, which are shown in the last two rows of Fig. 3. According to our observation, the incomplete segmentation results mainly occur in the following two cases. (1) In the early stage, the low-density infections with fuzzy boundaries are difficult to identify. (2) The appearance of consolidation may be similar to bones, which is also hard for the model to recognize. Therefore, it may be inaccurate to make severity predictions by directly calculating the volume ratio between the segmented infections and lung regions. Therefore, we make full use of the segmentation results to provide auxiliary information for the original CT scan inputs.

Table 3. The comparison results on the validation set of COVID-19 severity detection challenge (unit: %).

ID	Methods	Con	Mixup	Macro F1	F1			
					Mild	Moderate	Severe	Critical
1	CNN-RNN [20]	–	–	63.00	–	–	–	–
2	ResNeSt-50 [37]			65.47	82.05	61.54	68.29	50.00
3	ResNeSt-50 [37]		✓	67.30	71.79	75.00	60.87	61.54
4	CMC [11]	✓	✓	71.86	79.07	78.26	75.56	54.55
5	Lung-aware CMC	✓	✓	70.62	78.05	66.67	71.11	66.67
6	Infection-aware CMC	✓	✓	77.03	83.72	66.67	74.42	83.33

Table 4. The leaderboard on the COVID19 Severity Detection Challenge (unit: %).

Rank	Teams	Macro F1	F1			
			Mild	Moderate	Severe	Critical
1	FDVTS (Ours)	51.76	58.97	44.53	58.89	44.64
2	Code 1055	51.48	61.14	34.06	61.91	48.83
3	CNR-IEMN	47.11	55.67	37.88	55.46	39.46
4	MDAP	46.00	58.14	42.29	43.78	40.03
5	etro	44.92	57.69	29.57	62.85	29.57
6	Jovision-Deepcam	41.49	62.20	40.24	38.59	24.93
7	ResNet50-GRU [20]	40.30	61.10	39.47	37.07	23.54

5.4 Results on the COVID-19 Severity Detection Challenge

Results on the Validation Set. Table 3 shows the results of the baseline model and our methods on the validation set of COVID-19 severity detection challenge. The baseline model is based on a CNN-RNN architecture [20] that performs 3D CT scan analysis, following the work [22–24] on developing deep neural architectures for predicting COVID-19. It achieves 63.00% Marco F1 Score. In our experiments, a simple 3D ResNeSt-50 model reaches better performance with 65.47% Macro F1 compared to the baseline. The Mixup technique helps to boost the detection result by ∼2% improvement. When we adopt the CMC architecture with Contrastive learning and Mixup, it further improves the detection performance with 71.86% Macro F1 Score. In addition, the CMC model significantly increases the F1 Scores on Moderate and Severe categories, which are easily confused. The experimental results demonstrate the effectiveness of Contrastive learning and Mixup modules.

Furthermore, we report the performance of our proposed lung-aware CMC and infection-aware CMC in the 5th and 6th rows in Table 3. It can be seen that the combination of original images and lung region masks does not significantly benefit the severity detection results. However, the infection-aware CMC which takes the concatenation of original images and infection region masks as the input significantly boost the performance to 77.03% Macro F1 Score. Additionally, this model shows the best or comparable results on the Mild (83.72%), Moderate (66.67%), Severe (74.42%), and Critical (83.33%) categories.

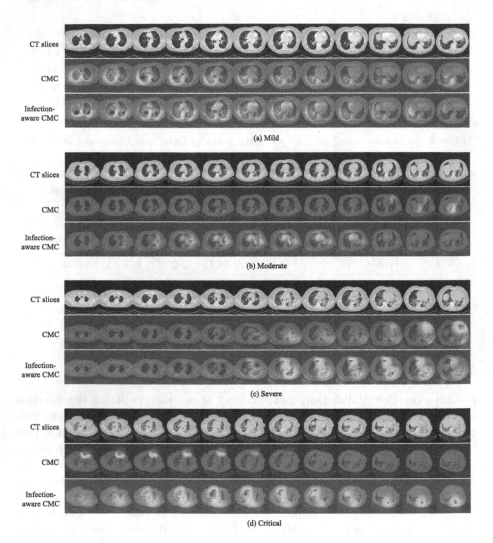

Fig. 4. The visualization of model outputs on CT scan of different severity categories.

The Leaderboard on the COVID-19 Severity Detection Challenge. In the COVID-19 severity detection challenge, we ensemble the CMC and infection-aware CMC models as our final model. The comparison results on the testing dataset are shown in Table 4. Our model ranks first in the challenge, surpassing the second team by 0.28% on the Macro F1 Score. Compared to other methods, our model shows relatively stable performance on each category, indicating that our model has learned the discriminative representations of different severity categories.

Visualization Results. We exploit the Class Activation Mapping (CAM) [40] to show the attention maps for COVID-19 severity detection. Four CT scan series of different categories are selected from the validation set, and the visualization results are shown in Fig. 4. For each group, the first row shows a series of CT slices, and the second and third rows present the CAMs from CMC and infection-aware CMC models, respectively.

Though the CMC model makes correct predictions, the attention maps just roughly indicate the suspicious area (see subfig. (a)), and it even over-focuses on irrelevant regions (e.g. bones) rather than the infections (see subfig. (b)(c)(d)). In contrast, our proposed infection-aware CMC model greatly improves the localization ability of suspicious lesions. Most infection areas can be located precisely, and the irrelevant regions are largely excluded. These highlighted areas in the attention maps demonstrate the reliability and interpretability of the detection results from our model.

6   Conclusions

In this paper, we present our solution for COVID-19 severity detection challenge in the 2nd COV19D Competition. Our proposed infection-aware 3D CMC network is composed of two stages, namely infection segmentation and COVID-19 severity classification. In the first stage, two segmentation networks are trained to extract lung fields as well as segment infected regions. The infection masks further provide auxiliary information for the original CT scans. In the second stage, the CMC network takes the CT scan and its lesion masks as the input and outputs the severity prediction. The experimental results on the competition leaderboard demonstrate the effectiveness of our infection-aware CMC network.

Acknowledgement. This work was supported by the Scientific & Technological Innovation 2030 - "New Generation AI" Key Project (No. 2021ZD0114001; No. 2021ZD0114000), and the Science and Technology Commission of Shanghai Municipality (No. 21511104502; No. 21511100500; No. 20DZ1100205). Yuejie Zhang, Xiaobo Zhang, and Rui Feng are corresponding authors.

References

1. Cao, Y., et al.: Longitudinal assessment of Covid-19 using a deep learning-based quantitative CT pipeline: illustration of two cases. Radiol.: Cardiothorac. Imaging **2**(2), e200082 (2020)
2. Chaganti, S., et al.: Automated quantification of CT patterns associated with Covid-19 from chest CT. Radiol.: Artif. Intell. **2**(4) (2020)

3. Chen, J., et al.: Deep learning-based model for detecting 2019 novel coronavirus pneumonia on high-resolution computed tomography. Sci. Rep. **10**(1), 1–11 (2020)

4. Chen, L.C., Zhu, Y., Papandreou, G., Schroff, F., Adam, H.: Encoder-decoder with atrous separable convolution for semantic image segmentation. In: Proceedings of the European Conference on Computer Vision (ECCV), pp. 801–818 (2018)

5. Chung, M., et al.: CT imaging features of 2019 novel coronavirus (2019-nCoV). Radiology (2020)

6. Fan, D.P., et al.: Inf-Net: automatic Covid-19 lung infection segmentation from CT images. IEEE Trans. Med. Imaging **39**(8), 2626–2637 (2020)

7. Fang, Y., et al.: Sensitivity of chest CT for Covid-19: comparison to RT-PCR. Radiology (2020)

8. Feng, Y.Z., et al.: Severity assessment and progression prediction of Covid-19 patients based on the lesionencoder framework and chest CT. Information **12**(11), 471 (2021)

9. He, K., Zhang, X., Ren, S., Sun, J.: Deep residual learning for image recognition. In: 2016 IEEE Conference on Computer Vision and Pattern Recognition, pp. 770–778 (2016)

10. He, K., et al.: Synergistic learning of lung lobe segmentation and hierarchical multi-instance classification for automated severity assessment of Covid-19 in CT images. Pattern Recogn. **113**, 107828 (2021)

11. Hou, J., Xu, J., Feng, R., Zhang, Y., Shan, F., Shi, W.: CMC-COV19D: contrastive mixup classification for Covid-19 diagnosis. In: Proceedings of the IEEE/CVF International Conference on Computer Vision, pp. 454–461 (2021)

12. Hou, J., et al.: Periphery-aware Covid-19 diagnosis with contrastive representation enhancement. Pattern Recogn. **118**, 108005 (2021)

13. Huang, G., Liu, Z., Van Der Maaten, L., Weinberger, K.Q.: Densely connected convolutional networks. In: 2017 IEEE Conference on Computer Vision and Pattern Recognition, pp. 2261–2269 (2017)

14. Huang, L., et al.: Serial quantitative chest CT assessment of Covid-19: a deep learning approach. Radiol.: Cardiothorac. Imaging **2**(2) (2020)

15. Ioannou, G., Papagiannis, T., Tagaris, T., Alexandridis, G., Stafylopatis, A.: Visual interpretability analysis of deep CNNs using an adaptive threshold method on diabetic retinopathy images. In: Proceedings of the IEEE/CVF International Conference on Computer Vision, pp. 480–486 (2021)

16. Javaheri, T., et al.: CovidCTNet: an open-source deep learning approach to identify Covid-19 using CT image. arXiv preprint arXiv:2005.03059 (2020)

17. Jin, S., et al.: AI-assisted CT imaging analysis for Covid-19 screening: building and deploying a medical AI system in four weeks. MedRxiv (2020)

18. Kendall, A., Gal, Y., Cipolla, R.: Multi-task learning using uncertainty to weigh losses for scene geometry and semantics. In: 2018 IEEE Conference on Computer Vision and Pattern Recognition, pp. 7482–7491 (2018)

19. Khosla, P., et al.: Supervised contrastive learning. In: Annual Conference on Neural Information Processing Systems 2020 (2020)

20. Kollias, D., Arsenos, A., Kollias, S.: AI-MIA: Covid-19 detection & severity analysis through medical imaging. arXiv preprint arXiv:2206.04732 (2022)

21. Kollias, D., Arsenos, A., Soukissian, L., Kollias, S.: MIA-COV19D: Covid-19 detection through 3-D chest CT image analysis. arXiv preprint arXiv:2106.07524 (2021)

22. Kollias, D., et al.: Deep transparent prediction through latent representation analysis. arXiv preprint arXiv:2009.07044 (2020)

23. Kollias, D., Tagaris, A., Stafylopatis, A., Kollias, S., Tagaris, G.: Deep neural architectures for prediction in healthcare. Complex Intell. Syst. **4**(2), 119–131 (2018)

24. Kollias, D., et al.: Transparent adaptation in deep medical image diagnosis. In: Heintz, F., Milano, M., O'Sullivan, B. (eds.) TAILOR 2020. LNCS (LNAI), vol. 12641, pp. 251–267. Springer, Cham (2021). https://doi.org/10.1007/978-3-030-73959-1_22

25. Lin, T.Y., Dollár, P., Girshick, R., He, K., Hariharan, B., Belongie, S.: Feature pyramid networks for object detection. In: Proceedings of the IEEE Conference on Computer Vision and Pattern Recognition, pp. 2117–2125 (2017)

26. Ma, J., et al.: Towards data-efficient learning: a benchmark for Covid-19 CT lung and infection segmentation. Med. Phys. **48**(3), 1197–1210 (2021). https://doi.org/10.1002/mp.14676

27. Milletari, F., Navab, N., Ahmadi, S.A.: V-Net: fully convolutional neural networks for volumetric medical image segmentation. In: 2016 Fourth International Conference on 3D Vision (3DV), pp. 565–571. IEEE (2016)

28. Pu, J., et al.: Automated quantification of Covid-19 severity and progression using chest CT images. Eur. Radiol. **31**(1), 436–446 (2021)

29. Qi, X., et al.: Machine learning-based CT radiomics model for predicting hospital stay in patients with pneumonia associated with SARS-CoV-2 infection: a multicenter study. medRxiv (2020)

30. Ronneberger, O., Fischer, P., Brox, T.: U-Net: convolutional networks for biomedical image segmentation. In: Navab, N., Hornegger, J., Wells, W.M., Frangi, A.F. (eds.) MICCAI 2015. LNCS, vol. 9351, pp. 234–241. Springer, Cham (2015). https://doi.org/10.1007/978-3-319-24574-4_28

31. Shan, F., et al.: Lung infection quantification of Covid-19 in CT images with deep learning. arXiv preprint arXiv:2003.04655 (2020)

32. Shi, F., et al.: Review of artificial intelligence techniques in imaging data acquisition, segmentation and diagnosis for Covid-19. IEEE Rev. Biomed. Eng. **15**, 4–15 (2020)

33. Tang, Z., et al.: Severity assessment of coronavirus disease 2019 (Covid-19) using quantitative features from chest CT images. arXiv preprint arXiv:2003.11988 (2020)

34. Wang, S., et al.: A deep learning algorithm using CT images to screen for corona virus disease (Covid-19). Eur. Radiol. 1–9 (2021)

35. WHO: Coronavirus disease (Covid-19) pandemic (2022). https://www.who.int/emergencies/diseases/novel-coronavirus-2019

36. Xu, X., et al.: A deep learning system to screen novel coronavirus disease 2019 pneumonia. Engineering **6**(10), 1122–1129 (2020)

37. Zhang, H., et al.: ResNeST: split-attention networks. arXiv preprint arXiv:2004.08955 (2020)

38. Zhang, H., Cisse, M., Dauphin, Y.N., Lopez-Paz, D.: mixup: Beyond empirical risk minimization. arXiv preprint arXiv:1710.09412 (2017)

39. Zheng, C., et al.: Deep learning-based detection for Covid-19 from chest CT using weak label. MedRxiv (2020)

40. Zhou, B., Khosla, A., Lapedriza, A., Oliva, A., Torralba, A.: Learning deep features for discriminative localization. In: Proceedings of the IEEE Conference on Computer Vision and Pattern Recognition, pp. 2921–2929 (2016)

41. Zhou, Z., Rahman Siddiquee, M.M., Tajbakhsh, N., Liang, J.: UNet++: a nested U-Net architecture for medical image segmentation. In: Stoyanov, D., et al. (eds.) DLMIA/ML-CDS -2018. LNCS, vol. 11045, pp. 3–11. Springer, Cham (2018). https://doi.org/10.1007/978-3-030-00889-5_1

Variability Matters: Evaluating Inter-Rater Variability in Histopathology for Robust Cell Detection

Cholmin Kang[(✉)] [iD], Chunggi Lee [iD], Heon Song [iD], Minuk Ma [iD], and Sérgio Pereira [iD]

Lunit, Seoul 06241, Republic of Korea
kcholmin@gmail.com, {cglee,heon.song,sergio}@lunit.io
http://lunit.io

Abstract. Large annotated datasets have been a key component in the success of deep learning. However, annotating medical images is challenging as it requires expertise and a large budget. In particular, annotating different types of cells in histopathology suffer from high inter- and intra-rater variability due to the ambiguity of the task. Under this setting, the relation between annotators' variability and model performance has received little attention. We present a large-scale study on the variability of cell annotations among 120 board-certified pathologists and how it affects the performance of a deep learning model. We propose a method to measure such variability, and by excluding those annotators with low variability, we verify the trade-off between the amount of data and its quality. We found that naively increasing the data size at the expense of inter-rater variability does not necessarily lead to better-performing models in cell detection. Instead, decreasing the inter-rater variability with the expense of decreasing dataset size increased the model performance. Furthermore, models trained from data annotated with lower inter-labeler variability outperform those from higher inter-labeler variability. These findings suggest that the evaluation of the annotators may help tackle the fundamental budget issues in the histopathology domain.

1 Introduction

Histopathology plays an important role in the diagnosis of cancer and its treatment planning. The process, typically, requires pathologists to localize cells and segment tissues in whole slide images (WSIs) [1]. However, with the advent of digital pathology, these tedious and error-prone tasks can be done by Deep Learning models. Still, these models are known to require the collection of large annotated datasets. Particularly, cell detection needs annotations from very large numbers of individually annotated cells.

The identification of cells in WSIs is challenging due to their diversity, subtle morphological differences, and the astounding amount of cells that exist in a

---

Supplementary Information The online version contains supplementary material available at https://doi.org/10.1007/978-3-031-25082-8_37.

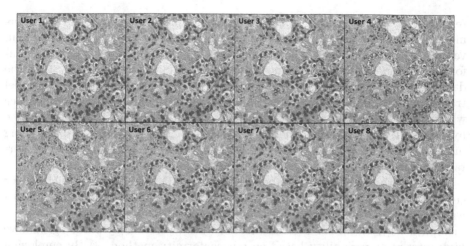

Fig. 1. Example of inter-rater variability in the annotation of cells. The colored circles represent: red - tumor cell, blue - lymphocytes, and green - other cells. (Color figure online)

single WSI. Therefore, when manually performed, this task is time-consuming, and suffers from high inter- and intra-rater variability among annotators [2, 3]. For example, Fig. 1 shows the large discrepancy in cell annotations among different annotators. The inconsistencies among the annotators negatively affect the performance of the model [4]. For example, when using such inconsistent data, the model may overfit to the noise and generalize poorly [5]. Therefore, some degree of consistency in the annotation must be guaranteed.

To improve the consistency among annotators and reduce the variability due to annotations in the dataset, we propose to exclude annotators with large variability from other annotators. This method is based on the existence of an anchor annotator who works as a reference point to which each annotator are compared. The concept of anchor has been recently proposed [6,7], where a cluster center is fixed for more efficient and stable training. We propose to have an anchor annotator that annotated a control set. In this way, we can measure the conformity of the other annotators in relation to the anchor one. During annotation, a given expert can receive a new image to annotate, or, instead, with some probability a control image. The set of annotated control images of an annotator is used to measure its conformity to the anchor annotator through a modified F1-score (mF1). In this way, annotators with high conformity would achieve lower variability, thus, more reliable annotations in relation to the anchor.

We collected and annotated a total of 29,387 patches from 12,326 WSIs. A total of 120 board-certified pathologists participated in the annotation process, which, to the best of our knowledge, is the largest-scale pathology annotation analysis reported in the literature. Based on our conformity, we divide the annotated data and perform an analysis on the discrepancy among annotators. Our results show the conformity of 0.70, with a standard deviation of 0.08. Furthermore, the kappa score showed a mean agreement of $\kappa = 0.43$ with a minimum value between $\kappa = -0.01$ and $\kappa = 0.87$. These results show relatively low conformity among most of the annotators. Furthermore, we also verify that the data with lower variability leads to better-performing models. Indeed, the model trained with a smaller dataset with lower variability showed better performance when compared to the model trained with the full dataset. This supports the hypothesis that a smaller higher quality dataset can lead to a well-performing model, at a reduced cost, when compared with a large dataset with high inter-rater variability.

The contributions of this paper are five-fold. 1) We propose a method for calculating the conformity between annotators and an anchor annotator. In this way, we can measure the variability among annotators in cell detection tasks in histopathology. 2) We observe that the annotations created by pathologists have high inter-rater variability based on our conformity method. 3) By employing the before-mentioned conformity method, we propose a process to reduce annotation variability within a dataset by excluding annotators with low conformity in relation to an anchor. This leads to more efficient usage of the annotation budget, without sacrificing performance. 4) We show that a dataset with low inter-rater variability leads to better model performance. 5) To the best of our knowledge, this is the first work that tackles the problem of measuring variability among annotators in a large-scale dataset for a cell detection task, with the participation of a large crowd of board-certified pathologists.

2   Related Work

Deep Learning techniques, particularly Convolutional Neural Networks (CNN) revolutionized the field of computer vision and image recognition [8,9] by achieving unprecedented performances. As such, it has also been extensively explored in the domain of digital pathology [10]. Cell detection tasks have received attention due to its laborious and error-prone nature [11–14]. However, there has been little attention to the effect of inter-rater variability in cell detection tasks, and how it affects the performance of the models.

In comparison to natural images, the annotation of medical images is a more challenging task. First, domain experts are needed, e.g., medical doctors. Second, due to the difficulty and ambiguity of the tasks, these tasks are more prone to subjectivity which leads to the disagreement between experts [15,16]. The inter-rater variability and reproducibility of annotations have been studied in some medical-related tasks, such as sleep pattern segmentation in EEG data [17], or segmentation in Computed Tomography images [18]. In the case of histopathology, some studies demonstrate that computer-assisted reading of images decreases the variability among observers [15,16].

To overcome the disagreements, some works attempt to utilize the annotations of multiple raters by learning several models from single annotations and from the agreement of multiple annotators [19]. In this way, the effect of disagreement can be mitigated, but it assumes multiple annotations for the same images.

Other approaches accept that inter-rater variability will exist, and explore crowds of non-experts to collect large amounts of data by crowd-sourcing, either with or without the assistance of experts. Nonetheless, the scopes were limited to tasks that do not require a high degree of expertise or the annotation of many instances [20,21]. Such assumptions do not hold in cell annotation in histopathology. Crowd-sourcing cell annotations were previously performed [22], but, the evaluation of the conformity of the annotators in this setting, where there is high inter-rater variability, remains unexplored. Inconsistencies among the annotators introduce noise to the annotations [4]. In turn, it can lead to models that overfit to the noise and generalize poorly [5]. Therefore, some degree of label variability must be guaranteed.

Other related prior work tackles the problem of lack of conformity between annotators from the perspective of a noisy label [23,24], or, tried to classify the reliability of annotators using the probabilistic model [25]. Other works tried to resolve large variability of annotator performance using the EM algorithm [26] or Gaussian process classifiers [27]. However, these approaches rely on the training dynamics, so, it is not possible to distinguish if a given sample is noisy or just difficult. Moreover, the noisy data is discarded, raising two concerns: 1) resources are used to gather unused data, and 2) the dataset needs to be sufficiently large to allow discarding data without impacting the performance of the model. Instead, we hypothesize that noise can be introduced by annotators with high variability to other annotators, and preventive selection of conforming annotators leads to higher quality annotations. This becomes even more necessary in large-scale settings, where a large pool of annotators needs to be sourced externally.

3   Evaluating Annotators and Variability

During the annotation process, the annotators receive unlabeled patches to annotate from the training set **T**. Control set **C** refers to the unlabeled dataset that will later be used to measure the variability among annotators. Images from the control set **C** are randomly, and blindly, assigned to the annotators. Since the annotators do not know when a control image is provided, this ensures that the annotation skill showed in **C** is similar to that of **T**. Similar methods are used in practice when evaluating crowd-sourced workers [28]. An *anchor annotator* is chosen as the reference annotator to perform analysis on discrepancy (i.e., variability) among annotators. The annotations of each annotator are compared with the annotations done by the *anchor annotator*. Conformity is measured between each annotated cell of an annotator and the annotated cells of an *anchor annotator*. The conformity is used to show the variability of annotators and annotators with large conformity will have small variability among them. Finally, annotators are divided into groups according to their conformity score. The model is, then, trained on the data of each group and aggregation of groups. The annotation, conformity measurement, and model development process are depicted in Fig. 2(a).

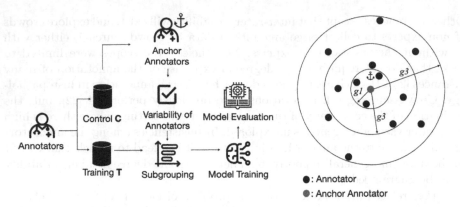

(a) An overview of annotator variability evaluation process

(b) An example of anchor annotator and Conformity measurement

Fig. 2. Evaluation of inter-rater variability. (a) The control set is created to blindly evaluate the annotator conformity during the annotation process. A score is computed per annotator, which is used to evaluate the impact of their annotations in the cell detection task. (b) Annotator selection method. Conformity is measured between each annotation done by an annotator and anchor annotator. In the annotator selection phase, annotations are divided into subgroups (such as g_1, g_2, or g_3 in the figure) according to annotator conformity percentiles.

3.1   Anchor Annotator

In order to perform an analysis of the variability among annotators, reference data should be set to measure how much each annotator is far from the reference point. In this work, one annotator is defined as the anchor annotator and is chosen as a criterion to evaluate variability among annotators. The rationale for setting an anchor annotator and evaluating variance is motivated by the concept of an anchor cluster center [6,7], where a cluster center is fixed for more efficient and stable training. The conceptual overview of anchor annotator selection is shown in Fig. 3. As shown in the Figure, an anchor annotator is the one with most number of images annotated so that the annotated images are the superset of the patches annotated by other annotators.

3.2   Annotator Variability and Conformity Calculation

Annotator conformity is calculated using Algorithm 1. To compute the annotators' conformity, a hit criterion needs to be defined. We consider a circular region of radius $\theta = 10$ micrometers centered in each cell of $\mathbf{C}$ in a given patch. Annotations from $\mathbf{C}$ that are located inside a given circle are True Positive (TP) candidates. Among these candidates, the closest point that matches the same class as the annotation from $\mathbf{C}$ is considered a TP. Given this criterion, we also compute False Positive (FP) and False Negative (FN) quantities. Based on

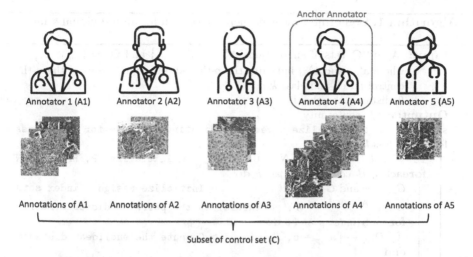

Fig. 3. A conceptual overview of anchor annotator selection. Anchor annotator is chosen among annotators with the most amount of annotations done in control set C. As shown in the figure, the patches annotated by A4 is the superset of patches annotated by A1, A2, A3, and A5. Thus, A4 is chosen as an anchor annotator.

the TP, FP, and FN, F1-score per class is computed and the conformity of the annotator is defined as mean F1-score over classes.

3.3 Annotator Grouping Based on the Conformity

Finally, annotators are divided into subgroups according to their conformity to the anchor annotator. Figure 2(b) shows the annotators' annotation distribution. First, an anchor annotator is selected. Then, conformity between the anchor annotator and other annotators is calculated. Finally, each annotator is evaluated by computing the conformity, as described in Algorithm 1. In this work, we consider this conformity as a variability. If all annotations done by all annotators are equal to those done by the anchor annotator, the conformity will be 1. However, if the annotations are different from the anchor annotator, the annotator conformity calculated by Algorithm 1 will decrease, leading to increased variability and decreasing conformity.

4 Experimental Results

4.1 Dataset

We collected a large-scale dataset for the task of tumor and lymphocyte cells detection in H&E-stained WSIs, from 16 primary origins cancers (e.g. lung, breast, kidney, etc.). As WSIs are typically large for exhaustive annotation, occupying several giga-pixels squared in area, it is a common practice to divide

Algorithm 1: Conformity measurement algorithm for an individual annotator

---

Input: **A** and **C**: sets of annotations $\mathbf{A}_n^k = \{\mathbf{a}_{n,1}^k, \ldots\}$ and $\mathbf{C}_n^k = \{\mathbf{c}_{n,1}^k, \ldots\}$
where $\mathbf{a}_{n,p}^k$ and $\mathbf{c}_{n,q}^k$ indicates the p-th and q-th annotations of the n-th
control patch and class k
θ: distance threshold

Output: $conf$: conformity

$s \leftarrow 0$;   ▷ Initialize a vector indicating F1-score for each class

foreach *class k* **do**

    $TP \leftarrow 0$, $FP \leftarrow 0$, and $FN \leftarrow 0$;   ▷ Initialize TP, FP, and FN

    foreach *n-th control data patch* **do**

        $G_A \leftarrow \emptyset$ and $G_M \leftarrow \emptyset$;   ▷ Initialize assigned index sets

        $D \in \mathbb{R}^{|\mathbf{A}_n^k| \times |\mathbf{C}_n^k|} \leftarrow 0$;   ▷ Initialize the pairwise distance matrix

        foreach *index p and q* **do**

            $D_{p,q} \leftarrow \|\mathbf{a}_{n,p}^k - \mathbf{c}_{n,q}^k\|_2$;   ▷ Compute the euclidean distance

        end

        foreach *i-th index pair* (p_i^*, q_i^*) *such that* $\forall j > i, D_{p_i^*, q_i^*} \leq D_{p_j^*, q_j^*}$ **do**

            if $D_{p_i^*, q_i^*} \leq \theta$, $p_i^* \notin G_A$, *and* $q_i^* \notin G_M$ **then**

                $G_A \leftarrow G_A \cup \{p_i^*\}$ and $G_M \leftarrow G_M \cup \{q_i^*\}$;   ▷ Assign indices

            end

        end

        $TP \leftarrow TP + |G_A|$;   ▷ Update TP

        $FP \leftarrow FP + |\mathbf{A}_n^k| - |G_A|$;   ▷ Update FP

        $FN \leftarrow FN + |\mathbf{C}_n^k| - |G_M|$;   ▷ Update FN

    end

    $s_k \leftarrow \frac{2TP}{2TP+FP+FN}$;   ▷ Compute the F1-score for class k

end

$conf \leftarrow \frac{1}{|s|}\sum_k s_k$;   ▷ Compute the conformity as mean F1-score

---

them into smaller patches [29]. We sampled patches of size 1024×1024 pixels
from a total of 12,326 WSIs. The annotations consist of point annotations that
locate the nuclei of cells, as well as their classes. We collect 3 sets at the WSI
level: control set (**C**), Training set (**T**) and Validation set (**V**). The validation
set is annotated by three annotators, and has no overlap with **C**. The number
of images and cell annotations in each set is presented in Table 1. A total of
29,387 unique patches are annotated. As can be seen in the Table, the average
number of annotated cells in a single image is similar among C, T, and V. A
large crowd of 120 board-certified pathologists participated in the annotation
process. During the annotation process, annotators are provided with guidelines
on how to annotate each cell. Annotators are guided to annotate each cell to
one of six classes: Tumor Cell, Lymphoplasma cells, Macrophage, Fibroblast,
Endothelial cell, and others. The class "other" include nucleated cells with vague
morphology that are not included in the specific cell types. Each cell is explained

Table 1. Description of the Control, Training, and Validation sets in terms of the number patches, annotated cells, and WSIs.

Dataset	Patches	Annotated cells	Avg annotated cells per patch	Number of WSI
Control set	150	32,746	218.31	141
Training set	21,795	3,727,343	171.02	8,875
Validation set	7,442	1,390,551	186.85	3,310

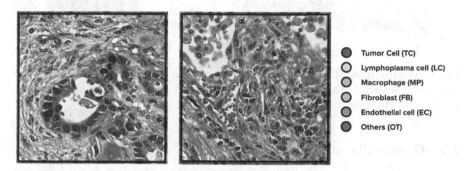

Fig. 4. Example of the reference images provided to annotators alongside the annotation guidelines.

with its definition and its clinical characteristics. Furthermore, visual examples were provided to exemplify the guides, such as in Fig. 4.

4.2 Cell Detection Method and Experimental Details

Having the conformity scores of the annotators, we evaluate its efficacy in filtering out annotations from annotators with higher variability by training a model in the downstream task of cell detection. Similar to [30], we pose the problem as a dense pixel classification task, but we use a DeepLabv3+ [31] architecture. The point annotation of each cell is converted to a circle of radius of 5 pixels, hence, resulting in a map similar to semantic segmentation tasks. The output is a dense prediction likelihood omap that, similar to [30], requires a post-processing stage to retrieve the unique locations of the cells[1]. In this experiment, we target Tumor-Infiltrating Lymphocyte (TIL) task which requires the number of tumor cells and lymphocytes in given WSI. Therefore, among the annotations, we regard macrophage, fibroblast, and endothelial cell as "others" class. All models are trained using the Adam [32] optimizer (learning rate of 1e-4) and the soft dice loss function [33]. Experiments are performed on 4 NVIDIA V100 GPUs, and implemented using PyTorch (version 1.7.1) [34].

---

[1] Post-processing consists of Gaussian filtering ($\sigma = 3$) the likelihood maps, followed by local maxima detection within a radius of 3 pixels.

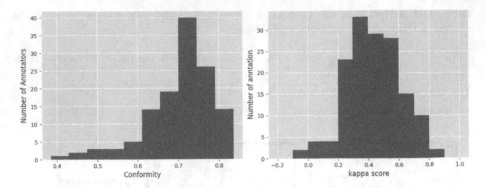

Fig. 5. Distributions of the variability of the external annotators as mF1-score (left), and the agreement of the annotations by the external annotators in relation to the set **C**, measured in terms of Fleiss' kappa (right).

4.3   Conformity of the Annotators

Following the proposed evaluation procedure of the annotators, we computed the individual annotator conformity. In this work, an annotator with the most number of annotations done from set C is chosen as an anchor annotator so that most annotators can have conformity calculated. The closer a score is to 1, the better the annotator conforms with an anchor annotator. Figure 5, left, shows the distribution of the conformity score. The median conformity is 0.70 (minimum $= 0.38$, and maximum $= 0.83$). We further conduct a Shapiro-Wilk Normality test [35] and verify that it does not follow a Normal distribution at a significance level $\alpha = 0.05$, with a p-value $= 9.427e - 8$. Moreover, we observe a negative skewness of -1.3158, which confirms the observation of the longer left tail. Varying expertise is a known problem when annotating cells in histopathology. Nonetheless, one would expect the conformity of the experts to reference annotation to be normally distributed. Instead, we observe that while the median annotator conformity is relatively high, there are some annotators on the left tail that perform significantly worse than the majority. This supports the observation of high inter-rater variability.

To further analyze the inter-rater variance, we measured the agreement among all annotators on each Control patch using Fleiss' kappa [36], which is a statistical measure of the reliability of agreement for categorical ratings with more than two raters. This test was run to show inter-rater variability rather than to show absolute quality of annotators. In this analysis, the inter-rater agreement is measured in relation to three categories: tumor cells, lymphocytes, and unmatched annotations. As previously described, the annotated cells from each pathologist have been matched to the nearest cell of the same class in **C**, following the hit criterion. On the other hand, if the hit criterion is not satisfied, we treat it as an unmatched annotation. We compute the Fleiss' kappa for all patches in **C**. The mean agreement among the annotations of the pathologists was $\kappa = 0.43$ with the minimum and maximum of $\kappa = -0.01$ and $\kappa = 0.87$,

Table 2. Model performance by each cancer primary origin and annotation conformity percentile range. For example, the column "p_{0-25}" denotes the training set with patches annotated by annotators with conformity in percentiles 0 to 25, i.e., the 25% annotators with the lowest conformity. At the bottom, it is presented the total annotation time in hours.

Primary origin	Percentile ranges					
	p_{0-25}	p_{75-100}	p_{0-50}	p_{50-100}	p_{25-100}	p_{0-100}
Biliary Tract	62.09	62.9	63.71	64.06	65.1	64.89
Breast	62.32	64.74	64.53	65.53	64.83	65.3
Colorectum	61.87	63.09	63.83	63.8	64.02	63.84
Esophagus	64.11	61.42	65.56	66.24	66.33	66.19
Head and Neck	59.07	61.68	61.76	62.54	62.63	62.61
Kidney	49.68	53.94	52.10	56.07	57.34	56.49
Liver	62.50	65.16	64.06	66.09	65.29	65.16
Lung	58.18	60.24	60.27	61.24	60.75	60.72
Melanoma	59.08	60.03	60.18	61.64	61.73	61.27
Ovary	53.92	55.64	56.57	57.56	57.79	57.32
Pan Urinary	62.71	64.82	65.33	65.45	65.02	64.8
Pancreas	56.68	58.49	58.62	59.93	61.16	60.06
Prostate	55.70	56.41	56.12	58.13	58.39	58.64
Stomach	59.74	61.07	62.13	62.79	63.54	63.10
Uterine Cervix	61.86	64.33	64.37	65.23	65.69	65.64
Uterine Endometrium	53.18	54.70	54.13	56.07	55.53	55.51
Average mF1	58.92	60.77	60.87	62.03	62.18	61.76
Annotation time (hours)	598	575	1,122	1,176	1,703	2,298

respectively. This range (-0.01 to 0.87) of kappa values suggest poor to an excellent agreement, but the mean is located in a fair to moderate agreement range. This further reinforces the observation that there exists high inter-rater variability. We observe that the classification of the cells is sensitive in specific patches and pathologists, as shown in Fig. 1.

4.4   Impact of the Conformity of the Annotators in Cell Detection

In this experiment, we divide the annotators by their conformity, from lower to higher. Following this conformity, we define 4 subsets of the data based on the percentile of the conformity of the annotators. Therefore, the interval p_{0-25} contains 25% of the patches in the dataset that were annotated by annotators with high variability. The intervals p_{25-50}, p_{50-75}, and p_{75-100} also contain 25% of the patches each, with decreasing annotator variability as compared to the anchor annotator. Therefore, following our hypothesis, the interval p_{75-100} must contain the annotations with the lowest variability. We hypothesize that data annotated by annotators with lower variability should lead to better-performing cell detection models. Table 2 shows the performance of the cell detection model in terms of conformity by the before-mentioned intervals of data across the 16 cancer primary origin organs. The results are the average of 3 models trained with different random seeds. We use the paired Wilcoxon Signed Rank test [37] at a significance level $\alpha = 0.05$ to compare results and test our hypothesis.

We compare the performance of models trained with similar amounts of data, but with different conformity in relation to the annotation. First, we contrast interval p_{0-25} and p_{75-100}, i.e., the data annotated by the highest variance and lowest variance annotators, respectively. Note that both sets contain 25% of patches of the full training dataset. We observe that the data from the best performing annotators (interval p_{75-100}) leads to better performance compared to interval p_{0-25}. This is true for all of cancer primary origins and is corroborated by an existing statistical difference between the results (p-value = 0.00614). Similarly, when we use 50% of the data to train the cell detection model (intervals p_{0-50} and p_{50-100}), a similar trend is observed. The model trained from the data from annotators with lower variability outperforms the model trained with data from interval p_{0-50}; again, a statistical difference was found (p-value = 0.00054). This result shows that given the same amount of data, the model trained with high conformity data outperforms that trained with low conformity data.

Additionally, we evaluate the performance of models trained from smaller datasets labeled by annotators with lower variability compared to the use of the full training set. The intervals p_{25-100} and p_{50-100} contain an upper 75% and 50% of the training dataset, respectively. Nonetheless, in both cases, they outperform the model trained with the full training set, with statistical significance in the case of the interval p_{25-100} (p-value = 0.01778). These results suggest that smaller but low inter-rater variability datasets can outperform models trained from the larger datasets, but with higher variability annotations.

The data from the interval p_{50-100} represent the upper 50% of the available training data, but the performance is on par with the variability setting of the interval p_{25-100}. Therefore, in retrospect, we could have collected 50% fewer data, which would have reduced the total amount of annotation time from 2298 h to 1176 h. This is especially significant as the annotation process requires expert pathologists. Thus, we believe that the proposed annotation scoring procedure can be helpful in using the available budget more effectively, by requesting annotations from the experts with low variability to anchor annotator and obtain either a smaller dataset with good model performance or a large-scale dataset with lower inter-rater variability annotations in the future data collection process.

5   Conclusions

In this paper, we proposed a simple, yet effective method to assess the variability of a crowd of 120 experts in the time-consuming and difficult task of cell annotation in histopathology images stained with H&E. Our findings support the previous observations that this task suffers from high inter-rater variability. Furthermore, we conducted a proof-of-concept experiment to show the effect of inter-rater variability on a machine learning model for the task of cell detection. From the experiment, we showed that with the same size of data, the data labeled by annotators with lower inter-rater variability led to better model performance than the data labeled by annotators with higher inter-rater variability. Furthermore, excluding the data with the most inter-rater variability from the full dataset was beneficial, despite resulting in a smaller dataset. Hence, contrary to common findings in deep learning research, the fewer amount of data with lower inter-rater variability resulted in better-performing machine learning model. These findings are useful in the medical domain where annotations are costly and laborious. With this finding, we can argue that collecting fewer data from low inter-rater variability is more beneficial than collecting data without considering the inter-rater variability. Despite the findings that increasing conformity leads to increasing model performance, we recognize that there are some limitations in our work. First, our work is based on the premise that the anchor annotator is well-performing. The work is based on the assumption that all annotators are sufficiently expert and have done their best to annotate. Further investigation is needed with iterated experiments on setting annotators as anchor annotators. Second, the generalizability of our work on stained WSI other than H&E stain is not guaranteed. Further investigation on other staining methods with more diverse skill backgrounds of digital pathologist annotators is needed. We hope the idea of scoring the annotators helps improve the annotation budget management in the histopathology domain.

References

1. Diao, J.A., et al.: Human-interpretable image features derived from densely mapped cancer pathology slides predict diverse molecular phenotypes. Nat. Commun. **12**(1), 1–15 (2021)
2. Lutnick, B., et al.: An integrated iterative annotation technique for easing neural network training in medical image analysis. Nat. Mach. Intell. **1**(2), 112–119 (2019)
3. Cheplygina, V., Perez-Rovira, A., Kuo, W., Tiddens, H.A.W.M., de Bruijne, M.: Early experiences with crowdsourcing airway annotations in chest CT. In: Carneiro, G., et al. (eds.) LABELS/DLMIA -2016. LNCS, vol. 10008, pp. 209–218. Springer, Cham (2016). https://doi.org/10.1007/978-3-319-46976-8_22
4. Guo, B., Chen, H., Yu, Z., Xie, X., Huangfu, S., Zhang, D.: FlierMeet: a mobile crowdsensing system for cross-space public information reposting, tagging, and sharing. Trans. Mob. Comput. **14**(10), 2020–2033 (2014)
5. Li, J., Wong, Y., Zhao, Q., Kankanhalli, M.S.: Learning to learn from noisy labeled data. In: CVPR (2019)
6. Wang, Y., Wang, D., Pang, W., Miao, C., Tan, A.H., Zhou, Y.: A systematic density-based clustering method using anchor points. Neurocomputing **400**, 352–370 (2020)
7. Miller, D., Sunderhauf, N., Milford, M., Dayoub, F.: Class anchor clustering: a loss for distance-based open set recognition. In: Proceedings of the IEEE/CVF Winter Conference on Applications of Computer Vision, pp. 3570–3578 (2021)

8. Krizhevsky, A., Sutskever, I., Hinton, G.E.: ImageNet classification with deep convolutional neural networks. In: Advances in Neural Information Processing Systems, vol. 25 (2012)

9. Girshick, R., Donahue, J., Darrell, T., Malik, J.: Rich feature hierarchies for accurate object detection and semantic segmentation. In: Proceedings of the IEEE Conference on Computer Vision and Pattern Recognition, pp. 580–587 (2014)

10. Meijering, E.: Cell segmentation: 50 years down the road [life sciences]. IEEE Signal Process. Mag. **29**(5), 140–145 (2012)

11. Chen, T., Chefd'hotel, C.: Deep learning based automatic immune cell detection for immunohistochemistry images. In: Wu, G., Zhang, D., Zhou, L. (eds.) MLMI 2014. LNCS, vol. 8679, pp. 17–24. Springer, Cham (2014). https://doi.org/10.1007/978-3-319-10581-9_3

12. Gao, Z., Wang, L., Zhou, L., Zhang, J.: Hep-2 cell image classification with deep convolutional neural networks. IEEE J. Biomed. Health Inform. **21**(2), 416–428 (2016)

13. Cireşan, D.C., Giusti, A., Gambardella, L.M., Schmidhuber, J.: Mitosis detection in breast cancer histology images with deep neural networks. In: Mori, K., Sakuma, I., Sato, Y., Barillot, C., Navab, N. (eds.) MICCAI 2013. LNCS, vol. 8150, pp. 411–418. Springer, Heidelberg (2013). https://doi.org/10.1007/978-3-642-40763-5_51

14. Xue, Y., Ray, N.: Cell detection in microscopy images with deep convolutional neural network and compressed sensing. arXiv preprint arXiv:1708.03307 (2017)

15. Hekler, A., et al.: Deep learning outperformed 11 pathologists in the classification of histopathological melanoma images. Eur. J. Cancer **118**, 91–96 (2019)

16. Bertram, C.A., et al.: Computer-assisted mitotic count using a deep learning-based algorithm improves interobserver reproducibility and accuracy. Vet. Pathol. **59**(2), 211–226 (2022)

17. Schaekermann, M., Beaton, G., Habib, M., Lim, A., Larson, K., Law, E.: Understanding expert disagreement in medical data analysis through structured adjudication. Proc. ACM Hum.-Comput. Interact. **3**(CSCW), 1–23 (2019)

18. Haarburger, C., Müller-Franzes, G., Weninger, L., Kuhl, C., Truhn, D., Merhof, D.: Radiomics feature reproducibility under inter-rater variability in segmentations of CT images. Sci. Rep. **10**(1), 1–10 (2020)

19. Sudre, C.H., et al.: Let's agree to disagree: learning highly debatable multirater labelling. In: Shen, D., et al. (eds.) MICCAI 2019. LNCS, vol. 11767, pp. 665–673. Springer, Cham (2019). https://doi.org/10.1007/978-3-030-32251-9_73

20. Maier-Hein, L., et al.: Can masses of non-experts train highly accurate image classifiers? In: Golland, P., Hata, N., Barillot, C., Hornegger, J., Howe, R. (eds.) MICCAI 2014. LNCS, vol. 8674, pp. 438–445. Springer, Cham (2014). https://doi.org/10.1007/978-3-319-10470-6_55

21. Kwitt, R., Hegenbart, S., Rasiwasia, N., Vécsei, A., Uhl, A.: Do we need annotation experts? A case study in celiac disease classification. In: Golland, P., Hata, N., Barillot, C., Hornegger, J., Howe, R. (eds.) MICCAI 2014. LNCS, vol. 8674, pp. 454–461. Springer, Cham (2014). https://doi.org/10.1007/978-3-319-10470-6_57

22. Amgad, M., et al.: NuCLS: a scalable crowdsourcing, deep learning approach and dataset for nucleus classification, localization and segmentation. arXiv:2102.09099 (2021)

23. Veit, A., Alldrin, N., Chechik, G., Krasin, I., Gupta, A., Belongie, S.: Learning from noisy large-scale datasets with minimal supervision. In: CVPR (2017)

24. Jiang, L., Zhou, Z., Leung, T., Li, L.J., Fei-Fei, L.: MentorNet: learning data-driven curriculum for very deep neural networks on corrupted labels. In: ICML (2018)

25. Yan, Y., Rosales, R., Fung, G., Subramanian, R., Dy, J.: Learning from multiple annotators with varying expertise. Mach. Learn. **95**(3), 291–327 (2014)
26. Raykar, V.C., et al.: Learning from crowds. J. Mach. Learn. Res. **11**(4) (2010)
27. Rodrigues, F., Pereira, F., Ribeiro, B.: Gaussian process classification and active learning with multiple annotators. In: International Conference on Machine Learning, pp. 433–441. PMLR (2014)
28. Bartolo, M., Thrush, T., Riedel, S., Stenetorp, P., Jia, R., Kiela, D.: Models in the loop: aiding crowdworkers with generative annotation assistants. arXiv:2112.09062 (2021)
29. Khened, M., Kori, A., Rajkumar, H., Krishnamurthi, G., Srinivasan, B.: A generalized deep learning framework for whole-slide image segmentation and analysis. Sci. Rep. **11**(1), 1–14 (2021)
30. Swiderska-Chadaj, Z., et al.: Learning to detect lymphocytes in immunohistochemistry with deep learning. MIA **58**, 101547 (2019)
31. Chen, L.-C., Zhu, Y., Papandreou, G., Schroff, F., Adam, H.: Encoder-decoder with atrous separable convolution for semantic image segmentation. In: Ferrari, V., Hebert, M., Sminchisescu, C., Weiss, Y. (eds.) ECCV 2018. LNCS, vol. 11211, pp. 833–851. Springer, Cham (2018). https://doi.org/10.1007/978-3-030-01234-2_49
32. Kingma, D.P., Ba, J.: Adam: a method for stochastic optimization. In: Bengio, Y., LeCun, Y. (eds.) ICLR (2015)
33. Milletari, F., Navab, N., Ahmadi, S.A.: V-Net: fully convolutional neural networks for volumetric medical image segmentation. In: 3DV (2016)
34. Paszke, A., et al.: Automatic differentiation in PyTorch (2017)
35. Shapiro, S.S., Wilk, M.B.: An analysis of variance test for normality (complete samples)†. Biometrika **52**(3–4), 591–611 (1965)
36. Falotico, R., Quatto, P.: Fleiss' kappa statistic without paradoxes. Qual. Quant. **49**(2), 463–470 (2015)
37. Woolson, R.F.: Wilcoxon signed-rank test. Wiley Encyclopedia of Clinical Trials, pp. 1–3 (2007)

FUSION: Fully Unsupervised Test-Time Stain Adaptation via Fused Normalization Statistics

Nilanjan Chattopadhyay⬡, Shiv Gehlot⬡, and Nitin Singhal[(✉)]

Advanced Technology Group, AIRA MATRIX, Thane, India
{nilanjan.chattopadhyay,shiv.gehlot,nitin.singhal}@airamatrix.com

Abstract. Staining reveals the micro-structure of the aspirate while creating histopathology slides. Stain variation, defined as a chromatic difference between the source and the target, is caused by varying characteristics during staining, resulting in a distribution shift and poor performance on the target. The goal of stain normalization is to match the target's chromatic distribution to that of the source. However, stain normalisation causes the underlying morphology to distort, resulting in an incorrect diagnosis. We propose FUSION, a new method for promoting stain-adaption by adjusting the model to the target in an unsupervised test-time scenario, eliminating the necessity for significant labelling at the target end. FUSION works by altering the target's batch-normalization statistics and fusing them with source statistics using a weighting factor. The algorithm reduces to one of two extremes based on the weighting factor. Despite the lack of training or supervision, FUSION surpasses existing equivalent algorithms for classification and dense predictions (segmentation), as demonstrated by comprehensive experiments on two public datasets.

Keywords: Stain variation · Unsupervised · Stain adaptation

1 Introduction

Staining is used in histopathology slide preparation to highlight the essential structure of the aspirate. Differences in stain chemicals, lighting conditions, or staining time may produce stain colour variances in images collected at the same or separate facilities. The inter-center chromatic difference results in sub-optimal performance on the target test set because of stain variation, limiting model deployment across the centres.

Transfer learning is a simple but effective strategy for dealing with distribution shifts, but it requires annotation efforts at the target end, which can be challenging, especially in medical image analysis. By training a network to learn stain variance via mapping color-augmented input images to the original images, self-supervised learning eliminates the need for annotation [3, 20]. However, the settings of the train-time augmentations influence the performance of

---

N. Chattopadhyay and S. Gehlot—These authors contributed equally.

© The Author(s), under exclusive license to Springer Nature Switzerland AG 2023
L. Karlinsky et al. (Eds.): ECCV 2022 Workshops, LNCS 13807, pp. 566–576, 2023.
https://doi.org/10.1007/978-3-031-25082-8_38

these algorithms on the target domain. *Stain normalisation* is a stain variation handling method that does not require any training at the source or target end. It aligns the source and target chromatic distributions using reference image(s) from the source domain [4,6,8–10,12–14,18]. However, depending on the reference image(s), its performance can vary greatly. Changes in the underlying structure of the images further degrade the performance. The use of generative modeling-based solutions eliminates the need for a reference image. However, it requires the data from target domain in addition to a larger training sample, which limits its application [16,25,26].

These limitations are overcome via domain adaptation [2,21] and test time training [17], but with a modified training process. The goal of adaptation is to generalise a model $f_\theta^s(x)$ trained on source data $\{x^s, y^s\}$ to target data x^t. Test time adaptation (TTA) modifies the testing procedure while preserving the initial training. Two TTA techniques explored in the literature are Entropy Minimization (EM) [24,27] and Normalization Statistics Adaptation (SNA) [11,15]. In EM, the backpropagation to update the adaptation parameters θ is driven by the entropy function formulated using the predictions of $f_\theta^s(x)$ on x^t. Batch normalisation statistics and parameters are controlled in [24] by entropy minimization of the target's prediction. For gradient update, [27] minimises the entropy of the marginal distribution computed from several augmented versions of a single test sample. SNA eliminates the need for backpropagation by focusing on batch normalization-specific statistics (mean and variance) and only requires forward steps. During the inference in [11], the batch normalisation statistics are updated with x^t. On the other hand, [15] updates statistics with x^t considering x^s as a prior.

FUSION, the proposed methodology, performs batch statistics adaptation as well. FUSION is a generic technique that combines source and target batch normalisation layers via a weighting factor that favors the source or the target end. The stain fluctuation between x^s and x^t is likewise connected to the weighting factor. FUSION outperforms the conventional SNA techniques because of its generic nature. The domain difference between x^s and x^t is represented in this study by stain variation; thus, the terms *"model adaptation"* and *"stain adaptation"* are used interchangeable.

2   FUSION

Batch normalisation (BN) is built into current Convolutional Neural Networks (CNNs) for stable and faster training. The second-order batch statistics (μ, σ) and two learnable parameters for scaling and shifting are used by BN layers to normalise the features of each batch.

$$\hat{g}(x_B^s) = \gamma^s \frac{g(x_B^s) - \mathbb{E}[g(x_B^s)]}{\sqrt{(Var[g(x_B^s)] + \epsilon}} + \alpha^s, \tag{1}$$

where, x_B^s is a training batch from x^s, and $g(x_B^s)$ are activations. The moving averages of μ and σ are also kept during the training phase to use during inference.

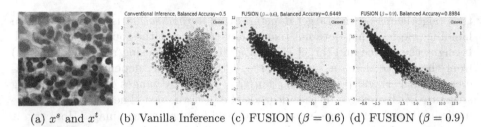

(a) x^s and x^t (b) Vanilla Inference (c) FUSION ($\beta = 0.6$) (d) FUSION ($\beta = 0.9$)

Fig. 1. The vanilla inference performs poorly on the target due to considerable stain variation between x^s (source) and x^t (target). FUSION may bypass this constraint by combining the x^s and x^t normalisation statistics, resulting in significant performance gains.

These statistics can differ greatly between two centres (source and target) in the case of stain variation (Fig. 1). As a result, in an *inter-center training-testing* system, the training statistics may lead to poor performance [11] during inference on test set from a different center. However, we believe that the batch's (x_B) requirement to represent the target data distribution $p_{test}(x)$ is important for classification because a batch may not contain samples from all classes. Accordingly, for inference on x^t, Eq. 1 is modified as:

$$\hat{g}(x_B^t) = \gamma^s \frac{g(x_B^t) - \mathbb{E}[g(x_B^t)]}{\sqrt{(Var[g(x_B^t)] + \epsilon}} + \alpha^s; \qquad x_B^t \in p_{test}(x) \qquad (2)$$

Even though each test batch represents the test data distribution, it is not sufficient to describe the statistics of the complete test set. Although all of the test samples can be pooled into a single batch, this has computational limits. We describe a computationally efficient strategy for leveraging the statistics of the complete test set during inference. The Eq. 2 is utilized for inference in the *first step*, and running estimates of μ and σ produced from the test-batches are kept at the same time (similar to training). In the *second step*, the collected running averages are utilized in the final inference, altering Eq. 2 to:

$$\hat{g}(x_B^t) = \gamma^s \frac{g(x_B^t) - \mathbb{M}^t}{\sqrt{(\mathbb{V}^t + \epsilon}} + \alpha^s, \qquad (3)$$

where, $\mathbb{M}^t$ and $\mathbb{V}^t$ represents the running statistics of mean and variance, respectively in the *first step*.

Fusing the Batch Normalizations. Only the statistics of the target are considered in Eq. 3, completely discarding the statistics of the source. Excessive perturbations of the source statistics may cause the performance to deteriorate. As a result, optimal performance can be achieved by combining both statistics. To this purpose, Eq. 3 is updated as follows (Fig. 2):

$$\hat{g}(x_B^t) = \gamma^s \left(\beta \left[\frac{g(x_B^t) - \mathbb{M}^t}{\sqrt{(\mathbb{V}^t + \epsilon}} \right] + (1 - \beta) \left[\frac{g(x_B^t) - \mathbb{M}^s}{\sqrt{(\mathbb{V}^s + \epsilon}} \right] \right) + \alpha^s, \qquad (4)$$

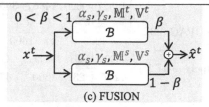

(a) Conventional Inference   (b) Modified Statistics Adaptation

$\mathcal{B}$: Batch Normalization

(c) FUSION

Fig. 2. Conventional inference (a) uses the running statistics of the source ($\mathbb{M}^s, \mathbb{V}^s$) for inference on target x^t. The modified statistics adaptation approach (b) replaces it with the target statistics, $\mathbb{M}^t, \mathbb{V}^t$. Unlike prior approaches to statistics adaptation, which only considered per-batch statistics, this method considers the complete target set. FUSION (c) is a generalized statistics adaptation approach that fuses the source and target statistics with a weighting factor β, reducing to (a) and (b) for $\beta = 0$ and 1, respectively.

$\beta \in [0, 1]$ is a hyperparameter, $\mathbb{M}^s$ and $\mathbb{V}^s$ represents moving averages for source. Equation 4 is a generalized test time statistic normalization adaptation. For $\beta = 1$, it reduces to Eq. 3, considering only the target statistics. While for $\beta = 0$, it focuses only on the source statistics. With $0 < \beta < 1$ as the weighting parameter, Eq. 4 exploits both the domains.

3   Experiments

Applications: FUSION is evaluated on *classification* and *segmentation* tasks using Camelyon -17 and TUPAC datasets. Classification and segmentation are performed on the five-center Camelyon-17 [1] dataset, but only classification is performed on the three-center TUPAC dataset. Sample images and a detailed dataset description may be found in Table 1 and Fig. 3, respectively. Datasets vary widely among locations, making it probable that a model developed for one location may perform badly when applied to other locations. The effectiveness of FUSION in preventing such performance decline is put to the test.

Implementation: ResNet-18 [5] and EfficientNet-B0 [19] are trained for classification with and without Train-time augmentations (TrTAug). Rotation-based augmentation is the default, whereas TrTAug also includes color-based augmentations such as hue, saturation, and value (HSV) variations to induce stain invariance. The Feature Pyramid Network (FPN) [7] with ResNet-34 [19] as an encoder is utilized for the segmentation task. Each network is trained with SGDM optimizer for 55 epochs and a step LR scheduler of 0.1 decay factor. The

Table 1. Multi centers (sources) datasets are used for classification and segmentation to analyze inter-center stain adaptation with FUSION. For each dataset, highlighted center is considered as source and remaining as the targets.

Application	Dataset (Centers)						Size
Classification	Camelyon-17 [1]	C0	C1	C2	C3	C4	256 × 256
		13527	9690	14867	30517	26967	
	TUPAC [23]	C0		C1		C2	128 × 128
		4260		1600		1344	
Segmentation	Camelyon-17 [1]	C0	C1	C2	C3	C4	1024 × 1024
		237	231	300	437	685	

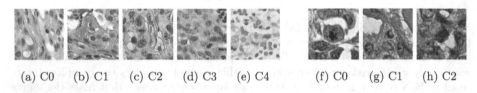

(a) C0   (b) C1   (c) C2   (d) C3   (e) C4   (f) C0   (g) C1   (h) C2

Fig. 3. Sample images from different centers of Camelyon17 (a-e) and TUPAC (f-h), highlighting the inter-center stain variation.

learning rate was set to .001, the batch size to 64, and the weight decay to .01. For FUSION, optimal β is selected through grid search from $[0.6, 0.7, 0.8, 0.9]$.

Baselines: We exhibit the influence of test-time augmentations (TTAug), stain normalisation [8, 22], and normalisation statistics adaption [11, 15] in addition to vanilla inference. The [11] and [15] are represented as vanilla batch normalization adaptation (Vanilla-BNA) and sample based BNA (SB-BNA), respectively. The influence of Eq. 2 in conjunction with Vanilla-BNA (Vanilla-BNA + Eq. 2) is also investigated. FUSION-full or FUSION with $\beta = 1$ (Eq. 3) is also analyzed. N is set to 20 for SB-BNA, as suggested in [15].

4   Results

FUSION Gives the Best Classification and Segmentation Results in Varied Test Setups. Table 2 indicates that conventional inference results in poor performance. This is attributed to drastic stain variation between C2 (source) and other centers (targets), implying the necessity of stain adaptation. FUSION outperforms the other approaches when it comes to classification on Camelyon-17 under stain-variation condition. The performance improvement for C0, C1, C3, and C4 in terms of balanced accuracy for EfficientNet-B0 (without TrTAug) is 43.26%, 22.89%, 38.85%, and 27.60%, respectively. Introducing TrTAug, as expected, introduces stain invariance in the network, resulting in increased performance across target centers. FUSION outperforms other methods and provides highest increment over the baseline with TrTAug. Similarly,

Table 2. *Balanced accuracy* for *classification* on Camelyon-17 [1] with C2 as the source. Best results are highlighted in bold. Maximum increment and minimum decrement are represented in blue and red, respectively. Vanilla BNA has poor performance without Eq. 2. Hence, for classification, each batch at inference time must represents the test set $(x_B^t \in p_{test}(x))$.

Source	Target			
C2	C0	C1	C3	C4
EfficientNet-B0 [19]	50.34 ± 0.57	49.43 ± 0.64	49.90 ± 0.13	63.99 ± 2.67
+ TTAug	51.07 ± 1.27 (↑ 0.73)	43.37 ± 1.08 (↓ 6.06)	44.73 ± 0.94 (↓ 5.17)	54.06 ± 0.71 (↓ 9.93)
+ Macenko [8]	50.21 ± 0.19 (↓ 0.13)	50.32 ± 2.84 (↑ 0.89)	49.83 ± 3.03 (↓ 0.07)	49.95 ± 0.05 (↓ 14.04)
+ Vahadane [22]	84.65 ± 2.20 (↑ 30.31)	76.29 ± 3.26 (↑ 26.86)	76.67 ± 3.13 (↑ 26.77)	46.08 ± 4.35 (↓ 17.91)
+ Vanilla BNA [11]	53.47 ± 0.19 (↑ 3.13)	53.27 ± 0.24 (↑ 3.84)	53.47 ± 0.24 (↑ 3.57)	51.55 ± 0.17 (↓ 12.44)
+ Vanilla BNA [11] +(2)	92.43 ± 0.24 (↑ 42.09)	72.11 ± 0.40 (↑ 22.69)	87.29 ± 0.52 (↑ 37.39)	91.44 ± 0.73 (↑ 27.45)
+ SB-BNA [15]	67.94 ± 5.33 (↑ 17.6)	62.32 ± 2.35 (↑ 12.98)	57.52 ± 2.57 (↑ 7.62)	85.67 ± 1.46 (↑ 21.68)
+ FUSION-full (3)	92.61 ± 0.12 (↑ 42.27)	71.27 ± 0.37 (↑ 21.84)	87.59 ± 0.31 (↑ 37.69)	92.14 ± 0.76 (↑ 28.15)
+ FUSION (4)	**93.59 ± 0.13** (↑ 43.25)	**72.32 ± 0.54** (↑ 22.89)	**88.75 ± 0.32** (↑ 38.85)	**91.59 ± 0.90** (↑ 27.60)
+ TrTAug	93.32 ± 1.11	75.28 ± 3.66	85.28 ± 1.51	82.01 ± 3.12
+ TTAug	92.25 ± 0.99 (↓ 1.07)	75.03 ± 2.66 (↓ 0.25)	82.57 ± 0.88 (↓ 2.71)	79.02 ± 2.42 (↓ 2.99)
+ Macenko [8]	92.72 ± 1.12 (↓ 0.60)	82.98 ± 3.09 (↑ 7.70)	83.80 ± 3.24 (↓ 1.48)	83.75 ± 1.83 (↑ 1.74)
+ Vahadane [22]	93.41 ± 0.78 (↑ 0.09)	80.53 ± 1.73 (↑ 5.25)	88.79 ± 0.59 (↑ 3.21)	71.37 ± 1.18 (↓ 10.64)
+ Vanilla BNA [11]	53.78 ± 0.18 (↓ 39.54)	53.03 ± 0.23 (↓ 22.25)	53.75 ± 0.13 (↓ 31.53)	51.86 ± 0.20 (↓ 30.15)
+ Vanilla BNA [11] +(2)	94.29 ± 0.16 (↑ 0.97)	83.41 ± 1.08 (↑ 8.13)	91.67 ± 0.22 (↑ 6.39)	94.81 ± 0.22 (↑ 12.8)
+ SB-BNA [15]	93.96 ± 0.79 (↑ 0.64)	83.87 ± 1.01 (↑ 8.59)	90.45 ± 0.74 (↑ 5.17)	93.22 ± 0.72 (↑ 11.21)
+ FUSION-full (3)	94.69 ± 0.22 (↑ 1.37)	83.83 ± 1.22 (↑ 8.55)	91.93 ± 0.29 (↑ 6.65)	95.38 ± 0.18 (↑ 13.38)
+ FUSION (4).	**95.28 ± 0.33** (↑ 1.96)	**88.68 ± 0.87** (↑ 13.4)	**93.02 ± 0.34** (↑ 7.74)	**95.58 ± 0.12** (↑ 13.57)

Table 3. *Dice Score* for *segmentation* on Camelyon-17 [1] with C1 as the source. FUSION has maximum increment for C0, C2, C3, and least decrement for C4. Due to dense predictions, Eq. 2 is inherently satisfied, making its combination with Vanilla BNA redundant.

Source	Target			
C1	C0	C2	C3	C4
ResNet-34 [5]	62.75 ± 7.95	53.49 ± 14.00	58.23 ± 8.87	50.52 + 12.61
+ TTA	56.23 ± 6.65 (↓ 6.52)	47.30 ± 8.87 (↓ 6.19)	50.46 ± 7.01 (↓ 7.77)	39.60 ± 8.91 (↓ 10.92)
+ Vanilla BNA [11]	70.56 ± 4.36 (↑ 7.81)	55.70 ± 3.11 (↑ 2.21)	59.80 ± 3.44 (↑ 1.57)	38.05± 4.80 (↓ 10.92)
+ Vanilla BNA [11]+(2)	70.47 ± 4.23 (↑ 7.72)	56.08 ± 3.57 (↑ 2.59)	59.85 ± 3.57 (↑ 1.62)	38.26 ± 4.59 (↓ 12.26)
+ FUSION-full (3)	70.36 ± 5.00 (↑ 7.61)	57.08 ± 6.17 (↑ 3.59)	60.62 ± 3.79 (↑ 2.39)	40.95 ± 6.73 (↓ 9.57)
+ FUSION (4)	**71.30 ± 5.95** (↑ 8.55)	**60.68 ± 9.66** (↑ 7.19)	**63.97 ± 4.81** (↑ 5.74)	**48.06 ± 7.40** (↓ 2.46)

FUSION is the optimal technique for TUPAC dataset, as shown in Fig. 4. It performs on par to FUSION-full in a few cases while superior in others.

FUSION is the best performing approach for segmentation (Table 3), with the highest gain of 8.55% in terms of dice score for C0. Performance on C4 shows decrease as seen in classification as well. However, FUSION results in least decrement in comparison to other methods. As can be seen, FUSION-full underperforms FUSION in most cases, implying that full adaptation is inefficient and that an optimal combination of x^s and x^t statistics is required.

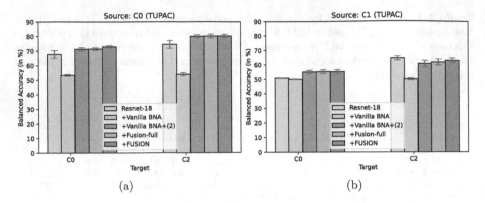

Fig. 4. *Balanced accuracy* for *classification* on TUPAC [23] with source C0 (a) and C1 (b). Vanilla BNA, like Camelyon-17 in Table 2, performs worse than vanilla inference in the absence of Eq. 2. In certain circumstances, FUSION-full and FUSION are comparable to Vanilla BNA, while in some cases, they are superior. Also, with C1 as the source and C2 as the target, FUSION produces the least decrement, proving the validity of statistics merging.

In Vanilla BNA, Each Batch Must be Representative of the Test Data Distribution. In the absence of Eq. 2, vanilla BNA performs worse than vanilla inference. Batch normalisation statistics update without Eq. 2 does not examine the impact of all classes in the dataset, resulting in poor performance. As pixels in a single image belong to many classes, this is satisfied by default for segmentation. Also, because Eq. 3 considers updates due to all test samples (statistical running averages) throughout inference, Eq. 2 is not required separately.

FUSION Restores Activation Distribution Shifts Caused by Stain Differences. BatchNorm normalizes the distribution of activations over a minibatch during training. By setting the first two moments (mean and variance) of the distribution of each activation to be zero and one, it tries to correct for the changes in distribution of activations of a layer in the network caused by update of parameters of the previous layers. This distributional changes in layer inputs after every minibatch is known as internal covariate shift.

By default, a running mean and variance of activations for all minibatches is tracked in BatchNorm layer to be used during test time such that the activation mean is 0 and variance is 1. But when the model is used for prediction on an unseen and different stain, the distribution of activations changes and behaves differently from training time. The existing BatchNorm statistics, since based on different distribution, can't correct for the new distribution shift. This causes the new mean and variance of activations after BatchNorm layer differ from the intended values of mean 0 and variance 1.

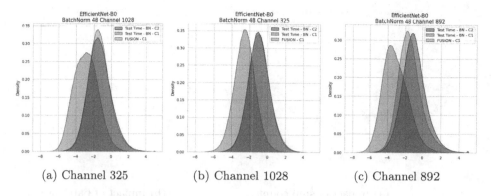

(a) Channel 325   (b) Channel 1028   (c) Channel 892

Fig. 5. (a-c) Due to a stain difference between the source and target, the feature map distribution shifts with EfficientNet-B0 after the BatchNorm layer. After the last BatchNorm layer, the activations are recovered from the EfficientNet-B0 model. C2 was used to train the model at the beginning. The blue line displays the density plot of a channel using source validation data, while the red line represents the shift owing to the target's different stain. FUSION aims to match the training distribution by correcting the covariate shift (green line). (Color figure online)

Let μ_c^l be the running mean and σ_c^l be the variance recorded during training at layer l for channel c. With stain variation between x^s and x^t:

$$\mathbb{E}[g_c^l(x^s)] \neq \mathbb{E}[g_c^l(x^t)] \tag{5}$$

$$\mathsf{Var}[g_c^l(x^s)] \neq \mathsf{Var}[g_c^l(x^t)] \tag{6}$$

After applying BatchNorm the above equations can be written as:

$$\mathbb{E}[\mathsf{BN}(g_c^l(x^s))] \neq \mathbb{E}[\mathsf{BN}(g_c^l(x^t))] \tag{7}$$

$$\mathsf{Var}[\mathsf{BN}(g_c^l(x^s))] \neq \mathsf{Var}[\mathsf{BN}(g_c^l(x^t))] \tag{8}$$

FUSION is designed to restore these activations in a way that:

$$\mathbb{E}[\mathsf{BN}(g_c^l(x^s))] \approx \mathbb{E}[\mathsf{FUSION}(g_c^l(x^t))] \tag{9}$$

$$\mathsf{Var}[\mathsf{BN}(g_c^l(x^s))] \approx \mathsf{Var}[\mathsf{FUSION}(g_c^l(x^t))] \tag{10}$$

where FUSION can be calculated using Eq. 4.

The shift in activation distribution (Fig. 5a, 5b, 5c) that explains the performance decrease with stain variation is correlated to the stain difference. With similar stains, the distribution shift is minor, and source batch normalisation statistics are sufficient. With substantial stain variation, however, the shift is significant with major reduction in test performance, and FUSION corrects it back to the source. As a result, FUSION's achieves higher performance improvement even on greater stain difference (higher distribution shift).

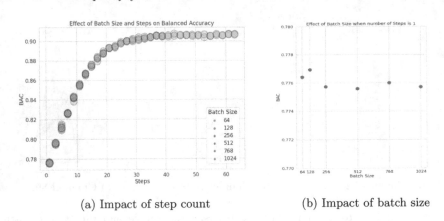

(a) Impact of step count   (b) Impact of batch size

Fig. 6. Impact of (a) steps count and (b) batch size on FUSION's performance in terms of balanced accuracy. The model is trained on *C2* with varying batch size and steps count, and mean balanced accuracy is reported on remaining test centers (centers *C0, C1, C3, C4*). (a) FUSION's performance is related to BatchNorm statistics which correlates to batch size and number of iterations/steps. Increased steps counts improves the performance irrespective of batch sizes. As moving average of mean and variance is used for updating BatchNorm statistics, large number of steps are required to overcome the affect of momentum. (b) When applying FUSION, batch size has negligible effect on performance when step count is fixed (set to 1). The difference in performance is not dependent on batch size. The similar observations can be made in (a) where points of different batch sizes overlap each other for same step count.

The Performance of FUSION Correlates to the Number of Steps. Batch size and step counts have an impact on batch normalisation statistic calculations. The performance of updating statistics from the target does not improve as batch size is increased. The amount of steps taken to generate the new statistics, on the other hand, is critical. As shown in Fig. 6a, the performance of different batch sizes converges roughly in the same way as step count increases. This could be related to the reduced influence of momentum while using step counts for batch statistics upgradation. Further evidence of importance of step count vs batch size can be seen in Fig. 6b where for the same step count, batch size has no correlation with performance.

FUSION is Generic and Provides Flexibility During Inference. Full adaptation is neither generic nor efficient, as it may result in substantial statistical deviation. Similarly, poor performance without adaptation is due to incompatibility of the source statistics with the target. FUSION's generalizability, which allows it to focus on any source or target via the weighting factor β, is a key advantage. FUSION provides more flexibility during inference because of its entirely unsupervised and test-time adaptive nature.

5   Conclusion

FUSION was proposed for entirely unsupervised test-time stain adaptation, and its performance was evaluated in various scenarios. Due to its unique qualities that combine source and target batch normalisation statistics, FUSION outperforms similar approaches in the literature. The rigorous testing with various datasets, applications, and architectures reveals that FUSION provides optimal performance. The characteristics of FUSION, such as its impact on countering the covariate shift, also justify the obtained higher performance. Analysis of batch size and the step counts impact on FUSION during inference highlights its dependency on step counts instead of batch size. However, as suspected, batch size of one is not sufficient for FUSION to work as batch normalization statistics calculation is required. Apart from its performance, FUSION provides inference time flexibility to steer it towards the source or target through a single hyperparameter. Hence, it allows no adaptation, full adaptation, or partial adaptation.

References

1. Bándi, P., et al.: From detection of individual metastases to classification of lymph node status at the patient level: the camelyon17 challenge. IEEE Trans. Med. Imaging **38**(2), 550–560 (2019)
2. Ganin, Y., Lempitsky, V.S.: Unsupervised domain adaptation by backpropagation. arXiv abs/1409.7495 (2015)
3. Gehlot, S., Gupta, A.: Self-supervision based dual-transformation learning for stain normalization, classification and segmentation. In: Lian, C., Cao, X., Rekik, I., Xu, X., Yan, P. (eds.) MLMI 2021. LNCS, vol. 12966, pp. 477–486. Springer, Cham (2021). https://doi.org/10.1007/978-3-030-87589-3_49
4. Gupta, A., et al.: GCTI-SN: geometry-inspired chemical and tissue invariant stain normalization of microscopic medical images. Med. Image Anal. **65**, 101788 (2020)
5. He, K., Zhang, X., Ren, S., Sun, J.: Deep residual learning for image recognition. CoRR abs/1512.03385 (2015)
6. Kothari, S., et al.: Automatic batch-invariant color segmentation of histological cancer images. In: From Nano to Macro, 2011 IEEE International Symposium on Biomedical Imaging, pp. 657–660 (2011)
7. Lin, T., Dollár, P., Girshick, R.B., He, K., Hariharan, B., Belongie, S.J.: Feature pyramid networks for object detection. CoRR abs/1612.03144 (2016)
8. Macenko, M., et al.: A method for normalizing histology slides for quantitative analysis. In: ISBI, pp. 1107–1110 (2009)
9. Magee, D., et al.: Colour normalisation in digital histopathology images. In: Proceedings of the Optical Tissue Image analysis in Microscopy, Histopathology and Endoscopy (MICCAI Workshop), vol. 100 (2009)
10. McCann, M.T., Majumdar, J., Peng, C., Castro, C.A., Kovačević, J.: Algorithm and benchmark dataset for stain separation in histology images. In: 2014 IEEE International Conference on Image Processing (ICIP), pp. 3953–3957 (2014)
11. Nado, Z., Padhy, S., Sculley, D., D'Amour, A., Lakshminarayanan, B., Snoek, J.: Evaluating prediction-time batch normalization for robustness under covariate shift. CoRR abs/2006.10963 (2020). https://arxiv.org/abs/2006.10963

12. Reinhard, E., Adhikhmin, M., Gooch, B., Shirley, P.: Color transfer between images. IEEE Comput. Graph. Appl. **5**, 34–41 (2001)
13. Ruderman, D.L., Cronin, T.W., Chiao, C.C.: Statistics of cone responses to natural images: implications for visual coding. JOSA A **15**(8), 2036–2045 (1998)
14. Ruifrok, A., Ruifrok, D.: Quantification of histochemical staining by color deconvolution. Anal. Quant. Cytol. Histol./Int. Acad. Cytol. [and] Am. Soc. Cytol. **23**(4), 291–299 (2001)
15. Schneider, S., Rusak, E., Eck, L., Bringmann, O., Brendel, W., Bethge, M.: Improving robustness against common corruptions by covariate shift adaptation. CoRR abs/2006.16971 (2020). https://arxiv.org/abs/2006.16971
16. Shaban, M.T., Baur, C., Navab, N., Albarqouni, S.: StainGAN: stain style transfer for digital histological images. arXiv preprint arXiv:1804.01601 (2018)
17. Sun, Y., Wang, X., Liu, Z., Miller, J., Efros, A.A., Hardt, M.: Test-time training for out-of-distribution generalization. CoRR abs/1909.13231 (2019)
18. Abe, T., Murakami, Y., Yamaguchi, M.: Color correction of pathological images based on dye amount quantification. Opt. Rev. **12**(4), 293–300 (2005)
19. Tan, M., Le, Q.V.: EfficientNet: rethinking model scaling for convolutional neural networks. CoRR abs/1905.11946 (2019)
20. Tellez, D., et al.: Quantifying the effects of data augmentation and stain color normalization in convolutional neural networks for computational pathology. Med. Image Anal. **58**, 101544 (2019)
21. Tzeng, E., Hoffman, J., Darrell, T., Saenko, K.: Simultaneous deep transfer across domains and tasks. CoRR abs/1510.02192 (2015)
22. Vahadane, A., et al.: Structure-preserved color normalization for histological images. In: 2015 IEEE 12th International Symposium on Biomedical Imaging (ISBI), pp. 1012–1015 (2015)
23. Veta, M., et al.: Predicting breast tumor proliferation from whole-slide images: The TUPAC16 challenge. Med. Image Anal. **54**, 111–121 (2019)
24. Wang, D., Shelhamer, E., Liu, S., Olshausen, B., Darrell, T.: Tent: fully test-time adaptation by entropy minimization. In: International Conference on Learning Representations (2021). https://openreview.net/forum?id=uXl3bZLkr3c
25. Zanjani, F.G., Zinger, S., Bejnordi, B.E., van der Laak, J.A.W.M.: Histopathology stain-color normalization using deep generative models. In: 1st Conference on Medical Imaging with Deep Learning (MIDL 2018), pp. 1–11 (2018)
26. Zanjani, F.G., Zinger, S., Bejnordi, B.E., van der Laak, J.A.W.M., de With, P.H.N.: Stain normalization of histopathology images using generative adversarial networks. In: 2018 IEEE 15th International Symposium on Biomedical Imaging (ISBI 2018), pp. 573–577 (2018). https://doi.org/10.1109/ISBI.2018.8363641
27. Zhang, M., Levine, S., Finn, C.: MEMO: test time robustness via adaptation and augmentation. CoRR abs/2110.09506 (2021). https://arxiv.org/abs/2110.09506

Relieving Pixel-Wise Labeling Effort for Pathology Image Segmentation with Self-training

Romain Mormont[1]($\boxtimes$) ⓘ, Mehdi Testouri[2], Raphaël Marée[1] ⓘ,
and Pierre Geurts[1] ⓘ

[1] University of Liège, Liège, Belgium
{r.mormont,raphael.maree,p.geurts}@uliege.be
[2] University of Luxembourg, Luxembourg City, Luxembourg
mehdi.testouri@uni.lu

Abstract. Data scarcity is a common issue when training deep learning models for digital pathology, as large exhaustively-annotated image datasets are difficult to obtain. In this paper, we propose a self-training based approach that can exploit both (few) exhaustively annotated images and (very) sparsely-annotated images to improve the training of deep learning models for image segmentation tasks. The approach is evaluated on three public and one in-house dataset, representing a diverse set of segmentation tasks in digital pathology. The experimental results show that self-training allows to bring significant model improvement by incorporating sparsely annotated images and proves to be a good strategy to relieve labeling effort in the digital pathology domain.

Keywords: Deep learning · Image segmentation · Self-training · Data scarcity · Digital pathology

1 Introduction

Computational pathology is on the brink of revolutionizing traditional pathology workflows by providing artificial intelligence-based tools that will help pathologists making their work faster and more accurate. However, the development of such tools is hindered by data scarcity as pathology training datasets are not as numerous and large as in other fields of research. This is particularly true for image segmentation as per-pixel delineation of objects is one of the most time-consuming types of annotation there is.

In this work, we investigate a method for training a binary image segmentation model in a context where the segmentation ground truth is sparse. More precisely, we focus on a setup where the training set is composed of an exhaustively labeled subset $\mathcal{D}_l$ and a sparsely labeled subset $\mathcal{D}_s$. In particular, the images in $\mathcal{D}_l$ come with exhaustively labeled segmentation masks (*i.e.* pixels

---

Supplementary Information The online version contains supplementary material available at https://doi.org/10.1007/978-3-031-25082-8_39.

of all objects of interest have been labeled as positive) whereas in $\mathcal{D}_s$, some (unknown) pixels belonging to objects of interest have not been labeled, hence the sparsity. Typically, image segmentation methods require that the training images come with exhaustive pixel-wise labels. In our setup, they would allow us to only use $\mathcal{D}_l$ and force us to ignore $\mathcal{D}_s$. We believe that it is possible to include the sparsely labeled images as well, and hopefully improve the performance over using only $\mathcal{D}_l$. One way of achieving this would be to somewhat "complete" the sparse segmentation masks and make them exhaustively labeled. Generating pseudo-labels is precisely what self-training approaches do, therefore, we devise an automated self-training workflow to exploit both sets.

Our self-training workflow consists of two separate phases. During the "*warm-up*" phase, we train a U-Net [23] model in a classical supervised manner on $\mathcal{D}_l$ for a few epochs. Then, during the "*self-training*" phase (sketched in Fig. 1), we repeat the following process for an arbitrary number of epochs: pseudo labels are generated for the unlabeled pixels from images in $\mathcal{D}_s$ using the currently trained model and the pseudo-labeled images are included in the training set for the next epoch. To control the impact of model uncertainty on pseudo-labeled pixels, we furthermore study different weighting schemes to tune their contributions to the training loss. We use binary ("hard") pseudo-labels and propose an auto-calibration approach for generating them. Experiments are carried out on three exhaustively labeled public datasets, namely MoNuSeg [12], GlaS [26] and SegPC-2021 [7], on which sparsity is artificially enforced for evaluation purpose. Through these experiments, we investigate the interest of our method and the impact of its hyperparameters in different scarcity conditions and in comparison with different baselines. In a second stage, we apply our method on an actual sparsely labeled dataset for cytology diagnosis.

Our main contributions and conclusions are as follows: **(1)** We design a self-training pipeline for binary segmentation to handle datasets composed on both exhaustively and sparsely annotated images (Sect. 3). **(2)** We show on three public datasets that this self-training approach is beneficial as, even in significant scarcity conditions, it improves over using only exhaustively labeled data (see Sect. 6.1). **(3)** We confirm the interest of the approach in a real-world scenario, where a significant improvement of performance is achieved by exploiting sparse annotations (see Sect. 6.3). **(4)** We show that, at fixed annotation budget, it is not necessarily better to focus the annotation effort on exhaustive labels rather than sparse labels (see Sect. 6.2).

2   Related Works

Self-training is not a new family of methods and has been applied more recently in the context of deep learning. A classical approach of a self-training process is the teacher-student paradigm where the teacher model is used to generate pseudo-labels that will be used to further train the student model. The most straightforward approach consists in using a single model alternatively as a teacher for pseudo-labeling and then as a student for the training step [32,33],

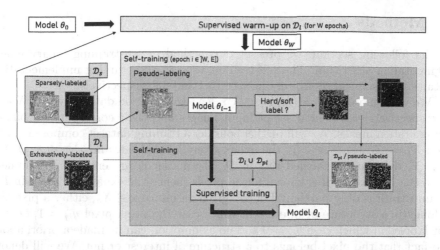

Fig. 1. Our approach illustrated. The model is first warmed-up on the exhaustively-labeled data then further trained by self-training on the combined sparsely- and exhaustively-labeled sets. A more formal definition is provided in Algorithm 1

but some approaches have used more complex interactions between the teacher and the student [22,30]. A critical design choice for a self-training algorithm is how it will make use of the pseudo-labels. Label uncertainty can be used to filter out the pseudo-labels not deemed reliable enough [6,15]. Coupled with aggressive data augmentation, consistency between pseudo-labels and model predictions is often formulated as a training loss [27,30,32,34].

It is not surprising that self-training has also been applied to medical image tasks to combat data scarcity [21,29] and to computational pathology in particular. Most works in this domain currently deal with image classification tasks [8,11,20,25,28] but detection and segmentation have also been explored. Jiayun Li *et al.* [17] combine weakly-supervised learning and self-training to predict per-pixel Gleason score across entire WSIs. Jiahui Li *et al.* [16] build a signet ring cell detector by training two models on pseudo-labels generated by one another. Segmentation architectures have also been trained with self-training for cardiac MRI [1] and COVID 19-related lung infection [5] segmentation. Bokhorst *et al.* [4] segment tissues from colorectal cancer patients into 13 classes representing different types of tissues. Their setup is similar to ours as their dataset is composed of both exhaustively- and sparsely-labeled images. They apply a weight map to tune and balance the contribution of individual pixels to the loss but they ignore unlabeled pixels entirely during training.

Beyond self-training, there exist methods, implemented in software such as QuPath [2], Ilastik [3] or Cytomine [19] and based on traditional computer vision or machine learning, that allow users to interactively complete a partial hand-drawn annotation of a given image. Among these methods are, for instance, Graph Cut [13] and GrabCut [24], or more recently DEXTRE [18] and NuClick [10]. While the self-training approach explored in this paper can certainly be exploited in an interactive mode, this question will be left as future work.

3   Methods

In the following section, we present our method, a self-training binary image segmentation algorithm. The self-training aspects and training implementation details are discussed separately in Sects. 3.1 and 3.2 respectively.

We will denote by $\mathcal{D} = (X, Y) \subset \mathcal{X} \times \mathcal{Y}$ a segmentation dataset, where X and Y respectively represent a set of input images and their corresponding binary segmentation masks. We will further consider a training dataset composed of two sets: $\mathcal{D}_l = (X_l, Y_l) \subset \mathcal{X} \times \mathcal{Y}$, the exhaustively labeled set, and $\mathcal{D}_s = (X_s, Y_s) \subset \mathcal{X} \times \mathcal{Y}$, the sparsely labeled set. In $\mathcal{D}_l$, the masks Y_l are entirely determined, since the ground truth is known for all pixels (hence the exhaustiveness). In $\mathcal{D}_s$, ground truth is only partially known: given an image $\mathbf{x} \in X_s$, either a pixel x_{ij} belongs to a structure of interest in which case the mask pixel $y_{ij} = 1$, or it is not labeled in which case $y_{ij} = 0$ and no assumption can be made a priori about the fact that the pixel belongs to a structure of interest or not. We will denote $n_l = |\mathcal{D}_l|$ and $n_s = |\mathcal{D}_s|$ the sizes of the sets. The total number of training images for a dataset will be denoted $n = n_l + n_s$.

3.1   Self-training

Our self-training algorithm is described in Algorithm 1 (and depicted in Fig. 1). It features a warm-up phase during which the model is classically trained on the set $\mathcal{D}_l$ (training implementation details are given in Sect. 3.2). The number of warm-up epochs $W > 0$ is fixed so that the model is able to converge on the labeled data. The warmed-up model is used as starting point for the self-training phase. Each self-training round e starts by pseudo-labeling $\mathcal{D}_s$ with the model θ_{e-1}. For an image $\mathbf{x} \in X_s$, the pseudo-label assigned to pixel x_{ij} is given by:

$$y_{ij}^{(pl)} = \begin{cases} 1, & \text{if } y_{ij} = 1 \\ g(\hat{y}_{ij}), & \text{otherwise} \end{cases} \tag{1}$$

where $\hat{y}_{ij}$ is the sigmoid output of model θ_{e-1} for pixel (i, j) given $\mathbf{x}$ as input and g is a function for generating the pseudo-label from $\hat{y}_{ij}$ (see below). In other words, we preserve the expert ground truth as pseudo-labels when available and use the model predictions for unlabeled pixels (this is the Combine step from Algorithm 1). With this strategy, entirely unlabeled images can also be included in $\mathcal{D}_s$. Our algorithm uses a single model (*i.e.* teacher = student) which is not reset between self-training rounds.

Soft and Hard Pseudo-labels. We considered two different pseudo-labeling strategies, or two different g functions (see Eq. 1). Initially, we decided to simply take g to be the identity function $g(x) = x$ in which case the sigmoid output of the model was used as pseudo-label. This strategy is commonly called "*soft*" labeling. During the next self-training round, this soft label will be compared to the network prediction which can be seen as a form of consistency constraint similar

---

Algorithm 1: Our self-training approach. The `Train` operation trains the given model on the provided dataset according to the protocol explained in Section 3.2. The `Predict` operation produces segmentation masks for a set of input images using the model. The `Combine` operation combines ground truth masks and pseudo labels from the given sets as explained in Section 3.1.

Data: The exhaustively- and sparsely labeled sets $\mathcal{D}_l$ and $\mathcal{D}_s$, a segmentation model θ_0, W and E respectively the number of warm up epochs and the total number of epochs.

Result: A self-trained segmentation model θ_E.

1 // *Warm up*
2 **for** $e \leftarrow 1$ **to** W **do**
3 $\quad | \quad \theta_e = \texttt{Train}(\theta_{e-1}, \mathcal{D}_l)$
4 **end**
5 **for** $e \leftarrow W+1$ **to** E **do**
6 $\quad |$ // *Pseudo labeling*
7 $\quad | \quad \hat{Y}_s = \texttt{Predict}(\theta_{e-1}, X_s)$
8 $\quad | \quad Y_{pl} = \texttt{Combine}(\hat{Y}_s, Y_s)$
9 $\quad | \quad \mathcal{D}_{pl} = (X_s, Y_{pl})$
10 $\quad |$ // *Self-training*
11 $\quad | \quad \theta_e = \texttt{Train}(\theta_{e-1}, \mathcal{D}_l \cup \mathcal{D}_{pl})$
12 **end**
13 **return** θ_E

---

to those in [14, 27, 30]. However, early experiments have shown that this approach causes training instabilities. Therefore, we investigated a second strategy where the sigmoid output is binarized using a threshold $T_e \in [0, 1]$:

$$g(x) = \begin{cases} 1, & \text{if } x > T_e \\ 0, & \text{otherwise} \end{cases} \tag{2}$$

where e is a self-training round. We call this strategy *"hard"* labeling as pseudo-labels are either 0 or 1. In addition to ensuring some sort of consistency between the pseudo-labels and the predictions, as in the *"soft"* approach, this threshold-ing also encourages the model to produce confident predictions (closer to 0 or 1). Because we want to avoid T_e to be an additional hyperparameter to tune, we propose an auto-calibration strategy based on the Dice score:

$$Dice_T(\mathbf{y}, \hat{\mathbf{y}}) = \frac{2 \times \sum_{i,j} \left[\mathbb{K}_{\hat{y}_{ij} \geq T} \times y_{ij} \right]}{\sum_{i,j} \mathbb{K}_{\hat{y}_{ij} \geq T} + \sum_{i,j} y_{ij}} \tag{3}$$

where T is the threshold applied to the model output to generate a binary prediction. The auto-calibration procedure selects T_e such that the Dice score in (3) is maximized for the images from an exhaustively labeled set $\mathcal{D}_a$:

$$T_e = \arg\max_T \sum_{(\mathbf{x},\mathbf{y}) \in \mathcal{D}_a} Dice_T(\mathbf{y}, h(\mathbf{x}; \theta_e)). \tag{4}$$

The ideal choice for $\mathcal{D}_a$ would be to use an external validation set but, in extreme data scarcity conditions, extracting such a validation set would penalize the performance of the algorithm by removing a significant amount of training data. Therefore, in this context, we consider using the training subset $\mathcal{D}_l$ as $\mathcal{D}_a$. This approach has the advantage of not requiring additional training data but also induces a risk of overfitting which might hurt generalization performance. We hope that the overfitting problem would be compensated by the improvement brought by hard labeling.

3.2 Training

In this section, we will provide more information about the `Train` procedure from Algorithm 1 which trains a model θ with a dataset $\mathcal{D}$. We use U-Net [23] as a segmentation architecture. We set the initial number of feature maps to 8 instead of 64 in the original article, with the rest of the network scaled accordingly. The main goal of this reduction of model capacity is to limit overfitting given the highly scarce data conditions explored in this work, but it would be worth exploring more complex architectures as future work.

The number of rounds W and E and the number of training iteration per round are chosen independently per dataset. Every training iteration, we build a mini-batch by sampling $B = 8$ images uniformly at random with replacement from $\mathcal{D}_l \cup \mathcal{D}_{pl}$ and by extracting one randomly located 512×512 patch and its corresponding mask from each of these images. The batch size was selected based on hardware memory constraints. We apply random data augmentation following best practices for machine learning in general and for self-training in particular [27,32]. We apply horizontal and vertical flips, color perturbation in the HED space [31] (bias and coefficient factors up to 2.5%), Gaussian noise (standard deviation up to 10%) and Gaussian blur (standard deviation up to 5%).

As a training loss, we average the per-pixel binary cross-entropy ℓ over all pixels of the batch, as defined in:

$$\ell(\hat{y}; y) = y \log \hat{y} + (1 - y) \log(1 - \hat{y}) \tag{5}$$

$$\mathcal{L} = -\frac{1}{B} \sum_{b=1}^{B} \frac{1}{|\mathbf{y}_b|} \sum_i \sum_j w_{ij,b} \ell(\hat{y}_{ij,b}; y_{ij,b}) \tag{6}$$

We multiply the per-pixel loss by a weight $w_{ij,b}$ for pixel (i,j) of the bth image of the batch in order to tune the contribution of this pixel to the loss (see below). We use Adam [9] as an optimizer with initial learning rate $\gamma = 0.001$ and default hyperparameters ($\beta_1 = 0.9$, $\beta_2 = 0.999$, no weight decay).

Weighting Schemes. Different strategies are evaluated for generating the per-pixel weight $w_{ij,b}$ in Eq. 6. For the sake of simplicity, we will drop the batch identifier b in the subsequent equations and denote this weight by w_{ij}. We introduced this weight term to have the possibility to tune the contribution of pseudo-labeled pixels when computing the loss. It is important to note that this weight only applies to pseudo-labeled pixels and, therefore, the ground truth pixels will always be attributed a weight of $w_{ij} = 1$. It is also important to note that the weight is inserted as a constant in our loss and the weight function is not differentiated during back-propagation.

We study four different weighting strategies each producing an intermediate weight w_{ij}^ξ where ξ is the strategy identifier. Because we want to avoid the amplitude of the loss and gradients to be impacted by the weighting strategy, we normalize it to obtain the final weight $w_{ij} = w_{ij}^\xi / \overline{w}^\xi$, where $\overline{w}^\xi$ is the average weight over all pixels of a patch. Our weighting strategies are as follows.

- "*constant*": weight w_{ij}^{cst} given by hyperparameter constant $C \in \mathbb{R}^+$,
- "*entropy*": weight w_{ij}^{ent} given by the entropy of the model prediction,
- "*consistency*": weight w_{ij}^{cty} based on model predictions spatial consistency,
- "*merged*": weight w_{ij}^{mgd} computed by combining the two previous strategies.

These four strategies are detailed further in Supplementary Section B.

4   Data

In this section, we describe the datasets we use to evaluate our method. It includes three public exhaustively labeled segmentation datasets: MoNuSeg [12], GlaS [26] and SegPC-2021 [7]. The datasets are described in Table 1 and illustrated in Supplementary Section D. MoNuSeg contains images of tissues coming from different organs, patients and hospitals where epithelial and stromal nuclei are annotated. Given the variety of sources, the images exhibit significant variations of staining and morphology. The density of annotations is also quite high compared to the other datasets. GlaS features images containing both benign tissues and malignant tissues with colonic carcinomas. The gland annotations vary greatly in shape and size. Originally, SegPC-2021 contains 3 classes: background, cytoplasm and nucleus. In this work, we merge the two latter classes as we focus on binary segmentation. One of the challenges of this dataset is the presence of non-plasma cells which should be ignored by the algorithm although they are very similar to plasma cells. Moreover, artefacts are present on the images (*e.g.* cracked scanner glass in foreground, scale reference or magnification written on the image).

Additionally, we use a dataset that actually motivated the development of our method, a sparsely labeled dataset for thyroid nodule malignancy assessment. Pathologists[1] sparsely annotated nuclei and cell aggregates with polygon annotations in 85 whole-slide images on Cytomine [19]. The training set is a set of 4742 crops, one for each polygon annotation. The test set is a set of 45 regions of interest (2000×2000 pixels each) with annotations highlighting structures of interest (nuclei features and architectural cell patterns) made by a computer science student. This dataset is illustrated in Supplementary Section E.

---

[1] Isabelle Salmon's team from Erasme Hospital, Brussels, Belgium.

Table 1. Summary statistics for the four datasets used in this work. An image is a region of interest extracted from a whole-slide image, except for the training set of Thyroid FNAB where it is a crop of an annotation.

Dataset	Training set		Test set	
	Images	Annots	Images	Annots
MoNuSeg	30	17362	14	11484
GlaS	85	763	80	781
SegPC-2021	298	1643	300	900
Thyroid FNAB	4742		45	307

5   Experimental Setup

In this section, we present context information for our experiments: what are our baselines, how we have simulated sparse datasets from the public datasets and what evaluation protocol we have applied.

Transforming the Datasets. In order to fit the sparsely labeled settings described in Sect. 3, we generate new datasets from SegPC-2021, GlaS and MoNuSeg. This generation is controlled by two parameters: n_l and ρ. The former is the number of images to keep in the exhaustively labeled set $\mathcal{D}_l$. These images are chosen at random without replacement in the original training set. The latter parameter ρ is the percentage of annotations to remove from the images to make the remaining images sparsely labeled.

Baselines. We compare our self-training approach to three baselines. The first one, referred to as "*upper*", consists in using the full dataset, without removing any ground truth (*i.e.* $|\mathcal{D}_l| = n$ and $|\mathcal{D}_s| = 0$). Since it has access to all the annotations, this baseline is expected to represent an upper bound for all other strategies.

The second baseline consists in using the sparsely-annotated set $\mathcal{D}_s$ as if it was exhaustively annotated ($\mathcal{D}_{train} = \mathcal{D}_l \cup \mathcal{D}_s$). This strategy makes sense especially for moderately sparse datasets. Indeed, convolutional layers (as in U-Net) are able to cope with a bit of label noise given that gradients are averaged over feature maps to update the model parameters. Therefore, a bit of noise in certain parts of the images can be compensated by the feedback of ground truth labels in other locations. This baseline will be referred to as "$\mathcal{D}_l \cup \mathcal{D}_s$".

The third and last baseline, referred to as "$\mathcal{D}_l$ *only*", consists in not using at all the sparsely-annotated images during the training process (*i.e.* $\mathcal{D}_{train} = \mathcal{D}_l$).

Our self-training approach is of practical interest if it outperforms the two latter baselines. Moreover, the closer to the "*upper*" baseline the better.

Evaluation. We have built an evaluation protocol in order to ensure a fair comparison between the different strategies and the baselines. Ultimately, all approaches produce a segmentation model θ that will be the one evaluated. As an evaluation metric, we use the Dice score introduced in Eq. 3 which requires a threshold T to turn the probability map produced by the model into a binary segmentation. To assess the performance of a model independently of the thresholding strategy, we pick T so that the Dice score is maximized on the test set. In other words, in order to determine the threshold, we apply the optimization procedure described in Eq. 4 where $\mathcal{D}_a$ is the test set $\mathcal{D}_{\text{test}}$. The Dice score resulting from this optimization will be referred to as "Dice*". Obviously, in a context of extreme data scarcity, it will not be possible to tune the threshold this way because of the lack of a sufficiently large validation or test set. We will leave the design of fully automatic strategies to tune such threshold as future work but we believe that visually tuning such threshold on new images would be feasible in interactive applications where the segmentation models would be mostly used to assist the labeling of new images.

Every experiment and hyperparameters combination we evaluate is run with ten random seeds to evaluate the variability. The seed affects the dataset sparsity (n_l and ρ), model initialization, mini-batch sampling and data augmentation. We report Dice average and standard deviation over these random seeds.

6 Results

6.1 Self-training Performance at Fixed n_l

In order to study how our self-training approach performs under different data scarcity conditions, we have generated several versions of our datasets by varying ρ with n_l fixed and have run the baselines and different hyperparameters combinations on the generated datasets. As discussed in Sect. 3.1, we have used hard labels exclusively. The detailed hyperparameter combinations used in this section are provided in Supplementary Section A.

Results are shown for all three datasets in Fig. 2. In general, self-training is always able to outperform significantly the "$\mathcal{D}_l$ only" and "$\mathcal{D}_l \cup \mathcal{D}_s$" baselines with a significantly reduced amount of sparse annotations (the exact value is dataset dependant, see below). Regarding the baselines, "upper" outperforms the two others. Moreover, using sparsely labeled images as if they were exhaustively labeled (i.e. $\mathcal{D}_l \cup \mathcal{D}_s$) appears not to be a good idea as it is outperformed by all self-training approaches and baselines in almost all scarcity conditions. The performance of this baseline increases as one adds more sparse annotations however and is able to catch up with the "$\mathcal{D}_l$ only" baseline in the lowest scarcity conditions validating the hypothesis presented in Sect. 5.

MoNuSeg. On this dataset, we can divide the analysis by differentiating three scarcity regimes: extreme ($\rho \in [95\%, 100\%]$), significant ($\rho \in [80\%, 90\%]$) and medium ($\rho \in [25\%, 75\%]$). Overall, most self-training approaches benefit from additional sparse annotations as their score increase when ρ decreases. This statement is true for all weighting strategies but the "*constant*" with $C = 0.1$ of which the performance plateau near $\rho = 85\%$, before decreasing as ρ decreases.

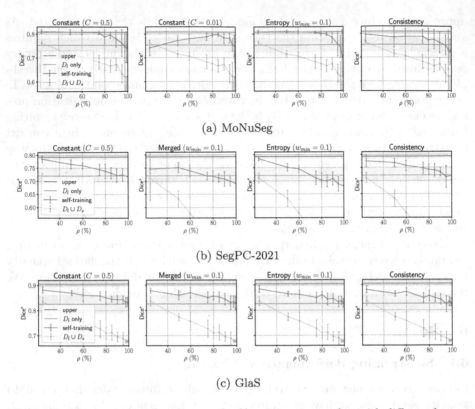

(a) MoNuSeg

(b) SegPC-2021

(c) GlaS

Fig. 2. Performance of our baselines and self-training approaches with different hyper-parameters combinations with a varying ρ and a fixed labeled set size n_l. We only report the weighting schemes that show representative behaviour. Other weighting strategies are evaluated and reported in Supplementary Section C.

In the extreme regime, all self-training approaches exhibit high variance and are outperformed by the "$\mathcal{D}_l$ only" baseline, or yield comparable performance. In this situation, it appears to be better to work in a fully supervised fashion using only images from $\mathcal{D}_l$ rather then using our self-training approach. Indeed, it seems that the noise brought by the extreme annotation sparsity (or complete lack of annotation when $\rho = 100\%$) degrades the model significantly which cannot even make efficient use of the exhaustively labeled images anymore. For $\rho = 95\%$, two self-training approaches ("*constant*" with $C = 0.1$ and "*entropy*") are on average better then the baseline but variance is still high making it difficult to really conclude that they are more efficient.

The situation is reversed in the significant regime where most self-training approaches (except the "*consistency*" weighting strategy) outperform the "$\mathcal{D}_l$ only" baseline and variance decreases significantly as well. As for the "*upper*" baseline, it remains more efficient than self-training. For $\rho = 90\%$, the most efficient weighting strategy on average is "*constant*" with $C = 0.1$ which also exhibits the smallest variance of all the self-training approaches. We believe that such a low constant is particularly helpful to combat the noise brought by the high sparsity as pseudo-labeled pixels contribute way less during training. For $\rho = 85\%$ and 90%, the *"constant"* with $C = 0.1$ strategy plateaus whereas others catch up in terms of performance and variance decrease with the *"entropy"* and *"merged"* (plot for this strategy can be found in Supplementary Fig. 1) approaches taking up the lead.

In the medium regime, three self-training approaches reach, and even slightly surpass, the upper baseline: *"constant"* with $C = 0.5$, *"entropy"* and *"merged"*. This result is interesting because it means that our self-training approach is able to reach the performance of a fully supervised approach but using only $\sim 30\%$ of the original annotations (*i.e.* $\rho = 75\%$, approximately 5k annotations instead of 17k) which is a significant annotation budget saving. The approach *"constant* $(C = 0.1)$"* decreases with ρ indicating that such a low C prevents the model to learn efficiently from the additional annotations (compared to the significant regime). This strategy even finished below the $\mathcal{D}_l \cup \mathcal{D}_s$ baseline at $\rho = 25\%$.

Overall, results on MoNuSeg are quite satisfactory. Although our approach is struggling in an extreme scarcity regime, it quickly catches up with the *"upper"* baseline as one adds more annotations to $\mathcal{D}_s$. In this case, the choice of weighting strategy matters and depends on the sparsity of the dataset.

SegPC-2021. Regarding the trend, our self-training approach behaves similarly on SegPC-2021 (see Fig. 2b) compared to MoNuSeg: all self-training approaches without exception seem to benefit from additional annotations in $\mathcal{D}_s$. Moreover, the $\mathcal{D}_l \cup \mathcal{D}_s$ baseline is particularly inefficient and finishes just below the *"$\mathcal{D}_l$ only"* baseline at $\rho = 25\%$. However, in the extreme regime, the gap between self-training and the *"$\mathcal{D}_l$ only"* baseline is less than on MoNuSeg. The rate at which our approach improves over the *"$\mathcal{D}_l$ only"* is also slower as it takes a larger ρ (around 75%) for the performance of self-training to become significantly better than this baseline. The best-performing weighting strategies also differ. The best strategies overall are *"constant"* (with $C = 0.5$ or 1) and *"consistency"*. The *"merged"* and *"entropy"* are worse than the others, although the latter catches up at $\rho = 25\%$. Only the "constant "and "entropy" strategies come close to catching up with the upper baseline but it takes proportionally more annotations compared to MoNuSeg as it happens around $\rho = 25\%$.

GlaS. On this dataset, all self-training approaches benefit from additional sparse annotations in $\mathcal{D}_s$. Compared to the *"$\mathcal{D}_l$ only"* baseline, the self-training approaches are never worse, even in the extreme scarcity regime, and it takes a ρ between 60% and 75% for self-training to become significantly better. Self-training is not able to catch up the *"upper"* baseline in this case.

6.2   Labeling a New Dataset: Sparsely or Exhaustively?

The fact that self-training is able to equal or outperform the *"$\mathcal{D}_l$ only"* and *"upper"* baselines suggests that it might be more interesting to consider an alternative

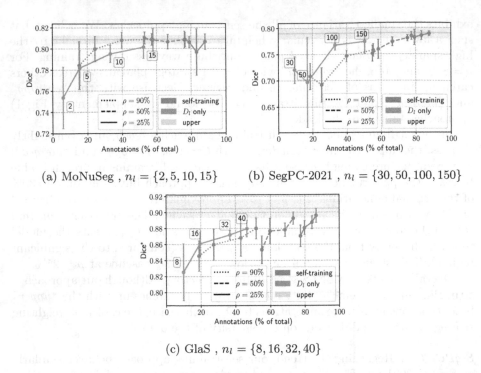

(a) MoNuSeg , $n_l = \{2, 5, 10, 15\}$ (b) SegPC-2021 , $n_l = \{30, 50, 100, 150\}$

(c) GlaS , $n_l = \{8, 16, 32, 40\}$

Fig. 3. Self-training *vs.* the baselines for varying ρ and n_l. A point corresponds to 10 runs of a given approach (see line color) and some sparsity parameters, ρ (see line style) and n_l (see subcaptions, increasing from left to right on a constant ρ curve, see "$\mathcal{D}_l$ *only*" for example). A new dataset is generated for each run, with the corresponding n_l and ρ values. The x value of a point corresponds to the average percentage of annotations used by the 10 datasets, or, alternatively, to the annotation budget dedicated for labeling the dataset.

annotation strategy to exhaustive labeling when annotating a new dataset. At fixed annotation budget, it might indeed be more interesting to combine sparse labeling and self-training rather than performing fully supervised training on an exhaustively labeled dataset (*i.e.* $\mathcal{D}_l$ only). To answer this question, we have conducted a set of experiments where we compare a self-training approach (entropy weighting strategy, $w_{min} = 0.1$) and the baselines all run against different sparsity regimes, varying both $\rho \in \{90\%, 50\%, 25\%\}$ and n_l (values are dataset specific). The results of these experiments are given in Fig. 3 where the values of n_l we have used are also specified. In these plots, the performance of all methods are reported over a common metric, the percentage of annotations used, which can be equated with the annotation budget for creating the dataset.

Our experiments show very dataset-dependent results. On MoNuSeg, we observe that self-training outperforms supervised training for all tested budgets. This indicates that it would have been more interesting to sparsely annotate this

Table 2. Experiment on the Thyroid FNAB dataset. The self-training approach uses the "*entropy*" weighting strategy and $w_{min} = 0.1$.

Method	Dice*
Self-training	89.05 ± 0.85
$\mathcal{D}_l$ only	80.30 ± 5.39
$\mathcal{D}_l \cup \mathcal{D}_s$	83.62 ± 3.52

dataset. However, this conclusion does not hold for the other datasets as, within the same annotation budget, using "$\mathcal{D}_l$ *only*" outperforms self-training.

This experiment also allows to compare which labeling scheme is better for self-training: for a given annotation budget, is it better to favor a larger set $\mathcal{D}_l$ or to add more sparse annotations in $\mathcal{D}_s$? For MoNuSeg and SegPC-2021, it appears that, for a similar annotation budget, self-training performance are comparable whatever the values of n_l and ρ. Therefore, for those datasets, it does not really matter if the annotation budget is spent for exhaustive or sparse labeling. For GlaS, however, there is a performance loss when switching from a lower ρ value to a higher (*e.g.* going from $(\rho, n_l) = (90\%, 40)$ to $(50\%, 8)$ in Fig. 3c). It indicates that, it is more interesting to label images exhaustively rather than sparsely for this dataset.

At this point, the experiments in this section do not allow us to provide definitive guidelines on how to focus annotation efforts to achieve optimal performance on a new dataset, but at least, they show that it can be beneficial to sparsely annotate more images than to exhaustively annotate fewer images.

6.3 Experiments on Thyroid CytologyFNAB

The Thyroid FNAB dataset presents a great opportunity to test our method on a real case of sparsely labeled dataset. It is also interesting to note that this dataset is larger than the three public datasets used in this study (almost 5k images in total). In order to fit the sparsely-labeled settings we have introduced, we split the dataset in two subsets based on the nature of the annotations. The set $\mathcal{D}_l$ is assigned annotations of cell aggregates and $\mathcal{D}_s$ is assigned annotations of individual cells and nuclei. The motivation for this split is presented in Supplementary Section E.

Based on the results of Sect. 6.1, we have chosen to use the "*entropy*" weighting strategy with $w_{min} = 0.1$ for our self-training approach, as it provides consistently good results across datasets. We compare this approach with two of our three baselines: "$\mathcal{D}_l$ *only*" and "$\mathcal{D}_l \cup \mathcal{D}_s$". The "*upper*" baseline obviously cannot be evaluated because we do not have access to the complete ground truth. The resulting performances are given in Table 2.

We observe that our self-training approach significantly outperforms the two baselines and remains quite stable as its standard deviation is below 1%. This confirms the interest of self-training when working with a sparse dataset.

7   Conclusion

In this work, we have introduced a method based on self-training for training a deep binary segmentation model with sparsely labeled data. Using 4 datasets, including an actual sparsely labeled one, we have shown that the method could indeed make use of sparse annotations to improve model performance over using only exhaustively labeled data. For one of our datasets, our self-training approach using only 30% of the original training annotations is even able to reach performance comparable to using all of them in a supervised way.

In the future, we want to extend the method to multi-class segmentation and further study the impact of various training choices and hyperparameters (model complexity, weighting strategies, soft pseudo-labeling, *etc.*) that we could not explore due to time and computing resources constraints. We also want to further study how the type of dataset (variability in images, density of ground truth, large or small annotations, *etc.*) impacts the performance margins of self-training. Moreover, in this work, we have removed annotation randomly from the datasets. In practice, it is unlikely that the existing annotations are really randomly chosen and it would be interesting to study the effect of the labeling process. This work has mostly been focused on high scarcity conditions but self-training methods have also shown to be beneficial with very large labeled and unlabeled sets in other contexts. In computational pathology, whole slide images usually offer a great potential for a large pool of unlabeled data. Therefore, we would like to study how our method would perform in such a context. Finally, we would also like to explore how our self-training algorithm could be used in an interactive mode to assist new dataset labeling. Eventually, we want to implement the algorithm on the Cytomine [19] platform.

Acknowledgements. Raphaël Marée is supported by the BigPicture EU Research and Innovation Action (Grant agreement number 945358). This work was partially supported by Service Public de Wallonie Recherche under Grant No. 2010235 - ARIAC by DIGITALWALLONIA4.AI. Computational infrastructure is partially supported by ULiège, Wallonia and Belspo funds. We would also like to thank Isabelle Salmon (ULB, Erasme Hospital, Belgium) and her team for providing the Thyroid FNAB dataset.

References

1. Bai, W., et al.: Semi-supervised learning for network-based cardiac MR image segmentation. In: Descoteaux, M., Maier-Hein, L., Franz, A., Jannin, P., Collins, D.L., Duchesne, S. (eds.) MICCAI 2017. LNCS, vol. 10434, pp. 253–260. Springer, Cham (2017). https://doi.org/10.1007/978-3-319-66185-8_29
2. Bankhead, P., et al.: QuPath: open source software for digital pathology image analysis. Sci. Rep. **7**(1), 1–7 (2017)
3. Berg, S., et al.: ilastik: interactive machine learning for (bio)image analysis. Nat. Methods (2019). https://doi.org/10.1038/s41592-019-0582-9
4. Bokhorst, J.M., Pinckaers, H., van Zwam, P., Nagtegaal, I., van der Laak, J., Ciompi, F.: Learning from sparsely annotated data for semantic segmentation in histopathology images. In: International Conference on Medical Imaging with Deep Learning-Full Paper Track (2018)

5. Fan, D.P., et al.: Inf-net: automatic Covid-19 lung infection segmentation from CT images. IEEE Trans. Med. Imaging **39**(8), 2626–2637 (2020)
6. Grandvalet, Y., Bengio, Y.: Semi-supervised learning by entropy minimization. In: Advances in Neural Information Processing Systems, vol. 17 (2004)
7. Gupta, A., Gupta, R., Gehlot, S., Goswami, S.: SegPC-2021: segmentation of multiple myeloma plasma cells in microscopic images. IEEE Dataport **1**(1), 1 (2021)
8. Jaiswal, A.K., Panshin, I., Shulkin, D., Aneja, N., Abramov, S.: Semi-supervised learning for cancer detection of lymph node metastases. arXiv preprint arXiv:1906.09587 (2019)
9. Kingma, D.P., Ba, J.: Adam: a method for stochastic optimization. arXiv preprint arXiv:1412.6980 (2014)
10. Koohbanani, N.A., Jahanifar, M., Tajadin, N.Z., Rajpoot, N.: NuClick: a deep learning framework for interactive segmentation of microscopic images. Med. Image Anal. **65**, 101771 (2020)
11. Koohbanani, N.A., Unnikrishnan, B., Khurram, S.A., Krishnaswamy, P., Rajpoot, N.: Self-path: self-supervision for classification of pathology images with limited annotations. IEEE Trans. Med. Imaging **40**(10), 2845–2856 (2021)
12. Kumar, N., et al.: A multi-organ nucleus segmentation challenge. IEEE Trans. Med. Imaging **39**(5), 1380–1391 (2019)
13. Kwatra, V., Schödl, A., Essa, I., Turk, G., Bobick, A.: Graphcut textures: image and video synthesis using graph cuts. ACM Trans. Graph. (TOG) **22**(3), 277–286 (2003)
14. Laine, S., Aila, T.: Temporal ensembling for semi-supervised learning. arXiv preprint arXiv:1610.02242 (2016)
15. Lee, D.H., et al.: Pseudo-label: the simple and efficient semi-supervised learning method for deep neural networks. In: Workshop on challenges in representation learning, ICML, vol. 3, p. 896 (2013)
16. Li, J., et al.: Signet ring cell detection with a semi-supervised learning framework. In: Chung, A.C.S., Gee, J.C., Yushkevich, P.A., Bao, S. (eds.) IPMI 2019. LNCS, vol. 11492, pp. 842–854. Springer, Cham (2019). https://doi.org/10.1007/978-3-030-20351-1_66
17. Li, J., et al.: An EM-based semi-supervised deep learning approach for semantic segmentation of histopathological images from radical prostatectomies. Comput. Med. Imaging Graph. **69**, 125–133 (2018)
18. Maninis, K.K., Caelles, S., Pont-Tuset, J., Van Gool, L.: Deep extreme cut: from extreme points to object segmentation. In: Proceedings of the IEEE Conference on Computer Vision and Pattern Recognition, pp. 616–625 (2018)
19. Marée, R., et al.: Collaborative analysis of multi-gigapixel imaging data using Cytomine. Bioinformatics **32**(9), 1395–1401 (2016)
20. Peikari, M., Salama, S., Nofech-Mozes, S., Martel, A.L.: A cluster-then-label semi-supervised learning approach for pathology image classification. Sci. Rep. **8**(1), 1–13 (2018)
21. Peng, J., Wang, Y.: Medical image segmentation with limited supervision: a review of deep network models. IEEE Access (2021)
22. Pham, H., Dai, Z., Xie, Q., Le, Q.V.: Meta pseudo labels. In: Proceedings of the IEEE/CVF Conference on Computer Vision and Pattern Recognition, pp. 11557–11568 (2021)
23. Ronneberger, O., Fischer, P., Brox, T.: U-Net: convolutional networks for biomedical image segmentation. In: Navab, N., Hornegger, J., Wells, W.M., Frangi, A.F. (eds.) MICCAI 2015. LNCS, vol. 9351, pp. 234–241. Springer, Cham (2015). https://doi.org/10.1007/978-3-319-24574-4_28

24. Rother, C., Kolmogorov, V., Blake, A.: "GrabCut" interactive foreground extraction using iterated graph cuts. ACM Trans. Graph. (TOG) **23**(3), 309–314 (2004)
25. Shaw, S., Pajak, M., Lisowska, A., Tsaftaris, S.A., O'Neil, A.Q.: Teacher-student chain for efficient semi-supervised histology image classification. arXiv preprint arXiv:2003.08797 (2020)
26. Sirinukunwattana, K., et al.: Gland segmentation in colon histology images: the GlaS challenge contest. Med. Image Anal. **35**, 489–502 (2017)
27. Sohn, K., et al.: FixMatch: simplifying semi-supervised learning with consistency and confidence. In: Advances in Neural Information Processing Systems, vol. 33, pp. 596–608 (2020)
28. Su, H., Shi, X., Cai, J., Yang, L.: Local and global consistency regularized mean teacher for semi-supervised nuclei classification. In: Shen, D., et al. (eds.) MICCAI 2019. LNCS, vol. 11764, pp. 559–567. Springer, Cham (2019). https://doi.org/10.1007/978-3-030-32239-7_62
29. Tajbakhsh, N., Jeyaseelan, L., Li, Q., Chiang, J.N., Wu, Z., Ding, X.: Embracing imperfect datasets: a review of deep learning solutions for medical image segmentation. Med. Image Anal. **63**, 101693 (2020)
30. Tarvainen, A., Valpola, H.: Mean teachers are better role models: weight-averaged consistency targets improve semi-supervised deep learning results. In: Advances in Neural Information Processing Systems, vol. 30 (2017)
31. Tellez, D., et al.: Whole-slide mitosis detection in H&E breast histology using PHH3 as a reference to train distilled stain-invariant convolutional networks. IEEE Trans. Med. Imaging **37**(9), 2126–2136 (2018)
32. Xie, Q., Luong, M.T., Hovy, E., Le, Q.V.: Self-training with noisy student improves imagenet classification. In: Proceedings of the IEEE/CVF Conference on Computer Vision and Pattern Recognition, pp. 10687–10698 (2020)
33. Yarowsky, D.: Unsupervised word sense disambiguation rivaling supervised methods. In: 33rd Annual Meeting of the Association for Computational Linguistics, pp. 189–196 (1995)
34. Zhu, Y., et al.: Improving semantic segmentation via self-training. arXiv preprint arXiv:2004.14960 (2020)

CNR-IEMN-CD and CNR-IEMN-CSD Approaches for Covid-19 Detection and Covid-19 Severity Detection from 3D CT-scans

Fares Bougourzi[1]([✉]) [iD], Cosimo Distante[1] [iD], Fadi Dornaika[2,3] [iD],
and Abdelmalik Taleb-Ahmed[4] [iD]

[1] Institute of Applied Sciences and Intelligent Systems, National Research Council
of Italy, 73100 Lecce, Italy
`fares.bougourzi@isasi.cnr.it, cosimo.distante@cnr.it`
[2] University of the Basque Country UPV/EHU, San Sebastian, Spain
`fadi.dornaika@ehu.eus`
[3] IKERBASQUE, Basque Foundation for Science, Bilbao, Spain
[4] Univ. Polytechnique Hauts-de-France, Univ. Lille, CNRS, Centrale Lille, UMR
8520 - IEMN, 59313 Valenciennes, France
`Abdelmalik.Taleb-Ahmed@uphf.fr`

Abstract. Since its appearance in late 2019, Covid-19 has become an active research topic for the artificial intelligence (AI) community. One of the most interesting AI topics is Covid-19 analysis from medical imaging. CT-scan imaging is the most informative tool about this disease.

This work is part of the 2nd COV19D competition, where two challenges are set: Covid-19 Detection and Covid-19 Severity Detection from the CT-scans. For Covid-19 detection from CT-scans, we proposed an ensemble of 2D Convolution blocks with Densenet-161 models (CNR-IEMN-CD). Here, each 2D convolutional block with Densenet-161 architecture is trained separately and in the testing phase, the ensemble model is based on the average of their probabilities. On the other hand, we proposed an ensemble of Convolutional Layers with Inception models for Covid-19 severity detection CNR-IEMN-CSD. In addition to the Convolutional Layers, three Inception variants were used, namely Inception-v3, Inception-v4 and Inception-Resnet.

Our proposed approaches outperformed the baseline approach in the validation data of the 2nd COV19D competition by 11% and 16% for Covid-19 detection and Covid-19 severity detection, respectively. In the testing phase, our proposed approach CNR-IEMN-CD ranked fifth and improved the baseline results by 18.37%. On the other hand, our proposed approach CNR-IEMN-CSD ranked third in the test data of the 2nd COV19D competition for Covid-19 severity detection, and improved the baseline results by 6.81%.

Keywords: Covid-19 · Deep leaning · CNNs · Recognition · Severity

© The Author(s), under exclusive license to Springer Nature Switzerland AG 2023
L. Karlinsky et al. (Eds.): ECCV 2022 Workshops, LNCS 13807, pp. 593–604, 2023.
https://doi.org/10.1007/978-3-031-25082-8_40

1 Introduction

Since the appearance of the Covid-19 pandemic in the late of 2019, reverse transcription-polymerase chain reaction (RT-PCR) has been considered as the golden standards for Covid-19 Detection. However, the RT-PCR test has many drawbacks including inadequate supply of RT-PCR kits, time-consuming consumption, and considerable false negative results [9,15,23]. To overcome these limitations, medical imaging techniques have been widely used as tools. These imaging techniques include X-rays and CT-scans [5,22]. In fact, CT-scans are not only used to detect Covid-19 infected cases, but they could also be used to track the patient's condition and predict the severity of the disease [4,7].

In the last decade, Deep Learning methods have become widely used in most computer vision tasks and have achieved high performance compared to traditional methods [2,3]. However, the main drawback of Deep Learning, especially the CNN architecture, is the need for huge labeled data, which is difficult to obtain in medical fields [7]. On the other hand, most of the proposed CNN architectures have been used for static images (single image input) [4].

In this work, we used pre-trained CNN architectures to detect Covid-19 infection and Covid-19 severity from 3D CT scans in the context of the 2nd COV19D challenge. We also trained a CNN model to filter out the sections that do not show a lung region. For Covid-19 severity detection, an Att-Unet model was trained to segment the lung regions and remove unimportant features. In addition, two volume input branches were used to reduce information loss due to resizing and varying the number of slices from one CT-scan to another. The main contributions of this work can be summarized as follows:

- For Covid-19 detection from CT scans, we proposed an ensemble of Convolution 2D blocks with Densenet-161 models, which we call CNR-IEMN-CD. Each convolution 2D block with densenet-161 architecture was trained separately, and the ensemble model is based on the average of their probabilities.
- To detect the severity of Covid-19 using CT scans, we proposed an ensemble of Convolutional Layer with Inception models, which we call CNR-IEMN-CSD. In addition to the Convolutional Layers, three Inception variants were used, namely Inception-v3, Inception-v4, and Inception-Resnet. The Inception variants were trained twice separately and the ensemble model is the average of the probabilities of the six models.
- Our proposed approaches outperformed the baseline approach in the validation data of the 2nd COV19D competition by 11% and 16% in Covid-19 detection and Covid-19 severity detection, respectively. Moreover, our proposed approach CNR-IEMN-CD ranked fifth and improved the results of the baseline approach by 18.37%. On the other hand, our proposed approach CNR-IEMN-CSD ranked third in the test data of the 2nd COV19D compe-

tition for Covid-19 severity detection and improved the baseline results by
6.81
– The codes used are publicly available at https://github.com/faresbougourzi/
2nd-COV19D-Competition. (Last accessed 28th June 2022).

This paper is organized as follows: Sect. 2 describes our proposed approaches
for Covid-19 detection and severity detection. The experiments and results are
described in Sect. 3. In Sect. 4, the obtained results of Covid-19 Detection Challenge are discussed. Finally, we conclude our paper in Sect. 5.

2   Our Approaches

In the following two sections, we will describe our proposed approaches for Covid-
19 detection and severity prediction, respectively.

2.1   Covid-19 Detection

Our proposed approach to Covid-19 detection for the 2nd COV19D competition
is summarized in Fig. 1. Because the 3D CT scans have a different number of
layers depending on the scanner and acquisition settings and contain multiple
slices with no lung regions, we trained a model to remove the non-lung slices. For
this purpose, we manually labelled 20 CT-scans from the training data as lung
and non-lung. In addition, we used the following datasets: COVID-19 CT segmentation [19], Segmentation dataset nr.2 [19], and COVID-19- CT Seg dataset
[16] to have more labelled data as lung and non-lung slices. After creating the
lung and non-lung sets, we trained the ResneXt-50 model [24] to remove the
slices that showed no lungs at all. Since the CT scans have different numbers of
slices, we concatenated all lung slices and then reduced them to $224 \times 224 \times 64$ as
shown in Fig. 1. To distinguish between Covid-19 and non-Covid-19 CT scans,
we separately trained three Densenet-161 [8] models. Since the input size of
the Densenet-161 model is an RGB image and we have a 3D volume of size
$224 \times 224 \times 64$, we added Convolution block with a 3×3 kernel, a stride of 1,
and a padding of 1. The 3×3 convolution block takes 64 input channels and produces 3 output channels. It should be noted that each CNN model was trained
separately. In the decision or test phase, we propose to form an ensemble of the
three trained CNN architectures by averaging their probabilities.

2.2   Covid-19 Severity Detection

Our proposed approach to predict Covid-19 severity for the 2nd COV19D competition is summarized in Fig. 2. Similar to the Covid-19 detection in Sect. 2.1,

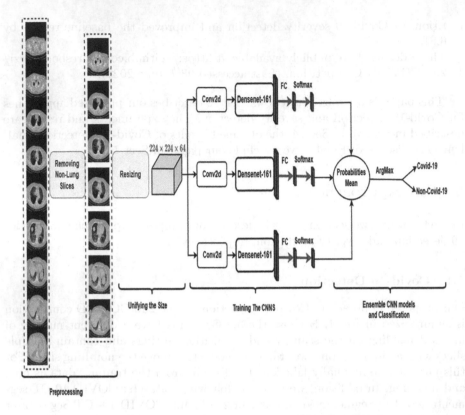

Fig. 1. Our proposed CNR-IEMN-CD approach for Covid-19 detection.

we used the trained ResneXt-50 model to remove non-lung slices. To segment the lungs, we used the Att-Unet [6,17] architecture trained with the following datasets COVID-19 CT segmentation [19], Segmentation dataset nr.2 [19], and COVID-19-CT-Seg [16]. After removing the non-lung slices and segmenting the lung regions to remove unnecessary features, we divided generated for each scan, from the obtained slices, two volumes $299 \times 299 \times 32$ and $299 \times 299 \times 16$. The goal of the two volumes is to reduce the information lost due to resizing and to obtain two views of each CT scan. The two input volumes were fed into convolution blocks. The convolution layer consists of three 3×3 convolution blocks with stride 1 and padding 1 (see Fig. 2). The first convolution block transforms the volume $299 \times 299 \times 32$ into $299 \times 299 \times 3$. Similarly, the second convolution block transforms the volume $299 \times 299 \times 16$ into $299 \times 299 \times 3$. The third convolution block transforms the concatenation of the outputs of the first and second convolution blocks ($299 \times 299 \times 3$) into $299 \times 299 \times 3$. The output volume of the third convolution block is used as input to a CNN backbone. In our experiments, we used Inception architectures as CNN backbones, namely, inception-V3, inception-V4, and Inception-Resnet [21]. Our approach to predict the severity of Covid-19 consists of six ConvLayers + CNN branches, as shown in Fig. 2, where each branch was trained separately. To create a more comprehensive ensemble model, we trained each CNN model twice. The final prediction of Covid-19 severity was obtained by averaging the prediction probabilities of the six models.

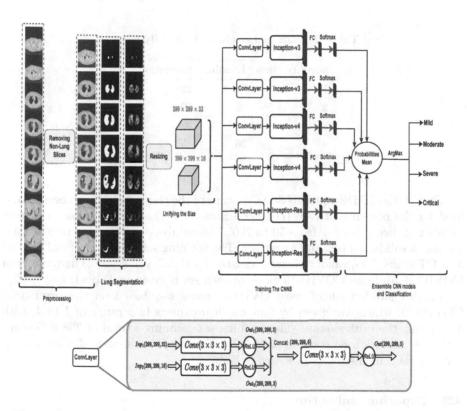

Fig. 2. Our proposed CNR-IEMN-CSD approach for Covid-19 severity detection.

3   Experiments and Results

3.1   The COV19-CT-DB Database

The COVID19-CT-Database (COV19-CT-DB) [1,10–14] consists of chest CT scans that are annotated for the existence of COVID-19.

Table 1. Data samples in each set.

Set	Training	Validation
Covid-19	882	215
Non-Covid-19	1110	289

Table 2. Data samples in each Severity Class

Severity class	Training	Validation
1	85	22
2	62	10
3	85	22
4	26	5

COVID19-CT-DB contains 3-D CT scans of the chest, which have been examined for the presence of COVID-19. Each of the 3-D scans contains a different number of slices, ranging from 50 to 700. The database was divided into a training set, a validation set, and a test set. The training set contains a total of 1992 3-D CT scans. The validation set consists of 494 3-D CT scans. The number of COVID-19 and non COVID-19 cases in each set is given in Table 1.

Further subdivision of the COVID-19 cases was based on the severity of COVID-19, which was given by four medical experts in a range of 1 to 4, with 4 denoting the critical state. The training set contains a total of 258 3-D scans CT. The validation set consists of 61 3-D CT scans. The number of scans in each severity class in these sets is shown in Table 2.

3.2   Experimental Setup

For Deep Learning training and testing, we used the Pytorch [18] Library with NVIDIA GPU Device GeForce TITAN RTX 24 GB. The batch size used consists of 16 CT-scan volumes for both tasks. We trained the networks for 40 epochs. The initial learning rate is 0.0001, which decreases by 0.1 after 15 epochs, followed by another 0.1 decrease after 30 epochs.

3.3   Covid-19 Recognition

Table 3 summarizes the Covid-19 detection results (F1 score) obtained with the validation data. As described in Sect. 2.1, we trained three Convolutional Block + Densenet-161 architectures. For simplicity, we refer to Convolutional Block + Densenet-161 with Densenet-161 and the training iteration. This table shows the results of the individual Densenet-161 models and the ensemble of 2 and three models. The ensemble of three models is denoted by (CNR-IEMN-CD). From these results, the ensemble of two models achieves better results than the individual models. The ensemble of three models, on the other hand, achieves the best performance. Compared to the baseline results, our proposal CNR-IEMN-CD improved the results by 13%.

Table 3. Covid-19 detection results on the 2nd COV19D competition validation data.

Model	Architecture	F1-Score
-	Baseline	77
1	Densenet-161 Model1	88.34
2	Densenet-161 Model2	88.83
3	Densenet-161 Model3	87.81
4	Ensemble (1&2)	89.61
5	Ensemble (2&3)	89.85
6	Ensemble (1&2&3)	90.07

Since five submissions were allowed in the 2nd COV19D competition, our submissions are 1, 2, 4, 5 and 6. The obtained results are summarized in Table 4. From these results, we notice that the ensemble of the three Densenet-161 models achieved the best performance than using single models and the ensemble of two models as well. Furthermore, our approach achieved higher performance than the baseline method by 18.37%. On the other hand, our approach ranked the fifth in the 2nd COV19D competition for Covid-19 Detection.

Table 4. Covid-19 Detection Results on the 2nd COV19D competition testing data.

Model	Architecture	F1-Score
-	Baseline	66.00
1	Densenet-161 Model1	81.91
2	Densenet-161 Model2	77.03
4	Ensemble (1&2)	83.12
5	Ensemble (2&3)	84.12
6	Ensemble (1&2&3)	84.37

3.4   Covid-19 Severity Detection

Table 5 summarizes the results (F1 score) of detecting the severity of Covid-19 using the validation data. As described in Sect. 2.2, we designed three models consisting of convolutional layers and CNN backbones. The CNN backbones used are Inception-V3, Inception-V4, and Inception-Resnet, and each model was trained twice to obtain a greater variety of predictors. In total, we have six models. For simplicity, we refer to ConvLayer + CNN backbone with the backbone architecture name. Table 5 shows the obtained results of the individual models and the ensemble of two models of the same architecture and the ensemble of the six models (CNR-IEMN-CSD). The obtained results show that the ensemble of two models improves the performance compared to the single models. Moreover, the ensemble of the six models achieved the best performance and was 17% better than the baseline.

Table 5. Covid-19 Severity Detection Results on the 2nd COV19D Severity Detection competition validation data.

Model	Architecture	F1-Score
-	Baseline	63.00
1	Inception-v3 Model1	74.86
2	Inception-v3 Model2	76.25
3	Inception-v4 Model1	75.60
4	Inception-v4 Model2	73.48
5	Inception-Res Model1	73.37
6	Inception-Res Model2	72.01
7	Ensemble Models(1-2)	79.10
8	Ensemble Models(3-4)	79.54
9	Ensemble Models(5-6)	79.10
10	Ensemble Models(1-6)	80.08

Since five submissions were allowed in the 2nd COV19D competition, our submissions are 1, 7, 8, 9 and 10. The obtained results on the testing data of the 2nd COV19D Severity Detection competition are summarized in Table 6. From these results, we notice our proposed CNR-IEMN-CSD approach achieved the best performance than using single models and the ensemble of two models as well. Furthermore, our approach achieved higher performance than the baseline method by 6.8%. On the other hand, our approach ranked the fifth in the 2nd COV19D competition.

4 Discusssion

To understand the behavior of our proposed approach for Covid-19 detection challenge, we visualize some slices from predicted CT-scans as TP, TN, FP, FN as shown in Fig. 3. As stated in [20], the COVID-19 lung imaging manifestations are highly overlapped with those of the viral pneumonia and this makes the diagnosis of Covid-19 challenging even for the radiologists. It can be noticed that in the 2nd COV19D competition for Covid-19 Detection, only two classes were defined, Covid-19 and Non-Covid. Thus, that Non-Covid class may presents No infection at all or other lung diseases.

Table 6. Covid-19 Severity Detection Results on the 2nd COV19D Severity Detection competition Testing data.

Model	Architecture	F1-Score	Mild	Moderate	Severe	Critical
-	Baseline	40.30	61.10	39.47	37.07	23.54
2	Inception-v3 Model2	39.89	34.86	35.44	58.00	31.25
7	Ensemble Models(1-2)	43.04	53.09	18.52	63.52	37.04
8	Ensemble Models(3-4)	45.49	54.66	26.79	63.48	37.04
9	Ensemble Models(5-6)	44.37	53.42	26.67	60.36	37.04
10	Ensemble Models(1-6)	47.11	55.67	37.88	55.46	39.46

The first row shows that many Non-Covid cases are classifies as Covid-19, these examples shows the appearance of infection manifestations, however, the signs are not related to Covid-19 infection. The second row shows similar infection signs and the CT-scans were correctly classified as Covid-19. From the third row shows that our approach can misclassify when the Covid-19 infection appears in small region of the lungs. The fourth row shows that our approach is able to classify Non-Covid CT-scans despite the appearance of the infection signs of the other diseases. Especially, with using the ensemble of multiple models. From these examples, we argue that separating other diseases as separate class can help the trained models to distinguish between the Covid-19 and Non-Covid cases.

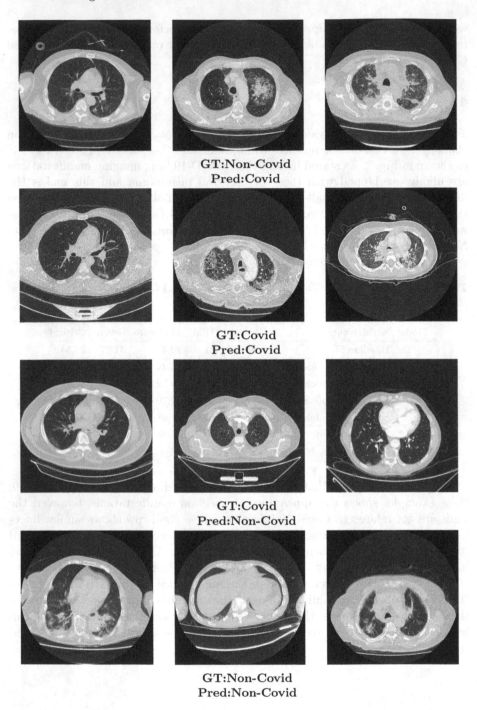

Fig. 3. Visualization illustration of some example slices from the predicted CT-scans. The first to the fourth rows show examples of wrongly predicted CT-scans as Covid-19, correctly predicted CT-scans as Covid-19, wrongly predicted CT-scans as Non-Covid-19, and CT-scans as Non-Covid-19, respectively.

5   Conclusion

In this work, we proposed two ensemble CNN-based approaches for the 2nd COV19D competition. For Covid-19 detection from CT scans, we proposed an ensemble of 2D convolution blocks with Densenet-161 models. Here, the individual convolution 2D blocks with Densenet-161 architecture are trained separately, and the ensemble model is based on the average of their probabilities at testing phase. On the other hand, we proposed an ensemble of Convolutional Layers with Inception models for Covid-19 severity detection. In addition to the Convolutional Layers, three Inception variants were used, namely Inception-v3, Inception-v4, and Inception-Resnet.

In the testing phase, our proposed approaches ranked fifth and third for Covid-19 detection and Covid-19 Severity detection, respectively. Furthermore, our approaches outperformed the baseline results by 18.37% and 6.81% for Covid-19 detection and Covid-19 Severity detection, respectively. In future work, we will evaluate our approaches in other datasets and use the attention mechanism to select the most representative slices for Covid-19 detection.

Acknowledgments. The authors would like to thank Arturo Argentieri from CNR-ISASI Italy for his support on the multi-GPU computing facilities.

References

1. Arsenos, A., Kollias, D., Kollias, S.: A large imaging database and novel deep neural architecture for covid-19 diagnosis. In: 2022 IEEE 14th Image, Video, and Multidimensional Signal Processing Workshop (IVMSP), pp. 1–5. IEEE (2022)
2. Bougourzi, F., Dornaika, F., Mokrani, K., Taleb-Ahmed, A., Ruichek, Y.: Fusing Transformed Deep and Shallow features (FTDS) for image-based facial expression recognition. Expert Syst. Appl. **156**, 113459 (2020). https://doi.org/10.1016/j.eswa.2020.113459
3. Bougourzi, F., Dornaika, F., Taleb-Ahmed, A.: Deep learning based face beauty prediction via dynamic robust losses and ensemble regression. Knowl.-Based Syst. **242**, 108246 (2022)
4. Bougourzi, F., Contino, R., Distante, C., Taleb-Ahmed, A.: CNR-IEMN: A Deep Learning Based Approach to Recognise Covid-19 from CT-Scan. In: ICASSP 2021–2021 IEEE International Conference on Acoustics, Speech and Signal Processing (ICASSP), pp. 8568–8572, June 2021. https://doi.org/10.1109/ICASSP39728.2021.9414185, iSSN: 2379-190X
5. Bougourzi, F., Contino, R., Distante, C., Taleb-Ahmed, A.: Recognition of COVID-19 from CT scans using two-stage deep-learning-based approach: CNR-IEMN. Sensors **21**(17), 5878 (2021). https://doi.org/10.3390/s21175878, number: 17 Publisher: Multidisciplinary Digital Publishing Institute
6. Bougourzi, F., Distante, C., Dornaika, F., Taleb-Ahmed, A., Hadid, A.: ILC-Unet++ for Covid-19 Infection Segmentation. In: Mazzeo, P.L., Frontoni, E., Sclaroff, S., Distante, C. (eds.) Image Analysis and Processing. ICIAP 2022 Workshops, LNCS, pp. 461–472. Springer, Cham (2022). https://doi.org/10.1007/978-3-031-13324-4_39

7. Bougourzi, F., Distante, C., Ouafi, A., Dornaika, F., Hadid, A., Taleb-Ahmed, A.: Per-COVID-19: a benchmark dataset for COVID-19 percentage estimation from CT-Scans. J. Imaging **7**(9), 189 (2021). https://doi.org/10.3390/jimaging7090189, number: 9 Publisher: Multidisciplinary Digital Publishing Institute

8. Huang, G., Liu, Z., Van Der Maaten, L., Weinberger, K.Q.: Densely connected convolutional networks. In: Proceedings of the IEEE Conference on Computer Vision and Pattern Recognition, pp. 4700–4708 (2017)

9. Jin, Y.H., Cai, L., Cheng, Z.S.e.a.: A rapid advice guideline for the diagnosis and treatment of 2019 novel coronavirus (2019-nCoV) infected pneumonia (standard version). Military Med. Res. **7**(1), 4 (2020). https://doi.org/10.1186/s40779-020-0233-6

10. Kollias, D., Arsenos, A., Kollias, S.: Ai-mia: Covid-19 detection & severity analysis through medical imaging. arXiv preprint arXiv:2206.04732 (2022)

11. Kollias, D., Arsenos, A., Soukissian, L., Kollias, S.: Mia-cov19d: Covid-19 detection through 3-d chest ct image analysis. In: Proceedings of the IEEE/CVF International Conference on Computer Vision, pp. 537–544 (2021)

12. Kollias, D., et al.: Deep transparent prediction through latent representation analysis. arXiv preprint arXiv:2009.07044 (2020)

13. Kollias, D., Tagaris, A., Stafylopatis, A., Kollias, S., Tagaris, G.: Deep neural architectures for prediction in healthcare. Complex Intell. Syst. **4**(2), 119–131 (2018)

14. Kollias, D., et al.: Transparent adaptation in deep medical image diagnosis. In: TAILOR, pp. 251–267 (2020)

15. Kucirka, L.M., Lauer, S.A., Laeyendecker, O., Boon, D., Lessler, J.: Variation in False-Negative rate of reverse transcriptase polymerase chain Reaction–Based SARS-CoV-2 Tests by time since exposure. Ann. Internal Med. **173**(4), 262–267 (2020). https://doi.org/10.7326/M20-1495, publisher: American College of Physicians

16. Ma, J., et al.: Toward data efficient learning: a benchmark for COVID 19 CT lung and infection segmentation. Medical Physics 48, 1197–1210 (2021). https://doi.org/10.1002/mp.14676. https://ui.adsabs.harvard.edu/abs/2021MedPh.48.1197M, aDS Bibcode: 2021MedPh.48.1197M

17. Oktay, O., Schlemper, J., Folgoc, L.L.e.a.: Attention U-Net: learning where to look for the pancreas. arXiv:1804.03999 [cs] (May 2018), arXiv: 1804.03999

18. Paszke, A., Gross, S., Massa, F., et al.: Pytorch: an imperative style, high-performance deep learning library. In: Advances in neural information processing systems, pp. 8026–8037 (2019)

19. RADIOLOGISTS: COVID-19 CT-scans segmentation datasets. http://medicalsegmentation.com/covid19/ (2019). Accessed 18 Aug 2021

20. Sun, Z., Zhang, N., Li, Y., Xu, X.: A systematic review of chest imaging findings in covid-19. Quant. Imaging Med. Surg. **10**(5), 1058 (2020)

21. Szegedy, C., Vanhoucke, V., Ioffe, S., Shlens, J., Wojna, Z.: Rethinking the inception architecture for computer vision (2015)

22. Vantaggiato, E., Paladini, E., Bougourzi, F., Distante, C., Hadid, A., Taleb-Ahmed, A.: Covid-19 recognition using ensemble-CNNs in two new chest x-ray databases. Sensors **21**(5), 1742 (2021)

23. Wu, Y.H., et al.: JCS: An Explainable COVID-19 diagnosis system by joint classification and segmentation. IEEE Trans. Image Process. **30**, 3113–3126 (2021). https://doi.org/10.1109/TIP.2021.3058783, conference Name: IEEE Transactions on Image Processing

24. Xie, S., Girshick, R., Dollár, P., Tu, Z., He, K.: Aggregated residual transformations for deep neural networks (2017)

Representation Learning with Information Theory to Detect COVID-19 and Its Severity

Abel Díaz Berenguer[1]([✉])(iD), Tanmoy Mukherjee[1,2](iD), Yifei Da[1](iD),
Matías Nicolás Bossa[1](iD), Maryna Kvasnytsia[1](iD), Jef Vandemeulebroucke[1,2](iD),
Nikos Deligiannis[1,2](iD), and Hichem Sahli[1,2](iD)

[1] Department of Electronics and Informatics (ETRO), Vrije Universiteit Brussel,
Pleinlaan 2, 1050 Brussels, Belgium
{aberengu,tmukherj,yda,mnbossa,mkvasnyt,jefvdmb,
ndeligia,hsahli}@etrovub.be
[2] Interuniversity Microelectronics Centre (IMEC), Kapeldreef 75,
3001 Heverlee, Belgium

Abstract. Successful data representation is a fundamental factor in machine learning based medical imaging analysis. Deep Learning (DL) has taken an essential role in robust representation learning. However, the inability of deep models to generalize to unseen data can quickly overfit intricate patterns. Thereby, the importance of implementing strategies to aid deep models in discovering useful priors from data to learn their intrinsic properties. Our model, which we call a dual role network (DRN), uses a dependency maximization approach based on Least Squared Mutual Information (LSMI). LSMI leverages dependency measures to ensure representation invariance and local smoothness. While prior works have used information theory dependency measures like mutual information, these are known to be computationally expensive due to the density estimation step. In contrast, our proposed DRN with LSMI formulation does not require the density estimation step and can be used as an alternative to approximate mutual information. Experiments on the CT based COVID-19 Detection and COVID-19 Severity Detection Challenges of the 2nd COV19D competition [24] demonstrate the effectiveness of our method compared to the baseline method of such competition.

Keywords: Representation learning · Mutual information ·
COVID-19 detection

1 Introduction

Reverse transcription polymerase chain reaction (RT-PCR) is a widely used screening method and currently adopted as the standard diagnostic method for suspected COVID-19 cases. Accounting for the high false positives in RT-PCR tests, readily available medical imaging methods, such as chest computed tomography (CT) have been advocated as a possibility to identify COVID-19 patients

L. Karlinsky et al. (Eds.): ECCV 2022 Workshops, LNCS 13807, pp. 605–620, 2023.
https://doi.org/10.1007/978-3-031-25082-8_41

[10,15,59]. Deep learning (DL) has demonstrated impressive results in COVID-19 detection from CT [17]. However, their performance generally depends on the prevalence of confounding lung pathologies within the patient cohort. Diseases such as community-acquired pneumonia or influenza have roughly similar visual manifestations in chest CT volumes, which increases the challenges for accurately detecting COVID-19. In addition, the COVID-19 Positives can have different degrees of infection severity. For example, higher severity is usually consistent with increased regions of multifocal and bilateral ground-glass opacities and consolidations [40]. Detecting different degrees of COVID-19 severity entails learning robust feature representations. This situation motivates researchers to apply strategies to enhance feature representations for effectively discriminating between COVID-19 Positives and COVID-19 Negatives, as well as for detecting the COVID-19 severity.

Tarvainen and Valpola [53] suggested that consistency regularization guides the two networks of the mean teacher model to adaptively correct their training errors and attain consistent predictions by improving each other feature representations. Consistency regularization has been widely used to learn meaningful feature representations [12,37,47]. For instance, in semi-supervised learning (SSL), consistency regularization based methods have proven useful in improving representation learning by harnessing unlabeled data [11,13,30,32,44,48]. However, while successful, consistency based methods ignore the structure across different volumes and might impose incoherent consistency [38].

Mutual information (MI) based representation learning [19,23] methods have been getting attention due to their ability to learn representations that can be employed for further downstream tasks. Though they have achieved promising results, MI estimation is computationally expensive due to the density estimation step. In this work, we argue that maximizing the mutual information on the latent space of DRN, which uses as input perturbated views of the same image, enhances the representations of intricate patterns to boost the ability of deep models to generalize to unseen data. In particular, we propose to use the least squared mutual information (LSMI) [50], a powerful density ratio method that has found applications across various domains in machine learning. Our contributions are:

- We propose to use a dependency estimation regularization on the latent space and call our model dual role network (DRN). Our proposal involves a loss function that combines the supervised objective with consistency regularization and LSMI to ensure consistency across predictions and MI maximization on nonlinear projections of the DRN latent space.
- We evaluate the proposed model on the COVID-19 detection and COVID19 severity detection tasks and achieved favorable results compared to the baseline of the 2nd COV19D competition [24][1]. Our proposal ranked ninth among the ten methods for COVID-19 detection and fifth for COVID-19 severity detection.

The remainder of this paper is organized as follows. In Sect. 2, we give an overview of the related work on COVID-19 detection. We describe the proposed

---

[1] https://mlearn.lincoln.ac.uk/eccv-2022-ai-mia/.

method, including the dependency measure in Sect. 3. Next, in Sect. 4, we discuss the experimental results. Finally, we draw our conclusions in Sect. 5.

2   Related Works

COVID-19 Detection methods can be roughly categorized into three categories. The first category involves methods that perform lung lesion segmentation or detection followed by classification [45,55]. The second category includes methods that simultaneously perform lesion segmentation and classification [16,58]. The major drawback of these two categories is the need of experts to annotate the lung lesions for segmentation.

More akin to our work is a third category of methods that perform direct CT detection. Miron *et al.* [35] proposed an slice-level and a volumetric approach using Convolutional Neural Network (CNN). Their volumetric approach employs a 3D ResNet50 [18] with inflated weights achieving better results than their slice-slice level approach. Wang *et al.* [56] proposed DeCoVNet, a 3D-CNN with residual blocks [18]. Zhang and Wen [62] adopted the Swin Transformer [33] to extract features from masked lung CT slices and aggregated the slice-level representations with global pooling to feed a fully connected layer and output the classification scores. Anwar [4] used AutoML to search for the most appropriate CNN architecture for COVID-19 detection from CT volumes. Hsu *et al.* [22] proposed 2D and 3D models to cope with COVID-19 detection. Their 2D model uses Swin Transformer [33] backbone followed by outlier removal and Wilcoxon signed-rank test to output the predictions. Their 3D proposal employs 2D Resnet50 [18] to extract features from the CT slices followed by context feature learning based on a self-attention mechanism to aggregate the feature representations across the volume for the final classification.

Tan *et al.* [52] adopted the 3D CNN-BERT with ResNet34 [18] backbone to extract features from a series of masked lung CT slices. Next, the authors proposed aggregating the outputs from 3D CNN-BERT via pooling and incorporated a Multilayer Perceptron to obtain the COVID-19 classification scores. Hou *et al.* [21] proposed the Contrastive Mixup Classification for COVID-19 Detection (CMC-COV19D). CMC-COV19D uses an adaptative training strategy combining a classification loss with Mixup and contrastive loss as an auxiliary task. The augmented views corresponding to Mixup and the negative and positive pairs pass through a shared encoder CNN to obtain the predictions. Kollias *et al.* [25] proposed a CNN plus Recurrent Neural Networks (RNN) to detect between COVID-19 and its severity from chest CT. Their CNN-RNN constitutes the baseline method to benchmark in the COV19-CT-DB [24].

Deviating from the described works, we propose dual role network combining consistency regularization with mutual information maximization via LSMI to learn useful data representations for detecting COVID-19 and its severity. In this work, we explore a 3D ConvNeXt [34] as backbone for the dual role network. To our knowledge, no previous works have attempted to incorporate the perspectives of contrastive learning using LSMI. Specifically, we use LSMI as a dependency measure to circumvent intractable mutual information estimation. Our model is generalizable to different medical imaging tasks and is backbone agnostic.

3   Background and Proposed Method

In this section, we provide an overview of our method. While we are motivated by the COVID-19 detection and COVID-19 severity detection problems, this method can be generalized across different medical imaging tasks.

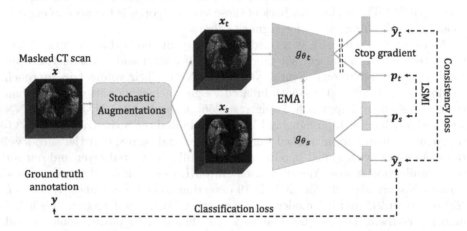

Fig. 1. Overview of the proposed method. The masked chest CT scan is stochastically augmented into two views, x_t and x_s, respectively. Both transformations pass feed-forward through the dual role network which output the d-dimensional features projection p and two-dimensional vector $\hat{y}$ for classification. We involve three loss terms to compute a total loss and optimize g_{θ_s}. First, the classification loss between ground truth annotation y and $\hat{y}_s$. Second, the consistency loss between $\hat{y}_t$ and $\hat{y}_s$. Third, the LSMI between both networks feature projection p_t and p_s, respectively. The total loss is backpropagated to optimize the network g_{θ_s} parameters, and we stop the gradient on the network g_{θ_t} to update its parameters using EMA.

Figure 1 provides a schema of the proposed method. Given an input masked chest CT scan $x \in \mathbb{R}^{h \times w \times c}$, and its corresponding annotation label y, we carry out a set of stochastic perturbations on x to obtain two different views of the input instance, denoted as x_s and x_t. Subsequently, we train two networks g_{θ_s} and g_{θ_t} and encourage their output to consistently match. Both the networks share the same backbone architecture (ConvNeXt [34] in this work), but these are parameterized by θ_s and θ_t, respectively. During training, we optimize the parameters θ_s via backpropagation while θ_t are updated using exponential moving average (EMA):

$$\theta_{t\tau} = \eta\theta_{t\tau-1} + (1 - \eta)\theta_{s\tau} \tag{1}$$

where $\eta \in [0,1]$ is the network decay rate coefficient and τ is the training step. One can notice that the teacher network operates like model ensembling with average exponential decay that have demonstrated usefulness to boost models performance [39].

To ensure learning robust feature representations, we equip both networks with a projection head on top of the last hidden layer that outputs the latent code used by the classification head. One head outputs $\hat{\boldsymbol{y}}$ for the mainstream predictions concerning the COVID-19 detection tasks (i.e., COVID-19 detection and COVID-19 severity detection). The second head outputs a d-dimensional projection of the feature representation resulting from the last hidden layer, defined as $p \in \mathbb{R}^d$. The projection heads are multilayer perceptrons with two linear layers followed by layer normalization and GELU.

To optimize g_{θ_s}, we use the following loss function:

$$J(\theta_s) = H(\hat{\boldsymbol{y}}_s, \boldsymbol{y}) + \lambda \underbrace{d_c(\hat{\boldsymbol{y}}_s, \hat{\boldsymbol{y}}_t)}_{\substack{\text{Consistency} \\ \text{regularization}}} + \beta \underbrace{d_l(\boldsymbol{p}_s, \boldsymbol{p}_t)}_{\text{LSMI}} \tag{2}$$

where H is the cross-entropy loss between the g_{θ_s} output, $\hat{\boldsymbol{y}}_s$, and the ground truth label $\boldsymbol{y}$. The output of the two classification heads are $\hat{\boldsymbol{y}}_s$ and $\hat{\boldsymbol{y}}_t$. d_c is the mean squared error (MSE) and d_l is the LSMI regularizer between the output from projection heads, $\boldsymbol{p}_s$ and $\boldsymbol{p}_t$. For d_l one can also use Kullback Leibler (KL) divergence, and related information theoretic measures. λ and β are hyperparameters to control the contribution of the different regularizers.

3.1 Dependence Measure

We use **least squared loss mutual information (LSMI)** as a dependence measure in this work. LSMI belongs to the family of f-divergences [3]. Thus, it is a non-negative measure that is zero only if the random variables are independent. Commonly used f-divergences, such as KL, require densities estimation to mutual information [49], yet density estimation is a known difficult problem [46,54]. Different to f-divergences requiring densities estimation, LSMI approximates mutual information from data via direct density ratio [51]. LSMI involves simple steps of using linear equations which makes it computationally efficient and numerically more stable compared to the f-divergences requiring density estimation [49]. It also does not assume data drawn from parametric distributions. Furthermore, to the best of our knowledge, there has not been previous work combining deep models with LSMI as a representation learning metric. To estimate LSMI we take a direct density ratio estimation approach [51]. This leads [60,61] to the estimator[2]:

---

[2] For the complete derivation of the LSMI estimator, readers are referred to [60].

$$\widehat{\mathrm{LSMI}}(\{(\boldsymbol{p}_{s_i}, \boldsymbol{p}_{t_i})\}_{i=1}^n) = \frac{1}{2n}\mathrm{tr}\left(\mathrm{diag}\left(\widehat{\boldsymbol{\alpha}}\right)\boldsymbol{K}\boldsymbol{L}\right) - \frac{1}{2},$$

where $\mathrm{tr}(\cdot)$ is the trace operator, $\boldsymbol{K}$ is the Gram matrix for $\boldsymbol{p}_s$ and $\boldsymbol{L}$ is the Gram matrix for $\boldsymbol{p}_t$, and $\widehat{\boldsymbol{\alpha}}$ is the model parameter written by [51]:

$$\widehat{\boldsymbol{\alpha}} = \left(\widehat{\boldsymbol{H}} + \delta\boldsymbol{I}_n\right)^{-1}\widehat{\boldsymbol{h}}, \quad \widehat{\boldsymbol{H}} = \frac{1}{n^2}(\boldsymbol{K}\boldsymbol{K}^\top)\circ(\boldsymbol{L}\boldsymbol{L}^\top), \quad \widehat{\boldsymbol{h}} = \frac{1}{n}(\boldsymbol{K}\circ\boldsymbol{L})\mathbf{1}_n.$$

where δ is a regularizer, and we use the Gaussian kernel:

$$\boldsymbol{K}_{ij}=\exp\left(-\frac{\|\boldsymbol{p}_{s_i} - \boldsymbol{p}_{s_j}\|_2^2}{2\sigma_{p_s}^2}\right), \quad \boldsymbol{L}_{ij}=\exp\left(-\frac{\|\boldsymbol{p}_{t_i} - \boldsymbol{p}_{t_j}\|_2^2}{2\sigma_{p_t}^2}\right),$$

where $\sigma_{p_s} > 0$ and $\sigma_{p_t} > 0$ are the Gaussian width.

4   Experimental Results

4.1   Datasets Description

We conducted the experiments to validate the effectiveness of our model on the COVID19-CT-Database (COV19-CT-DB) [5,24–29]. This dataset is provided to benchmark the models on the 2nd COV19D Competition of the ECCV2022 Workshop: AI-enabled Medical Image Analysis - Digital Pathology & Radiology/COVID19[3].

COVID-19 Detection Challenge. The training set involves 1992 chest Computer Tomography (CT) scans: 882 are COVID-19 Positives cases, and 1110 are COVID-19 Negatives cases. The validation set has a total of 504 CT scans: 215 COVID-19 Positives, and 289 COVID-19 Negatives cases, respectively. Experienced radiologists made the annotations of the CT scans with about 98% degree of agreement [24–26], which makes the COV19-CT-DB appealing to validate our methods because it was not built-up using RT-PCR results as ground truth labels. Further information on the dataset is available in [24].

COVID-19 Severity Detection Challenge. The training set for severity detection comprises 258 CT scans: 85 are Mild severity, 62 are Moderate severity, 83 are Severe severity, and 26 are Critical severity. The validation set has a total of 59 CT scans: with 22 Mild severity, 10 Moderate severity, 22 Severe severity, and 5 as Critical severity. For more information about this dataset description, readers are referred to [24].

In addition to the data provided for the 2nd COV19D Competition, **we used the publicly available[4] STOIC dataset to pretrain the backbone used in our model.** The STOIC dataset consists of 2000 CT scans randomly selected from a cohort of 10000 subjects suspected of being COVID-19 Positives. The

---

[3] https://vcmi.inesctec.pt/aimia_eccv/.
[4] https://stoic2021.grand-challenge.org/.

PCR of these 2000 subjects is provided, enabling us with 1051 COVID-19 Positives and 949 COVID-19 Negatives subjects. We randomly split the data with stratification into training and validation. Finally, to pretrain our ConvNeXt backbone, we employed 930 COVID-19 Positives and 870 COVID-19 Negatives subjects in training, and for validation we used 121 COVID-19 Positives and 79 COVID-19 Negatives.

4.2 Implementation and Training Details

We implemented our proposal using Pytorch[5] upon the PyTorch Image Models [57]. We employed the U-net [20] to extract the lung masks from the CT scans to remove non-lung information. Hounsfield units were clipped to the intensity window $[-1150, 350]$ and normalized to the range $[0, 1]$. The masked CT scans were subsequently resized to $224 \times 224 \times 112$. Although our method is backbone agnostic, in this implementation we adopt ConvNeXt [34]. To cope with the CT scans for COVID-19 detection and severity, we extended the ConvNeXt [34] to 3D ConvNeXt to harness the spatial information within adjacent slices. Besides, we inflated the weights from the 2D ConvNeXt-B [34] pre-trained on ImageNet to initialize the learning parameters.

We trained the models with AdamW during 300 epoch and early stopping equal 100 epochs. Following [34], we also used layer scale with initial value of $1e - 6$. We linearly ramped up the learning rate during 20 epochs to $4e - 5$, subsequently cosine decaying schedule is used. The weight decay is also scheduled to go from $2e - 5$ to $2e - 2$ following a cosine schedule. The EMA decay, η, is equal to 0.9998. We also used stochastic drop length equal to 0.4. Besides, we employed label smoothing equal to 0.4. The consistency λ and contrastive β weights were set equal to 0.1 and 0.5, with a linear ramp up during the first 30 epochs. To find these values we used the Tree-structured Parzen Estimator algorithms [8,9] from Optuna[6] [2]. The values for λ and β proved to have high impact on the performance of the proposed model. As a general rule, we found that good combinations are with $\beta \leq \lambda$, $0 < \beta \leq 0.3$, and $0.3 \leq \lambda \leq 0.6$. The dimension, d, of the projection vectors p_s and p_t is set to 256. We used Monai[7] to apply images transformations and the stochastic augmentations consisting of Gaussian smoothness, contrast adjustment, random zooming and intensity shifting. To select the network for inference at test time, we validated g_{θ_s} and g_{θ_t} at each validation step and kept the one with the best performance in terms of macro F1-score. Our implementation is available at https://gitlab.com/aberenguer/MT-ConvNeXt-COVID-19.

4.3 COVID-19 Detection

We summarize the results of our DRN with consistency regularization and various information theory loss terms in Table 1. We use InfoMax [19], KL,Jensen

---

[5] https://pytorch.org/.

[6] https://optuna.readthedocs.io/en/stable/index.html.

[7] https://monai.io/.

Shannon (JS) divergence, and LSMI to assess their impact on the effectiveness of our model in coping with the COVID-19 detection task. We decided to employ InfoMax, KL, and JS because previous experimental evidence [6,19,42] has demonstrated these f-divergences are also suitable for deep mutual information estimation. From Table 1, one can notice that our DRN combining consistency regularization and MI maximization is advantageous to boosting the performance of deep models in COVID-19 detection. Furthermore, LSMI regularization (DRN-MSE-LSMI) yielded the best macro F1-score compared to InfoMax, KL and JS. All the variants summarized in Table 1 outperformed the baseline of the COVID-19 detection benchmark [24] whose macro F1-score was reported equal to 0.77. The results suggest that LSMI, used as a representation learning metric, has appealing advantages over other f-divergences [49] to reduce the discrepancy between two probability distributions with similar underlying factors while keeping their "relevant" information to improve deep models' discriminative ability.

Table 1. The effects of various dependency measures (KL, InfoMax, JS, and LSMI) on the performance of our DRN with consistency regularization for COVID-19 detection. We report the macro F1-score with a threshold equal to 0.5 on the validation set of the COVID-19 Detection Challenge.

Model	Classification loss	Consistency loss (d_c)	Dependency measure (d_l)				Macro F1-score
	CE	MSE	InfoMax	KL	JS	LSMI	
DRN-MSE-InfoMax	✓	✓	✓	✗	✗	✗	0.86
DRN-MSE-KL	✓	✓	✗	✓	✗	✗	0.88
DRN-MSE-JS	✓	✓	✗	✗	✓	✗	0.89
DRN-MSE-LSMI	✓	✓	✗	✗	✗	✓	0.90

Figure 2 portrays examples of CT slices corresponding to correct and incorrect predictions yielded by DRN-MSE-LSMI for COVID-19 detection. One can notice from Fig. 2 (A–B) that DRN-MSE-LSMI adequately detected COVID-19 positives. On the other hand, in Fig. 2 (C–D), one can see that DRN-MSE-LSMI yielded false positives for those CT scans corresponding to subjects with confounding pathologies that exhibit multifocal abnormalities such as lobar consolidation and tree-in-bud that are more related to bacterial infections. It reveals that DRN-MSE-LSMI still lacks the capabilities to cope with all those COVID-19 negatives with lung abnormalities indicating infection but not corresponding to COVID-19. Analyzing the COVID-19 negatives, one can observe in Fig. 2 (E–F) that DRN-MSE-LSMI correctly detected subjects without findings (E) and those with anomalies corresponding to other infections (F) but not COVID-19. The DRN-MSE-LSMI also produced some false negatives, mainly for subjects with unifocal ground glass opacities, as shown in Fig. 2 (G–H).

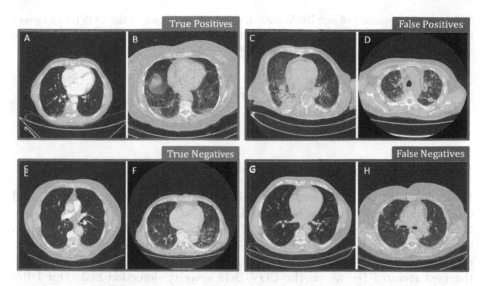

Fig. 2. Examples of predictions for COVID-19 detection. From A to B: true positives, C to D: false positives, E to F: true negatives, and G to H: false negatives. CT slices extracted from the validation set of the COV19-CT-DB.

Based on the results reported in Table 1, we selected our DRN-MSE-LSMI for final submission to the COVID-19 Detection Challenge of the 2nd COV19D Competition [24]. In Table 2, we report the performance of our DRN-MSE-LSMI on the COVID-19 Detection test set, which comprised more than 5000 chest CT scans. Compared to the CNN-RNN baseline, our DRN-MSE-LSMI achieved favorable results. However, considering the performance yielded by DRN-MSE-LSMI on validation, we noticed that our model sharply dropped in performance. Although similar results have been observable in the medical imaging community [14,41], the outcomes draw our attention to improving the model's trustworthiness at test time. We conjecture that using cross-entropy in our loss function (Eq. 2) is perhaps inadequate as it can guide the models to yield overconfident predictions [36]. An alternative would be the focal loss [31] that has proven effective in increasing the model robustness at test time [36]. Sangalli *et al.* [43] also suggested a constrained optimization to emphasize reduction of the false positives while keeping an adequate rate of true positives. Besides, reducing the bias effects [1] could also improve model reliability at test time. Furthermore, the choices of random augmentations constitute a critical aspect of our model that remained unexplored in this work. We believe our proposal could benefit more from stochastic transformations learned directly from data to improve discriminative ability at test time. Thereby, despite our promising results, several research questions remain to be investigated in our further work.

Table 2. Performance of our DRN-MSE-LSMI on the test set of the COVID-19 Detection Challenge and comparison to the competition baseline [24]. Our predictions were made by thresholding the output probabilities at 0.5.

Model	Macro F1-score
DRN-MSE-LSMI	0.78
CNN-RNN [24]	0.69

4.4 COVID-19 Severity Detection

In Table 3 we report the results of DRN with consistency regularization and different dependency measures, i.e., KL, InfoMax, JS, and LSMI, to evaluate their impact on the performance for COVID-19 severity detection. Similar to the trend exhibited by DRN in the above section, combining the consistency regularization and the LSMI (DRN-MSE-LSMI) for MI maximization achieved the best macro F1-score on the COVID-19 severity detection task. Our DRN also attained superior results compared to the macro F1-score of 0.63 corresponding to the baseline for COVID-19 Severity Detection Challenge [24]. The experimental results on the COVID-19 severity detection task also support the advantages of using LSMI for deep representation learning based on information theory. While KL and LSMI share similar properties [49], the better performance of LSMI could be explained by its numerical stability compared to KL which is rather susceptible to outliers.

Table 3. The effects of various dependency measures (KL, InfoMax, JS, and LSMI) on the performance of our DRN with consistency regularization for COVID-19 severity detection. We report the macro F1-score on the validation set of the COVID-19 Severity Detection Challenge.

Model	Classification loss	Consitency loss (d_c)	Dependency measure (d_l)				Macro F1-score
	CE	MSE	InfoMax	KL	JS	LSMI	
DRN-MSE-InfoMax	✓	✓	✓	✗	✗	✗	0.69
DRN-MSE-KL	✓	✓	✗	✓	✗	✗	0.71
DRN-MSE-JS	✓	✓	✗	✗	✓	✗	0.72
DRN-MSE-LSMI	✓	✓	✗	✗	✗	✓	0.73

We illustrate different examples of correct and incorrect predictions for COVID-19 severity detection in Fig. 3. One can observe that DRN-MSE-LSMI also yielded adequate results. In Fig. 3 (A–B), (E–F), (I–J), and (M–N), one can see that, for all severity categories, DRN-MSE-LSMI could detect the degree of severity regardless of the difficulties in distinguishing between them. However,

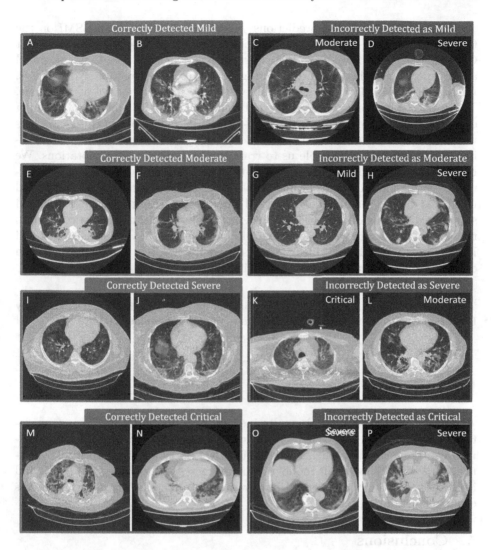

Fig. 3. Examples of correct and incorrect predictions for COVID-19 severity detection. From A to B: correctly detected Mild, C to D: wrongly detected as Mild, E to F: accurately detected Moderate, G to H: incorrectly detected Moderate, I to J: properly detected as Severe, K to L mistakenly detected as Severe, M to N: correctly detected Critical, and O to P: erroneously detected as Critical. CT slices extracted from the validation set of the COV19-CT-DB.

in Fig. 3 (C–D) (G–H), (K–L), and (O–P), one can also observe that DRN-MSE-LSMI yielded several erroneous predictions. Most erroneous predictions corresponded to subjects that were Mild or Critical that DRN-MSE-LSMI detected as Moderate or Severe, respectively. Such incorrect predictions might be attributed to the underrepresentation of the Moderate and Critical categories in the training set.

We also submitted the predictions yielded by our DRN-MSE-LSMI on the test set of the COVID-19 Severity Detection Challenge [24] that had 265 chest CT scans. In Table 4, we provide its performance. One can note that our model also outperformed the baseline, encouraging our research direction. Nonetheless, DRN-MSE-LSMI also had a noticeable decrease in performance at test time. These outcomes are not unexpected if one considers the amount of training data for the challenging task of COVID-19 severity detection. In this sense, it is worth noting that our model is theoretically suitable for Semi-supervised learning (SSL), which might enable us to train it with limited data annotations. We believe that leveraging unannotated data would help scale the model's performance and reliability at test time. We could feed our model with labeled and unlabeled data to combine the statistical strengths between the unsupervised and supervised learning objectives [7]. When an unannotated sample data comes into the training, the supervised objective will become zero, but the consistency and LSMI terms can still act to learn representations. SSL is a compelling research direction to cope with COVID-19 severity detection.

Table 4. Performance of our DRN-MSE-LSMI on the test set of the COVID-19 Severity Detection Challenge and comparison to the competition baseline [24].

Model	Macro F1-score
DRN-MSE-LSMI	0.44
CNN-RNN [24]	0.40

Overall, the experimental results indicate that our DRN combining consistency regularization and contrastive learning via mutual information maximization is a promising approach for detecting COVID-19 and its severity. In addition, the evaluations of our model on the test sets for detecting COVID-19 and its severity highlighted potential improvements to increase the trustworthiness of our model during testing.

5   Conclusions

We presented a representation learning method using information theory for COVID-19 Detection and COVID-19 Severity Detection. The proposed estimator is non-parametric and does not require density estimation, unlike traditional MI based methods. Experiments demonstrate the effectiveness of our proposal compared to the baseline of the 2nd COV19D competition of the ECCV2022 Workshop: AI-enabled Medical Image Analysis - Digital Pathology & Radiology/COVID19. We found that DRN with LSMI for enhancing representation learning encourages the dual role network to achieve adequate results. Our future work will look into enhancing our model's trustworthiness. We will study the effects of different supervised terms, our choices for stochastic augmentations and leverage unlabeled data to attempt scaling the model performance with limited annotations.

Acknowledgement. We want to thank the organizers of the 2nd COV19D Competition occurring in the ECCV 2022 Workshop: AI-enabled Medical Image Analysis - Digital Pathology & Radiology/COVID19 for providing access to extensive and high-quality data to benchmark our model. This research has been partially financed by the European Union under the Horizon 2020 Research and Innovation programme under grant agreement 101016131 (ICOVID).

References

1. Abrahamyan, L., Ziatchin, V., Chen, Y., Deligiannis, N.: Bias loss for mobile neural networks. In: Proceedings of the IEEE/CVF International Conference on Computer Vision, pp. 6556–6566 (2021)
2. Akiba, T., Sano, S., Yanase, T., Ohta, T., Koyama, M.: Optuna: a next-generation hyperparameter optimization framework. In: Proceedings of the 25th ACM SIGKDD International Conference on Knowledge Discovery and Data Mining, pp. 2623–2631 (2019)
3. Ali, S.M., Silvey, S.D.: A general class of coefficients of divergence of one distribution from another. J. Roy. Stat. Soc. Ser. B (Methodol.) **28**(1), 131–142 (1966)
4. Anwar, T.: Covid19 diagnosis using autoML from 3D CT scans. In: Proceedings of the IEEE/CVF International Conference on Computer Vision Workshops, pp. 503–507, October 2021
5. Arsenos, A., Kollias, D., Kollias, S.: A large imaging database and novel deep neural architecture for Covid-19 diagnosis. In: 2022 IEEE 14th Image, Video, and Multidimensional Signal Processing Workshop (IVMSP), pp. 1–5. IEEE (2022)
6. Belghazi, M.I., et al.: Mutual information neural estimation. In: International Conference on Machine Learning, pp. 531–540. PMLR (2018)
7. Bengio, Y., Courville, A., Vincent, P.: Representation learning: a review and new perspectives. IEEE Trans. Pattern Anal. Mach. Intell. **35**(8), 1798–1828 (2013)
8. Bergstra, J., Bardenet, R., Bengio, Y., Kégl, B.: Algorithms for hyper-parameter optimization. In: Advances in Neural Information Processing Systems, vol. 24 (2011)
9. Bergstra, J., Yamins, D., Cox, D.: Making a science of model search: hyperparameter optimization in hundreds of dimensions for vision architectures. In: International Conference on Machine Learning, pp. 115–123. PMLR (2013)
10. Bernheim, A., et al.: Chest CT findings in coronavirus disease-19 (Covid-19): relationship to duration of infection. Radiology **295**(3), 200463 (2020)
11. Bortsova, G., Dubost, F., Hogeweg, L., Katramados, I., de Bruijne, M.: Semi-supervised medical image segmentation via learning consistency under transformations. In: Shen, D., et al. (eds.) MICCAI 2019. LNCS, vol. 11769, pp. 810–818. Springer, Cham (2019). https://doi.org/10.1007/978-3-030-32226-7_90
12. Chen, X., et al.: Anatomy-regularized representation learning for cross-modality medical image segmentation. IEEE Trans. Med. Imaging **40**(1), 274–285 (2020)
13. Cui, W., et al.: Semi-supervised brain lesion segmentation with an adapted mean teacher model. In: Chung, A.C.S., Gee, J.C., Yushkevich, P.A., Bao, S. (eds.) IPMI 2019. LNCS, vol. 11492, pp. 554–565. Springer, Cham (2019). https://doi.org/10.1007/978-3-030-20351-1_43
14. Cutillo, C.M., Sharma, K.R., Foschini, L., Kundu, S., Mackintosh, M., Mandl, K.D.: Machine intelligence in healthcare-perspectives on trustworthiness, explainability, usability, and transparency. NPJ Digit. Med. **3**(1), 1–5 (2020)

15. Dong, D., et al.: The role of imaging in the detection and management of Covid-19: a review. IEEE Rev. Biomed. Eng. **14**, 16–29 (2021)

16. Goncharov, M., et al.: CT-based Covid-19 triage: deep multitask learning improves joint identification and severity quantification. Med. Image Anal. **71**, 102054 (2021)

17. Greenspan, H., San José Estépar, R., Niessen, W.J., Siegel, E., Nielsen, M.: Position paper on COVID-19 imaging and AI: from the clinical needs and technological challenges to initial AI solutions at the lab and national level towards a new era for AI in healthcare. Med. Image Anal. **66**, 101800 (2020)

18. He, K., Zhang, X., Ren, S., Sun, J.: Deep residual learning for image recognition. In: Proceedings of the IEEE Conference on Computer Vision and Pattern Recognition, pp. 770–778 (2016)

19. Hjelm, D., et al.: Learning deep representations by mutual information estimation and maximization. In: International Conference on Learning Representations (ICLR), April 2019

20. Hofmanninger, J., Prayer, F., Pan, J., Röhrich, S., Prosch, H., Langs, G.: Automatic lung segmentation in routine imaging is primarily a data diversity problem, not a methodology problem. Eur. Radiol. Exp. **4**(1), 1–13 (2020). https://doi.org/10.1186/s41747-020-00173-2

21. Hou, J., Xu, J., Feng, R., Zhang, Y., Shan, F., Shi, W.: CMC-Cov19D: contrastive mixup classification for Covid-19 diagnosis. In: Proceedings of the IEEE/CVF International Conference on Computer Vision, pp. 454–461 (2021)

22. Hsu, C.C., Chen, G.L., Wu, M.H.: Visual transformer with statistical test for Covid-19 classification. arXiv preprint arXiv:2107.05334 (2021)

23. Hu, W., Miyato, T., Tokui, S., Matsumoto, E., Sugiyama, M.: Learning discrete representations via information maximizing self-augmented training. In: International Conference on Machine Learning, pp. 1558–1567. PMLR (2017)

24. Kollias, D., Arsenos, A., Kollias, S.: AI-MIA: Covid-19 detection & severity analysis through medical imaging. arXiv preprint arXiv:2206.04732 (2022)

25. Kollias, D., Arsenos, A., Soukissian, L., Kollias, S.: MIA-Cov19d: Covid-19 detection through 3-D chest CT image analysis. arXiv preprint arXiv:2106.07524 (2021)

26. Kollias, D., Arsenos, A., Soukissian, L., Kollias, S.: MIA-Cov19d: Covid-19 detection through 3-D chest CT image analysis. In: Proceedings of the IEEE/CVF International Conference on Computer Vision, pp. 537–544 (2021)

27. Kollias, D., et al.: Deep transparent prediction through latent representation analysis. arXiv preprint arXiv:2009.07044 (2020)

28. Kollias, D., Tagaris, A., Stafylopatis, A., Kollias, S., Tagaris, G.: Deep neural architectures for prediction in healthcare. Complex Intell. Syst. **4**(2), 119–131 (2018)

29. Kollias, D., et al.: Transparent adaptation in deep medical image diagnosis. In: TAILOR, p. 251–267 (2020)

30. Li, X., Yu, L., Chen, H., Fu, C.W., Xing, L., Heng, P.A.: Transformation-consistent self-ensembling model for semisupervised medical image segmentation. IEEE Trans. Neural Netw. Learn. Syst. **32**(2), 523–534 (2021)

31. Lin, T., Goyal, P., Girshick, R., He, K., Dollár, P.: Focal loss for dense object detection. IEEE Trans. Pattern Anal. Mach. Intell. **42**(2), 318–327 (2020)

32. Liu, Q., Yu, L., Luo, L., Dou, Q., Heng, P.A.: Semi-supervised medical image classification with relation-driven self-ensembling model. IEEE Trans. Med. Imaging **39**(11), 3429–3440 (2020)

33. Liu, Z., et al.: Swin transformer: hierarchical vision transformer using shifted windows. In: Proceedings of the IEEE/CVF International Conference on Computer Vision, pp. 10012–10022 (2021)

34. Liu, Z., Mao, H., Wu, C.Y., Feichtenhofer, C., Darrell, T., Xie, S.: A convnet for the 2020s. In: Proceedings of the IEEE/CVF Conference on Computer Vision and Pattern Recognition, pp. 11976–11986 (2022)
35. Miron, R., Moisii, C., Dinu, S., Breaban, M.E.: Evaluating volumetric and slice-based approaches for Covid-19 detection in chest CTs. In: Proceedings of the IEEE/CVF International Conference on Computer Vision, pp. 529–536 (2021)
36. Mukhoti, J., Kulharia, V., Sanyal, A., Golodetz, S., Torr, P., Dokania, P.: Calibrating deep neural networks using focal loss. Adv. Neural. Inf. Process. Syst. **33**, 15288–15299 (2020)
37. Ning, Z., Tu, C., Di, X., Feng, Q., Zhang, Y.: Deep cross-view co-regularized representation learning for glioma subtype identification. Med. Image Anal. **73**, 102160 (2021)
38. Peng, J., Pedersoli, M., Desrosiers, C.: Boosting semi-supervised image segmentation with global and local mutual information regularization. CoRR abs/2103.04813 (2021)
39. Polyak, B.T., Juditsky, A.B.: Acceleration of stochastic approximation by averaging. SIAM J. Control. Optim. **30**(4), 838–855 (1992)
40. Prokop, M., Van Everdingen, W., van Rees, V., et al.: CoRads: a categorical CT assessment scheme for patients suspected of having Covid-19-definition and evaluation. Radiology **296**, E97–E104 (2020)
41. Ranschaert, E.R., Morozov, S., Algra, P.R.: Artificial Intelligence in Medical Imaging: Opportunities. Applications and Risks. Springer, Cham (2019). https://doi.org/10.1007/978-3-319-94878-2
42. Rubenstein, P., Bousquet, O., Djolonga, J., Riquelme, C., Tolstikhin, I.O.: Practical and consistent estimation of f-divergences. In: Wallach, H., Larochelle, H., Beygelzimer, A., d'Alché-Buc, F., Fox, E., Garnett, R. (eds.) Advances in Neural Information Processing Systems, vol. 32. Curran Associates, Inc. (2019)
43. Sangalli, S., Erdil, E., Hötker, A., Donati, O., Konukoglu, E.: Constrained optimization to train neural networks on critical and under-represented classes. Adv. Neural. Inf. Process. Syst. **34**, 25400–25411 (2021)
44. Seo, H., Yu, L., Ren, H., Li, X., Shen, L., Xing, L.: Deep neural network with consistency regularization of multi-output channels for improved tumor detection and delineation. IEEE Trans. Med. Imaging **40**(12), 3369–3378 (2021)
45. Shi, F., et al.: Large-scale screening to distinguish between Covid-19 and community-acquired pneumonia using infection size-aware classification. Phys. Med. Biol. **66**(6), 065031 (2021)
46. Silverman, B.W.: Density Estimation for Statistics and Data Analysis. Routledge, Abingdon (2018)
47. Sinha, S., Dieng, A.B.: Consistency regularization for variational auto-encoders. Adv. Neural. Inf. Process. Syst. **34**, 12943–12954 (2021)
48. Su, H., Shi, X., Cai, J., Yang, L.: Local and global consistency regularized mean teacher for semi-supervised nuclei classification. In: Shen, D., et al. (eds.) MICCAI 2019. LNCS, vol. 11764, pp. 559–567. Springer, Cham (2019). https://doi.org/10.1007/978-3-030-32239-7_62
49. Sugiyama, M.: Machine learning with squared-loss mutual information. Entropy **15**(1), 80–112 (2013)
50. Sugiyama, M., Suzuki, T., Kanamori, T.: Density Ratio Estimation in Machine Learning, 1st edn. Cambridge University Press, New York (2012)
51. Suzuki, T., Sugiyama, M.: Sufficient dimension reduction via squared-loss mutual information estimation. In: AISTATS (2010)

52. Tan, W., Liu, J.: A 3D CNN network with BERT for automatic Covid-19 diagnosis from CT-scan images. In: Proceedings of the IEEE/CVF International Conference on Computer Vision, pp. 439–445 (2021)
53. Tarvainen, A., Valpola, H.: Mean teachers are better role models: weight-averaged consistency targets improve semi-supervised deep learning results. In: Proceedings of the 31st International Conference on Neural Information Processing Systems, pp. 1195–1204 (2017)
54. Vapnik, V.: The Nature of Statistical Learning Theory. Springer, New York (1999). https://doi.org/10.1007/978-1-4757-3264-1
55. Wang, B., et al.: AI-assisted CT imaging analysis for Covid-19 screening: building and deploying a medical AI system. Appl. Soft Comput. **98**, 106897 (2021)
56. Wang, X., et al.: A weakly-supervised framework for Covid-19 classification and lesion localization from chest CT. IEEE Trans. Med. Imaging **39**(8), 2615–2625 (2020)
57. Wightman, R.: Pytorch image models (2019). https://github.com/rwightman/pytorch-image-models, https://doi.org/10.5281/zenodo.4414861
58. Wu, Y.H., et al.: JCS: an explainable Covid-19 diagnosis system by joint classification and segmentation. IEEE Trans. Image Process. **30**, 3113–3126 (2021)
59. Xie, X., Zhong, Z., Zhao, W., Zheng, C., Wang, F., Liu, J.: Chest CT for typical coronavirus disease 2019 (Covid-19) pneumonia: relationship to negative RT-PCR testing. Radiology **296**(2), E41–E45 (2020)
60. Yamada, M., Sigal, L., Raptis, M., Toyoda, M., Chang, Y., Sugiyama, M.: Cross-domain matching with squared-loss mutual information. IEEE TPAMI **37**(9), 1764–1776 (2015)
61. Yamada, M., Sugiyama, M.: Cross-domain object matching with model selection. In: AISTATS (2011)
62. Zhang, L., Wen, Y.: A transformer-based framework for automatic Covid19 diagnosis in chest CTs. In: Proceedings of the IEEE/CVF International Conference on Computer Vision, pp. 513–518 (2021)

Spatial-Slice Feature Learning Using Visual Transformer and Essential Slices Selection Module for COVID-19 Detection of CT Scans in the Wild

Chih-Chung Hsu(✉) ⓘ, Chi-Han Tsai, Guan-Lin Chen, Sin-Di Ma, and Shen-Chieh Tai

Institute of Data Science, National Cheng Kung University, No. 1, University Road, Tainan City, Taiwan
cchsu@gs.ncku.edu.tw

Abstract. Computed tomography (CT) imaging could be convenient for diagnosing various diseases. However, the CT images could be diverse since their resolution and number of slices are determined by the machine and its settings. Conventional deep learning models are hard to tickle such diverse data since the essential requirement of the deep neural network is the consistent shape of the input data in each dimension. A way to overcome this issue is based on the slice-level classifier and aggregating the predictions for each slice to make the final result. However, it lacks slice-wise feature learning, leading to suppressed performance. This paper proposes an effective spatial-slice feature learning (SSFL) to tickle this issue for COVID-19 symptom classification. First, the semantic feature embedding of each slice for a CT scan is extracted by a conventional 2D convolutional neural network (CNN) and followed by using the visual Transformer-based sub-network to deal with feature learning between slices, leading to joint feature representation. Then, an essential slices set algorithm is proposed to automatically select a subset of the CT scan, which could effectively remove the uncertain slices as well as improve the performance of our SSFL. Comprehensive experiments reveal that the proposed SSFL method shows not only excellent performance but also achieves stable detection results.

Keywords: Spatial-slice correlation · COVID-19 classification · Convolutional neural networks · Visual transformer · Computed tomography

1 Introduction

SARS-CoV-2 (COVID-19), a contagious disease that has caused an extremely high infection rate in the world, can cause many infections in a short time. Doctors need to use the image to confirm the severity of the infection in the lung medical imaging. However, human judgment resources are limited, and a large amount of data is impossibly labeled on time. Therefore, many studies were proposed to solve the problem of the insufficient workforce via deep learning methods and provide auxiliary systems for preliminary marking or judgment. The lack of a workforce leads to several problems. First, the difference in image resources caused by the types of image settings is mostly

© The Author(s), under exclusive license to Springer Nature Switzerland AG 2023
L. Karlinsky et al. (Eds.): ECCV 2022 Workshops, LNCS 13807, pp. 621–634, 2023.
https://doi.org/10.1007/978-3-031-25082-8_42

obtained by X-ray or computed tomography (CT) imaging schemes, in which the settings of the equipment and resolution are usually diverse, causing poor generalizability of the model. Second, unlike X-ray, CT scan produces more detailed and complete imaging information, so it is possible to see the inside of solid organs. However, rare studies focus on the CT scan due to the high diversity of the CT slices, compared to X-ray-based approaches [1,4,12,14,15].

There are many real-world challenges in computer vision, aiming to solve the problems of the COVID-19 dataset described above and classify slices. For example, Fang et al. [7] proposed enhanced local features based on convolution and deconvolution in CT scans to improve the classification accuracy. Singh et al. [6] use multi-objective differential evolution-based convolutional neural networks to resolve the difference between infected patients. Pathak et al. [28] proposed a transfer learning model to classify COVID-19 patients via the semantic features extracted by ResNet [9], in which the weights in the ResNet were pretrained by the ImageNet dataset.

In addition, Transformer has significantly contributed to current research in image classification or segmentation. For example, Jiang et al. [15] proposed a method that combined Swin-Transformer and Transformer in Transformer module to classify chest X-ray images into three classes: COVID-19, Pneumonia, and Normal (healthy). As reported by [15], the accuracy could achieve 0.9475. Le Dinh et al. [22] also have done some model experiments with chest X-ray images. They found that the Transformer-based models performed better than convolution-based models on all three metrics: precision, recall, and F1-Score. However, they mentioned that the datasets used in their work were from many open-source datasets. To some extent, this might affect the model's accuracy. Furthermore, X-ray images obtained from different machines have various image qualities, image color channels as well as resolutions. These factors have a significant impact on the model training pipeline. Although this uses the data proposed by the official competition, which contains a large number of thoracic CT images, the slice positions in the data are different, and the arrows manually marked by the hospital are included, which leads to the high complexity of the data and the difficulty of classification. CT scans results in medical image analysis are usually regarded as three-dimensional data, but in most traditional convolutional neural network (CNN) models, only two-dimensional data operations can be performed, and due to the data set provided by the competition, each case's total number of slices varies.

COV19-CT-DB dataset proposed by Kollias et al. [2,17–21] has large annotated COVID-19 and non-COVID-19 CT scans, which can improve the learning capability for deep learning methods. However, the resolution and length of CT scans in COV19-CT-DB were inconsistent between each CT scan; thus, deep neural networks trained by this dataset were difficult to achieve promising results. To deal with this issue, Chen et al. [3] proposed the 2D and 3D methods to explore the importance of slices of a CT scans. The 2D method, adaptive distribution learning with statistical testing (ADLeaST), combined statistical analysis with deep learning to determine whether the input CT scans is COVID-19 or non-COVID-19 by hypothesis testing. The COVID-19 and non-COVID-19 slices in a CT scans were mapped to follow specific distributions via a deep neural network that the slices without lung tissue can be excluded, thus increasing the stability and explainability of deep learning methods. However, the 2D method could be influenced by implicit slices (i.e., positive slices without obvious symptoms), which were randomly sampled from the COVID-19 CT scans for training. In the 3D method, a 3D volume-based CNN was proposed to automatically capture the relationship of slices via the self-attention mechanism in the Within-Slice-Transformer (WST) and Between-Slice-Transformer (BST) modules. However, the 3D methods suffered from a fatal overfitting issue due to the insufficient training samples and the complex architecture of Vision Transformer.

Although Chen et al. [3] made an improvement on the performance by center slice sampling on a CT scans with fixed percentages, we argue that the important slices could also be excluded since the sampling range was imprecise. This research proposed a novel two-step method for the challenge dataset, COV19-CT-DB [2, 17–21], to break the limitation. Firstly, a new strategy based on lung area estimating for slice selection was proposed, termed the essential slices selection algorithm. In a CT scans, the slice with a larger area of lung tissue is more likely to contain symptoms of COVID-19 since the lungs are the organs most affected by COVID-19. The spatial feature learning model (i.e., conventional convolutional neural network) will be trained by the slices consisting of the most lung tissue, thus making the embedding features more representative. Secondly, the embedding features from the spatial model pretrained on the selected slice are used to explore the context feature between slices in a CT scans. The two-step models considered the temporal variation of a CT scans for classifying without a limited range of slices.

2 Related Work

Due to the rapid increase in the number of new and suspected COVID-19 cases, there is an urgent need to detect COVID-19 from CT scans automatically using deep learning approaches. As Hou et al. [11] mentioned, the existing methods for COVID-19 diagnosis with CT scans can be roughly divided into two categories:

- Slice-level based approach. First, the infection regions are selected by semantic segmentation and followed by utilizing them for further COVID-19 classification.
- Training the classification model to diagnose COVID-19 directly with an end-to-end model. A 3D-CNN-based approach such as [11]. It explores the spatial-slice context feature representation via 3D convolution. In [11], the contrastive representation learning (CRL) and mixup pneumonia classification were adopted to further improve the performance. However, it is well-known that the computational complexity is relatively high.

Radu et al. [26] apply two ways to improve the performance from the baseline of the MIA-COV19D Competition within ICCV 2021 [18]. They addressed the challenge in two different ways: 1) treating the CT as a volume and thus using 3D convolutions, and 2) treating the CT as a set of slices. They rescale the given images to deal with a large number of slices per volume and large dimensions of an individual slice. The second approach adopts 2D convolutions and predictions at the slice level, followed by an aggregation step where the outputs at the slice level are aggregated into a response at the CT level. The best result of their work is 3D convolutions with label smoothing [27] with cross-entropy and Sharpness Aware Minimization [8] on top of stochastic gradient descent (SGD) as loss function and optimizer, respectively.

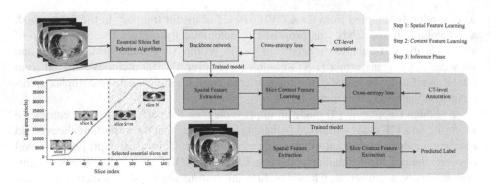

Fig. 1. Flowchart of the proposed two-step model based on spatial-slice feature learning module.

Weijun Tan et al. [30] proposed a segmentation-aware COVID-19 classification, enabling the unnecessary parts of the slices could be removed further. As the processed slices, a lung area estimation algorithm was adopted to eliminate the non-lung slices to form a subset for better feature representation. Then, the BERT-enabled 3D-CNN model was adopted to learn the context feature from the selected subset. However, directly learning the context features from selected raw data (i.e., slices) could be ineffective. Furthermore, the 3D convolution requires high computational resources, making it hard to deploy it to real-world applications.

Most of the peer methods show great performance on the public benchmark dataset. However, resolving the diverse issue caused by the varying number of slices and resolution over different CT scans is still hard. In this paper, we take advantage of peer methods to form the novel spatial-slice feature learning methods for effectively addressing this problem.

3   Spatial-Slice Feature Learning

In this paper, we mainly focus on improving the spatial-level model by the proposed essential slices set selection algorithm, as well as enhancing the stability of learning the context features. The selected slice set will be different for each CT scan and thus should be able to improve the training quality of the context feature learning model. The proposed slice set selection algorithm and two-step-aware models are introduced in Sect. 3.1 and Sect. 3.3, respectively. The model flowchart is shown in Fig. 1. First, the proposed essential slices set selection algorithm is adopted to find the most important slices to form a CT subset, followed by adopting them to learn a spatial model (traditional CNN models). In this way, the feature representation of each slice could be effective since the model is trained from the clean and high-quality slices. Then, the second-step model, termed slice context feature learning, is used to extract the relationship between two or more successive slices. Since the number of slices of different CT scans could vary, as the requirement of most CNNs requires a fixed shape, randomly sampling with a fixed number of slices is adopted in the second model to form the input data of the context feature learning model. Once the spatial and slice context feature learning models are trained, the identical producer is adopted to find the predicted result.

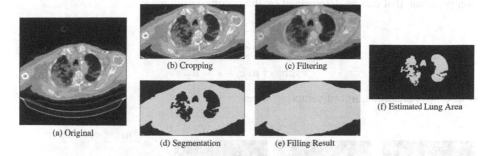

Fig. 2. Preprocessing slice for lung area estimating. From (a) to (f) are original slice, cropped slice, filtered slice, segmentation map, filling result obtained by binary dilation algorithm, lung area estimated by the difference between (d) and (e), respectively.

3.1 Essential Slices Set Selection Algorithm

Given a CT scan $C = \{\mathbf{X}_i\}_{i=1}^n$ with label y and n slices, we aim to discover the range of slices $C^* = \{\mathbf{X}_i\}_{i=s}^e$ which contains the largest area of lung tissue with start index s and end index e. Firstly, a cropping algorithm was employed to exclude the background of a CT scan slice. The cropped region $C_{crop} = \{(\mathbf{X}_{crop}^{(i)})\}_{i=s}^e$ can reduce the computational complexity and increase the estimating precision for all slices in a CT scan, as shown in Fig. 2(b). Then the minimum filtering operator with window size $k \times k$ is adopted to eliminate the noise caused by the CT scanner, as shown in Fig. 2(c). The filtering process can be defined as:

$$\mathbf{X}_{filter} = Filter(\mathbf{X}_{crop}, k). \tag{1}$$

Afterwards, the segmentation map $\mathbf{M}$ of the filtering slice can be determined by a specific threshold t, as shown in Fig. 2(d). The process can be written as:

$$\mathbf{M}[i,j] = \begin{cases} 0, & \text{if } \mathbf{X}_{filter}[i,j] < t \\ 1, & \text{if } \mathbf{X}_{filter}[i,j] >= t \end{cases} \tag{2}$$

where i is the width index and j is the height index of a slice. In order to discover the lung area, the binary dilation algorithm [32] was adopted to calculate the filling result $\mathbf{M}_{filling}$. The lung region can be obtained by the difference of the segmentation map and its filling map, as illustrated in Fig. 2(d)(e)(f). The lung area of a slice can be defined as:

$$Area(\mathbf{X}) = \sum_i \sum_j \mathbf{M}_{filling}(i,j) - \mathbf{M}(i,j). \tag{3}$$

Finally, the best selection for start index s and end index e is formulated as a optimization problem, that can be determined by the accumulating area:

$$\arg\max_{s,e} \sum_{i=s}^{e} Area(\mathbf{X}_i), \tag{4}$$

$$\text{subject to } e - s \leq n_c,$$

where n_c is the constraint of sampling slices.

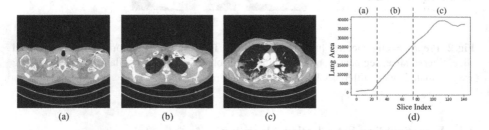

Fig. 3. Estimated chest area of COVID-19 CT scan. The curve in orange denotes the selected range of slices determined by the proposed method. (Color figure online)

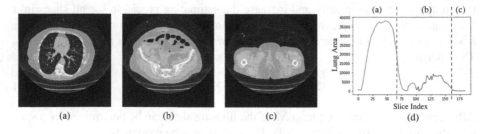

Fig. 4. Estimated chest area of non-COVID-19 CT scan. The curve in orange denotes the selected range of slices determined by the proposed method. (Color figure online)

A schematic of selection of slices of CT scan is shown in Fig. 3 and Fig. 4. After calculating the slice range C^*, we use the sub-dataset as the training data to train the spatial model by the binary cross-entropy loss, which can be defined as:

$$L_{bce} = -\frac{1}{N} \sum_{i=1}^{N} y_i \cdot \log(p_i) + (1 - y_i) \log(1 - p_i), \tag{5}$$

where $p_i = F_{2d}(\mathbf{X}_i; \theta_{2d})$, θ_{2d} is the trainable parameters of the spatial model, and N is the total number of selected slices.

3.2 Spatial Feature Learning

It is essential to select a powerful and effective backbone network to have a promising result. There are many state-of-the-art CNNs proposed recently, such as Efficient-Net [29], ConvNeXt [25], and Swin-Transformer [24]. Since EfficientNet provides relatively more pre-trained models using ImageNet [5], it benefits model development because it is unnecessary to train the backbone from scartch. As a result, EfficientNet-b3a [29, 31] was adopted as the backbone for semantic feature extraction. Since the feature dimension of the EfficientNet is high, an fully connected layer with embedding size 224 is concatenated to the last layer of EfficientNet to reduce the computational complexity for the proposed spatial-slice feature learning (SSFL) model. In the training phase, we randomly sampled eight CT scans and followed them by truncating the CT scans into their subset based on the proposed essential slices set selection algorithm. Then, sixteen slices were randomly sampled from the selected subset as a batch.

Although EfficientNet could effectively extract the spatial features, the context feature between slices is hard to discover yet. As stated previously, we discuss two different context feature mining strategies based on Visual Transformer in the following Subsections.

3.3 Slice Context Feature Learning

Swin-Transformer was proposed by Liu et al. [24]. Compared with conventional visual Transformer, the computational complexity of Swin-Transformer is greatly reduced. As the depth increases, the Swin-Transformer gradually merges image blocks to construct a hierarchical Transformer, which can be used as a general visual backbone network for tasks such as image classification, object detection, and semantic segmentation. Recently, Swin-Transformer-V2 was updated by Liu et al. [23], setting new performance records on four representative vision tasks, including ImageNet-V2 image classification, COCO object detection, ADE20K semantic segmentation, and Kinetics-400 video action classification.

To deal with the context feature learning, the semantic features extracted from the essential slices set via spatial model are re-arranged into a pseudo-image. This pseudo-image contains the relationship between slices' features at different locations, which Swin-Transformer could judiciously capture their context feature. It is well-known that the Transformer-based approach could capture an even longer dependency correlation than traditional recurrent-based networks.

It is also essential to verify that the long-range dependency correlation benefits the COVID-19 classification for diverse CT scans. In this way, a Long Short-Term Memory (LSTM) [10] Network is adopted as the context feature learning module. Unlike traditional recurrent neural networks (RNN), the training stability of LSTM is relatively high and effective since it can handle the vanishing gradient problem faced in RNN.

More specifically, since the slices context feature could be essential, it is crucial to exploit the correlation between two or even more successive slices. As the significant advantages of capturing long-range dependency brought by visual Transformers, Swin-Transformer is adopted in this paper to capture the context features from the extracted feature embedding of the individual slices. To verify the effectiveness of the visual Transformer, an LSTM is also adopted to capture the context feature as the baseline model.

First, all the slices of the input CT will be fed into the backbone network to extract their semantic features. A randomly sampling strategy is designed to form the features set. Specifically, we randomly sample k_s features from the CT scan as well as form them into the 2D representation, leading to a $k_s \times 224$ matrix. In this paper, the $k_s = 224$. Then, the Swin-Transformer is adopted to learn the context feature representation. Unlike the conventional visual Transformer, the local shifted window will restrict the range of self-attention within the window, making the computation more efficient. On the other hand, since the number of CT scans is diverse, some of the CT scans could be hard to meet the requirement of $k_s = 224$. In the case of an insufficient number of slices in a CT scan, we circularly sample the slices' features until the number of the features satisfies the predefined k_s. The version of the Swin-Transformer is the base-version, where the embedding dimension, depths of each block, and the number of the head are 128, (2,2,18,2), and (4,8,16,32), respectively.

We also developed the context feature learning based on LSTM to verify the effectiveness of the Transformer. Similarly, the input data of the LSTM is formed by aggregating k_l features randomly sampled from all slices. Then, we establish a four-layer Bidirectional LSTM with the 128 hidden units, followed by concatenating a linear layer with 64 neurons to reduce the dimensionality, as our base context feature learning module. Finally, the prediction could be made by the classification head with a single neuron. In this paper, $k_l = 128$.

4 Experiments

4.1 Experimental Settings

Implementation Details. The dataset used in this paper for performance evaluation is COV19-CT-DB [2, 17–21] consisting of 1993, 484, 5281 CT scans for training, validation, and testing, respectively. In the experiments, the performance is evaluated in the validation set only since the ground truth in the testing set of COV19-CT-DB is inaccessible. The size of slices is inconsistent in some CT scans that only the last slice is acquired in the horizontal plane. To deal with this issue, the CT scans with an uncommon scanned axis (other than the horizontal plane) for the training manually. In the testing dataset, we simply verify the shape consistency of each slice in a CT scan to remove the abnormal slices. If the shape of the slices is inconsistent in a CT scan, we keep the last slice only. The rest of the CT scans were selected by the proposed method as described in Sect. 3.1. The constraint size n_c was set to half of the total sizes for each CT scan.

It is well-known that the sensitivity and specificity are the essential quality indices for evaluating the accuracy and reliability of a diagnostic test. In this experiments, three well-known metrics, sensitivity(SE), specificity(SP), and Macro-F1-Score, are used to evaluate the performance of the proposed SSFL.

The experiments of the hyper-parameters selection of the proposed essential slices set selection algorithm is conducted in Fig. 5 and table 1. The different parameters in the proposed selection algorithm show quite different results. Specifically, the parameters with $k = 5$ and $t = 50$ can estimate the lung area more accurately without ground glass opacity, while it is bad at ground glass opacity(e.g., Fig. 5(b), middle one). In contrast, the parameters with $k = 5$, $t = 100$, $k = 10$, and $t = 100$, are more resistant to ground glass opacity symptoms but at poor resolution (e.g., Fig. 5(c)(d), top one), leading to miscalculate the area of some human tissue as the area of the lung.

In Table 1, we can conclude that Macro-F1-Score evaluated in the validation set based on parameters $k = 5$ and $t = 100$ achieves the state-of-the-art performance, compared to other settings. Also, the impact of the performance to the length of the sampled slices n_c is also conducted in Table 1 to verify the best n_c for SSFL. It is clear that $n_c = 50\%$ is significantly better than $n_c = 40\%$, at parameters $k = 5$ and $t = 100$. In summary, the length of the sampled slices could be suggested to $n_c = 50$ for COVID-19 classification in SSFL.

Hyper-Parameters of Spatial Model. Adam [16] was used as the optimizer with learning rate 0.0001, $\beta_1 = 0.9$, $\beta_2 = 0.999$ and the weight decay is 0.0005. The total epoch is 50, each image is resized to 256×256, and data augmentation, including random horizontal flip, random shift, scale, rotation, hue, saturation, brightness, contrast, and normalization are used in the training phase. The loss function is binary cross-entropy loss. For the testing phase, each image is resized to 256×256 and the standard normalization is adopted. We also verified the performance based on the spatial feature only. In this way, we use the average of the predicted probabilities of sixteen slices sampled from a CT scan as a well-known testing-time augmentation strategy.

Hyper-Parameters of Slice Context Model. For training Swin-Transformer model, Adam [16] was used as the optimizer with learning rate 0.0003, $\beta_1 = 0.9$, $\beta_2 = 0.999$, and the weight decay is $1e - 4$. The total epoch is 30, batch size is 16, and the loss function is binary cross-entropy loss.

For training LSTM model, AdamW [13] was used as the optimizer with learning rate $3e - 5$, $\beta_1 = 0.9$, $\beta_2 = 0.999$, and the weight decay is $1e - 4$. The total epoch is 50, the batch size is 32, and the loss function is binary cross-entropy loss.

4.2 Performance Evaluation

The experimental results are depicted in Table 2. It is clear that the proposed method with context feature learning significantly outperforms other state-of-the-art models. As we may know, the spatial model could be even more effective if the training slices are high-quality enough. It is interesting to verify the effectiveness of the spatial model used in this paper (i.e., EfficientNet). The proposed spatial model trained on the essential slices set shows great performance compared to the traditional state-of-the-art ADLeasT [3] in terms of Macro-F1-Score measurement. Furthermore, the standard testing-time augmentation (TTA) is used to improve the performance and stability of the proposed spatial models, as shown in rows 4 and 5 in Table 2. Although the ensemble strategy

Table 1. Comparing the proposed slice selection and fixed percentage sampling. The setting (5, 50, 40%) in slice selection denotes the tuning parameters window size $k = 5$, threshold $t = 50$, and sampling slice $n_c = 40\%$.

Method	Setting	SE (mean 10)	SP (mean 10)	Macro-F1 (mean 10)
Fixed percentage	30%–70%	0.888	0.951	0.922
	25%–75%	0.883	0.962	0.926
Slice Selection	(5, 50, 40%)	0.879	0.955	0.919
	(5, 50, 50%)	**0.902**	0.947	0.926
	(5, 100, 40%)	0.897	0.955	0.928
	(5, 100, 50%)	0.893	**0.973**	**0.936**
	(5, 100, 60%)	0.883	0.955	0.922
	(10, 100, 40%)	0.897	0.940	0.920
	(10, 100, 50%)	0.869	0.970	0.923

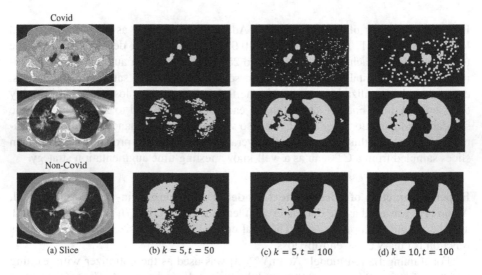

Covid

Non-Covid

(a) Slice   (b) $k = 5, t = 50$ (c) $k = 5, t = 100$ (d) $k = 10, t = 100$

Fig. 5. Visualized result of estimated lung area with different tuning parameters. k and t denote the window size of a filter and threshold for obtaining the segmentation map, respectively.

could be used to improve the performance in many tasks, it is essential that the overfitting still remains.

In contrast, the proposed slice context feature learning could be used to automatically aggregate the spatial features by exploiting their correlation to obtain more stable results. As a result, the proposed method shows even better specificity and sensitivity when context feature learning is applied. Although the performance gap between the proposed method with Swin-Transformer and LSTM is closer to each other, we claim that the Swin-Transformer-based approach shows relatively computational complexity since the self-attention calculation is restricted in the windows. Furthermore, a recurrent-based approach is hard to parallelize, increasing the inference time; thus, real-world applications could be harmful. In summary, the proposed spatial-slice feature learning framework achieves state-of-the-art performance for the COVID-19 classification task in the diverse CT dataset.

Table 2. Performance evaluation of validation set of the proposed methods and other peer methods in terms sensitivity (SE), specificity (SP), and Macro-F1-Score.

	SE	SP	Macro-F1
Baseline [17]	–	–	0.77
ADLeasT [3]	0.833	0.978	0.911
2D CNN	0.897	0.959	0.930
2D CNN (Ensemble-10)	0.893	0.973	0.936
2D CNN (Ensemble-50)	0.893	0.974	**0.937**
SSFL-LSTM	**0.900**	0.959	0.932
SSFL-ST	0.879	**0.981**	0.934

Furthermore, the performance evaluation of the COV19D Challenge [17] is conducted in Table 3. Although the validation Macro-F1-Score of the SSFL with Swin-Transformer (SSFL-ST) model is not the best, its performance achieved state-of-the-art in the private testing set of the COV19D challenge. The proposed SSFL model can take advantage of the relationship between slices and slices, while the weights of the spatial model are determined by validation loss during training, implying that the slice context features obtained by the spatial model might have a data leakage issue. Compared with the SSFL with LSTM (SSFL-LSTM) model trained with the complete feature embedding, the SSFL-ST model only learn the partial slice context features with window-based self-attention calculation, suppressing the issue of data leakage.

Table 3. Performance comparison of the private testing set of COV19D Challenge in terms of Macro-F1-Score.

Submission #	Model variants	Validation	Testing
1	2D CNN (Ensemble-10)	0.936	0.868
2	SSFL-LSTM	0.932	0.857
3	SSFL-ST	0.934	**0.891**
4	2D CNN (Ensemble-50)	**0.937**	0.867
5	2D CNN (Ensemble-10)+SSFL-LSTM+SSFL-ST	–	0.867

5   Conclusions

In this paper, an effectively and efficiently spatial-slice feature learning (SSFL) framework has been proposed to resolve the problems of the inconsistent resolution and number of slices between different CT scans for COVID-19 classification. The visual Transformer is adopted to capture the long-range dependence correlation of the slices' features extracted by conventional convolutional neural networks. Furthermore, the proposed essential slices set selection algorithm based on lung-area estimation could be adopted to eliminate the outlier slices from the CT scan, making the SSFL could be learned more stable and efficient. By randomly selecting slices from the essential slices set of the CT scan, the proposed SSFL could prevent the overfitting due to this randomness. Furthermore, the proposed SSFL shows a relatively lower false-positive rate on the COVID-19 detection, implying that the extra costs due to false-positive samples could be maximumly reduced. Comprehensive experiments also demonstrated the effectiveness and superiority of the proposed SSFL over the other peer methods.

Acknowledgement. This study was supported in part by the National Science and Technology Council, Taiwan, under Grants 110-2222-E-006 -012, 111-2221-E-006 -210, 111-2221-E-001-002, 111-2634-F-007-002. We thank to National Center for High-performance Computing (NCHC) for providing computational and storage resources.

References

1. Abbas, A., Abdelsamea, M.M., Gaber, M.M.: Classification of Covid-19 in chest x-ray images using Detrac deep convolutional neural network. Appl. Intell. **51**(2), 854–864 (2021)
2. Arsenos, A., Kollias, D., Kollias, S.: A large imaging database and novel deep neural architecture for Covid-19 diagnosis. In: 2022 IEEE 14th Image, Video, and Multidimensional Signal Processing Workshop (IVMSP), pp. 1–5. IEEE (2022)
3. Chen, G.L., Hsu, C.C., Wu, M.H.: Adaptive distribution learning with statistical hypothesis testing for Covid-19 CT scan classification. In: 2021 IEEE/CVF International Conference on Computer Vision Workshops (ICCVW), pp. 471–479 (2021). https://doi.org/10.1109/ICCVW54120.2021.00057
4. Chen, J.: Design of accurate classification of Covid-19 disease in x-ray images using deep learning approach. J. ISMAC **2**, 132–148 (2021). https://doi.org/10.36548/jismac.2021.2.006
5. Deng, J., Dong, W., Socher, R., Li, L.J., Li, K., Fei-Fei, L.: Imagenet: a large-scale hierarchical image database. In: 2009 IEEE Conference on Computer Vision and Pattern Recognition, pp. 248–255. IEEE (2009)
6. Singh, D., Kumar, V., Kaur, M.: Classification of Covid-19 patients from chest CT images using multi-objective differential evolution-based convolutional neural networks. Eur. J. Clin. Microbiol. Infect. Diseases (2020)
7. Fang, L., Wang, X.: Covid-19 deep classification network based on convolution and deconvolution local enhancement. Comput. Biol. Med. **135**, 104588 (2021)
8. Foret, P., Kleiner, A., Mobahi, H., Neyshabur, B.: Sharpness-aware minimization for efficiently improving generalization. arXiv preprint arXiv:2010.01412 (2021)

9. He, K., Zhang, X., Ren, S., Sun, J.: Deep residual learning for image recognition. In: Proceedings of the IEEE Conference on Computer Vision and Pattern Recognition, pp. 770–778 (2016)
10. Hochreiter, S., Schmidhuber, J.: Long short-term memory. Neural Comput. **9**(8), 1735–1780 (1997)
11. Hou, J., Xu, J., Feng, R., Zhang, Y., Shan, F., Shi, W.: CMC-Cov19d: contrastive mixup classification for Covid-19 diagnosis, pp. 454–461 (2021). https://doi.org/10.1109/ICCVW54120.2021.00055
12. Hussain, E., Hasan, M., Rahman, M.A., Lee, I., Tamanna, T., Parvez, M.Z.: Corodet: a deep learning based classification for Covid-19 detection using chest x-ray images. Chaos Solit. Fract. **142**, 110495 (2021)
13. Ilya, L., Frank, H.: Decoupled weight decay regularization. arXiv preprint arXiv:1711.05101 (2017)
14. Ismael, A.M., Şengür, A.: Deep learning approaches for Covid-19 detection based on chest x-ray images. Expert Syst. Appl. **164**, 114054 (2021)
15. Jiang, J., Lin, S.: Covid-19 detection in chest x-ray images using swin-transformer and transformer in transformer. arXiv preprint arXiv:2110.08427 (2021)
16. Kingma, D.P., Ba, J.: Adam: a method for stochastic optimization. arXiv preprint arXiv:1412.6980 (2014)
17. Kollias, D., Arsenos, A., Kollias, S.: AI-MIA: Covid-19 detection and severity analysis through medical imaging. arXiv preprint arXiv:2206.04732 (2022)
18. Kollias, D., Arsenos, A., Soukissian, L., Kollias, S.: MIA-Cov19d: Covid-19 detection through 3-D chest CT image analysis. arXiv preprint arXiv:2106.07524 (2021)
19. Kollias, D., et al.: Deep transparent prediction through latent representation analysis. arXiv preprint arXiv:2009.07044 (2020)
20. Kollias, D., Tagaris, A., Stafylopatis, A., Kollias, S., Tagaris, G.: Deep neural architectures for prediction in healthcare. Complex Intell. Syst. **4**(2), 119–131 (2018)
21. Kollias, D., et al.: Transparent adaptation in deep medical image diagnosis. In: TAILOR, pp. 251–267 (2020)
22. Le Dinh, T., Lee, S.H., Kwon, S.G., Kwon, K.R.: Covid-19 chest x-ray classification and severity assessment using convolutional and transformer neural networks. Appl. Sci. **12**(10) (2022). https://doi.org/10.3390/app12104861, https://www.mdpi.com/2076-3417/12/10/4861
23. Liu, Z., et al.: Swin transformer v2: scaling up capacity and resolution. arXiv preprint arXiv:2111.09883 (2022)
24. Liu, Z., et al.: Swin transformer: hierarchical vision transformer using shifted windows. arXiv preprint arXiv:2103.14030 (2021)
25. Liu, Z., Mao, H., Wu, C.Y., Feichtenhofer, C., Darrell, T., Xie, S.: A convnet for the 2020s. In: Proceedings of the IEEE/CVF Conference on Computer Vision and Pattern Recognition, pp. 11976–11986 (2022)
26. Miron, R., Moisii, C., Dinu, S., Breaban, M.: Covid detection in chest CTs: improving the baseline on Cov19-CT-DB. arXiv preprint arXiv:2107.04808 (2021)
27. Müller, R., Kornblith, S., Hinton, G.: When does label smoothing help? arXiv preprint arXiv:1906.02629 (2020)
28. Pathak, Y., Shukla, P.K., Tiwari, A., Stalin, S., Singh, S., Shukla, P.: Deep transfer learning based classification model for Covid-19 disease. IRBM **43**(2) (2020)
29. Tan, M., Le, Q.: Efficientnet: rethinking model scaling for convolutional neural networks. In: International Conference on Machine Learning, pp. 6105–6114. PMLR (2019)

30. Tan, W., Liu, J.: A 3D CNN network with BERT for automatic Covid-19 diagnosis from CT-scan images. arXiv preprint arXiv:2106.14403 (2021)
31. Wightman, R.: PyTorch image models (2019). https://github.com/rwightman/pytorch-ima ge-models. https://doi.org/10.5281/zenodo.4414861
32. Wikipedia contributors: Mathematical morphology—Wikipedia, the free encyclopedia (2022). https://en.wikipedia.org/w/index.php?title=Mathematical_morphology&oldid=1082 436538. Accessed 2 July 2022

Multi-scale Attention-Based Multiple Instance Learning for Classification of Multi-gigapixel Histology Images

Made Satria Wibawa[1(✉)], Kwok-Wai Lo[2], Lawrence S. Young[3], and Nasir Rajpoot[1,4]

[1] Department of Computer Science, Tissue Image Analytics Centre, University of Warwick, Coventry, UK
{made-satria.wibawa,n.m.rajpoot}@warwick.ac.uk
[2] Department of Anatomical and Cellular Pathology, The Chinese University of Hong Kong, Hong Kong, China
kwlo@cuhk.edu.hk
[3] Warwick Medical School, University of Warwick, Coventry, UK
l.s.young@warwick.ac.uk
[4] The Alan Turing Institute, London, UK

Abstract. Histology images with multi-gigapixel of resolution yield rich information for cancer diagnosis and prognosis. Most of the time, only slide-level label is available because pixel-wise annotation is labour intensive task. In this paper, we propose a deep learning pipeline for classification in histology images. Using multiple instance learning, we attempt to predict the latent membrane protein 1 (LMP1) status of nasopharyngeal carcinoma (NPC) based on haematoxylin and eosin-stain (H&E) histology images. We utilised attention mechanism with residual connection for our aggregation layers. In our 3-fold cross-validation experiment, we achieved average accuracy, AUC and F1-score 0.936, 0.995 and 0.862, respectively. This method also allows us to examine the model interpretability by visualising attention scores. To the best of our knowledge, this is the first attempt to predict LMP1 status on NPC using deep learning.

Keywords: Deep learning · Multiple instance learning · Attention · H&E · LMP1 · NPC

1 Introduction

Analysing histology images for cancer diagnosis and the prognosis is not a task without challenges. A Whole Slide Image (WSI) in 40× magnification with an average resolution of $200,000 \times 150,000$ contains 30 billion pixels and a size of ∼90GB in the uncompressed state. Moreover, histology image mainly has a noisy/ambiguous label. Most of the time label of the WSI is assigned on the slide level, for example, the label for cancer stage. This may create ambiguous interpretations by the machine learning/deep learning model, because in such

© The Author(s), under exclusive license to Springer Nature Switzerland AG 2023
L. Karlinsky et al. (Eds.): ECCV 2022 Workshops, LNCS 13807, pp. 635–647, 2023.
https://doi.org/10.1007/978-3-031-25082-8_43

a large region, not all regions correspond to the slide label. Some regions may contain tumours, and the rest is non-tumour regions e.g. background, stroma, and lymphocyte.

The standard approach to handling multi-gigapixel WSIs is by extracting its region into smaller patches that the machine can process. The noisy label problem also occurs in this approach since we do not have a label for each patch. The multiple instance learning (MIL) paradigm is generally used to overcome this problem [3]. In MIL, each patch is represented as an instance in a bag. Since WSIs have more than one patch, the bag contains multiple instances, hence the name 'multiple' instances learning. During training, only global (slide-level) image labels are required for supervision. Aggregation mechanism is then utilised to summarise all information in instances to make a final prediction.

The selection of aggregation mechanism is one of the vital parameter in MIL. Several aggregation mechanisms have been introduced in MIL for computational histopathology, such as the max and the median score [10] or the average of all patches scores [4,16]. The limitation of such mechanisms is they are not trainable. Therefore, attention mechanism [1] was utilised in the recent MIL model on computational pathology [7,15]. Attention mechanism is trainable, it computes the weights of the instances representation. Strong weights/scores indicate instances are more important to final prediction than instances with low scores.

Several studies have already utilised MIL for cancer diagnosis and prognosis. For example, Lu and colleagues [13] proposed attention-based multiple instance learning for subtyping renal cell carcinoma and non-small-cell lung carcinoma. Campanella and colleagues [2] used multiple instance learning with RNN-based aggregation to discriminate tumour regions in breast, skin, and prostate cancer. No current study examines the use of multiple instance learning in nasopharyngeal carcinoma.

Nasopharyngeal carcinoma (NPC) is a malignancy that develops from epithelial cells within lymphocyte-rich nasopharyngeal mucosa. The high incidence rate of NPC mainly occurs in southern China, southeast Asia and north Africa, which are 50–100 times greater than the rates in other regions of the world. NPC has unique pathogenesis involving genetic, lifestyle and viral (Epstein-Barr virus or EBV) cofactors [19]. EBV infection on NPC encodes several viral oncoprotein, one of the most important ones is latent membrane protein (LMP1). LMP1 is the predominant oncogenic driver of NPC [9]. Overexpression of LMP1 promotes NPC progression by inducing invasive growth of human epithelial and nasopharyngeal cells. Due its nature, LMP1 is an excellent therapeutic target for EBV-associated NPC [11]. However, detection of LMP1 on NPC with standard testing such as immunohistochemistry may cause additional costs and delays the diagnosis. On the other hand, histopathology images are widely available and provide rich information of the cancer on the tissue level. Deep learning-based algorithms have been applied for several tasks and deliver promising results in EBV-related cancer in histopathology. Deep learning has been used to predict EBV status in gastric cancer [21], predict EBV integration sites [12] and predict microsatellite instability status in gastric cancer [14]. However, despite the advance in computational histopathology and the significant importance of LMP1 in NPC progression, this topic is still an understudy area.

In this paper, we proposed a deep learning pipeline based on the MIL paradigm to predict LMP1 status in NPC. Our main contributions in this paper are: (1)To the best of our knowledge, this is the first study that attempts to utilise deep learning on LMP1 status prediction based on histology images of NPC. (2)We propose multi-scale attention-based with residual connection[1] for aggregation layer in MIL. (3)Our proposed MIL pipeline outperformed other known MIL methods for predicting LMP1 status in NPC.

2   Materials and Methods

2.1   Dataset

This study used two kinds of datasets: the first for the tissue classification task and the second is the LMP1 dataset.

– **Tissue classification dataset.** For this task, we combined two datasets from Kather100K [8] dataset and an internal dataset. Kather100K contains 100,000 non-overlapping image patches from H&E stained histological images of colorectal cancer. The number of tissue classes in this dataset is nine, we exclude the normal colon mucosa class (NORM) from the dataset. We argued that the NORM class is irrelevant in head and neck tissue. Thus, only eight classes from the Kather100K dataset were utilised in this study. The internal dataset consists of three cohorts. The number of images patches in this dataset is 180,344 with seven classes which taken at 20x magnification. We combined images with the same classes from Kather100K and internal datasets in the final dataset.
– **LMP1 dataset.** LMP1 data was collected from an University Hospital. The dataset comprises 101 NPC cases with a H&E stained whole slide images (WSIs) for each case. All WSIs were taken by an Aperio scanner at 40x magnification and 0.250μ per pixel resolution. Expression of LMP1 was determined by immunohistochemical staining. The proportion score was according to the percentage of tumour cells with positive membrane and cytoplasmic staining (0–100). The intensity score was assigned for the average intensity of positive tumour cells (0,none; 1,weak; 2,intermediate; 3,strong). The LMP1 staining score was the product of proportion and intensity scores, ranging from 0 to 300. The LMP1 expression was categorized into absence/low/negative (score 0–100) and high/positive (score 101–300). Details of the class distribution of the dataset in this study can be seen in Table 1.

---

[1] Model code: https://github.com/mdsatria/MultiAttentionMIL.

Table 1. Details of the dataset

Label	Number of cases
LMP1 positive	25
LMP1 negative	76
Total cases	101

2.2 Tissue Classification

We utilised a pretrained ResNet-50 [6] with ImageNet [5] to classify images into tissue and non-tissue classes. Dataset for tissue classification was split into three parts, i.e. training, validation and testing with the data distribution of 80%, 10% and 10%, respectively. Training images were augmented in several methods, including colour augmentation, kernel filter (sharpening and blurring) and geometric transformation. Before images used in training, we normalised the value of the pixel within the range of -1 and 1.

We trained all layers in the ResNet50 model, using an Adam optimizer with starting learning rate of 1×10^{-3} and weight decay of 5×10^{-4}. Training was conducted on 20 epochs and every 10 epochs, we decreased learning rate by ten times. To regularize the model, we apply a dropout layer with the probability of 0.5 before the fully connected layer. We use cross entropy to calculate the loss of the network which is defined as:

$$CE = \sum_{i=1}^{C} t_i \log(f(s)_i) \tag{1}$$

where

$$f(s)_i = \frac{e_i^s}{\sum_{j=1}^{C} e_j^s} \tag{2}$$

t_i is the ground truth label and s_i is the predicted label and $f(s)_i$ is a softmax activation function. In our case, there were eight classes, thus $C \in [0..7]$.

2.3 LMP1 Prediction

The main pipeline of LMP1 prediction is illustrated in Fig. 1. After tumour patches were identified in the tissue classification stage, we encoded all the tumour patches into feature vectors. We experimented with two backbone networks for encoding patches into feature vectors, namely ResNet-18 and EfficientNet-B0 [17].

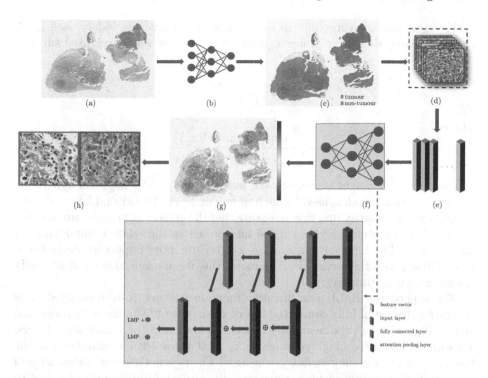

Fig. 1. The workflow of LMP1 prediction. Patches from WSIs (a) were classified into the tumour and non-tumour regions by the tumour classification model based on ResNet50 (b). The tumour regions of WSIs can be seen as the red area of the heatmap (c). All tumour patches were then used as instances of the bag (d) and encoded into feature vectors (e) with backbone networks. Our MIL model which consists of four fully connected layers, and three attention-based aggregation layers with the residual connection (f) then trained with these feature vectors. The model will generate attention scores for each patch. The red area in the heatmap shows a high attention score (g&h).

Let B denote the bag, x is the feature vector extracted from backbone network and K is the number of patches in the slide, bag of instances will be $B = \{x_1, ..., x_K\}$. Within the ResNet-18 and EfficientNet-B0 as the backbone networks, each patch is represented by a 512 and 1280-dimensional features vector, respectively. The number of K will vary depending from the number of tumour patches in their corresponding slide. Slide labels will be inherited into their corresponding patches. Thus, for all bags of instances $\{B_1, B_2, ..., B_N\}$ with N as the number of slide, every x in B will have same label as their corresponding slide $Y_n \in \{0, 1\}, n = 1...N$.

We used attention mechanism as an aggregation layer as widely used in [7, 15, 20]. An attention layer in aggregation layer is defined as a weighted sum:

$$z = \sum_{k=1}^{K} a_k x_k, \tag{3}$$

where
$$a_k = \frac{\exp\{w^{\intercal} \tanh (V x_k^{\intercal})\}}{\sum_{j=1}^{K} \exp\{w^{\intercal} \tanh (V x_j^{\intercal})\}} \tag{4}$$

and $w \in \mathbb{R}^{L \times 1}$ is weight vector of each instance, $V \in \mathbb{R}^{L \times M}$. tanh is hyperbolic tangent function which is used in the first hidden layer. The second layer employs a softmax non-linearity function to ensure that the attention weights sum is equal to one. The attention weights can be interpreted as the relative importance of the instance. The higher the attention weight, the more important an instance in the final prediction score. Furthermore, this mechanism also creates a fully trainable aggregation layer.

We applied the ReLU non-linearity function for each fully connected layer except for the last of fully connected layers. Each output from the fully connected layer will go through the next fully connected layer and the aggregation layers. This model has three weighted sums z_1, z_2, z_3 that are then accumulated in the last aggregation via the residual connection. The accumulation of weighted sum from each aggregation layer is then used in the last fully connected layers to predict label of the bag.

We trained our model with learning rate of 1×10^{-4} and with Adam optimization algorithm. All the experiments were implemented on Python language and Pytorch framework. We conducted our experiment on workstation with Intel(R) i5-10500 CPU (3.1 GHz), 64 GB RAM and single GPU NVIDIA RTX 3090.

2.4   Metrics for Evaluation

We use the accuracy score to measure our model performance. Due to the imbalanced class distribution in our dataset, we also use F1-score and Area Under the Curve and Receiver Operating Characteristic (AUROC) curve. F1-score is defined as follow:
$$F1 = \frac{TP}{TP + \frac{1}{2}(FP + FN)} \tag{5}$$

where TP, TN, FN, and FP are true positive, true negative, false negative and false positive, respectively. All the experiments were conducted on three-folds cross validation with stratified label. Due to the usage of cross-validation in our experiment, the average and standard deviation of all metrics were also reported.

3 Results

3.1 Tissue Classification

Tissue classifier model achieved accuracy of 0.990 and F1-score of 0.970 on the test set. This model then used on the LMP1 cohort to stratify tumour patches. We extracted non-overlapping patches with the size of 224×224 at a 20x magnification level from LMP1 cohort. We conducted a segmentation on lower-resolution image to find foreground and background region in image. We use the segmentation result as a guidance to segregate patches with tissue and background patches. Only patches contains tissue were inferred in the tumour classification process.

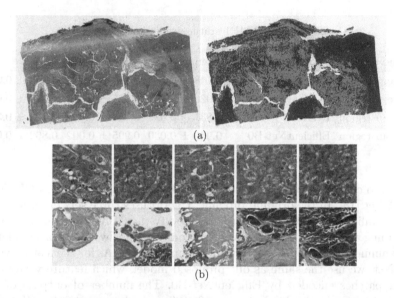

Fig. 2. Tissue classification result. (a) Top left image is a thumbnail of one of slide in the dataset and top right image is the corresponding tumour region heatmap. (b) Samples of tumour patches is depicted in the top row and non-tumour region in the bottom row

As can be seen in Fig. 2, tumour patches were separated nicely in the WSIs. There was some minor misclassification in some slides due domain shift problem, but the rate was too low and could be ignored. Backbone networks encoded these tumour patches into vector representation and used them as bag of instances for LMP1 status prediction.

3.2 LMP1 Prediction

We argue that the tumour region is more important than any other tissue region for predicting LMP1. The non-tumour region may hinder the performance of

LMP1 prediction and weighs irrelevant regions. To this end, we compared the performance of model with all tissue region as bag of instance vs model which only employs tumour patches as bags of instance. We also compared two backbone networks, namely ResNet-18 and EfficientNet-B0 in this experiment. The result of this experiment can be seen in Table 2. There were differences in performance between the same backbone networks. The model that trained with only the tumour region delivered better performance predicting LMP1 status than the model that trained with all tissue regions. Furthermore, EfficienetNet-B0 was better than ResNet-18 in terms of the backbone network. The best performance was achieved by using tumour regions as instances and EfficientNet-B0. This model achieved average accuracy, AUC and F1-score of 0.936, 0.995 and 0.862, respectively.

Table 2. LMP1 status prediction results

Tissue region	Backbone network	Accuracy	AUC	F1-score
All region	ResNet-18	0.748 ± 0.035	0.877 ± 0.013	0.310 ± 0.257
All region	EfficientNet-B0	0.926 ± 0.032	0.978 ± 0.014	0.825 ± 0.095
Tumour region	ResNet-18	0.831 ± 0.082	0.954 ± 0.023	0.423 ± 0.364
Tumour region	EfficientNet-B0	0.936 ± 0.030	0.995 ± 0.002	0.862 ± 0.076

We also compared our result with other known methods such as CLAM [13] and MI-Net [18]. In the original pipeline, CLAM uses three sets of data, namely training, validation and testing sets. We modify the CLAM pipeline only to use the training and testing set to allow a fair comparison. Two aggregation methods were examined for MI-Net: the max and mean methods. As for the feature vectors in MI-Net, we use the same as our proposed model, which feature vectors from tumour patches encoded by EfficientNet-B0. The number of outputs of fully-connected layers in the MI-Net are 256, 128 and 64, respectively. MI-Net was trained with Adam optimizer and learning rate of 1×10^{-4}. We also ensure CLAM and MI-Net use the same data as ours in the cross-validation training.

Table 3. Performance comparison of our model

Model	Tissue region	Accuracy	AUC	F1-score
CLAM [13]	All region	0.723 ± 0.022	0.548 ± 0.092	–
MI-Net Max [18]	All region	0.852 ± 0.011	0.700 ± 0.018	0.571 ± 0.037
MI-Net Mean [18]	All region	0.762 ± 0.002	0.541 ± 0.016	0.165 ± 0.076
MI-Net Max [18]	Tumour region	0.832 ± 0.031	0.661 ± 0.054	0.477 ± 0.130
MI-Net Mean [18]	Tumour region	0.767 ± 0.006	0.559 ± 0.041	0.207 ± 0.136
Our model	Tumour region	**0.936** ±0.030	**0.995** ± 0.002	**0.862** ± 0.076

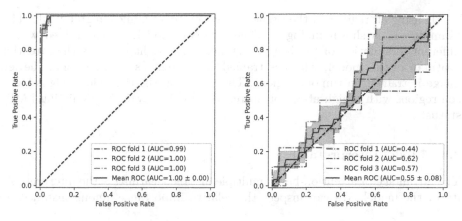

Fig. 3. AUROC curve. Left image is AUROC curve for our model and right image is AUROC curve for CLAM

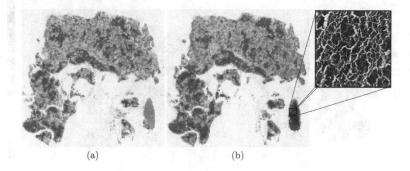

(a)   (b)

Fig. 4. Prediction comparison on different instance types. LMP1 heatmap prediction with tumour patches as instances(a) and all patches as instances(b). Region with high attention score in(b) was blood cell, which irrelevant region regarding to LMP1.

Based on the performance comparison in Table 3 and AUROC in Fig. 3, CLAM performed poorly on this particular task. The average accuracy score achieved was only 0.723, with average AUC of 0.548. On the other hand, MI-Net performed well in all experiments compared to the CLAM. Both max and mean aggregation methods surpass CLAM performance in accuracy and AUC. However, aggregation with the max method was better than the mean method in this case. This may occur because the mean method averages from many other irrelevant instances while the max method selects only relevant instances. Our model achieved the best accuracy, AUC and F1-score compared to other models. The second-best model was MI-Net with max-aggregation that trained on tumour regions of the tissue. Both of MI-Net and our models performed better when only tumour regions was used as bag of instances. This indicates that tumour region is more relevant to predicting LMP1 in our case.

Utilising all regions in the tissue may generate an unmeaningful result. This example can is shown in Fig. 4. There are two models which trained with a different formulation of the bag of instances. The right image is the attention heatmap of the model, which is trained on all regions of tissue, and the left image is trained on tumour regions. The model on the right side weighs blood cell regions with a high attention score, which is irrelevant in predicting LMP1 status.

3.3   Model Interpretability

The advantage of attention-based multiple instance learning is the interpretability of the model. We can inspect the relative importance among instances in

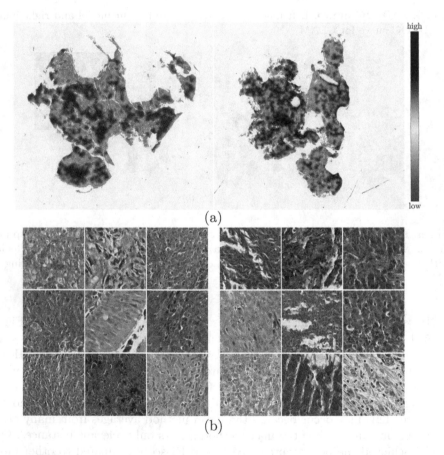

Fig. 5. Attention heatmap of LMP1 prediction model. (a) is attention heatmap, the left image is LMP1 negative and the right image is LMP1 positive. (b) is randomly selected patches from dataset with highest attention score, left images is true negative cases and right images is true positive cases

regards to bag labels. Attention heatmaps from both of LMP1 negative and positive can be seen in Fig. 5. This figure also visualised the patches with top 10% attention scores from both LMP1 classes. Based on the colour heterogeneity of patches, colour variation from the staining method was uncorrelated to LMP1 status.

We also conducted a simple test to prove that the multiple attention layer chooses the most relevant instances in regards to the bag label. Currently, there is no well-defined explanation regarding LMP1 effect on tumour micro-environment and its morphology. Therefore, we use MNIST as an example. We trained the same MIL architecture as the LMP1 status prediction task on the MNIST dataset. We defined the task into a binary classification. Bags which contained at least one image of digit 1 were defined as the positive class, and bags without image of digit 1 were defined as the negative class. We generated 5000 bags with a balanced class distribution. Each bags contain ten images/instances. The number of digits 1 was randomly selected in the positive class. We trained our model with the same learning rate 1×10^{-4} and optimized it with the Adam

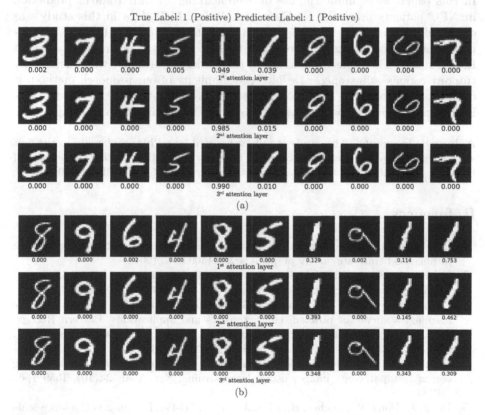

Fig. 6. Attention scores from three aggregation layers. These are two examples of bags in MNIST (a,b) with their corresponding attention scores in three aggregation layers. Number below each digits is the attention scores.

algorithm. To encode the image digits in MNIST into a features vector, we flatten the image of 28×28 into a 1-dimensional array of 784 elements. We then changed the scale of the feature vector from 0–255 to 0–1.

Figure 6 depicts an example of inferred results from a positive bag. Bag (a) contains two positive instances/two images of digit 1. These instances have the highest attention scores among other instances in the first aggregation layers. In the second attention layer, the attention scores for negative instances were reduced while the scores for positive instances were increased. This trend continues until the last aggregation layer. Thus all negative instances have zero value of attention. In another case, in bag (b), the attention scores in the last layers were averaged among the positive instances. This experiment proves that multiple attention-based of aggregation layers help the model to select more relevant instances.

4   Conclusion

In this paper, we examine the use of deep learning for LMP1 status prediction in NPC patients using the H&E-stained WSIs. LMP1 data in this study was collected from a University Hospital with a total number of 101 cases. We proposed multi attention aggregation layer for MIL in the tumour region to predict LMP1 status. There was an increase in the model performance when using only tumour regions as instances. Despite the simplicity of our proposed method, it outperformed the other known MIL models. To the best of our knowledge, this is the first attempt to predict LMP1 status on NPC using deep learning.

Acknowledgements. This study is fully supported by a PhD scholarship to the first author funded by Indonesia Endowment Fund for Education (LPDP), Ministry of Finance, Republic of Indonesia under grant number Ref: S-575/LPDP.4/2020.

References

1. Bahdanau, D., Cho, K., Bengio, Y.: Neural machine translation by jointly learning to align and translate. arXiv preprint arXiv:1409.0473 (2014)
2. Campanella, G., et al.: Clinical-grade computational pathology using weakly supervised deep learning on whole slide images. Nat. Med. **25**(8), 1301–1309 (2019)
3. Carbonneau, M.A., Cheplygina, V., Granger, E., Gagnon, G.: Multiple instance learning: a survey of problem characteristics and applications. Pattern Recogn. **77**, 329–353 (2018). https://doi.org/10.1016/j.patcog.2017.10.009, https://www.sciencedirect.com/science/article/pii/S0031320317304065
4. Coudray, N., et al.: Classification and mutation prediction from non-small cell lung cancer histopathology images using deep learning. Nat. Med. **24**(10), 1559–1567 (2018)
5. Deng, J., Dong, W., Socher, R., Li, L.J., Li, K., Fei-Fei, L.: Imagenet: a large-scale hierarchical image database. In: IEEE Conference on Computer Vision and Pattern Recognition, pp. 248–255. IEEE (2009)

6. He, K., Zhang, X., Ren, S., Sun, J.: Deep residual learning for image recognition. In: IEEE Conference on Computer Vision and Pattern Recognition, pp. 770–778 (2016)

7. Ilse, M., Tomczak, J., Welling, M.: Attention-based deep multiple instance learning. In: International Conference on Machine Learning, pp. 2127–2136. PMLR (2018)

8. Kather, J.N., et al.: Predicting survival from colorectal cancer histology slides using deep learning: a retrospective multicenter study. PLoS Med. **16**(1), e1002730 (2019)

9. Kieser, A., Sterz, K.R.: The latent membrane protein 1 (LMP1). Epstein Barr Virus **2**, 119–149 (2015)

10. Klein, S., et al.: Deep learning predicts HPV association in oropharyngeal squamous cell carcinomas and identifies patients with a favorable prognosis using regular H&E stainsdeep learning predicts HPV association in OPSCC. Clin. Cancer Res. **27**(4), 1131–1138 (2021)

11. Lee, A.W., Lung, M.L., Ng, W.T.: Nasopharyngeal Carcinoma: From Etiology to Clinical Practice. Academic Press, Cambridge (2019)

12. Liang, J., et al.: DeepeBV: a deep learning model to predict Epstein-Barr Virus (EBV) integration sites. Bioinformatics (2021). https://doi.org/10.1093/bioinformatics/btab388

13. Lu, M.Y., Williamson, D.F., Chen, T.Y., Chen, R.J., Barbieri, M., Mahmood, F.: Data-efficient and weakly supervised computational pathology on whole-slide images. Nat. Biomed. Eng. **5**(6), 555–570 (2021)

14. Muti, H.S., et al.: Development and validation of deep learning classifiers to detect Epstein-Barr Virus and microsatellite instability status in gastric cancer: a retrospective multicentre cohort study. Lancet Digit. Health **3**(10), e654–e664 (2021)

15. Qiu, S., Guo, Y., Zhu, C., Zhou, W., Chen, H.: Attention based multi-instance thyroid cytopathological diagnosis with multi-scale feature fusion. In: 2020 25th International Conference on Pattern Recognition (ICPR), pp. 3536–3541. IEEE (2021)

16. Schaumberg, A.J., Rubin, M.A., Fuchs, T.J.: H&E-stained whole slide image deep learning predicts SPOP mutation state in prostate cancer. BioRxiv p. 064279 (2018)

17. Tan, M., Le, Q.: Efficientnet: rethinking model scaling for convolutional neural networks. In: International Conference on Machine Learning, pp. 6105–6114. PMLR (2019)

18. Wang, X., Yan, Y., Tang, P., Bai, X., Liu, W.: Revisiting multiple instance neural networks. Pattern Recogn. **74**, 15–24 (2018)

19. Wong, K.C., et al.: Nasopharyngeal carcinoma: an evolving paradigm. Nat. Rev. Clin. Oncol. **18**(11), 679–695 (2021)

20. Zhang, H., et al.: DTFD-Mil: double-tier feature distillation multiple instance learning for histopathology whole slide image classification. In: Proceedings of the IEEE/CVF Conference on Computer Vision and Pattern Recognition, pp. 18802–18812 (2022)

21. Zheng, X., et al.: A deep learning model and human-machine fusion for prediction of EBV-associated gastric cancer from histopathology. Nat. Commun. **13**(1), 1–12 (2022)

A Deep Wavelet Network
for High-Resolution Microscopy
Hyperspectral Image Reconstruction

Qian Wang[1,2] and Zhao Chen[1,2]([✉])

[1] School of Computer Science and Technology, Donghua University, Shanghai, China
chenzhao@dhu.edu.cn
[2] Shanghai Key Laboratory of Multidimensional Information Processing,
East China Normal University, Shanghai, China

Abstract. Microscopy hyperspectral imaging (MHSI) integrates conventional imaging with spectroscopy to capture images through numbers of narrow spectral bands, and has attracted much attention in histopathology image analysis. However, due to the current hardware limitation, the observed microscopy hyperspectral images (MHSIs) has much lower spatial resolution than the coupled RGB images. High-resolution MHSI (HR-MHSI) reconstruction by fusing low-resolution MHSIs (LR-MHSIs) with high-resolution (HR) RGB images can compensate for the information loss. Unfortunately, radiometrical inconsistency between the two modalities often causes spectral and spatial distortions which degrade quality of the reconstructed images. To address these issues, we propose a Deep Wavelet Network (WaveNet) based on spectral grouping and feature adaptation for HR-MHSI reconstruction. For preservation of spectral information, the band-group-wise adaptation framework exploits intrinsic spectral correlation and decompose MHSI reconstruction in separate spectral groups. In spatial domain, we design the mutual adaptation block to jointly fuse the wavelet and convolution features by wavelet-attention for high-frequency information extraction. Specifically, to effectively inject spatial details from HR-RGB to LR-MHSI, radiometry-aware alignment is conducted between the two modalities. We compare WaveNet with several state-of-the-art methods by their performance on a public pathological MHSI dataset. Experimental results demonstrate its efficacy and reliability under different settings.

Keywords: Microscopy hyperspectral imaging · Super-resolution · Wavelet network · Mutual adaptation · Fusion

1 Introduction

Microscopy hyperspectral imaging(MHSI) is a promising technique in the field of histopathology image analysis [6]. The spectrum signatures of various cells and tissues can provide extra information and aid in revealing disease-related pathology changes. The MHSI system [17] and the acquired image sets are illustrated

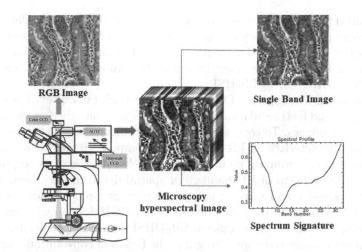

Fig. 1. Illustration of the MHSI system [17] with acquired image sets. The MHSI consists of an AOTF (Acousto-Optic Tunable Filter), a grayscale CCD and a color CCD. During image acquisition, AOTF modulates the wavelength of one path of light to image numerous spectral bands of the hyperspctral image on the grayscale CCD and the other path of light enters the color CCD to capture the RGB image.

in Fig. 1. The system can capture microscopy hyperspectral images (MHSIs) coupled with RGB images. Therefore, these two sets of images reflect the same slide with different spectral and spatial resolutions. Usually, there are numerous spectral bands in one set of MHSIs. However, the high spectral resolution of MHSI limits radiant energy collected per band, resulting in low spatial resolution in the observed MHSIs [3]. Thus, MHSI is prone to suffer from inadequate details, which hinders its application to pathology analysis. Therefore, high-resolution microscopy hyperspectral image (HR-MHSI) reconstruction is an essential step before exploiting MHSI to the greatest extent in histopathological diagnosis. Single image HR reconstruction from low-resolution hyperspectral images (LR-HSIs) and fusion method based on LR-HSIs and HR-RGB images are two main directions for high-resolution hyperspctral image (HR-HSI) reconstruction. Inspired by deep learning models in natural image HR reconstruction, many works achieve reasonable HR-HSI quality from single LR-HSI with CNN architecture. However, its spatial resolution is much lower than that of HR-RGB image. Moreover, the coupled HR-RGB image can be easily collected in routine pathology examination. To this end, an increasing number of deep learning-based methods of fusing LR-HSI with HR-RGB are proposed [14] [1] [26], designing CNN models by the strategy such as multi-level and multi-scale fusion [10], spectral-spatial residual fusion [3], iterative optimization for observation model and fusion process [22]. However, most CNN-based methods directly concatenate LR-HSI and HR-RGB bands in spatial or feature domain, which fail to exploit spectral physical properties and lead to spectral distortion. Meanwhile, the spatial radiometry between the two modalities is aligned mainly via high-pass detailed features, but recent research [18] finds that CNN models are prone to first pick up low frequency components than high frequency, especially at the early stage of the training. The spatial-detail information and radiometry transformation of high frequency may no longer be maintained by the model when passing from HR-RGB to LR-HSI.

In this work, we propose a Deep Wavelet Network (WaveNet) based on spectral grouping and feature adaptation for HR-MHSI reconstruction for microscopy hyperspectral images. To cope with the spectral distortion, we leverage spectral correlation characteristic of LR-MHSIs to decompose HR-MHSI reconstruction task in a spectral grouping manner, where each subnetwork is responsible for reconstructing a group of HR bands. For spatial-detail preservation, we jointly exploit the wavelet and convolution in the proposed mutual adaptation block where LR-MHSIs adapt to spatial details in HR-RGB domain whereas HR-RGB images adapts to spectral response in LR-MHSI domain. Specifically, a wavelet-attention mechanism is designed to guide the fusion of convolution features with abundant frequency information in wavelet domain. In addition, the radiometrical inconsistency problem is dedicatedly addressed by aligning radiance offset between LR-MHSI and HR-RGB features in mutual adaptation, which allows radiometry-free detailed features to be further injected into LR-HSIs.

The main contributions of this work are as follows. 1) The proposed WaveNet is a band-group-wise fusion framework for HR-MHSI reconstruction with strong physical interpretability. It is designed based on spectral correlation and band grouping. It can be applied to enhancement of MHSIs with an arbitrary number of bands or spectral groups. 2) The main component of WaveNet, namely, the mutual adaptation block, fuses convolutional feature maps and wavelet transformation results of multimodal pathology images by a series of cross-connected wavelet attention modules. 3) Particularly, the radiometry-aware modules (RAM) in each mutual adaptation block are designed for radiometry alignment of LR-MHSI and HR-RGB features, effectively injecting spatial details from HR-RGB images to LR-MHSIs. The experimental results on a real world histopathology dataset validates the efficacy of the group-wise fusion mechanism as well as the ability of spatial detail preservation of the wavelet-convolution adaptation modules.

2   Related Works

Most fusion-based HR-HSI reconstruction methods fall into two categories: the model-based HR-HSI reconstruction and the deep-learning based HR-HSI reconstruction. Model-based methods are developed upon the observation model between LR-HSI, HR-RGB and the desired HR-HSI. A bunch of works has been proposed, including spectral unmixing [12], tensor factorization [3,11], and sparse representation [7,9] They exploit the prior knowledge of HSIs [4] to design objective functions, since the number of observation data are much fewer than the unknowns. However, the pre-defined constraints which are sometimes inconsistent with the real-world scenarios, limits its performance. The deep-learning based methods could directly learn the mapping function between LR-HSI, HR-RGB and HR-HSI by nonlinear representation power, thus manifesting large improvement [19,22,23,28]. By utilizing the advanced networks in CNN with carefully designed fusion strategy, state-of-the-art HR-HSI reconstruction results have been achieved. For instance, Zhu et al. designed a lightweight progressive zero-centric residual network (PZRes-Net) to learn a high resolution and zero-centric residual image based on the spectral-spatial separable convolution with dense connections [30]. Wang et al. proposed to optimize the observation model and fusion processes which is implemented by invertible neural network iteratively and alternatively and guarantee spatial and spectral accuracy by enforcing bidirectional data consistency [21]. Dong et al. proposed a model-guided deep convolutional network (MoG-DCN) based on a U-net denoiser, leveraging both the observation matrix of HSI and deep image prior [5].

However, most CNN-based fusion networks inherently extract high-level features instead of high-pass information when injecting details from HR-RGB images to LR-HSIs, leading to inevitable spectral and spatial distortion. In the field of, wavelet transform (WT) has exhibited its strength in high-frequency information preservation. Many works utilize CNN to jointly learn sub-bands decomposition by WT and reconstruction results by inverse WT [8,24,29]. For example, Liu et al. introduced WT into a modified U-Net architecture to replace downsampling and upsampling operations to retain multi-scale detailed features [13]. Some works construct additional wavelet stage to enhance the reconstructed images [2,15]. Given the concerns of radiometrical inconsistency between LR-HSI and RGB, WT has not been well applied in fusion-based HR-HSI reconstruction.

3   Method

As shown in Fig. 2, the proposed Deep Wavelet Network consists of two phases: spectral grouping and feature adaptation. At first, the whole spectral-band image fusion is divided into groups of fusion subnetworks given the intrinsic spectral correlation. It solves the space-varying radiance offset between LR-MHSI and HR-RGB in a group-wise manner, allowing for better spectral preservation. Then, within each subnetwork, we design several modules which delicately fuse convolutional feature maps and wavelet decomposition components to effectively inject spatial details from HR-RGB images to LR-MHSIs.

3.1   Spectral Grouping

Let $\mathbf{Z} \in \mathbb{R}^{N \times P} (N = R \times C)$ denotes the desired HR-MHSI with N pixels (R rows and C columns) and B bands, to be recovered from its degraded measurement $\mathbf{X} \in \mathbb{R}^{M \times P} (M = R_L \times C_L)$ with M pixels (R_L rows and C_L columns) and P bands and HR RGB $\mathbf{Y} \in \mathbb{R}^{N \times Q}$ with N pixels and Q bands ($M \ll N, Q \ll P$). In this work, we use the strategy of spectral grouping to learn group-wise mapping functions from $\mathbf{X}, \mathbf{Y}$ to $\mathbf{Z}$. The grouping process is obtained by clustering

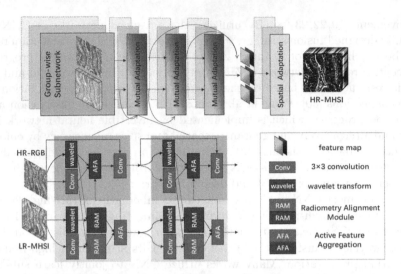

Fig. 2. Illustration of the proposed WaveNet framework

adjacent bands into groups based on the correlation matrix $\mathbf{R} \in \mathbb{R}^{P \times P}$. In our case, Pearson correlation is adopted and each coefficient is defined by

$$\mathbf{R}_{i,j} = \frac{\text{Cov}\left(\mathbf{z}_i, \mathbf{z}_j\right)}{\sigma_{\mathbf{z}_i} \sigma_{\mathbf{z}_j}}, \tag{1}$$

where $\mathbf{z}_i \in \mathbb{R}^{N \times 1}$ and $\mathbf{z}_j \in \mathbb{R}^{N \times 1}$ are i-th and j-th spectral bands $(1 \leq i, j \leq P)$, $Cov(\cdot; \cdot)$denotes the covariance between two variables and σ denotes the standard deviation. To determine the group number, a traditional method is using one empirical fixed threshold. However, it assumes that different subgroups have similar correlation typology which is not the case in some complicated datasets. Figure 3 shows the averaged correlation matrix of training samples in the dataset, where the first several spectral bands have weaker coefficients than others and cannot be grouped together if using a fixed threshold. Therefore, we propose to use multiple thresholds when the dataset has diverse correlation typology. In this way, the WaveNet learns the group-wise mapping function $f_k(\cdot)$ via minimizing

$$\mathcal{L} = \sum_{k=1}^{K} \|f_k(\mathbf{X}(:, \lambda_k), \mathbf{Y}; \theta_k) - \mathbf{Z}(:, \lambda_k)\|_F^2, \tag{2}$$

where λ_k denotes the k-th group of spectral bands, k is the number of groups and $k = 1, 2, \ldots K$.

3.2 Mutual Adaptation

The module contains two vital sub-blocks, Radiometry Alignment Module (RAM) and Active Feature Aggregation (AFA). The former is designed to

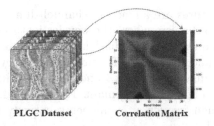

PLGC Dataset   **Correlation Matrix**

Fig. 3. Illustration of the averaged correlation matrix derived from training samples in PLGC dataset

address the radiance offset by local-global self-attention, which transforms high-spatial information from HR-RGB to LR-MHSI without spectral distortion. The latter applies wavelet-attention to aggregate the radiometry-free features from convolution and wavelet domain, highlighting the fusion features with details and textures.

Radiometry Alignment Module. Radiometrical inconsistency between LR-MHSI and HR-RGB is usually addressed in local and global perspective. But most of the methods work in a sequential way, where the former part solves the inconsistency locally with the later to refine it globally. The sequential manner limits the exploitation of underlying relations between local and global radiometry variation. In this work, we apply local and global radiometry transformation in a parallel fashion. Traditional convolution aggregate features over a localized receptive field, making it useful for local radiometry alignment. Comparably, self-attention module adaptively obtains a similarity attention distribution over the other related pixels so as to be suitable for global radiometry alignment. As shown in Fig. 4(a), the alignment module first crosses the feature maps to preserve the spectral-spatial position, then integrates a parallel convolution and self-attention layer for radiometry transformation by the ACmix module, finally followed by a nonlinear activation function LeakyReLu and 3×3 convolution. We refer readers to [16] for the detailed structure of ACmix.

Active Feature Aggregation. To benefit from WT for information preservation, we apply WT on feature maps in both LR-MHSI and HR-RGB modalities, and for each modality, attention mechanism is used to integrate WT with convolution. A dual-path feature extraction module is constructed with one path for 3×3 convolution and the other for WT. We adopt the widely used Haar wavelet for simplicity and efficiency [24]. It decomposes each feature map by four filters, $\mathbf{f_{LL}}$, $\mathbf{f_{LH}}$, $\mathbf{f_{HL}}$ and $\mathbf{f_{HH}}$ to convolve with feature maps to obtain four frequency sub-bands. Take the LR-MHSI for example, $\{\mathbf{x}_{LL}, \mathbf{x}_{LH}, \mathbf{x}_{HL}, \mathbf{x}_{HH}\} \in \mathbb{R}^{W/2 \times H/2 \times C_1}$ denotes average pooling channel as well as vertical, horizontal and diagonal derivatives, respectively, then a pixel-shuffle operation is used to upsample the sub-band feature maps. To this end, convolution and wavelet features are aligned in channel dimension. Furthermore, we adopt the attention mechanism where the wavelet features act as "query value" to emphasize high-frequency "key value" in convolution domain. As shown in Fig. 4(b), the wavelet and convolution features are fused by the pixel-wise manipulation, followed by a dual-layer of 1×1 and 3×3 convolution to reduce feature discrepancy in spectral and spatial domain. Thus, the domain-relevant features are extracted with high-frequency components actively emphasized and then added to the wavelet features.

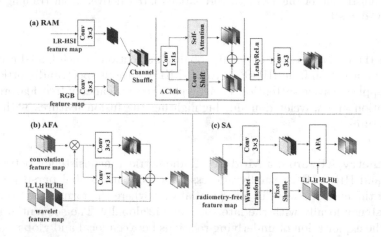

Fig. 4. Illustration of the proposed specific module RAM, AFA and SA

3.3 Multi-scale Spatial Adaptation

Let $\{\mathbf{X}_m^w(:, \lambda_k), \mathbf{X}_m^c(:, \lambda_k)\} \in \mathbb{R}^{W \times H \times C_1}$ denote the wavelet and convolution LR-MHSI feature maps of the m-th mutual adaptation block, respectively, and $\{\mathbf{Y}_m^w, \mathbf{Y}_m^c\} \in \mathbb{R}^{W \times H \times C_2}$ denote the HR-RGB feature maps. Through the cascade mutual adaptation blocks (in our case is three), the spatial-detail information of HR-RGB has been transformed into LR-MHSI progressively. The output feature set of LR-MHSI and HR-RGB denotes as $\left\{\hat{\mathbf{X}}_m^c(:, \lambda_k), \hat{\mathbf{Y}}_m^c\right\}$, then the outputs from each block are concatenated in channel dimension and fed into the spatial adaptation (SA) module. In this way, we exploit WT to depict the contextual and textural information in multi-scale high-resolution features. The structure of SA is shown in Fig. 4(c) The HR-MHSI is reconstructed by combing outputs of grouping subnetworks. Therefore, the loss function is formulated as

$$\mathcal{L}(\mathbf{Z}, \hat{\mathbf{Z}}) = \sum_{k=1}^{K} \|\mathbf{Z}(:, \lambda_k) - \hat{\mathbf{Z}}(:, \lambda_k)\|_1, \tag{3}$$

where $\| \cdot \|_1$ is the $\ell 1$ norm. The m-th mutual adaptation and spatial adaptation are substituted for $\mathbf{MA_k}(\cdot)$, $\mathbf{SA}(\cdot)$ respectively, and the implementation is detailed in Algorithm 1.

Algorithm 1. The WaveNet for HR-MHSI reconstruction

Input: X(LR-MHSI) and **Y**(HR-RGB);
Step1: Spectral Grouping
  1a) Calculate correlation matrix **R** based on Eq.(2);
  1b) Set group number K and initialize input set $\{\mathbf{X}(:, \lambda_k), \mathbf{Y}\}_{k=1}^K$ for subnetworks;
Step2: Mutual Adaptation
Set module number M
for $k = 1$ to K **do**
    for $m = 1$ to M **do**
      2a) Obtain convolution and wavelet features by
          $\{\mathbf{X}_m^c(:, \lambda_k), \mathbf{X}_m^w(:, \lambda_k)\}$ and $\{\mathbf{Y}_m^c, \mathbf{Y}_m^w\}$;
      2b) Update LR-MHSI features by
          $\hat{\mathbf{X}}_m(:, \lambda_k) \leftarrow \mathbf{MA}_k(\mathbf{X}_m^c(:, \lambda_k), \mathbf{Y}_m^c, \mathbf{X}_m^w(:, \lambda_k), \mathbf{Y}_m^w)$;
      2c) Update HR-RGB features by $\hat{\mathbf{Y}}_m \leftarrow \mathbf{MA}_k(\mathbf{Y}_m^c, \mathbf{Y}_m^w, \mathbf{Y}_m)$;
    end for
end for
Step3: Spatial Adaptation
for $k = 1$ to K **do**
  3a) Let $\hat{\mathbf{X}}(:, \lambda_k) \leftarrow concatenate(\hat{\mathbf{X}}_1(:, \lambda_k), \hat{\mathbf{X}}_2(:, \lambda_k) \cdots \hat{\mathbf{X}}_m(:, \lambda_k))$;
  3b) Reconstruct HR-MHSI by $\mathbf{Z}(:, \lambda_k) \leftarrow \mathbf{SA}(\hat{\mathbf{X}}(:, \lambda_k)) + \mathbf{X}(:, \lambda_k)$;
end for
Output: Z(HR-MHSI).

---

4   Experiments

4.1 Dataset

We evaluate the fusion-based HR-MHSI reconstruction model on a public pathological HSI dataset – PLGC (precancerous lesions of gastric cancer) dataset [27] which is composed of three sub-category datasets, namely NT (normal gastric tissue), IM (intestinal metaplasia) and GIN (gastric intraepithelial neoplasia). Each MHSI data is captured under 10x magnification with size of $2048 \times 2448 \times 40$ ranging from 450–700 nm at an interval of 6.25 nm, whereas a HR-RGB counterpart is acquired with the same field of view. We adopt the preprocessing method in our previous work [20] to calibrate the raw data due to optical path impurities, artifacts and inhomogeneous spectral response. Following the similar protocol of [22], the original MHSIs are used as ground truths, which are then blurred by a $r \times r$ Gaussian filter and downsampled with the factor of 2 to generate the LR-MHSIs. However, the conventional strategy, which fixes the kernel r and standard deviation std of the Gaussian filter, may not generalize to real case. Therefore, we randomly select kernel std in the range of [0.5, 1.5].

4.2   Implementation Detail

We conduct comparative experiments with several state-of-the-art HSI fusion method which are totally deep learning based: Two-stream Fusion Network (TFNet) [14], Spatial-Spectral Reconstruction Network (SSR-NET) [26], two versions of multispectral and hyperspectral image fusion network (MHF)-blind MHF (BMHF-net) and consistent MHF (CMHF-net) [23], residual learning network based on gradient transformation prior (GTP-PNet) [25], progressive zero-centric residual network (PZRes-Net) [30] and model-guided deep convolutional network (MoG-DCN) [5]. For fair comparisons, all the above-mentioned models were trained with suggested parameters and the remaining experimental settings followed the same protocol.

We select five widely used image quality metrics for quantitative evaluation, including peak signal-to-noise ratio (PSNR), spectral angle mapper (SAM), relative dimensionless global error in synthesis (ERGAS), structure similarity (SSIM), and universal image quality index (UIQI). All the metrics are averaged over the bands except the SAM. The smaller ERGAS and SAM are and the closer they are to zero, whereas the larger PSNR, SSIM and UIQI are and the closer they are to one, the better the HR-MHSI is.

The network is implemented in PyTorch framework on Linux 18.04 with a single NVIDIA GeForce GTX 3090Ti GPU. A total of 14768 pairs of LR-MHSI of size $128 \times 128 \times 32$ and HR-RGB of size $256 \times 256 \times 3$ are generated, with the pixel value normalized in $[0, 1]$. For data partition, the percentage of training, validation and testing is 6:2:2 and no data augmentation is used for all the experiments. We train the model by Adam optimizer with the learning rate initialized as 3×10^{-4}. When validation loss hasnt changed for 5 epochs, the cosine annealing decay strategy is employed to decrease learning rate to 8×10^{-5}.

4.3   Experimental Analysis

Fusion Results on Real-World Dataset. As mentioned above, to deal with the uncertainty in the parameters of blur kernels, we trained the models in the mix manner with the standard deviation of the Gaussian blur kernel randomly varying from 0.5–1.5. To testify the generalization, the models are evaluated under two different degradation settings: Gaussian blur kernel with fixed 1.0 standard deviation and mix standard deviation as the training manner. Table 1 and Table 2 give the comparison among deep learning-based methods with best performance highlighted in bold. It shows that the proposed method outperforms all the comparative methods in terms of all the five metrics under both degradation settings, except the SSIM of which is second only to the PZRes-Net. Our WaveNet model achieves highest PSNR 35.30 dB (34.96 dB) and lowest ERGAS 0.4090 (0.4268). It should be attributed to high-frequency components in wavelet domain and active feature aggregation in convolution domain. The lowest SAM of 0.0140 (0.0143) verifies superior ability of WaveNet in spectral preservation. This is strong evidence for the efficacy of radiometry-aware feature fusion scheme. Though the PZRes-Net achieves highest SSIM in both settings, which may be caused by the zero-mean normalization as it coincides with the calculation of SSIM, the model suffers a lot from distortion in the fused image with much higher ERGAS value. Comparing the two tables, we can see the performances of all the methods drop for different blur kernels, but the proposed WaveNet network still maintains its edge in overall performances. Specifically, GTP, PZRes, MoG and our WaveNet produce desirable performance with PSNR higher than 34 dB, which exceeds other four methods significantly. Those methods all integrate prior knowledge of MHSIs into deep learning models, such as the gradient transformation prior in GTP, zero-centric high-frequency details in PZRes, domain knowledge in MoG, and spectral correlation in WaveNet.

The per-band average PSNR values of test images in the two settings are plotted in Fig. 5. Since different tissues such as precancerous cells, acinar and lymphocyte exhibit unique spectra characteristics, the spatial properties per band vary significantly along the spectral dimension. Our method consistently produces the highest PSNR over all bands, which indicates the efficacy of the group-wise adaptation for taking the spectral correlation into consideration.

Table 1. Evaluation of the averaged performance on the test methods for scale factor 2 and gaussian blur with std 1 (best results are highlighted in bold)

Metric	SSRNET	CMHF	BMHF	TFNet	GTP	PZRes	MoG	WaveNet
ERGAS	0.7023	0.6760	0.5572	0.4875	0.4548	0.5176	0.4379	**0.4090**
SAM	0.0213	0.0220	0.0188	0.0167	0.0154	0.0148	0.0149	**0.0140**
PSNR	30.87	31.25	32.79	33.79	34.44	34.70	34.80	**35.30**
SSIM	0.9120	0.9133	0.9175	0.9202	0.9203	**0.9994**	0.9223	0.9271
UIQI	0.7050	0.7529	0.7600	0.7723	0.7787	0.7838	0.7835	**0.8066**

Table 2. Evaluation of the averaged performance on the test methods for scale factor 2 and gaussian blur with std in [0.5, 1.5] (best results are highlighted in bold)

Metric	SSRNET	CMHF	BMHF	TFNet	GTP	PZRes	MoG	WaveNet
ERGAS	0.7524	0.6912	0.5644	0.4973	0.4709	0.5413	0.4448	**0.4268**
SAM	0.0216	0.0223	0.0189	0.0168	0.0156	0.0150	0.0149	**0.0143**
PSNR	30.35	31.08	32.70	33.63	34.18	34.35	34.67	**34.96**
SSIM	0.9104	0.9110	0.9163	0.9185	0.9181	**0.9996**	0.9209	0.9228
UIQI	0.7066	0.7471	0.7586	0.7677	0.7745	0.7790	0.7802	**0.7965**

Visualization on Subcategory. This section is devoted to evaluate different models on the subcategory data since NT, IM and GIN hold intrinsic features of microstructure according to the stage of gastric carcinogenesis. Figures 6, 7 and 8 show the pseudo-color image of a test sample from NT, IM and GIN subcategory respectively, with bands 28(623 nm)–15(540 nm)–4(469 nm) as R-G-B. Due to

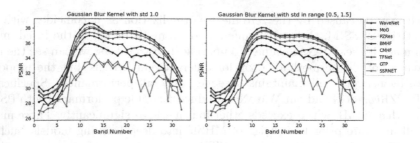

Fig. 5. Average PSNR of each band in PLGC dataset

the scale factor and gaussian blur, LR-MHSI suffers from the over-smoothed contents of tissues and blurred edges of cells. The proposed method works well on recovering the contextual and textural features in all the three subcategories. It is credited to its advantage on injecting spatial-detail information from HR-RGB to LR-MHSI.

Ablation Study. We verify the effect of three key components of the WaveNet model, including the group number, RAM and AFA. To support the ablation study, we have implemented several variants of the proposed network. Firstly, we changed the threshold of the correlation matrix to divide the spectral bands into 2, 4, 8 groups with 32, 16, 8 feature channels per group. Table 3 shows that WaveNet with two group-wise subnetworks lead to better reconstruction results in the PLGC dataset, which is in line with topology of the spectral correlation matrix in Fig. 3. It also worth noting that 8 groups with 4 spectral bands per group achieve competitive quality, this could be explained that the band number per group is small enough to have similar properties. Next, we investigated the contribution of the RAM and AFA module by training the network without the

Table 3. Comparisons of WaveNet model with different group number

Model	ERGAS	SAM	PSNR	SSIM	UIQI
Group 2	**0.4268**	**0.0143**	**34.96**	**0.9228**	**0.7965**
Group 4	0.4412	0.0146	34.67	0.9214	0.7940
Group 8	0.4352	0.0145	34.80	0.9160	0.7977

Table 4. Ablation studies towards the RAM and AFA module

Model	ERGAS	SAM	PSNR	SSIM	UIQI
w/o RAM	0.4496	0.0146	34.57	0.9192	0.7957
w/o AFA	0.4595	0.0149	34.48	0.9153	0.7917
Full WaveNet	**0.4268**	**0.0143**	**34.96**	**0.9228**	**0.7965**

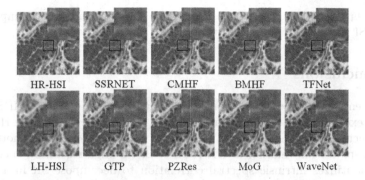

Fig. 6. The pseudo-color image of ground-truth HR-MHSI, simulated LH-MHSI of a test sample in NT category and results from 8 test methods with bands 28(623 nm)–15(540 nm)–4(469 nm).

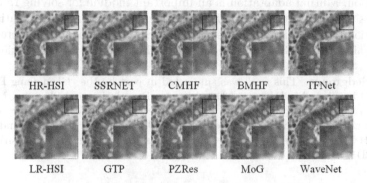

Fig. 7. The pseudo-color image of ground-truth HR-MHSI, simulated LH-MHSI of a test sample in IM category and results from 8 test methods with bands 28(623 nm)–15(540 nm)–4(469 nm).

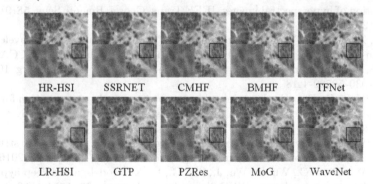

Fig. 8. The pseudo-color image of ground-truth HR-MHSI, simulated LH-MHSI of a test sample in GIN category and results from 8 test methods with bands 28(623 nm)–15(540 nm)–4(469 nm).

module. From Table 4, we can see that both the two components improve the HR MHSI reconstruction performance.

5   Conclusions

To compensate for the information loss in LR-MHSIs and exploit MHSI to the greatest extent in histopathological diagnosis, we propose an effective HR-MHSI reconstruction network named WaveNet. It is a fusion-based method to fuse the high-spatial information in HR-RGB with the high-resolution spectrum in MHSI. It utilizes intrinsic spectral correlation to decompose SR in a spectral grouping manner. It can adjust to any group number and achieve high spectral preservation. It also has high spatial preservation ability since the wavelet transform is incorporated with convolution to blend abundant high-frequency information. Mutual adaptation is an important module for solving radiometrical inconsistency between two modalities, which allows for injecting spatial-detail information from HR-RGB images to LR-MHSIs. Experiments conducted on the public real-world PLGC dataset validates its high-quality reconstruction.

Acknowledgment. This work was supported in part by the "Chenguang Program" supported by Shanghai Education Development Foundation and Shanghai Municipal Education Commission under Grant 18CG38, the Shanghai Key Laboratory of Multidimensional Information Processing of the East China Normal University under Grant MIP20224 and the Fundamental Research Funds for the Central Universities under Grant 22D111211.

References

1. Arun, P.V., Buddhiraju, K.M., Porwal, A., Chanussot, J.: CNN-based super-resolution of hyperspectral images. IEEE Trans. Geosci. Remote Sens. **58**(9), 6106–6121 (2020)
2. Chen, Z., Guo, X., Yang, C., Ibragimov, B., Yuan, Y.: Joint spatial-wavelet dual-stream network for super-resolution. In: Martel, A.L., et al. (eds.) MICCAI 2020. LNCS, vol. 12265, pp. 184–193. Springer, Cham (2020). https://doi.org/10.1007/978-3-030-59722-1_18
3. Dian, R., Li, S., Fang, L.: Learning a low tensor-train rank representation for hyperspectral image super-resolution. IEEE Trans. Neural Netw. Learn. Syst. **30**(9), 2672–2683 (2019)
4. Dong, W., et al.: Hyperspectral image super-resolution via non-negative structured sparse representation. IEEE Trans. Image Process. **25**(5), 2337–2352 (2016)
5. Dong, W., Zhou, C., Wu, F., Wu, J., Shi, G., Li, X.: Model-guided deep hyperspectral image super-resolution. IEEE Trans. Image Process. **30**, 5754–5768 (2021)
6. Dremin, V.: Skin complications of diabetes mellitus revealed by polarized hyperspectral imaging and machine learning. IEEE Trans. Med. Imaging **40**(4), 1207–1216 (2021)
7. Fang, L., Zhuo, H., Li, S.: Super-resolution of hyperspectral image via superpixel-based sparse representation. Neurocomputing **273**, 171–177 (2018)

8. Guo, T., Seyed Mousavi, H., Huu Vu, T., Monga, V.: Deep wavelet prediction for image super-resolution. In: Proceedings of the IEEE Conference on Computer Vision and Pattern Recognition Workshops, pp. 104–113 (2017)

9. Han, X.H., Shi, B., Zheng, Y.: Self-similarity constrained sparse representation for hyperspectral image super-resolution. IEEE Trans. Image Process. **27**(11), 5625–5637 (2018)

10. Han, X.H., Zheng, Y., Chen, Y.W.: Multi-level and multi-scale spatial and spectral fusion CNN for hyperspectral image super-resolution. In: Proceedings of the IEEE/CVF International Conference on Computer Vision Workshops (2019)

11. Kanatsoulis, C.I., Fu, X., Sidiropoulos, N.D., Ma, W.K.: Hyperspectral super-resolution: a coupled tensor factorization approach. IEEE Trans. Sig. Process. **66**(24), 6503–6517 (2018)

12. Lanaras, C., Baltsavias, E.: Hyperspectral super-resolution by coupled spectral unmixing, pp. 3586–3594 (2015). https://doi.org/10.1109/ICCV.2015.409

13. Liu, P., Zhang, H., Zhang, K., Lin, L., Zuo, W.: Multi-level wavelet-CNN for image restoration. In: Proceedings of the IEEE Conference on Computer Vision and Pattern Recognition Workshops, pp. 773–782 (2018)

14. Liu, X., Liu, Q., Wang, Y.: Remote sensing image fusion based on two-stream fusion network. Inf. Fusion **55**, 1–15 (2020)

15. Ma, W., Pan, Z., Guo, J., Lei, B.: Achieving super-resolution remote sensing images via the wavelet transform combined with the recursive res-net. IEEE Trans. Geosci. Remote Sens. **57**(6), 3512–3527 (2019)

16. Pan, X.,et al.: On the integration of self-attention and convolution. In: Proceedings of the IEEE/CVF Conference on Computer Vision and Pattern Recognition, pp. 815–825 (2022)

17. Sun, L., et al.: Diagnosis of cholangiocarcinoma from microscopic hyperspectral pathological dataset by deep convolution neural networks. Methods **202**, 22–30 (2022)

18. Wang, H., Wu, X., Huang, Z., Xing, E.P.: High-frequency component helps explain the generalization of convolutional neural networks. In: Proceedings of the IEEE/CVF Conference on Computer Vision and Pattern Recognition, pp. 8684–8694 (2020)

19. Wang, Q., Li, Q., Li, X.: Hyperspectral image superresolution using spectrum and feature context. IEEE Trans. Industr. Electron. **68**(11), 11276–11285 (2020)

20. Wang, Q., et al.: Identification of melanoma from hyperspectral pathology image using 3D convolutional networks. IEEE Trans. Med. Imaging **40**(1), 218–227 (2020)

21. Wang, W., et al.: Enhanced deep blind hyperspectral image fusion. IEEE Trans. Neural Netw. Learn. Syst.(2021)

22. Wang, W., Zeng, W., Huang, Y., Ding, X., Paisley, J.: Deep blind hyperspectral image fusion. In: Proceedings of the IEEE/CVF International Conference on Computer Vision, pp. 4150–4159 (2019)

23. Xie, Q., Zhou, M., Zhao, Q., Xu, Z., Meng, D.: MHF-net: an interpretable deep network for multispectral and hyperspectral image fusion. IEEE Trans. Pattern Anal. Mach. Intell. (2020)

24. Xin, J., Li, J., Jiang, X., Wang, N., Huang, H., Gao, X.: Wavelet-based dual recursive network for image super-resolution. IEEE Trans. Neural Netw. Learn. Syst. (2020)

25. Zhang, H., Ma, J.: GTP-Pnet: a residual learning network based on gradient transformation prior for pansharpening. ISPRS J. Photogramm. Remote. Sens. **172**, 223–239 (2021)

26. Zhang, X., Huang, W., Wang, Q., Li, X.: SSR-net: spatial-spectral reconstruction network for hyperspectral and multispectral image fusion. IEEE Trans. Geosci. Remote Sens. **59**(7), 5953–5965 (2020)
27. Zhang, Y., Wang, Y., Zhang, B., Li, Q.: A hyperspectral dataset of precancerous lesions in gastric cancer and benchmarkers for pathological diagnosis. J. Biophotonics (2022, in press). https://doi.org/10.1002/jbio.202200163
28. Zheng, Y., Li, J., Li, Y., Guo, J., Wu, X., Chanussot, J.: Hyperspectral pansharpening using deep prior and dual attention residual network. IEEE Trans. Geosci. Remote Sens. **58**(11), 8059–8076 (2020)
29. Zhong, Z., Shen, T., Yang, Y., Lin, Z., Zhang, C.: Joint sub-bands learning with clique structures for wavelet domain super-resolution. In: Advances in Neural Information Processing Systems, vol. 31 (2018)
30. Zhu, Z., Hou, J., Chen, J., Zeng, H., Zhou, J.: Hyperspectral image super-resolution via deep progressive zero-centric residual learning. IEEE Trans. Image Process. **30**, 1423–1438 (2020)

Using a 3D ResNet for Detecting the Presence and Severity of COVID-19 from CT Scans

Robert Turnbull[✉]

Melbourne Data Analytics Platform, University of Melbourne,
Melbourne, VIC 3060, Australia
robert.turnbull@unimelb.edu.au
https://mdap.unimelb.edu.au

Abstract. Deep learning has been used to assist in the analysis of medical imaging. One use is the classification of Computed Tomography (CT) scans for detecting COVID-19 in subjects. This paper presents Cov3d, a three dimensional convolutional neural network for detecting the presence and severity of COVID-19 from chest CT scans. Trained on the COV19-CT-DB dataset with human expert annotations, it achieves a macro f1 score of 87.87 on the test set for the task of detecting the presence of COVID-19. This was the 'runner-up' for this task in the 'AI-enabled Medical Image Analysis Workshop and Covid-19 Diagnosis Competition' (MIA-COV19D). It achieved a macro f1 score of 46.00 for the task of classifying the severity of COVID-19 and was ranked in fourth place.

Keywords: Deep learning · Computer vision · Medical imaging · Computed Tomography (CT) · 3D ResNet · COVID-19

1 Introduction

Medical professionals need fast and accurate methods for detecting the presence and severity of COVID-19 in patients. The World Health Organisation (WHO) recommends nucleic acid amplification tests (NAAT), such as real-time reverse transcription polymerase chain reaction (rRT-PCR), because they are highly specific and sensitive [24]. Though NAAT tests remain the gold-standard for diagnostic testing for COVID-19, these kinds of tests have the disadvantage of taking a long time and not occurring at the point of need [19]. Medical imaging can be used to complement these diagnostic strategies [12]. Computed Tomography (CT) scans use x-rays to reconstruct cross-sectional images to produce a three dimensional representation of the internals of the body [20]. Thoracic radiologists have used chest CT scans to correctly diagnose patients with COVID-19, including cases where rRT-PCR gives a negative result [25]. But this technique requires a human interpreter with sufficient knowledge and experience to provide reliable results. Advances in deep learning (DL) for computer vision offers the

L. Karlinsky et al. (Eds.): ECCV 2022 Workshops, LNCS 13807, pp. 663–676, 2023.
https://doi.org/10.1007/978-3-031-25082-8_45

possibility of using artificial intelligence for the task of CT scan image analysis [15]. Early in the pandemic, attempts to use deep learning showed significant promise for accurate detection of COVID-19 [3,14,16]. We can anticipate that refining techniques in artificial intelligence for medical imaging analysis will assist in improved clinical results in the future, not only for COVID-19, but for other illnesses as well. To further this area of research, the 'AI-enabled Medical Image Analysis Workshop and Covid-19 Diagnosis Competition' (MIA-COV19D) was created as part of the International Conference on Computer Vision (ICCV) in 2021 [13]. This competition sought submissions to predict the presence of COVID-19 in a dataset of CT scan cross-sectional images [1]. The winning submission produced a macro F1 score of 90.43 on the test partition of the dataset [7]. In 2022, the AI-enabled Medical Image Analysis Workshop issued a second competition with an enlarged dataset and a new task which was to also predict the severity of the disease in scans with positive cases [12]. This article presents an approach to the tasks of this competition by using a three dimensional convolutional neural network called Cov3d.[1]

2   The COV19-CT-DB Database

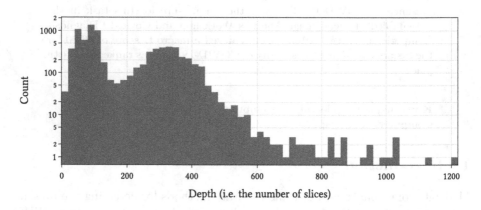

Fig. 1. A histogram of the number of slices NB. The vertical axis uses a logarithmic scale.

The database (COV19-CT-DB) comprises more than 7,700 chest CT scans from more than 3,700 subjects and was annotated to record whether or not the subject had COVID-19 [12]. Each CT scan contains 2D slices perpendicular to the long axis of the subject. The number of slices (hereafter referred to as the 'depth') of the scan ranges from 1 to 1201. It has a bimodal distribution (see Fig. 1) with more then 58% of scans have fewer than 160 slices and almost 99% having fewer

---

[1] The code for this project has been released under the Apache-2.0 license and is publicly available at https://github.com/rbturnbull/cov3d/.

than 500 slices. The slices are provided as sequentially numbered JPEG files with a typical resolution of 512×512 pixels. It has been divided into training, validation and test partitions (Table 1).

Table 1. The number of CT scans in the partitions of the database

COVID-19	Training	Validation	Test
Positive	882	215	–
Negative	1,110	269	–
Total	2,292	484	5,281

Table 2. The number of CT scans with severity annotations in the partitions of the database.

Index	Severity	Training	Validation	Test
1	Mild	85	22	–
2	Moderate	62	10	–
3	Severe	85	24	–
4	Critical	26	5	–
	Total	258	61	265

For a minority of the CT scans where COVID-19 was present, four experts annotated the severity of disease in the patient in four categories: mild, moderate, severe and critical (Fig. 2). These categories correspond with greater degrees of pulmonary parenchymal involvement. These annotations are indicated in CSV files accompanying the database. The number of scans with these annotations for the three partitions in the database are given in Table 2.

3   Methods

This paper discusses Cov3d, a classifier of CT scan images using three dimensional convolutional neural networks. Cov3d uses the deep learning framework PyTorch [17]. Cov3d also uses the fastai library which adds higher-level components [8]. It is built using the FastApp framework for packaging deep learning models and providing an auto-generated command-line interface (CLI).[2] The major components of the training pipeline are shown in Fig. 3. FastApp provides a training command in the CLI where the user provides the paths to the

---

[2] TorchApp Framework is available as an alpha release at: https://github.com/rbturnbull/torchapp/.

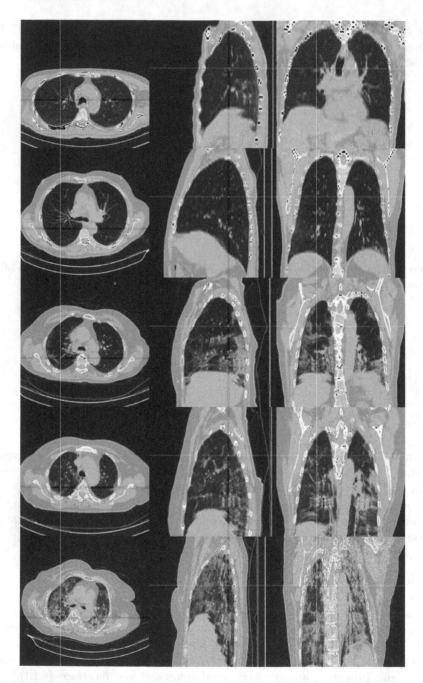

Fig. 2. Cross-sections from the five CT scans in the axial (left/blue), sagittal (middle/green), and coronal (right/red) planes. The top row displays a subject without COVID-19 (ct_scan_65 in the non-COVID-19 validation set). The following four rows show subjects having COVID-19 with severity categorizes as (in order): mild, moderate, severe and critical (ct_scan_97, ct_scan_0, ct_scan_93, ct_scan_19 respectively in the COVID-19 validation set). These five scans serve as examples of correct predictions for the best single models for the two tasks discussed in the results section. NB. The aspect ratios for the images in the sagittal and coronal planes have been distorted to present the all the slices fitted into a square image. (Color figure online)

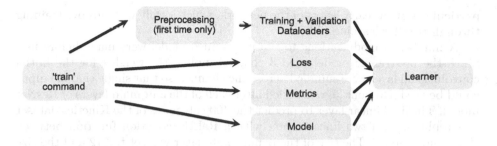

Fig. 3. The major components of the Cov3d training pipeline.

data and any hyperparameters for the run. The data files are first preprocessed for the first run of any particular resolution. FastApp then takes the training and validation dataloaders, the loss, metrics and model defined in the Cov3d app and creates a fastai 'Learner' which controls the training loop. At the end of training, the 'Learner' is saved for later inference.

3.1 Preprocessing

As seen in Fig. 1, the number of slices in each CT scan is significantly varied. To regularize these so that multiple scans could be processed in batches, and to reduce the size of the inputs to fit within the memory of the hardware, each scan went through a preprocessing step. First the 2D slices were reduced from a resolution of 512^2 pixels to 128^2, 256^2 or 320^2 pixels through bicubic interpolation. Then the slices were interpolated using 1D cubic interpolation along the axis perpendicular to the cross sections so that the depth was exactly half of the width/height of the processed scan. This gave three resolution sizes: small ($64{\times}128{\times}128$), medium ($128{\times}256{\times}256$) and large ($160{\times}320{\times}320$). These processed 3D single-channel volumes were saved as PyTorch tensor files and are loaded from disk during the training cycle.

3.2 Neural Network Model

ResNet models have been hugely popular for computer vision tasks on 2D images [4]. The residual 'shortcut' connections between the layers address the problem of vanishing gradients to allow for models with a greater number of layers. Cov3d uses a neural network model analogous to a two dimensional ResNet-18 model, but with 3D convolutional and pooling operations instead of their 2D equivalents. In particular, Cov3d uses the 3D ResNet-18 model included in the Torchvision library [23]. This model has been pre-trained on the Kinetics-400 dataset which classifies short video clips [10]. The time dimension in the pre-trained model was used for the depth dimension of the CT scans. Though the CT scans are quite different to video clip data in Kinetics-400, we can anticipate that the pre-trained network will have learned to identify shapes and patterns, particularly at the early layers of the network, which ideally will improve training through transfer learning (TL).

A number of modifications to the pre-trained model were made before use. Since the pre-trained model used 3 channel inputs, the weights for the initial convolutional layer were summed across the channels so that single channel input could be used. Dropout [5] was added after each of the four main layers of ResNet model. The final linear layer to predict the 400 categories of the Kinetics dataset was replaced with two linear layers with a ReLU activation function between them and dropout. The size of the penultimate layer was set to 512 and the size of the final layer was dependent on the loss function discussed below. An outline of the modified 3D ResNet-18 architecture is given in Table 3.

Table 3. The neural network architecture based on the ResNet-18. Square brackets represent convolutional residual blocks. 3D Convolutional layers are represented by the kernel size in the order of depth × width × height and then the number of filters after a comma.

Layer name	Description
Stem	1×7×7, 64, stride 1×2×2
Layer 1	$\begin{bmatrix} 3 \times 3 \times 3, 64 \\ 3 \times 3 \times 3, 64 \end{bmatrix}$ × 2, Dropout(0.5)
Layer 2	$\begin{bmatrix} 3 \times 3 \times 3, 128 \\ 3 \times 3 \times 3, 128 \end{bmatrix}$ × 2, Dropout(0.5)
Layer 3	$\begin{bmatrix} 3 \times 3 \times 3, 256 \\ 3 \times 3 \times 3, 256 \end{bmatrix}$ × 2, Dropout(0.5)
Layer 4	$\begin{bmatrix} 3 \times 3 \times 3, 512 \\ 3 \times 3 \times 3, 512 \end{bmatrix}$ × 2, Dropout(0.5)
Global Average Pooling	$1 \times 1 \times 1$
Penultimate Layer	512 dimensions, Dropout(0.5)
Final Layer	$1 + 5$ dimensions

3.3   Loss

For detection of the presence of COVID-19, the penultimate layer connects to a single logit x_i which models the log-odds that the subject has the disease, where i refers to the index of the input. This can be converted to a probability p_i using the sigmoid function:

$$p_i = \frac{1}{1 + e^{-x_i}} \tag{1}$$

The loss for task 1 (i.e. detection of the presence of COVID-19) is denoted ℓ_{covid} and is given by the weighted binary cross entropy:

$$\ell_{covid}(p_i, y_i) = -w_i \left(y_i \log p_i + (1 - y_i) \log(1 - p_i) \right) \tag{2}$$

where w_i is the weight of the input item and y_i is the target probability. The weights are set to compensate for the different proportion of COVID-19 and non-COVID-19 scans in the training partition (see Table 1).

Szegedy et al. introduced a regularization mechanism called label-smoothing which modifies the target probabilities to mitigate against the network becoming overconfident in the result [22]. This technique was employed here and therefore the target probability for positive COVID-19 scans is set to $1 - \epsilon_p$ and the target probability for negative COVID-19 scans was ϵ_p. The hyperparameter ϵ_p was set to either 0.0 or 0.1.

Task 2 requires classifying the severity of COVID-19 in patients into one of the four categories. The dataset for this task is quite small (Table 2) which makes training a model challenging. We can expand this dataset by treating the COVID-19 negative scans as belonging to a fifth category. Furthermore, the COVID-19 positive scans where the severity has not been annotated may be regarded as belonging to a superset of those four categories. This may be modelled by connecting the penultimate layer to a vector of five dimensions ($z_{i,c}$) corresponding to the four severity categories ($c = 1, 2, 3, 4$) and an additional category for being COVID-19 negative ($c = 0$). The probability distribution over the categories ($s_{i,c}$) is given by the softmax function:

$$s_{i,c} = \frac{e^{z_{i,c}}}{\sum_{j=0}^{4} e^{z_{i,j}}} \tag{3}$$

The probability that the scan is merely COVID-19 positive is given by summing over the four severity categories ($\sum_{c=1}^{4} s_{i,c}$). If the input is in one of the four annotated categories, or is classed as non-COVID-19, then the loss is thus given with the cross entropy:

$$\ell_{severity}(s_{i,c}, y_i) = -w_i \sum_{c=0}^{4} y_{i,c} \log(s_{i,c}) \tag{4}$$

where w_i is the weight of the input item and y_i is the target probability for class c. If the input is annotated as COVID-19 positive but without a severity category then the loss is given by:

$$\ell_{severity}(s_{i,c}, y_i) = -w_i \log(\sum_{c=1}^{4} s_{i,c}) \tag{5}$$

In this way, the second task to predict the severity can be trained using the entire database. This assists in the severity task because it requires the network to learn that features in the inputs to differentiate the severity categories should be absent in scans where the patient does not have COVID-19. Accuracy metrics for this task are restricted to the instances in the validation set where the severity was annotated.

As above, label smoothing was used as a regularization technique. But unlike in Szegedy where the target probabilities were combined with a uniform distribution [22], in this case the target probability for each class was reduced to

$1 - \epsilon_s$ and the remaining probability of ϵ_s is divided between neighboring categories while non-neighboring categories remain with a target probability of zero. Non-COVID-19 scans were considered to be neighboring to 'mild' severity scans.

The two loss functions discussed above could be used independently to train in separate models or they can be combined in a linear combination to train a single model to perform both tasks simultaneously:

$$\ell_{combined} = (1 - \lambda)\ell_{covid} + \lambda\ell_{severity} \tag{6}$$

where $0.0 \leq \lambda \leq 1.0$.

3.4   Training Procedure

The models were trained for 30, 40 or 50 epochs through the training dataset. The batch size was limited to two because of memory constraints. The Adam optimization method was used for updating the training parameters [11]. The maximum learning rate was set to 10^{-4} and this, together with the momentum for the optimizer, were scheduled according to the '1cycle' policy outlined by Smith [21]. Scikit-learn was used to calculate the macro f1 scores on the validation set every epoch [18]. The weights yielding the highest macro f1 score on the validation dataset for the two tasks were saved for later inference.

3.5   Regularization and Data Augmentation

To mitigate against overfitting on the training dataset, there is the option to randomly reflect the input scans through the sagittal plane each training epoch. At inference, the reflection transformation was then applied to the input and the final probability predictions were taken from the mean. Weight decay of 10^{-5} was applied.

4   Results

The models were trained using NVIDIA Tesla V100-SXM2-32 GB GPUs on the University of Melbourne high-performance computing system Spartan. The results were logged using the 'Weights and Biases' platform for experiment tracking [2].

The results of experiments with different hyperparameter settings are shown in Table 4. No one set of hyperparameters achieved the highest result in both tasks so two separate models are stored for inference on the two tasks. The highest macro f1 score for task 1 was 94.76 which is significantly above the baseline of 77. This model used the highest resolution (160×320×320) and had a λ value of 0.1 which weighed the loss heavily to ℓ_{covid}. The model correctly predicted the presence or absence of COVID-19 for the five scans displayed in Fig. 3. The confusion matrix for making predictions with this model on the validation set is shown in Table 5. The model produced eight false positives and

Table 4. Results of experiments. The best result for each task is highlighted in bold. In descending order of task 1 macro f1.

ID	Depth	Width/Height	Reflection	λ	Epochs	ϵ_p	ϵ_s	Task 1 macro f1	Task 2 macro f1
1 [P1]	160	320	Yes	0.1	40	0.1	0.0	**94.76**	47.64
2 [P2]	160	320	Yes	0.1	50	0.1	0.0	94.12	44.17
3 [P3]	160	320	No	0.1	30	0.1	0.1	93.94	45.08
4 [P4]	160	320	Yes	0.0	40	0.0	0.1	93.72	–
5	160	320	Yes	0.5	30	0.1	0.1	93.72	59.52
6	160	320	Yes	0.9	50	0.1	0.0	93.71	63.69
7	128	256	Yes	0.1	40	0.1	0.0	93.51	44.24
8	128	256	Yes	0.1	50	0.1	0.0	93.28	48.12
9	128	256	Yes	0.1	40	0.0	0.0	93.11	38.8
10	160	320	Yes	0.1	30	0.1	0.1	92.87	41.53
11	160	320	Yes	0.9	40	0.1	0.0	92.77	66.23
12	64	128	Yes	0.1	40	0.1	0.1	92.68	39.88
13	128	256	No	0.5	30	0.1	0.1	92.64	59.09
14 [S1]	128	256	Yes	0.9	40	0.1	0.0	92.56	**75.46**
15	128	256	No	0.1	30	0.1	0.1	92.49	44.16
16 [S3]	160	320	Yes	0.9	40	0.1	0.1	92.37	67.66
17	128	256	Yes	0.0	40	0.0	0.1	92.26	–
18	64	128	Yes	0.2	30	0.1	0.0	92.22	47.45
19	64	128	Yes	0.1	50	0.1	0.0	92.04	47.25
20	64	128	Yes	0.0	40	0.1	0.0	91.83	–
21	64	128	Yes	0.1	40	0.1	0.0	91.83	43.76
22	64	128	Yes	0.0	40	0.0	0.1	91.83	–
23	128	256	Yes	0.1	30	0.1	0.1	91.67	43.54
24	64	128	No	0.1	30	0.1	0.0	90.93	38.73
25 [S4]	64	128	Yes	0.9	40	0.1	0.0	90.62	67.54
26	64	128	No	0.5	30	0.1	0.0	89.73	61.43
27	128	256	Yes	1.0	40	0.0	0.0	–	67.12
28 [S2]	64	128	Yes	1.0	40	0.0	0.0	–	68.11

seventeen false negatives (see examples in Fig. 5). Of the seventeen scans that were wrongly classified as not having COVID-19, eight were also part of the validation set in the severity task. These included three scans labelled 'mild', four labelled 'severe' and one labelled 'critical'. The latter example is shown at the bottom of Fig. 5. This particular instance includes a high degree of background which may have contributed to the erroneous prediction.

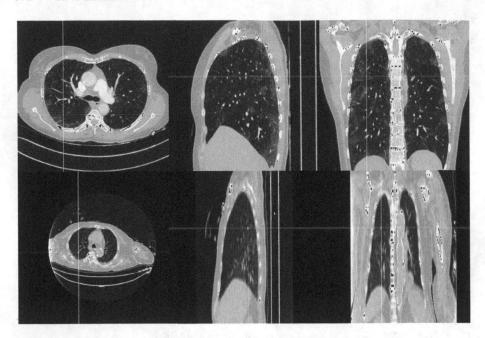

Fig. 4. Cross-sections for incorrect predictions from experiment #14 in the axial (left/blue), sagittal (middle/green), and coronal (right/red) planes. Top: a 'mild' case that was predicted as 'critical': ct_scan_159 from the COVID-19 validation set. Bottom: a false negative: ct scan 156 from the COVID-19 validation set. This scan was labelled as 'critical' severity. (Color figure online)

The highest macro f1 score for task 2 was 75.52 for experiment 14. This is significantly above the baseline of 63. The confusion matrix for predictions on the validation set for this experiment is shown in Table 6. The model correctly predicted the severity for the scans in the bottom four rows of Fig. 3. Two examples of incorrect predictions from this experiment are shown in Fig. 4. The top row of this image shows a 'mild' case that was predicted as 'critical'. As in the case in Fig. 5, this scan has a significant amount of background. These background regions may have confused the model because the majority of training instances did not include this. This problem may have been allayed through automated cropping as part of the preprocessing step.

Table 5. Confusion matrix for predictions on the validation set for experiment #1 for the task of predicting the presence of COVID-19.

		Ground truth		
		Positive	Negative	Total
Predictions	Positive	198	8	206
	Negative	17	261	278
Total		215	269	484

Table 6. Confusion matrix for predictions on the validation set for experiment #14 for the task of predicting the severity of COVID-19.

		Ground truth				
		Mild	Moderate	Severe	Critical	Total
Predictions	Mild	18	2	2	0	22
	Moderate	0	7	1	0	8
	Severe	3	1	17	0	21
	Critical	1	0	4	5	10
Total		22	10	24	5	61

The submission for the test set included the four highest performing models for each task (marked with submission IDs P1–P4 and S1–S4) and also a basic ensemble of the four averaging the probability predictions from each model. The results on the test set for tasks 1 and 2 are present in Tables 7 and 8.

Table 7. Results on the test set for task 1. Best result in bold.

Submission	Experiment	Macro f1	f1 NON-COVID	f1 COVID
P1	1	**87.87**	96.95	78.80
P2	2	87.60	96.88	78.31
P3	3	85.92	96.26	75.59
P4	4	87.12	96.76	77.47
P5	Ensemble	85.97	96.32	75.61

The highest performing model on the test set for task 1 was submission [P1] (i.e. experiment 1 in Table 4). It achieved a macro f1 score of 87.87. This was the 'runner-up' submission for the competition, behind the two teams which were awarded shared first place with a macro f1 score of 89.11 [6,9].

The highest performing model on the test set for task 2 was the ensemble of submissions S1–S4 (i.e. experiments 14, 28, 16 25). It achieved a macro f1 score of 46.00. This achieved the rank of fourth place amongst the six teams which achieved an f1 score higher than the baseline of 40.3.

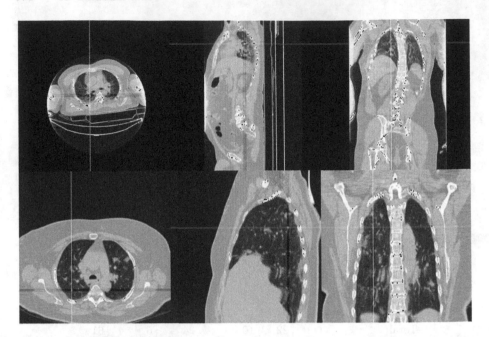

Fig. 5. Cross-sections for incorrect predictions from experiment 1 in the axial (left/blue), sagittal (middle/green), and coronal (right/red) planes. Top: a 'mild' case that was predicted as 'critical': ct scan 159 from the COVID-19 validation set. Bottom: a 'severe' case that was predicted as 'mild': ct scan 72 from the COVID-19 validation set. (Color figure online)

Table 8. Results on the test set for task 2. Best result in bold.

Submission	Experiment	Macro f1	f1 Mild	f1 Moderate	f1 Severe	f1 Critical
S1	14	37.57	62.25	36.64	51.38	0.00
S2	28	33.94	55.48	17.65	55.51	7.14
S3	16	39.84	64.12	40.56	50.24	4.45
S4	25	42.31	66.23	45.75	57.28	0.00
S5	**Ensemble**	**46.00**	58.14	42.29	43.78	40.03

5   Conclusion

Cov3d is a three dimensional model for detecting the presence and severity of COVID-19 in subjects from chest CT scans. It is based on a 3D ResNet pre-trained on video data. The model was trained using a customized loss function to simultaneously predict the dual tasks in the 'AI-enabled Medical Image Analysis Workshop and Covid-19 Diagnosis Competition'. The results for both tasks improve upon the baseline results for the challenge. Cov3d and the other submissions to the competition demonstrate that AI holds real promise in interpreting medical imaging such as CT scans. As these methods improve, they can be used to complement an array of other diagnostic methods to provide better patient care.

Acknowledgements. This research was supported by The University of Melbourne's Research Computing Services and the Petascale Campus. Sean Crosby and Naren Chinnam were instrumental in arranging the computational resources necessary. I thank Patricia Desmond and David Turnbull for providing feedback on earlier versions of this article.

References

1. Arsenos, A., Kollias, D., Kollias, S.: A large imaging database and novel deep neural architecture for Covid-19 diagnosis. In: 2022 IEEE 14th Image, Video, and Multidimensional Signal Processing Workshop (IVMSP), pp. 1–5 (2022). https://doi.org/10.1109/IVMSP54334.2022.9816321
2. Biewald, L.: Experiment tracking with weights and biases (2020). Software available: https://www.wandb.com/
3. Harmon, S.A., et al.: Artificial intelligence for the detection of Covid-19 pneumonia on chest CT using multinational datasets. Nat. Commun. **11**(1), 4080 (2020). https://doi.org/10.1038/s41467-020-17971-2
4. He, K., Zhang, X., Ren, S., Sun, J.: Deep residual learning for image recognition. In: 2016 IEEE Conference on Computer Vision and Pattern Recognition (CVPR), pp. 770–778 (2016). https://doi.org/10.1109/CVPR.2016.90
5. Hinton, G.E., Srivastava, N., Krizhevsky, A., Sutskever, I., Salakhutdinov, R.R.: Improving neural networks by preventing co-adaptation of feature detectors (2012). https://doi.org/10.48550/ARXIV.1207.0580, https://arxiv.org/abs/1207.0580
6. Hou, J., Xu, J., Feng, R., Zhang, Y.: FDVTS's solution for 2nd Cov19d competition on Covid-19 detection and severity analysis (2022). https://doi.org/10.48550/ARXIV.2207.01758, https://arxiv.org/abs/2207.01758
7. Hou, J., Xu, J., Feng, R., Zhang, Y., Shan, F., Shi, W.: CMC-Cov19d: Contrastive mixup classification for Covid-19 diagnosis. In: 2021 IEEE/CVF International Conference on Computer Vision Workshops (ICCVW), pp. 454–461 (2021). https://doi.org/10.1109/ICCVW54120.2021.00055
8. Howard, J., Gugger, S.: Fastai: A layered API for deep learning. Information **11**(2) (2020). https://doi.org/10.3390/info11020108, https://www.mdpi.com/2078-2489/11/2/108
9. Hsu, C.C., Tsai, C.H., Chen, G.L., Ma, S.D., Tai, S.C.: Spatiotemporal feature learning based on two-step LSTM and transformer for CT scans (2022). https://doi.org/10.48550/ARXIV.2207.01579, https://arxiv.org/abs/2207.01579
10. Kay, W., et al.: The kinetics human action video dataset (2017). https://doi.org/10.48550/ARXIV.1705.06950, https://arxiv.org/abs/1705.06950
11. Kingma, D.P., Ba, J.: Adam: a method for stochastic optimization (2014). https://doi.org/10.48550/ARXIV.1412.6980, https://arxiv.org/abs/1412.6980
12. Kollias, D., Arsenos, A., Kollias, S.: AI-MIA: Covid-19 detection & severity analysis through medical imaging. arXiv preprint arXiv:2206.04732 (2022)
13. Kollias, D., Arsenos, A., Soukissian, L., Kollias, S.: MIA-COV19D: Covid-19 detection through 3-D chest CT image analysis. arXiv preprint arXiv:2106.07524 (2021)
14. Kollias, D., et al.: Deep transparent prediction through latent representation analysis. arXiv preprint arXiv:2009.07044 (2020)

15. Kollias, D., Tagaris, A., Stafylopatis, A., Kollias, S., Tagaris, G.: Deep neural architectures for prediction in healthcare. Complex Intell. Syst. **4**(2), 119–131 (2018)
16. Kollias, D., et al.: Transparent adaptation in deep medical image diagnosis. In: TAILOR, pp. 251–267 (2020)
17. Paszke, A., et al.: PyTorch: an imperative style, high-performance deep learning library. In: Wallach, H., Larochelle, H., Beygelzimer, A., d'Alché-Buc, F., Fox, E., Garnett, R. (eds.) Advances in Neural Information Processing Systems, vol. 32, pp. 8024–8035. Curran Associates, Inc. (2019). http://papers.neurips.cc/paper/9015-pytorch-an-imperative-style-high-performance-deep-learning-library.pdf
18. Pedregosa, F., et al.: Scikit-learn: machine learning in Python. J. Mach. Learn. Res. **12**, 2825–2830 (2011)
19. Peeling, R.W., Heymann, D.L., Teo, Y.Y., Garcia, P.J.: Diagnostics for Covid-19: moving from pandemic response to control. Lancet (London, England) **399**, 757–768 (2022). https://doi.org/10.1016/S0140-6736(21)02346-1
20. Seeram, E.: Computed tomography: a technical review. Radiol. Technol. **89**, 279CT–302CT (2018)
21. Smith, L.N.: A disciplined approach to neural network hyper-parameters: part 1 - learning rate, batch size, momentum, and weight decay (2018). https://doi.org/10.48550/ARXIV.1803.09820, https://arxiv.org/abs/1803.09820
22. Szegedy, C., Vanhoucke, V., Ioffe, S., Shlens, J., Wojna, Z.: Rethinking the inception architecture for computer vision. In: 2016 IEEE Conference on Computer Vision and Pattern Recognition (CVPR), pp. 2818–2826. IEEE Computer Society, Los Alamitos, June 2016. https://doi.org/10.1109/CVPR.2016.308, https://doi.ieeecomputersociety.org/10.1109/CVPR.2016.308
23. Tran, D., Wang, H., Torresani, L., Ray, J., LeCun, Y., Paluri, M.: A closer look at spatiotemporal convolutions for action recognition. In: 2018 IEEE/CVF Conference on Computer Vision and Pattern Recognition, pp. 6450–6459 (2018). https://doi.org/10.1109/CVPR.2018.00675
24. World Health Organization: Recommendations for national SARS-CoV-2 testing strategies and diagnostic capacities. https://www.who.int/publications/i/item/WHO-2019-nCoV-lab-testing-2021.1-eng. Accessed 27 June 2022
25. Xie, X., Zhong, Z., Zhao, W., Zheng, C., Wang, F., Liu, J.: Chest CT for typical coronavirus disease 2019 (COVID-19) Pneumonia: relationship to negative RT-PCR testing. Radiology **296**(2), E41–E45 (2020). https://doi.org/10.1148/radiol.2020200343, pMID: 32049601

AI-MIA: COVID-19 Detection and Severity Analysis Through Medical Imaging

Dimitrios Kollias[1(✉)], Anastasios Arsenos[2], and Stefanos Kollias[2]

[1] Queen Mary University of London, London, UK
d.kollias@qmul.ac.uk
[2] National Technical University of Athens, Athens, Greece

Abstract. This paper presents the baseline approach for the organized 2nd Covid-19 Competition, occurring in the framework of the AIMIA Workshop in the European Conference on Computer Vision (ECCV 2022). It presents the COV19-CT-DB database which is annotated for COVID-19 detection, consisting of about 7,700 3-D CT scans. Part of the database consisting of Covid-19 cases is further annotated in terms of four Covid-19 severity conditions. We have split the database and the latter part of it in training, validation and test datasets. The former two datasets are used for training and validation of machine learning models, while the latter is used for evaluation of the developed models. The baseline approach consists of a deep learning approach, based on a CNN-RNN network and report its performance on the COVID19-CT-DB database. The paper presents the results of both Challenges organised in the framework of the Competition, also compared to the performance of the baseline scheme.

Keywords: AIMIA · COVID19-CT-DB · COVID19 detection · COVID19 severity detection · Deep learning · CT scans · Medical imaging

1 Introduction

Medical imaging seeks to reveal internal structures of human body, as well as to diagnose and treat diseases. Radiological quantitative image analysis constitutes a significant prognostic factor for diagnosis of diseases, such as lung cancer and Covid-19. For example, CT scan is an essential test for the diagnosis, staging and follow-up of patients suffering from lung cancer in its various histologies. Lung cancer constitutes a top public health issue, with the tendency being towards patient-centered control strategies. Especially in the chest imaging field, there has been a large effort to develop and apply computer-aided diagnosis systems for the detection of lung lesions patterns.

Deep learning (DL) approaches applied to the quantitative analysis of multimodal medical data offer very promising results. Current targets include building robust and interpretable AI and DL models for analyzing CT scans for the discrimination of immune-inflammatory patterns and their differential diagnosis with SARS-CoV-2-related or other findings. Moreover, detection of the new coronavirus (Covid-19) is of extreme significance for our Society. Currently, the gold standard reverse transcription polymerase chain reaction (RT-PCR) test is identified as the primary diagnostic tool for the Covid-19; however, it is time consuming with a certain number of false positive

© The Author(s), under exclusive license to Springer Nature Switzerland AG 2023
L. Karlinsky et al. (Eds.): ECCV 2022 Workshops, LNCS 13807, pp. 677–690, 2023.
https://doi.org/10.1007/978-3-031-25082-8_46

results. The chest CT has been identified as an important diagnostic tool to complement the RT-PCR test, assisting the diagnosis accuracy and helping people take the right action so as to quickly get appropriate treatment. That is why there is on-going research worldwide to create automated systems that can predict Covid-19 from chest CT scans and x-rays.

Recent approaches use Deep Neural Networks for medical imaging problems, including segmentation and automatic detection of the pneumonia region in lungs. The public datasets of RSNA pneumonia detection and LUNA have been used as training data, as well as other Covid-19 datasets. Then a classification network is generally used to discriminate different viral pneumonia & Covid-19 cases, based on analysis of chest x-rays, or CT scans.

Various projects are implemented in this framework in Europe, US, China and worldwide. However, a rather fragmented approach is followed: projects are based on specific datasets, selected from smaller, or larger numbers of hospitals, with no proof of good performance generalization over different datasets and clinical environments. A DL methodology which we have developed using latent information extraction from trained DNNs has been very successful for diagnosis of Covid-19 through unified chest CT scans and x-rays analysis provided by different medical centers and medical environments.

At the time of CT scan recording, several slices are captured from each person suspected of COVID-19. The large volume of CT scan images calls for a high workload on physicians and radiologists to diagnose COVID-19. Taking this into account and also the rapid increase in number of new and suspected COVID-19 cases, it is evident that there is a need for using machine and deep learning for detecting COVID-19 in CT scans. Such approaches require data to be trained on. Therefore, a few databases have been developed consisting of CT scans. However, new data sets with large numbers of 3-D CT scans are needed, so that researchers can train and develop COVID-19 diagnosis systems and trustfully evaluate their performance.

The current paper presents a baseline approach for the Competition which is a part of the Workshop "AI-enabled Medical Image Analysis - Digital Pathology & Radiology/COVID19 (AIMIA)" which occurs in conjunction with the European Conference on Computer Vision (ECCV) 2022 in Tel-Aviv, Israel, October 23- 24, 2022.

The AIMIA-Workshop context is as follows. Deep Learning has made rapid advances in the performance of medical image analysis challenging physicians in their traditional fields. In the pathology and radiology fields, in particular, automated procedures can help to reduce the workload of pathologists and radiologists and increase the accuracy and precision of medical image assessment, which is often considered subjective and not optimally reproducible. In addition, Deep Learning and Computer Vision demonstrate the ability/potential to extract more clinically relevant information from medical images than what is possible in current routine clinical practice by human assessors. Nevertheless, considerable development and validation work lie ahead before AI-based methods can be fully ready for integrated into medical departments.

The workshop on AI-enabled medical image analysis (AIMIA) at ECCV 2022 aims to foster discussion and presentation of ideas to tackle the challenges of whole slide image and CT/MRI/X-ray analysis/processing and identify research opportunities in the context of Digital Pathology and Radiology/COVID19.

The COV19D Competition is based on a new large database of chest CT scan series that is manually annotated for Covid-19/non-Covid-19, as well as Covid-19 severity diagnosis. The training and validation partitions along with their annotations are provided to the participating teams to develop AI/ML/DL models for Covid-19/non-Covid-19 and Covid-19 severity prediction. Performance of approaches will be evaluated on the test sets. The COV19-CT-DB is a new large database with about 7,700 3-D CT scans, annotated for COVID-19 infection.

The rest of the paper is organized as follows. Section 2 presents former work on which the presented baseline has been based. Section 3 presents the database created and used in the Competition. The ML approach and the pre-processing steps are described in Sect. 4. The obtained results, as well as the performance of the winning approaches in both Challenges are presented in Sect. 5. Conclusions and future work are described in Sect. 6.

2   Related Work

In [9] a CNN plus RNN network was used, taking as input CT scan images and discriminating between COVID-19 and non-COVID-19 cases. In [16], the authors employed a variety of 3-D ResNet models for detecting COVID-19 and distinguishing it from other common pneumonia (CP) and normal cases, using volumetric 3-D CT scans. In [22], a weakly supervised deep learning framework was suggested using 3-D CT volumes for COVID-19 classification and lesion localization. A pre-trained UNet was utilized for segmenting the lung region of each CT scan slice; the latter was fed into a 3-D DNN that provided the classification outputs. The presented approach is based on a CNN-RNN architecture that performs 3-D CT scan analysis. The method follows our previous work [12–14, 19] on developing deep neural architectures for predicting COVID-19, as well as neurodegenerative and other [13, 15] diseases and abnormal situations [4].

These architectures have been applied for: a) prediction of Parkinson's, based on datasets of MRI and DaTScans, either created in collaboration with the Georgios Gennimatas Hospital (GGH) in Athens [13, 19], or provided by the PPMI study sponsored by M. J. Fox for Parkinson's Research [15], b) prediction of COVID-19, based on CT chest scans, scan series, or x-rays, either collected from the public domain, or aggregated in collaboration with the Hellenic Ministry of Health and The Greek National Infrastructures for Research and Technology [12]. Other architectures, based on capsule nets [5, 18] can also be used in this framework.

The current Competition follows and extends the MIA-COV19D Workshop we successfully organized virtually in ICCV 2021, on October 11, 2021. 35 Teams participated in the COV19D Competition; 18 Teams submitted their results and 12 Teams scored higher than the baseline. 21 presentations were made in the Workshop, including 3 invited talks [11].

In general, some workshops on medical image analysis have taken place in the framework of CVPR/ICCV/ECCV Conferences in the last three years. In particular: in ICCV 2019 the 'Visual Recognition for Medical Images' Workshop took place, including papers on the use of deep learning in medical images for diagnosis of diseases. Similarly, in CVPR 2019 and 2020 the following series of Workshops took place: 'Medical

Computer Vision' and 'Computer Vision for Microscopy Image Analysis', dealing with general medical analysis topics. In addition, the 'Bioimage Computing' took place in ECCV 2020. Moreover, the issue of explainable DL and AI has been a topic of other Workshops (e.g., CVPR 2019; ICPR 2020 and 2022; ECAI Tailor 2020).

3   The COV19-CT-DB Database

The COVID19-CT-Database (COV19-CT-DB) consists of chest CT scans that are annotated for the existence of COVID-19. COV19-CT-DB includes 3-D chest CT scans annotated for existence of COVID-19. Data collection was conducted in the period from September 1 2020 to November 30 2021. It consists of 1,650 COVID and 6,100 non-COVID chest CT scan series, which correspond to a high number of patients (more than 1150) and subjects (more than 2600). In total, 724,273 slices correspond to the CT scans of the COVID-19 category and 1,775,727 slices correspond to the non COVID-19 category [1].

Annotation of each CT slice has been performed by 4 very experienced (each with over 20 years of experience) medical experts; two radiologists and two pulmonologists. Labels provided by the 4 experts showed a high degree of agreement (around 98%). Each of the 3-D scans includes different number of slices, ranging from 50 to 700. This variation in number of slices is due to context of CT scanning. The context is defined in terms of various factors, such as the accuracy asked by the doctor who ordered the scan, the characteristics of the CT scanner that is used, or specific subject's features, e.g., weight and age.

Figure 1 shows four CT scan slices, two from a non-COVID-19 CT scan (on the top row) and two from a COVID-19 scan (on the bottom row). Bilateral ground glass

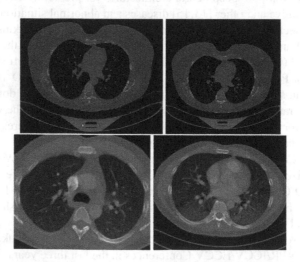

Fig. 1. Four CT scan slices, two from a non-COVID-19 CT scan (on the top row) and two from a COVID-19 scan (on the bottom row) including bilateral ground glass regions in lower lung lobes.

regions are seen especially in lower lung lobes in the COVID-19 slices. The database has been split in training, validation and testing sets. The training set contains, in total, 1992 3-D CT scans. The validation set consists of 494 3-D CT scans. The number of COVID-19 and of Non-COVID-19 cases in each set are shown in Table 1. There are different numbers of CT slices per CT scan, ranging from 50 to 700.

Table 1. Data samples in each set

Set	Training	Validation
COVID-19	882	215
Non-COVID-19	1110	289

A further split of the COVID-19 cases has been implemented, based on the severity of COVID-19, as annotated by four medical experts, in the range from 1 to 4, with 4 denoting the critical status. Table 2 describes each of these categories [17].

Table 2. Description of the severity categories

Category	Severity	Description
1	Mild	Few or no Ground glass opacities. Pulmonary parenchymal involvement $\leq 25\%$ or absence
2	Moderate	Ground glass opacities. Pulmonary parenchymal involvement 25–50%
3	Severe	Ground glass opacities. Pulmonary parenchymal involvement 50–75%
4	Critical	Ground glass opacities. Pulmonary parenchymal involvement $\geq 75\%$

In particular, parts of the COV19-CT-DB COVID-19 training and validation datasets have been accordingly split for severity classification in training, validation and testing sets.

The training set contains, in total, 258 3-D CT scans. The validation set consists of 61 3-D CT scans. The numbers of scans in each severity COVID-19 class in these sets are shown in Table 3.

Table 3. Data samples in each severity class

Severity class	Training	Validation
1	85	22
2	62	10
3	85	22
3	26	5

4 The Deep Learning Approach

4.1 3-D Analysis and COVID-19 Diagnosis

The input sequence is a 3-D signal, consisting of a series of chest CT slices, i.e., 2-D images, the number of which is varying, depending on the context of CT scanning. The context is defined in terms of various requirements, such as the accuracy asked by the doctor who ordered the scan, the characteristics of the CT scanner that is used, or the specific subject's features, e.g., weight and age.

The baseline approach is a CNN-RNN architecture, as shown in Fig. 2. At first all input CT scans are padded to have length t (i.e., consist of t slices). The whole (unsegmented) sequence of 2-D slices of a CT-scan are fed as input to the CNN part. Thus the CNN part performs local, per 2-D slice, analysis, extracting features mainly from the lung regions. The target is to make diagnosis using the whole 3-D CT scan series, similarly to the annotations provided by the medical experts. The RNN part provides this decision, analyzing the CNN features of the whole 3-D CT scan, sequentially moving from slice 0 to slice $t - 1$.

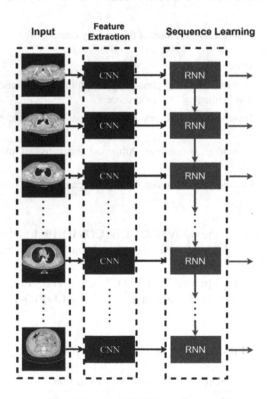

Fig. 2. The CNN-RNN baseline model

In the case of the COVID19 Detection Challenge, the outputs of the RNN part feed the output layer -which consists of 2 units- that uses a softmax activation function and provides the final COVID-19 diagnosis. In the case of the COVID19 Severity Detection Challenge, the outputs of the RNN part feed the output layer -which consists of 4 units- that uses a softmax activation function and provides the final COVID-19 severity diagnosis.

In this way, the CNN-RNN network outputs a probability for each CT scan slice; the CNN-RNN is followed by a voting scheme that makes the final decision. In the case of the COVID19 Detection Challenge the voting scheme can be either a majority voting or an at-least one voting (i.e., if at least one slice in the scan is predicted as COVID-19, then the whole CT scan is diagnosed as COVID-19, and if all slices in the scan are predicted as non-COVID-19, then the whole CT scan is diagnosed as non-COVID-19). In the case of the COVID19 Severity Detection Challenge, the voting scheme is a majority voting.

The main downside of this model is that there exists only one label for the whole CT scan and there are no labels for each CT scan slice. Thus, the presented model analyzes the whole CT scan, based on information extracted from each slice.

4.2 Pre-processing, Implementation Details and Evaluation Metrics

At first, CT images were extracted from DICOM files. Then, the voxel intensity values were clipped using a window/level of 350 Hounsfield units (HU)/ -1150 HU and normalized to the range of $[0, 1]$.

Regarding implementation of the proposed methodology: i) we utilized ResNet50 as CNN model, stacking on top of it a global average pooling layer, a batch normalization layer and dropout (with keep probability 0.8); ii) we used a single one-directional GRU layer consisting of 128 units as RNN model. The model was fed with 3-D CT scans composed of the CT slices; each slice was resized from its original size of $512 \times 512 \times 3$ to $224 \times 224 \times 3$. As a voting scheme for the COVID19 Detection Challenge, we used the at-least one.

Batch size was equal to 5 (i.e., at each iteration our model processed 5 CT scans) and the input length 't' was 700 (the maximum number of slices found across all CT scans). Softmax cross entropy was the utilized loss function for training the model. Adam optimizer was used with learning rate 10^{-4}. Training was performed on a Tesla V100 32 GB GPU.

Evaluation of the achieved performance in both Challenges was made in terms of macro F1 score. The macro F1 score is defined as the unweighted average of the class-wise/label-wise F1-scores. In the first Challenge this is the unweighted average of the COVID-19 class F1 score and of the non-COVID-19 class F1 score. In the second Challenge this was similarly defined as the unweighted average of the F1 scores of the severity categories.

5   The COV19D Competition Results

COV19D Competition consisted of two sub-challenges, namely, COVID19 Detection Challenge and COVID19 Severity Challenge. Training and validation data from the COVID-19 category (as indicatively shown in Figs. 3 and 4) and the non COVID-19

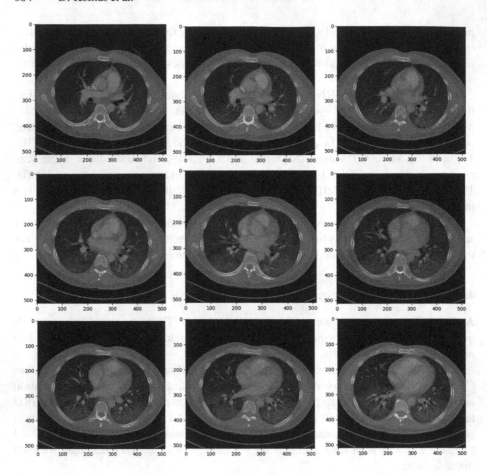

Fig. 3. 9 slices belonging to a CT-scan of the COVID-19 category, showing bilateral ground glass areas in lower lung lobes.

category (as indicatively shown in Fig. 5) were provided and used by the participating teams for training their models.

Nine teams that submitted results scored higher than the baseline and made valid submissions regarding the COVID19 Detection Challenge. The achieved performance by each team is shown in Table 4. Six teams made valid submissions and surpassed the baseline for COVID19 Severity Challenge. The corresponding performance of each team is shown in Table 5.

5.1 COVID19 Detection Challenge Results

The winners of this Competition were **ACVLab** from National Cheng Kung University, Taiwan and **FDVTS** from Fudan University, China. The runner-up was **MDAP** from the University of Melbourne. All results provided by the 9 teams are shown in Table 4; links of the related Github codes and arXiv papers are also provided. In the following we give

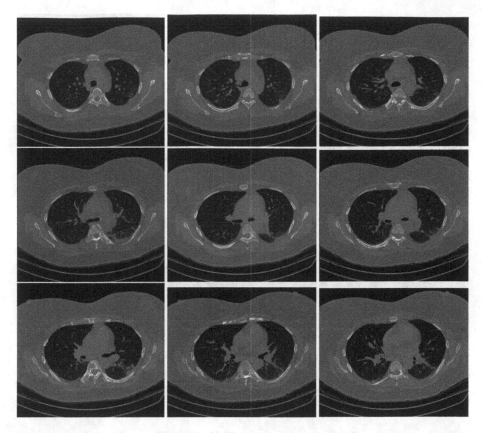

Fig. 4. 9 slices belonging to a CT-scan of the COVID-19 type, showing pneumonia bilateral thickening filtrates.

a short description of the best methods that the top three teams implemented and used. The index on the right of each team represents their official rank in the classification task.

Team **ACVLab** 1 proposed two spatiotemporal feature learning methodologies based on 2D-CNN + LSTM and 2D-CNN + Swin Transformer. At first, they trained the 2D CNN network which consisted of EfficientNet-b3a as the backbone and a fully connected layer with embedding size 224. Then they used the resulted embeddings from the 2D CNN network as input to the LSTM model which consisted of four Bi-LSTM layers with 128 hidden units. For the swin model they followed the same procedure and used the slice's embeddings from the 2D CNN model as input. In order to deal with the varying size of slices of the CT-scans the randomly selected 120 slices for the LSTM model and 224 slices for the swin model.

Team **FDVTS** 1 achieved the top performance in this task along with ACVLab team. In their approach, they employed a 3D Contrastive Mixup Classification network for COVID-19 diagnosis on chest CT images. They used the same network for

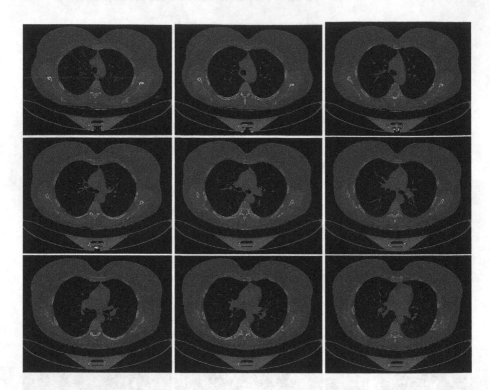

Fig. 5. 9 slices belonging to a CT scan of the non COVID-19 category

both sub-challenges. Their network named CMC-COV19D employed the contrastive representation learning (CRL) as an auxiliary task to learn more discriminative representations of COVID-19. The CRL was comprised of the following components. 1) A stochastic data augmentation module, which transformed an input CT into a randomly augmented sample. They generated two augmented volumes from each input CT. 2) A base encoder, which mapped the augmented CT sample to a representation vector. 3) A projection network, which was used to map the representation vector to a relative low-dimension vector. 4) A classifier, which classified the vector to the final prediction.

Team **MDAP** 2 was the runner-up of the COVID19 Detection Challenge. Their proposed methodology named Cov3d, was a three dimensional convolutional neural network able to detect the presence and severity of COVID19 from chest CT scans. In order to cope with the varying input size problem they used 1D cubic interpolation to set the depth of the CT-scans to a fixed number. Cov3d used a neural network model analogous to a two dimensional ResNet-18 model but with 3D convolutional and pooling operations instead of their 2D equivalents. In particular, Cov3d used the ResNet-3D-18 model. This model has been pre-trained on the Kinetics-400 dataset which classifies short video clips.

Table 4. COVID19 detection challenge results: F1 Score in %; the best performing submission is in bold

Teams	Submission #	Macro F1	F1 (NON-COVID)	F1 (COVID)	Github
ACVLab [7]	1	86.8	96.66	76.95	link
	2	85.75	96.26	75.22	
	3	**89.11**	**97.45**	**80.78**	
	4	86.72	96.64	76.80	
	5	86.75	96.61	76.89	
FDVTS [6]	1	88.61	97.13	80.09	link
	2	**89.11**	**97.31**	**80.92**	
	3	88.40	97.08	79.73	
	4	88.65	97.14	80.15	
	5	88.70	97.15	80.24	
MDAP [21]	**1**	**87.87**	**96.95**	**78.80**	link
	2	87.60	96.88	78.31	
	3	85.92	96.26	75.59	
	4	87.12	96.76	77.47	
	5	85.97	96.32	75.61	
Code 1055 [10]	1	82.13	95.04	69.22	link
	2	86.02	96.47	75.57	
	3	83.63	95.47	71.78	
	4	83.93	95.57	72.29	
	5	**86.18**	**96.37**	**76.00**	
CNR-IEMN [3]	1	81.91	95.34	68.48	link
	2	77.03	92.94	61.12	
	3	83.12	95.55	70.69	
	4	84.12	95.90	72.34	
	5	**84.37**	**95.98**	**72.76**	
Dslab [23]	1	81.69	95.74	67.64	link
	2	81.04	95.47	66.61	
	3	82.38	95.85	68.92	
	4	**83.78**	**96.22**	**71.33**	
	5	83.53	96.19	70.87	
Jovision-Deepcam [20]	1	77.71	93.13	62.29	link
	2	80.13	94.24	66.01	
	3	**80.82**	**94.56**	**67.07**	
	4	77.66	93.01	62.31	
	5	75.51	92.36	58.65	
ICL [8]	**1**	**79.55**	**93.77**	**65.34**	link
etro [2]	**1**	**78.72**	**93.48**	**63.95**	link
Baseline	1	69	83.62	54.38	

5.2 COVID19 Severity Detection Challenge Results

The winner of this Competition was **FDVTS** from Fudan University, China. The runner-up (with a slight difference from the winning teams) was **Code 1055** from Augsburg University, Germany. The team that ranked third was **CNR-IEMN**.

All results of the 6 teams are shown in Table 5; links are also provided to both, of the related Github codes and arXiv papers. In the following we provide a short description of the best methods that the top three teams implemented and used. The index on the right of each team represents their official rank in the classification task.

Team **FDVTS** 1 achieved the top performance in this task. They used the same methodology for both sub-challenges.

Table 5. COVID19 severity detection challenge results: F1 Score in %; the best performing submission is in bold

Teams	Submission #	Macro F1	F1 (MILD)	F1 (MODERATE)	F1 (SEVERE)	F1 (CRITICAL)	Github
FDVTS [6]	1	46.21	53.90	35.66	56.57	38.71	link
	2	41.15	62.50	35.30	51.81	15.00	
	3	47.14	62.94	35.21	60.61	29.79	
	4	45.96	53.33	39.16	52.63	38.71	
	5	**51.76**	**58.97**	**44.53**	**58.89**	**44.64**	
Code 1055 [10]	1	46.67	58.16	40.85	55.24	32.43	link
	2	49.36	56.95	30.14	52.48	57.90	
	3	**51.48**	**61.14**	**34.06**	**61.91**	**48.83**	
	4	46.01	60.00	27.78	52.78	43.48	
	5	49.90	63.01	28.99	48.52	59.09	
CNR-IEMN [3]	1	39.89	34.86	35.44	58.00	31.25	link
	2	43.04	53.09	18.52	63.52	37.04	
	3	**47.11**	**55.67**	**37.88**	**55.46**	**39.46**	
	4	44.37	53.42	26.67	60.36	37.04	
	5	45.49	54.66	26.79	63.48	37.04	
MDAP [21]	1	37.57	62.25	36.64	51.38	0.00	link
	2	33.94	55.48	17.65	55.51	7.14	
	3	39.84	64.12	40.56	50.24	4.45	
	4	42.31	66.23	45.75	57.28	0.00	
	5	**46.00**	**58.14**	**42.29**	**43.78**	**40.03**	
etro [2]	1	**44.92**	**57.69**	**29.57**	**62.85**	**29.57**	link
Jovision-Deepcam [20]	1	**41.49**	**62.20**	**40.24**	**38.59**	**24.93**	link
Baseline	1	**40.3**	**61.10**	**39.47**	**37.07**	**23.54**	

Team **Code 1055** 2 was the runner-up of COVID19 Severity Detection Challenge. The architecture they utilized was a three dimensional version of the recently published ConvNeXt architecture and was modified in order to be able to process the three-dimensional input tensors. This architecture was especially characterized by multiple alterations already proven to be useful in the context of Vision Transformers that are applied to the standard ResNet architecture. To make use of the ConvNeXt architecture, they replaced all 2D components with corresponding 3D components.

Team **CNR-IEMN** 3 developed an ensemble of convolutional layers with Inception models for COVID19 Severity Detection Challenge. In addition to the convolutional layers, three Inception variants were used, namely Inception-v3, Inception-v4 and Inception-Resnet. Regarding the pre-processing of the CT-scans they did the following steps: removed the non-lung slices and segmented the lung regions to remove unnecessary features, they resized the obtained slices into two volumes of size 224 × 224 × 32 and 224 × 224 × 16. The goal of the two volumes was to reduce the information lost due to resizing and to obtain two views of each CT scan. The two input volumes were fed into the convolution block.

6   Conclusions and Future Work

In this paper we have introduced a new large database of chest 3-D CT scans, obtained in various contexts and consisting of different numbers of CT scan slices. We have also developed a deep neural network, based on a CNN-RNN architecture and used it for COVID-19 diagnosis on this database.

The scope of the paper has been to present a baseline scheme regarding the performance that can be achieved based on analysis of the COV19-CT-DB database. This was used in the Competition part of the ECCV 2022 AIMIA Workshop, which included two Challenges: one targeting COVID-19 detection and another targeting COVID-19 severity detection.

The paper also presented the results of the methods developed in the framework of both Challenges and briefly described the main methodology followed by each of the three top ranking teams in the Challenges. By examining these results, it can be concluded that the detection of COVID-19 severity, based on a rather small of cases for each severity category, has been a harder problem to tackle. More data will be released in the future for each examined category, so that better performance can be reached.

The model presented as baseline in the paper will be the basis for future expansion towards more transparent modelling of COVID-19 diagnosis.

Acknowledgments. We would like to thank GRNET, the Greek National Infrastructures for Research and Technology, for supporting the presented work, through Project "Machaon - Advanced networking and computational services to hospital units in public cloud environment".

References

1. Arsenos, A., Kollias, D., Kollias, S.: A large imaging database and novel deep neural architecture for Covid-19 diagnosis. In: 2022 IEEE 14th Image, Video, and Multidimensional Signal Processing Workshop (IVMSP), pp. 1–5. IEEE (2022)
2. Berenguer, A.D., Mukherjee, T., Bossa, M., Deligiannis, N., Sahli, H.: Representation learning with information theory for Covid-19 detection (2022). https://doi.org/10.48550/ARXIV.2207.01437, https://arxiv.org/abs/2207.01437
3. Bougourzi, F., Distante, C., Dornaika, F., Taleb-Ahmed, A.: Ensemble CNN models for Covid-19 recognition and severity perdition from 3D CT-scan (2022). https://doi.org/10.48550/ARXIV.2206.15431, https://arxiv.org/abs/2206.15431
4. Caliva, F., et al.: A deep learning approach to anomaly detection in nuclear reactors. In: 2018 International Joint Conference on Neural Networks (IJCNN), pp. 1–8. IEEE (2018)
5. De Sousa Ribeiro, F., Leontidis, G., Kollias, S.: Introducing routing uncertainty in capsule networks. Adv. Neural. Inf. Process. Syst. **33**, 6490–6502 (2020)
6. Hou, J., Xu, J., Feng, R., Zhang, Y.: Fdvts's solution for 2nd Cov19d competition on Covid-19 detection and severity analysis (2022). https://doi.org/10.48550/ARXIV.2207.01758, https://arxiv.org/abs/2207.01758
7. Hsu, C.C., Tsai, C.H., Chen, G.L., Ma, S.D., Tai, S.C.: Spatiotemporal feature learning based on two-step LSTM and transformer for CT scans (2022). https://doi.org/10.48550/ARXIV.2207.01579, https://arxiv.org/abs/2207.01579
8. Jung, O., Kang, D.U., Kim, G., Chun, S.Y.: Adaptive GLCM sampling for transformer-based Covid-19 detection on CT (2022). https://doi.org/10.48550/ARXIV.2207.01520, https://arxiv.org/abs/2207.01520
9. Khadidos, A., Khadidos, A.O., Kannan, S., Natarajan, Y., Mohanty, S.N., Tsaramirsis, G.: Analysis of Covid-19 infections on a CT image using deepsense model. Front. Publ. Health **8** (2020)
10. Kienzle, D., Lorenz, J., Schön, R., Ludwig, K., Lienhart, R.: Custom pretrainings and adapted 3D-convnext architecture for Covid detection and severity prediction (2022). https://doi.org/10.48550/ARXIV.2206.15073, https://arxiv.org/abs/2206.15073
11. Kollias, D., Arsenos, A., Soukissian, L., Kollias, S.: Mia-Cov19d: Covid-19 detection through 3-D chest CT image analysis. In: Proceedings of the IEEE/CVF International Conference on Computer Vision, pp. 537–544 (2021)

12. Kollias, D., et al.: Deep transparent prediction through latent representation analysis. arXiv preprint arXiv:2009.07044 (2020)
13. Kollias, D., Tagaris, A., Stafylopatis, A., Kollias, S., Tagaris, G.: Deep neural architectures for prediction in healthcare. Complex Intell. Syst. **4**(2), 119–131 (2018)
14. Kollias, D., et al.: Transparent adaptation in deep medical image diagnosis. In: TAILOR, pp. 251–267 (2020)
15. Kollias, S., Bidaut, L., Wingate, J., Kollia, I., et al.: A unified deep learning approach for prediction of Parkinson's disease. IET Image Process. (2020)
16. Li, Y., Xia, L.: Coronavirus disease 2019 (Covid-19): role of chest CT in diagnosis and management. Am. J. Roentgenol. **214**(6), 1280–1286 (2020)
17. Morozov, S.P., et al.: Mosmeddata: data set of 1110 chest CT scans performed during the Covid-19 epidemic. Digit. Diagn. **1**(1), 49–59 (2020)
18. Ribeiro, F.D.S., Leontidis, G., Kollias, S.: Capsule routing via variational Bayes. In: Proceedings of the AAAI Conference on Artificial Intelligence, vol. 34, pp. 3749–3756 (2020)
19. Tagaris, A., Kollias, D., Stafylopatis, A., Tagaris, G., Kollias, S.: Machine learning for neurodegenerative disorder diagnosis-survey of practices and launch of benchmark dataset. Int. J. Artif. Intell. Tools **27**(03), 1850011 (2018)
20. Tan, W., Yao, Q., Liu, J.: Two-stage Covid19 classification using BERT features (2022). https://doi.org/10.48550/ARXIV.2206.14861, https://arxiv.org/abs/2206.14861
21. Turnbull, R.: Cov3d: detection of the presence and severity of COVID-19 from CT scans using 3D ResNets [Preliminary Preprint], July 2022. https://doi.org/10.26188/20226087. v1, https://melbourne.figshare.com/articles/preprint/Cov3d_Detection_of_the_presence_and_severity_of_COVID-19_from_CT_scans_using_3D_ResNets_Preliminary_Preprint_/20226087
22. Wang, X., et al.: A weakly-supervised framework for Covid-19 classification and lesion localization from chest CT. IEEE Trans. Med. Imaging **39**(8), 2615–2625 (2020)
23. Zheng, L., et al.: PVT-Cov19d: pyramid vision transformer for Covid-19 diagnosis (2022). https://doi.org/10.48550/ARXIV.2206.15069, https://arxiv.org/abs/2206.15069

Medical Image Segmentation: A Review of Modern Architectures

Natalia Salpea[1]([⊠]) [ID], Paraskevi Tzouveli[1] [ID], and Dimitrios Kollias[2] [ID]

[1] National Technical University of Athens, Athens, Greece
natalia.e.salpea@gmail.com
[2] Queen Mary University of London, London, UK

Abstract. Medical image segmentation involves identifying regions of interest in medical images. In modern times, there is a great need to develop robust computer vision algorithms to perform this task in order to reduce the time and cost of diagnosis and thus to aid quicker prevention and treatment of a variety of diseases. The approaches presented so far, mainly follow the U-type architecture proposed along with the UNet model, they implement encoder-decoder type architectures with fully convolutional networks, and also transformer architectures, exploiting both attention mechanisms and residual learning, and emphasizing information gathering at different resolution scales. Many of these architectural variants achieve significant improvements in quantitative and qualitative results in comparison to the pioneer UNet, while some fail to outperform it. In this work, 11 models designed for medical image segmentation, as well as other types of segmentation, are trained, tested and evaluated on specific evaluation metrics, on four publicly available datasets related to gastric polyps and cell nuclei, which are first augmented to increase their size in an attempt to address the problem of the lack of a large amount of medical data. In addition, their generalizability and the effect of data augmentation on the scores of the experiments are also examined. Finally, conclusions on the performance of the models are provided and future extensions that can improve their performance in the task of medical image segmentation are discussed.

Keywords: Medical image segmentation · Polyps · Cell nuclei · UNet · Vnet · Attention UNet · TransUNet · SwinUNet · ResUNet · ResUNet++ · R2U-Net · MSRF-Net · DeepLabv3+ · ResUNet-a · CVC-ClinicDB · Kvasir-SEG · 2018 Data Science Bowl · SegPC

1 Introduction

Medical image segmentation is an essential process for faster and better diagnosis of diseases and for ensuring faster and more effective treatment of a large number of patients simultaneously, reducing the need for human intervention as well as limiting human errors, time and cost. It belongs to the field of semantic segmentation and involves the identification of regions of interest in medical data

L. Karlinsky et al. (Eds.): ECCV 2022 Workshops, LNCS 13807, pp. 691–708, 2023.
https://doi.org/10.1007/978-3-031-25082-8_47

such as 2D images, 3D brain MRIs [26,42,46], CT scans [27,28], ultrasounds and X-rays [31], among others. Segmentation results can help identify areas associated with diseases, such as gastrointestinal polyps, in order to analyze them and decide if they are cancerous and need to be removed. This diagnosis can be life-critical for patients as it can help prevent serious forms of cancer. Due to the slow and tedious process of manual segmentation [43], there is a great need for computer vision algorithms that can perform segmentation quickly and accurately without the requirement for human involvement. Other challenges in the field of medical imaging are the small amount of available data, and the need for professionals specialized in this field when generating the data. When annotating polyps in images for example, protocols and guidelines are defined by the expert performing the annotation at that time. However, there are discrepancies between experts, for example in deciding whether a region is cancerous or not. Furthermore, due to the quite significant variation between data due to different imaging quality, the software used to perform the annotation, the expert performing the annotation, the brightness of the images and general biological variations, segmentation algorithms need to be robust and have high generalizability.

In this paper we perform a thorough study of the most important architectures presented in the field of medical image segmentation, we test and evaluate them on four publicly available and widely used datasets and we compare the results to draw conclusions on which architectures are most favorable to the task at hand and which are most promising for future developments. The process followed is illustrated in Fig. 1. Specifically, after the data were collected, split into training, test and validation sets, they were preprocessed and augmented to increase their size and introduce an additional diversity factor. The 11 models, discussed in a following section, are then trained on the augmented data, on the non augmented data and on supersets resulting from combining sets, with appropriately selected hyperparameters and loss functions. The models are then tested over the test subsets, the segmentation masks are generated and qualitatively compared with the actual masks, and the selected evaluation metrics are calculated. Once all the data are collected, a qualitative and quantitative comparison of the architectures follows. In addition, experiments and comparisons are performed to test the effect of data augmentation on the performance of the models, as well as testing the generalization ability of the models through training and evaluation on different datasets.

2   Related Work

In recent years, Convolutional Neural Networks (CNN) have dominated the field of segmentation in various types of medical data, but also in general [6,29,30]. The most modern architectures for semantic and instance segmentation are usually encoder-decoder networks, whose success is due to skip-connections, which allow semantically important and dense feature maps to be forwarded from the encoder to the decoder and reduce the semantic gap between the two subnetworks. The implementation parameters of this type of models differ depending

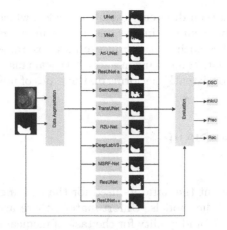

Fig. 1. Entire work pipeline of the essay

on the biomedical application. Moreover, the small amount of publicly available data makes the models' task more difficult and requires specialized design in order to be addressed as a problem. One of the first and most popular approaches to the task of semantic medical image segmentation that addresses the above problem is UNet [39], which converted the Fully Convolutional Network (FCN) of [32], which has only convolutional layers, into a U-type encoder-decoder architecture in which low- and high-level features are combined via skip-connections. However, a semantic gap between these features may occur. In order to bridge the gap, it is proposed in [22] to add convolutional units to the skip-connections. In [35], a version of UNet with an attention mechanism is proposed which helps to implement a pruning mechanism to skip unnecessary spatial features passing through the skip-connections. UNet was not only limited to the medical image domain but versions of UNet for different types of segmentation appeared, such as [13]. The 3D version of UNet, VNet [34] performs segmentation on sparsely annotated volumetric images, and consists of an FCN with skip-connections. Residual links improve information flow while reducing the convergence time, which is why they were preferred in several approaches, such as in [49]. In [25], Squeeze and Excitation blocks, attention mechanisms as well as Atrous Spatial Pyramid Pooling (ASPP) are added to the ResUNet model. In [2], residual connections are combined into a recurrent network which leads to better feature representations. The ASPP mechanism appeared in [9] to pool global features, and evolved in [10] into an architecture that uses skip-connections between the encoder and decoder. Squeeze and Excitation blocks were proposed in [20]. These blocks model interdependencies between channels and extract global maps that emphasize important features and ignore less relevant features. The concept of gated shape stream appeared in [44] and served to produce more detailed segmentation maps by taking into account the shape of the region of interest. To maintain high resolution representations, it was proposed to exchange low and high level features between different resolution scales, which is shown in [45] to lead to more spatially accurate final segmentation maps. Finally, in [41] many of the above techniques such as the Squeeze and Excitation blocks, the gated shape stream, the attention mechanism and multi-scale fusion are combined into a robust model that consistently performs on the types of medical data contained in the present work.

3   Experimental Setup

3.1   Datasets

In this section we present the datasets used for the training, the validation and the testing process of the models. Unfortunately, there are not enough public datasets of sufficient size and quality for the task of medical image segmentation. To address the first problem of insufficient sample number, data augmentation works as a catalyst. However, the second problem, that of quality, is not considered an easy challenge to address because the quality of the images can only be improved when there is a significant breakthrough in the sample acquisition technology. The sets we used belong to two broad categories involving polyp segmentation and cell nuclei segmentation respectively. Different types of sets were chosen in order to test the robustness and the generalizability of the models. The numbers of samples of each dataset is shown in Table 1(a), and examples from each dataset are shown in Fig. 2.

One of the datasets used in this paper is CVC-ClinicDB [4], which is an open access database of colonoscopy video frames that contain various examples of polyps. The dataset consists of images with a resolution of 384×288 pixels and their binary segmentation masks corresponding to the areas covered by polyps in the images. The images are property of Hospital Clinic, Barcelona, Spain and the masks are property of the Computer Vision Center, Barcelona, Spain.

Another dataset used in this work is Kvasir-SEG [24] which is an open access dataset with images of gastrointestinal polyps and corresponding segmentation masks, with manual annotation and verification by an experienced gastroenterologist. The resolution of the images varies from 332×487 to 1920×1072 pixels. The image and the mask files are encoded with JPEG compression. Data were collected using endoscopic equipment at Vestre Viken Health Trust (VV) in Norway.

The 2018 Data Science Bowl dataset [5] contains a large number of segmented images of cell nuclei. The images were acquired under a variety of conditions and vary in terms of cell type, magnification and imaging mode. Each segmentation mask contains one nucleus, so each image can have more than one mask assigned to it. This dataset contains images from more than 30 biological experiments on different samples, cell strains, microscopy instruments, imaging conditions, operators, research facilities and staining protocols. The annotations were manually created by a team of expert biologists at the Broad Institute.

For the SegPC dataset [14,17–19], microscopic images were recorded from bone marrow aspirate slides of patients diagnosed with Multiple Myeloma, a type of white blood cancer. The slides were stained and plasma cells, which are cells of interest, were segmented. Images were recorded in raw BMP format. This dataset aims at a robust cell segmentation which is the first step in creating a computer-assisted diagnostic tool for plasma cell cancer, namely Multiple Myeloma (MM).

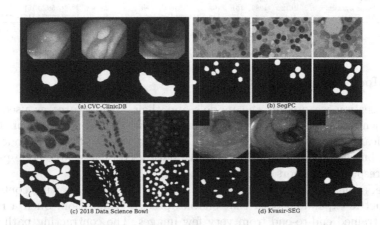

Fig. 2. Examples of image-mask pairs from all datasets [4,5,14,17–19,24]

Some of the datasets were originally not in the format dictated by the experimental procedure, so a preprocessing procedure was needed. In particular, all the datasets were divided into training, validation and test folders in the ratio of 80%, 10% and 10% respectively. For the 2018 Data Science Bowl and SegPC datasets there was additional preprocessing. In the 2018 Data Science Bowl a separate mask is given for each nucleus identified in the image. These masks were merged so that for each image there is a unique binary mask that includes all nuclei. Similarly for the SegPC all the masks corresponding to one image are merged into a final binary mask containing all the nuclei (ignoring the cytoplasm to remain consistent with the 2018 Data Science Bowl dataset). Finally, joining the Kvasir-SEG and CVC-CLinicDB datasets led to the creation of a new larger dataset that was named "Polyps dataset", and similarly joining the 2018 Data Science Bowl and the SegPC datasets led to the "Cells Dataset". The creation of these sets was deemed necessary for some experiments to be reported in later sections concerning the generalizability of the models.

Due to the relatively small number of samples that are publicly available for the task of medical image segmentation in the areas considered in this paper, it is necessary to augment the data [37]. Techniques such as random rotation, shear, zoom, horizontal and vertical flip, were applied to the training sets resulting in a large increase in the number of samples, not exceeding the memory and runtime limits of the available equipment used in this work. The updated numbers of samples are shown on Table 1(b).

Table 1. (a) Number of samples in all datasets before augmentation (b) Number of samples in all datasets after augmentation

Dataset	Images	Train	Valid	Test
CVC-ClinicDB	612	489	61	62
Kvasir-SEG	1000	800	100	100
2018 Data Science Bowl	670	536	67	67
SegPC	497	397	50	50
Polyps Dataset	1612	1289	161	162
Cells Dataset	1167	933	117	117

(a)

Dataset	Images	Train	Valid	Test
CVC-ClinicDB	860	737(+50%)	61	62
Kvasir-SEG	1496	1296(+62%)	100	100
2018 Data Science Bowl	918	784(+46%)	67	67
SegPC	745	645(+62%)	50	50
Polyps Dataset	2356	2033(+58%)	161	162
Cells Dataset	1664	1429(+60%)	117	117

(b)

3.2 Model Architectures

In this section we present the models and their architectures as used in the experiments of this paper. Several of the most widely used models in the field of medical image segmentation as well as some State-of-the-Art models were chosen, but not only limited to that. We have used open-source code provided by the respective authors for all the models.

UNet's [39] architecture consists of a contracting path to capture context and a symmetric expanding path that enables precise localization. Such a network can be trained end-to-end from very few images. The contracting path follows the typical architecture of a convolutional network whereas the expansive path consists of upsampling of the feature map, convolutions and a concatenation with the corresponding cropped feature map from the contracting path.

VNet [34] utilises the power of fully convolutional neural networks to process MRI volumes. It also introduces residual connections along with a novel objective function. In this work, a 2D version of VNet is used.

Attention UNet [35] is an attention gate model for medical imaging that can focus on target objects of variable shape and size. It learns to ignore non relevant features and to highlight the ones that are important to a specific task, while also eliminating the need for external localization modules.

ResUNet [49] is a network which combines residual learning with U-Net, and is designed for road area extraction. The network consists of residual units and has the extra advantages of easy training and need for fewer parameters.

ResUNet++ [25] is an improved ResUNet architecture designed for segmentation of colonoscopic images, it takes advantage of residual blocks, squeeze and excitation blocks, Atrous Spatial Pyramidal Pooling (ASPP) and attention blocks.

ResUNet-a [13] is a deep learning architecture designed for the semantic segmentation of high resolution aerial images, that uses a UNet backbone along with residual connections, atrous convolutions, pyramid scene parsing pooling and multi-tasking inference.

R2U-Net [2] is a Recurrent Residual Convolutional Neural Network inspired by UNet, which leverages the power of UNet, residual networks, as well as Recurrent CNNs. It succeeds in easy training and more effective feature representation for segmentation tasks.

SwinUNet [7] is a pure Transformer UNet inspired architecture for medical image segmentation. The tokenized image patches are fed into the Transformer-based U-shaped Encoder-Decoder architecture with skip-connections for semantic feature learning.

TransUNet [8] combines the strengths of both Transformers and U-Net and is presented as a strong alternative for medical image segmentation. The Transformer encodes tokenized image patches from a convolutional neural network (CNN) feature map as the input sequence and the decoder upsamples the encoded features which are then combined with the high-resolution CNN feature maps to enable precise localization.

DeepLabv3+ [10] is a novel architecture which combines the multi-scale fusion advantages of spatial pyramid pooling modules and the sharp object boundaries of encoder-decoder networks. It extends DeepLabv3 [9] by adding a decoder module to refine the segmentation results especially along object boundaries.

MSRF-Net [41] is a Multi-Scale Residual Fusion Network specially designed for medical image segmentation. It is able to exchange multi-scale features of varying receptive fields using Dual-Scale Dense Fusion (DSDF) blocks which can exchange information rigorously across two different resolution scales, allowing the preservation of resolution.

3.3 Metrics

Loss Functions. One of the loss functions used in this work is Binary Cross Entropy which calculates the performance of a binary classification model (the type of semantic segmentation in this task is equivalent to classification with two classes, 0 and 1) that produces as output values of probability 0 to 1. It is described by Formula 1.

$$\mathcal{L}_{BCE} = (1 - y)log(1 - \hat{y}) + ylog\hat{y} \tag{1}$$

where y is the real value and $\hat{y}$ is the predicted value. In addition to Binary Cross Entropy, there is also Dice Loss which is a measure of the overlap between the real image and the model-computed image and is described by Formula 2.

$$\mathcal{L}_{DSC} = 1 - \frac{2y\hat{y} + 1}{y + \hat{y} + 1} \tag{2}$$

where y is the real value and $\hat{y}$ is the predicted value. In all of the models presented above, both loss functions were used in example experiments and for each one of them, the loss function that led to the best results was eventually selected. In the MSRF-Net model a combination of loss functions defined in Formula 3 was used:

$$\mathcal{L}_{MSRF} = \mathcal{L}_{comb} + \mathcal{L}_{comb}^{DS^0} + \mathcal{L}_{comb}^{DS^1} + \mathcal{L}_{BCE}^{SS} \tag{3}$$

where $\mathcal{L}_{comb} = \mathcal{L}_{DSC} + \mathcal{L}_{BCE}$ is the combination of Dice and Binary Cross Entropy losses, $\mathcal{L}_{comb}^{DS^0}$ and $\mathcal{L}_{comb}^{DS^1}$ are the two deep supervision output losses of the model, and $\mathcal{L}_{BCE}^{SS}$ is the Binary Cross Entropy Loss computed for the shape stream of the MSRF-Net model [41].

Evaluation Metrics. For the evaluation of the models' performance we used standard medical image segmentation evaluation metrics, such as the (mean) Dice Coefficient described in Formula 4 and (mean) Intersection Over Union described in Formula 5, which measure the overlap between the predicted and the true images, as well as Precision described in Formula 6 and Recall described in Formula 7, which measure the ratio of correct predictions in similar ways.

$$DSC = \frac{2 A \cap B|}{|A| + |B|} \tag{4}$$

$$IoU = \frac{|A \cap B|}{|A \cup B|} \tag{5}$$

$$Precision = \frac{TruePositives}{TruePositives + FalsePositives} \tag{6}$$

$$Recall = \frac{TruePositives}{TruePositives + FalseNegatives} \tag{7}$$

where $|A|$ and $|B|$ are the sets of the activations in the predicted mask and the ground truth mask respectively.

3.4   Implementation Details

All the augmentation, training and evaluation code has been implemented using the Keras [11] framework with TensorFlow [1] as the backend. All experiments were conducted on ARIS [16], the Greek supercomputer developed and maintained by GRNET in Athens, and more specifically on 2 x NVIDIA Tesla k40m accelerated GPU nodes. All the images and masks from the four datasets were resized to 256×256, the number of epochs is set to 200. The optimizer for all models was chosen to be Adam, with a learning rate of 0.0001. We used a batch size of 8 for CVC-ClinicDB and SegPC and a batch size of 16 for Kvasir-SEG and 2018 Data Science Bowl.

4   Results

Table 2(a) shows the results of the experiments on the CVC-ClinicDB dataset which, as previously analysed, includes images of polyps derived from colonoscopy videos. The best performing model in terms of Dice Coefficient and mean Intersection over Union metrics is DeepLabv3+ followed by MSRF-Net, R2U-net, TransUnet and UNet. The MSRF-Net and R2U-Net achieve the highest Precision, while the highest Recall is achieved by DeepLabv3+. The good performance of these models is visually confirmed by Fig. 3 and the masks produced by them resemble the real masks to a good extent. Models such as SwinUNet, VNet and ResUNet appear to fail on this particular dataset. The masks generated by them show significant imperfections compared to the real masks.

Table 2(b) shows the results of the experiments on the Kvasir-SEG dataset which also includes images of gastrointestinal polyps. The most robust models

Table 2. Quantitative results of our experiments on the four original augmented datasets

Method	DSC	mIoU	Precision	Recall	Method	DSC	mIoU	Precision	Recall
MSRF-Net	0.87	0.84	**0.95**	0.89	MSRF-Net	0.86	**0.73**	**0.89**	0.80
UNet	0.92	0.80	0.91	0.87	UNet	0.90	0.71	**0.89**	0.78
VNet	0.81	0.64	0.79	0.76	VNet	0.76	0.49	0.73	0.60
Att-UNet	0.87	0.69	0.92	0.73	Att-UNet	0.86	0.62	0.82	0.72
ResUNet-a	0.86	0.71	0.85	0.81	ResUNet-a	0.78	0.49	0.73	0.60
SwinUNet	0.79	0.56	0.70	0.71	SwinUNet	0.77	0.50	0.79	0.66
TransUNet	0.94	0.80	0.92	0.86	TransUNet	0.90	0.70	0.87	0.79
R2U-Net	0.94	0.83	**0.95**	0.88	R2U-Net	0.88	0.69	0.85	0.79
DeepLabv3+	**0.99**	**0.89**	0.94	**0.94**	DeepLabv3+	**0.96**	**0.73**	0.87	**0.82**
ResUNet	0.70	0.64	0.80	0.77	ResUNet	0.60	0.49	0.66	0.66
ResUNet++	0.78	0.72	0.90	0.78	ResUNet++	0.65	0.55	0.74	0.69

(a) CVC-ClinicDB (b) Kvasir-SEG

Method	DSC	mIoU	Precision	Recall	Method	DSC	mIoU	Precision	Recall
MSRF-Net	0.91	0.86	0.91	**0.95**	MSRF-Net	0.80	0.68	0.82	0.79
UNet	0.95	**0.87**	0.92	0.94	UNet	0.88	0.66	0.81	0.78
VNet	0.94	0.85	0.91	0.93	VNet	0.85	0.57	0.75	0.71
Att-UNet	0.95	0.86	0.92	0.93	Att-UNet	0.89	0.66	0.80	0.79
ResUNet-a	0.91	0.76	0.90	0.83	ResUNet-a	0.77	0.41	0.60	0.57
SwinUNet	0.94	0.85	0.92	0.92	SwinUNet	0.80	0.45	0.63	0.62
TransUNet	0.95	**0.87**	**0.93**	0.93	TransUNet	0.90	0.68	0.81	0.80
R2U-Net	0.95	0.86	0.90	**0.95**	R2U-Net	0.89	**0.69**	**0.85**	0.78
DeepLabv3+	**0.98**	0.85	0.91	0.93	DeepLabv3+	**0.98**	0.64	0.78	0.78
ResUNet	0.88	0.84	0.89	0.94	ResUNet	0.77	0.64	0.74	**0.82**
ResUNet++	0.89	0.83	0.89	0.93	ResUNet++	0.76	0.62	0.76	0.77

(c) 2018 Data Science Bowl (d) SegPC

with respect to the mIoU metric are MSRF-Net and DeepLabv3+ followed by UNet and TransUNet, while with respect to the DSC metric the most robust model is DeepLabv3+ followed by R2U-Net, TransUNet and UNet. The highest Precision is achieved by MSRF-Net and UNet, and the highest Recall is achieved by DeepLabv3+. Observing the visual results in Fig. 3, the strong models produce fairly qualitative masks, in which the target polyp is detected, however, additional small regions may be misidentified as polyps. The models that seem to fail are VNet, ResUNet-a and SwinUNet, which is visually confirmed by the inaccurate generated masks.

Table 2(c) lists the results on the 2018 Data Science Bowl dataset which includes images of nuclei of different cell types. An initial observation is that all models achieve significantly higher performance compared to the other datasets. According to the DSC metric the strongest models are DeepLabv3+, R2U-Net, TransUNet, Attention UNet, SwinUNet, UNet and VNet while according to mIoU the strongest are TransUNet, UNet followed by MSRF-Net, Attention UNet, R2U-Net, SwinUNet, VNet and DeepLabv3+. The highest Precision is achieved by TransUNet and the highest Recall is achieved by MSRF-Net and R2U-Net. The only model that shows significantly lower performance than the others is ResUNet-a. Figure 3 shows the very high similarity between the masks of the models and the actual mask in terms of the grayscale image, while in terms of the color image the masks of the models that achieve the highest scores in the mIoU metric are clearly of higher quality.

Finally, Table 2(d) lists the results on the SegPC dataset which consists of plasma cell images from Multiple Myeloma patients. The model performances for this dataset are lower than the above datasets, which makes SegPC the more challenging dataset. The strongest model based on the DSC metric is DeepLabv3+ followed by TransUNet,R2U-Net, Attention UNet and UNet, while based on the mIoU metric the strongest model is R2U-Net followed by MSRF-Net and TransUNet. The highest Precision is achieved by the R2U-Net model and the highest Recall is achieved by the ResUNet model. These results are confirmed visually in Fig. 3. The models that appear to fail based on both the quantitative results in Table 2(d) and the visual results in Fig. 3 are SwinUNet and ResUNet-a.

We also examine the effect of the data augmentation on the quantitative results of the experiments. The greater the increase in scores observed with the increasing size of the datasets, the more the power of the models is highlighted, as it is shown that their performance will improve in the event that more and larger medical image datasets are produced in the future.

Table 3 presents the results without data augmentation. It is observed that data augmentation causes an approximate 4–5% increase in the mIoU metric, a 2–3% increase in the DSC metric, a 1–2% increase in the Precision metric and a 3–4% increase in the Recall metric. Some models benefit more from the larger data volume than others, for example DeepLabv3+ achieves a 2% increase in the mIoU metric while MSRF-Net achieves a 4% increase in the Kvasir-SEG dataset. Other models seem to not achieve an increase through data augmentation, for example ResUNet, ResUNet++ in the CVC-ClinicDB dataset. In general, the increasing trend in model performance as a function of increasing data volume is confirmed. Models such as MSRF-Net, R2U-Net, TransUNet seem to scale faster and to a greater extent with increasing data size.

Table 3. Quantitative results of our experiments on the four original non-augmented datasets

Method	DSC	mIoU	Precision	Recall	Method	DSC	mIoU	Precision	Recall
MSRF-Net	0.89	0.83	0.94	0.80	MSRF-Net	0.81	0.69	0.84	0.79
UNet	0.90	0.78	0.91	0.85	UNet	0.87	0.64	0.83	0.74
VNet	0.79	0.59	0.78	0.71	VNet	0.75	0.44	0.69	0.55
Att-UNet	0.83	0.67	0.86	0.75	Att-UNet	0.84	0.58	0.84	0.66
ResUNet-a	0.84	0.70	0.89	0.76	ResUNet-a	0.75	0.46	0.64	0.62
SwinUNet	0.78	0.54	0.70	0.71	SwinUNet	0.75	0.45	0.65	0.60
TransUNet	0.91	0.76	0.90	0.83	TransUNet	0.87	0.66	0.80	0.80
R2U-Net	0.90	0.76	0.92	0.81	R2U-Net	0.87	0.66	0.85	0.75
DeepLabv3+	**0.99**	**0.87**	**0.96**	**0.91**	DeepLabv3+	**0.96**	**0.75**	**0.89**	**0.83**
ResUNet	0.66	0.64	0.81	0.76	ResUNet	0.54	0.45	0.58	0.67
ResUNet++	0.77	0.72	0.94	0.76	ResUNet++	0.60	0.50	0.70	0.63
(a) CVC-ClinicDB					(b) Kvasir-SEG				
Method	**DSC**	**mIoU**	**Precision**	**Recall**	**Method**	**DSC**	**mIoU**	**Precision**	**Recall**
MSRF-Net	0.91	**0.86**	0.91	0.94	MSRF-Net	0.79	**0.67**	**0.81**	0.78
UNet	0.95	**0.86**	0.91	**0.95**	UNet	0.86	0.63	0.78	0.77
VNet	0.94	0.85	0.91	0.93	VNet	0.84	0.57	0.75	0.71
Att-UNet	0.95	**0.86**	0.91	0.94	Att-UNet	0.88	0.65	0.78	**0.80**
ResUNet-a	0.93	0.83	0.88	0.94	ResUNet-a	0.74	0.39	0.52	0.39
SwinUNet	0.94	0.85	0.92	0.92	SwinUNet	0.77	0.43	0.56	0.65
TransUNet	0.95	**0.86**	**0.95**	0.90	TransUNet	0.88	0.63	0.75	**0.80**
R2U-Net	0.95	**0.86**	0.92	0.93	R2U-Net	0.85	0.63	0.77	0.78
DeepLabv3+	**0.98**	0.84	0.92	0.91	DeepLabv3+	**0.97**	0.61	0.76	0.76
ResUNet	0.89	0.85	0.90	0.93	ResUNet	0.70	0.56	0.69	0.76
ResUNet++	0.89	0.84	0.89	0.94	ResUNet++	0.72	0.60	0.79	0.72
(c) 2018 Data Science Bowl					(d) SegPC				

The generalizability is an important characteristic of a model and thus considered necessary to evaluate. In order to evaluate the generalization ability [23], models are trained on one of the two datasets including gastrointestinal polyps (CVC-ClinicDB, Kvasir-SEG) and evaluated on the other dataset. Similarly for

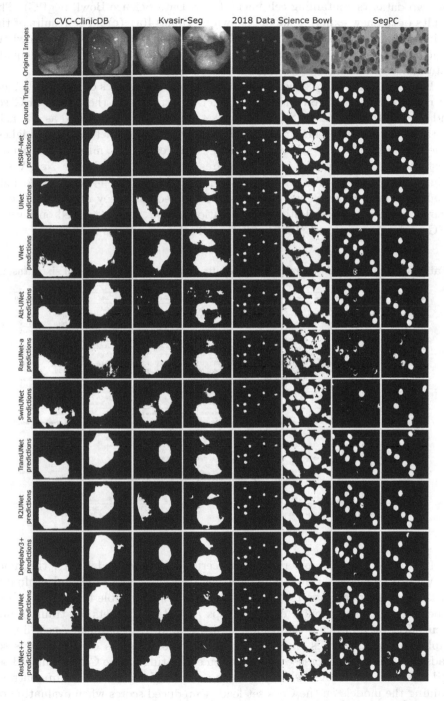

Fig. 3. Qualitative results of our experiments on the four original augmented datasets

the two datasets containing cell nuclei (2018 Data Science Bowl, SegPC). The results of the new experiments are shown in Tables 4(a)–(c) . The results of the training experiment on SegPC and evaluation on the 2018 Data Science Bowl are omitted because they are of extremely low quality, making SegPC an inadequate dataset for training, in terms of generalizability.

While the results are generally unsatisfactory, with scores on the mIoU metric not exceeding 53% on the CVC-ClinicDB set, 66% on the Kvasir-SEG set and 30% on the 2018 Data Science Bowl set, the results can be used as a benchmark for the models' generalizability. The DeepLabv3+ model shows the highest generalization ability by achieving the highest scores in DCS, mIoU, while in contrast the ResUNet-a model fails in this area. Models such as MSRF-Net, VNet, TransUNet, R2U-Net, Attention UNet, UNet and ResUNet++ show quite good results in these experiments, therefore it is judged that they can generalize to a sufficient degree. An additional observation is that models trained on Kvasir-SEG perform better on CVC-ClinicDB while the opposite is not true.

Table 4. Quantitative generalizability results of our experiments on all six datasets

(a) Trained on CVC-ClinicDB Tested on Kvasir-SEG

Method	DSC	mIoU	Precision	Recall
MSRF-Net	0.41	0.30	0.42	0.52
UNet	0.53	0.23	0.24	0.84
VNet	0.59	0.26	0.28	0.80
Att-UNet	0.56	0.24	0.26	0.80
ResUNet-a	0.46	0.10	0.25	0.10
SwinUNet	0.55	0.24	0.25	0.76
TransUNet	0.57	0.25	0.27	0.82
R2U-Net	0.55	0.24	0.25	0.78
DeepLabv3+	**0.91**	**0.53**	**0.66**	0.72
ResUNet	0.35	0.22	0.23	**0.85**
ResUNet++	0.30	0.23	0.34	0.40

(b) Trained on Kvasir-SEG Tested on CVC-ClinicDB

Method	DSC	mIoU	Precision	Recall
MSRF-Net	0.52	0.42	0.53	0.67
UNet	0.75	0.45	0.69	0.56
VNet	0.66	0.30	0.44	0.49
Att-UNet	0.72	0.42	0.78	0.47
ResUNet-a	0.54	0.14	0.19	0.38
SwinUNet	0.63	0.26	0.30	0.63
TransUNet	0.76	0.47	0.70	0.59
R2U-Net	0.74	0.46	0.77	0.53
DeepLabv3+	**0.97**	**0.66**	**0.93**	**0.70**
ResUNet	0.38	0.30	0.50	0.43
ResUNet++	0.51	0.37	0.52	0.57

(c) Trained on 2018 Data Science Bowl Tested on SegPC

Method	DSC	mIoU	Precision	Recall
MSRF-Net	0.42	0.27	0.27	0.96
UNet	0.69	0.29	0.29	0.94
VNet	0.69	0.29	**0.30**	0.93
Att-UNet	0.68	0.28	0.29	0.95
ResUNet-a	0.54	0.11	0.23	0.14
SwinUNet	0.66	0.26	0.27	0.95
TransUNet	0.68	0.28	0.28	0.96
R2U-Net	0.66	0.26	0.27	**0.97**
DeepLabv3+	**0.87**	0.28	0.29	0.94
ResUNet	0.39	0.25	0.25	**0.97**
ResUNet++	0.45	**0.30**	**0.30**	0.96

(d) Trained on Polyps Dataset Tested on CVC-ClinicDB and Kvasir-SEG

Tested on	CVC-ClinicDB				Kvasir-SEG			
Method	DSC	mIoU	Precision	Recall	DSC	mIoU	Precision	Recall
MSRF-Net	0.90	**0.86**	**0.94**	0.91	0.85	0.75	0.88	0.84
UNet	0.90	0.77	**0.94**	0.80	0.80	0.69	0.88	0.76
VNet	0.78	0.54	0.77	0.64	0.83	0.58	0.82	0.66
Att-UNet	0.91	0.74	**0.94**	0.78	0.87	0.64	0.82	0.75
ResUNet-a	0.81	0.61	0.92	0.65	0.79	0.53	0.77	0.63
SwinUNet	0.78	0.48	0.66	0.64	0.75	0.44	0.56	0.68
TransUNet	0.92	0.74	0.88	0.82	0.89	0.69	0.83	0.81
R2U-Net	0.89	0.75	**0.94**	0.79	0.90	0.73	**0.92**	0.78
DeepLabv3+	**0.99**	**0.86**	0.93	**0.92**	**0.96**	**0.76**	0.86	**0.87**
ResUNet	0.55	0.50	0.58	0.77	0.58	0.46	0.54	0.75
ResUNet++	0.76	0.68	0.93	0.71	0.68	0.59	0.83	0.68

(e) Trained on Cells Dataset Tested on 2018 Data Science Bowl and SegPC

Tested on	2018 Data Science Bowl				SegPC			
Method	DSC	mIoU	Precision	Recall	DSC	mIoU	Precision	Recall
MSRF-Net	0.91	**0.87**	0.93	**0.93**	0.81	**0.70**	**0.84**	0.81
UNet	0.95	**0.87**	0.93	**0.93**	0.87	0.63	0.75	0.70
VNet	0.92	0.80	**0.94**	0.84	0.76	0.39	0.45	0.74
Att-UNet	0.94	0.86	**0.94**	0.91	0.87	0.63	0.70	0.86
ResUNet-a	0.94	0.84	0.91	0.92	0.75	0.37	0.43	0.73
SwinUNet	0.93	0.82	0.90	0.91	0.77	0.43	0.54	0.69
TransUNet	0.95	**0.87**	**0.94**	**0.93**	0.87	0.64	0.70	**0.88**
R2U-Net	0.94	**0.87**	0.92	**0.93**	0.87	0.64	0.74	0.83
DeepLabv3+	**0.98**	0.85	0.92	0.91	**0.97**	0.61	0.75	0.77
ResUNet	0.85	0.72	0.90	0.78	0.57	0.43	0.49	0.79
ResUNet++	0.86	0.74	0.82	0.89	0.69	0.54	0.71	0.69

In addition, in the experiments concerning the larger datasets, the models were trained on the new Polyps, Cells datasets and tested and evaluated on their constituent subsets. The results are shown in Table 4(d)–(e). In these experiments, the suitability of the Polyps and Cells datasets for training and their effect on improving the results is also examined in parallel. According to Table 4(d) and (e), it is shown that training most models on the Polyps set leads to significantly increased scores when evaluating on the CVC-ClinicDB set and less increased scores when evaluating on the Kvasir-SEG set. In contrast, training the models on the Cells set leads to reduced scores when evaluating on the 2018 Data Science Bowl and SegPC sets. Models that respond positively to training on the Polyps and Cells sets compared to training on the individual sets that comprise them have higher generalizability. As demonstrated in the above experiments, models such as DeepLabv3+, MSRF-Net, VNet, Attention UNet, TransUNet, UNet, and R2U-Net generalize better than the other models.

5 Discussion

Models such as ResUNet and ResUNet-a, designed for domains other than medical image segmentation, do not perform well enough on datasets involving polyps but achieve satisfactory scores on datasets with cell nuclei, indicating that this project is quite specialized and thus requires specialized models to achieve state-of-the-art results. The utilization of residual connections in combination with recurrent networks is found to be very effective in the R2U-Net while their combination with Squeeze and excitation blocks, Atrous Spatial Pyramid Pooling [9,15,33,36,40] and attention does not lead to any improvement over UNet and at the same time complicates the architecture a lot. The utilization of the attention mechanism does not perform on the polyp datasets however it leads to better identification of cells with disease in the SegPC set and their separation from healthy cells, as is the case with Attention UNet. The appearance of transformers in U-type architectures, while not promising as an idea at first because by their nature transformers lack localization capability, in combination with convolutional networks, as is the case in TransUNet, they are very efficient encoders bringing a small improvement in UNet's scores. In contrast, without using convolutional networks together with transformers, the final results are insufficient, as is the case with SwinUNet. Finally, models that focus more on combining information from different scales and therefore manage to reduce the semantic gap between the low-level encoder features and the high-level decoder features, such as DeepLabv3+ and MSRF-Net, seem to achieve significant improvement in scores on datasets with polyps that are considered more challenging, as well as SegPC. It is important to note that the MSRF-Net model is considerably more demanding to train in terms of computational resources and time than the other models that perform equally well. Moreover, DeepLabv3+'s ResNet50 backbone which is pretrained on the large ImageNet [12] dataset makes it more powerful and contributes to its fast training.

In realistic scenarios where these models would perform polyp segmentation on real images from different laboratories, with different technologies, from different patients, the generalizability is considered an essential feature [41]. The models must be able to perform adequately regardless of parameters such as the equipment with which the images were captured or the patient whom the images were obtained from. It was shown from the study in the previous section that the models with the highest scores also have a high generalization ability.

As for the datasets, those related to polyps are challenging and carry some peculiarities, such as some black boxes that show high contrast in the images and can confuse the models as to the important areas of the image, and little general contrast in the images especially in the polyp locations which makes the task of detecting polyps more difficult. In the datasets associated with cell nuclei the task of identifying nuclei is considered to be quite easy because there is considerable contrast between them and the plain monochrome background, with difficulties being encountered in identifying specific cells affected by Multiple Myeloma (in SegPC) and in images where nuclei are depicted more faintly or there is significant overlap between them.

6   Conclusions

The models tested in the experiments of this work generally follow the U-architecture with variations such as the use of transformers, the existence of an attention mechanism, the exploitation of residual learning and the exchange of features at different scales. While the original model that inspired the creation of almost all the others, UNet, seems to consistently perform quite well on all datasets, some of the variations presented lead to improved results while others are not as successful. We observe that multi-scale feature fusion via atrous spatial pyramid pooling or information exchange between scales, along with residual connections and attention lead to the most noticeable improvements. The combination of transformers with convolutional networks is also proven to be a promising approach. Data augmentation is beneficial to all the favoured proposed architectures.

While several models with different architectures were considered in this paper and therefore led to a global study of the medical image segmentation task, there are additional models that could be tested, such as DCSAU-Net [48], MCGU-Net [3] but also UNet++ [50], U^2-Net [38], UNet3+ [21], which add new details and changes that can significantly favor the results of the experiments.

Moreover, in Fig. 3 it is easily observed that in some examples in the CVC-ClinicDB and Kvasir-SEG datasets, the more efficient models produce masks that misidentify as polyps, regions that are not polyps which can reduce their scores. Combining the above models with bounding boxes from a sufficiently powerful Object Detection model, such as Detectron2 [47], either during the evaluation or during training, could lead to the elimination of the incorrectly detected regions.

Finally, a meticulous ablation study could be conducted focused on specific aspects of the mentioned architectures so that more detailed conclusions can be drawn. Also a larger and perhaps more intense data augmentation could certainly bring about an improvement in the results, since an increasing trend in scores as a function of data volume was already observed.

Acknowledgements. This work was conducted in the Artificial Intelligence and Learning Systems Laboratory of the School of Electrical and Computer Engineering of the National Technical University of Athens. The computations in this paper were performed on equipment provided by the Greek Research and Technology Network.

References

1. Abadi, M., et al.: TensorFlow: large-scale machine learning on heterogeneous systems (2015). Software available: https://www.tensorflow.org/
2. Alom, M.Z., Hasan, M., Yakopcic, C., Taha, T.M., Asari, V.K.: Recurrent residual convolutional neural network based on u-net (R2U-Net) for medical image segmentation (2018). https://doi.org/10.48550/ARXIV.1802.06955, https://arxiv.org/abs/1802.06955
3. Asadi-Aghbolaghi, M., Azad, R., Fathy, M., Escalera, S.: Multi-level context gating of embedded collective knowledge for medical image segmentation (2020). https://doi.org/10.48550/ARXIV.2003.05056, https://arxiv.org/abs/2003.05056
4. Bernal, J., Sánchez, F.J., Fernández-Esparrach, G., Gil, D., Rodríguez, C., Vilariño, F.: WM-DOVA maps for accurate polyp highlighting in colonoscopy: validation vs. saliency maps from physicians. Comput. Med. Imaging Graph. **43**, 99–111 (2015). https://doi.org/10.1016/j.compmedimag.2015.02.007, https://doi.org/10.1016/j.compmedimag.2015.02.007
5. Caicedo, J.C., et al.: Nucleus segmentation across imaging experiments: the 2018 data science bowl. Nat. Methods **16**(12), 1247–1253 (2019). https://doi.org/10.1038/s41592-019-0612-7
6. Caliva, F., et al.: A deep learning approach to anomaly detection in nuclear reactors. In: 2018 International Joint Conference on Neural Networks (IJCNN), pp. 1–8. IEEE (2018)
7. Cao, H., et al.: Swin-unet: unet-like pure transformer for medical image segmentation (2021). https://doi.org/10.48550/ARXIV.2105.05537, https://arxiv.org/abs/2105.05537
8. Chen, J., et al.: Transunet: transformers make strong encoders for medical image segmentation (2021). https://doi.org/10.48550/ARXIV.2102.04306, https://arxiv.org/abs/2102.04306
9. Chen, L.C., Papandreou, G., Kokkinos, I., Murphy, K., Yuille, A.L.: DeepLab: semantic image segmentation with deep convolutional nets, atrous convolution, and fully connected CRFs (2016). https://doi.org/10.48550/ARXIV.1606.00915, https://arxiv.org/abs/1606.00915
10. Chen, L.C., Zhu, Y., Papandreou, G., Schroff, F., Adam, H.: Encoder-decoder with atrous separable convolution for semantic image segmentation (2018). https://doi.org/10.48550/ARXIV.1802.02611, https://arxiv.org/abs/1802.02611
11. Chollet, F., et al.: Keras (2015). https://github.com/fchollet/keras
12. Deng, J., Dong, W., Socher, R., Li, L.J., Li, K., Fei-Fei, L.: Imagenet: a large-scale hierarchical image database. In: 2009 IEEE Conference on Computer Vision and Pattern Recognition, pp. 248–255. IEEE (2009)
13. Diakogiannis, F.I., Waldner, F., Caccetta, P., Wu, C.: Resunet-A: a deep learning framework for semantic segmentation of remotely sensed data (2019). https://doi.org/10.48550/ARXIV.1904.00592, https://arxiv.org/abs/1904.00592
14. Gehlot, S., Gupta, A., Gupta, R.: EDNFC-net: convolutional neural network with nested feature concatenation for nuclei-instance segmentation. In: ICASSP 2020–2020 IEEE International Conference on Acoustics, Speech and Signal Processing (ICASSP). IEEE, May 2020. https://doi.org/10.1109/icassp40776.2020.9053633
15. Giusti, A., Cireşan, D.C., Masci, J., Gambardella, L.M., Schmidhuber, J.: Fast image scanning with deep max-pooling convolutional neural networks (2013). https://doi.org/10.48550/ARXIV.1302.1700, https://arxiv.org/abs/1302.1700

16. GRNET: Aris documentation - hardware overview. https://doc.aris.grnet.gr/system/hardware/

17. Gupta, A., et al.: GCTI-SN: geometry-inspired chemical and tissue invariant stain normalization of microscopic medical images. Med. Image Anal. **65**, 101788 (2020). https://doi.org/10.1016/j.media.2020.101788

18. Gupta, A., Gupta, R., Gehlot, S., Gehlot, S.: Segpc-2021: segmentation of multiple myeloma plasma cells in microscopic images (2021). https://doi.org/10.21227/7NP1-2Q42, https://ieee-dataport.org/open-access/segpc-2021-segmentation-multiple-myeloma-plasma-cells-microscopic-images

19. Gupta, A., Mallick, P., Sharma, O., Gupta, R., Duggal, R.: PCSeg: color model driven probabilistic multiphase level set based tool for plasma cell segmentation in multiple myeloma. PLOS ONE **13**(12), e0207908 (2018). https://doi.org/10.1371/journal.pone.0207908

20. Hu, J., Shen, L., Albanie, S., Sun, G., Wu, E.: Squeeze-and-excitation networks (2017). https://doi.org/10.48550/ARXIV.1709.01507, https://arxiv.org/abs/1709.01507

21. Huang, H., et al.: Unet 3+: a full-scale connected unet for medical image segmentation (2020). https://doi.org/10.48550/ARXIV.2004.08790, https://arxiv.org/abs/2004.08790

22. Ibtehaz, N., Rahman, M.S.: Multiresunet: rethinking the u-net architecture for multimodal biomedical image segmentation (2019). https://doi.org/10.48550/ARXIV.1902.04049, https://arxiv.org/abs/1902.04049

23. Jha, D., et al.: A comprehensive study on colorectal polyp segmentation with resunet, conditional random field and test-time augmentation. IEEE J. Biomed. Health Inform. **25**(6), 2029–2040 (2021). https://doi.org/10.1109/jbhi.2021.3049304

24. Jha, D., et al.: Kvasir-SEG: a segmented polyp dataset. In: Ro, Y.M., et al. (eds.) MMM 2020. LNCS, vol. 11962, pp. 451–462. Springer, Cham (2020). https://doi.org/10.1007/978-3-030-37734-2_37

25. Jha, D., et al.: Resunet++: an advanced architecture for medical image segmentation (2019). https://doi.org/10.48550/ARXIV.1911.07067, https://arxiv.org/abs/1911.07067

26. Kollia, I., Stafylopatis, A.G., Kollias, S.: Predicting Parkinson's disease using latent information extracted from deep neural networks. In: 2019 International Joint Conference on Neural Networks (IJCNN), pp. 1–8. IEEE (2019)

27. Kollias, D., Arsenos, A., Kollias, S.: AI-Mia: Covid-19 detection & severity analysis through medical imaging. arXiv preprint arXiv:2206.04732 (2022)

28. Kollias, D., Arsenos, A., Soukissian, L., Kollias, S.: Mia-Cov19d: Covid-19 detection through 3-D chest CT image analysis. In: Proceedings of the IEEE/CVF International Conference on Computer Vision, pp. 537–544 (2021)

29. Kollias, D., Tagaris, A., Stafylopatis, A., Kollias, S., Tagaris, G.: Deep neural architectures for prediction in healthcare. Complex Intell. Syst. **4**(2), 119–131 (2018)

30. Kollias, D., Yu, M., Tagaris, A., Leontidis, G., Stafylopatis, A., Kollias, S.: Adaptation and contextualization of deep neural network models. In: 2017 IEEE Symposium Series on Computational Intelligence (SSCI), pp. 1–8. IEEE (2017)

31. Kollias, D., et al.: Transparent adaptation in deep medical image diagnosis. In: Heintz, F., Milano, M., O'Sullivan, B. (eds.) TAILOR 2020. LNCS (LNAI), vol. 12641, pp. 251–267. Springer, Cham (2021). https://doi.org/10.1007/978-3-030-73959-1_22

32. Long, J., Shelhamer, E., Darrell, T.: Fully convolutional networks for semantic segmentation (2014). https://doi.org/10.48550/ARXIV.1411.4038, https://arxiv.org/abs/1411.4038

33. Holschneider, M., Kronland-Martinet, R., Morlet, J., Tchamitchian, P.: A real-time algorithm for signal analysis with the help of the wavelet transform. In: Combes, J.M., Grossmann, A., Tchamitchian, P. (eds.) Wavelets. Inverse Problems and Theoretical Imaging, pp. 289–297. Springer, Heidelberg (1990). http://kronland.fr/wp-content/uploads/2015/05/RealTimeAlgo_Springer89.pdf. https://doi.org/10.1007/978-3-642-75988-8_28

34. Milletari, F., Navab, N., Ahmadi, S.A.: V-net: fully convolutional neural networks for volumetric medical image segmentation (2016). https://doi.org/10.48550/ARXIV.1606.04797, https://arxiv.org/abs/1606.04797

35. Oktay, O., et al.: Attention U-net: learning where to look for the pancreas (2018). https://doi.org/10.48550/ARXIV.1804.03999, https://arxiv.org/abs/1804.03999

36. Papandreou, G., Kokkinos, I., Savalle, P.A.: Modeling local and global deformations in deep learning: epitomic convolution, multiple instance learning, and sliding window detection. In: 2015 IEEE Conference on Computer Vision and Pattern Recognition (CVPR). IEEE, June 2015. https://doi.org/10.1109/cvpr.2015.7298636

37. Psaroudakis, A., Kollias, D.: Mixaugment & mixup: augmentation methods for facial expression recognition. In: Proceedings of the IEEE/CVF Conference on Computer Vision and Pattern Recognition, pp. 2367–2375 (2022)

38. Qin, X., Zhang, Z., Huang, C., Dehghan, M., Zaiane, O.R., Jagersand, M.: U^2-net: going deeper with nested U-structure for salient object detection (2020). https://doi.org/10.48550/ARXIV.2005.09007, https://arxiv.org/abs/2005.09007

39. Ronneberger, O., Fischer, P., Brox, T.: U-net: convolutional networks for biomedical image segmentation. In: Navab, N., Hornegger, J., Wells, W.M., Frangi, A.F. (eds.) MICCAI 2015. LNCS, vol. 9351, pp. 234–241. Springer, Cham (2015). https://doi.org/10.1007/978-3-319-24574-4_28

40. Sermanet, P., Eigen, D., Zhang, X., Mathieu, M., Fergus, R., LeCun, Y.: Overfeat: integrated recognition, localization and detection using convolutional networks (2013). https://doi.org/10.48550/ARXIV.1312.6229, https://arxiv.org/abs/1312.6229

41. Srivastava, A., et al.: MSRF-net: a multi-scale residual fusion network for biomedical image segmentation (2021). https://doi.org/10.48550/ARXIV.2105.07451, https://arxiv.org/abs/2105.07451

42. Tagaris, A., Kollias, D., Stafylopatis, A.: Assessment of Parkinson's disease based on deep neural networks. In: Boracchi, G., Iliadis, L., Jayne, C., Likas, A. (eds.) EANN 2017. CCIS, vol. 744, pp. 391–403. Springer, Cham (2017). https://doi.org/10.1007/978-3-319-65172-9_33

43. Tagaris, A., Kollias, D., Stafylopatis, A., Tagaris, G., Kollias, S.: Machine learning for neurodegenerative disorder diagnosis-survey of practices and launch of benchmark dataset. Int. J. Artif. Intell. Tools **27**(03), 1850011 (2018)

44. Takikawa, T., Acuna, D., Jampani, V., Fidler, S.: Gated-SCNN: gated shape CNNs for semantic segmentation. In: 2019 IEEE/CVF International Conference on Computer Vision (ICCV). IEEE, October 2019. https://doi.org/10.1109/iccv.2019.00533

45. Wang, J., et al.: Deep high-resolution representation learning for visual recognition (2019). https://doi.org/10.48550/ARXIV.1908.07919, https://arxiv.org/abs/1908.07919

46. Wingate, J., Kollia, I., Bidaut, L., Kollias, S.: Unified deep learning approach for prediction of Parkinson's disease. IET Image Proc. **14**(10), 1980–1989 (2020)
47. Wu, Y., Kirillov, A., Massa, F., Lo, W.Y., Girshick, R.: Detectron2 (2019). https://github.com/facebookresearch/detectron2
48. Xu, Q., Duan, W., He, N.: DCSAU-net: a deeper and more compact split-attention u-net for medical image segmentation (2022). https://doi.org/10.48550/ARXIV.2202.00972, https://arxiv.org/abs/2202.00972
49. Zhang, Z., Liu, Q., Wang, Y.: Road extraction by deep residual u-net (2017). https://doi.org/10.48550/ARXIV.1711.10684, https://arxiv.org/abs/1711.10684
50. Zhou, Z., Rahman Siddiquee, M.M., Tajbakhsh, N., Liang, J.: UNet++: a nested U-net architecture for medical image segmentation. In: Stoyanov, D., et al. (eds.) DLMIA/ML-CDS -2018. LNCS, vol. 11045, pp. 3–11. Springer, Cham (2018). https://doi.org/10.1007/978-3-030-00889-5_1

Medical Image Super Resolution by Preserving Interpretable and Disentangled Features

Dwarikanath Mahapatra[1]([envelope]), Behzad Bozorgtabar[2], and Mauricio Reyes[3]

[1] Inception Institute of Artificial Intelligence, Abu Dhabi, United Arab Emirates
dwarikanath.mahapatra@inceptioniai.org
[2] Ecole Polytechnique Fédérale de Lausanne, Écublens, Switzerland
Behzad.bozorgtabar@epfl.ch
[3] ARTORG Center for Biomedical Engineering Research, University of Bern,
Bern, Switzerland
mauricio.reyes@med.unibe.ch

Abstract. State of the art image super resolution (ISR) methods use generative networks to produce high resolution (HR) images from their low resolution (LR) counterparts. In this paper we show with the help of interpretable saliency maps that generative approaches to ISR can introduce undesirable artefacts which can adversely affect model performance in downstream tasks such as disease classification. We propose a novel loss term that aims to maximize the similarity of interpretable class activation maps between the HR and LR images. This not only preserves the images' explainable information but also leads to improved performance in terms of super resolution output and downstream classification accuracy. We also incorporate feature disentanglement that plays an important role in our method's superior performance for super resolution and downstream classification task.

Keywords: Interpretability · Image super resolution · Class activation maps · Classification

1 Introduction

Image super-resolution (ISR) generates high resolution (HR) images from low resolution (LR) medical images and enables a closer examination of pathological regions. While state of the art generative ISR methods produce images consistent with the dataset's global statistics, they may introduce locally inconsistent noise or artefacts, which becomes visible in frequency domain analysis [24]. Although super-resolved images might be perceptually similar to the original images, changes to a single pixel can fool the CNN model [2,15,18]. Such adversarial behaviour can be dangerous in medical image applications since inaccurate diagnosis can have serious consequences. Although CNN based classification

© The Author(s), under exclusive license to Springer Nature Switzerland AG 2023
L. Karlinsky et al. (Eds.): ECCV 2022 Workshops, LNCS 13807, pp. 709–721, 2023.
https://doi.org/10.1007/978-3-031-25082-8_48

methods demonstrate state of the art performance, their black box nature makes it difficult to explain the series of intermediate steps. In this paper we show that interpretable saliency maps can identify regions having novel artifacts or spurious information in the ISR process, and demonstrate that enforcing semantic similarity of saliency maps between HR and LR images improves the quality and reliability of super resolution outputs. This improvement is shown not only through classical metrics, such as peak signal-to-noise ratio and others, but also through better performance in downstream classification tasks.

1.1 Related Work:

ISR has been widely studied for MR image analysis applications [23], as well as MR spectroscopy [11]. Some of the early works of ISR for MRI performed multiple-frame ISR (i.e., slice-wise)) by registering multiple noisy LR images which proved to be very challenging [32]. Recent deep learning based ISR approaches have shown superior performance for MR image super resolution [4,32]. However they use large models that pose challenges in real world settings. Zhang et al. in [29] propose a squeeze and excitation reasoning attention network as part of a lightweight model for ISR. [6] achieve multi contrast MRI super resolution using multi stage networks. The approach in [30] leverages hierarchical features in residual deep networks (RDN), while [3] combines 3D dense networks and adversarial learning for MRI super resolution.

Interpretability is an important component of medical image diagnosis systems, and there are many works that leverage interpretable activation maps or attention maps for tasks such as active learning [19], and multiple medical image analysis tasks [8,22]. Our work aims to preserve interpretable features in the super resolution process.

1.2 Relevance of Interpretable Class Activation Maps

Figure 1 illustrates the motivation of our method. Figure 1 (a) shows an original image and its corresponding class activation maps (for ground truth class 'Atelectasis') are shown in Fig. 1 (f), (with Deep Taylor [20]) and Fig. 1 (i) (GradCam [25]). This image is downsampled by 2× and Fig. 1 (c) shows a super resolved image obtained using [29] and the corresponding activation maps in Figs. 1 (h,k). The red arrows show regions having new activation areas in the saliency map, introduced in the generation stage. Figure 1 (b) shows the super resolved HR image obtained using our approach and the corresponding saliency maps are shown in Fig. 1 (g,j). Our method generates saliency maps similar to the original image.

We train a DenseNet-121 [10] classifier (DenseNet$_{Ref}$) on the training set of original images, and use this to generate activation maps for every label. Note that the back propagation based approaches generate saliency maps for each disease label, and our proposed loss function aims to maximize the similarity of activation maps for all labels. We also train DenseNet$_{Our}$ and DenseNet-[29] using super resolved images from the corresponding methods. Figure 1 (a)

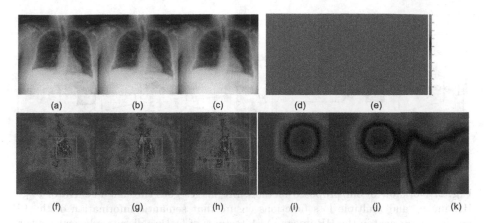

Fig. 1. Illustrating the benefits of using interpretable saliency maps. (a) original image and its corresponding class activation maps (for ground truth class 'Atelectasis') are shown in (f), (with Deep Taylor [20]) and (i) (GradCam [25]). Original image is downsampled by 2×. (b) Super resolved HR image obtained using our approach, the difference image w.r.t to (a) is shown in (d), and the corresponding saliency maps are shown in (g,j). (c) Super resolved image obtained using [29], the difference image w.r.t to (a) is shown in (e) and the corresponding activation maps in (h,k). Red arrows show regions having new activation areas in the saliency map, introduced in the generation stage. (Color figure online)

has disease label 'Atelectasis' which is correctly identified by DenseNet$_{Ref}$ and DenseNet$_{Our}$, but DenseNet-[29] classifies it as 'Pleural Effusion'. Clearly the new activation area in the saliency map influences the classification decision. Most super resolution methods do well when trained on healthy images, but tend to introduce artefacts for input disease images. By preserving the interpretable features we are able to overcome this limitation.

1.3 Our Contribution

We propose to preserve interpretable image components between low and high resolution images to obtain better quality super resolved images, which leads to better performance in downstream tasks. We also show interpretable saliency maps can be used to identify regions where additional information is introduced by the ISR process, and how such information can be used to minimize artefacts and hence provide improved super-resolved images.

2 Method

2.1 Overview

Figure 2 shows the workflow of our proposed method. A LR image $x \in \mathscr{R}^{N \times N}$ goes through a generator network consisting of a series of convolution blocks

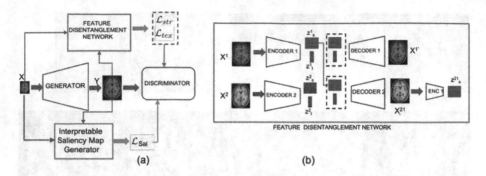

Fig. 2. Workflow of our proposed method. LR images goes through a generator to get HR image, and multiple loss functions ensure that semantic information of the LR image is preserved in the HR image. Architectures of feature disentanglement network using swapped autoencoders is shown.

and an upsampler to obtain a high resolution (HR) image $y \in \mathscr{R}^{M \times M}$, where $M > N$. The discriminator module ensures y satisfies the following constraints (Fig. 3):

1. The HR and LR image should have similar semantic characteristics since a higher resolution version should not alter image semantics. For this purpose we disentangle the image into structure and texture features, and ensure their respective semantic information is consistent across both images.
2. The HR image should preserve the interpretable (or explainable) features of the original LR image. To achieve it we enforce similarity of interpretable class activation maps for the HR and LR images.

2.2 Feature Disentanglement

In order to separate the images into structure and texture components we train an autoencoder (AE) shown in Fig. 2 (b). In a classic AE, the encoder E and generator G form a mapping between image x and latent code z using an image reconstruction loss

$$\mathscr{L}_{rec}(E, G) = \mathbb{E}_{x \sim X} \left[\|x - G(E(x))\|_1 \right] \qquad (1)$$

To ensure that the generated image is realistic we have discriminator D that calculates the adversarial loss for generator G and encoder E as:

$$\mathscr{L}_{adv,Dis}(E, G, D) = \mathbb{E}_{x \sim X} \left[-\log(D(G(E(x)))) \right] \qquad (2)$$

As shown in Fig. 2 (b) we divide the latent code into two components - a texture component z_t and a structural component z_s. Note that there are two encoders, E^1, E^2 for the two images. For the decoder stage, image X^1 is reconstructed by combining the corresponding z_s^1, z_t^1. However for image X^2 if

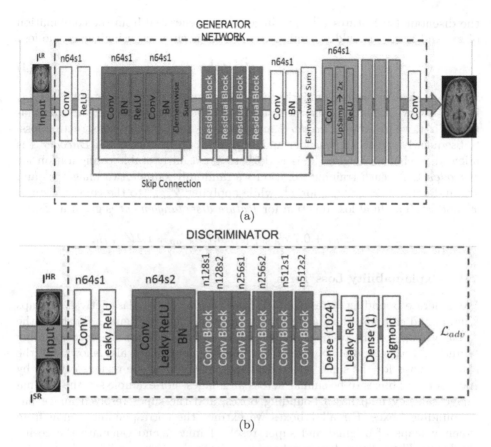

Fig. 3. (a) Generator network; (b) Discriminator network. $n64s1$ denotes 64 feature maps (n) and stride (s) 1 for each convolutional layer.

we replace z_s^2 with z_s^1 and combine with z_t^2, then we expect to get the structure of X^1 and the texture of X^2, and we term the new image as X^{21}. This is achieved using the 'swapped-GAN' loss [21]:

$$\mathscr{L}_{swap}(E, G, D) = \mathbb{E}_{x^1, x^2 \sim X, x^1 \neq x^2} \left[-\log(D(G(z_s^1, z_t^2))) \right] \tag{3}$$

In the original work of [21] the authors use co-occurrent patch statistics from X^2 and X^{21} such that any patch from X^{21} cannot be distinguished from a group of patches from X^2. This constraint is feasible in natural images since different images have unique texture, but is challenging to implement in medical images since images from the same modality (showing the same organ) have similar texture. Thus limited unique features are learnt under such a constraint. Moreover, in medical images the image areas driving the diagnosis are the most important ones to consider, and any ISR approach should not negatively affect the diagnosis. To overcome this limitation we introduce certain novel components into the loss function. The output X^{21} when passed through encoder E^1 outputs the disentangled features z_s^{21}, z_t^{21}. Since X^{21} is generated from the combination of z_s^1, and z_t^2, z_s^{21} should be similar to z_s^1. This is enforced using the cosine loss

$$\mathcal{L}_{str,Dis} = 1 - \langle z_s^1, z_s^{21} \rangle \tag{4}$$

where $\langle . \rangle$ denotes cosine similarity and Dis denotes disentanglement stage. z_s is a tensor with spatial dimensions and z_t is vector encoding texture information. First, we pre-train $Encoder$ 1 and $Decoder$ 1 using the reconstruction loss. Subsequently the second encoder and decoder are introduced, and $Encoder$ 1 is allowed to change weights. This is done to get a fairly stable configuration for $Encoder$ 1. At each training iteration we randomly sample two images x^1 and x^2, and enforce $\mathcal{L}_{rec}, \mathcal{L}_{adv}$ for x^1, while applying $\mathcal{L}_{swap}$ to the combination of x^1 and x^2. The final loss function for *feature disentanglement* is given in Eq. 5,

$$\mathcal{L}_{Dis} = \mathcal{L}_{Rec} + 0.7\mathcal{L}_{Adv,Dis} + 0.7\mathcal{L}_{swap} + 1.2\mathcal{L}_{str,Dis} \tag{5}$$

2.3   Explainability Loss Term

We generate bounding boxes at all distinct regions of the class activation maps (Fig. 1) and compare them across the two saliency maps. Figure 1 (f) shows the saliency map for the original image (for class label 1) and the corresponding bounding boxes in orange. Similarly, bounding boxes are also shown for the saliency maps for the super resolved images. We enforce the regions covered by the bounding boxes to be similar across all images. For example say the original image saliency map has 3 bounding boxes, and the super resolved image has 4 bounding boxes. For all 7 boxes we extract the corresponding region from saliency maps of original and super resolved image, and calculate the cosine similarity. The saliency loss is defined as

$$\mathcal{L}_{Sal} = \frac{1}{K \times I} \sum_{k=1}^{K} \sum_{i} 1 - \langle f_{k,i}^{Org}, f_{k,i}^{SR} \rangle \tag{6}$$

where $\langle . \rangle$ denotes cosine similarity. f denotes the feature which is the flattened bounding box patch of intensity values, i denotes the index of the bounding box and K denotes the saliency maps corresponding to K disease labels. I denotes the total number of bounding boxes across all saliency maps and varies with the images. Thus we calculate the average loss across all the bounding boxes as well as across all saliency maps corresponding to the different labels. Note we also experimented with a global saliency loss where we calculate the saliency loss for the entire image, but the results were not optimal since the sparsity of the saliency maps does not give accurate features at the global level.

Training the Super Resolution Network: The architecture of the generator and discriminator for super resolution are shown in the Supplementary material. Given the LR image x and the intermediate generated HR image y, we obtain their respective disentangled latent feature representations as z_s^x, z_t^x and z_s^y, z_t^y.

Thereafter we calculate the semantic similarity between them using the cosine similarity loss as

$$\mathscr{L}_{str,SR} = 1 - \langle z_s^x, z_s^y \rangle$$
$$\mathscr{L}_{tex} = 1 - \langle z_t^x, z_t^y \rangle. \tag{7}$$

Additionally we obtain the respective class activation maps and get the saliency loss. Once the above loss terms are obtained, we train the whole super resolution network in an end to end manner using the following loss function.

$$\mathscr{L}_{SR}(X,Y) = \mathscr{L}_{adv,SR} + \lambda_1 \mathscr{L}_{Sal}(X,Y) + \lambda_2 \mathscr{L}_{str,SR}(X,Y) + \lambda_3 \mathscr{L}_{tex}(X,Y). \tag{8}$$

3 Experiments and Results

3.1 Dataset Description

We use two datasets to show super resolution results: 1) **fastMRI** [28] - following [27], we filter out 227 and 24 pairs of proton density (PD) and fat suppressed proton density weighted images (FS-PDWI) volumes for training and validation. 2) The **IXI dataset**: Three types of MR images are included in the datasets (i.e., PD, T1, and T2)[1]. Each of them has $500, 70$, and 6 MR volumes for training, testing, and validation respectively. Subvolumes of size $240 \times 240 \times 96$ are used and since 2D images are used, we get $500 \times 96 = 48,000$ training samples. For the downstream **Classification** task we use two datasets: 1) NIH ChestXray14 dataset [26] having $112,120$ expert-annotated frontal-view X-rays from $30,805$ unique patients and 14 disease labels. Original images were resized to 256×256. 2) The **CheXpert** Dataset [12] consisting of $224,316$ chest radiographs of $65,240$ patients labeled for the presence of 14 common chest conditions. Original images were resized to 256×256. The values of λ's for each dataset is given in the Tables.

3.2 Implementation Details

AE Network: The encoder consists of 4 convolution blocks followed by max pooling after each step. The decoder is also symmetrically designed. 3×3 convolution filters are used and $64, 64, 32, 32$ filters are used in each conv layer. The input to the AE is 256×256 and dimension of z_{tex} is 256, while z_{str} is 64×64. **Super Resolution Network:** We train our model using Adam [14]) with $\beta_1 = 0.9, \beta_2 = 0.999$, a batch size of 256 and a weight decay of 0.1, for 100 epochs. We implement all models in PyTorch and train them using one NVIDIA Tesla V100 GPU with 32 GB of memory. **Saliency Map Calculation:** To generate interpretability saliency maps for Deep Taylor [20] we use the iNNvestigate library [1][2], which implements several known interpretability approaches. We also show results using GradCAM [25].

---

[1] http://brain-development.org/ixi-dataset/.
[2] https://github.com/albermax/innvestigate.

Table 1. Quantitative results for IXI and fastMRI dataset. Higher values of PSNR and SSIM, and lower value of NMSE indicate better results. GC denotes use of GradCam saliency maps while DT denotes use of Deep Taylor saliency maps.

	IXI-T2 Images			Fast MRI	
	$2\times;\ \lambda_1,\lambda_2,\lambda_3 = 1.5, 1.2, 1.1$	$4\times;\ \lambda_1,\lambda_2,\lambda_3 = 1.8, 1.1, 1.0$		$2\times;\ \lambda_1,\lambda_2,\lambda_3 = 1.4, 1.3, 1.2$	$4\times;\ \lambda_1,\lambda_2,\lambda_3 = 1.8, 1.3, 1.3$
	PSNR/SSIM/NMSE	PSNR/SSIM/NMSE		PSNR/SSIM/NMSE	PSNR/SSIM/NMSE
Bicubic	28.5/0.878/0.038	27.1/0.864/0.036	Bicubic	25.2/0.487/0.067	17.1/0.198/0.084
[7]	30.2/0.891/0.034	28.4/0.878/0.033	[16]	26.66/0.512/0.063	18.363/0.208/0.082
[5]	37.32/0.9796/0.027	29.69/0.9052/0.031	[31]	28.27/0.667/0.051	21.81/0.476/0.067
[30]	38.75/0.9838/0.026	31.45/0.9324/0.029	[17]	28.870/0.670/.048	23.255/0.507/0.062
[32]	39.71/0.9863/0.027	32.05/0.9413/0.031	[13]	29.484/0.682/0.049	28.219/0.574/0.059
[29]	40.30/0.9874/0.022	32.62/0.9472/0.029	[6]	31.769/0.709/0.045	29.819/0.601/0.054
[9]	41.9/0.991/0.020	34.2/0.951/0.027	–	–	–
Our$_{GC}$	44.7/0.9965/0.016	35.8/0.967/0.023	Our$_{GC}$	35.2/0.743/0.038	33.4/0.69/0.044
Our$_{DT}$	46.8/0.9979/0.013	37.5/0.977/0.02	Our$_{GT}$	36.9/0.757/0.034	35.2/0.73/0.040
Ablation studies using deep Taylor maps					
$\mathscr{L}_{tex} + \mathscr{L}_{Sal}$	41.3/0.982/0.02	34.2/0.948/0.025	$\mathscr{L}_{tex} + \mathscr{L}_{Sal}$	32.8/0.724/0.042	31.1/0.66/0.051
$\mathscr{L}_{str} + \mathscr{L}_{Sal}$	43.7/0.989/0.019	34.8/0.952/0.024	$\mathscr{L}_{str} + \mathscr{L}_{Sal}$	33.9/0.729/0.041	32.7/0.67/0.051
$\mathscr{L}_{Sal}$	37.4/0.978/0.026	34.6/0.942/0.026	$\mathscr{L}_{Sal}$	30.6/0.703/0.046	29.5/0.62/0.052
$\mathscr{L}_{tex} + \mathscr{L}_{str}$	37.4/0.987/0.024	35.0/0.956/0.025	$\mathscr{L}_{tex} + \mathscr{L}_{str}$	31.8/0.711/0.043	30.6/0.66/0.051

3.3 Quantitative Results

For a given upscaling factor we first downsample the original image by that factor and recover the original size using different super resolution methods, and compare the performance using different metrics such as peak signal to noise ratio (PSNR), Structural Similarity Index Metric (SSIM), and Normalized Mean Square Error (NMSE). Table 1 shows the average values of different methods for IXI-T2 and fastMRI datasets at upscaling factors of 2× and 4×. Our method shows the best performance for both datasets and outperforms the next best method by a significant margin. While there is an expected noticeable performance drop for higher scaling factors, our method still outperforms other methods significantly. Our proposed method's advantage is the combination of structure, texture and interpretable features that improve the image quality significantly. The table shows results for Deep Taylor (Our$_{DT}$) and GradCAM (Our$_{GC}$). Our$_{DT}$ gives better performance since Deep Taylor gives more informative maps, as is obvious from the two sets of maps shown in Fig. 1. In the Supplementary material we show results for IXI-T1 and IXI-PD images.

3.4 Ablation Studies

Table 1 also shows ablation study outcomes where different loss terms are excluded during training. Excluding the saliency features results in significantly reduced performance, and excluding the structure and texture losses leads to worse performance. On the other hand excluding only one or more of structure and texture features leads to poor performance despite including Saliency features. Thus we conclude that disentangled features and explainable information are important for accurate super resolution (Table 2).

Table 2. Quantitative results for IXI Dataset. Higher values of PSNR and SSIM, and lower value of NMSE indicate better results.$_{GC}$ denotes use of GradCam saliency maps while $_{DT}$ denotes use of Deep Taylor saliency maps.

	IXI - PD images		IXI - T1 images	
	2×	4×	2×	4×
	PSNR/SSIM/NMSE	PSNR/SSIM/NMSE	PSNR/SSIM/NMSE	PSNR/SSIM/NMSE
Bicubic	30.4/0.9531/.042	29.13/0.8799/0.048	33.80/0.9525/0.030	28.28/0.8312/0.051
[7]	31.7/ 0.892/ 0.035	29.5/ 0.870/ 0.033	30.7/ 0.883 /0.032	28.5/ 0.861/ 0.037
[5]	38.96/0.9836/0.022	31.10/0.9181/0.030	37.12/0.9761/ 0.26	29.90/0.8796/ 0.034
[30]	40.31/0.9870/0.021	32.73/0.9387/ 0.029	37.95/0.9795/ 0.028	31.05/0.9042/ 0.031
[32]	41.28/0.9895/0.02	33.40/0.9486/ 0.027	38.27/0.9810/ 0.025	31.23/0.9093/ 0.032
[29]	41.66 /0.9902/0.019	33.97/0.9542/0.024	38.74/0.9824/0.021	32.03/0.9219/0.026
[9]	42.9/0.9936/0.018	35.3/0.962/0.023	39.9/0.989/0.021	33.6/0.927/0.024
Our$_{GC}$	45.4/0.9975/0.014	37.9/0.973/0.019	42.2/0.995/0.017	36.1/0.945/0.02
Our$_{DT}$	47.7/0.9983/0.011	39.6/0.986/0.015	44.3/0.998/0.015	37.8/0.959/0.018
Ablation studies				
$\mathscr{L}_{tex} + \mathscr{L}_{Sal}$	41.5/0.983/0.02	35.7/0.962/0.023	39.9/0.984/0.02	33.9/0.937/0.022
$\mathscr{L}_{str} + \mathscr{L}_{Sal}$	43.6/0.992/0.017	36.3/0.965/0.021	40.8/0.989/0.019	35.1/0.936/0.022
$\mathscr{L}_{Sal}$	37.3/0.9762/0.025	34.6/0.949/0.025	37.8/0.971/0.024	31.8/0.911/0.026
$\mathscr{L}_{tex} + \mathscr{L}_{str}$	38.1/0.9831/0.024	35.4/0.975/0.024	39.3/0.979/0.022	32.9/0.929/0.025

3.5   Classification Results on ChestXrays

We train a Baseline Densenet-121 classifier [10] on the original HR images. The values obtained for cheXpert and NIH datasets indicate the upper limit of performance values for the particular classifier. We then take the super resolved images from different methods and train different DenseNet-121 classifiers. The performance for each ISR method is summarized in Table 3. The closer the values to 'Baseline', the better the super-resolution output is. Our proposed method comes the closest to the Baseline with no statistical significant difference ($p = 0.072$, with $p < 0.03$ for other methods). Note that if a different classifier is used the results will vary, but we expect that the relative performance of different super resolution methods will be similar.

Table 3. AUC values for NIH and CheXpert datasets. Baseline DenseNet-121 classifiers and Deep Taylor maps for ablation are used.

	Baseline	Our$_{GC}$	Our$_{DT}$	[9]	[29]	[32]	[30]	[5]	Bicubic	$\mathscr{L}_{str} + \mathscr{L}_{Sal}$	$\mathscr{L}_{tex} + \mathscr{L}_{Sal}$	$\mathscr{L}_{Sal}$	$\mathscr{L}_{tex} + \mathscr{L}_{str}$
CheXpert	92.3	90.4	91.6	89.4	88.8	88.0	85.8	83.7	81.5	91.1	90.5	89.2	88.6
NIH	90.2	88.1	89.4	87.8	86.1	84.9	83.2	82.4	80.1	88.8	88.0	87.2	85.7

3.6   Qualitative Results

Figure 4 shows the recovered images and the difference image with the original image. Our method shows a very accurate reconstruction with minimal regions in the error map, while the recovered images from other methods are blurred and of poor quality. These results demonstrate the effectiveness of our approach.

Figure 5 shows ablation study results on the IXI and fastMRI datasets for different superresolution factors. Our method demonstrates superior performance of our original proposed method and the importance of different terms in our loss function.

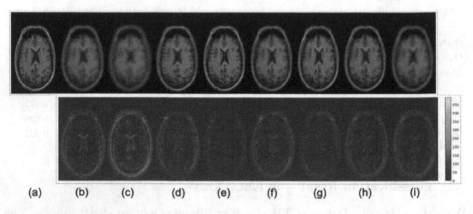

Fig. 4. Visualization of super resolution results at 2× factor for the IXI dataset. The top row shows the original image and the super resolved images and the bottom row shows the corresponding difference images. (a) Original image; Super-resolved images obtained using: b) [30]; (c) [32]; (d) [29]; (e) Our proposed method; Using (f) $\mathscr{L}_{tex} + \mathscr{L}_{Sal}$; (g) $\mathscr{L}_{str} + \mathscr{L}_{Sal}$; (h) $\mathscr{L}_{Sal}$; (i) $\mathscr{L}_{tex} + \mathscr{L}_{str}$.

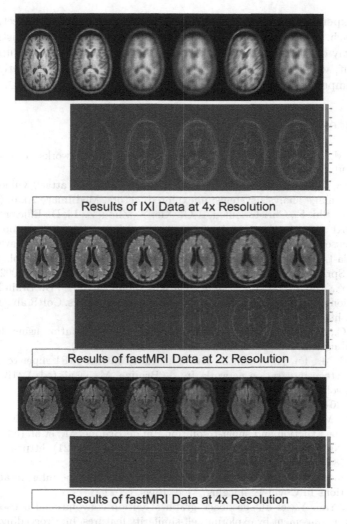

Fig. 5. Visualization of superresolution results. Each figure the top row is the original image followed by the difference image in the bottom row. Column 1 - original image; Reconstructed Image using: Column 2 - Our Proposed method; Column 3 - $\mathscr{L}_{str}+\mathscr{L}_{Sal}$; Column 4 - $\mathscr{L}_{tex} + \mathscr{L}_{Sal}$; Column 5 - $\mathscr{L}_{tex} + \mathscr{L}_{str}$; Column 6 - $\mathscr{L}_{Sal}$.

4   Conclusions

In this paper we have shown the effectiveness of including interpretable information from saliency maps in the super resolution process. By preserving interpretable information across the lower and higher resolution images we improve the super resolution output. Another factor contributing to improved super resolution output is the feature disentanglement step where separate structure and texture features are preserved between the input and output super resolved images. Experimental results on brain MRI images show the effectiveness of our approach. We also demonstrate our method's benefits in a downstream task of chest xray classification. Our added loss functions give superior super resolution output, which leads to better classification performance. Ablation studies show the importance of each of the proposed loss term.

References

1. Alber, M., Lapuschkin, S., et al.: innvestigate neural networks. J. Mach. Learn. Res. **20**(93), 1–8 (2019)
2. Bortsova, G., González-Gonzalo, C., et. al.: Adversarial attack vulnerability of medical image analysis systems: unexplored factors. Med Image Anal. (2021)
3. Chen, Y., Shi, F., Christodoulou, A.G., Xie, Y., Zhou, Z., Li, D.: Efficient and accurate MRI super-resolution using a generative adversarial network and 3D multi-level densely connected network. In: Frangi, A.F., Schnabel, J.A., Davatzikos, C., Alberola-López, C., Fichtinger, G. (eds.) MICCAI 2018. LNCS, vol. 11070, pp. 91–99. Springer, Cham (2018). https://doi.org/10.1007/978-3-030-00928-1_11
4. Chen, Y., Xie, Y., Zhou, Z., Shi, F., Christodoulou, A.G., Li, D.: Brain MRI super resolution using 3D deep densely connected neural networks. CoRR abs/1801.02728 (2018). http://arxiv.org/abs/1801.02728
5. Dong, C., Loy, C.C., He, K., Tang, X.: Image super-resolution using deep convolutional networks. IEEE Trans. PAMI **38**, 295–307 (2016)
6. Feng, C.-M., Fu, H., Yuan, S., Xu, Y.: Multi-contrast MRI super-resolution via a multi-stage integration network. In: de Bruijne, M., et al. (eds.) MICCAI 2021. LNCS, vol. 12906, pp. 140–149. Springer, Cham (2021). https://doi.org/10.1007/978-3-030-87231-1_14
7. Feng, C.-M., Yan, Y., Fu, H., Chen, L., Xu, Y.: Task transformer network for joint MRI reconstruction and super-resolution. In: de Bruijne, M., et al. (eds.) MICCAI 2021. LNCS, vol. 12906, pp. 307–317. Springer, Cham (2021). https://doi.org/10.1007/978-3-030-87231-1_30
8. Fuhrman, J.D., Gorre, N., et al.: A review of explainable and interpretable AI with applications in Covid-19 imaging. Med. Phys. (2021)
9. Hu, X., Yan, Y., et al.: Feedback graph attention convolutional network for MR images enhancement by exploring self-similarity features. In: Proceedings of MIDL, vol. 143, pp. 327–337 (2021)
10. Huang, G., Liu, Z., van der Maaten, L., Weinberger, K.: Densely connected convolutional networks. In: https://arxiv.org/abs/1608.06993, (2016)
11. Iqbal, Z., Nguyen, D., et al.: Super-resolution in magnetic resonance spectroscopic imaging utilizing deep learning. Front. Oncol. **9** (2019)
12. Irvin, J., Rajpurkar, P., et al.: Chexpert: a large chest radiograph dataset with uncertainty labels and expert comparison. In: arXiv preprint 1901.07031 (2019)
13. Kim, J., Lee, J.K., Lee, K.M.: Accurate image super-resolution using very deep convolutional networks. In: IEEE CVPR, June 2016
14. Kingma, D., Ba, J.: Adam: a method for stochastic optimization. arXiv:1412.6980 (2014)
15. Li, X., Zhu, D.: Robust detection of adversarial attacks on medical images. In: IEEE ISBI, pp. 1154–1158 (2020)
16. Lim, B., Son, S., et al.: Enhanced deep residual networks for single image super-resolution (2017)

17. Lyu, Q., Shan, H., et al.: Multi-contrast super-resolution MRI through a progressive network. IEEE Trans. Med. Imag. **39**(9), 2738–2749 (2020)
18. Ma, X., et al.: Understanding adversarial attacks on deep learning based medical image analysis systems. Pattern Recogn. **110**, 107332 (2021)
19. Mahapatra, D., Poellinger, A., Shao, L., Reyes, M.: Interpretability-driven sample selection using self supervised learning for disease classification and segmentation. IEEE Trans. Med. Imaging **40**(10), 2548–2562 (2021)
20. Montavon, G., Lapuschkin, S., et al.: Explaining nonlinear classification decisions with deep Taylor decomposition. Pattern Recogn. **65**, 211–222 (2017)
21. Park, T., et al.: Swapping autoencoder for deep image manipulation. In: Advances in Neural Information Processing Systems (2020)
22. Reyes, M., Meier, R., et al.: On the interpretability of artificial intelligence in radiology: challenges and opportunities. Radiol. Artif. Intell. **2**(3), e190043 (2020)
23. Scherrer, B., Gholipour, A., Warfield, S.K.: Super-resolution reconstruction to increase the spatial resolution of diffusion weighted images from orthogonal anisotropic acquisitions. Med. Image Anal. **16**(7), 1465–1476 (2012)
24. Schwarz, K., Liao, Y., Geiger, A.: On the frequency bias of generative models. In: Advances in Neural Information Processing Systems (NeurIPS) (2021)
25. Selvaraju, R.R., Cogswell, M., Das, A., Vedantam, R., Parikh, D., Batra, D.: Grad-CAM: visual explanations from deep networks via gradient-based localization. In: Proceedings of ICCV, pp. 618–626 (2017)
26. Wang, X., Peng, Y., Lu, L., Lu, Z., Bagheri, M., Summers, R.: Chestx-ray8: hospital-scale chest x-ray database and benchmarks on weakly-supervised classification and localization of common thorax diseases. In: In Proceedings of CVPR (2017)
27. Xuan, K., Sun, S., Xue, Z., Wang, Q., Liao, S.: Learning MRI k-space subsampling pattern using progressive weight pruning. In: Martel, A.L., et al. (eds.) MICCAI 2020. LNCS, vol. 12262, pp. 178–187. Springer, Cham (2020). https://doi.org/10.1007/978-3-030-59713-9_18
28. Zbontar, J., Knoll, F., et al.: FastMRI: an open dataset and benchmarks for accelerated MRI (2019)
29. Zhang, Y., Li, K., et al.: MR image super-resolution with squeeze and excitation reasoning attention network. In: IEEE CVPR, pp. 13420–13429 (2021)
30. Zhang, Y., Tian, Y., Kong, Y., Zhong, B., Fu, Y.: Residual dense network for image super-resolution. In: CVPR (2018)
31. Zhao, C., et al.: A deep learning based anti-aliasing self super-resolution algorithm for MRI. In: Frangi, A.F., Schnabel, J.A., Davatzikos, C., Alberola-López, C., Fichtinger, G. (eds.) MICCAI 2018. LNCS, vol. 11070, pp. 100–108. Springer, Cham (2018). https://doi.org/10.1007/978-3-030-00928-1_12
32. Zhao, X., Zhang, Y., Zhang, T., Zou, X.: Channel splitting network for single MR image super-resolution. IEEE Trans. Image Process. **28**(11), 5649–5662 (2019)

Multi-label Attention Map Assisted Deep Feature Learning for Medical Image Classification

Dwarikanath Mahapatra[1(✉)] and Mauricio Reyes[2]

[1] Inception Institute of Artificial Intelligence, Abu Dhabi, UAE
dwarikanath.mahapatra@inceptioniai.org
[2] ARTORG Center for Biomedical Engineering Research, University of Bern,
Bern, Switzerland
mauricio.reyes@med.unibe.ch

Abstract. Effective deep feature learning and representation is essential in training deep learning models for medical image analysis. Recent developments proposing attention mechanisms have shown promising results in improving model performance. However, their mechanism is difficult to interpret and requires architectural modifications, further reducing benefits for transfer learning and using pre-trained models. We propose an interpretability-guided attention mechanism, formulated as inductive bias operating on the loss function of the trained model. It encourages the learned deep features to yield more distinctive attention maps for multilabel classification problems. Experimental results for medical image classification show our proposed approach outperforms conventional methods utilizing different attention mechanisms while yielding class attention maps in higher agreement with clinical experts. We show how information from unlabeled images can be used directly without any model modification. The proposed approach is modular, applicable to existing network architectures used for medical imaging applications, and yields improved model performance and class attention maps.

Keywords: Interpretability · Multiclass attention · Inductive bias · Medical image classification

1 Introduction

Machine learning models for clinical applications should extract clinically relevant features for downstream tasks such that model performance is optimal while ensuring the extracted features do not stem from spurious correlations [9]. Originally proposed in the context of neural language translation, self-attention based approaches have gained popularity for medical image computing. They generate, without need of supervision, a context or weight vector characterizing relationships among learned feature components to better guide the learning process. Recent works using attention mechanisms include Squeeze and Excitation (SENet) that models channel interdependencies and adaptively calibrates

© The Author(s), under exclusive license to Springer Nature Switzerland AG 2023
L. Karlinsky et al. (Eds.): ECCV 2022 Workshops, LNCS 13807, pp. 722–734, 2023.
https://doi.org/10.1007/978-3-031-25082-8_49

channel-wise feature responses [11], or [2] which uses 2D relative self-attention to replace convolution in image classification. In [27], Convolutional Block Attention Module (CBAM) is proposed for feed-forward convolutional neural networks. In CBAM intermediate attention maps along the channel and spatial dimensions are inferred separately, and multiplied to the input feature map for adaptive feature refinement. Other attention based approaches include self-supervised contrastive learning [5], where targeted data augmentations along with a contrastive loss is used to disregard data variations for feature learning, However, this assumes implicit knowledge of downstream task invariances, which is challenging for medical applications.

Despite performance improvement, prominent disadvantages of previously proposed methods are worth mentioning: 1) They require modification of existing architectures to include additional attention layers (or at its extreme, they build on fully attention models as in [20]). This hinders utilization of existing pre-trained models that have been well studied and proven to be accurate for different tasks, as well as related transfer learning strategies shown to be effective in medical imaging applications [6,10,25]. 2) Lower interpretability of learned features: Despite demonstrations of improved activation maps via post-hoc visualizations, such visualizations describe only local or partial features within the layers of a model which is difficult to assess for a clinical end-user. This is particularly important in medical imaging applications where interpretability of deep learning models is a crucial aspect to ensure trustability, auditability, and ultimately a safe deployment of these technologies [3,8,16,21].

Motivated by these aspects and building on findings from the area of interpretability, we propose a novel approach for deep feature learning, which is characterized by: (i) Improved intelligibility of learned features, as by design it incorporates an explicit and interpretable attention mechanism via a novel multi-label activation map modelling; and (ii) modularity, so it can be applied to any existing architecture without need of modifications. To this end, our proposed deep feature learning approach is cast as an inductive bias loss term (i.e., it encodes desired characteristics of learned features), allowing a very modular utilization in conjunction with other previously studied loss functions.

Our proposed approach is based on the following basic rationale: A trained radiologist learns to perform differential diagnosis (i.e., being able to distinguish among potential diagnoses) on medical images based on *distinguishable label-specific* (e.g. disease/condition) image patterns or characteristics. Through experience radiologists improve their levels of distinctiveness of specific image patterns attributed to potential diagnoses. Consequently, during model training we propose to incorporate a novel inductive bias in the loss term of the model such that learned features enhance label distinctiveness of multi-label interpretability activation maps. We term the proposed approach as MultiLAM-Net for multi-label attention map assisted deep feature learning.

We exploit the fact that for a single sample, multiple label-specific activation maps can be generated, mimicking pseudo-conterfactual[1] attribution maps (i.e., how would the activation map look like should the sample be of a given label), that are used here to enhance label distinctiveness. We propose three different variants to aggregate sample-wise multi-label activation maps information via a graph representation. We further describe how the proposed approach enables use of unlabeled samples for an unsupervised deep feature learning setup. In summary, the main contributions of the study are:

- Interpretability guided deep feature learning approach for medical image classification.
- Improved model performance and quality of label-specific activation maps.
- Modular and easy to implement, can be utilized in conjunction with other existing loss functions, different interpretability approaches, and existing classification model architectures without modifications.
- Our approach can use unlabeled datasets for an unsupervised deep feature learning setup, which is not possible for attention-based approaches necessitating ground-truth information to back propagate gradients.

We present results on multi-label classification on publicly available chest X-ray dataset, and compare the proposed approach to different state-of-the-art methods, and several ablation studies. We demonstrate the added value of the proposed approach in terms of classification performance and improved attention maps, reflected through a better agreement with expert-annotated attention maps, than with previous attention-based approaches.

2   Methods

2.1   Multi-label Activation Maps

Given a set of N labeled training images $\{(\mathbf{x}_i, y_i) : 1 \leq i \leq N\}$, with $(\mathbf{x}_i, \mathbf{y}_i) \in \mathcal{X} \times \mathcal{Y}$ and $\mathcal{X} \subseteq \mathbb{R}^d$ and $\mathcal{Y} \in \{1, \ldots, K\}$ denotes image and label domains, respectively. A deep learning model $\mathbf{M} : \mathcal{X} \mapsto \mathcal{Y}$ is commonly updated by minimizing a standard loss term, such as the weighted cross entropy loss (L_{WCE}). We aim at complementing commonly used loss terms with an interpretability-guided scheme building in multi-label activation maps. Activation maps (also known as saliency maps or heatmaps) are a popular interpretability approach developed initially for natural images [24], and later further investigated for medical image applications [4]. Activation maps highlight image areas of importance (also called pixel attribution) for model's predictions and for a given label.

Given an input sample $\mathbf{x} \in \mathbb{R}^d$, and classification model $\mathbf{M}$, a label-specific activation map $\mathbf{A}_{x,c} \in \mathbb{R}^d$ identifies relevant regions of interest in the image to be classified as label $c \in \{1, \ldots, K\}$. In multi-label classification problems, a given sample can be classified not only into one label, but multiple ones (i.e.,

---

[1] called here pseudo-counterfactuals for the sake of explanations, but we note there is no casualty modeling involved.

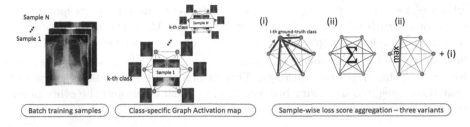

Fig. 1. Conceptual description of proposed MultiLAMNet. Given a batch of training images, a graph representation of multi-label activation maps is produced per sample, where nodes correspond to latent representation of label-specific activation maps and edge weights represent the similarity between nodes. Three variants to perform sample-wise loss aggregation are proposed.

co-occurrence of diseases/conditions). Here we exploit the fact that for a single training sample, **multiple** activation maps can be generated by selection of the targeted label (effectively changing the network's backward gradient trace). Hence, the proposed MultiLAMNet approach aims at enhancing the distinctiveness across all activation maps $\mathbf{A}_{x,c=i}$ and $\mathbf{A}_{x,c=j}(i \neq j)(\forall i, j \in \{1, \ldots, K\})$.

To this end, and to collectively account for all class labels and derive a sample-wise loss score, we propose a graph representation as follows: (i) Each training sample is represented as a graph with K nodes. (ii) Each node in the graph corresponds to the latent representation of the label-specific activation map, $\mathbf{Z}(A_{x,c})$, which is generated via forward pass of the activation maps for label c till the second-to-last layer of the network. This follows findings regarding the effectiveness of deep representations [28]. The set of latent representations for a given sample is $\{\mathbf{Z}(A_{x,c=i})\}(i \in \{1, \ldots, K\})$ and describes label-specific representations of potential labels. Please note the term "potential" used here, as this derivation does not necessarily require prior knowledge of the ground-truth label(s). The presence/absence of this information will be used below to derive variants of the proposed approach. (iii) Each edge weight corresponds to the similarity between two given labels. Our undirected graph has $(K \times (K-1))/2$ edge weights representing all pairs of label combinations. As similarity, in this work we use the cosine similarity between latent representations, denoted as $\langle \mathbf{Z_i}, \mathbf{Z_j} \rangle$. Figure 1 illustrates the proposed graph representation.

2.2   Sample-wise Multi-label Loss Scores

In this section we describe three variants to derive sample-wise multi-label loss scores based on the graph representation described above.

Ground-Truth Node Aggregation: When ground truth label information is available, we define the multi-label loss score as:

$$L_{node} = \frac{1}{M_x} \sum_{\substack{i=1 \\ y(i)=1}}^{K} \sum_{\substack{j=1 \\ j \neq i}}^{K} \langle \mathbf{Z}(A_{x,c=i}), \mathbf{Z}(A_{x,c=j}) \rangle, \tag{1}$$

where $y(i) = 1$ indicates aggregating similarity between nodes corresponding to the ground-truth label(s) and all other nodes, and M_x is the number of labels present in sample $\mathbf{x}$. If a given sample has only one class label (single label classification case), Eq. (1) reduces to $L_{node} = \sum_{j=1;j\neq y(x)}^{K} \langle \mathbf{Z}(A_{x,c=y(I)}, \mathbf{Z}(A_{x,c=j}) \rangle$. Figure 1-(i) illustrates this strategy. This variant emphasizes loss changes such that the aggregated similarity between ground truth labels and the other potential labels is minimized.

Overall Graph Aggregation: In this second variant, the loss score does not require knowledge of the ground-truth label(s) and hence, given a pre-trained model, it can be used within an unsupervised setup where unlabeled data is available. Figure 1-(ii) illustrates this second strategy. The loss term describes the *overall* graph node interactions:

$$L_{over} = \frac{2}{K(K-1)} \sum_{i=1}^{K-1} \sum_{j=i+1}^{K} \langle \mathbf{Z}(A_{x,c=i}), \mathbf{Z}(A_{x,c=j}) \rangle \quad (2)$$

Ground-Truth Node Aggregation with Maximum Penalization: In this third variant we studied a mechanism to further reduce large intra-sample similarities. It extends L_{node} in Eq. 1 as:

$$L_{node\_max} = L_{node} + max\{\langle \mathbf{Z}(\mathbf{A}_{\mathbf{x},\mathbf{c}=\{\forall i|\mathbf{y}(i)=1\}}), \mathbf{Z}(\mathbf{A}_{\mathbf{x},\mathbf{c}=\{\forall j|j\neq i\}}) \rangle\} \quad (3)$$

The intuition is to focus on pairs of labels of particular difficulty for the classification. Figure 1-(iii) illustrates this third strategy. The total loss is:

$$L_{Total} = L_{WCE} + \lambda L_{MultiLAM} \quad (4)$$

where $L_{MultiLAM}$ can be any of the variants $\{L_{node}, L_{over}, L_{node\_max}\}$ presented above, and λ is a weighting factor (more details below and sensitivity analysis in supplementary).

2.3 Implementation Details

The proposed MultiLAMNet was implemented in PyTorch. For the selected use-case of multi-label lung disease classification from X-ray images we trained a DenseNet-121 architecture since it has been effectively used for this task before [12,19]. However, we note that any deep learning classification model can be used with MultiLAMNet. We used Adam optimizer [13] with $\beta_1 = 0.93$, $\beta_2 = 0.999$, and batch normalization, with learning rate of 10^{-3}, and 10^5 update iterations, and early stopping based on the validation accuracy.

To generate interpretability saliency maps, we used default parameters of the iNNvestigate implementation of DeepTaylor [1] and of GradCAM [22], to demonstrate modularity. DeepTaylor operates by running a backward pass on the network in order to produce a decomposition of the neural network's output for the specified class label. Each neuron of a deep network is viewed as a function that can be expanded and decomposed on its input variables. The decompositions of multiple neurons are then aggregated or propagated backwards, resulting in a class activation map. In this study we selected DeepTaylor decomposition [17] due to its popularity and previous uses in other medical image applications [7,15,23]. For all experiment we empirically set $\lambda = 1.4$ in Eq. 4 to weigh the loss terms. We show sensitivity analysis for this parameter in the supplementary. Training and test was performed on a NVIDIA Titan X GPU with 12 GB RAM. The input images were all of size 320×320 pixels. Training the baseline DenseNet-121 with L_{WCE} for 50 epochs took 13 h, and for our method it took 15.1 h (extra 16.1% time). All the reported results are on the test set and an average of 3 different runs was conducted to ensure a fair and robust assessment.

3 Results

We demonstrate the benefits of MultiLAMNet for multi-label lung disease classification from chest X-rays (a multi-label classification problem). We used the CheXpert dataset [12] consisting of 224, 316 chest X-ray images of 65, 240 patients labeled for the presence of common chest conditions. The training set has 223, 414 images while validation and test set have 200 and 500 images, respectively. The validation ground-truth was obtained using majority voting from annotations of three board-certified radiologists. Test images were labeled by consensus of five board-certified radiologists. The test set evaluation protocol is based on five disease labels: *Atelectasis, Cardiomegaly, Consolidation, Edema,* and *Pleural Effusion*, which were selected in order to compare to the work of Pham et al. [18], which is the 2nd ranked method in the CheXpert challenge (same AUC as 1st ranked approach and with an available implementation).

Additionally, in order to show the ability of the proposed approach to use unlabeled data (Eq. 2), we present results obtained when further training a pre-trained model with only 75% of training data, using the remaining 25% as unlabeled data and assuming no knowledge about the ground-truth labels. Furthermore, we evaluated the activation maps yielded by every benchmarked model. An experienced lung radiologist with over 15 years of experience manually annotated salient regions describing diagnosed conditions on a subset of 25 randomly selected cases.

Comparison Methods and Ablation Experiments:

1. Our proposed method MultiLAMNet, using each variant at a time, referred to as MultiLAMNet-{Node,Over,Node_Max}, following Eqs. (1) (2) (3).
2. Pham et al. [18], which is the 2nd ranked method in the CheXpert challenge (same AUC as 1st ranked approach and with an available implementation), and uses Directed Acyclic Graphs (DAGs) for improved performance.
3. Squeeze & Excitation (SE) [11], using a channel-wise attention mechanism.
4. Convolutional Block Attention module (CBAM) [27], which combines channel, and spatial-wise attention mechanisms.

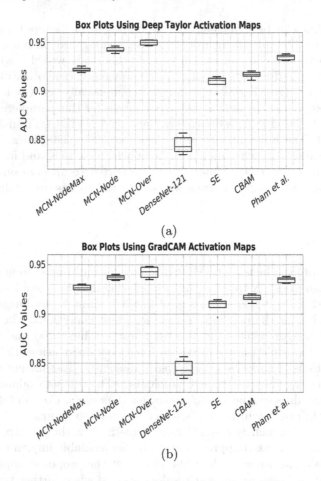

Fig. 2. Average AUC_{PR} values for each pathology from the 5-fold evaluation CheXpert validation set using (a) Deep Taylor Maps; (b)GradCAM Maps. MCN: MultiLAMNet; SE: Squeeze & Excitation [11]; CBAM: Convolutional Block Attention module (CBAM) [27]; Pham et al.: Directed Acyclic Graph [18].

The above-mentioned approaches are based on attention mechanisms. In addition, we include further baselines and variations of the proposed approach: (i) *DenseNet*: Uses only L_{WCE} for training, and (ii) MultiLAMNet-GradCAM: Same as MultiLAMNet but using GradCAM [22] to yield activation maps. This experiment aims to show the modularity and effectiveness of MultiLAMNet using alternative interpretability methods.

3.1   Evaluation Metrics

The trained classification models were assessed via the Area Under the Precision Recall (AUC-PR) curve measures. To evaluate activation maps produced by

Table 1. Similarity of activation maps with expert maps. Mean (standard deviation), using SSIM-structural similarity index [26]; DSC-Dice Metric; HD$_{95}$-95th percentile Hausdorff Distance (mm). Bold values indicate best performance.

	Baselines				Proposed MultiLAMNet		
	DenseNet	SE	CBAM	Pham	Variants		
	-121	[11]	[27]	[18]	Node	Overall	NodeMax
SSIM	52.1(4.1)	62.2(4.0)	64.3(4.1)	68.5(4.2)	73.4(3.5)	**75.8(3.1)**	71.8(4.0)
DSC	69.1(4.4)	72.3(3.9)	77.4(3.7)	80.2(3.7)	87.2(3.1)	**89.9(3.1)**	84.8(3.5)
HD$_{95}$	14.7(3.5)	12.4(3.4)	11.8(3.1)	10.9(3.2)	10.0(2.7)	**9.4(2.5)**	10.2(2.7)

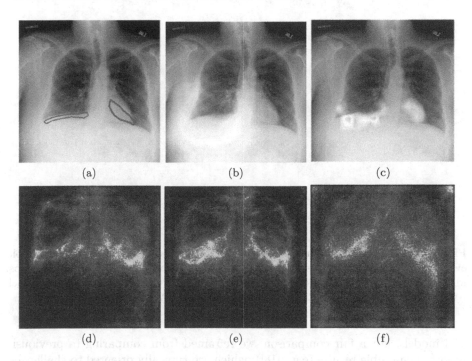

(a)   (b)   (c)

(d)   (e)   (f)

Fig. 3. (a) Salient regions drawn by radiologist; GradCam maps by: (b) DenseNet-121; (c) MultiLAMNet. Deep Taylor maps by: (d) DenseNet-121, and (e) MultiLAMNet; (f) Activation map by Pham et al. [18].

the different benchmarked models, we compared them with the manually-drawn regions of interest by the experienced lung radiologist using Dice (DC), Hausdorff Distance 95%(HD_{95}), and Structural Similarity Index Measure (SSIM) metrics. To calculate these metrics, we binarized activation maps using the ConvexHull function of SciPy with default parameters. We show distribution of classification results for all labels in the CheXpert validation set of 200 images, using a 5-fold validation in order to show consistency of the performance across different folds

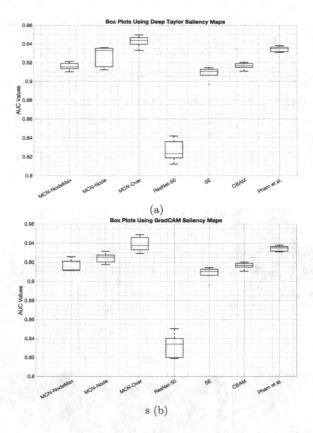

Fig. 4. Average AUC$_{PR}$ values **using ResNet50** for each pathology from the 5-fold evaluation CheXpert validation set using (a) Deep Taylor Maps; (b)GradCAM Maps. MCN: MultiLAMNet; SE: Squeeze & Excitation [11]; CBAM: Convolutional Block Attention module (CBAM) [27]; Pham et al.: Directed Acyclic Graph [18].

and models. For a fair comparison, we refrained from comparing to previously reported ensemble results (e.g., [18]), which are typically oriented to challenges.

3.2   Improved Classification Results

Figure 2 shows AUC-PR results for all benchmarked methods. MultiLAMNet yielded the highest mean AUC-PR for all conditions in the CheXpert 5-fold validation set, followed by Pham et al. [18], CBAM [27] and SE [11]. Same pattern was found for GradCAM, Fig. 2(b).

3.3   Comparison with Radiologist's Saliency Maps

We compared activation maps from all benchmarked approaches with manually annotated salient regions generated by a radiologist with over 15 years of expe-

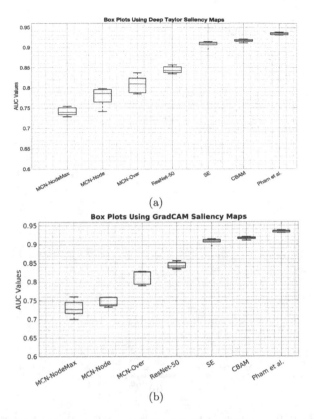

(a)

(b)

Fig. 5. Average AUC_{PR} values **using DenseNet-121 and inverse proposed loss (i.e., 1-cosine similarity)** for each pathology from the 5-fold evaluation CheXpert validation set using (a) Deep Taylor Maps; (b) GradCAM Maps. MCN: MultiLAMNet; SE: Squeeze & Excitation [11]; CBAM: Convolutional Block Attention module (CBAM) [27]; Pham et al.: Directed Acyclic Graph [18].

rience. Quantitative measures in Table 1 highlight MultiLAMNet's superior performance across metrics indicating a better alignment with the expert. Although we do not expect a perfect alignment, a good activation map should highlight a majority of the regions of interest characterizing the lung condition. Figure 3 shows saliency maps drawn by an radiologist and those obtained with and without using MultiLAMNet. Our method generates maps more closely aligned to the expert annotations.

3.4 Additional Results

Figure 4 shows the average AUC_{PR} values **using ResNet50** for each pathology from the 5-fold evaluation CheXpert validation set using Deep Taylor and GradCAM. Figure 5 shows the corresponding plots **using DenseNet-121 and inverse proposed loss (i.e., 1-cosine similarity)** (Fig. 6).

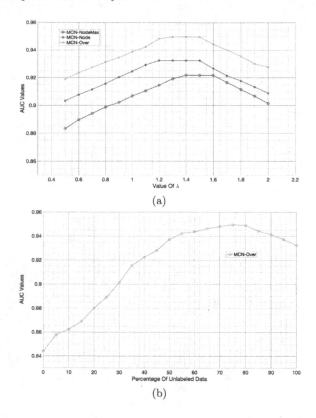

(a)

(b)

Fig. 6. (a) Sensitivity analysis of hyperparameter λ (Eq. 4). (b) Variation of AUC at different percentages of used unlabeled data and proposed loss L_{over} in Eq. 2. Example: 70% refers to training with 30% labeled data using cross entropy loss, and remaining 70% unlabeled samples using loss L_{over} in Eq. 2. Baseline DenseNet-121 corresponds to point at 0% (AUC = 0.844).

4   Conclusion

In this study we propose a novel deep feature learning inductive bias assisted by a multi-class activation map graph modelling. Beyond the originally defined objective of model interpretation, in this study we demonstrate how interpretability can be used to produce an effective deep feature learning inductive bias mechanism, simultaneously leading to improved model performance and interpretability. MultiLAMNet is easy to implement, modular and different from other attention-based methods it does not require model architectural changes, and can be used in an unsupervised manner. Results on multi-class lung disease classification shows superior performance along with a better agreement between class activation maps and expert-generated maps. In comparison to contrastive learning [5], the proposed approach does not require artificial construction of contrastive samples via data augmentation, but naturally utilizes the expected intra-sample class-distinctiveness of activation maps to drive the learning process. We believe these results align with recent findings regarding the importance of effective inductive bias [14], and it can be extended for multi-omics problems where different data types (e.g. imaging, text, clinical data, etc.) can be handled by existing interpretability approaches, leading to a flexible multi-omics interpretability-guided deep feature learning framework.

References

1. Alber, M., et al.: innvestigate neural networks. J. Mach. Learn. Res. **20**(93), 1–8 (2019)
2. Bello, I., Zoph, B., Vaswani, A., Shlens, J., Le, Q.V.: Attention augmented convolutional networks. In: Proceedings of the IEEE/CVF International Conference on Computer Vision, pp. 3286–3295 (2019)
3. Budd, S., Robinson, E.C., Kainz, B.: A survey on active learning and human-in-the-loop deep learning for medical image analysis. Med. Image Anal. **71**, 102062 (2021)
4. Cardoso, J., et al.: Interpretable and Annotation-Efficient Learning for Medical Image Computing Third International Workshop, iMIMIC 2020, Second International Workshop, MIL3ID 2020, and 5th International Workshop, Labels 2020, Held in Conjunction with MICCAI 2020, Lima, Peru, 4–8 October 2020, Proceedings (2020)
5. Chen, T., Kornblith, S., Norouzi, M., Hinton, G.: A simple framework for contrastive learning of visual representations. In: International Conference on Machine Learning, pp. 1597–1607. PMLR (2020)
6. Cheplygina, V., de Bruijne, M., Pluim, J.P.: Not-so-supervised: a survey of semi-supervised, multi-instance, and transfer learning in medical image analysis. Med. Image Anal. **54**, 280–296 (2019)
7. Eitel, F., Ritter, K.: Testing the robustness of attribution methods for convolutional neural networks in MRI-based Alzheimer's disease classification. In: Suzuki, K., et al. (eds.) ML-CDS/IMIMIC -2019. LNCS, vol. 11797, pp. 3–11. Springer, Cham (2019). https://doi.org/10.1007/978-3-030-33850-3_1
8. Fuhrman, J.D., Gorre, N., Hu, Q., Li, H., El Naqa, I., Giger, M.L.: A review of explainable and interpretable AI with applications in COVID-19 imaging. Med. Phys. **49**, 1–14 (2021)
9. Geirhos, R., et al.: Shortcut learning in deep neural networks. Nat. Mach. Intell. **2**(11), 665–673 (2020)
10. Ghafoorian, M., et al.: Transfer learning for domain adaptation in mri: application in brain lesion segmentation. In: Descoteaux, M., Maier-Hein, L., Franz, A., Jannin, P., Collins, D.L., Duchesne, S. (eds.) MICCAI 2017. LNCS, vol. 10435, pp. 516–524. Springer, Cham (2017). https://doi.org/10.1007/978-3-319-66179-7_59
11. Hu, J., Shen, L., Sun, G.: Squeeze-and-excitation networks. In: Proceedings of the IEEE Conference on Computer Vision and Pattern Recognition, pp. 7132–7141 (2018)
12. Irvin, J., Rajpurkar, P., et al.: Chexpert: a large chest radiograph dataset with uncertainty labels and expert comparison. In: arXiv preprint arXiv:1901.07031 (2019)
13. Kingma, D., Ba, J.: Adam: a method for stochastic optimization. In: arXiv preprint arXiv:1412.6980 (2014)

14. Locatello, F., et al.: Challenging common assumptions in the unsupervised learning of disentangled representations. In: International Conference on Machine Learning, pp. 4114–4124. PMLR (2019)

15. Mahapatra, D., Poellinger, A., Shao, L., Reyes, M.: Interpretability-driven sample selection using self supervised learning for disease classification and segmentation. IEEE Trans. Med. Imaging 1–1 (2021). https://doi.org/10.1109/TMI.2021.3061724

16. McCrindle, B., Zukotynski, K., Doyle, T.E., Noseworthy, M.D.: A radiology-focused review of predictive uncertainty for AI interpretability in computer-assisted segmentation. Radiol. Artif. Intell. 3(6), e210031 (2021)

17. Montavon, G., Lapuschkin, S., Binder, A., Samek, W., Muller, K.R.: Explaining nonlinear classification decisions with deep Taylor decomposition. Pattern Recogn. 65, 211–222 (2017)

18. Pham, H.H., Le, T.T., Tran, D.Q., Ngo, D.T., Nguyen, H.Q.: Interpreting chest x-rays via CNNs that exploit hierarchical disease dependencies and uncertainty labels. In: arXiv preprint arXiv:1911.06475 (2020)

19. Rajpurkar, P., et al.: Chexnet: radiologist-level pneumonia detection on chest x-rays with deep learning. In: arXiv preprint arXiv:1711.05225 (2017)

20. Ramachandran, P., Parmar, N., Vaswani, A., Bello, I., Levskaya, A., Shlens, J.: Stand-alone self-attention in vision models. arXiv preprint arXiv:1906.05909 (2019)

21. Reyes, M., et al..: On the interpretability of artificial intelligence in radiology: challenges and opportunities. Radiol. Artif. Intell. 2(3), e190043 (2020). https://doi.org/10.1148/ryai.2020190043

22. Selvaraju, R.R., Cogswell, M., Das, A., Vedantam, R., Parikh, D., Batra, D.: Grad-CAM: visual explanations from deep networks via gradient-based localization. In: Proceedings of the ICCV, pp. 618–626 (2017)

23. Silva, W., Poellinger, A., Cardoso, J.S., Reyes, M.: Interpretability-guided content-based medical image retrieval. In: Martel, A.L., et al. (eds.) MICCAI 2020. LNCS, vol. 12261, pp. 305–314. Springer, Cham (2020). https://doi.org/10.1007/978-3-030-59710-8_30

24. Simonyan, K., Vedaldi, A., Zisserman, A.: Deep inside convolutional networks: visualising image classification models and saliency maps. CoRR, December 2013. http://arxiv.org/abs/1312.6034

25. Tajbakhsh, N., et al.: Convolutional neural networks for medical image analysis: full training or fine tuning? IEEE Trans. Med. Imaging 35(5), 1299–1312 (2016)

26. Wang, Z., et al.: Image quality assessment: from error visibility to structural similarity. IEEE Trans. Imag. Proc. 13(4), 600–612 (2004)

27. Woo, S., Park, J., Lee, J.-Y., Kweon, I.S.: CBAM: convolutional block attention module. In: Ferrari, V., Hebert, M., Sminchisescu, C., Weiss, Y. (eds.) ECCV 2018. LNCS, vol. 11211, pp. 3–19. Springer, Cham (2018). https://doi.org/10.1007/978-3-030-01234-2_1

28. Zhang, R., Isola, P., Efros, A.A., Shechtman, E., Wang, O.: The unreasonable effectiveness of deep features as a perceptual metric. In: Proceedings of the IEEE Conference on Computer Vision and Pattern Recognition, pp. 586–595 (2018)

Unsupervised Domain Adaptation Using Feature Disentanglement and GCNs for Medical Image Classification

Dwarikanath Mahapatra[1]([✉]), Steven Korevaar[2], Behzad Bozorgtabar[3], and Ruwan Tennakoon[2]

[1] Inception Institute of Artificial Intelligence, Abu Dhabi, UAE
dwarikanath.mahapatra@inceptioniai.org
[2] Royal Melbourne Institute of Technology, Melbourne, Australia
s3544280@student.edu.au, ruwan.tennakoon@rmit.edu.au
[3] Ecole Polytechnique Fédérale de Lausanne, Lausanne, Switzerland
Behzad.bozorgtabar@epfl.ch

Abstract. The success of deep learning has set new benchmarks for many medical image analysis tasks. However, deep models often fail to generalize in the presence of distribution shifts between training (source) data and test (target) data. One method commonly employed to counter distribution shifts is domain adaptation: using samples from the target domain to learn to account for shifted distributions. In this work we propose an unsupervised domain adaptation approach that uses graph neural networks and, disentangled semantic and domain invariant structural features, allowing for better performance across distribution shifts. We propose an extension to swapped autoencoders to obtain more discriminative features. We test the proposed method for classification on two challenging medical image datasets with distribution shifts - multi center chest Xray images and histopathology images. Experiments show our method achieves state-of-the-art results compared to other domain adaptation methods.

Keywords: Unsupervised domain adaptation · Graph convolution networks · Camelyon17 · CheXpert · NIH Xray

1 Introduction

The success of convolutional neural networks (CNNs) has set new benchmarks for many medical image classification tasks such as diabetic retinopathy grading [9], digital pathology image classification [18] and chest X-ray images [14], as well as segmentation tasks [24]. Clinical adoption of algorithms has many challenges due to the domain shift problem: where the target dataset has different characteristics than the source dataset on which the model was trained. These differences are often a result of different image capturing protocols, parameters, devices, scanner manufacturers, etc. Since annotating samples from hospitals

L. Karlinský et al. (Eds.): ECCV 2022 Workshops, LNCS 13807, pp. 735–748, 2023.
https://doi.org/10.1007/978-3-031-25082-8_50

and domains is challenging due to scarcity of experts and the resource intensive nature of labeling, it is essential to design models from the beginning that perform consistently on images acquired from multiple domains.

Most approaches to the domain shift problem can be categorized based on the amount of data from the target domain. Fully-supervised domain adaptation (DA) and semi-supervised DA assume availability of a large amount and a small amount of fully-labeled instances from the target domain, respectively, along with data from a source domain [26]. Unsupervised domain adaptation (UDA) techniques [5,23] use only unlabeled data from the target domain and labeled source domain data. UDA methods aim to learn a domain-invariant representation by enforcing some constraint (e.g. Maximum Mean Discrepancy [16]) that brings the latent space distributions, z, of distinct domains closer; ideally leading to more comparable performance in classification/segmentation.

Convolutional neural network (CNN) based state-of-the-art (SOTA) UDA methods often obtain impressive results but typically only enforce alignment of global domain statistics [31] resulting in loss of semantic class label information. Semantic transfer methods [19,22] address this by propagating class label information into deep adversarial adaptation networks. Unfortunately, it is difficult to model and integrate semantic label transfer into existing deep networks. To deal with the above limitations [20] propose a Graph Convolutional Adversarial Network (GCAN) for unsupervised domain adaptation. Graph based methods better exploit global relationship between different nodes (or samples) than CNNs and learn both global and local information beneficial for DA.

One way to address domain shift is with feature disentanglement [17] by separating latent feature representations into domain invariant (often called structure or content) and domain variant (texture or style) components, thus minimizing the impact of domain variation. In this work we combine feature disentanglement with graph convolutional networks (GCN) for unsupervised domain adaptation and apply it to two different standard medical imaging datasets for classification and compare it to SOTA methods.

Related Work: Prior works in UDA focused on medical image classification [2], object localisation, and lesion segmentation [11,15], and histopathology stain normalization [4]. Heimann et al. [11] used GANs to increase the size of training data and demonstrated improved localisation in X-ray fluoroscopy images. Likewise, Kamnitsas et al. [15] used GANs for improved lesion segmentation in magnetic resonance imaging (MRI). Ahn et al. [2] use a hierarchical unsupervised feature extractor to reduce reliance on annotated training data. Chang et al. [4] propose a novel stain mix-up for histopathology stain normalization and subsequent UDA for classification. Graph networks for UDA [20,30] have been used in medical imaging applications [1] such as brain surface segmentation [8] and brain image classification [12,13]. However, none of them combine feature disentanglement with GCNs for UDA.

Our Contributions: Although previous works used feature disentanglement and graph based domain adaptation separately, they did not combine them for medical image classification. In this work: 1) we propose a feature disentanglement module trained to align structural components coupled with a Graph Convolutional Adversarial Network (GCAN) for unsupervised domain adaptation in medical images. 2) We perform feature disentanglement using swapped autoencoders to obtain texture and structural features, which are used for graph construction and defining generator losses. Since, in medical imaging applications, images from the same modality (showing the same organ) have similar texture, we introduce a novel cosine similarity loss components into the loss function that enforces the preservation of the structure component in the latent representation; 3) We demonstrate our method's effectiveness on multiple medical image datasets such as chest xrays and histopathology images.

2   Method

Given a source data set, $\mathcal{D}_S = \{(x_i^S, y_i^S)\}_{i=1}^{n_s}$, consisting of n_s labeled samples, and a target data set, $\mathcal{D}_T = \{(x_i^T)\}_{i=1}^{n_t}$ consisting of n_t unlabeled target samples, unsupervised domain adaptation aims to learn a classifier that can reliably classify unseen target samples. Here, $x_i^S \sim p_S$ is a source data point sampled from source distribution p_S, $y_i^S \in \mathcal{Y}_S$ is the label, and $x_i^T \sim p_T$ is a target data point sampled from target distribution p_T. As per the covariate shift assumption, we assume that $p_S(y|x) = p_T(y|x) \ \forall x$. Thus the only thing that changes between source and target is the distribution of the input samples, x.

The overall block diagram of our system is depicted in Fig. 1. Our proposed approach consists of a feature disentanglement module (Fig. 1-b) which separates semantic textural and structural features (often called style and content respectively). The output of the feature disentanglement module is constructed into a graph and then fed into an adversarial generator based graph neural network (Fig. 1-a) that generates features which are domain invariant. The graph neural network learns more global relationships between samples, which in turn leads to more discriminative and domain invariant feature learning.

2.1   Feature Disentanglement Network

Figure 1(b) shows the architecture of our feature disentanglement network (FDN). The FDN consists of two encoders $(E_S(\cdot), E_T(\cdot))$ and two decoder (generator) networks $(G_S(\cdot), G_T(\cdot))$, for the source and target domains respectively. Similar to a classic autoencoder, each encoder, $E_\bullet(\cdot)$, produces a latent code z_i for image $x_i^\bullet \sim p_\bullet$. Each decoder, $G_\bullet(\cdot)$, reconstructs the original image from z_i. Furthermore, we divide the latent code, z_i, into two components: a texture component, z_i^{tex}, and a structural component, z_i^{str}. Structural information is the local information that describes an image's underlying structure and visible organs or parts, and is expected to be invariant across similar images from different domains. The disentanglement network is trained using the following loss function:

$$\mathcal{L}_{Disent} = \mathcal{L}_{Rec} + \lambda_1 \mathcal{L}_{swap} + \lambda_2 \mathcal{L}_{str} \tag{1}$$

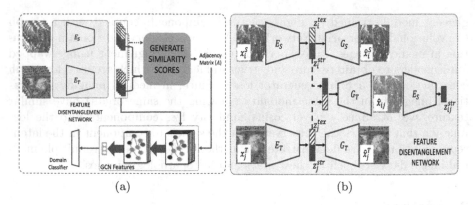

Fig. 1. (a) Workflow of our proposed method. Given images from source and target domains we disentangle features into texture and structure features, and generate similarity scores to construct an adjacency matrix. Generated features by the GCN are domain invariant and along with the latent features are used to train a domain classifier. (b) Architecture of feature disentanglement network with swapped structure features

$\mathcal{L}_{Rec}$, is the commonly used image reconstruction loss and is defined as:

$$\mathcal{L}_{Rec} = \mathbb{E}_{x_i^S \sim p_S}\left[\left\|\left|x_i^S - G_S(E_S(x_i^S))\right|\right\|\right] + \mathbb{E}_{x_j^T \sim p_T}\left[\left\|\left|x_j^T - G_T(E_T(x_j^T))\right|\right\|\right] \quad (2)$$

As shown in Fig. 1(b), we also combine the structure of source domain images (z_i^{str}) with the texture of target images (z_j^{tex}) and send it though G_T. the reconstructed images, $\hat{x}_{ij}$, should contain the structure of the original source domain image but, should look like an image from the target domain. To enforce this property, we use the adversarial loss term $\mathcal{L}_{swap}$ defined as:

$$\mathcal{L}_{swap} = \mathbb{E}_{x_j \sim p_T}[\log(1 - D_{FD}(x_j))] + \mathbb{E}_{x_i \sim p^S, x_j \sim p^T}\left[\log(D_{FD}(G_T([z_i^{str}, z_j^{tex}])))\right]. \quad (3)$$

Here, D is a discriminator trained with target domain images, thus the generator should learn to produce images that appear to be from the target domain.

Furthermore, the newly generated image, $\hat{x}_{ij}$, should retain the structure of the original source domain image x_i^S from where the original structure was obtained. In [25] the authors use co-occurrent patch statistics from $\hat{x}_{ij}$ and x_i^S such that any patch from x_i^S cannot be distinguished from a group of patches from $\hat{x}_{ij}$. This constraint is feasible in natural images since different images have far more varied style and content, but is challenging to implement in medical images since images from the same modality (showing the same organ) appear similar. Thus limited unique features are learnt under such a constraint. To overcome this limitation we introduce a novel component into the loss function. The output, $\hat{x}_{ij}$, when passed through encoder E_S produces the disentangled features $z_{ij}^{str}, z_{ij}^{tex}$. Since $\hat{x}_{ij}$ is generated from the combination of z_i^{str} and z_j^{tex}, z_{ij}^{str} should be similar to z_i^{str} (i.e. the structure should not change). This is enforced using the cosine similarity loss

$$\mathcal{L}_{str} = 1 - \langle z_i^{str}, z_{ij}^{str} \rangle \quad (4)$$

where $\langle \cdot \rangle$ denotes cosine similarity.

2.2 Graph Convolutional Adversarial Network

To learn more global relationships between samples, the features from the feature disentanglement module is used to construct an instance graph and which is processed with a Graph Convolutional Network (GCN). The GCN learns layerwise propagation operations that can be applied directly on graphs. Given an undirected graph with m nodes, the set of edges between nodes, and an adjacency matrix $\mathbf{A} \in R^{m \times m}$, a linear transformation of graph convolution is defined as the multiplication of a graph signal $\mathbf{X} \in R^{k \times m}$ with a filter $\mathbf{W} \in R^{k \times c}$:

$$\mathbf{Z} = \hat{\mathbf{D}}^{-\frac{1}{2}} \hat{\mathbf{A}} \hat{\mathbf{D}}^{-\frac{1}{2}} \mathbf{X}^T \mathbf{W}, \tag{5}$$

where $\hat{\mathbf{A}} = \mathbf{A} + \mathbf{I}$, $\mathbf{I}$ being the identity matrix, and $\hat{\mathbf{D}}_{ii} = \sum_j \hat{\mathbf{A}}_{ij}$. The output is a $c \times m$ matrix $\mathbf{Z}$. The GCN can be constructed by stacking multiple graph convolutional layers followed by a non-linear operation (such as ReLU).

In our implementation, each node in the instance graph represents a latent feature of a sample (either source or target domain) $[z^{tex}, z^{str}]$. The adjacency matrix is constructed using the cosine similarity score as $\hat{\mathbf{A}} = \mathbf{X}_{sc}\mathbf{X}_{sc}^T$. Note that the cosine similarity score quantifies the semantic similarity between two samples. Here $\mathbf{X}_{sc} \in R^{w \times h}$ is the matrix of cosine similarity scores, w is the batch size, and $h = 2$ is the dimension of the similarity scores of each sample obtained from corresponding $[z_{tex}, z_{str}]$.

The flow of our proposed approach is shown in Fig. 1(a). Our network is trained by minimizing the following objective function:

$$\mathcal{L}(\mathcal{X}_S, \mathcal{Y}_S, \mathcal{X}_T) = \mathcal{L}_C(\mathcal{X}_S, \mathcal{Y}_S) + \lambda_{adv}\mathcal{L}_{Adv}(\mathcal{X}_S, \mathcal{X}_T) \tag{6}$$

The classification loss $\mathcal{L}_C$ is a cross entropy loss $\mathcal{L}_C(\mathcal{X}_S, \mathcal{Y}_S) = -\sum_{c=1}^{n_s} y_c \log(p_c)$, y_c is the indicator variable and p_c is the class probability.

The second loss component is an adversarial loss function defined in Eq. 7. A domain classifier D identifies whether features from the graph neural network G_{GNN} are from the source or target domain. Conversely, G_{GNN}, is being trained to produce samples that can fool D. This two-player minimax game will reach an equilibrium state when features from G_{GNN} are domain-invariant.

$$\mathcal{L}_{Adv}(\mathcal{X}_S, \mathcal{X}_T) = \mathbb{E}_{x \in D_S}[\log(1 - D(G_{GNN}(x)))] + \mathbb{E}_{x \in D_T}[\log(D(G_{GNN}(x)))] \tag{7}$$

Training and Implementation: Given $\mathbf{X}$ and $\mathbf{A}$ the GCN features are obtained according to Eq. 5. Source and target domain graphs are individually constructed and fed into the parameters-shared GCNs to learn representations. The dimension of z_{tex} is 256, while z_{str} is 64×64. **VAE Network:** The encoder consists of 3 convolution blocks followed by max pooling. The decoder is symmetrically designed to the encoder. 3×3 convolution filters are used and $64, 32, 32$ filters are used in each conv layer. The input to the VAE is 256×256. For the domain classifier we use DenseNet-121 network with pre-trained weights from ImageNet and fine-tuned using self supervised learning. As a pre-text task we use the classifier to predict the intensity values of masked regions.

3 Experiments and Results

3.1 Results for CAMELYON17 Dataset

Dataset Description: We use the CAMELYON17 dataset [3] to evaluate the performance of the proposed method on tumor/normal classification. In this dataset, a total of 500 $H\&E$ stained WSIs are collected from five medical centers (denoted as by $C1_{17}, C2_{17}, C3_{17}, C4_{17}, C5_{17}$ respectively). 50 of these WSIs include lesion-level annotations. All positive and negative WSIs are randomly split into training/validation/test sets and provided by the organizers in a 50/30/20% split for the individual medical centers to obtain the following split: $C1_{17}$:37/22/15, $C2_{17}$: 34/20/14, $C3_{17}$: 43/24/18, $C4_{17}$: 35/20/15, $C5_{17}$: 36/20/15. 256×256 image patches are extracted from the annotated tumors for positive patches and from tissue regions of WSIs without tumors for negative patches. We use $\lambda_1 = 0.9, \lambda_2 = 1.2, \lambda_{adv} = 1.3$.

Implementation Details. Since the images have been taken from different medical centers their appearance varies despite sharing the same disease labels. This is due to slightly different protocols of $H\&E$ staining. Stain normalization has been a widely explored topic which aims to standardize the appearance of images across all centers, which is equivalent to domain adaptation. Recent approaches to stain normalization/domain adaptation favour use of GANs and other deep learning methods. We compare our approach to recent approaches and also with [4] which explicitly performs UDA using MixUp.

To evaluate our method's performance: 1) We use $C1_{17}$ as the source dataset and train a ResNet-50 classifier [10] (ResNet$_{C1}$). Each remaining dataset from the other centers are, separately, taken as the target dataset, the corresponding domain adapted images are generated, and classified using ResNet$_{C1}$. As a baseline, we perform the experiment without domain adaptation denoted as $No - UDA$ where ResNet$_{C1}$ is used to classify images from other centers. We report results for a network trained in a fully-supervised manner on the training set from the same domain ($FSL - SameDomain$) to give an upper-bound expectation for a UDA model trained on other domain's data, where a ResNet-50 is trained on the training images and used to classify test images, all from the same hospital. This approach will give the best results for a given classifier (in our case ResNet-50). All the above experiments are repeated using each of $C2_{17}, C3_{17}, C4_{17}, C5_{17}$ as the source dataset. Table 1 summarizes our results. PyTorch was used in all implementations.

The results in Table 1 show that UDA methods are better than conventional stain normalization approaches as evidenced by the superior performance of our proposed method and [4]. Our method performs the best amongst all the methods, and approaches $FSL - SameDomain$, the theoretical maximum performance a domain adaptation method can achieve. This shows that our proposed GCN based approach performs better than other UDA methods. The ablation studies also show that our proposed individual loss term $\mathcal{L}_{Str}$ has a significant contribution to the overall performance of our method and excluding it significantly degrades the performance.

Table 1. Classification results in terms of AUC measures for different domain adaptation methods on the CAMELYON17 dataset. Note: $FSL - SD$ is a fully-supervised model trained on target domain data.

	Comparison and proposed method							Proposed	Ablation study results		
	No UDA	MMD	CycleGAN	[28]	[6]	[21]	[4]	Method	FSL-Same	$\mathcal{L}_{Str}$	$\mathcal{L}_{Swap}$
Center 1	0.8068	0.8742	0.9010	0.9123	0.9487	0.9668	0.979	**0.988**	0.991	0.979	0.953
Center 2	0.7203	0.6926	0.7173	0.7347	0.8115	0.8537	0.948	**0.963**	0.972	0.951	0.941
Center 3	0.7027	0.8711	0.8914	0.9063	0.8727	0.9385	0.946	**0.958**	0.965	0.948	0.932
Center 4	0.8289	0.8578	0.8811	0.8949	0.9235	0.9548	0.965	**0.979**	0.986	0.967	0.949
Center 5	0.8203	0.7854	0.8102	0.8223	0.9351	0.9462	0.942	**0.949**	0.957	0.934	0.921
Average	0.7758	0.8162	0.8402	0.8541	0.8983	0.9320	0.956	**0.967**	0.974	0.956	0.939
p	0.0001	0.0001	0.002	0.003	0.013	0.024	0.017	-	0.07	0.01	0.001

3.2 Results on Chest Xray Dataset

We use the following chest Xray datasets: **NIH Chest Xray** Dataset: The NIH ChestXray14 dataset [29] has $112,120$ expert-annotated frontal-view X-rays from $30,805$ unique patients and has 14 disease labels. Original images were resized to 256×256, and $\lambda_1 = 0.9, \lambda_2 = 1.2, \lambda_{adv} = 1.2$. **CheXpert** Dataset: This datset [14] has $224,316$ chest radiographs of $65,240$ patients labeled for the presence of 14 common chest conditions. The validation ground-truth is obtained using majority voting from annotations of 3 board-certified radiologists. Original images were resized to 256×256, and $\lambda_1 = 0.95, \lambda_2 = 1.1, \lambda_{adv} = 1.3$. These two datasets have the same set of disease labels.

We divide both datasets into train/validation/test splits on the patient level at $70/10/20$ ratio, such that images from one patient are in only one of the splits. Then we train a DenseNet-121 [27] classifier on one dataset (say NIH's train split). Here the NIH dataset serves as the source data and CheXpert is the target dataset. We then apply the trained model on the training split of the NIH dataset and tested on the test split of the same domain the results are denoted as $FSL - Same$. When we apply this model to the test split of the CheXpert data without domain adaptation the results are reported under No-UDA.

Tables 2,3 show classification results for different DA techniques where, respectively, the NIH and CheXpert dataset were the source domain and the performance metrics are for, respectively, CheXpert and NIH dataset's *test split*. We observe that UDA methods perform worse than $FSL - Same$. This is expected since it is very challenging to perfectly account for domain shift. However all UDA methods perform better than fully supervised methods trained on one domain and applied on another without domain adaptation.

The DANN architecture [7] outperforms MMD and cycleGANs, and is on par with graph convolutional methods GCAN [20] and GCN2 [13]. However our method outperforms all compared methods which can be attributed to the combination of GCNs, which learn more useful global relationships between different samples, and feature disentanglement which in turn leads to more discriminative feature learning.

Table 2. Classification results on the CheXpert dataset's test split using NIH data as the source domain. Note: $FSL - SD$ is a fully-supervised model trained on target domain data.

	Atel.	Card.	Eff.	Infil.	Mass	Nodule	Pneu.	Pneumot.	Consol.	Edema	Emphy.	Fibr.	PT.	Hernia
No UDA	0.697	0.814	0.761	0.652	0.739	0.694	0.703	0.781	0.704	0.792	0.815	0.719	0.728	0.811
MMD	0.741	0.851	0.801	0.699	0.785	0.738	0.748	0.807	0.724	0.816	0.831	0.745	0.754	0.846
CycleGANs	0.765	0.874	0.824	0.736	0.817	0.758	0.769	0.832	0.742	0.838	0.865	0.762	0.773	0.864
DANN	0.792	0.902	0.851	0.761	0.849	0.791	0.802	0.869	0.783	0.862	0.894	0.797	0.804	0.892
FSL−Same	*0.849*	*0.954*	*0.903*	*0.814*	*0.907*	*0.825*	*0.844*	*0.928*	*0.835*	*0.928*	*0.951*	*0.847*	*0.842*	*0.941*
GCAN	0.798	0.908	0.862	0.757	0.858	0.803	0.800	0.867	0.776	0.865	0.908	0.811	0.799	0.904
GCN2	0.809	0.919	0.870	0.765	0.871	0.807	0.810	0.882	0.792	0.883	0.921	0.817	0.812	0.914
Proposed	**0.825**	**0.931**	**0.884**	**0.788**	**0.890**	**0.818**	**0.828**	**0.903**	**0.818**	**0.910**	**0.934**	**0.828**	**0.830**	**0.923**
Ablation study results														
$\mathcal{L}_{Str}$	0.811	0.914	0.870	0.771	0.872	0.808	0.819	0.884	0.803	0.887	0.916	0.809	0.812	0.907
$\mathcal{L}_{Swap}$	0.787	0.885	0.845	0.741	0.843	0.780	0.791	0.856	0.776	0.859	0.887	0.785	0.786	0.883

Table 3. Classification results on the NIH Xray dataset's test split using CheXpert data as the source domain. Note: $FSL - SD$ is a fully-supervised model trained on target domain data.

	Atel.	Card.	Eff.	Infil.	Mass	Nodule	Pneu.	Pneumot.	Consol.	Edema	Emphy.	Fibr.	PT.	Hernia
No UDA	0.718	0.823	0.744	0.730	0.739	0.694	0.683	0.771	0.712	0.783	0.803	0.711	0.710	0.785
MMD	0.734	0.846	0.762	0.741	0.785	0.738	0.709	0.793	0.731	0.801	0.821	0.726	0.721	0.816
CycleGANs	0.751	0.861	0.785	0.761	0.817	0.758	0.726	0.814	0.746	0.818	0.837	0.741	0.737	0.836
DANN	0.773	0.882	0.819	0.785	0.837	0.791	0.759	0.838	0.770	0.836	0.863	0.766	0.762	0.861
FSL−Same	*0.814*	*0.929*	*0.863*	*0.821*	*0.869*	*0.825*	*0.798*	*0.863*	*0.805*	*0.872*	*0.904*	*0.802*	*0.798*	*0.892*
GCAN	0.78	0.895	0.811	0.777	0.828	0.782	0.751	0.832	0.765	0.828	0.857	0.761	0.756	0.851
GCN2	0.786	0.906	0.833	0.789	0.831	0.802	0.763	0.835	0.774	0.837	0.868	0.768	0.763	0.860
Proposed	**0.798**	**0.931**	**0.884**	**0.799**	**0.843**	**0.818**	**0.773**	**0.847**	**0.784**	**0.849**	**0.879**	**0.779**	**0.774**	**0.868**
Ablation study results														
$\mathcal{L}_{Str}$	0.788	0.919	0.875	0.786	0.832	0.804	0.768	0.834	0.778	0.835	0.863	0.766	0.765	0.853
$\mathcal{L}_{Swap}$	0.770	0.903	0.860	0.771	0.811	0.788	0.748	0.814	0.761	0.816	0.845	0.746	0.749	0.836

3.3 T-SNE Visualizations

In Figs. 2, 3, we show the T-sne plots for our Proposed method, for GCN2 [13], DANN [7] and our method without L_{Str}. The features are obtained from the final layer, and for all methods the output feature vector is 512 dimensional, and the corresponding parameter values are $\lambda_1 = 0.95, \lambda_2 = 1.1$. The plots show visualizations for 5 disease labels for the CheXpert dataset. We show data from the source (marked with dots) and target domain (marked with '+') for these labels. For our proposed method (Fig. 2 (a)), the source and target domain data of the same label map to nearby areas, which indicates that the domain adaptation step successfully generates domain invariant features. Additionally, the clusters for different labels is fairly well separated. These two characteristics are not observed for the other methods. GCN2 has some level of separability but there is undesirable overlap across different classes. The overlap across classes is even worse when we exclude the feature disentanglement components.

The t-sne visualizations clearly indicates that our proposed approach using feature disentanglement is highly effective in learning domain invariant representations for different disease labels. It also highlights the importance of feature disentanglement step. By separating the images into structure and texture features we are able to learn representations that have better discriminative power.

3.4 Results with L_1 and L_2 Loss

Tables 4, 5, 6 show results, similar to previous tables, but using L_1, L_2 and compared with our propsed cosine loss. These results demonstrate the effectiveness of our approach using different loss functions.

Table 4. Classification results in terms of AUC measures on the CAMELYON17 dataset using **L1, L2 loss instead of Cosine Loss**.

	Proposed method- loss variants			Ablation study results		
	L_1	L_2	Cosine	FSL-Same	$\mathcal{L}_{Str}$	$\mathcal{L}_{Swap}$
Center 1	0.971	0.979	**0.988**	0.991	0.979	0.953
Center 2	0.946	0.954	**0.963**	0.972	0.951	0.941
Center 3	0.943	0.95	**0.958**	0.965	0.948	0.932
Center 4	0.96	0.967	**0.979**	0.986	0.967	0.949
Center 5	0.932	0.939	**0.949**	0.957	0.934	0.921
Average	0.950	0.958	**0.967**	0.974	0.956	0.939
p	0.03	0.04	-	0.07	0.01	0.001

Table 5. Classification results on the CheXpert dataset's test split using **L1, L2 loss instead of Cosine Loss**.

	Atel.	Card.	Eff.	Infil.	Mass	Nodule	Pneu.	Pneumot.	Consol.	Edema	Emphy.	Fibr.	PT.	Hernia
FSL−Same	*0.849*	*0.954*	*0.903*	*0.814*	*0.907*	*0.825*	*0.844*	*0.928*	*0.835*	*0.928*	*0.951*	*0.847*	*0.842*	*0.941*
Proposed	0.825	0.931	0.884	0.788	0.890	0.818	0.828	0.903	0.818	0.910	0.934	0.828	0.830	0.923
Proposed-L2	0.814	0.920	0.872	0.775	0.879	0.806	0.814	0.892	0.801	0.901	0.922	0.817	0.819	0.911
Proposed-L1	0.806	0.911	0.863	0.767	0.870	0.797	0.803	0.885	0.792	0.892	0.913	0.808	0.81	0.904
Ablation study results														
$\mathcal{L}_{Str}$	0.811	0.914	0.870	0.771	0.872	0.808	0.819	0.884	0.803	0.887	0.916	0.809	0.812	0.907
$\mathcal{L}_{Swap}$	0.787	0.885	0.845	0.741	0.843	0.780	0.791	0.856	0.776	0.859	0.887	0.785	0.786	0.883

Table 6. Classification results on the NIH Xray dataset's test split using **L1,L2 loss instead of Cosine Loss**.

FSL−Same	*0.814*	*0.929*	*0.863*	*0.821*	*0.869*	*0.825*	*0.798*	*0.863*	*0.805*	*0.872*	*0.904*	*0.802*	*0.798*	*0.892*
Proposed	0.798	0.931	0.884	0.799	0.843	0.818	0.773	0.847	0.784	0.849	0.879	0.779	0.774	0.868
Proposed-L2	0.789	0.92	0.873	0.787	0.831	0.804	0.761	0.834	0.772	0.835	0.864	0.763	0.761	0.851
Proposed-L1	0.781	0.909	0.865	0.779	0.825	0.795	0.75	0.825	0.764	0.826	0.857	0.756	0.753	0.842
Ablation study results														
$\mathcal{L}_{Str}$	0.788	0.919	0.875	0.786	0.832	0.804	0.768	0.834	0.778	0.835	0.863	0.766	0.765	0.853
$\mathcal{L}_{Swap}$	0.770	0.903	0.860	0.771	0.811	0.788	0.748	0.814	0.761	0.816	0.845	0.746	0.749	0.836

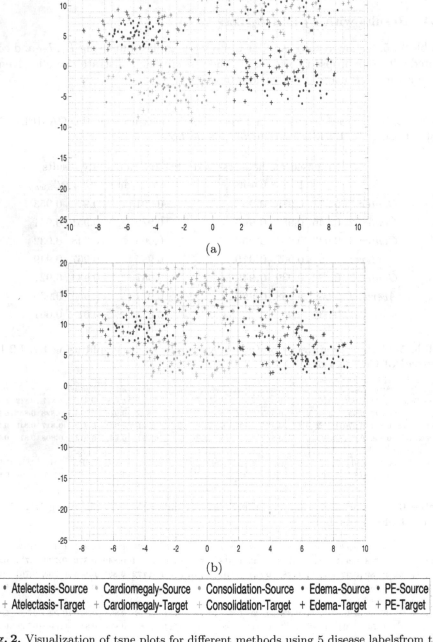

Fig. 2. Visualization of tsne plots for different methods using 5 disease labelsfrom the CheXpert Dataset: (a) Our proposed method; (b) Our method without using structure loss.

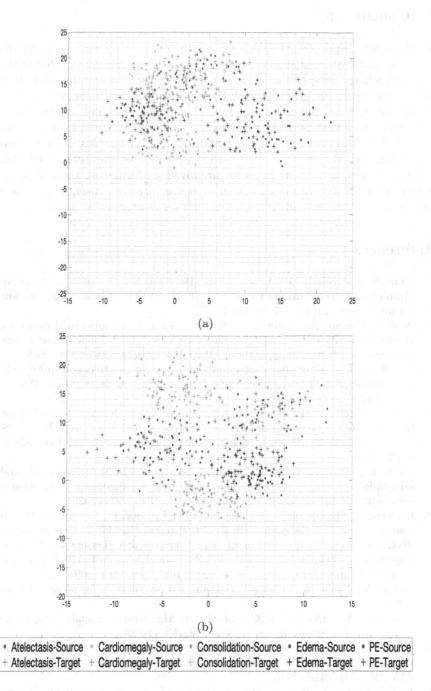

Fig. 3. Visualization of tsne plots for different methods using 5 disease labelsfrom the CheXpert Dataset: (a) using DANN; (b) the GCN2 method of [13].

4   Conclusion

In this paper we propose an UDA method using graph adversarial convolutional networks and feature disentanglement using a novel loss term. Our novel loss term extends the swapped autoencoder architecture thus learning more discriminative features. Our graph convolutional network based unsupervised domain adaptation method outperforms conventional CNN methods as graphs better learn the interaction between samples by focusing on more global interactions while CNNs focus on the local neighborhood. This enables GCN to perform better UDA as demonstrated by results on multiple datasets. While feature disentanglement also contributes to improved performance, there is scope for improvement. In future work we wish to explore stronger disentanglement techniques, and aim to extend this approach for segmentation-based tasks.

References

1. Ahmedt-Aristizabal, D., Armin, M.A., Denman, S., Fookes, C., Petersson, L.: Graph-based deep learning for medical diagnosis and analysis: past, present and future. Sensors **21**(14), 47–58 (2021)
2. Ahn, E., Kumar, A., Fulham, M., Feng, D., Kim, J.: Unsupervised domain adaptation to classify medical images using zero-bias convolutional auto-encoders and context-based feature augmentation. IEEE TMI **39**(7), 2385–2394 (2020)
3. Bándi, P., et al.: From detection of individual metastases to classification of lymph node status at the patient level: the CAMELYON17 challenge. IEEE Trans. Med. Imag. **38**(2), 550–560 (2019)
4. Chang, J.-R., et al.: Stain mix-up: unsupervised domain generalization for histopathology images. In: de Bruijne, M., et al. (eds.) MICCAI 2021. LNCS, vol. 12903, pp. 117–126. Springer, Cham (2021). https://doi.org/10.1007/978-3-030-87199-4_11
5. Chen, C., Dou, Q., Chen, H., et al.: Unsupervised bidirectional cross-modality adaptation via deeply synergistic image and feature alignment for medical image segmentation. IEEE Trans. Med. Imag. **39**(7), 2494–2505 (2020)
6. Gadermayr, M., Appel, V., Klinkhammer, B.M., Boor, P., Merhof, D.: Which way round? A study on the performance of stain-translation for segmenting arbitrarily dyed histological images. In: Frangi, A.F., Schnabel, J.A., Davatzikos, C., Alberola-López, C., Fichtinger, G. (eds.) MICCAI 2018. LNCS, vol. 11071, pp. 165–173. Springer, Cham (2018). https://doi.org/10.1007/978-3-030-00934-2_19
7. Ganin, Y., Ustinova, E., et al.: Domain-adversarial training of neural networks (2016)
8. Gopinath, K., Desrosiers, C., Lombaert, H.: Graph domain adaptation for alignment-invariant brain surface segmentation (2020)
9. Gulshan, V., Peng, L., et al.: Development and validation of a deep learning algorithm for detection of diabetic retinopathy in retinal fundus photographs. JAMA **316**(22), 2402–2410 (2016)
10. He, K., Zhang, X., Ren, S., Sun, J.: Deep residual learning for image recognition. In: Proceedings of the CVPR (2016)

11. Heimann, T., Mountney, P., John, M., Ionasec, R.: Learning without labeling: domain adaptation for ultrasound transducer localization. In: Mori, K., Sakuma, I., Sato, Y., Barillot, C., Navab, N. (eds.) MICCAI 2013. LNCS, vol. 8151, pp. 49–56. Springer, Heidelberg (2013). https://doi.org/10.1007/978-3-642-40760-4_7

12. Hong, Y., Chen, G., Yap, P.T., Shen, D.: Multifold acceleration of diffusion MRI via deep learning reconstruction from slice-undersampled data. In: IPMI, pp. 530–541 (2019)

13. Hong, Y., Kim, J., Chen, G., Lin, W., Yap, P.T., Shen, D.: Longitudinal prediction of infant diffusion MRI data via graph convolutional adversarial networks. IEEE Trans. Med. Imaging 38(12), 2717–2725 (2019)

14. Irvin, J., Rajpurkar, P., et. al.: CheXpert: a large chest radiograph dataset with uncertainty labels and expert comparison. In: arXiv preprint arXiv:1901.07031 (2017)

15. Kamnitsas, K., Baumgartner, C., et al.: Unsupervised domain adaptation in brain lesion segmentation with adversarial networks. In: IPMI, pp. 597–609 (2017)

16. Kumagai, A., Iwata, T.: Unsupervised domain adaptation by matching distributions based on the maximum mean discrepancy via unilateral transformations 33, 4106–4113 (2019)

17. Liu, A.H., Liu, Y.C., Yeh, Y.Y., Wang, Y.C.F.: A unified feature disentangler for multi-domain image translation and manipulation. Advances in Neural Information Processing Systems, vol. 31 (2018)

18. Liu, Y., Gadepalli, K., et al.: Detecting cancer metastases on gigapixel pathology images. In: arXiv preprint arXiv:1703.02442 (2017)

19. Luo, Z., Zou, Y., et al.: Label efficient learning of transferable representations across domains and tasks. In: Proceedings of NeuRIPS, pp. 164–176 (2017)

20. Ma, X., Zhang, T., Xu, C.: GCAN: Graph convolutional adversarial network for unsupervised domain adaptation. In: IEEE CVPR, pp. 8258–8268 (2019)

21. Mahapatra, D., Bozorgtabar, B., Thiran, J.-P., Shao, L.: Structure preserving stain normalization of histopathology images using self supervised semantic guidance. In: Martel, A.L., et al. (eds.) MICCAI 2020. LNCS, vol. 12265, pp. 309–319. Springer, Cham (2020). https://doi.org/10.1007/978-3-030-59722-1_30

22. Motiian, S., Jones, Q., et al.: Few-shot adversarial domain adaptation. In: Proceedings of the NeurIPS, pp. 6673–6683 (2017)

23. Ouyang, C., Kamnitsas, K., Biffi, C., Duan, J., Rueckert, D.: Data efficient unsupervised domain adaptation for cross-modality image segmentation. In: Shen, D., et al. (eds.) MICCAI 2019. LNCS, vol. 11765, pp. 669–677. Springer, Cham (2019). https://doi.org/10.1007/978-3-030-32245-8_74

24. Painchaud, N., Skandarani, Y., et al.: Cardiac segmentation with strong anatomical guarantees. IEEE Trans. Med. Imag. 39(11), 3703–3713 (2020)

25. Park, T., et al.: Swapping autoencoder for deep image manipulation. In: Advances in Neural Information Processing Systems (2020)

26. Puybareau, É., Zhao, Z., et al.: Left atrial segmentation in a few seconds using fully convolutional network and transfer learning. In: STACOM Atrial Segmentation and LV Quantification Challenges, pp. 339–347 (2019)

27. Rajpurkar, P., Irvin, J., et al.: Chexnet: radiologist-level pneumonia detection on chest x-rays with deep learning. In: arXiv preprint arXiv:1711.05225 (2017)

28. Vahadane, A., Peng, T., et al.: Structure-preserving color normalization and sparse stain separation for histological images. IEEE Trans. Med. Imag. 35(8), 1962–1971 (2016)

29. Wang, X., Peng, Y., et al.: Chestx-ray8: Hospital-scale chest x-ray database and benchmarks on weakly-supervised classification and localization of common thorax diseases. In: Proceedings of the CVPR (2017)

30. Wu, M., Pan, S., Zhou, C., Chang, X., Zhu, X.: Unsupervised domain adaptive graph convolutional networks. In: Proceedings of The Web Conference 2020, pp. 1457–1467 (2020)

31. Xie, S., Zheng, Z., Chen, L., Chen, C.: Learning semantic representations for unsupervised domain adaptation. In: Proceedings of the ICML, vol. 80, pp. 5423–5432 (2018)

Author Index

Printed in the United States
by Baker & Taylor Publisher Services

Printed in the United States
by Baker & Taylor Publisher Services